Principles of School Business Management

SECOND EDITION

Association of School Business Officials International

R. Craig Wood, Ed.D.
University of Florida

David C. Thompson, Ed.D.
Kansas State University

Lawrence O. Picus, Ph.D.
University of Southern California

Don I. Tharpe, Ed.D.
Association of School Business Officials International

Principals of School Business Management

Printed in the United States of America

For more information or to purchase additional copies, contact:
ASBO International
11401 North Shore Drive
Reston, VA 22090-4232
703/478-0405

ISBN #0-910170-70-3

Contents

CHAPTER 3

CHAPTER 4

CHAPTER 5

Chapter 6

Chapter 7

Chapter 8

Chapter 9

Chapter 10

Chapter 11

CHAPTER 12

CHAPTER 13

CHAPTER 14

CHAPTER 15

CHAPTER 16

CHAPTER 17

CHAPTER 18

CHAPTER 19

Chapter 23

Chapter 24

Index

Preface

As in the previous edition, this second edition is designed to serve as a basic reference for practicing school business administrators as well as a text for graduate courses in school business administration. Hence, the second edition is a reflection of these dual purposes and offers a comprehensive overview of school business management. The text is designed to be used in sections as each chapter may be viewed in its own right. As each task area varies due to statutory and regulatory requirements, the practitioner will be able to accept what is applicable within his or her own state statutory guidelines and local board policies. By design, the second edition continues in the tradition of not being a quick, how-to-guide in becoming a school business administrator. The text, rather, reflects the view that school business administration is an area that is large and diverse. Unlike many texts in the field, this text offers the reader a detailed display of the task areas.

The second edition was an ambitious undertaking, as school business management is a broad, complex, and dynamic area of educational leadership. The authors embrace the concept that school business management is not a narrow discipline. The contemporary school business administrator must be knowledgeable concerning a wide variety of areas. The text has again taken the approach that is far more realistic and reflective of modern practice, that is, the school business administrator's acting and being responsible for a wide variety of activities designed to organize and maintain the school district in order to ensure the enhancement of educational opportunity. Thus, practitioners may find certain chapters to be of primary interest.

The authors wish to thank the Association of School Business Administrators, International, and the Board of Directors for their support, vision, and willingness to again undertake such a massive and significant project. Don I. Tharpe, the Executive Director of ASBO, has always been highly supportive of the text. Additionally, Peg Kirkpatrick, of the ASBO staff deserves special recognition for bringing this text to fruition. Most importantly, the membership of the Association of School Business Officials, International are to be thanked for having the most outstanding, viable, and professional organizations dedicated to the excellence of educational administration.

Gainesville, Florida R. Craig Wood

Authors

R. Craig Wood is Professor, The Graduate School, and Director UCEA Center for Education Finance, University of Florida. Dr. Wood's professional career has spanned that of classroom teacher, school district business manager, assistant superintendent for finance, assistant professor, associate professor, and currently Professor, Department of Educational Leadership of the College of Education at the University of Florida in Gainesville, Florida. Previous to his present position, he was a professor at Purdue University. He currently holds a Doctoral Research Faculty appointment with the Graduate School of the University of Florida. He holds his doctorate in educational administration from the Virginia Polytechnic Institute and State University. His post doctoral study was done at Harvard University.

A specialist in education finance and law, he is one of the most prolific authors in education finance and law in America. Dr. Wood's publications record consists of nearly 200 publications in numerous book chapters, monographs, and journal articles. He has also has published numerous times in the *American Education Finance Association's Annual Yearbooks,* the *Journal of Education Finance,* the *NOLPE Handbook of School Law,* West's *Education Law Reporter, School Business Affairs*, *Government Union Review,* and the *Journal of Collective Bargaining in the Public Sector.* Additionally he has published numerous chapters in books by NOLPE and ASBO as well as many others. He is a coauthor of the texts, *Education Finance Law: Constitutional Challenges to State Aid Plans* (NOLPE, 1993) and *Fiscal Leadership for Schools: Concepts and Practices* (Longman, 1994).

He presently serves as the chair of the Editorial Advisory Board of the Association of School Business Officials as well as on the Board of Directors of the American Education Finance Association. He serves on the editorial review boards of *Education Law Reporter* and the finance law editor for the *Journal of Education Finance.*

He has delivered numerous general sessions and keynote addresses as well as his research to such national audiences as the American Education Finance Association, the National Organization on Legal Problems of Education, the Association of School Business Officials, the American Association of School Administrators, the University Council for Educational Administration, the National Center for Educational Statistics, the National Conference of State Legislatures, and the American Education Research Association. He has conducted national workshops for the National Association of Attorneys General.

His consulting activities have served numerous public and private agencies including the states of Delaware, South Dakota, and Montana and numerous school districts as well as law firms regarding education finance litigation. He has served as the lead expert witness for state defendants and plaintiffs while conducted education finance research reports involving constitutional challenges to education finance distribution formulas in nearly a dozen states. Dr. Wood is also the Director of the University Council for Educational Administration's Center for Education Finance at the University of Florida. He is a Director of Wood, Thompson & Associates Inc.

David C. Thompson's professional career has spanned classroom teacher, elementary principal, high school principal, superintendent of schools, and the professoriate. He has coauthored several textbooks, *including Fiscal Leadership for Schools: Concepts and Practices* (Longman, 1994) and *Education Finance Law: Constitutional Challenges to State School Finance Plans—An Analysis of Strategy* (NOLPE, 1993). He has also published in other respected books and journals, including the yearbooks, of the American Education Finance Association, the *Journal of Education Finance*, West's *Education Law Reporter*, and many others. He is also Editor of the journal *Educational Considerations*. A specialist in education finance litigation, his work has included contracts with public and private firms including state legislatures and law firms throughout the nation, involving him in school finance lawsuits as chief analyst, expert, or advisor representing thousands of school districts and several million students. His research as been present to nearly every major national organization including the National Association of Attorneys General, the American Education Research Association, The American Education Finance Association, National Organization Legal Problems of Education, National Center for Education Statistics, National Conference of Professors of Educational Administration , and many keynote addresses to large legislative audiences. Dr. Thompson is Professor in the University Graduate School and Chair of the Department of Educational Administration and founding Co-Director of the University Council of Educational Administration's (UCEA) Center for Education Finance at Kansas State University and Director of the firm of Wood, Thompson & Associates, Inc.

Lawrence O. Picus is an Associate Professor in the School of Education at the University of Southern California, specializing in school finance, school business administration and budgeting, and the economics of education. He is the Director of the Center for Research in Education Finance, a School of Education research center which studies issues of school finance and productivity as they relate to school reform. He is also a senior research fellow with the Finance Center of the Consortium for Policy Research in Education, and coordinator of the USC School of Education's Annual Summer Superintendents' Conference. Picus serves as a member of ASBO's Editorial Advisory Committee, and is the 1995 President-elect of the American Education

Finance Association (AEFA). Picus is the coauthor of *School Finance: A Policy Perspective*, and *Administrative Uses of Microcomputers*, and the editor of the 1995 AEFA Yearbook, *Where Does the Money Go? Resource Allocation in Elementary and Secondary Schools.* Picus has authored numerous articles published in leading academic journals including *Educational Administration Quarterly, Education Evaluation and Policy Analysis*, *Educational Policy*, the *Journal of Education Finance* and *the Journal of Public Policy Analysis and Management.* Picus earned a Ph.D. in Public Policy Analysis from the RAND Graduate School, and holds a Masters Degree from the University of Chicago and a Bachelors Degree from Reed College.

Don I. Tharpe is executive director of the Association of School Business Officials International. His career in education and association management spans more than 20 years. Prior to coming to ASBO in 1985 as Director of Professional Programs, he held various senior-level positions at the American Vocational Association, including membership, administration, and government relations. In 1990, he was appointed to lead ASBO. Tharpe has also been a classroom teacher and supervisor of Trade and Industrial Education in Jefferson City, Missouri.

He earned his doctorate in educational administration from Virginia Polytechnic Institute and State University, Blacksburg, Virginia. Tharpe is an active member of the American Society of Association Executives and the Greater Washington Society of Association Executives.

Acknowledgments

The authors wish to express their sincere appreciation to the current ASBO Board of Directors for their support of this publication:

President:	Andre J. Montagne, RSBA
President-Elect:	Denny G. Bolton, Ph.D., RSBA
Vice President:	Clair O. Belsher, CSBO
Past President:	William E. Gall, RSBA
Directors:	Roger M. Giroux, Ph.D., RSBA
	David B. Puthoff, RSBO
	Sam MacKinlay, CA, CGA
	Glenn A. Nienhuis, RSBO
	Pam Deering
	Douglas L. Johnson, RSBA

Chapter 1

Introduction to School Business Management

Introduction

To the graduate student, as well as the educational practitioner, the realm of public elementary and secondary education business management is an exciting and challenging area of public educational leadership.[1] Everything within the contemporary public school district revolves around the efficient and proper functioning of the organization to enhance learning. School business management is vital in order for the teaching and learning process to occur.

School business management involves the proper financial operations of all the public school districts of the nation. Every public school district in America has a chief fiscal officer. This person, regardless of title, is ultimately responsible for the efficient fiscal operation of the school district. The titles for such a person vary from district to district and from state to state. In many states, the position requires state certification along with a graduate degree. In other states no such requirement exists. Thus, the requirements and academic background for the chief fiscal officer vary greatly among school districts and states.

Historical Overview

The importance of planning for the management of school systems was recognized in colonial statutes. The original colonies required that a committee of freemen establish revenue, collection, and expenditure procedures. The statutes called for a clerk or treasurer to be the fiscal manager, and no provision for compensation was provided. According to chapter 26 of the Province Law of the Bay Colony, 1692-93, an educational system with its various managerial components was to be established. The chapter stated:

> That the inhabitants of each town within this province, shall take due care, from time to time, to be constantly provided of an able, learned orthodox minister, or ministers, (or school masters) shall be

> suitably encouraged and sufficiently supported and maintained by the inhabitants of such town. And all contracts, agreements and order heretofore made, or that shall hereafter be made by the inhabitants of any town within this province, respecting their ministers or school masters, as to their settlement or maintenance, shall remain good and valid according to the true intent hereof, the whole time for which they were or shall be made, in all the particulars thereof; and shall accordingly be pursued, put in execution and fulfilled. And where there is no contract and agreement made in any town respecting the support and maintenance of the ministry (or school), or when the same happens to be expired, and the inhabitants of such town shall neglect to make suitable provision therein, upon complaint thereof made unto the quarter sessions of the peace for the county where such town lies, the said court of quarter sessions shall and hereby are empowered to order a competent allowance unto such minister (or master) according to the estate and ability of the town; the same to be assessed upon the inhabitants by warrant from the court, directed to the selectmen, who are thereupon to proceed to make and proportion such assessment in manner as is directed for other public charges, and to cause the same to be levied by the constables of such town, by warrant under the hands of the selectmen, or of the town clerk by their order.[2]

The act further stated that if the town failed to provide for the school, the court was then empowered to establish one and to order levies to support the school. Fiscal responsibilities would be removed from the treasurer and assumed by a court-appointed agent.

As public schools grew in size and in responsibility, the unpaid clerk's time was severely overextended. However, 150 years passed before the first full-time school business administrator was employed by a public school system. The city council of Cleveland, Ohio, hired an "acting manager" in 1841 to keep the books, prepare the payroll, and care for the school building.[3] Cleveland added "superintendent" to its administrative staff in 1853.

As local school districts began to flourish, state legislatures enacted laws requiring that school districts designate a chief school fiscal officer. State departments of education have, in many states, followed with certification requirements. Today, the position of school business manager has become an accepted administrative position in the great majority of school districts. Job titles for the manager range from administrative assistant to superintendent for business services. While each school district is organized differently, the task of fiscal management generally falls to a person with vast training and experience.

The Growth of a Profession

The first school business manager was employed to "keep the books and care for the building."[4] This pattern continued; as school boards expressed interest in hiring administrators prepared in bookkeeping and business functions rather than teaching. Early school business managers were not expected to participate in educational decision making, and educators were to leave the business aspects of the school district to the school business manager. The growth of school districts and the acceptance of the management team idea have greatly changed the concept and expectations of school business officials' duties, title, and responsibilities.

Several factors have been in the forefront of the changing concept of school business management. The most significant of these concepts developed from societal and governmental change.

The Decline of Rural Population

Movement from the farms to the cities resulted in an increase in school business administrators. Generally, rural and small school districts did not employ school business administrators. The decline of the family farm led to the school district consolidation movement resulting in the merger of small schools throughout the nation. School districts of larger geographical size with more complex educational programs, transportation needs, and administrative reorganizations opened rural schools to the position of school business administrator.

Diversity of Funding and Accountability

Historically, public schools were supported through local taxation. State governments regulated the educational programs, length of the school year, attendance, and in a few instances taxation limits. However, little attention was given to local school district fiscal management practices.

This lack of supervision changed in the mid-1960s when the federal government began a massive infusion of capital and accompanying regulations. The state governments in the 1970s, following California's lead, began a series of cost-cutting measures through revenue reduction legislation. More complex accounting systems to track revenue and expenditures from categorical grants, block grants, rental property, leasing contracts, intergovernmental agreements, self insurance, and so on were adopted by school districts. The issuance of *Federal Handbook II* accounting procedures caused concern among school administrators and hastened the employment of school business administrators in many small school districts. Today, the role of the school business administrator is complex, highly technical, and one that calls for a fiscal, as well as an educational, vision of the school district.

Increasing Expenditures

The 1970s produced a $23 billion growth in educational expenditures despite a declining enrollment for the corresponding period. As was suggested in the first edition of this text, per-pupil costs will continue to rise with the "search for excellence" and with the increased competition to increase teacher salaries and extend the school year. These reasons, along with the movement toward national, state, and local school reform, continue to exert fiscal forces upon public schools. For a variety of reasons, ranging from simple economic pressures to having the next generation competitive with the world marketplace, public educational expenditures are expected to rise for the foreseeable future. Additionally, the long-postponed renovation of school buildings cannot be postponed indefinitely, and new buildings are being constructed in the high population growth states.

While it is sometimes difficult to grasp, our nation is a country of vast change that continues every day within every generation. Public education is highly reflective of these changes. It is important to note that throughout the United States the demographic changes are vast and complex. Briefly, the school age population data of our nation reflect continued growth. Table 1.1 reflects the overall general trends for the near future.

Table 1.1
Population Projections 1981-82 to 2003-04*

Group	1981-82	1992-93 est.	2003-04 projected	% Change 1981-82 to 1992-93	% Change 1992-93 to 2003-04
5 -13 yrs old	30,711	33,005	37,107	7	12
14-17 yrs old	15,610	13,649	16,277	-13	19
18 yrs old	4,186	3,307	3,918	-21	18
18-24 yrs old	30,509	26,063	27,563	-15	6
25-29 yrs old	20,275	20,278	17,431	(**)	-14
30-34 yrs old	18,798	22,343	19,684	19	-12
35-44 yrs old	26,454	39,973	44,336	51	11

*Population numbers are reported in thousands.
** represents between -.05 and 0.5. All population projections are from the middle series of population projections of the Bureau of the Census.

The current and projected expenditure pattern for public elementary and secondary education are expected to continue to rise, but at a decreasing rate, for the near future. Overall expenditure projections are shown in table 1.2.

Table 1.2
Public Elementary and Secondary Education Expenditures*

	1981-82	1992-93 estimate	2003-04 projected	Change 1981-82 to 1992-93	Change 1992-93 to 2003-04
**Current expenditures	$148.5	$211.9	$284.4	43%	34%
Current expenditures per pupil in average daily attendance	$4,003	$5,352	$6,218	34%	16%
Average annual salaries for teachers	$28,306	$28,306	$34,316	21%	12%

*All figures in billions of constant 1991-92 dollars, figures taken from Projections of Education Statistics to 2004, (U.S. Dept. of Educ., National Center for Education Statistics..
**Excludes capital outlay and interest.

Because of these economic pressures and tax resistance movements, school districts are often forced to seek alternative funding to maintain programs and facilities. For example, school districts are engaging in refinancing revenue bonds and lease/purchase agreements for school buildings. These are complex methods of extending limited resources and continue to grow in acceptance and popularity among educational fiscal leaders. Additionally, innovative fund-raising techniques, long employed by private schools, are being adopted by public school districts.[5] As the pressures on public education continue to mount, the role and leadership of the school business administrator continues to grow in importance.

The Role of the School Business Administrator

Administrative Organization

The role of the school business administrator is determined by the organizational pattern of the school district. The general classification is that of unit control. In a few states, there is an organizational structure of dual control. In school districts that have adopted a dual-control administrative pattern, the school business administrator reports directly to the school board and is not responsible to the superintendent. The theory behind this plan is that the superintendent is responsible for educational matters and the school business administrator is responsible for financial and management matters. This type of administrative organization provides for the educational and business functions of a school district to be brought together at the board level for decision making.

The great majority of public school districts have adopted unit control which places the superintendent of schools at the head of the administrative structure. All administrators report to the superintendent, and the superintendent reports to the school board. Although the superintendent is the head of the organization, this structure does not rule out diversified decision making. The school business administrator is a member of the administrative cabinet. All members are advisers to the superintendent and are often called on to represent the superintendent in their areas of expertise. The business administrator's operations are affected by every educational decision, and business decisions affect the educational operation. It is imperative that a team concept be maintained by the members of the cabinet.

As the operation of public schools become more complex, the need for specialization within the area of school business administration becomes more critical. The school business administrator becomes the cabinet representative for the areas of accounting, finance, transportation, food service, data processing, purchasing, maintenance, and so on. A general administrative knowledge of all the preceding areas is imperative.

Basic Principles

Frederick Hill, and many other authors, has discussed a number of basic principles that serve as guidelines to school business administration.[6] These principles include:

- School business administration operates as a service area to the administrative organization. Its purpose is to support and enhance the educational programs of the school district.

- For effective operation of the school district, the administrative organization should consist of the unit control plan with one chief administrator reporting to the board of education. The school business administrator should be a highly trained member of the superintendent's cabinet qualified to act on the superintendent's behalf.

- Greater attention should be given to the preparation and in-service components of the school business administrator's education as the operation of schools becomes more complex.

- Sound business practices must be observed in all school business activities. The articulation of these practices to cabinet members and the public is the responsibility of the school business administrator.

- Ethical codes of conduct should be observed at all times. No appearance of impropriety can be tolerated. Schuyler Joyner, past president of the Association of School Business Officials International (ASBO), placed ethical conduct in clear perspective when he said, "If you wouldn't want to see it on the front page of your local paper, don't do it."[7]

Areas of Responsibility

No uniform pattern of responsibilities exists for the school business administrator. Various texts list similar responsibilities but do not agree on a common list. Regardless of the responsibilities, personal leadership is required if responsibility is to be carried out. In discussing leadership for the chief fiscal officer more than thirty years ago, A.W. Flowers defined leadership as:

- Personality in *ACTION.*
- Incisive *JUDGMENT* in all situations.
- Genuine *ENTHUSIASM* for work and for people.
- Contagious *VITALITY* which encourages followers.[8]

As a member of the superintendent's cabinet, the school business administrator is responsible for providing leadership and support for the school district's programs. C.W. McGuffey found seven basic functions of management that were rated as important by school business administrators.[9] Tasks of the school business operation fell under these general responsibilities. McGuffey concluded that a successful school business administrator must possess or develop management skills in these general areas of responsibility:

Planning: Planning is the process of looking ahead and preparing for the future by identifying resources and needs and developing a master plan to follow. The recent rise in public interest in the educational process has raised questions concerning resource allocation and the value of current practices. The school business administrator has been given the responsibility of providing the database for resource allocation and unit costs. This greatly expanded role has forced the academic and business operations to cooperatively develop projections of educational needs. At a time when the role of the school business administrator is becoming more technical, the need to expand and become more of a generalist is also present.

Organizing: A manager must create and maintain the conditions and relationships necessary for the business office to function efficiently. The organizational structure should assure that appropriate procedures for accomplishing the work are established and remain in place. Requirements of manpower, materials, finances, and other resources must be determined and assigned at the needed time. Lines of authority and a clear communication process must be understood by all agencies and levels.

Staffing: The staffing responsibility is much broader than the title implies. It assumes the ability to correctly and clearly state job requirements in an advertisement, and screen, employ, and provide the job introduction to the personnel needed to achieve the organization's goals. Staffing also involves producing programs for the professional development of the staff and advancement opportunities within the organization. This process requires an evaluation procedure and a reward and remedial system.

Directing: Direction is related to control, but is different from it. Direction involves guidance and leadership. Individuals must be motivated to achieve the organization's objectives. A feeling of "ownership" of the organization must be established within the personnel. The function of directing includes the components of communication, morale, job satisfaction, productivity, and loyalty.

Controlling: Once goals have been established, all action is regulated in relation to the goals. "Checks" are established to measure accomplishments, and actual results are compared with the criteria and needed corrections are made. This function is based on the manager's ability to foresee potential conflicts and correct course before the conflict arises. Trouble spots are avoided or tackled in such a way to assure project success.

Decision making: The decision-making process involves three steps: identification of the problem, review of alternative solutions, and the choice of the appropriate solution. Decision making is the initial element in the administrative process. Decisions, as a general rule, should be made as close as possible to the point of implementation. If the manager delegates this authority, it must be understood that the power will fall on personnel who have the clinical analysis skills and technical knowledge to solve the problem. Decision making should be a deliberate process that examines all possible solutions and their consequences before the choice is made.

Evaluating: Evaluation is the process of measuring the organization's activities against the achievement of basic goals. Evaluation is a continuous process that provides for examination of ongoing activities. Decisions to alter the program are often possible on an in-progress basis. The ultimate challenge in evaluating educational programs is to translate goals into operational terms that can be used to establish future needs in the areas of staff and budget.

The school business administrator is expected to manage operations of the office within the areas of responsibility. Delegation of authority is required if the day-to-day tasks are to be accomplished.

Professional Preparation and Certification

The debate concerning the preparation of the school business administrator has gone on since the first school business administrator was employed. Whether a business background is preferable to an educational background may never be settled. It is generally agreed that the school business administrator needs an understanding of both areas if he or she is to function as a contributing member of the superintendent's cabinet. The business administrator must possess the financial and business management skills needed by any manager in the private sector. In addition, the business administrator must perceive and understand the probable effects of business application and functions on the educational programs of the schools.

Several states have adopted certification requirements for the school business administrator. Generally, these requirements include the possession of at least a B.A. degree, more typically a M.A. degree, and the presentation of competency mastery in selected areas. Education and business administration programs should prepare graduates for the entry into the area of school business administration. Reflecting the importance of this field, the National Council for the Accreditation of Teacher Education (NCATE) recognizes the area of school business administration.

Research was done by C.W. McGuffey to determine the relative importance of skills needed by the school business administrator. The skills (task clusters) were ranked by practicing school business administrators. Of the twenty-eight skills, only warehousing and school supplies management were ranked as of low importance. Any preparation program must certainly include these task clusters in addition to addressing the seven areas of responsibility discussed earlier.

McGuffey's task clusters follow in order of importance.

High Importance

- Financial planning and budgeting.
- Fiscal accounting and financial reporting.
- Cash management.
- Fiscal audits and reports.
- General management.
- Payroll management.
- Purchasing.

Average Importance

- School insurance and risk management.
- Debt service and capital fund management.
- Legal control.
- Office management.
- Educational resources management.
- School activity and student body funds.
- Personnel management.
- School plant maintenance.
- School property management.
- School plant operations.
- School community relations.
- Collective negotiations.
- Plant security and property protection.
- Data processing.
- School transportation services.
- School construction management.
- School food services.
- Staff development.
- Grantsmanship.
- Educational facilities planning.[10]

The preparation for these positions will always be a combination of academic graduate school programs accompanied by intensive in-service programs offered by such organizations as the Association of School Business Officials International and the appropriate state organizations, as well as the classic on-the-job training that exists in nearly every public school district.

School business administrator certification is required in nearly two dozen states. The Professional Development Research Committee of ASBO recommends that certification of the school business administrator be adopted in all states. It further recommends that provisions include professional certification standards recognizing the business background as well as the educational background.

In addition to professional preparation, the following desirable personal characteristics have been suggested in ASBO's *School Business Administrator:*

- ***The School Business Administrator should possess plenty of directed drive.*** The true leader has more drive than his or her peers. The school business administrator has more energy, more staying power, more confidence.

- ***The School Business Administrator should like people and show it.*** A good administrator really likes people. A sincere liking for people as unique personalities, coupled with a bright faith in the future, make a winning combination.

- ***The School Business Administrator should be able to get along well with others.*** Without this characteristic, hard sledding will interfere with his or her achievements and the achievements of the group of which the school business administrator is the leader.

- ***The School Business Administrator should be a good listener.*** One of the best instruments of the administrator is a pair of good ears.

- ***The School Business Administrator should be a person of integrity.*** Honesty, sincerity and truthfulness still win respect and instill like traits in others.

- ***The School Business Administrator should be fair in his or her dealings with all.*** The administrative menu offers a steady diet of "hot potatoes." The good administrator does not pass the buck, but accepts the blame and builds a deserved reputation for fairness.

- ***The School Business Administrator should establish a good work climate.*** The true leader will be patient, be firm, have a sense of humor—and the staff will produce results. By laying a foundation of trust, sincerity, and mutual respect, a business manager can lay to rest many staff anxieties.

- ***The School Business Administrator should be a self-starter.*** A leader sets the pace for new ideas and picks up the ball and runs.

- ***The School Business Administrator should possess a high threshold of annoyance and a sense of values.*** The wise one rarely "blows off steam." He or she is no victim of his or her emotions. The school business administrator is bigger than the problems that he or she confronts. The school business administrator's perspective prevents him or her from taking himself too seriously, and has inner resources to deal with stress.

- ***The School Business Administrator should develop high articulation and writing skills and watch his or her language.*** The ability to draw persuasive "word pictures" will help others to see and to accept group objectives. His or her language should be above reproach, and he or she should carefully avoid offense to anyone.

• ***The School Business Administrator should possess the ability to make the problems of others his or her own.*** The staff will come to the school business administrator if they receive help when they ask for it. The willingness to accept the problems of others and to share in the solution is evidence of leadership.

• ***The School Business Administrator should be able to accept criticism.*** He or she needs to guard against creating a "yes" person attitude on the part of his or her associates. A good administrator knows how to take criticism and how to use it constructively.

• ***The School Business Administrator should be able to render decisions.*** Frequently, more grief among staff members is created by no decision than by a poor one.

• ***The School Business Administrator should be willing to "bask in reflected glory:"*** The real business manager sees him or herself as a member of a team rather than the star player. The good leader sees that good work is recognized and proper credit given to all who have shared in it.[11]

Personal Attitudes and Values

The emphasis on the importance of relationships with people suggests that attention must be given to the kind of person the school business administrator is expected to be. Certain criteria for good administrators have been further suggested by Frederick Hill:

• They are first of all a good people.

• They want to be part of a strong team rather than be strong standing alone. They select able and creative persons as co-workers, cooperate with them in a give-and-take relationship, and desire the efficiency and broad viewpoint that is possible when several first-rate minds are analyzing a situation.

• They have respect for all those with whom they deal. They respect them, first, because they believe in the fundamental dignity and worth of every human being; and second, because they believe that a person should be treated with consideration, not only for what he or she is, but for what he or she might become.

• They select qualified assistants and delegate as much routine to them as possible.

• They are compassionate.

• They never forget that their power is delegated to them by others. They make use of it for the purpose for which it was put in their hands. One's authority is . . . limited on every hand by the objectives one is seeking and the interests and desires of those who are affected by the actions.[12]

When is a School Business Administrator Needed?

No set formula concerning school expenditures, number of employees, enrollment or average daily attendance is used to determine when a school business administrator is needed. The need for such a position depends on several factors relating tasks to be performed to the number and skills of staff currently employed. Factors that should be seriously considered by the school board before the final decision is reached include:

- ***Growth rate of the school district.*** Is the school district stable or is it undergoing a period of acceleration or shifts in enrollment?

- ***Extent of local programs and services.*** Are such services as transportation programs and school lunches provided for students on a broad or limited basis? To what degree does the school district participate in preschool, adult, or post-high-school programs?

- ***Pattern of organization for the school district.*** Does the school district have several buildings on scattered sites or are all facilities located close to each other, thereby reducing logistical problems?

- ***Condition of existing facilities.*** Are maintenance problems and new facility needs minimal or will these require considerable attention by central office administrators?

- ***Interests and qualifications of other central office staff members.*** Does the school district employ sufficient central office staff members so that the superintendent can assume responsibility for school business affairs? Have assurances that adequate leadership attention will be given to the development of other facets of the total school program?

- ***Role of nonprofessional staff members.*** Does the school district have experienced noncertified staff members who require minimum supervision as they assume responsibility for certain phases of the school business operation?

- ***Magnitude and scope of the total school program.*** To what extent does the school district provide community educational and recreational services and participate in vocational programs?

Whenever the demands of the business area grow to such a magnitude that it taxes the present staff, a business assistant or an administrative assistant in charge of business should be hired. Frederick Hill recommends that an assistant be employed when the average daily attendance reaches 1,000, and that a full-time school business administrator be hired when the enrollment is between 1,500 and 2,000 pupils.[13]

The US. National Education Goals[14]

Since the first edition of this work, our society has become even more concerned regarding the quality of public education in the United States. Perhaps the most notable example of this concern was expressed in March of 1994 when the United States Congress passed Goals 2000: Educate America Act. This federal legislation establishes educational goals for our nation. The Goals 2000 Act encourages all states to set world-class educational standards and establish high expectations for all students. At the same time, the act provides states with a great deal of latitude to link the national goals with each state's reform strategy. Moreover, one of the act's objectives is to streamline reporting requirements, hopefully freeing up time and personnel for school improvement activities.

While it is impossible to predict what this means for school business administrators, the tasks and goals of the office may be highly affected. However, it is safe to conclude that the role and tasks of the school business administrator will become more complex.

One of the more significant areas of interest to school business administrators is that of linking public schools with social service agencies. Goals 2000 attempts to improve links between the public schools and other social service agencies. Goals 2000 encourages state plans to provide incentives for such collaborations. For example, working with other social agencies means different demands may be made on the way school facilities are used. In particular, schools have more people coming in and out of them, and they may be required to stay open longer to provide additional services to children and their families.

The reform movement has placed, and will most likely increase, the pressure to engage in decentralized budgeting for many school districts throughout the nation. It has been suggested that under the decentralized schools envisioned by the Goals 2000 Act, more budget authority would be devoted to the school site. Thus, the budget office function may change to one of supporting decision making rather than serving as a mediator among competing requests. More particularly, as school site officials begin to make more decisions locally, they will need assistance in estimating the costs of resources, conducting cost analyses, and finding the most cost-effective ways to implement educational programs.

Furthermore, it has been suggested that the entire purchasing process may change significantly in many school districts. If school sites attempt to make different programmatic decisions, the benefits of the traditional centralized purchasing process could very well be significantly diminished.

Within the movement toward the Goals 2000 issues, several clear tasks for school business administrators emerge. The school business administrator must determine, along with the assistance of others, the best fit between centralized and decentralized fiscal operations. This task is complicated by the observation that nearly every school district will determine these issues differently given the uniqueness of each school district within the nation. Notwithstanding these differences, there are certain functions that benefit from economies of scale if managed from the central office. Such things as employee payroll and benefits, as well as risk management functions and transportation, most likely should be managed from the centralized office. Centralization of these functions should have little impact on school-based decisions.

However, particular issues affecting transportation, for example, may have substantial cost implications when giving control to school sites. If schools are free to choose when to start and end the school day, considerable inefficiencies may be introduced into the transportation operation. School districts will have to decide whether to require all schools to operate on schedules that allow for greater efficiencies, or find a way to charge back costs to schools that elect "inefficient" school times.

Regardless of such issues, such programs as Goals 2000 depend on strong coordination between the business office and other employees of a school district. The challenge for school business administrators will be to find ways to ensure proper accountability while providing support for more decentralized decision making.

Goals 2000 Act

The Goals 2000 Act states that the purpose is to "improve learning and teaching by providing a national framework of education reform; to promote the research, consensus building, and systemic changes needed to ensure equitable educational opportunities and high levels of educational achievement for all students . . . " The act provides nearly $90 million to support state efforts to attain the eight national goals and establishes procedures for each of the states to submit improvement plans to the federal government to receive those moneys. Notwithstanding the fact that such an appropriation is largely insignificant, the program may provide further impetus for systemic change in the schools of the United States. Briefly, the goals are as follows:

- **Goal 1: School Readiness.**

By the year 2000, all children in America will start school ready to learn.

- **Goal 2: School Completion.**

By the year 2000, the high school graduate rate will increase to at least 90 percent.

- **Goal 3: Student Achievement and Citizenship.**

By the year 2,000 all students will leave grades 4, 8 and 12 having demonstrated competency over challenging subject matter including English, mathematics, science, foreign languages, civics and government, economics, arts, history, and geography; and every school in America will ensure that all students learn to use their minds well, so they may be prepared for responsible citizenship, further learning, and productive employment in our nation's modern economy.

- **Goal 4: Teacher Education and Professional Development.**

By the year 2000, the nation's teaching force will have access to programs for the continued improvement of their professional skills and the opportunity to acquire the knowledge and skills needed to instruct and prepare all American students for the next century.

- **Goal 5: Mathematics and Science.**

By the year 2000, United States' students will be first in the world in mathematics and science achievement.

- **Goal 6: Adult Literacy and Lifelong Learning.**

By the year 2000, every adult American will be literate and will possess the knowledge and skills necessary to compete in a global economy and to exercise the rights and responsibilities of citizenship.

- **Goal 7: Safe, Disciplined, and Alcohol- and Drug-Free Schools.**

By the year 2000, every school in the United States will be free of drugs, violence, and the unauthorized presence of firearms and alcohol, and will offer a disciplined environment conducive to learning.

- **Goal 8: Parental Participation.**

By the year 2000, every school will promote partnerships that will increase parental involvement and participation in promoting the social, emotional, and academic growth of children.

Elements of Future Educational Reform

Regardless of various reforms, school business administrators will continue to be the individuals who are ultimately responsible for the efficient fiscal operation of public education. Important tasks will continue in a variety of ways. Future tasks will continue to grow more complicated and require even greater educational and fiscal management vision in order for public education to serve the next generation within our society.

Essential elements of successful educational improvement which must be adapted for every school district as outlined in Section 306 of the Goals 2000 Act include:

- Teaching and learning standards and assessments.
- Opportunity-to-learn standards or strategies.
- Use of technology.
- Governance, accountability, and management.
- Parent and community support and involvement.
- Making improvements system-wide.
- Promoting grassroots efforts.
- Dropout prevention strategies.
- Program improvement and accountability.
- Coordination with school-to-work programs and vocational instruction.
- Milestones and time lines.
- Regular review.[15]

The changing nature of public education and the various organizational structures form an opportunity for central office leaders, including school business administrators, to work more closely with school site leaders. Ways to provide assistance to schools include the following:

- Research potential for grant opportunities.
- Form collaboratives with other school districts.
- Pool financial resources to support professional development training opportunities.
- Identify key community members to serve on steering committees and task forces.
- Assist in the forging of school-business partnerships throughout the local community.
- Serve as liaisons to local government officials, state officials, town agencies, service agencies, civic organizations, and churches.
- Network local school districts with higher education in the community.
- Support the purchase and use of technology as an integral part of the school's learning environment. Tools that can help create high-performance learning environments where all children reach high standards include voice

mail, interactive satellite links, electronic bulletin boards, on-line computer networks, software, and equipment.[16]

Thus, the school business administrator should view the reforms as an opportunity to achieve greater efficiency while having the opportunity to serve the school district in a better fashion. Notwithstanding how a school district is organized, the school business administrator must be a person of leadership and vision.

Code of Ethics

The ancient Greeks used to state that character determines destiny. Never has this concept been more true as society grows in complexity and has higher expectations of its governmental agencies. Ethical behavior is a facet of all professions. In accepting membership in the Association of School Business Officials International, the school business administrator accepts the Code of Ethics established by the organization. An individual accepts the obligation to exhibit high standards of personal and professional conduct. It must be noted that the Board of Directors has withdrawn membership to individuals who have violated the Code of Ethics.

Summary

The school business administrator has become an important part of the superintendent's cabinet. It is generally accepted that a school business administrator can contribute to the team concept in greater magnitude if he or she has a strong background in business applications as well as an understanding of the educational organization. This chapter introduces this fascinating and varied area of public educational leadership. It is hoped that the student and practitioner of this vital area will continue to learn and build competencies and skills beyond what any textbook can possibly offer. The professional organization of all school business administrators is the Association of School Business Officials, International, located in Reston, Virginia. The Association provides many outstanding inservice programs to practicing school business administrators. Thus, it is hoped that the reader will always be reminded that learning based on a textbook is merely the beginning of building one's professional competency.

Endnotes

[1]Portions of this chapter are based on the previous edition written by F.M. Miles.

[2]Massachusetts, *The Acts and Resolves, Public and Private, of the Province of the Massachusetts Bay*, Vol. 1 (Boston: Wright and Potter, Printers of the State, 1869), 62.

[3]F.W. Hill, *The School Business Administrator* (Reston VA: Association of School Business Officials, 1960), 10.

[4]*IBID.*

[5] See, R.C. Wood, "New Revenues For Financing Education at the Local Level," *The Impacts of Litigation and Legislation on Public School Finance.* Tenth Ann. American Educ. Finance Ass'n Yearbook, Chapter Three (New York: Harper & Row, 1989) 59-75.

[6]F.W. Hill, *The School Business Administrator*, 3d, (Reston, VA: Association of School Business Officials, 1982).

[7]S. Joyner, "The School Business Administrator," (paper presented at the School Finance Workshop, Flagstaff, AZ, 1971).

[8]A.W. Flowers, "The Role of the School Business Official in the Management Team," (paper presented the School Finance Workshop, Flagstaff, AZ, 1971).

[9]C.W. McGuffey, "Competencies Needed by the Chief School Business Official," (University Contacts Committee manuscript, Association of School Business Officials, 1978).

[10]McGuffey, *A Handbook on Professional Certification of School Business Officials*, (Reston, VA: Association of School Business Officials, 1980).

[11]*Ibid.*

[12]Hill, *The School Business Administrator*, 3d, 51-52.

[13]Hill, *The School Business Administrator.*

[14]This section is based on L.O. Picus, "Education's Systemic Reform: Implications for School Business Administrators," *School Business Affairs*, (January 1995).

[15]See, T.K. Peterson, "Navigating Goals 2000: Sorting Out Funding and Programs, *School Business Affairs*, (January 1995), 12.

[16]J.R. Belair, "Goals 2000: A School's Perspective," *School Business Affairs*, (January 1995) 22.

Chapter 2
Legal Context of the Public School District

Introduction

Many individuals assume that school business management is primarily and exclusively dominated by such areas of expertise as accounting, budgeting, insurance management, and purchasing, as well as a host of other repetitive task areas.[1] The aspiring school business administrator may not consider legal knowledge an important area of cognitive information. Legal knowledge or expertise is not typically part of the school business official's job description. This view is completely erroneous and based on a total lack of understanding of the role and tasks of the contemporary school business administrator.

Legal knowledge is very specialized. Usually, one hires an attorney only when in need of such specialized help. There are certainly special times when school officials must have access to attorneys and the specialized knowledge they provide. However, viewing legal knowledge in this light oversimplifies the status and role of the school business official. This view is compatible with that of the person who is a mere functionary, the person who does what one is told or one who does only that which his or her predecessor did ten, fifteen, or twenty years ago. The functionary need not worry about the legal context of the school district or his or her position within it. Such a view is unrealistic for the contemporary school business official. In this time of rapid change and litigiousness, school business officials, in particular, need to understand the legal context in which they operate.

Two levels are involved. First, there is the matter of power and authority. Most simply stated, what does the school district, and the school business administrator who acts on its behalf, have the power to do? With the variety of participants acting in the public schools, i.e., teachers, administrators, parents, students, and taxpayers, there is a certain degree of confusion concerning authority. A basic understanding of the governmental organization of the school district and an acquaintance with the variety of legal sources which can control a particular relationship are fundamental in terms of defining one's authority. In certain circumstances, the relationship between parties will be proscribed by statutory authority granted to the school district by the state legislature; this is authority limited by the state and federal constitutions. Additionally, considerable latitude exists for the school district to bind itself and other persons and entities to contracts. The district also has responsibilities to protect the health, safety, and welfare of its students. The scope of the relationship is determined by a legal source: statutory, regulatory, constitutional, contractual, or tortuous. These relationships are significant because they have implications for potential liability and because they define the scope of authority entrusted to the school district.

A second level important in understanding the legal context of the financial functions of the school district is an appreciation of the legal process. At a very simplistic level, a difference between statutory law and judge-made case law exists. A thorough understanding of the controlling statutes by the school business administrator is vitally necessary for the successful function of the fiscal operations. There is a skill involved in reading a case, relating the facts of the dispute to the conclusion or holding of the court, and then relating this case to other similar cases so as to articulate a general standard of acceptable behavior or action. Knowledge of the legal process involves learning the importance of several types of authorities.

Sources of legal authority range from written federal and state constitutions, statutes, administrative rules and regulations, and judge-made law. State constitutions have the highest priority followed by statutes. Judicial decisions involve interpretations of constitutions and statutes and the development of legal principles. Part of the Anglo-American jurisprudential heritage has been this deference to judge-made law, known as common law. This common law has been particularly significant in the development of contracts and torts principles. Although more and more of this common law area is being controlled by statute, it is still an area in which case law can make for enforceable standards of expected behavior. Finally, administrative rules and regulations explicate the obligations and standards of administrative agencies. These rules and regulations have the full force of law so long as they are authorized by legislation and are not in conflict with governing constitutions, statutes, or court decisions.

The priority of time refers to reading the most recent decision. Although the principle of *stare decisis,* or precedent, is fundamental to the operation of American jurisprudence, there are times when the highest courts at the federal and state level reverse earlier decisions. These are times when, in trying to understand what a series of cases mean, attention to the date of a particular decision is critical.

As one becomes more knowledgeable concerning the legal process one becomes increasingly attentive to the facts of a particular matter and to the jurisdictional constraints that surround judicial decision making. Consider, for example, the not unusual circumstances, where a school business administrator from state A describes for an administrator in state B a particular judicial decision which has the potential for changing the way he or she performs a certain task. Before he or she changes the practice, he or she will want to know the facts of the case, compare these to the way procedures operate in his or her district, know the source of law applied to the facts, and determine whether the same or similar law applies in his or her jurisdiction.

Increased legal knowledge by school business officials is not proposed as a substitute for legal advice from attorneys. Legal expertise is necessary. This chapter is premised upon the belief that knowledge of the legal process makes one a more competent public school administrator. One will have a better sense of knowing when to ask for advice and what type of questions to ask. One can then better practice preventive law by shaping policies and practices to avoid litigation while accomplishing the educational and fiscal objectives of the school district.

The cases cited herein provide examples of the kinds of disputes that are being litigated which involve school business officials, and the cases assist in explaining certain legal principles. The nature of the disputes will change, but the underlying question of authority or power will most likely be at the center of much future litigation. Does the school district have the authority to engage in a given activity in such a manner? Does a given procedure violate a constitutional limitation or a federal or state statutory provision? Attention to the existence of appropriate authority is necessary when a school district tries to respond to different circumstances and different needs. And there are times when the district is simply challenged with litigation by a disgruntled parent or citizen. Thus, whether one is a creative administrator attempting to develop new responses for educating children or an administrator faced with taking an action that a certain constituency disapproves of, the administrator will want to be attentive to matters of power and authority.

The Local School District as a Legal Entity

In this federalist system, the state and national governments have financial resources which can be made available to public schools. Education finance authorities have long viewed that public schools are primarily, but not exclusively, the responsibility of the state.[2] The Federal Constitution does not specifically mention education as one of the powers of the federal government, and therefore the language of the Tenth Amendment which specifies that "powers not delegated to the United States by the Constitution . . . are reserved to the States . . ." gives the states primary responsibility for public education. The federal government has provided some resources for public schools under authority from the general welfare clause of Article I, but these moneys, generally made available to further specific social policies, require public school districts to meet certain guidelines to qualify.[3] Although the federal government has become financially involved in education, there is no question about the primary responsibility for education falling on the states and school districts.

Authority to Act

The state legislature has plenary power concerning education and the establishment of educational policy, and this power is limited only by the federal constitution and the state's own constitution. This plenary power allows the state legislature to delegate authority, normally to state administrative agencies or local boards of education, in the area of education and to alter or revise its delegation. Except for Hawaii, which operates only one state-wide school district, the states have delegated major authority to local school districts for the provision of public education. The operation of public school districts is generally enough of a routine that there is seldom a significant question of authority or power. When a school district faces a new type of problem and wants to develop a policy concerning the problem which goes beyond anything that has been done before, there can be a significant question concerning whether authority exists for such actions.[4] Or, when a school district takes a certain action which proves to be of concern to a certain constituency of the school district, this constituency will consider legal challenges to the proposed school district action. One of the most common challenges will be to the school district's authority to undertake such actions.

Consider the following example of this second type of question concerning authority. Local boards of education deal with a variety of educational decisions every year. Each decision has the potential to result in a suit challenging the board of education's authority under a variety of different legal theories. Consider, for example, a legal challenge made by a disgruntled group of parents upset by a particular decision similar to one that many boards of education are being forced to make: the closing of a school. In this example, due to declining enrollments, the board of education in a school district of eight high schools foresaw the probability of having to close one high school. Utilizing recommended practices for such a school-closing decision, the board appointed a citizens' advisory committee to develop criteria to guide the decision. The school board adopted the criteria, which included retention of the schools that accommodated the most complete selection of programs for students, had low operating and maintenance costs, minimized student busing, accommodated the physically handicapped, and were located in areas with the largest potential for growth in student enrollment. The board of education directed the administration, in conjunction with a consultant, to apply the criteria. The board of education received a report ranking the schools for closing according to the stated criteria. The board continued to reassess the rankings and sought more information until a mid-May meeting when it finally approved the closing of a high school which had appeared third on the original list of schools reported to the school board.

Suit was brought against the school district seeking an injunction thus halting an implementation of the school closing order on the grounds that the decision was arbitrary and capricious because the board of education failed to comply with the criteria it had adopted in making the final closing decision. The trial court found that such a decision was arbitrary, however the appellate court reversed on appeal. The appellate court held that the state legislature delegated authority to close schools to the local school board. Since the board of education had no authority to delegate its responsibilities and duties in administering the school closing to an advisory committee, the court held, *inter alia*, that, "the board acted within its discretion in the (school closing)" and concluded that to require a strict application of the criteria prepared by the committees would impermissibly subordinate or limit the discretionary powers vested in the board by the legislature.[5]

The question of appropriate power or authority can be critical in determining what actions a particular board of education or state agency can make. The following sections consider several important themes that are interwoven through this area of the law and provide several examples of litigation where these themes have been utilized by the courts in deciding a case.

Legislative Delegation of Authority

The concept applying to the state legislature's authority to delegate is relatively straight forward. However, court determinations of such delegation are difficult to project in that each is based on specific factual circumstances. The general concept is that the legislature may delegate by statute administrative powers to administrative agencies in the executive branch of government, but may not delegate lawmaking or legislative powers. The rationale for this distinction between legislative and administrative delegation is drawn from the constitutional distinction between legislative and executive powers. The transfer of legislative authority to administrative agencies would dilute legislative autonomy and threaten the basic principle of separation of powers central to the constitutional system of government. Additionally, this would tend to put the decision maker out of reach of the electorate, the important political check that exists in our democratic system. As matters become more complicated it is impossible for the legislature to administer vast and varied administrative agencies which are ultimately under its authority. Therefore, courts, in distinguishing delegable administrative power from nondelegable lawmaking power, are often persuaded by the enabling statutes for reasonably precise guidelines which might describe the policy objective(s) of the legislature or guide an administrator's discretion in implementing the legislative goal. The general legal theory is that the delegation is valid so long as the statutory guidelines limit the discretion of the administrators and do not allow for administrative judgment to be substituted for that of the elected legislators.

As discussed, the application of this rule is much more difficult to predict. Particular cases include passages that describe some of the elements that go into the balance. For example, consider the following:

> The line separating that which is purely regulation, and that which is purely legislation, is necessarily indistinct, and becomes more so as the line separating such authority is approached. Therefore, courts . . . will resolve the doubt in favor of the validity of the [delegating] act rather than holding it invalid . . .which is especially true when the [administrative] act is essential and necessary for carrying out the broad purpose and intent of the Legislature.[6]

As this suggests, it is important to identify the subject matter of the legislation, the specificity of guidelines surrounding the legislation and the extent to which other interests (either individual interests which may be protected by the constitution or group interests that may be dealt with in other legislation) are present. Finally, there are significant differences from state to state concerning how state courts will treat this distinction.

One vivid example of this difference between states is in the area of collective bargaining. Confronted with the question of whether local school boards, absent express statutory authority, could recognize a labor organization as an exclusive representative of a group of public employees and negotiate and enter into binding contracts with the organization regarding terms and conditions of employee employment, the courts of Illinois and Virginia, for example, reached opposite conclusions. An Illinois appellate court held that local school boards had such authority.[7] Yet, the Virginia Supreme Court applied a strict construction that allows local public agencies to exercise only those powers conferred expressly or by necessary implication by the state legislature to conclude that collective bargaining was not authorized.[8] Presumably, the labor history and political philosophy of a state is not lost on the judiciary's decision concerning what test to apply.

Even when powers are appropriately delegated, it can be difficult to know how these powers will extend. Consequently, courts are often called upon to determine whether the administrative body has acted beyond its legally delegated authority. Beyond those issues that are expressly stated, many statutes also provide broad discretionary power (which may be necessary for the proper and efficient management of public education) as the courts will recognize the implied powers to act in certain ways (which may be necessary to accomplish the desired legislative policy during changing times and conditions). Yet, it is possible that the authority goes beyond the intent of the statute, when it moves past being an implied power to being a nondelegated activity which is illegal.

Because these general principles of legislative authority and proper delegation of power are widely appealed to, it is instructive to consider some selected cases involving these issues. The Michigan Supreme Court held that early retirement benefits, although not salaries, were working conditions within the meaning of the school code.[9] Consequently, the board of education had authority to provide for such benefits in its collective bargaining agreement with teachers, an agreement that did not violate the state constitution.[10]

With the financial health and subsequent vitality of local school districts so dependent upon the state aid distribution formula, it is not surprising that local districts have attempted various ways of influencing this formula.[11] Due to the significance of this one area, chapter 6 discusses these issues in greater detail.

A North Carolina case involved a challenge concerning the appropriate authority of a school district to initiate an extended day-care program to more adequately satisfy the different needs of many of its latchkey students. The case provides a good example of what the court found important in deciding whether such action was authorized under appropriate state statute and philosophy.

The school district initiated an extended day program which, for a fee, provided activities after school for children that might not have parents at home upon completion of the normal school day at 2:30 p.m. The only costs to the school district were fuel and lighting costs associated with use of the building. A coalition of day-care center operators challenged the authority of the school district to initiate such a program. The North Carolina Court of Appeals held that the school district had the authority under state statute to operate such a program for latchkey children.[12] Additionally, the court upheld the authority of the school district to expend moneys for heating and lighting the school building for the benefit of these extended-day pupils as a public benefit. The program provided academic improvement for many students. Because the school board had the authority to absorb the fuel and electricity costs of the program (it was for a public purpose, to improve the educational achievements of latchkey children) the court rejected the contention that the expenditure must be approved by the voters in a referendum.

As public school districts are financially strapped, they seek alternative avenues for funding. One apparently attractive source is the generation of new, different taxes. These newly imposed taxes seem to be challenged often, and as a generalization, they are not popular in the courts.

A Pennsylvania school district imposed a tax of 1 percent of the construction cost for the "erection, alteration, repair, renovation, extension or replacement of any building or improvement to real property." The state statute which authorized such taxing authority granted the court the authority to invalidate such a tax if it were excessive or unreasonable. The Pennsylvania Commonwealth Court held this tax was unreasonable because it singled out new homeowners and remodelers to bear the cost of additional school support.[13]

The tax was held to be an unreasonable alternative to raising additional moneys for supporting the public schools through an increase in the real estate tax millage.

A board of education in Florida imposed a discretionary 2 mill tax levy allowed by law. A Florida district court of appeals upheld a challenge to this levy, overturning it on the basis that the levy was illegal and therefore void, because the school district had not satisfied the statutory requirements in publicizing the intended levy and allowing an opportunity for community response.[14] For example, one of the statutory requirements involved specific notice about intended use of the money. "Such notice shall specify the projects or number of school buses anticipated to be funded by such additional taxes . . . ."[15] The district court of appeals rejected the school district's claim that there was substantial compliance with the statute. When the taxing power is exercised by a tax authority that does not have inherent power to tax, generally, the courts strictly interpret the statutes granting the tax power. Such failure to comply with a statute authorizing a levy is generally considered not just an irregularity, but an omission which invalidates the tax.

A Texas court of appeals held that a school district has no interest or implied authority to levy taxes for maintenance of schools.[16] Authority to levy taxes must arise from some affirmative grant of power from the constitution or the legislators. In Texas, a public school district that had validly split from another school district had no power to assess or collect ad valorem taxes, since qualified voters of the district had never voted to approve the tax.

A school district passed a resolution to recommend a tax exemption for a proposed hotel resort, in return for the hotel owner's contractual offer to pay all of the assessed taxes in lieu of actual taxes if gambling were legalized in the state and if the owner were to institute gambling in the hotel. This resolution was declared null and void because, as a matter of law in New York, the legislative body could not bargain away or limit its future powers by a contract which is based on speculative future events.[17]

Appropriate Administrative Delegation of Authority

The law is clear that an administrative agency may not redelegate those discretionary powers conferred upon it to another body or a subordinate employee. In the public school context, this standard is most significant in limiting the actions to which an employee of the public school district can legally bind the school district when the board of education has not officially taken action. Some cases distinguish between the ministerial actions of employees, those that are purely mechanical and therefore legitimate, from those that are discretionary and are therefore not proper. Once again, it is instructive to consider several cases where the delegation principles have been utilized to resolve a dispute, usually concerning whether the public school district is bound to an agreement that an administrator made with a third party on behalf of the school district.

A North Carolina appellate court articulated who had the authority to make contracts. "Under the system of public education in this state, local school boards alone have the duty or authority to enter into or authorize purchases or supplies and equipment for the respective local school systems."[18] A supplier of merchandise for sale by students could not sue the school district for the value of unsold/unreturned merchandise when the school board had not entered into the contract. The action of the high school principal could not obligate the school board. In addition, the law recognized that public officials are deemed to have notice of the limited nature and extent of the authority of principals, or other school officials, to bind the board of education. Consequently, an argument that the building principal had apparent authority to bind the board of education to the contract was rejected.[19]

An Ohio court of appeals reversed a trial court verdict that had awarded almost $20,000 to a supplier of word-processing equipment.[20] The Cleveland School Board had approved purchase of the equipment, it had been ordered by the director of purchasing, and a certificate has been signed by the treasurer stating that unencumbered funds were available for the expenditure. Approximately two months later, the business manager learned of the contract and determined that the equipment would not meet the educational needs of the district. The contract was canceled and the school board passed a resolution rescinding the purchase. The state of Ohio had two applicable statutes. One applied generally to any subdivision of the state and required a certificate of the fiscal officer of the subdivision. A second, applied specifically to school districts, required a certificate signed by the treasurer, president of the board, and superintendent. The court of appeals held that this contract for the word-processing equipment was void because the certificate did not contain the requisite three signatures, even though the Cleveland school district did not follow this procedure with most of their expenditures. The vendor was held responsible for knowing the provisions of the statute, and as in this case where the purchase was not routine and might involve the purchase of further equipment, it was appropriate for the school district to require compliance with the more specific statute. Therefore, the court ruled that the contract was void and unenforceable.

A student in the junior class of a Louisiana high school decided to rent a sound system and have a disc jockey play records for the junior-senior prom, rather than employ the more traditional band. A disc jockey was selected and a contract signed by one of the two class sponsors after discussing the matter with the building principal. When the building principal later discovered that he was hiring a sound system rather than a band, something he apparently did not understand at the time of the signing, he tried to terminate the contract because he did not think it appropriate for the prom. In an action for damages for breach of contract, a Louisiana appeals court held that the school district was liable for the amount of the liquidated damages clause stated in the contract.[21] The court held that the principal had implied authority to represent the school board and that this authority was properly delegated to the faculty sponsor who actually signed the contract. Because the principal was properly exercising his authority as the building principal, the school board was a party to the contract and could be held liable for its breach. In addition, there was no personal liability on the part of the building principal, even though he broke the contract, because he was acting on the implied authority given him by the local school board. There was no Louisiana statute limiting the authority of the school district to be bound by contract only upon certification approval and signature by the school board, a statute that generally exists in other states.

A coalition of parents, students, and residents of the "west side" of Kansas City protested the closing of the local high school. Not receiving the desired response concerning the reopening of the high school with higher quality and with more community control, the coalition occupied the school to try to force an accommodation from the board. The superintendent met with the group and reached an agreement to establish a community-sponsored experimental high school in the building, to start in approximately one year. About two months before the experimental high school was to open, and after considerable preparation had gone into the planning for the school, the superintendent decided that financial constraints would not allow the opening of the experimental school as planned. The coalition of parents sought specific performance of the agreement, which had been adopted, to open the experimental school. The Missouri court of appeals affirmed the trial court in holding that the agreement was an invalid contract.[22] The central issue was whether this agreement was an exercise of the school district's proprietary functions or governmental functions. The general rule, in these types of issues, is that a municipal corporation may, by contract, limit the exercise of its proprietary functions but not its governmental functions. The appellate court opined:

> This is the very function for which school districts exist—not building buildings or operating playgrounds or even hiring teachers—but deciding where and if schools are needed to educate our children. This is not to say that a district may not be able to contract for advice on this question, but to contract away the power ultimately to decide whether certain educational facilities should remain open to serve the district's children is to give up the most basic function of all.[23]

Consequently, the agreement was not enforceable because the public school district was not empowered to contract away this authority to close a public school.

The delegation doctrine can also be important in hiring and firing decisions. Illustrating this is a South Dakota school district that experienced an unexpected resignation in late August. The position was announced, and a candidate visited the school district and interviewed for the position. The candidate was told that he would have a two-week trial period, since the school board was not scheduled to meet before then. The candidate refused the position on a trial basis, so the contract was prepared, and he accepted by signing the contract. Before the contract was signed by any board members, the new teacher had trouble with his class, and the district refused to approve the contract and terminated him after nine days of teaching. The South Dakota Supreme Court held that no contract existed because the contract had not been signed by the appropriate school district representative as required by law. "A teacher shall be employed *only* upon written contract signed by the teacher and by the president of the school board and business manager of the school district."[24] The court did award compensation for moving expenses and housing costs which were incurred because of detrimental reliance on the teaching position. Once again, contracts can only be formed according to the specific standards stated in the applicable state statutes.

An Illinois appellate court held that a decision not to renew a superintendent's contract, and to determine that notice of the school board's intent not to renew the superintendent's contract should be given, are nondelegable and must be done by the school board.[25] Yet the functions of drafting the written notice and the reasons for nonrenewal and of delivering that notice are ministerial, and therefore, delegable. The hiring and firing of instructional personnel and determination not to renew contracts are not delegable because they are discretionary.

Shared Authority with Another Administrative Agency

Occasionally a conflict will arise when one administrative agency seeks to take action which impinges upon the authority of another administrative agency. Normally, courts resolve these disputes by identifying the particular activity and determining which administrative agency has more central responsibility over this problem.

Once again, several cases provide examples of the types of disputes that arise and how the courts, in general, resolve them. One type of conflict is between a municipality operating under home-rule authority and a public school district. The general rule is that education is a matter of statewide concern and does not come under the control of a municipality operating under a home-rule charter. A similar rationale was used, for example, when an Illinois appellate court held that a public school district, in the process of erecting a school building within a municipality, is exempted from the building code of the municipality.[26] On the other hand, many states have specific controlling state statutes, or state board of education rules and regulations, outlining the responsibilities of school districts in these instances.

A California court considered this relationship in greater detail. It stated that local authorities have the power to regulate a municipal affair. Yet state law prevails over local ordinances with regard to local matters if the subject matter were also of statewide concern. In the event of conflict between regulations of state and local governments, if the state legislation discloses an interest to preempt the field to the exclusion of local regulation, the question becomes one of predominance or superiority, which the courts will decide. Again, it must be noted that controlling state statutes must always be followed. Such controlling state statutes, particularly in this arena, may be contrary to this discussion. Again, it must be stressed that school business administrators must be thoroughly familiar with controlling statutory guidance in these areas.

In this particular example, the court determined that state legislation created a scenario for developing new and improved schemes for financing interim school facilities. This is a matter of statewide interest, and the court held it to preempt the authority of the school district to limit or control development in any additional way.

> The express authority to impose school impact fees and dedication upon developers under [the California statute] is broad in its scope, applies to all schools, but narrowly restricts the amount and use of such fees. To permit local entities to impose further broader fees or further burdens on developers for new school construction would be to write into this very comprehensive plan for school finance of new school construction a complicating feature, having statewide ramifications never suggested. Indeed, it could have extremely adverse effect on urban development generally. There is preemption in the legislative scheme we consider here, precluding the county, district, and board from collecting the fees from developers for permanent school facilities.[27]
>
> As further example, a California court of appeals ruled that school districts could impose a fee on new residential and commercial developments for the purposes of needed school construction. The court ruled that state statutes empowered school districts to impose such fees.[28] Thus, once again, it can be seen that enabling statutes permit a wide range of specific activities and powers to school districts among the states.

An Ohio board of education voted to purchase electrical energy from a different company than the one supplying electricity to the rest of the municipality. The municipality challenged the authority of the school district to make such a purchase. The Ohio Supreme Court interpreted the Ohio Constitution to grant exclusive authority for the contracting of public utility services to municipalities.[29] Consequently, although there is no explicit limit upon the school district's power to purchase electrical energy, the court held that the grant of authority to the municipality means that school districts are prohibited from contracting for public utility services, absent the express consent of the municipality. To do otherwise would limit the ability of the municipality to obtain the best utility rates for its residents. This case illustrates the complexity of each state's statutory control over public education as well as what can be, in many instances, the interdependency of public school districts with other governmental agencies.

In many states, an area of recurring conflict exists between public school districts and other governmental bodies responsible for collecting and distributing tax moneys. An issue in several states involved allocation of interest earned on tax moneys. A Missouri case provides one example. The allocation of interest moneys earned between the county collector of taxes and the various governmental bodies is a matter of state law. The Missouri Supreme Court, *en banc,* reaffirmed its position that interest on deposited school funds was payable to the treasurer of the school district, and were not to be credited to the general revenue fund of the county.[30] The court also refused to soften the application of the statute by allowing the county to pay all of its interest expenses out of interest revenue earned on all moneys being held by the county before distribution. No legislative authority existed to charge the interest cost of borrowing tax anticipation warrants back against the districts; therefore, the county collector of revenue was ordered in a mandamas action to reimburse the school district this charge for interest.

Two Arkansas cases involved the allocation of the collector's commission and the allocation of certain fees incurred by the county assessor and collector. In one case involving the appropriate allocation of the collector's commission, the Arkansas Supreme Court held that the commission must be based upon expenses related to the collection of tax revenue (not for expenses related to the accompanying sheriff's office).[31] Any extra moneys would be returned to the local school district. In a second case, the Arkansas Supreme Court held that the office rental charge for the county assessor and collector was one for the county rather than for assessor and collector, and could not be recovered from the school district through the county board of education's school tax collection account.[32]

Again, illustrating the unique specifics of conflict often found within the various states was that of the Massachusetts Supreme Court ruling that the mayor of Boston was responsible for keeping the school committee within its authorized expenditure.[33] If the mayor did not, the city was responsible for the additional moneys necessary to keep the schools open for the statutorily mandated minima. The state was not financially responsible.

Other Loci of Authority or Power

Before World War II, generally school administrators did not need to be knowledgeable concerning legal matters much beyond what was written in the school code. These statutes were considerably simpler than they are today. If the actions of these administrators were challenged, they would likely have been based on the question of proper delegation of authority and the general authority found under the *in loco parentis* doctrine.

The explosion of education law in the last generation is really a story about the expansion of different sources of authority. During this period there has been a marked increase in the applicability of constitutional standards to school matters.[34] Additionally, the amount of federal and state legislation affecting schools has increased dramatically.[35] This legislation provides avenues for new fiscal resources to encourage equity, and thereby also tends to limit the discretion of school boards and administrators in the operation of certain aspects of the public schools. State departments of education generally reflect more aggressive actions in shaping local and state school policies. A major addition to this area has been the expansion of public sector collective bargaining which has caused the contract to be an important legal source in matters of teacher and administrator employment. This topic is discussed in detail in chapter 9 of the text.

To the uninitiated, this cataloging of many different sources of legal authority may appear irrelevant to what a public educational administrator does in the fiscal management of the school district. In reality, the variety and range of these legal sources are quite important. Different sources of authority carry different degrees of weight. For example, because of the supremacy clause of the Federal Constitution,[36] it takes priority if it should come into conflict with state or federal statutes, state constitutions, or administrative rules and regulations. These various legal authorities may also be significant because the range of protection can differ widely.

Written Constitutions

Constitutions are the highest form of law, and take priority over conflicting legal sources such as statutes or administrative rules or regulations. A South Carolina Supreme Court decision is instructive on this relationship between the constitution and statutes.[37] The case dealt with the unconstitutionality of appointed school boards determining the amount of tax to be levied for school operations as not satisfying the constitutional standard of no taxation without representation. The supreme court stated several principles of law regarding constitutional challenges to statutory enactments that apply to federal as well as state constitutions.

> Our Constitution is not a grant of power, but a limitation on what, absent limitations therein, would be a plenary power in the people or their elected representatives. Accordingly it is not sufficient to find that an act is offensive to what may be prevailing notions of the proper sphere for state governments. It is necessary, in order for us to strike down an act of the General Assembly, to find that it offends specific provisions of the state constitution which have limited and circumscribed legislative action in that area. [A]ll reasonable doubt must be resolved in favor of the constitutionality of the act. If a constitutional construction of a statute is possible, that construction should be followed in lieu of an unconstitutional construction.[38]

Constitutions are usually sparsely written, with few words used to articulate broad principles, thus allowing for considerable latitude for judicial interpretation. In this federalist system, both the federal Constitution and state constitutions provide important sources of authority in the operations of public education.

Federal Constitution

The Fourteenth Amendment of the federal Constitution has had a major impact upon public education because of the significance of both the due process and equal protection clauses, as well as the amount of litigation, spawned through the interpretation of these Constitutional concepts. In addition, the Fourteenth Amendment has been the concept utilized to apply the protections of the Bill of Rights, which was approved to limit the authority of the federal government over individuals, and of the state and the administrative agencies of the states, such as public school districts. The operative language of the Fourteenth Amendment limits the actions of states: "No state shall deprive any person of life, liberty or property without due process of law." And the same limitation upon denial of equal protection under the law is made upon the state. The significance of this limitation is considerable. Constitutional rights under the first ten amendments and the Fourteenth Amendment, in particular, are rights against the intrusion of governmental power, not against individuals; that is, it is necessary for an aggrieved party to allege that the constitutional violation occurred under state action. If a private individual committed the wrong, no matter how grievous, there is no cause of action under the Fourteenth Amendment.

An elaborate analysis of what constitutes state action is beyond the scope of this chapter. It is sufficient to understand that public schools, which by definition, receive revenue sources from public moneys, do engage in state actions and are therefore responsible for complying with the federal Constitution. Private schools are not involved in state actions and do not have to meet equal protection or due process standards. Private schools may well decide to meet these standards, but such compliance is not a matter of constitutional compunction.

A few examples of disputes that involve constitutional questions may be instructive. All cited disputes involved statutes which were challenged as being unconstitutional.

A Pennsylvania case involved a challenge to an amendment of the state retirement code. The amendment required members of the state retirement system to contribute an additional percentage of their salary to the state retirement fund. A Pennsylvania Commonwealth Court held that this amendment was an unconstitutional impairment of the employee's right to contract.[39] A similar statute in Maryland, when challenged, was upheld as constitutional.[40] Again, this illustrates the complexity of state statutory and constitutional interpretations.

In *Plyler v. Doe,* a majority of the Supreme Court held that a Texas statute that withheld from local school districts any state funds for the education of children who were not ''legally admitted" into the United States and that authorized local school districts to deny enrollment to such school children violated the equal protection clause. [41] The majority of the Supreme Court stated that this type of discrimination, i.e., the denial of an education, is not rational unless it furthers some substantial goal of the state. A special concern was voiced specifically concerning the denial of education rather than other types of governmental benefits. The majority opinion held that the reasons given by the state were not compelling in this instance. At the time of this writing, Proposition 187 in California will offer the courts a major policy issue regarding the potential enforcement or reshaping of *Plyler v. Doe*. Several observations from the original *Plyler* are noted:

- The exclusion of children from school represents a major cost to the children and to the nation. "Whatever savings might be achieved by denying these children an education, they are wholly insubstantial in light of the costs involved to these children, the state, and the nation."[42]

- The denial of public school access to children of undocumented aliens will not have the alleged effect of discouraging illegal aliens from entering the country.

- The savings from a denial of this education will likely improve quality of education for remaining students. "In terms of educational cost and need...undocumented children are 'basically indistinguishable' from legally resident alien children."[43]

Representation on a New Jersey board of education for a regional high school was determined by the population of the underlying boroughs. When representation was recalculated according to the 1980 census figures, the borough was slated to decrease from two to one board member. The constitutionality of the state statute was challenged. The statute in question removed military personnel and civilians residing within military installments from being counted in the apportionment. A New Jersey superior court held this exclusion violated the equal protection clause, and therefore voided this particular reapportionment.[44]

A bond election in Texas, passed by the electorate, was challenged on several grounds: the unconstitutionality of the election statute, the misleading information given concerning the tax rate, and the legal limitation to be placed upon the interest rate. The Texas court of appeals held that the election was constitutional because the district did not enforce the unconstitutional property ownership and rendering requirements of the pertinent statutes when it held the election.[45] The actual rate of interest for the bonds was legal because it fell within the limits of the proposed referendum, even though the rate turned out to be higher than school officials predicted at the time of the election. Finally, the court held that the referendum established that the maximum rate allowed by law for the bonds would be determined at the time of issuance, not at the time of referendum.

State Constitutional Sources

The cases described herein demonstrate how significant federal Constitutional standards can be in limiting state statutes, and how state court judges are authorized, as courts of general jurisdictions, to interpret the federal Constitution just as federal judges. When there is disagreement, the U.S. Supreme Court controls. A parallel does not exist for state constitutions. The state constitution is the supreme law of the state, and federal judges do not have authority to interpret state constitutions. Certainly, the major issue before the state supreme courts is the determination of the constitutionality of the state aid distribution formula.[46] Given the importance and significance of this issue, the topic is discussed separately in chapter 6 within this text.

Another area concerning state constitutions is the meaning of providing a "free public education." The operational interpretation given to this or similar language, which appears in many state constitutions, varies widely from state to state. Some states interpret these clauses to mean that textbooks are included and must be provided free of charge,[47] while other states take the opposite view.[48] Litigation has expanded beyond textbooks. Illinois courts allow charging a rental fee for optional towel use in a high school physical education class[49] and for assessing a lunchroom supervision fee to parents who live close to school yet whose children bring lunches to eat at school.[50] The California Supreme Court, in contrast, held that school districts were constitutionally prohibited under the free-school guarantee from charging fees for students' participation in dramatic productions, musical performances, and athletic competition.[51] This broad state constitutional language can have a significant impact upon the financial management of a district; this situation varies widely throughout the United States.

State Boards of Education

State boards of education are important sources of authority. These boards differ on where they draw legitimacy in that about half are created by statute and half by state constitution.[52] Even when they are created by state constitutional authority, the state board of education is subject to the commands of the legislature. Consequently, unless charged with constitutional responsibility, state boards of education derive authority from the state legislatures and have authority to do only what is expressly or implicitly delegated to the board. Thus, it is argued that public school districts have only the authority that is expressly or implicitly delegated by the state board of education and the legislature.[53] On the other hand, a more expansive interpretation can be argued in that such state boards of education and state superintendents must have broad powers over public elementary and secondary education and that these powers must be sufficiently expansive to correspond with the high responsibilities of such offices.[54]

As state boards of education carry out the perceived responsibilities, disagreements can arise concerning whether the state board of education does enjoy such power.[55] Often, such disagreements occur between a state board of education and a local school district.[56] Generally, public school districts are subject to the powers of the state board of education. Both, the state board of education as well as all local school districts, within constitutional guidelines, are subject to the legislature.[57] Despite these broad powers, all such actions of the state board of education must comply with applicable state statutes.[58] Two examples provide an insight into how courts analyze such disagreements.

As illustration of this point, the Illinois Supreme Court was called upon to decide whether the Illinois State Board of Education had the specific authority to promulgate and enforce rules designed to prevent racial segregation.[59] The Illinois legislature passed a statute which provided, in part: "As soon as practicable, and from time to time thereafter, the [local school] board shall change or revise existing [attendance centers] or create new [attendance centers] in a manner which will take into consideration the prevention of segregation and the elimination of separation of children in public schools because of color, race, or nationality."[60] Pursuant to this statute, the state board developed rules to eliminate and prevent racial segregation in schools; these rules were officially adopted by the board in 1976. The most difficult requirement of these rules was that each attendance center in a district not vary by more than 15 percent from the minority racial composition of the pupils in all attendance centers. This requirement, in effect, defined nonconformance as a quota of 15 percent, plus or minus, from the school district-wide composition of minorities. The state board also identified compliance standards, and noncompliance could lead to nonrecognition and possible loss of state funding. Two school districts could not meet the 15 percent quota in individual attendance centers without instituting busing of students outside the neighborhood schools; the local school board refused to do this, and in fear of losing state aid for noncompliance, the districts challenged the authority of the state board to promulgate such rules.

The state board of education argued that it enjoyed statutory authority to promulgate such rules because of general language in the Illinois School Code. The Illinois Supreme Court rejected this claim.

> Nowhere in [the statute] is the board granted expressed authority to determine standards for racial desegregation. And the fact that the board may set standards for the 'operation, maintenance, administration, and supervision' of schools does not imply the authority now sought. If we were to hold otherwise, it would be difficult to conceive of any regulation which could not be justified under [this statute].[61]

The court also noted that a different statutory section provided a procedure that the board could use in combating segregation. The statute did not provide any standard or guidelines governing the state board's discretion to enforce the statute. Examining another section of the School Code, the Supreme Court was persuaded that the local school district had the authority to decide how to comply with the statute. The rules of the state board of education in this area were unenforceable.

A second case involved an attempt by the California State Department of Education to reclaim funds from a high school district. The court ruled that this action was contrary to the statutory authorization.[62] The high school had been legitimately paid for the vocational students. While the pertinent statutory section prohibited future filing or amended claim for vocational "average daily attendance" funds, it did not authorize the Department of Education to retroactively recapture funds paid.

A number of other levels of educational agencies exist in various states. The powers of these agencies are, like those of the state or local boards of education, only those which are delegated to them. A New Jersey case provides an example of this principle in application.

A county educational services commission was created in New Jersey " . . . for the purpose of carrying on programs of educational research and development and providing to public school districts such educational and administrative services as may be authorized pursuant to rules of the state board of education." This commission had contracted with a private corporation to furnish the services and programs that the commission was obligated to supply to member school districts. The superior court of New Jersey affirmed the trial court in holding that the contract was void as the court opined that the county educational services commission had no authority to enter into this contact.[63] The New Jersey legislature gave statutory authority to boards of education to contract with private corporations, but no such authority was granted to the commission. The court applied the general rule that "municipal bodies in the state have no powers other than those delegated by the legislature, and must perform their prescribed activities within the statutory ambit."[64] Thus, a school district could make contracts only within its legislatively expressed or implied authority; therefore, the contract was void.

Federal Statutes

Although public elementary and secondary education is predominantly supported by state and local taxing sources, the federal government does channel moneys to public elementary and secondary education. Since the passage of the Elementary and Secondary Education Act in the mid 1960s, the federal government has supported a number of programs; many are aimed at supporting the education of needy or handicapped students.

Congress derives authority for passing legislation affecting schools under the Fourteenth Amendment. Consequently, Congress could utilize this authority to require school districts to utilize the due process or equal protection guarantees. If Congress were to use this source of authority, such authority would need to be clearly understood as the basis of the legislation.[65] The more common source of authority for federal financial support is Article I, Section 8 of the U.S. Constitution: "The Congress shall have power to lay and collect Taxes, Duties, Imposts and Excises, to pay the Debts and provide for the common Defense and general Welfare of the United States . . .." Congress decides what expenditures further the general welfare, and the Supreme Court will not interfere with this discretion unless it is ". . . clearly wrong, a display of arbitrary power, not an exercise of judgment."[66] It is currently accepted that improving public elementary and secondary education is providing for the general welfare. Thus, under the current congressional political thinking, the broad utilization of block grants to the states to enhance and further public elementary and secondary education is consistent with the United States Constitution.

One contemporary issue which focuses on the federal government involves its attempts to be repaid some of its money, allegedly misspent by the recipient state agency. This issue can best be considered by closely examining two illustrative cases.

The Supreme Court articulated the statutory framework for the federal government recovering moneys received under Title I and improperly spent.[67] Title I of the Elementary and Secondary Education Act of 1965 created a program designed to improve educational opportunities available to disadvantaged children. The moneys are channeled from the federal government, to the state, to the local school districts. States give assurance to the Department of Education that local school districts will spend the funds only on qualifying programs. If federal auditors determine that a state has misapplied funds, a deficiency can be assessed against the state for repayment. Consequently, the determination of the existence and amount of liability is the responsibility of the Department of Education. The states can seek a review of the auditor's recommendation by the education appeal board. If dissatisfied with the decision by this board, the state can appeal the matter to the appropriate United States court of appeals; the court may decide whether the findings of the Secretary of Education are supported by substantial evidence and reflect application of the proper legal standards. Regardless of the usage of block grants, the receiving agencies must conform with the rules and regulations of the appropriation.

In a case typifying many controversies over the years, the U.S. Secretary of Education directed that the state of New York repay more than $20 million granted previously under Title I of the Education Consolidated and Improvement Act. The state utilized these moneys to supplant rather than to supplement nonfederal funding for selected special education programs. The court of appeals declined to review agreeing that the Secretary had reasonably determined when such formulas were required according to federal rules and regulations.[68]

Much of federal moneys that go to public schools has been tied to furtherance of civil rights. Title VI[69] prohibits racial discrimination in schools receiving federal money; Title IX[70] prohibits sexual discrimination and harassment,[71] ADA[72] prevents discrimination against the those with disabilities in public schools receiving federal moneys. The legislation links a penalty of loss of federal moneys to violation of the protected civil rights. Litigation involving these statutes has involved two important points. First, a question can be raised about who has authority to sue under a particular statute. The U.S. Supreme Court recognized an individual right of action to sue under Title IX, allowing individuals the opportunity to sue on their own behalf and not be reliant solely on the will of the administrative agency to press a violation claim.[73]

A second question involves the linkage between the civil rights violation and the federal moneys involved. In a 1984 case, the U.S. Supreme Court held that violation under Title IX must be program specific to be actionable for the withholding of federal moneys.[74] A violation is not sufficient to reach any moneys which the government body may receive from the federal government; for federal moneys to be subject to penalty, the violation must occur in the same area in which the moneys are being received.

Contracts

A contract is a significant vehicle that a school district can utilize to obligate itself to another party in exchange for some benefit. These contracts have the authority of law so long as the subject of the contract falls within the power of the school district to undertake, and the contract has been properly approved. These issues were raised in the first part of this chapter, and are taken up in subsequent chapters of this text that focus on particular activities of the school business official.

An Ohio case, once again, illustrates the significance of obtaining official school board approval in order for a contract to be binding. An Ohio board of education formally adopted a tuition policy for nonresident students and formally accepted two sisters pursuant to the policy. The girls continued as tuition students for that year and the next three academic years, even though the school board did not formally approve this arrangement for any of these three years. The school board revised its tuition policy that eliminated tuition students for the next year; the sisters obtained an injunction from the trial court on the basis that the superintendent had told them that, when the district entered into the tuition student relationship, it would allow the students to continue until they graduated. The superintendent of schools had come to this conclusion after individually talking with several board members. The appellate court reversed the judgment and dissolved the injunction on the basis that no enforceable contract existed beyond the first year. Pertinent state statute stated "[n]o contract shall be binding upon any board [of education] unless it is made or authorized at a regular or special meeting of such board." The representations of the superintendent, even though based on conversations with individual school board members, did not, as a matter of law, meet the statutory requirements. Thus, no binding contract existed.[75]

As discussed in later chapters, contracts between school districts and employees are very important; and with more states authorizing collective bargaining, these contracts become especially important. Increasingly, the employment relationship will be spelled out through the master collective bargaining contract rather than through board policy. Thus, the nature of collective bargaining is discussed separately in chapter 9 of this text.

An interesting issue involved in teacher contracts is the interplay between the contract and board policy. To what extent can school board policy supplement contractual provisions? Two cases provide examples of this type of dispute. An Arizona district had a collectively bargained contract with teachers that contained leave provisions and required submission of a "cause of absence" form by the teachers in order to obtain paid sick leave. The contract negotiations were not going well, and the teachers engaged in a "sick-in." Many absences were due to alleged illness. The superintendent initiated a policy, without prior formal board approval, which required teachers to provide a physician's certificate stating that the teacher was ill; if no such statement were made, the teacher's pay would be docked. The board of education approved this policy a week later. An Arizona appellate court held that the superintendent's administrative order had no force or effect without prior board approval.[76] The retroactive validation by the school board was not effective. The court suggested that the contract controls on this matter, and that the board of education action has no authority either for retroactive or future application.[77]

In a second case, the Illinois Supreme Court had to decide, *inter alia,* the arbitrability of a claim that a board of education had violated a term of the teachers' employment contract.[78] The board of education had contractually agreed to a sick-leave provision which did not require a physician's certificate after an absence of three days for personal illness. The board of education subsequently passed a policy which required a physician's statement for any absence of three days for personal illness pursuant to state law. "The school board may require a physician's certificate . . . as a basis for pay during leave after an absence of three days for personal illness." The court held that the state statute did not give the school board nondelegable authority to decide on matters of sick leave and thereby make the contractual provision unenforceable. Rather, the statute gave the school district an option, but the question of whether this policy violated the contractual section on sick leave was properly arbitrable under the grievance procedure.[79]

This interface between school board policy and contractual language depends on the specifics of the applicable state collective bargaining statutes, the scope of the management rights clause, mandatory bargaining requirements, and the content of the contractual language at issue. On matters involving employees, it is imperative to consider pertinent contractual provisions. Again, chapter 9 of this text discusses these issues at length.

Liability

It is important for public school administrators to be conscious of various types of authority, or limitations upon authority; not only because they may help explain the scope of the authority or limitation, but also because they have a large impact upon the potential liability of the school district. Potential liability varies considerably across several different causes of action. It is instructive to suggest a number of dimensions of which the public school administrator should be aware.

First, the availability of certain defenses depend upon the legal cause of action under highly specific state statutes. For example, sovereign immunity may act as a bar against suing the state or state employees in matters of tort. Even though the existence of absolute sovereign immunity is quite rare and differs widely among states, the doctrine is generally limited to tort liability.[80] In certain states, liability claims are limited to a set amount of damages. Further, damages can only be claimed against a school board, as the individual who is acting within the scope of his or her employment cannot be held personally liable. In fact, many issues regarding sovereign immunity questions tend to be highly factually specific as to the applicability under state statute.[81] It must be stressed that such claims are highly state statutory specific and generalizations beyond applicable state statutes should be very limited.

Generally, boards of education are agents of the state and are responsible for education. A local board of public education is bound by, and may sue or be sued on contract, in the same manner as municipal corporations. Employment contracts remain the function of local communities, and damages from breach of these contracts would be paid by the local taxpayers, not the state.

Second, based on highly specific state statutes, the measure of damages varies from one cause of action to another. Punitive damages are available in torts, not available in contracts, and available in those constitutional torts actions against individuals;[82] but they are not generally available in those constitutional torts actions against municipalities, including school districts.[83]

Third, there is important variation regarding personal liability and district liability. Many states have statutes which assume the cost of financial damages and costs of defending the suit if a district employee should be sued.[84] Generally, these protections are for actions while employees are acting within the scope of their employment. And, in a constitutional tort, a different standard of liability attaches to an individual than to a municipality. Individuals, in certain limited instances, enjoy a good faith, qualified immunity[85] while municipalities do not.[86] This area is highly state statutory specific and public school administrators and legal counsels should be very aware of the controlling state statutes.

Summary

This chapter has explored the variety of legal sources that interface in the public education context. The origin of these various sources can be important to the public school administrator who wants to understand the scope and boundaries of authority and potential personal and professional liability. This conceptual framework is important to the school business official in trying to shape school district policy and practice to be responsive to the educational needs of the future. As discussed herein, there are many issues that the successful school business administrator must be highly knowledgeable about, as well as being aware of all the ramifications. Many of the issues are highly state statutory specific. The successful public school administrator must be knowledgeable of these controlling statutes as well as the applicable local school board policy and regulations. In this manner, the school business manager may best be prepared to interact with school district legal counsel so as to best protect the interests of the taxpayers and the public school district.

Endnotes

[1]Portions of this chapter are based on the previous edition written by P.W. Thurston.

[2]See, R.C. Wood and D.C. Thompson, *Education Finance Law: Constitutional Challenges to State Aid Plans, An Analysis of Strategy,* (Topeka, KS: National Organization on Legal Problems of Education), 1993; and D.C. Thompson, R.C. Wood, and D.S. Honeyman, *Fiscal Leadership for Schools: Concepts and Practices:* (White Plains, NY: Longman Publishing Co.), 1994.

[3]Historically, federal lunch money, impact aid, and support for educating handicapped students are just a few examples of this federal involvement in education.

[4]See, e.g., State v. Whittle Communications, 402 S.E.2d 556 (N.C. 1991).

[5]Tyska v. Board of Educ., 117 Ill. App. 3d 917, 923, 453 N.E.2d 1344, 1351 (Ill. App. 1st Dist. 1983).

[6]Dicken v. Kentucky State Board of Educ., 199 S.W.2d 977, 981 (Ky. 1947).

[7]Chicago Div., Ill. Educ. Ass 'n v. Board of Educ., 76 Ill. App. 2d 456, 222 N.E.2d 243 (Ill. App. 1 Dist. 1966).

[8]Virginia v. County Bd. of Arlington County, 217 Va. 558, 232 S.E.2d 30 (Va. 1977).

[9]Jurva v. Attorney Gen. of Michigan, 351 N.W.2d 813 (Mich. 1984).

[10]For a similar case see, Robinson v. Joint School Dist., 670 P.2d 894 (Idaho. 1983).

[11]R.C. Wood and D.C. Thompson, *Education Finance Law.*

[12]Kiddie Korner Day Schools v. Charlotte-Mecklenburg Bd. of Educ., 285 S.E.2d 110 (N.C. App. 1981).

[13]Heisey v. Elizabethtown Area School Dist., 445 A.2d 1344 (Pa. Cmwlth. 1982): on a related manner see, San Van v. School Dist. of Derry Township, 635 A.2d 254

(Pa. Commw. Ct. 1993, see also, Tredyffrin-Easttwosn School Dist. v. Valley Forge Music Fair, 627 A.2d 814 (Pa. Commw. Ct. 1993).

[14]Wilson v. School Bd. of Marion City., 424 So. 2d 16 (Fla. App. 5 Dist. 1982).

[15]Ibid., 19.

[16]Manges v. Freer lnd. School Dist., 653 S.W.2d 553 (Tex. App. 4 Dist. 1983).

[17]Citizens to Save Minnewaka v. New Paltz Central School Dist. 468 N.Y.W.2d 920 (N.Y., A.D. 3 Dept. 1983).

[18]Community Projects For Students v. Wilder, 298 S.E.2d 434, 435 (N.C. App. 1982).

[19]Ibid.

[20]Cado Business Systems of Ohio v. Board of Educ., 457 N.E.2d 939 (Ohio App. 1983).

[21]Hebert v. Livingston Parish School Bd., 438 So. 2d 1141 (La. App. 1 Cir. 1983).

[22]Coalition to Preserve Educ. v. School Dist. of Kansas City, 649 S.W.2d 533 (Mo. App. 1983).

[23]Ibid., 537.

[24]Minor v. Sally Buttes School Dist. 58-2, 345 N.W.2d 48 (S.D. 1984), See also, DeMarr v. Kansas City School Dist, 802 S.W.2d 537 (Mo. Ct. App. 1991).

[25]Daleanes v. Board of Educ., 457 N.E.2d 1382 (Ill. App. 2 Dist. 1983).

[26]Board of Educ. v. West Chicago, 55 Ill. App. 2d 401, 205 N.E.2d 63 (Ill. App. 2 Dist. 1965).

[27]Candid Enterprises v. Grossmont Union H.S. Dist., 197 Cal. Rptr. 429 (Cal. App. 4 Dist. 1983).

[28]North State Dev. Co. v. Pittsburg School Dist., 270 Cal. Rptr. (Ct. App. 1990).

[29]Village of Lucas v. Lucas Local School Dist., 442 N.E.2d 449 (Ohio 1982).

[30]State ex. rel. School Dist. of Springfield v. Wickliffe, 650 S.W.2d 623 (Mo. 1983).

[31]Special School Dist. v. Sebastian County, 641 S.W.2d 702 (Ark. 1982).

[32]Venhaus v. Board of Educ., 659 S.W.2d 179 (Ark. 1983).

[33]Board of Educ. v. City of Boston, 434 N.E.2d 1224 (Mass. 1982).

[34]Since 1950, the U.S. Supreme Court has decided an increasingly large number of cases which affect schools. In the three decades between 1920 and 1949, the Supreme Court decided 15 cases affecting schools while in the decade between 1950-1959, 13 cases affecting schools were decided. During the 1960-1969 decade this rose to 22 cases, and then jumped to 73 for the decade of the 1970s, 84 cases were decided affecting education for the first half decade of the 1990.

[35]The Supreme Court has decided a sizable number of cases affecting schools which are based on statutory claims rather than constitutional claims. These cases were evenly split between those founded on constitutional claims and those based on statutory claims.

[36]Art. VI, U.S. Const. "This Constitution, and the Laws of the United States which shall be made in pursuance thereof and of all Treaties made, or which shall be made, under the authority of the United States, shall be the Supreme Law of the land; and the Judges in every State shall be bound thereby, anything in the Constitution or Laws of any State to the contrary notwithstanding."

[37]Crow v. McAlpine, 285 S.E.2d 355 (S. Car. 1981).

[38]Ibid., 356, quoting from Bauer v. South Carolina State Housing Auth., 271 S.C. 219, 226 S.E.2d 869 (S.Car. 1978).

[39]AFSCME v. Commonwealth of Penn., 472 A. 2d 746 (PA. Cmwlth. 1984). Also see Pennsylvania Fed. of Teachers v. School Dist. of Philadelphia, 472 A.2d 479 (Pa. Cmwlth. 1981).

[40]A. Tugend. "Federal Judge Clears Way For Changes In Maryland Pension System," *Educ. Week,* Oct. 10, 1984, at 5.

[41]102 S.Ct. 2382 (1982).

[42]Ibid. at 2402.

[43]Ibid. at 2401 .

[44]Borough of Oceanport v. Hughes, 451 A.2d 966 (N.J. Super. Ct. 1982).

[45]Ex Parte Progreso Indep. School Dist., 650 S.W.2d 158 (Tex. App. 13 Dist. 1983).

[46]See, R.C. Wood and D.C. Thompson, *Education Finance Law.*

[47]See, e.g., Paulson v. Minidoka County School Dist., 93 Idaho 469, 463 P.2d 935 (Idaho 1970), Bond v. Ann Arbor School Dist., 383 Mich. 693, 178 N.W.2d 484 (Mich. 1970).

[48]See, e.g., Hamer v. Board of Educ., 47 Ill. 2d 480, 265 N.E.2d 616 (Ill. 1970).

[49]Hamer v. Board of Educ., 9 Ill. App. 3d 663, 292 N.E.2d 569 (Ill. App. 2 Dist. 1973).

[50]Ambroiggio v. Board of Educ., 101 Ill. App. 3d 187, 427 N.E.2d 1027 (Ill. App. 2 Dist. 1981).

[51]Hartzell v. Connell, 201 Cal. Rptr. 601, 679 P.2d 35: Notwithstanding this issue, school districts, under specific guidelines, may charge fees for pupil transportation, see, Arcadia Unif. School Dist. V. State Dept. of Educ., 280 Cal. Rptr. 463 (Ct. Aoo. 1991). In general, see R. C. Wood, "The Supreme Court Rules on User Fee: Transportation," *NOLPE NOTES*, Oct. 1989, 1-2 see also, R. C. Wood, "Kadrmas

v. Dickinson Public Schools: A Further Retreat From Equality of Educational Opportunity," *J. of Educ. Finance,* v. 15, n. 3, Winter, 1990, 429-436.

[52]E.C. Bolmeier, School and the State Structure 116 (2d ed., 1973).

[53]See, R.C. Wood and J. Maiden, "Florida Department of Education v. Glaser: Local Control of Property Taxes Under the State Uniformity Clause," *NOLPE Notes,* May, 1995.

[54]Board of Educ. of Englewood Cliffs et. al, 608 A.2d 914 (N.J. Super. Ct. App. Div. 1992).

[55]For an example examining conflict between a state board of education and the state superintendent, see, State Board of Educ. v. Honig, 16 Cal. Rptr. 2d 727 (Cal. Ct. App. 1993).

[56]See, e.g., State Board for Elem. and Sec. Educ. v. Ball, 847 S.W.2d 743 (Ky, 1993).

[57]See, Eiche v. Louisiana Board of Elem. and Sec. Educ., 582 So. 2d 186 (La. 1991); State ex rel. Marokie v. Wagoner, 446 S.E.2d 680 (W Va. 1994).

[58]Fairbanks North Star Borough Sch. Dist.. v. NEA-Alaska, 817 P.2d 923 (Alaska 1991).

[59]Aurora East Public School Dist. v. Cronin, 442 N.E.2d 571 (Ill. 1982).

[60]Ill. Rev. Stat. 1977, ch. 122, par. 10-21.3.

[61]Aurora East Public School Dist. v. Cronin, 442 N. E.2d 511, 517 .

[62]Fullerton Union High School Dist. v. Riles, 188 Cal. Rptr. 897 (Cal. App. 1983).

[63]Remedial Educ. and Diagnostic Services v. Essex County Educ. Services Comm., 468 A . 2d 253 (N.J. Super. 1983).

[64]Ibid. at 254.

[65]For a case which elaborates the significance of the source of authority behind the statutory requirement see Pennhurst State School and Hospital v. Halderman, 451 U.S. 1, 101 S.Ct. 1531 (1981).

[66]Helvering v. Davis, 301 U.S. 6i9 (1937).

[67]Bell v. New Jersey, 461 U.S. 773 (1983).

[68]New York v. United States Dept. of Educ., 903 F.2d 930 (2d Cir. 1990).

[69]42 U.S. Code, Sect. 2000(d), et. seq. See also, 59 Fed. Reg. 11, 448 (1994).

[70]20 U.S. Code, Sect. 1681, et. seq.

[71]Franklin v. Gwinett County Public Schools, 112 S. Ct. 1028 (1992).

[72]42 U.S.C. § 12101 (1990).

[73]Cannon v. University of Chicago, 441 U.S. 677 (1979).

[74]Grove City College v. Bell, 104 S.Ct. 1211(1984).

[75]Walher v. Lochland City School Dist., 429 N.E.2d 1179 (Ohio App. 1980); See also, Lake Erie Inst. of Rehabilitation v. Marion County Bd. of Educ., 798 F. Supp. 262 (W.S. Pa. 1992); Mercantile Bank of Illinois v. School Dist. of Osceola, 834 S.W.2d 737 (Mo. 1992).

[76]Godbey v. Roosevelt School Dist., 638 P.2d 235 (Ariz. App. 1981).

[77]See, Board of Educ. v City of New Haven, 602 A.2d 1018 (Conn, 1992).

[78]Board of Educ. of Arbor Park School Dist. v. Ballweber, 451 N.E.2d 858 (Ill. 1983).

[79]It is not surprising that the arbitrator found this board policy in violation of the contract.

[80]See, e.g., Clevinger v. Board of Educ. of Pike County, 789 S.W.2d 5 (Ky. 1990); Johnson v. Washington County, 518 N.W.2d 594 (Minn. 1994); Murray v. Zarger, 642 A.2d 575 (Pa. Commw. Ct. 1994).

[81]See, e.g., Smith v. City of Chester, 842 F. Supp. 147 (E.D. Pa. 1994).

[82]Smith v. Wade, 103 S.Ct. 1625 (1983).

[83]City of Newport v. Fact Concerts, 453 U.S. 247 (1981).

[84]See, e.g., Ill. Rev. Stat., § 122:10-20.20.

[85]Wood v .Strickland, 420 U.S. 308 (1975).

[86]Owen v. City of Independence, 445 U.S. 622 (1980).

CHAPTER 3
Taxation

Systems of Taxation

The complexity of financing public education in the United States was introduced in the first two chapters of this text. This complexity readily extends to the focus of this chapter on taxation, wherein the sources of revenue for conducting the educational enterprise are examined in some detail. As was clearly noted in the first edition of this text, school business administrators have experienced a significant change in their jobs due in large part to a reversal in economic fortunes. This reversal has been historic in proportion, as education's revenue grew dramatically for nearly half a century. But revenue has declined for about the last twenty years. As a result, educational leaders and other policymakers at the national, state, and local levels have been faced with a reality demanding that they develop expertise not only in spending money, but also in understanding and managing scarce revenue.

Although it will be seen later that taxes evoke strong responses from citizens, it is clear that schools and forms of government in the United States rely almost exclusively on tax moneys to carry out legal and educational obligations. These taxes are of a varied nature, resulting in a mixed revenue system that has both advantages and disadvantages. The mixture forms a tax structure that sometimes appears complicated to lay people, thereby increasing the difficulty for school administrators who must face a constant need for increasing revenue while seeking to help patrons understand and support schools. It is thus critical for effective school business administrators to have a clear understanding of revenue sources for schools, to understand how these sources interrelate, to understand the fundamental mechanics of taxation, and to gather some skill in assessing the relative benefits and drawbacks of each existing or proposed tax.

Tax Systems

Although nearly everyone recognizes that government in a democracy is an expression of the will of the people, the reaction of citizens to taxes is remarkable. Despite the fact that taxes pay for services enjoyed by all people, there is a significant level of resentment that is apparent even at the mention of taxes. Although there are likely many good explanations of why people react negatively to taxes, it is reasonable to believe that their response stems from the ever-present visibility of the various tax systems that raise revenue for public purposes.

Despite the havoc that would be created without tax moneys, few people pay their taxes eagerly and even fewer would do so without force. As noted elsewhere, it is a basic tenet of public policy that taxation is the removal of private wealth for a public purpose sufficiently compelling to forcibly deny someone the absolute ownership of private wealth.[1] In a nation like the United States, which was built in part on rebellion against abusive taxation, taxation for any purpose can be expected to naturally engender some negative reaction.

Yet it is clear that although taxes are grudgingly paid, there is a legitimate purpose for raising moneys for public use. Every civilized nation, almost from the beginning of time, has devised some means to derive funds to support government or to make purchases from a collective purse that are not practicably available to private citizens. In the United States, tax systems have emerged on multiple levels that coincide with the nation's development and that reflect the thinking of citizens about how collective decision making should occur.

In this nation, tax systems have been built around three main levels of governance: federal tax system, individual state tax systems, and local tax systems. Each level has emerged for sensible reasons and has had to learn to coexist, albeit uneasily at times. As a result of overlapping units of government, which also result in overlapping demands for revenue, it is important to briefly trace this reality as a preface to understanding the relationship of tax systems and school revenue.

The Federal Tax System

Although there is dispute among scholars over the importance one may assign to federal, state, and local tax systems in the grand mix of the total tax structure, it stands without question that the federal tax system has become an omnipresent and powerful force in the world of taxation. In the modern era, federal taxes on the American people are staggering, as in 1994, total revenue from all sources was well in excess of $1 trillion. These moneys, gathered in vast amounts, were spent on thousands of programs affecting millions of people, a number that grows exponentially when it is realized that taxes are not exclusively reserved for domestic concerns. Aid to foreign countries and a global economy result in American tax dollars also touching hundreds of millions of people around the world. Yet the strength and power of the federal tax system has not always been so vast or so complicated. It can be shown that the federal system has only recently reached its present role, including its impact on public education.

Although it would be quite reasonable in the modern era to ask why citizens were not taxed on their incomes in support of federal revenue needs, the concept of individual obligation to government must be seen as part of the developing struggle to establish a federal tax base. Income as a basis for taxation had been proposed many times, including the Bay Colony's attempt in 1634 at taxing people on their income potential. But proposing a tax is a far cry from enacting or sustaining it. The federal income tax was proposed early in the new nation, as in 1862 Congress called for a tax of 3 percent on a variety of income sources. At first the tax appeared to be a good idea as it generated $73 million by 1866, but by 1870 revenues had fallen by nearly half. The tax died out in 1872 and did not arise again as a serious issue of debate until 1894, when the fiscal problems of the nation followed a financial panic caused Congress to enact a 2 percent tax on earnings above a $4,000 personal exemption. Yet even in severe economic times, lawmakers had problems convincing taxpayers of their duty, as a lawsuit seeking to overturn the 1894 law was successful on grounds that the U.S. Constitution did not permit direct federal taxation. In desperation, however, Congress was successful after contentious debate in ratifying the Sixteenth Amendment in 1913, giving Congress "...power to lay and collect taxes on incomes, from whatever source derived..."[2]

Dizzied by triumph, Congress began enacting new tax laws that have evolved into the massive federal tax system now known to all Americans. Beginning in 1913, individual incomes were taxed at differentiated rates and introducing familiar concepts like progressive taxation under which higher incomes pay higher overall percentages, personal exemptions for marital status, and so forth. Congress quickly found the revenue stream to be unpredictable, and acted variously under its new authority to pass laws such as the Current Tax Payment Act of 1943 requiring payroll deduction for federal taxes owed. Corporations and other entities were held liable for taxes, and the revenue base accelerated at a rapid pace. In contrast to the pathetic Revolutionary War days and meager revenues at the start of the Civil War, the new tax laws grew to the point where, today, federal revenue is more than $1 trillion annually.

Yet while the federal tax system has become vast with massive revenue and the broadest base it has ever known, it is important to realize that all revenue gathered by the federal government does not result in surplus. As seen in table 3.1, the needs and desires of the American people are not totally consonant with their historic attitude about taxation. While revenues are indeed vast, the data show that Americans expect world military superiority, leadership in international charity, personal safety in many forms such as social security and unemployment insurance, massive support for health care and research, veterans' benefits and other educational scholarship support, transportation and highway assistance, and many other costly services, including the high cost of government itself.

But such services are not free, and in return for making demands, Americans pay a host of federal taxes including tax on individual income, corporate income, various social taxes including social security and disability, estate and gift taxes, and many steep special excise taxes on commodities such as alcohol, tobacco, and fuel. The revenue has indeed become massive at $1.25 trillion in 1994, but demands are even greater as Congress overspent its revenue in the same year by incurring a federal deficit of $234.8 billion.[3] Overall, expenditures are nonetheless remarkable, as the first federal aid to education in the amount of $2 million appeared in 1919, while in 1994 Congress again approved reauthorization of the Elementary and Secondary Education Act of 1965 for more than $12 billion.

Table 3.1
Federal Revenues and Expenditures 1945-1994

(In Millions of Dollars)

YEAR	Receipts [1]	OUTLAYS [1]				Surplus or deficit (-)
		Total	Human resources	National defense	Percent of GDP [2]	
1945	45,159	92,712	1,859	82,965	43.7	-47,553
1950	39,443	42,562	14,221	13,724	16.0	-3,119
1955	65,451	68,444	14,908	42,729	17.8	-2,993
1960	92,492	92,191	26,184	48,130	18.3	301
1965	116,817	118,228	36,576	50,620	17.6	-1,411
1970	192,807	195,649	75,349	81,692	19.9	-2,842
1972	207,309	230,681	107,211	79,174	20.1	-23,373
1973	230,799	245,707	119,522	76,681	19.3	-14,908
1974	263,224	269,359	135,783	79,347	19.2	-6,135
1975	279,090	332,332	173,245	86,509	22.0	-53,242
1976	298,060	371,792	203,594	89,619	22.1	-73,732
1976 [5]	81,232	95,975	52,065	22,269	21.6	-14,744
1977	355,559	409,218	221,895	97,241	21.3	-53,659
1978	399,561	458,746	242,329	104,495	21.3	-59,186
1979	463,302	503,485	267,574	116,342	20.7	-40,183
1980	517,112	590,947	313,374	133,995	22.3	-73,835
1981	599,272	678,249	362,022	157,513	22.9	-78,976
1982	617,766	745,755	388,681	185,309	23.9	-127,989
1983	600,562	808,380	426,003	209,903	24.4	-207,818
1984	666,457	851,846	432,042	227,413	23.1	-185,388
1985	734,057	946,391	471,822	252,748	23.9	-212,334
1986	769,091	990,336	481,594	273,375	23.5	-221,245
1987	854,143	1,003,911	502,196	281,999	22.5	-149,769
1988	908,954	1,064,140	533,404	290,361	22.1	-155,187
1989	990,691	1,143,172	568,668	303,559	22.1	-152,481
1990	1,031,321	1,252,705	619,327	299,331	22.9	-221,384
1991	1,054,272	1,323,793	689,691	273,292	23.3	-269,521
1992	1,090,453	1,380,856	772,450	298,350	23.3	-290,403
1993	1,153,535	1,408,205	827,541	281,086	22.4	-254,670
1994, est.	1,249,071	1,483,829	879,911	279,824	22.3	-234,758

Source: U.S. Department of Commerce, Bureau of the Census, *Statistical Abstracts of the United States 1994* (Washington, D.C.: U.S. Government Printing Office, 1994), 330.

State Tax Systems

Although one might suggest that the federal tax system has both captured a meaningful revenue stream and stretched taxpayers to their limits, it is important to recognize that other parallel tax systems also exist. As noted previously, while scholars may argue about the relative stature of various tax systems in the grand tax scheme, it is accurate to say that state tax systems have added yet another vast layer of revenue-generating government. If federal revenue is huge, it is also true that the fifty states' collective revenue totaling more than $328 billion in 1992, is also astounding.

Like their federal counterparts, states spend money on programs affecting millions of people. Unlike federal moneys, however, state dollars are spent almost exclusively inside the borders of the nation, as federal and state governments have been assigned very different roles. Yet while state resources are large, it can be shown that state tax systems have only recently reached the present role, including a role in education.

While tax systems in the United States could be studied from a variety of perspectives, it is important to have an understanding of the unique way in which each unit of government has emerged. Most people today, if setting out to establish a new nation, would tend to look toward a strong central government to help frame identity and structure, given the prevailing global village concept. Of course, it would be a mistake to impose a modern mentality on history, because the culture of an era must be recognized if one is to understand how long-standing institutions have managed to evolve. This is particularly true when seeking to understand the development of state tax systems because there has been both parallelism and isolation in how state tax systems have grown alongside, yet distinctly different from, the federal system.

A critical concept in state tax system development is recognition of the fact that there has always been a tense rivalry between the federal government and the individual states. In fact, it can be argued that due to the fierce independence of the colonies, the nation itself was developed around the concept of *united* states' sovereignty. If tax ideology were a defining aspect of the new nation, it can easily be seen that the individual colonies would have already mirrored the same behavior because the separate behavior would be now joined in a collective national entity.

State autonomy in all matters, including financial, was further aided by westward expansion. Congress began granting aid to states that were eager to build roads and transportation systems across the land. Although some prosperity had come to the nation, there were crushing federal debts stemming from the War of 1812, the Louisiana Purchase in 1803, and the Civil War. Yet debt began to overtake the states too, as many states had borrowed huge sums for expansionist purposes. In 1840, no state forbade debt, but with the varying federal fortunes, which caused aid to states to both ebb and flow, an anti-debt mood took hold in many states. By 1855 nearly two-thirds of the thirty-two states had adopted constitutional prohibitions against deficit spending. Sadly for both federal and state governments alike, the revenue systems that had supported government's limited needs in the past were now barely adequate.

While a lengthy history could be traced, it is more important to note a changing perspective in the nation toward the subject of taxation for public purposes. Growth of the nation, combined with resignation to the fact that taxes were becoming a permanent reality, led to a more rational view that aggressiveness ought not to be directed against taxation itself, but rather toward assuring fundamental tax fairness. Beginning in about 1880, study commissions in the states began looking at ways to improve tax systems, taking interest in how taxation of property might be better administered. States had tried various taxes but had become dependent on the property tax because of problems in collecting other kinds of taxes. Yet states had also had difficulty administering the property tax because of a lack of uniformity in valuing property for tax purposes and because politics had severely hampered fair application of property tax laws.

By the turn of the twentieth century, state taxation had become a permanent reality. Critics were left wishing for fairer tax systems and for relief from property tax as the dominant source of revenue. By 1920, Wisconsin's success in implementing an income tax had spread to other states, leading the way for adoption of new taxes such as sales and excise taxes.

Although state tax systems have experienced the same resistance as was seen in federal history, the path has been somewhat easier due to both the times and society's advancement. Already more disposed toward state taxes under the home rule concept, states have been able to convince citizens that taxes were being spent on issues affecting their personal welfare. Similarly, development of society itself has aided tax schemes, as prosperity causes less resistance to taxes and as technology itself creates new objects to be taxed. Much of the twentieth century has been a boom economy, and technology has provided a vast array of objects that can be taxed. Continued growth in intergovernmental aid has added huge sums to state coffers, as both the federal and local governments contribute to state concerns. Modern state tax systems are often vast, collecting tax on income, sales, real and personal property, and so on.

While states originally had nearly as much difficulty raising revenue as did the federal government, the phenomenal growth of modern state taxation has resulted in an increase in state revenue from only $192 million in 1902 to more than $328 billion in 1992. The tax workhorses in states are clearly the individual income tax, which produced about $104.6 billion in 1992 and general sales tax, which produced $107.8 billion in the same year, Technology has also contributed a new tax base, such as $6.7 billion in taxes on automobiles in 1992. In contrast to the early days when states were weakly attempting to levy property tax on the few people who could not escape, state tax systems have become a vast and powerful revenue source.

Yet while state tax systems have become massive and broadly based, it is again important to see that all revenue gathered by states does not result in a surplus. As seen in table 3.2, the needs and desires of citizens are not always in keeping with historic attitudes about taxation. While state revenue is in fact vast, data in table 3.2 show that citizens expect still another spate of expensive services beyond those provided at the federal level. Although calculated differently from the data in table 3.2, in 1992 state expenditures actually exceeded revenue by $7 billion, with the biggest spending occurring in the areas of education, welfare, highways, health, and natural resources. Education is often the largest state cost. Similarly, welfare is an enormous state

Table 3.2
State Revenues and Expenditures 1992

(In Millions of Dollars)

ITEM	State (mil. dol.)
Revenue	**744,232**
Intergovernmental revenue	169,902
Revenue from own sources	574,330
General revenue from own sources	435,980
Taxes [3]	328,370
Property	6,673
Individual income	104,609
Corporation income	21,566
Sales and gross receipts	162,721
Customs duties	(X)
General	107,757
Selective [3]	54,964
Motor fuel	22,250
Alcoholic beverages	3,599
Tobacco products	6,119
Public utilities	7,762
Motor vehicle and operators' licenses	11,771
Death and gift	4,456
Charges and miscellaneous [3]	107,610
Current charges [3]	52,941
National defense and international relations	(X)
Postal service	(X)
Education [3]	28,901
School lunch sales	-
Higher education	28,434
Natural resources	1,396
Hospitals	12,729
Sewerage and sanitation	287
Parks and recreation	788
Housing and community development	213
Airports	650
Water transport and terminals	393
Special assessments	127
Sale of property	183
Interest earnings	26,603
Utility and liquor store revenue	6,579
Insurance trust revenue	131,771
Expenditure	**699,954**
Intergovernmental expenditure	201,314
Direct expenditure	498,640
General expenditure [3]	409,132
Education [3]	86,650
Elementary and secondary education	2,221
Higher education	70,904
Public welfare	125,500
Health and hospitals	41,643
Highways	40,266
Police protection	4,863
Fire protection	(X)
Corrections	18,306
Natural resources	9,022
Sewerage and sanitation	2,235
Housing and community development	1,606
Governmental administration	19,847
Parks and recreation	2,688
Interest on general debt	24,622
Utility and liquor store expenditure	9,613
Insurance trust expenditure	79,895
By character and object:	
Current operation	322,353
Capital outlay	50,126
Construction	39,001
Equip., land, and existing structures	11,125
Assistance and subsidies	20,784
Interest on debt (general and utility)	25,482
Insurance benefits and repayments	79,895
Expenditure for salaries and wages [4]	*112,685*

Source: U.S. Department of Commerce, Bureau of the Census, *Statistical Abstracts of the United States 1994* (Washington, D.C.: U.S. Government Printing Office, 1994), 298.

cost. Highways are yet another major cost, as states repair and improve the vast network of state roads. While these services were among the most costly, many others were provided to citizens in the various states.

Local Tax Systems

Because of the size and cost of both federal and state tax systems, it would be reasonable to wonder whether any room remains for the third layer known as local tax systems. Such speculation, however, is quickly defied as citizens are well aware of local government since its presence is often even more visible in their lives. If federal and state units have been assigned specific roles in society, it is true that local governments and their tax structures have also been assigned a very specific task. As with both federal and state government, these roles have resulted from parallel development with the other tax systems and from the isolation unique to American historical progression. These events have had a tremendous impact on the nature of the local tax system.

Development of the nation has followed a unique path of three strong levels of government. As stated earlier, the federal system has accepted responsibility for defense, international affairs, interstate concerns, and general welfare. This limited role stands in contrast to the states, which in a sovereign manner laid claim under the Constitution to all matters not reserved to higher government. These powers were so broad that states were able to delegate many responsibilities to a system of local government under a view that Americans prefer government that assigns issues of local concern to correspondingly local units of government. The logic was popular in a naturally rebellious nation, as it tended to make government more accountable on a daily basis and had the effect of removing higher government from daily invasion in the lives of people.

As it became more difficult for one local body to govern all local affairs, this concept further evolved into many units of local government such as counties, cities, and even township boards in some parts of the nation. In a nation generally fearful of strong central government, the third layer of localism was appealing and consequently not difficult to establish, especially given isolation on the frontier where federal and state governments had little visible influence in daily affairs.

While the logic was straightforward and defensible, development of fiscal support systems was severely hindered by the growth of multiple governmental layers. Although an argument could be held about which level of government was established first, the reality has been that federal and state governments have seized control of larger affairs, while local units care for a huge array of daily problems. While clearly overlapping with higher levels of government, local units have had to take broad responsibility for a host of concerns including education, welfare, health, roads, police and fire services, corrections, community development, and so on. For example, although state police exist in every state, this service overlaps with county sheriffs and local police units. Likewise, despite federal and state aid to education, in many states the largest share of education's resources are derived locally. The problem, however, is that given the enormous task assigned to local government and given tax systems already in place for support of higher government, local units are not financially well positioned to provide all the services that are expected. In part, this is an effect of trickle-down taxation.

It was seen earlier that the federal government has utilized the income tax for its primary revenue source, so much so that income overshadows any other source of federal revenue. States have also seized on the income tax, generating massive amounts of moneys from taxpayers within each respective state's borders. But because the federal unit has used income so heavily, practical limits to state utilization of income are often thought to exist. As a result, states have also tapped other revenue sources, particularly sales tax on a wide variety of retail exchanges. Under this scenario, one unit of government (usually a lower one) picks up what is left of a tax that has been dominated by some other unit and then further seeks alternative revenue sources.[4] As each unit of government takes primary control of a tax, each successive unit is left with fewer choices or the daunting task of finding new taxes. For local units, the principal unused tax source has been the property tax because, as noted earlier, no federal property tax exists and states have preferred to tax property only lightly, if at all. Since there is little room left to tax income at the local level and local sales tax is either marginal or illegal in many states, only the property tax is available to support the staggering array of services assigned to the single unit of government that is most visible in the lives of citizens.

The overall result has been a very narrow local tax base, with a challenge to be responsive to citizens' needs. Local governments primarily derive revenue from state and federal intergovernmental transfers and from their own local sources. Transfers from higher units, usually in the form of revenue sharing, have become an important source of income for local governments, as in 1992 local units received $20 billion from the federal government and about $195.5 billion from the states. Significantly, local tax revenue yielded $226.7 billion, an amount largely attributed to property taxes. As data in table 3.3 show, property tax revenue in the states amounted to $171.9 billion, of which $168 billion were collected locally: more than 75 percent of all taxes collected at the local level. Even more powerful, however, is the knowledge that all revenue derives from one source—local taxpayers—in that local citizens pay the federal and state tax from which intergovernmental transfers are added back to local revenue. In other words, all taxes come from the local level, no matter how money is sifted back to communities. A summary of all levels of government revenue and expenditure is seen in table 3.4.

Despite a narrow local tax base, expenditures for local services are significant by any standard. In 1992, localities led all units of government in spending for education, putting $226.7 billion into children's education needs. Similarly, local spending outpaced other units for essential services, spending

$46.5 billion on health care, $29.7 billion on police and fire service, $30.2 billion on sanitation, and $30.5 billion on the cost of administering thousands of local governmental jurisdictions. Perhaps not too surprisingly, local expectations also caused this level of government to overspend its revenue by $7.5 billion.

Under these conditions, it is not surprising that local government is often fiscally stressed. Neither is it surprising that the local unit is a first target for unhappy taxpayers who continue an American tradition of tax dissidence. As state and federal government grow and seem to become more inaccessible to ordinary citizens, it is reasonable that these people should cast about for ways to "get at big government." Since local government is in fact most accessible and since the property tax is the primary local revenue source, it becomes sensible to disgruntled citizens to attack the property tax in an attempt to relieve their tax burdens. This phenomenon has been repeated across the country over the last decade, with recent signs of renewal in some quarters of the nation.

The classic instance of overburdened taxpayers upset with property taxes appeared in California's Proposition 13, which was approved by state voters in 1978. Known as a taxpayer initiative, Proposition 13 called for a state constitutional amendment to limit government revenues. Rather than seeking a statute to achieve the same end, the initiative wanted to amend the state constitution permanently so that any effort to change property tax levels would require another amendment. Although highly complex in its operation, Proposition 13 essentially attempted to roll back property taxes to the 1975 level. The proposed law also sought to limit the permissible tax on property to 1 percent of a property's assessed value. Further, Proposition 13 sought to restrict future increases by limiting growth in assessed valuation to 2 percent annually unless a property changed owners, and forbade levying any new taxes on real property. Proposition 13 passed by a large majority, and was followed only one year later by Proposition 4, further restricting state revenues by limiting tax increases to the cost of living.

With passage of these laws, two events followed. The first event came about when Proposition 13 dramatically reduced property tax revenue in California by about $7 billion, a reduction of more than one half its former revenue level. Another 16 percent was lost to Proposition 4, with a grim outlook as the state's revenue became permanently tied to the economy. Schools were nearly ruined, as the loss of local property tax reduced school revenue by 29 percent, requiring the state to effectively take over funding of education. The second event was the national tidal wave of property tax reform triggered by California's tax rebellion. In almost no time, other states joined the tax initiative bandwagon, such as Massachusetts where taxpayers passed Proposition 2-1/2, rolling back property taxes to no more than 2.5 percent of market value and limiting annual revenue growth to that same percentage. In only a short

Table 3.3
Local Revenue and Expenditures 1992

(In Millions of Dollars)

ITEM	Local (mil. dol.)
Revenue	**647,872**
Intergovernmental revenue	216,305
Revenue from own sources	431,567
General revenue from own sources	357,419
Taxes [3]	227,240
Property	171,863
Individual income	10,562
Corporation income	2,029
Sales and gross receipts	33,391
Customs duties	(X)
General	23,073
Selective [3]	10,318
Motor fuel	694
Alcoholic beverages	281
Tobacco products	191
Public utilities	5,725
Motor vehicle and operators' licenses	830
Death and gift	30
Charges and miscellaneous [3]	130,179
Current charges [3]	83,959
National defense and international relations	(X)
Postal service	(X)
Education [3]	10,201
School lunch sales	3,737
Higher education	3,593
Natural resources	426
Hospitals	25,023
Sewerage and sanitation	2,828
Parks and recreation	3,001
Housing and community development	2,828
Airports	5,217
Water transport and terminals	1,231
Special assessments	241
Sale of property	539
Interest earnings	28,746
Utility and liquor store revenue	55,852
Insurance trust revenue	18,296
Expenditure	**655,325**
Intergovernmental expenditure	7,355
Direct expenditure	647,970
General expenditure [3]	562,842
Education [3]	240,120
Elementary and secondary education	226,696
Higher education	13,424
Public welfare	28,734
Health and hospitals	46,469
Highways	26,211
Police protection	29,682
Fire protection	14,358
Corrections	10,300
Natural resources	3,127
Sewerage and sanitation	30,163
Housing and community development	15,461
Governmental administration	30,488
Parks and recreation	13,040
Interest on general debt	30,633
Utility and liquor store expenditure	74,748
Insurance trust expenditure	10,381
By character and object:	
Current operation	501,141
Capital outlay	84,586
Construction	49,876
Equip., land, and existing structures	19,133
Assistance and subsidies	12,446
Interest on debt (general and utility)	39,416
Insurance benefits and repayments	10,381
Expenditure for salaries and wages [4]	*266,119*

Source: U.S. Department of Commerce, Bureau of the Census, *Statistical Abstracts of the United States 1994* (Washington, D.C.: U.S. Government Printing Office, 1994), 298.

Table 3.4
Revenue and Expenditures for all Government Units
1992

(In Millions of Dollars)

ITEM	All governments (mil. dol.)	Federal (mil. dol.)	State (mil. dol.)	Local (mil. dol.)
Revenue	**2,261,849**	**1,259,383**	**744,232**	**647,872**
Intergovernmental revenue	([2])	3,431	169,902	216,305
Revenue from own sources	2,261,849	1,255,952	574,330	431,567
General revenue from own sources	1,644,789	851,390	435,980	357,419
Taxes [3]	1,214,651	659,041	328,370	227,240
Property	178,536	(X)	6,673	171,863
Individual income	591,636	476,465	104,609	10,562
Corporation income	123,865	100,270	21,566	2,029
Sales and gross receipts	260,394	64,282	162,721	33,391
Customs duties	17,480	17,480	(X)	(X)
General	130,830	(X)	107,757	23,073
Selective [3]	112,084	46,802	54,964	10,318
Motor fuel	42,809	19,865	22,250	694
Alcoholic beverages	11,787	7,907	3,599	281
Tobacco products	11,500	5,190	6,119	191
Public utilities	21,338	7,851	7,762	5,725
Motor vehicle and operators' licenses	12,601	(X)	11,771	830
Death and gift	15,629	11,143	4,456	30
Charges and miscellaneous [3]	430,138	192,349	107,610	130,179
Current charges [3]	226,382	89,482	52,941	83,959
National defense and international relations	8,677	8,677	(X)	(X)
Postal service	45,158	45,158	(X)	(X)
Education [3]	39,102	-	28,901	10,201
School lunch sales	3,737	-	-	3,737
Higher education	32,027	-	28,434	3,593
Natural resources	16,970	15,148	1,396	426
Hospitals	37,876	124	12,729	25,023
Sewerage and sanitation	3,115	(X)	287	2,828
Parks and recreation	3,899	110	788	3,001
Housing and community development	5,749	2,708	213	2,828
Airports	5,888	21	650	5,217
Water transport and terminals	2,889	1,265	393	1,231
Special assessments	368	(X)	127	241
Sale of property	4,445	3,723	183	539
Interest earnings	65,746	10,397	26,603	28,746
Utility and liquor store revenue	62,431	(X)	6,579	55,852
Insurance trust revenue	554,629	404,562	131,771	18,296
Expenditure	**2,487,885**	**1,527,311**	**699,954**	**655,325**
Intergovernmental expenditure	([2])	186,036	201,314	7,355
Direct expenditure	2,487,885	1,341,275	498,640	647,970
General expenditure [3]	1,858,519	886,545	409,132	562,842
Education [3]	353,591	26,821	86,650	240,120
Elementary and secondary education	228,917	-	2,221	226,696
Higher education	84,328	-	70,904	13,424
Public welfare	201,956	47,722	125,500	28,734
Health and hospitals	118,791	30,679	41,643	46,469
Highways	67,290	813	40,266	26,211
Police protection	41,248	6,703	4,863	29,682
Fire protection	14,358	(X)	(X)	14,358
Corrections	31,017	2,411	18,306	10,300
Natural resources	63,650	51,501	9,022	3,127
Sewerage and sanitation	32,398	(X)	2,235	30,163
Housing and community development	32,549	15,482	1,606	15,461
Governmental administration	67,180	16,845	19,847	30,488
Parks and recreation	17,712	1,984	2,688	13,040
Interest on general debt	254,968	199,713	24,622	30,633
Utility and liquor store expenditure	84,361	(X)	9,613	74,748
Insurance trust expenditure	545,006	454,730	79,895	10,381
By character and object:				
Current operation	1,323,835	500,341	322,353	501,141
Capital outlay	227,807	93,095	50,126	84,586
Construction	99,797	10,920	39,001	49,876
Equip., land, and existing structures	112,433	82,175	11,125	19,133
Assistance and subsidies	129,852	96,622	20,784	12,446
Interest on debt (general and utility)	264,611	199,713	25,482	39,416
Insurance benefits and repayments	545,006	454,730	79,895	10,381
Expenditure for salaries and wages [4]	*540,355*	*161,551*	*112,685*	*266,119*

Source: U.S. Department of Commerce, Bureau of the Census, *Statistical Abstracts of the United States 1994* (Washington, D.C.: U.S. Government Printing Office, 1994), 298.

while, an additional seventeen states joined in the tax initiative race, creating financial crises for local governments that depend so heavily on the property tax base.

Although tax reform waned in the following years, it never died out completely and tends to revive during times of fiscal stress. States that are heavily property-tax dependent for support of governmental services are generally at greater risk of revolt, principally because taxpayers know they can still have a greater impact on local tax issues and because the narrowness of local tax bases requires heavy reliance on the property tax.

Although some critics were relieved when tax initiatives slackened in the late 1980s, there are signs that taxpayers have not forgotten the power they wield at the local level. New Jersey's troubled school finance system has been repeatedly struck down by courts as inequitable, yet taxpayers successfully forced a state legislative bill in 1991, rolling back the state's largest tax increase in history and simultaneously reducing property tax by $360 million, with another $469 million cut in 1994 under a new antitax governor. Michigan's abandonment of the property tax in 1993 resulted in dramatic reform of school funding, moving the state from a highly local-tax-base to a sales tax-dependent system. Likewise, Colorado's voters approved a strict constitutional amendment in 1992, limiting property tax increases and creating fiscal stress in rapidly growing districts. Scores of other cases abound, and many states are ripe for tax initiative. For example, New Hampshire's local taxpayers pay 90 percent of educational costs, while at the same time the state pays only 7 percent. Similarly, South Dakota's voters pushed a tax initiative onto the ballot in 1994 in a move to limit property tax to 1 percent of assessed value and to limit growth to 1.25 percent. Florida, in 1994, passed a severe limitation to government revenue via a constitutional amendment. In sum, tax unrest continues to be a serious issue, as local government has almost no other option for obtaining revenue.

Data on tax systems in the United States suggest a mixture of revenue sources for all units of government. This text's interest is related to tax systems as they affect public education, but it is critical that school business administrators comprehend the nature and magnitude of taxation at all levels. While education is only one responsibility of government, it is among the most important because schools prepare future citizens, who in turn will become taxpayers. While schools receive a mix of revenue like all other public sector units, there are issues relating to federal, state, and local tax systems. These issues become evident in table 3.5 which summarizes the amounts and sources of revenue to schools in the fifty states. These data show that tax systems yield immense school revenue, and that these funds come from all levels of government. Federal aid ranges from 3.1 percent in New Hampshire to 17 percent in Mississippi. State aid ranges from 8.5 percent in New Hampshire to 90.3 percent in Hawaii, depending on how each state meets its obligation to education. Strikingly, local aid ranges from .05 percent in Hawaii to 91.2 percent in Delaware, making clear that local taxes still support a large proportion of education's costs in most areas. While dramatic shifts have occurred in the relationship between federal, state, and local tax systems in their fiscal support of schools, it remains true that school business administrators primarily depend on a mixture of local property taxes and state aid for the operation of schools, with only minor assistance from the federal government.

Tax Concepts and Fundamentals

As noted in the first edition of this text, the field of taxation has its share of specialized vocabulary, some of which has been gradually introduced throughout the preceding discussion. Likewise, the field of taxation has also developed some principles by which new and existing taxes can be evaluated when determining whether they represent good or bad public policy. Because school business administrators are often called upon to discuss taxes both generally and specifically in their local communities or in the legislative decision-making process, it becomes important to extend the historical information in this chapter to include a better grasp of the elements of effective and desirable taxation.

The discussion thus centers on analysis of the elements thought to make up good tax systems, with discussion of principal forms of taxes woven into the analysis. These fundamental tax concepts make it possible in the last section of the chapter to critically examine income, sales, and property taxes from a perspective of how well these taxes fit the definition of an equitable, effective, and desirable system of taxation.

Although schools often do not benefit directly from various other types of taxes, brief mention is made of other tax instruments since it is common to hear educators and policymakers talk of searching for other revenue sources.[5]

Good Tax Systems

It was established earlier in the chapter that many people would not voluntarily choose to pay taxes. However, it was contended that the history of the nation has been such that people pay taxes more willingly if they believe that taxes are fair. Most experienced educational administrators, however, will quickly point out that the task of convincing people that tax systems are fundamentally fair is a tenuous situation in light of the many tax initiatives that have marked the recent past. Yet the task is no less critical because of its difficulty, making it important that educational administrators be conversant in the language and operation of tax systems.

Table 3.5
Federal, State, and Local Revenue by Dollars and Percentages of Totals FY 1992

(In Thousands of Dollars)

State or other area	Total	Federal		State		Local and intermediate	
		Amount	Percent of total	Amount	Percent of total	Amount	Percent of total
1	2	3	4	5	6	7	8
United States	**$234,485,729**	**$15,493,330**	**6.6**	**$108,792,779**	**46.4**	**$103,975,705**	**44.3**
Alabama	2,823,340	322,576	11.4	1,659,018	58.8	611,248	21.6
Alaska	1,120,970	128,612	11.5	762,663	68.0	205,165	18.3
Arizona	3,226,760	284,615	8.8	1,366,934	42.4	1,510,219	46.8
Arkansas	1,828,439	197,915	10.8	1,095,488	59.9	478,138	26.2
California	26,868,216	2,027,474	7.5	17,696,851	65.9	6,830,548	25.4
Colorado	3,058,633	152,090	5.0	1,307,982	42.8	1,510,328	49.4
Connecticut	3,891,217	126,225	3.2	1,583,668	40.7	2,063,543	53.0
Delaware	608,015	46,144	7.6	400,819	65.9	150,409	24.7
District of Columbia	711,172	66,508	9.4	—	—	641,350	90.2
Florida	10,810,522	788,420	7.3	5,227,256	48.4	4,350,167	40.2
Georgia	5,332,428	409,741	7.7	2,545,306	47.7	2,255,693	42.3
Hawaii	1,000,848	75,310	7.5	903,444	90.3	4,893	0.5
Idaho	861,955	69,859	8.1	532,475	61.8	242,120	28.1
Illinois	9,959,661	680,351	6.8	2,881,367	28.9	6,177,317	62.0
Indiana	5,127,888	272,355	5.3	2,710,144	52.9	1,975,429	38.5
Iowa	2,486,610	132,718	5.3	1,176,197	47.3	1,025,899	41.3
Kansas	2,264,365	123,564	5.5	959,173	42.4	1,112,810	49.1
Kentucky	2,939,351	296,573	10.1	1,969,899	67.0	651,896	22.2
Louisiana	3,377,064	363,958	10.8	1,848,734	54.7	1,068,290	31.6
Maine	1,246,798	73,876	5.9	621,026	49.8	548,461	44.0
Maryland	4,692,155	238,573	5.1	1,792,755	38.2	2,511,988	53.5
Massachusetts	5,621,629	296,702	5.3	1,728,360	30.7	3,483,002	62.0
Michigan	9,659,095	599,076	6.2	2,566,851	26.6	6,289,097	65.1
Minnesota	4,512,902	200,853	4.5	2,327,594	51.6	1,817,120	40.3
Mississippi	1,701,274	289,302	17.0	910,068	53.5	436,000	25.6
Missouri	4,053,529	258,032	6.4	1,538,752	38.0	2,088,076	51.5
Montana	821,111	72,483	8.8	343,293	41.8	373,016	45.4
Nebraska	1,506,050	93,705	6.2	517,098	34.3	761,716	50.6
Nevada	1,122,853	46,957	4.2	434,762	38.7	601,857	53.6
New Hampshire	1,015,187	31,098	3.1	86,597	8.5	871,238	85.8
New Jersey	10,523,002	436,024	4.1	4,438,939	42.2	5,451,200	51.8
New Mexico	1,368,013	169,616	12.4	1,009,593	73.8	154,408	11.3
New York	21,573,865	1,210,481	5.6	8,696,709	40.3	11,447,389	53.1
North Carolina	5,067,118	364,253	7.2	3,274,259	64.6	1,218,261	24.0
North Dakota	539,184	59,909	11.1	241,401	44.8	207,434	38.5
Ohio	9,736,287	571,416	5.9	3,974,682	40.8	4,797,389	49.3
Oklahoma	2,541,025	117,060	4.6	1,580,811	62.2	749,822	29.5
Oregon	2,869,231	183,784	6.4	877,897	30.6	1,722,487	60.0
Pennsylvania	11,561,337	664,767	5.7	4,788,825	41.4	5,874,822	50.8
Rhode Island	896,056	53,653	6.0	344,820	38.5	486,720	54.3
South Carolina	2,914,730	262,740	9.0	1,409,019	48.3	1,119,150	38.4
South Dakota	559,944	61,986	11.1	151,173	27.0	327,868	58.6
Tennessee	3,093,743	324,252	10.5	1,305,270	42.2	1,225,443	39.6
Texas	16,891,646	1,120,400	6.6	7,326,385	43.4	7,975,106	47.2
Utah	1,527,561	106,069	6.9	874,332	57.2	493,354	32.3
Vermont	645,751	32,761	5.1	204,369	31.6	395,643	61.3
Virginia	5,560,451	322,156	5.8	1,729,400	31.1	3,340,445	60.1
Washington	5,086,074	288,382	5.7	3,644,053	71.6	998,770	19.6
West Virginia	1,715,747	129,763	7.6	1,153,764	67.2	406,703	23.7
Wisconsin	4,966,200	216,430	4.4	1,958,288	39.4	2,693,730	54.2
Wyoming	598,728	31,762	5.3	314,216	52.5	242,527	40.5

Source: U.S. Department of Education, National Center for Education Statistics, Digest of Education Statistics 1994 (Washington, D.C.: National Center for Education Statistics, 1994), 153.

Establishing the philosophical basis for fair taxation is beyond the scope of this book.[6] But it is important to recognize that Americans expect fairness, even when they are perhaps not schooled in its classical theory. Although inarticulate in such matters, taxpayers widely concur with two postulates laid down by John Locke (1632-1704) and Adam Smith (1723-1790) who saw persons as rational creatures deserving dignity. Locke argued that democracy was a natural state and that democracy did not permit the unreasonable taking of a person's property, which he defined to include wealth. Smith carried this to its next level, arguing for representation in government, wherein he noted that a tax should be sensitive to the notions of equity, certainty, convenience, and economy. According to Smith, equity demanded sensitivity to one's ability to pay a tax. Certainty required that tax owed must be clear in the amount due, the manner in which it must be paid, and the time it becomes due. Economy referred to the concept that a tax should not result in government storing up unneeded reserves. While these concepts were more complex than presented here, it is easy to see that given a cultural heritage of tax protest in the United States, such thinking was appealing and could become institutionalized in the law.

The views of Locke and Smith seem very much alive in the modern context. Taxpayers want taxes to be fair and with the least possible ambiguity. A good tax system would then be one which takes only what is needed and falls equitably across the full range of taxpayers. Of course, any such discussion would first include concern about whether everyone is paying and would then turn to whether everyone contributes fairly. These concepts raise a whole host of other special concepts or terms, such as whether a tax must benefit taxpayers directly, how ability to pay should factor into a tax, the extent to which tax equity must be required, and a number of other concepts such as tax neutrality and whether a tax merits the financial and political cost of enforcement. Consequently, it is useful to discuss the fundamental concepts of tax systems using a framework of tax basis, tax yield, and tax equity.

Tax Basis

It is most reasonable to begin a discussion of tax principles with a clear recognition that taxation is an empty word unless there is some object to be taxed. Thus tax basis refers to any object against which a proposed tax could be levied. While there is much discussion on finding new revenue sources, it is important to make several observations about such talk. The first observation is that any new source of revenue would require identifying a new object to be taxed. The second observation is that unless the difficult decision to implement a new tax is made, such talk is intrinsically useless. New taxes are seldom welcomed unless citizens can be firmly convinced that some other tax will be reduced.

As discussed earlier, there have been many attempts at identifying new tax bases. In spite of this, only four major bases exist. Despite all the historical difficulties, wealth has been the first and most important basis for taxation. As seen earlier, all three levels of government have tried at various times to tax wealth, and all three levels have experienced difficulty in achieving these goals. Without doubt, real property has been the most important form of wealth. It is easy to see why this has been true, as real estate in the form of land and improvements located on the land suggest wealth in a very visible fashion. Wealth can actually be a much broader term, however, as there are other types of wealth that could be taxed. For example, wealth could be defined to include such things as personal property or money in the form of cash or stocks or bonds. Personal property can be seen to include any number of physical items not otherwise described as real property and is sometimes referred to as tangible wealth. Negotiable instruments such as currency, certificates of deposit, or stocks and bonds are often referred to as intangible wealth. The point is that wealth in its various forms is best represented by possessions, in that people who hold such items do so as a consequence of financial stability and prosperity.

Yet there is another side to wealth as a basis for taxation that must be noted. While it is true that intangible wealth represents the cash position of an individual, other forms of wealth may not be a ready asset and might not even accurately represent a person's financial state. For example, ownership of land could be interpreted differently when looking at income realized from the land. A farmer owning hundreds or thousands of acres might make a very poor living after the cost of real estate taxes and the expense of machinery and crop production. In fact, a farmer could regard land as a liability if it could not be sold in a stagnant market, particularly if the land were mortgaged. Similarly, a retired person's complete ownership of a home could give the appearance of a great asset, while analysis might show the person struggling to meet the cost of taxes and home maintenance. Likewise, buying a home and owning it outright are two different matters, in that a mortgage represents a liability against the homeowner's wealth. In a similar vein, other types of tangible wealth present their own problems as an indication of wealth because some are easy to hide and can even misrepresent one's personal worth. Finally, intangible wealth can also be misleading. For example, it is easy to hide wealth in tax-sheltered investments; in contrast, intangibles may not be a vigorous asset if the economy is struggling. Obviously, the point is that wealth takes several forms that could be the basis of taxation; that wealth as conceptualized is not a flawless picture of true wealth; and, that the high visibility of some kinds of wealth have made it the backbone of tax systems, particularly at the local level. As seen earlier, this has been especially true for schools.

A second basis for taxation that is often used is income. While it might be argued that every kind of wealth was income at the time it was accumulated, the concept of income as a basis for taxation is generally limited to changes in current financial position. Because people think about income on an annual basis, income as a basis for taxation most often refers to familiar concepts like wages, salary, bonuses, and other earned or unearned income represented by cash payments. Unlike other forms of wealth, income is unique because income tax laws allow taxpayers to adjust gross income by special deductions and credits that reduce tax liability. While the conceptual underpinnings of deductions may be subject to question, the differentiating factor seems to be that wealth is a more static notion unrelated to current income, while income is active in a single year. Income is combined with the belief that it costs money to produce income, thereby legitimizing deductions from gross income to produce a net amount on which taxes should be paid.

As discussed earlier, income as a basis for taxation has been significantly captured by the federal government, with nearly every state also taxing income to some smaller extent. To an even lesser degree, some local units of government, such as cities, have also managed to access this tax base, particularly in more urban areas. While schools seldom have direct access to this tax base, state and federal revenue from this source provides aid to schools.

A third basis for taxation is consumption, which is more commonly referred to as retail sales. Sales taxes actually take several forms, depending on whether a tax is general or specialized. A specialized tax is often called an excise tax. Many examples of excise taxes exist, such as taxes paid on motor fuels, car tires, liquor, minerals extracted from the earth, and similar types of objects that are singled out for taxation. In some cases, social goals may be part of the decision to levy excise taxes. For example, there is general consensus that the use of alcohol and tobacco add nothing helpful to society and that these items should be singled out and heavily taxed. Likewise, taxes on gambling in various forms is a type of selective tax that targets an undesirable social behavior. In these cases, the logic of levying taxes is to discourage the behavior and to simultaneously generate significant revenue, since these behaviors are so culturally ingrained as to be practicably ineradicable. In a somewhat different vein, other excises fuel and car tires to a fund for road maintenance and repair. States have often engaged in such taxes on fuel and minerals and gaming, while the federal government has long engaged in steep taxes on many kinds of commodities in lieu of enacting a general sales tax.

In contrast to excise taxes, general sales tax affects almost every retail transaction at the state and local level. While each state has its own laws that determine what is taxed and what is exempt from general sales tax, taxing consumption is a significant revenue source for intermediate and local governmental units. A general sales tax taxes consumption of goods and services under the theory that the tax is voluntary in that consumers willingly choose to pay when they decide to make a purchase. State laws make it difficult to generalize about what is actually taxed, because some states choose to exempt some items from the tax. Exemptions tend to be of three kinds. One kind of exemption pays tribute to an argument against the voluntary nature of sales tax when considering the impact of tax on lower income groups. For example, some states exempt food and medicine in the belief that these items are not purchased voluntarily. Another exemption may stem from efforts to stimulate the economy or from how 'retail' is defined. For example, a state may charge sales tax to a citizen buying lumber, but may exempt the same item when used by a contractor building a house. Thirdly, the completed house may be exempt from sales tax since real estate might not be subject to sales tax given the goal of promoting home ownership or encouraging job growth in the construction industry.

Regardless of how decisions to implement sales tax are made, this basis for taxation is a significant one that yields substantial revenue, for states and also for local governments that convince voters to permit "add on" sales taxes. When legally permissible, cities and counties sometimes ask voters to add a retail sales tax onto the existing state sales tax. For example, a community in a state with a 6 percent retail sales tax might pass a half-cent county sales tax and another half-cent city sales tax, making a total of 7 percent total sales tax. If that community experienced total sales of $100 million in a year, the state would gather $6 million in revenue, while the county and city would each capture an additional $500,000. In sum, general and specialized sales taxes have become a significant basis for taxation. While schools again seldom have direct access to this basis, sales tax often provides a significant part of the money sent to schools in the form of general aid, especially at the state level.

Finally, other bases for taxation exist that are commonly referred to as minor taxes. These generally merit little discussion, since it can be inferred from the name that they generate little revenue when compared to the three principal sources of wealth, income, and sales taxes. Another distinction also applies to many minor taxes. Several such taxes are quite different from the broad scope of other taxes in that they serve a regulatory role wherein revenue actually pays for some service rendered by government that benefits society as a whole. Fees and licenses fall into this category, as such taxes serve the public by taxing people who choose to engage in some regulated activity, often of a commercial nature or one that requires some type of monitoring. Good examples of commercial activities monitored by government might include licensing of businesses where public health or safety is involved. Examples of noncommercial activities could include the operation of national, state, or local parks and public recreation services that need services such as facilities and police patrols. Another obvious example is vehicle licensing, where both cars and drivers must be monitored for safety reasons. While it is clear that these taxes are necessary and useful and can even result in substantial money, it should also be apparent that the revenue they generate is for a purpose quite different than supporting general government and that the amount collected is understandably far less than is true for other tax bases since the user base is also smaller.

It was noted at the beginning of this section that taxation is meaningless unless there is some object to be taxed. While there is much talk of finding new tax bases, the discussion has shown that existing bases of wealth, income, and consumption effectively cover most of the conceivable objects that could be taxed. True, some people have argued that sales taxes could be broadened to take in currently untaxed items or to include taxing each level of production (i.e., the value added tax), or that property could be taxed differently or more comprehensively than it presently is taxed. Still others have argued for a different and more intensive system of income taxation. But in all these instances, critics are not actually proposing new tax bases, but rather they are arguing for a different method of pursuing established tax sources. Thus the search for new taxes by default takes either of two forms. The first form seeks to identify previously unidentified bases for taxes, a marginal activity at best. The second form seeks merely to increase taxes on some group, an activity promising little success unless citizens are convinced that some other tax will be reduced. If the first form is marginal, the second one is limited as well, since it is nearly self-defeating to offset an increase in one kind of tax by decreasing another tax. More appropriately, the search should focus on evaluating the fairness of existing taxes by seeking to make each tax as broad as possible so as to include the largest number of payers, and by making sure that the tax is structured as fairly as possible while generating enough yield to make the tax system worthwhile.

Tax Yield

Discussion of the three principal tax bases of wealth, sales, and income implied rather pointedly that these sources of revenue have been worth pursuing. One of the reasons that these bases have persisted is that each has generated significant tax yield. As implied in the discussion of minor tax sources, a general tax should provide adequate revenue or should be abandoned because it does not make sense to chase something that is not worth catching. A first rule in evaluating a tax would thus be that it should generate a satisfactory amount of revenue in order to justify its existence. While a grossly unfair tax yielding vast revenues would not be justified, a tax regarded as completely fair would not be worth pursuing if it yielded few dollars. In effect, a tax can be judged in part by whether it deserves to be levied on the basis of yield. Two terms useful in evaluating the a tax on the criterion of yield are elasticity and stability.

Elasticity of yield is a concept that compares how the yield of a tax responds to incremental changes in income. If a tax increases faster than income, the yield is said to be elastic. For example, in periods of inflation it is common to hear people complain that taxes go up faster than income. On first consideration, this does not seem to make sense because income had to go up first, with changes in taxes paid following close behind. If tax systems were perfectly tuned, the point might be well taken. However, income tax laws do not usually change as fast as the economy, wherein an income that barely keeps up with inflation may move the taxpayer into a higher tax bracket, while income itself did not result in a higher standard of living. Income is one of the few taxes described by this concept, since income tax in the United States is structured progressively.

In contrast, a tax that increases slower than changes in income is said to be inelastic. Many taxes fit this description, simply because it takes time for tax laws to respond to economic conditions. For example, property taxes are totally disconnected from income except by virtue of the type of housing chosen by different income levels—yet even here, income at all levels can go up (or down) sharply, while property taxes remain unchanged. Finally, a tax that responds to change in income in a 1:1 ratio is said to demonstrate elasticity of unity. It is difficult to think of such a tax, since the income tax is elastic while most other taxes are not related to income at all. In most instances, elasticity of unity can be found only by coincidence.

Performance of a tax on the concept of elasticity is important to its successful implementation. Taxes that are too elastic are resented. A common indicator of this problem is analyses of consumer purchasing power where, despite income growth, consumers are told they have gained little or no advantage on the consumer price index (CPI). Similarly, inelastic taxes can result in unfair benefit or disadvantage to taxpayers. For example, taxpayers with rapidly escalating incomes will not see a commensurate increase in sales or property taxes, and taxpayers faced with declining incomes will not see a drop in sales or property tax rates. Further, while there is a common belief that elasticity of unity is quite fair, it may not be fair at all given different income levels. The sales tax provides a classic example. As income declines, an increasingly larger proportion of income goes toward purchasing basics such as food, clothing, and housing while the sales tax rate does not change in response.

Oddly, any of the three conditions relating to elasticity results in disadvantage for government. If elasticity is significantly present in a tax, government is seen to be taking advantage. If inelasticity is present, policymakers are faced with raising taxes to combat the effect of growth in the costs of government. If elasticity of unity is observed, critics will complain about tax unfairness to lower income taxpayers. The challenge is to create taxes that respond to impossible conditions, in that increased tax revenue is demanded if services are to be maintained, while the need for services increases sharply at the very time when tax revenues may be decreasing—or alternatively, revenues may be responding elastically to favorable economic conditions but are nonetheless subject to criticism that they are preventing taxpayers from getting ahead of the tax curve.

These are formidable problems that represent yet another aspect of tax yield via the concept of stability. Actually, the elasticity of a tax and its stability are variants of the same concept, but they can be approached from slightly different angles to produce thoughtful evaluation. Elasticity describes one perspective on tax behavior, while stability of yield examines the same tax behavior more from the viewpoint of government in need of revenue. A stable tax is simply one which demonstrates constancy of yield, despite changes in the economy. As might be anticipated, a stable tax will not respond immediately to downward changes in income, as in the discussion of property taxes. Alternatively, a stable tax will not respond to upward economic shifts either, again as in the property tax discussion. Under these conditions, a tax may be favored or disfavored by different audiences, depending on whose needs are being considered. In other words, elasticity interests taxpayers most when, from a selfish viewpoint, they will want to see a lack of elasticity in inflationary times and immediate elasticity during economic hardship. Stability interests government most from an equally selfish view in that government's revenue needs almost never decline regardless of boom or bust economic conditions. The difference, of course, is that government must be concerned with both elasticity and stability, since it must placate its patrons while satisfying its own revenue needs.

Tax Equity

The issues of appropriate tax basis and tax yield finally emerge as a tough policy dilemma in a discussion that climaxes by considering some principles of tax equity. All the foregoing discussion has been sensitive to different views on how tax systems can be made equitable. Indeed it was stated that the task for government is to focus on evaluating the fairness of existing or new taxes by seeking to make each tax as broad as possible so as to include the largest possible number of taxpayers and by making sure that any tax is structured as fairly as possible while generating enough yield to make the tax system worthwhile. In the United States, issues of fairness have been of paramount and long-standing concern to taxpayers.

Tax equity is admittedly a difficult concept for many reasons. The first reason has been waggishly stated to be that the only fair tax is a tax someone else pays. Although humorous, the sentiment is also true. The second reason is that not enough is known about tax equity to implement it fully. The third reason is more daunting, however, in that even if enough were known about genuinely equitable tax systems, implementation would still likely be impossible due entirely to the first reason. In other words, a fair tax to most people is fairest when it can be avoided.

While it is true that knowledge about fair taxation is incomplete, it is equally true that much is known on the topic. While disagreement on implementation of any given tax is prevalent, there is general agreement on several concepts believed to make up a good and fair tax system. Once it is established that a tax is broadly based and can generate sufficient yield, the final and ultimate test is to evaluate that tax on the basis of central equity principles. These areas of agreement relate to concepts known as the benefit principle, the ability to pay principle, concepts of horizontal and vertical equity, tax incidence and tax impact, tax neutrality, and directness and certainty of taxes. These equity concepts are described next and, when taken with yield and tax basis considerations and issues about administrative ease and cost of a tax, form the structure of the last part of this chapter which looks carefully at the performance of various major taxes, with implications for school business leaders.

Benefit Principle

The first edition of this text began the discussion of tax basics by describing a tax as having two fundamental elements: a tax is compulsory and unrequited. These elements of a tax become very clear when restated in ordinary terms. The compulsory nature of taxation was described earlier in this chapter when it was noted that few, if any, people would pay taxes if participation were not required. The statement that a tax is unrequited requires some further conceptual development, but the notion is that under the present tax system in the United States, citizens have no right to expect a return on taxes paid either per se for having paid a tax or a return commensurate to the amount of contribution. This text introduces a different term to describe the same concept by discussing what has become widely known as the benefit principle.

As noted elsewhere, many equity decisions are heavily value-laden.[7] As implied earlier, equity often depends on the perspective each party brings to a discussion. Nowhere is equity more affected by such differing views than in a discussion of taxation, where one such value judgment that has been made by policymakers relates to the benefit principle, or the concept of unrequitedness. Yet it is not enough to say that this principle dominates by majority rule and that taxpayers are never requited for their tax payments. In fact, many taxes seem to defy this established equity principle. For example, real property taxes are paid only by property owners,[8] and personal property vehicle taxes are paid only by people who own cars. Even more directly, property owners pay special assessments, which in turn bring personal benefits directly to their homes such as water, sewer, and paving improvements. Likewise, only people who drive pay gasoline excise tax which is often earmarked for road improvements. While these analogies can be picked apart to show that property owners who do not drive also pay general road taxes or that nonproperty owners can use streets paved with someone else's money, there is nonetheless enough evidence of proportionality of private benefit for taxes paid to show at least some degree of requitedness since, for example, one citizen is not usually expected to pay another citizen's special assessment taxes.[9]

Yet, despite the inability to show that not every tax paid is unrequited, there is consensus that a common purse for the general public welfare has sufficient merit to deny anyone the right of a direct return. For example, few people would argue that requiring all taxpayers to support public education is a bad idea, even if some taxpayers have no direct involvement with schools. Similarly, people who never have reason to use fire or police services must still pay because these facilities make a better and safer community. An important derivation of benefit is thus obtained, in that taxes levied for the common benefit result in an indirect benefit to everyone, whether it be safer streets or a better economy resulting from good schools. While it is clearly a value judgment to say that taxpayers have no right to expect a return on their tax payments, it is a principle of equity that rejecting the benefit principle results in greater benefit to everyone, especially those who could not individually pay for the types of services often purchased with collective tax dollars.

Ability to Pay

The previous statement regarding benefit to people who could not afford to pay for services absent tax dollars raises a second criterion of equity. Having broken down the thinking that benefits must be proportional to taxes paid, modern equity theory then goes an additional step by arguing that the amount of tax should not be unreasonable in relation to one's ability to pay.

Application of the ability concept, however, has been incomplete and uneven to the point of confusion. Probably the truest attempt to consider ability to pay has been in structuring the income tax, wherein higher incomes are taxed not only on higher dollar amounts but also progressively higher percentages—hence, derivation of the term progressive taxation. Other taxes, however, fit this criterion very poorly. As noted earlier, property taxes bear no relationship to income (ability to pay) since such taxes are based on all taxpayers paying a uniform tax rate. Under this condition, taxes vary only as property values differ, thereby deriving the term proportional taxation because the tax is a flat proportion of the property's value—a condition that also describes sales tax since these taxes are based on a flat rate and are inconsiderate of ability to pay. Nonetheless, it must also be realized that those who have income generally have greater valued real estate. In fact, the disconnectedness of ability to pay and tax rates is a source of great criticism since it is easily shown that lower incomes pay higher percentages of most forms of tax in relation to income—a condition known as regressive taxation. Ability to pay is therefore a criterion of tax equity, but it is often absent in the real world.

Horizontal and Vertical Equity

A third condition of tax theory is a concept known as horizontal and vertical equity. If achieving equity has seemed formidable up to this point, these terms dramatically complicate the concept of equity. It must be recognized that the conceptualization of fairness up to this point has focused on equality—the equal treatment of equals—which is the concept of horizontality. Under these conditions, equity is satisfied when taxpayers in like circumstances pay like taxes. One such example would be when two persons of identical incomes pay identical amounts of income tax. Another example would find owners of two cars of identical make and age paying the same personal property tax. Still a third example would find all taxpayers in a given community paying the same sales tax rate. Under these conditions, American tax equity seems satisfied by equal treatment of equals.

In reality, however, horizontal equity is satisfied only after equality is assured—the unequal treatment of unequals—which is the notion of verticality. In other words, horizontal equity is found only after vertical equity is satisfied. Using the same examples as before, identical incomes likely should not result in identical taxes, unless income is adjusted for inequalities among taxpayers. This attempt has been made in the United States through the use of adjusted gross income, allowing for deductions such as medical expenses and so forth that effectively make identical incomes very unequal. Apart from income, however, vertical equity (and consequently horizontal equity) is very poorly applied. Using the previous examples, there is no effort made to assure that vehicle taxes have any relationship to ability to pay, and sales tax is often regressive as noted earlier. In other words, horizontal and vertical equity are interdependent of each other and are derivatives of the ability to pay criterion. It is clear that many taxes have serious problems, despite the fact that such equity is a mandatory part of modern tax theory.

Incidence and Impact

As noted earlier in text and footnote, there is an argument to be made about who actually pays a tax. The comment earlier was that only real property owners pay real estate taxes and that only vehicle owners pay personal property taxes on cars. This can obviously be extended to say that only people with incomes pay income taxes, and so on. The truth, however, is that different taxes have different characteristics in this regard, making who actually pays the tax a point of discussion.

The terms tax incidence and tax impact address this problem rather clearly, along with a related concept of tax shifting. If tax systems were concerned about horizontality and verticality, they must be even more concerned about who pays the tax in the end because this is ultimately a vertical equity question. The impact of a tax can be easily discovered, as this is known by simply observing who actually makes a tax payment. For example, real property owners do not really want to pay the tax, but they strive to capture the deduction to gross income that it represents. Thus in all instances, determining impact presents no problem. Tax incidence, however, is a point of much greater hesitation and may imply a vertical equity concern. The difficulty arises with certain types of taxes because one person may be seen as paying the tax (impact), while others actually pay the tax (incidence). For example, the income tax is not susceptible to this problem, as taxes on income are taxes on individuals.[10] Some other taxes, however, may be successfully shifted. For example, the owner of an apartment complex may simply raise rent to cover any tax increase, while simultaneously enjoying an increased tax deduction. In this case, renters bear the incidence of the property tax, while impact falls to the owner as a result of tax shifting. Under these conditions, no form of tax equity is well served, as it may be assumed that most people rent because they cannot afford to buy—a circumstance that does not argue for the renter's superior ability to pay the property tax.

Neutrality

Although building a fair tax system is exceedingly difficult under the conditions laid down thus far, another major tenet is central to tax equity. Modern tax theory demands that any tax levied exhibit neutrality, which requires that a tax leave citizens in the same relative position to one another after the tax as was true before the tax. This means that wealthy people should still be wealthier than poor people after each has paid a fair amount of tax. At first glance, this seems contradictory to the progressive system of taxation discussed earlier, because progressivity demands taking more from wealthy people based on ability to pay. To some extent, the incompatibility noted here is correct. These concepts are fundamentally incompatible unless a very delicate balance is observed which sets off ability to pay and progressivity against neutrality.

Tax structures in the United States often have a difficult time with the concept of neutrality, primarily because of vertical equity concerns discussed previously. Two alternative arguments can be constructed in this regard. One argument states that neutrality is not well observed in most taxes collected in this country because there is no deliberate effort made to assure that each type of tax is sensitive to equity. This argument would see the property tax as not neutral, for example, since there is no relationship between property tax and changes in income. Under this view, only the income tax meets the neutrality criterion. The other argument states that because rich people are still rich after taxes and poor people are still poor after taxes, neutrality is at least marginally observed, even though some attention needs to be paid to whether the rich become richer or the poor become poorer. The latter argument is not callous when calmly considered in light of the original definition of neutrality. Indeed, neutrality demands a lack of movement in relative wealth, rather than social mobility via taxing the poor so lightly that their condition is improved. Nonetheless, neutrality is difficult to achieve, if for no other reason than the inability of critics on both sides to agree on a definition.

Directness and Certainty

A final condition of tax equity relates to perceptions about taxes. This chapter has maintained that a principal reason for tax resistance has been failure to understand that how people perceive a tax. If people perceive a tax to be unfair, it is unfair for all practical purposes. Hence tax equity demands that taxes meet the twin criteria of directness and certainty.

Directness in taxation is a simple concept that appeals to virtually everyone by requiring that a tax not be hidden and that the use of the tax be apparent to everyone. The reality of taxation is such that most taxes are at least not hidden, although some actually suffer from overexposure. Taxes that trickle out gradually, such as sales tax, are less visible, while other taxes like the property tax occur in large and painful lump sums. Neither tax is hidden, although the requirement that their use be apparent may be less uniform. On the one hand, use of the property tax is generally quite apparent. Sales tax, on the other hand, may be less clear to people since the bulk of sales taxes are usually state taxes that disappear into the state's general fund. The other requirement of certainty simply demands that the type of tax and the amount be known in advance. On the whole, these equity criteria are not overly difficult to meet in the modern context of taxation.

Administrative Ease and Costs

A final element in considering the nature of a good tax system is to establish some criteria concerning how readily a tax can be applied. The history of taxation in the United States has demonstrated considerable difficulty in applying many kinds of taxes, and it was seen that productive taxation has been meaningful only in the twentieth century. As a result, consideration of any tax must give thought to ease and economy of administration.

American history and the discussion in this chapter have suggested that some taxes are much more difficult to administer than others. Some taxes are more easily evaded, leaving a commensurately higher burden to those who cannot evade them in order to generate the same revenue as if everyone had paid a fair share. For example, the federal government has devoted great energy to detecting income tax evasion, knowing that taxes are higher on law abiding citizens as a result of tax shirkers. Likewise, some types of taxes are more costly to administer given the method of evaluating the taxable object. For example, it is more difficult to administer real property taxes than it is to collect sales tax. Still other costs must be considered in nonfinancial terms, as some taxes are more invasive than others. For example, a tax on household items would be very difficult to implement in the United States because of the inviolate nature of the home, although such a tax is common in several other countries. Although admittedly difficult to achieve, a good tax system should demonstrate the greatest possible efficiency, taking into account both fiscal and political costs.

Under all the conditions discussed in this section, a good tax system can finally be defined. A good tax system should have as broad a basis as possible, taking into account a number of objects to be taxed so as to spread the tax burden across the greatest number of citizens. Theorists argue that everyone owes some tax, making it necessary that the basis for taxation be so broad as to capture everyone as equally as possible. A good tax system should also focus only on tax bases that can provide an adequate yield, abandoning nuisance taxes, or those taxes that are unprofitable when the fiscal or political costs of administration are taken into account. Finally, a good tax system should be highly sensitive to equity, although definition of an equitable tax is as varied as those who pay the tax. Yet regardless of whether citizens want to pay taxes or not, they will be more supportive of a tax that visibly demonstrates the desirable qualities of an equitable tax system. Accordingly, tax systems should be based on ability to pay insofar as possible, and should be sensitive to both horizontal and vertical equity. Likewise, tax shifting should be eliminated so that tax impact and incidence are identical. Finally any tax should seek a high level of neutrality, and directness and certainty should never be clouded.

Evaluating Tax Bases

Knowledge of the principles of good taxation is prerequisite to informed school business management. School leaders who are knowledgeable about the strengths and weaknesses of various existing taxes are well positioned to contribute to the ongoing debate about new and existing revenue sources for schools. Accordingly, this section looks at the characteristics of wealth, income, and sales tax bases in greater detail in order to aid administrators who become involved in state and local politics about school revenue. As indicated earlier, evaluation of each tax follows a framework of looking at the nature of its basis, tax yield, equity compared to other taxes, and administrative characteristics. While discussion is broad, the implications are directed toward adequate and equitable financial support for schools.

Wealth

As stated earlier, the most important basis for taxation has been wealth. The first edition of this book called the wealth tax "the third workhorse in the tax stable."[11] While not too much was made of chronological order of tax basis discussion in the earlier edition, wealth has been deliberately placed first in this new edition simply because, of all taxes available to schools, the property tax (i.e., wealth) has almost without exception been the only tax base directly accessible to local school districts. Thus, wealth is the logical starting place when evaluating taxes, as all three levels of government have attempted to tax wealth in some form, with highly different methods and degrees of success.

Basis

As alluded to in earlier discussion, the basis of wealth is any one of three kinds of property. Real property has been defined as land or physical improvements attached to the land. Tangible personal property is defined as any other physical property, although for tax purposes this has been practicably limited to items of significant value, rather than mere personal possessions. Personal property qualifying for taxation has thus generally been such items as automobiles, business inventories, and so forth, although it should be noted that items such as trailer homes are considered to be personal property in many states. Intangible property then encompasses other types of property that have no physical presence—rather, they have value only in that they can be exchanged for something else in a market economy. While this concept is vague, the difference is understood by naming such intangibles as stocks, bonds, and other negotiable interests that have cash value. In other words, intangibles are meaningless except for the intrinsic value attached to them in an exchange transaction. For reasons discussed earlier about the high visibility of real property, tangible and intangible wealth as bases for taxation have played far lesser roles in property tax schemes.

Simply agreeing that wealth is a legitimate basis for taxation, however, is insufficient. Wealth must be valued, or assessed to determine its worth. This presents problems, particularly in that different kinds of wealth must be assessed in appropriate ways. Intangible wealth is easy to evaluate because its exchange rate is known at any time, although it can present many tax evasion problems. Both real and tangible wealth, however, are more complex due to the nature and uses. For example, tangible wealth among individuals is different from tangible wealth in businesses. A person may own a car or boat that represents nothing beyond its use for pleasure, while some businesses have massive inventories representing most of their assets. Likewise, real property owned by individuals does not generate income, while realty owned by businesses creates customer traffic that represent the firm's survival. Still another group could be represented by individuals who own investment property, such as apartments or other rental units.

As noted in the first edition, each kind of wealth presents problems requiring different methods of valuing. In the case of commercial property, two common methods of assessing worth prevail in most states. One method requires that property be taxed on its income capacity or rental value—the amount of money such property would bring to the owner in a specified period of time such as one year. In effect, the tax becomes some portion of real or expected rents. The other method usually requires basing tax assessment on the original or replacement cost of a property, less some calculated depreciation amount. The first method works in almost any scenario involving property that produces income. The second method is more useful in valuing property that rarely changes hands and in which continual investment occurs. For example, factories are seldom sold and investment in machinery and equipment for modernization purposes is substantial. Under these conditions, businesses can write off a portion of investment on the logic that equipment is used up and reinvesting in the business benefits the economy, while also representing adjusted income.

In all other types of real property valuation, a different assessment or appraisal method must be used since noncommercial property cannot effectively be valued in relation to income potential. For thousands of communities, noncommercial property is the most important portion of the tax base, since private homes are included in this category. For school revenues, the property tax relating to homes and businesses is a critical aspect, as more than 50 percent of all property taxes collected at the local level are earmarked for public education. The process of appraising property for tax purposes is worth describing, since it is a process with which taxpayers are at least marginally familiar and is the source of many complaints and disputes about tax fairness.

As a general rule, trained appraisers assess all types of property in the various states. Training ranges from very little to extended study, but the purpose of an appraiser is to estimate the value of a property using standard procedures to assure that all properties are valued by the same criteria. First, an appraiser generally conducts a physical inspection of a property to determine its features. In the case of homes, facts such as the number of bedrooms and baths, type of air movement, size of living area, type of construction such as brick or frame, and special features such as swimming pools or fireplaces are recorded. The second step usually looks back at the previous appraisal to determine whether improvements have been made to the property. Still a third step may involve comparing the property's features to recent area home sales in a process often known as market comparables. This step seeks to adjust the new assessed value of the home so that it is valued as closely as possible to the market price of similar homes. This step is a very important one, since the selling price is the best reflection of a home's worth—in other words, it is only worth what a willing buyer will pay. Under these conditions, adjusting for recent sales may result in a higher or lower appraisal by responding to actual real estate market conditions. For new homes, the process is easier by simply noting the selling price of the house. In communities where home sales are slow or nonexistent, however, appraisal is even more difficult.

When the appraised value is finally established, state laws usually govern what happens next. Depending on the state, the process generally calls for applying a statutory tax rate against either the full market value of the property or a fractional value. In either case, the result uses the taxable portion of the home's value. Full market value simply says that the full appraisal price will be used in figuring the property tax owed. States using fractional assessment exempt a portion of the home's value for tax purposes, giving the appearance of not taxing the difference. For example, a full market value state would tax 100 percent of a home appraised at $100,000. In contrast, a fractional assessment state might require that 10 percent of a property's market value be subjected to taxation, wherein tax liability would be figured against $10,000 of the property's true market value. Regardless of which method is used, the process requires the tax rate (also called mill rate) to be multiplied against the assessed value as defined by state law. Using a full market value approach, a home appraised at $100,000 under a school tax rate of 10 mills[12] would yield $1,000 tax revenue. Under a 10 percent fractional assessment scheme, deriving the same revenue would require 100 mills tax effort. Graphically, this is seen as:

[MV x RA x TR]= Tax Due

where:

MV= market value ($100,000)

RA= assessment rate (100% or 10% as alternately shown)

TR= tax rate (10 mills or .010)

Full Market Value Approach	Fractional Assessment Approach
[($100,000 x 100%=$100,00)]	[($100,000 x 10%=$10,000)]
then:	then:
[($100,0000 x .010=$1,000)]	[($1,000 x .100 =.$1,000)]

Whichever process is applicable is repeated for each property in the taxing unit. In the vast majority of states, tax bills are calculated by the legally appointed tax office and sent to property owners. Further, schools and other subunits do not get involved in any aspect of appraisal except to keep an anxious watch on whether valuations are going up or down. The district has no interest in taxes on individual properties except as owners may protest their taxes, but the district is vitally interested in the aggregate assessed valuation, which in most states is the controlling factor in setting the tax rate for schools. For example, a district with a budget of $3 million and an assessed valuation of $ 100 million would require a tax rate of 30 mills ($3,000,000 ÷ $100,000,000= .030). If assessed valuation declines the next year, tax rates must rise to generate the same revenue or the district must reduce its costs. Conversely, increasing assessed valuations results either in extra revenue or reduced tax rates. Districts take particular interest in tax rate changes, as they know that school districts are only one of many taxing units, each with its own tax rate that joins with other tax rates to make a total tax rate paid by local taxpayers. The many differences among states make it difficult to generalize, but this example serves to indicate how individual taxing units are affected by the appraisal process.[13]

The first edition of this book noted that American education has largely been financed by the property tax since its beginning in New England. Given the complexity of property as a basis for taxation and given the historic reliance on property as the principal local expression of wealth, the pressure on school districts and local governments to reduce taxes is understandable. But it is also clear that any reduction in property tax results in one of two realities: a corresponding reduction in revenue to property tax-dependent entities such as schools, or an increase in some other tax to recapture lost revenue. Thus in states, such as Michigan, that have effectively eliminated the property tax for public education, the bulk of these moneys have been replaced with state income and sales tax revenues. Tax increases of any kind have been difficult in recent years, but property tax reduction in the form of tax reform and new inventions such as circuit breakers, e.g. age qualifications, disabilities, etc., and homestead exemptions have grown in popularity. Likewise, tax exemptions for various properties such as charities and tax abatements that supposedly create economic development have added pressure to the revenue crisis. Yet, as will be seen, the tremendous yield of the property tax has institutionalized it as the primary revenue source for local government.

Yield

The earlier discussion of federal, state, and local tax systems gave clear indication of the tremendous revenue yielded by taxing wealth. While wealth taxes take different forms at each level of government, the property tax at the local level is no exception to the concept of wealth as a powerful tax base—in 1992 the property tax yielded more than $ 1.22 trillion. Domination of this tax base by local governments was very clear, as the federal government captured no property taxes and as states gathered only a fraction of the total amount collected, raising just $328.4 billion. While these two levels of government were greatly different in the reliance on this tax base, total yield indicated vast profitability of taxing property in support of government operations.

For purposes of this discussion, yield of the property tax also needs to be considered on the criteria of elasticity and stability. While these data are mixed on whether the tax is elastic, the intermediate conclusion has been that it shows at least elasticity of unity. A few studies have suggested that it may be slightly inelastic, but there is an argument to be made that data have not been reliable due to the number of tax units that have experienced tax limitation movements—a phenomenon that could affect the ability of the property tax to keep pace with changes in income. Similarly, the many exceptions built into various states' property tax laws such as circuit breakers and homestead exemptions have an impact on these data by reducing tax revenue in ways that are not commensurately reflected in the income data against which elasticity is measured. As a general rule, however, the property tax has not been demonstrated to be either highly elastic or inelastic, leaving the tax in a relatively neutral position on this dimension.

On the other hand, the property tax has demonstrated every good feature sought in a stable tax. Since it is not linked to income, changes in the economy brought about by income shifts are very slow to show up in property tax data. In an unfortunate sense, the era of the Great Depression illustrates this point—many home owners experienced tax foreclosures because the property tax did not respond quickly enough to fiscal distress. In more recent times, the stability of this tax might have driven some portions of the tax revolt movement typified by Proposition 13 that caused significant shortfall in many other states as well. Short of disastrous conditions, however, stability of the property tax can be said to result in smooth revenue streams for government. Stability, coupled with yield, has been the primary force for using the property tax.

Equity

Heavy use of a tax, however, does not always indicate whether equity is present in the tax. As seen throughout this chapter, the property tax has been subjected to serious criticism by citizens who, rightly or wrongly, perceive the tax as unfair. As a result, some consideration of how the property tax performs on the dimensions of ability to pay, its relationship to progressive structure, horizontal and vertical load, impact and incidence, neutrality, and directness and certainty is important to continued reliance on this tax in support of government.

As noted elsewhere, the property tax exhibits some elements of both the benefit principle and the equity standard of ability to pay.[14] This dual tendency refers back to earlier discussion relating to the exclusionary benefit of some special assessment taxes wherein property owners are the primary beneficiaries of taxes paid. For example, some neighborhoods pay higher taxes than others, with resulting improvements that benefit local residents. These improvements are almost never exclusive of nonresidents, however, in that streets and parks and other improvements are generally open to everyone. Benefits spill over to the general public by reducing the number of improvements that would have to be publicly funded if those private communities were not to provide separate facilities for themselves. This does not mean, however, that property taxes are based on ability to pay except in the sense of higher priced properties having higher tax bills.

Depending on one's view of equity, a further criterion is whether the tax demonstrates progressive, proportional, or regressive tendencies. For reasons discussed earlier, many critics argue that the best taxes are substantially progressive in structure. As briefly noted, property taxes are alternately shown to exhibit all three behaviors, although they are shown never to be simultaneous. Although tax rates are applied uniformly, proportionality is offered by people choosing the rates they pay by their choice of residence. This view argues that wealthy people pay more tax, and poorer people pay less tax by virtue of residence choices. Regressivity is offered by the reality that lower income people pay a higher percentage of income as property taxes, despite the lower tax bill resulting from lower housing choices. Progressivity is shown by the lowest income group who may pay shifted property tax using public assistance dollars. This view argues that the larger society is paying all or part of a tax that would otherwise be inequitable. The result is that the theory base is not well founded and that assessment of progressivity cannot be accurately made until greater agreement is reached on how to determine equity of the property tax.

Still a third criterion of equity is consideration of horizontal and vertical performance of any tax. Horizontal equity requires equality among equals, and preferential treatment for unequals to better produce end equality. It is entirely reasonable to believe that owners of identical properties paying identical tax rates are recipients of horizontal equity. In the case of the property tax, however, several dilemmas arise that cast doubt on that belief. It is difficult to identify taxpayers in exactly equal circumstances for several reasons. One reason is that outward appearances may be deceiving, as in the case of a mortgaged property versus a debt-free title. A similar case can be made for retired people struggling to meet taxes and maintenance costs. Another reason is that property taxes can be shifted, particularly as in the case of renters who may end up paying a tax while the owner takes a double tax benefit by deducting taxes as an expense in addition to not actually paying the tax. The problem, of course, is that the property tax is unrelated to the income from which taxes are eventually paid. If horizontal equity performance were thus difficult to test, it becomes clear that vertical equity performance is in serious question. Still, other devices such as circuit breakers have introduced some degree of verticality to the system, making it even more difficult to confidently assess equity of property taxes on this criterion. Little progress is likely on this front because of the disconnectedness between the property tax and income, making such issues and incidence and impact difficult to disentangle cleanly.

A fourth equity consideration of property tax is the concept of neutrality. Depending on one's equity definition, property tax can be seen as neutral in that it falls equally to everyone owning property within a tax jurisdiction by virtue of uniform tax rates and conscientious attempts to assess properties on an equal footing. The property tax falls suspect to the more direct definition of neutrality, however, in that it should leave a taxpayer in the same financial position in relation to other taxpayers as was true before the tax was paid. Absent a link between income and property taxes, neutrality cannot be certain since it is possible to lose ground compared to other taxpayers after the property tax is paid. Again, the best example is a person on a fixed income whose property tax continues to rise without corresponding increases in income. Only in the narrowest sense can it be claimed that the relationship is constant by inferring that the rich will still be richer than the poor after property taxes are paid. These and other realities such as tax abatement trends cast doubt on the neutrality of the property tax.

The final equity criterion against which to judge a tax is directness and certainty of tax payments. In this regard, it is required that the tax be known in amounts and purpose. The property tax performs better on this dimension than on any other criterion except yield, in that the exceedingly local nature of the tax places it in high visibility for expenditure on local needs. Indeed, it can be argued that one of the best reasons for persistence of this tax is that people identify it with projects that benefit them personally, such as schools, hospitals, police, and fire protection. Accordingly, a high degree of local accountability for this tax glosses over a multitude of other sins represented by the property tax. As a result, property taxes are both reviled and valued at the local level, while performing in an uneven fashion on other equity criteria.

Administration and Economy

A final characteristic of a good tax system is economy and ease of administration. If a tax is expensive to apply and difficult to administer, it will be resented and result in low yield and evasion. It should be obvious that some taxes fare better when evaluated for administration and economy, while others perform poorly because of inherent problems.

The property tax presents a mixed picture on these criteria of good tax systems. Obviously, the recordkeeping requirements involved in creating the appraisal system are enormous, and maintaining the system is difficult since the data base must be kept as current as possible. The sheer number of parcels of property creates a severe workload for an appraiser's office because each property is unique and government offices are seldom overstaffed. Assuring that appraisers do a good job and that multiple appraisers are consistent in valuations is also a difficult task. In many states, appraisers either have not done a good job by virtue of ignorance or politics or as a consequence of simply too much to do. The result has been that numerous states have had to undertake massive statewide reappraisal, either under court order or just ahead of legal action because appraisals have been so uneven as to result in claims of unconstitutional taxation. Even in states where no such threats have emerged, appraisers are overworked because keeping an appraisal system current demands frequent physical reinspection of properties and sophisticated computer analyses to track and adjust property values on a methodical basis.

Yet despite the high cost of administering the property tax from these perspectives, it is a highly efficient and economical tax that is also relatively easy to administer. Ease of administration comes from inability of taxpayers to avoid this tax, since nonpayment will result in a tax lien and eventual tax foreclosure. Likewise, the extraordinary yield of this tax far outweighs its administrative cost. Low levels of noncompliance and high return on administrative costs make it both efficient and economical.

Conclusions About Wealth

Evaluation of wealth as a basis for taxation yields several valuable observations. First, it is commonly believed that wealth represents the basis on which people should pay tax. Second, it is easy to see why people have long believed that wealth is best represented by real property since it is so visible. While this does not preclude other forms of wealth, such as tangible or intangible property, it is even more understandable why real property is so heavily taxed when considering the difficulty of implementing and enforcing other wealth taxes. Third, it is also understandable that the massive yield of the property tax has led governments to overlook some of its weaknesses. Fourth, government has been allowed to continue the property tax despite its flaws, primarily because it has been viewed somewhat tolerantly as a local tax with direct local returns. But most importantly, the property tax has endured because local units of government have had no other meaningful choices.

For schools, this reality has caused the property tax to be overused as states have been slow to respond to the stress that local governments have experienced.[15] It is often forgotten that local government has many roles, of which education is only one. Although state aid to education has grown, the local share has declined as a percentage but not in total dollars raised. As a result, citizens have seen federal, state, and local taxes rise in tandem, despite increases in other types of intergovernmental aid. With the property tax being the most visible and most local of all taxes paid, it is not surprising that citizens have responded by seeking to reduce revenue at the one level of government most closely associated with their daily lives—an event that threatens many school districts all too often.

Income

A second major basis for taxation has been income. Although it was pointed out earlier that the income tax has been captured in large part by federal and state governments, income is nonetheless important to school funding for either of two reasons. The first reason is that state aid to schools in certain states includes a significant income component, even though schools themselves may not have had direct access to the income tax. The second reason is that in any discussion of modified school tax bases, income is likely to play a significant role in that newly discovered tax bases are only a remote possibility. But like other taxes, income can be found to have strengths and weaknesses when evaluated by the criteria of a good tax system as proposed in this chapter.

Basis

Although the basis of the income tax is obvious by its name, it is important to remember an earlier distinction made between wealth and income. While some people might correctly see income as intangible wealth, income as a basis for taxation is different in that society's rules for conceptualizing and taxing income have been structured quite uniquely. Unlike other forms of wealth, income is defined to measure the change in financial position of individuals or businesses over a limited time period. It is further differentiated in that income is made up primarily of cash payments related to earned or unearned income.

The basis of income is also quite different by virtue of how it is treated for tax purposes. As noted, the inclination to view income in relation to the cost of living has resulted in adjustments to income, usually in the form of deductions for certain allowable expenses of a personal or business nature. For individuals, common deductions from gross income are permitted for such things as mortgage interest, medical expenses, and dependents. The logic is that there is a minimum threshold for the cost of surviving which ought to be untaxed and that social policy can be enhanced through tax concessions. The latter case is particularly evident in deductions of mortgage interest and medical expenses: i.e., encouraging home ownership is a core American value, as is the vertical equity notion that good health should be government-subsidized via tax deductions. For businesses, allowing deductions for expenses is a recognition that it costs money to realize earnings, a concept that takes in the notions of gross income and net income. The result is that the basis of income is quite different from other wealth by limiting the period of tax liability and by adjusting gross income in recognition of certain costs.

Unlike other forms of wealth, however, it presents no problem to determine the value of income. No appraisal system is needed, and no problems arise as a result of different opinions about what an asset is worth. The problems confronting income as a tax base relate more to capturing all income so that no income goes untaxed, and to resolving difficult discussions about how the income tax should be structured. As with other forms of taxes, citizens want to pay the least amount possible—a reality resulting in both tax evasion and great disagreement about whether existing tax policy takes too much tax and whether the tax is properly structured to distribute the tax burden fairly. As a result, the income tax must be subjected to scrutiny for its yield and equity.

Yield

Earlier discussion about federal, state, and local tax systems gave clear indication of the tremendous revenue yielded from taxing income. While each level of government has varied access to the income tax, the total amount of tax collected in 1992 was truly astounding. Total income tax paid by Americans in that year was $591.6 billion, with the federal government taking 80 percent of all receipts at $476.5 billion. As noted, most states have joined in taxing income, collecting $104.6 billion in that same year for a share equaling 17.7 percent of all income tax receipts in the nation. Although local governments have not widely made a case for taxing income, local governments in states permitting a local income tax still managed to gather $10.6 billion in 1992, primarily in cities desperately seeking to make up for other declining tax bases. These data suggest that despite the difficulty in establishing income as a basis for taxation, income has resulted in a highly profitable source of revenue for supporting government services. Yet despite the fact that people have come to accept income tax as a permanent reality, the tax is not without merit or criticism on other dimensions making up a good tax.

It may be correctly observed that the tremendous yield of the income tax is due entirely to conditions relating to the structure of the tax itself. A good tax system was argued earlier to exhibit high yield, combined with other properties such as elasticity and stability. The income tax has been so tightly structured that its yield has been great by virtue of inability to escape tax liability through devices such as payroll withholding of taxes owed. This reality has been greatly aided by the progressive structure of the tax, particularly at the federal level. Federal income tax code follows two kinds of certain logic, wherein tax revenues increase as income rises and, by virtue of a progressive rate structure, the tax rises again with each successively higher income level. For example, in 1992 initial tax rates required a married taxpayer making $35,800 to pay 15 percent in the form of federal income tax. If income rose to $35,801, however, the tax rate threshold changed to 28 percent, based on the belief that the taxpayer was better able to support society's needs. Likewise, incomes above $86,500 paid 31 percent. Conversely, low income taxpayers might pay no taxes after adjustment for dependents and other allowable deductions. To a lesser extent, the same progressive rate structure has crept into state income tax codes.

The scenario just described has been judged to attribute a fairly high degree of elasticity to the income tax. Progressive tax systems are fundamentally elastic, since revenues will increase faster than incomes so long as income growth is sufficient enough to push taxpayers into successively higher tax brackets. During periods of normal or vigorous inflation, elasticity is quite evident, although this creates some problems as discussed. During periods of economic downturn, however, a different set of problems may be encountered, in that the stability of the income tax is weakened to a greater extent by changes in taxpayers' economic condition than is true of the property tax, for example. Since both federal and state governments are generally quite dependent on the income tax for general fund revenue that pay for a host of government programs, periods of significant economic recession could produce lagging income tax receipts, which in turn could severely restrict services.

Equity

Given the tremendous criticism of property taxes on the criterion of equity, many reformers have turned to the income tax in search of a more equitable method of exacting revenue from taxpayers. The search for a more equitable tax has been perennial, as each tax has shown itself to have less than perfect scores on several of the many characteristics of desirable tax systems. With the tremendous yield of the income tax and promotion by advocates of this tax, income as a basis for taxation has been closely evaluated for its equity performance.

While the property tax has shown mixed behavior on the benefit principle, there is little doubt that the income tax is disconnected from any expectation of receiving direct benefits from the tax, particularly at the federal level. In part, this is due to the fact that income taxes at the state and federal level are viewed as swallowed up by some distant treasury. At the same time the progressive rate structure, first built into the federal system and then adopted into many state tax codes, aids in denying the benefit principle in that progressivity holds the concept of ability to pay as its fundamental root. While people do not like progressive taxation because they argue it is a disincentive for personal industry in a capitalist society, they have nonetheless had to accept progressive taxation because tax laws are so punitive as to discourage nonconformity. The current progressive rate structure of 15 percent, 28 percent, and 31 percent is based on what the first edition of this book called marginality, or the concept that the last income dollar is less needed by a taxpayer than the first dollar. Marxian in principle, the theory argues from each according to ability, and to each according to need. Marginality is aided by the operation of adjusted gross (or net) income, wherein deductions for expenses is allowed to individuals and businesses. If ability to pay were a valid criterion for tax systems, design of the income tax theoretically fares well.

Of particular interest to reformers is performance on horizontal and vertical equity dimensions. This is inextricably entwined with how well the tax does on the impact and incidence criterion, in that a tax that can be shifted is not viewed as very equitable. The progressive structure of the income tax contributes favorably to evaluation on these criteria, as the unequal tax burden among income classes is intended to weigh against higher income groups by forcing them to pay a larger percentage of income in support of government. Horizontality is aided because everyone, regardless of income, pays the same tax rate on the first dollar earned, while rates increase only when income bracket thresholds are exceeded. Verticality is thus aided by the exact same design, wherein lower income groups avoid the tax burden paid by higher income groups. At the same time, the concept of adjusted gross income further aids verticality, because it is an attempt to mitigate income inequality on a personal level by subtracting subsistence allowances for dependents and other approved expenses. This performance is strengthened by evaluation of the impact and incidence of the tax, wherein most observers agree that impact and incidence are identical in that income tax cannot be shifted.

On the basis of structure alone, the income tax seems to fare well on horizontal and vertical dimensions. Operationally, however, critics are prone to question its performance as a result of features commonly built into the income tax code. Progressivity, for example, can be hindered by other well-intentioned policies such as tax exempt instruments that make it possible for very wealthy people to earn tax-free income that may result in less tax due than is true for low income wage earners. Similarly, taking advantage of tax deductions such as personal business expenses or home mortgage interest requires a taxpayer to first qualify for itemizing deductions when filing an income tax return, thereby likely disqualifying truly low income people from the advantages of these vertical aids. Likewise, the criteria relating to progressivity and neutrality is somewhat in conflict if it is accepted that neutrality leaves everyone exactly positioned after tax than before the tax: i.e., a delicate balancing act is involved to assure that progressivity does not leave the wealthy person even slightly less wealthy or the poor person even slightly more wealthy. Regardless of how finely one wishes to tune the equity equation and regardless of one's predisposition to believe the tax is progressive or regressive, there is evidence as seen in table 3.6 suggesting that the income tax falls in higher proportion to wealthier citizens.

The final equity criteria of directness and certainty also provide mixed performance data on the income tax, depending on one's viewpoint. From one view it can be argued that, if the tax lacks directness and certainty, it is because citizens fail to appreciate the vast interlinkages of federal and state governments in their daily lives. As seen earlier, federal revenues provide many forms of aid to citizens, ranging in a wide spectrum to include services such as welfare and Aid for Dependent Children (AFDC), aid to public and higher education, national defense, health services, roads and highways, federal justice, social security, and many other services and intergovernmental aids to states and localities. State income taxes also fund the majority of state program spending in many states, bringing the directness of this tax close enough to allow people to see and appreciate its use. Local governments that have managed to access the income tax are likely benefited even more in this

Table 3.6
Income Tax Burden by Income Class 1985

Income Decile	Effective Individual Income Tax Rate 1985
Progressive Assumptions	
Lowest	3.8%
Highest	13.5%
Regressive Assumptions	
Lowest	4.0%
Highest	14.5%

Source: Joseph Pechman, *Who Paid the Taxes in 1966-86.* Washington, D.C., Brookings Institution, 1986 (revised).

regard, as taxes become more visible as the distance to a taxing unit decreases.

On the other hand, performance on certainty and directness is less than adequate because people do not naturally favor the compulsory nature of taxation. The tax is certainly not hidden—although as noted elsewhere the income tax trickles out unnoticed by virtue of payroll withholding, and the annual accounting to state and federal internal revenue services is disliked by citizens who must focus once a year on the complexities of taxes. This accounting draws greater attention than would otherwise be warranted and results in an entire accounting and legal industry whose existence is based in helping citizens reduce tax burdens and defending taxpayers charged with wrongdoing.[16] Similarly, uses of tax moneys are not always understood or appreciated by taxpayers, many of whom see state and federal government as remote bureaucracies with unclear objectives and wasteful spending proclivities. Finally, the criterion of certainty is often poorly met, as citizens either pay additional tax at year-end or receive refunds for overpayment.

Administration and Economy

A final characteristic of a good tax system is economy and ease of administration. Taxes requiring great expense to assure compliance are not viewed favorably by anyone involved. Despite bad publicity about income tax evasion and an underground economy that thrives without a tax burden, the income tax presents a favorable picture on these criteria. Payroll withholding of taxes owed virtually assures that legitimate workers pay their taxes, and computer network systems have greatly increased the ability of government to note unreported income deposited to banks and to have banks report interests and dividends paid to depositors. Similarly, constant fear of audit and tax seizure for back taxes keeps many taxpayers more honest than might otherwise be true. Because of payroll accounting and compliance measures that are in place, administration of the tax is simple in comparison to the compliance system that would be required if each taxpayer were independently responsible for payment. Similarly, the high rate of compliance makes the tax quite efficient because the considerable cost of compliance is offset by the phenomenal yield of this tax.

Conclusions About Income

Evaluation of income as a basis for taxation yields several valuable observations. First, it is true that income represents the basis on which people should pay tax, i.e., people pay taxes using income, and it represents their cash position at any given moment, assuming wise consumer behavior. Second, it is also true to say that when evaluated on the criteria of a good tax system, the income tax performs better than some other taxes. Particularly, its basis is broad by capturing everyone legitimately earning income. Further, its structure is designed to provide high yield, elasticity, and a reasonable degree of stability. In addition, the tax conforms to many principles of an equitable tax system, particularly demonstrating the concepts of ability to pay and vertical loading. Third, although the tax does have weaknesses, these seem to be functions of operation, rather than flaws in the system itself. For example, mixed performance on neutrality and directness and certainty are either missteps on the part of government or misperceptions on the part of taxpayers. Fourth, it is clear that government has found a powerful revenue base in the income tax, although as a general rule lower units have successively less access to it. Finally, the massiveness of the income tax base has made intergovernmental aid possible in a way that was not previously available under the historic tax systems in this nation.

Income as a basis for taxation has had a greater impact on public education than is commonly realized. While admittedly most states do not permit schools to access the income tax directly, schools nonetheless benefit immensely as state coffers providing aid to schools were likely partially filled with income tax revenues. In 1990, 32 percent of all state revenues were derived from the personal income tax, with the remainder coming from various other sources including sales tax, corporation income tax, and excise taxes.[17] Similarly, federal aid to education comes in significant part from the federal income tax, since this tax represents the federal government's main source of revenue. It is accurate to say that if it were not for the income tax, schools would be desperately looking for other ways to regain lost aid, which unfortunately would cause them to turn in most instances to the property tax. Because the income tax has many of the attributes of good tax systems, effective school business administrators should work with policymakers to assess and expand the role of income in school funding schemes.

Sales

The third major basis for taxation has been a tax on consumption or, as more commonly named, the sales tax. If the property tax has been the mainstay of local government and the income tax the mainstay of federal government, sales tax has been a powerful force at the state level with considerable impact at the local level as well. In a limited fashion, a variant on sales tax has been a revenue source for the federal government as well. For schools, sales tax has been such an important indirect source of revenue that it behooves school business administrators to understand its behavior and potential as a continued revenue source. But like other taxes, the sales tax has strengths and weaknesses when evaluated by the criteria of a good tax system.

Basis

It seems obvious to say that the basis of a sales tax is sales. The concept is somewhat akin to people paying tax for the privilege of buying some commodity in the marketplace. The sales tax can be broader than just retail sales, however, as there has been talk of new taxes that would essentially be based on sales defined differently than the historic view of sales tax in this country. Conceptually, sales taxes as used in the United States can be divided into two main groups. The first group of taxes are the traditional retail sales wherein an item is taxed at the point of sale. The second group can be identified as selective sales taxes that were referred to as excise taxes earlier in this chapter. Although not used in this country, still a third level of taxation can be described wherein sales tax is charged at one or more levels of production, distribution, or wholesale exchange. This latter type of tax has long been referred to as a value added tax (VAT) and is common in some other countries, although lack of interest in this country has caused it to be virtually ignored.

As described earlier, the federal government financed its operations for many years using only excise taxes, primarily in the form of customs duties and import/export excise taxes. While the federal government is now primarily dependent on the income tax, it continues to raise millions of dollars in excise revenue by taxing such commodities as gasoline, tobacco, liquor, automobile tires, and the like. In fact, federal excise revenues in 1992 reached $46.8 billion. States have also been engaged in the excise tax business, as sales of gasoline and other selective commodities are often heavily taxed at the state level. In fact, in 1992 state excise taxes gathered a total of $54.9 billion. The excise tax has traditionally served one of several purposes. In the case of the federal government, it once served the role of being the only practicable tax that could be levied. A second role has been to control the balance of trade, as in tariffs designed to discourage excessive imports or exports. A third role has been to influence behaviors believed to be undesirable. For example, excise taxes on liquor and tobacco have been intended to discourage consumption, while at the same time capitalizing on the lack of success experienced by such taxes.

Although excises generate significant revenue, the real workhorse in the sales tax arena is the general sales tax at the retail level. This tax is levied at the point of sale on all nonexempted items within a given state. Because the federal government has never enacted a general sales tax, states have been quick to seize on this tax base, with city and county governments also levying sales tax to a lesser but meaningful extent. The revenue generated by state sales taxes is phenomenal, as in 1992 when retail sales tax generated $162.7 billion in those states with such taxes. The rush to employ the sales tax base is understandable, as states have been relegated to second place behind the federal government in the race for income tax revenue, and have likewise permitted local communities to dominate the property tax. Local governments have also joined the rush, generating another $33.4 billion in the same year. Most states have moved to adopt a sales tax, with many states also permitting lower units of government to adopt add-on retail sales taxes at the local level. As of 1994, only a very few states levied no general retail sales tax.

Yield

If one criterion of a good tax system were that a tax must generate significant yield, the sales tax clearly meets that criterion. Judging from the $196.1 billion received from general sales taxes at the state and local levels in 1992, it is clear that lawmakers have understood just how well this tax conforms to the productive yield criterion. Massive yield should not be surprising, however, because several realities preclude any other result. First, it is axiomatic that people cannot survive without market activity. In any civilized society, a system of exchange between producers and consumers exists so that a ready base for implementing a sales tax is inherent. Second, domination of most other tax bases by competing levels of government has forced aggressive utilization of the sales tax, particularly by state governments for whom the sales tax base often represents its greatest revenue source. Third, a general sales tax is by definition broad, taking in a vast array of items that, by small and constant increments, collectively generate huge revenue. The result becomes that an aggressive sales tax trickles simultaneously from so many sources that it ends up being a constant and heavy revenue stream.

Inherent in evaluating tax yield are concerns about whether a tax demonstrates elasticity and stability. Like other taxes, the sales tax has its own set of peculiar characteristics relating to these criteria. There is some disagreement about whether the sales tax is elastic. One school of thought suggests that the sales tax is elastic because it can be shown that tax revenue has had a steeper rate of increase than income over the same period of time. Another school of thought argues that this is true for undesirable reasons, in that revenue growth is a function of inflation driving up the price of goods, while incomes have not always kept pace. Critics also point out the recent propensity of revenue-hungry governments to raise sales tax rates, a reality that has been seen in many states in recent years. Still a third perspective on sales tax elasticity has been advanced, arguing that people buy more goods and services when their incomes are robust.

While sales tax elasticity is perhaps not resolved, the evidence is clearer regarding stability of the sales tax. In contrast to the highly stable property tax and the moderately stable income tax, the sales tax shows considerable instability for several reasons. The first reason is that sales of many goods indeed track the economy, as people put off major purchases until consumer confidence rises. In this scenario, sales tax revenue may drop sharply if people are not buying automobiles, televisions, refrigerators, and so forth. Second, consumers watching their expenses will be cautious even in areas of subsistence, buying hamburger instead of steak or buying generic labels instead of more costly name brands. While these may seem like small differences, millions of consumers saving pennies results in many sales tax dollars lost at the cash register. Third, consumer pessimism may stimulate growth in savings accounts. While this may seem to be a desirable goal, a negative result is lack of consumer activity. The sum of elasticity and stability suggests that the sales tax is a consumer-oriented tax that demonstrates significant volatility.

Equity

As seen in discussion of earlier tax bases, good tax systems must demonstrate a high level of equity if continued use is to be promoted. As might be expected, performance of the sales tax on the various equity criteria is better than for some taxes, but worse than others.

A central criterion of an equitable tax is whether it conforms to the ability principle. The sales tax is a poor performer in this regard, as the only link between income and sales tax is the taxpayer's financial ability to absorb the sales tax. The closest link that can be made in that instance is that income can be seen to drive consumer decisions, especially in discretionary buying where the price of the object plus tax might make it available only to higher income earners. Clearly there is no attempt at progressivity where the higher income taxpayer would be required to pay a higher sales tax rate. In fact, the reverse analogy is often used to suggest that the sales tax is regressive by arguing that the flat sales tax rate is a heavier burden on lower income groups who pay a higher proportion of their incomes in sales tax. Critics are particularly hard on the sales tax in this regard, arguing that food, clothing, medicine, and so forth should be exempt from sales tax since these are basic subsistence requirements.

The sales tax also scores fairly low when evaluated for the third criterion of horizontal and vertical equity. Part of the problem in this test rests in the fact that the sales tax is both compulsory and voluntary. Obviously the tax is compulsory in that any qualifying item requires the tax to be paid, and taxpayers are effectively compelled to buy such items as food, clothing, and so on. But the tax is also voluntary in that the consumer has a choice in whether to buy many items, and these choices generally correspond to financial well-being. For example, prosperous people can choose to dine out more often and in expensive locations. They can also choose to buy more expensive vehicles or travel by other costly means. Wealthy people are also free to buy more and better clothing. Importantly, wealthy people are free not to tax themselves by simply not buying such things, thereby adding the savings to their existing resources.

Prior to introducing such concepts, it would have been reasonable to see the sales tax as horizontally equitable since the sales tax affects everyone at the same rate. On second thought, however, considerable vertical inequality must be seen in the sales tax, primarily due to the unfavorable relationship between percentage of income paid as sales taxes for high income groups compared to low income groups. As seen in table 3.7, estimates of horizontal and vertical equity flowing from sales tax in relation to income show the lowest 10 percent of incomes paying between 7 percent and 7.2 percent of their earnings in the form of sales tax, while the highest 10 percent of incomes pay only 1.9 percent of their earnings as sales tax. Since vertical equity is the condition that produces the end equality sought by horizontal equity, it is reasonable to conclude that the sales tax provides neither.

A third equity trait sought by good tax systems relates to impact and incidence of a tax. Similar to income, the sales tax is quite clear in its performance on these criteria. Unlike income, however, the sales tax is a poor performer for very sensible reasons. Impact is simple to determine because the point of sale defines impact, i.e., the retailer. In contrast, it would be naïve to think that the retailer is actually paying the tax. Instead, it is quite clear that the tax is being paid by the consumer, who absorbs the tax incidence. The tax is thus shifted forward to the consumer, while the retailer merely acts to collect and remit the tax to the taxing unit. This concept serves to illustrate two additional points. First, impact and incidence can never be the same, as any effort to put the tax on anyone other than the consumer would merely increase the price at some pre-retail level, such as the wholesale price. Second, this effectively denies any breakthrough on a new sales tax such as a VAT because any pre-retail tax would simply be shifted forward to the consumer as many times as necessary—conditions that clearly go back to poor horizontal and vertical equity and would only increase prices.

A fourth equity criterion demands neutrality. If we accept the definition that a tax should not affect taxpayer behavior, then the sales tax fails this test. As discussed earlier, people may postpone purchases due to recent sales tax in-

Table 3.7
Sales Tax Burden by Income Class 1985

Income Decile	Tax as Income Proportion 1985
Progressive Assumptions	
Lowest	7.0%
Highest	1.9%
Regressive Assumptions	
Lowest	7.2%
Highest	1.9%

Source: Joseph Pechman, *Who Paid the Taxes in 1966-86.* Washington, D.C., Brookings Institution, 1986 (revised).

creases, or alternatively may rush to buy some item before tax rates increase. Likewise, consumer behavior is affected by the variable nature of sales taxes, wherein one might drive further to reach a taxing jurisdiction with a lower sales tax rate. Unfortunately, excise taxes seem to be the instances in which consumer behavior is not changed by either the tax or the economic conditions. For example, tax data on tobacco and alcohol and gambling show that these behaviors are not changed by any tax.

Performance of the sales tax on the final criteria of directness and certainty is somewhat better than for the preceding equity conditions. The tax is direct and obvious at the point of sale, and it has the added benefit of being incremental and gradual in its effect. Certainty depends on how informed each taxpayer is. Excise taxes, however, are somewhat less visible in that they are often built into the price of the commodity. But since the consumer pays the stated price of an item subject to excise tax with no other tax added, there are usually no adverse implications.

Administration and Economy

If taxes were judged solely by the criterion of administrative ease and economy, the sales tax would be a clear winner. While other taxes have been seen to require considerable effort to monitor and enforce collection, the sales tax is virtually effortless and cost-free. This is made possible by the method of collection, wherein the retailer serves as the assessor and remitter of these taxes, with no effort on the part of the taxpayer. In fact, retailers perform a function worth millions of dollars because their service is without charge to the taxing entity. The only cost to government for sales tax administration comes in checking on retailers and administering the tax proceeds. Since these services benefit citizens, the cost of the sales tax can be seen as insignificant.

Conclusions About Sales

Like income, sales as a basis for taxation has had a greater impact on schools than is commonly realized. While most states do not permit schools to access the sales tax directly, schools still benefit greatly as state aid to schools generally contains significant sales tax revenue. In 1990, 33.2 percent of all state revenues were derived from the sales tax, an amount slightly greater than income tax revenue.[18] In fact, the combined total of sales and income taxes resulted in more than two-thirds of all state revenue coming from these two important sources. Similarly, some federal aid to education no doubt came from federal excise taxes, since this tax represents one of the federal government's sources of revenue. It is accurate to say that if it were not for the sales tax, school revenue would be negatively affected. But its yield should not forgive its weaknesses, and recent trends showing increased reliance on sales taxes and increased sales tax rates in some states are alarming. For these reasons, school business administrators should work with policymakers to carefully assess the continued role of sales tax in education funding schemes.

Schools and Tax Systems: Realities and New Directions

This chapter began by acknowledging that discussion of tax systems would necessarily need to be broader than how taxes and schools relate on a daily basis. It was simultaneously noted, however, that many types of taxes constitute fiscal support schemes for schools and that new ways of funding education are constantly being sought. The argument was thus advanced that effective school business administrators should be knowledgeable about tax systems generally and the impact they have on schools. School leaders are increasingly involved in state and local policy decisions about how to fund education. In an era of growing government responsibility for an ever-increasing array of public concerns, it is only prudent for administrators to anticipate pressure to reduce taxes and to increase services. Since these concepts are fundamentally incompatible, good administrators must be prepared to offer support or amendment to legislative proposals on new ways to approach school funding. Since all school funding is derived from taxes, fair and productive taxation should be a central concern for school business leaders.

A key concept from this chapter has been that Americans now expect to pay taxes, but they are watchful in regard to how much they pay and whether a tax is perceived to be fair. In the current national climate, taxpayers claim that taxes are too high and too unfair. To some extent, it must be recognized that these critics make a good point. Schools have been almost exclusively dependent on the property tax—a tax that has been shown to continually increase and to show mediocre performance on the criteria of a good tax system. States have tried to relieve some property tax pressure by introducing state aid, but in many states the percentage of aid is still rather low, while state sales and income taxes can be quite steep. In states where aid ratios are high, property taxes have still not really decreased in most cases, giving taxpayers a fairly accurate impression that tax burdens are rapidly increasing. There has been sizable federal aid to education, but it has come at a time when federal income taxes are seen as oppressive. Other local governments have also increased tax rates, instilling the view of critics that costs escalate while services shrink.

A further concept in this chapter has been that tax systems need significant revision, both generally and as they relate to schools. Most of that criticism stems from performance of the various taxes on the elements demanded of good tax systems. Judged by these criteria, only the federal tax system demonstrates fundamental fairness by focusing on the income tax as the yardstick for individual responsibility to support the aims of government. Whether one approves or disapproves of how government spends money should not affect the structure of tax systems, as all taxes should be based in fairness and be related to ability to pay. As currently envisioned, only the income tax demonstrates these qualities, although it clearly does so in a contentious and incomplete manner due to the many ways in which the federal income tax code can be manipulated to personal or corporate advantage. Clearly the sales and property tax are dismal performers in this regard, as neither is linked in any way to ability to pay or to many other of the elements of a good tax system, except high yield.

Another point has been that the dependency of local governments, including schools, on the property tax demands meaningful reform. While debate rages over local control of education versus the ever-increasing hand of state and federal government, public education is disadvantaged to an enormous extent by the uniquely local nature of the property tax. Arguments favoring local control of education can be made absent the property tax, but no argument for fundamental fairness of the property tax as presently structured can be built. Property tax bases drawn along school district boundaries are presently exclusionary, making ability to pay for education a function of residence and resulting in uneven tax rates across states and the nation. Attempts to redraw boundaries to better equalize taxable wealth are met with outrage, while increased income taxes to fund equalization aid formulas are resisted at all levels. Likewise, increasing sales taxes often depends on convincing voters of the need for additional moneys, while at the same time this tax performs poorly on equity criteria and is both volatile and exclusionary unless controlled by central government. Due to the fact that increased external aid never comes without new controls, schools are faced with a real dilemma because the need for revenue is insatiable while external local, social, and legal environments are pulling in opposite directions.

Positive reform of the property tax, however, seems uncertain and even unlikely for several important reasons. First, this chapter has shown that new tax bases would already be in place if such were available. Unfortunately, sales, income, and property taxes are the three main workhorses in the tax stable, drawing a government wagon of burgeoning weight. Second, there is likely no way to restructure the local property tax so as to conform to the elements of a good tax system, particularly without radical redesign that would produce tremendous anxiety. The property tax system is already highly complicated, and attempting to make it responsive to ability to pay and vertical equity issues would require development of new theory and new technology that show little promise of emerging at the present time. Third, notwithstanding the state of Michigan, the enormous yield of the property tax suggests that it will continue to be used until real revolt occurs. Yet even the history of revolt cannot be predicted confidently, because the social nature of this country is changing so rapidly. While revolt might once have been predictable, there is a mood in the nation suggesting that people of wealth should pay all the costs of government, making them an ineffectual minority who could not successfully carry out a tax revolt even if it were warranted.

A danger greater than a tax revolt that could emerge from continued reliance on property tax would be a movement away from unity of public purpose and toward a class fragmentation of society. When social groups no longer identify with a common cause or with each other, psychological withdrawal from society can result in worse conditions than actual physical withdrawal. This was first illustrated in California's Proposition 13 which devastated that state's public schools, dropping the ranking of that state's educational system from one of the best in the country to one of the worst. This phenomenon was repeated in varying degrees across the nation, spawning taxpayer initiatives in many states. Beginning roughly with California's revolt, headlines across the country also began predicting the total demise of public education and have never ceased their grim forecasts for the fate of public schools. Although tax initiatives were declared no longer en vogue by the mid-1980s, the sudden surge of education reform in the states, the appearance of tax credits for private schools, takeovers by states of educationally bankrupt schools, schools for profit, and many other such events were not accidental. When people lose faith in public schools, a pillar of democracy is also lost, and it is natural for disaffected people to withdraw physical, emotional, psychological, and financial support from institutions they reject.

Although advocates of public education might charge that this view is too harsh, there is current evidence to support the claim. The recent rise in schools for profit and declarations of educational bankruptcy in states has created a widespread crisis of confidence in education, even greater than the panic surrounding the *Nation at Risk* report[19] over a decade ago. Takeovers of schools by states have become common, as in New Jersey where the state has commandeered control of three school systems, charging in the instance of the Newark schools "...misdirected instruction, badly neglected buildings, inefficient practices, and inequitable distribution of even the most basic resources."[20] Similar charges have been leveled against the New York City public school system, with four major reports since 1991 calling for total overhaul of the system. The author of the most recent report, prepared for the mayor's office, claimed that the system is "...perhaps the worst governmental structure I have ever studied."[21] Radical school reform seems to be gripping the nation, with for-profit school management companies spreading across the land. For example, schools in Colorado, Florida, Kansas, Massachusetts, Michigan, Ohio, and Texas have joined the Edison project, with many more under serious study. Charter schools have also taken the nation by storm, raising a fury of claims on both sides of the issue as advocates for equity and excellence square off in opposition.

Still other indicators of faltering support for public schools can be found, with implications for how schools are funded. California's voters, in 1994, passed a constitutional amendment in the form of Proposition 187 that would relieve a tremendous tax burden of more than \$2.4 billion by denying educational, medical, and other benefits to illegal aliens. Although tax initiatives died out in the 1980s, they have reemerged in the mid-1990s as voters in South Dakota have considered a tax limitation measure called Initiative One which would limit property taxes to 1 percent of assessed value and further limit property tax increases to 1.25 percent. At the same time, fate of the state's video lottery was in question, with the total revenue at risk set at \$850 million—an amount that could reduce school revenues by 45 percent. In 1994, Michigan's legislature abolished \$5.6 billion in school property taxes without a plan in place to replace lost revenue. At the same time, at least one estimate of the future finds school spending will increase by another 50 percent by the year 2005, bringing total spending to \$386 billion or an increase in average expenditure per pupil to \$7,942, up \$1,908 in only eleven years.[22] Some states are calling for reduced taxes, as in New Jersey where in 1991 taxpayers rolled back the state's largest tax increase in history and simultaneously injected \$360 million into property tax relief, with another \$469 million cut in 1994 under a new antitax governor. While many more instances could be cited, the property tax seems to be face with continued trouble at the voting booth.

As might be expected, solutions to these thorny issues are unclear because the roots are so deep and the nation is so enmeshed in social and economic turmoil. One of the basic complexities of modern life is that no issue is disconnected from all other issues, making even a seemingly straightforward problem extraordinarily entangled. For education, the entanglement is between social issues, economics, and highly politicized factions where truth from either side is ridiculed as rhetoric, or worse, labeled as malintentioned propaganda. Reformers continue to hope for the best, and policymakers continue to struggle with how to appease the courts, a growing and needy clientele, and shaky support for government's goals. Schools continue to lobby for increased support, while continuing to squeeze the property tax. Still others turn to novel ways to raise money, as in some states where school districts have turned to corporate sponsors to help raise funds. For example, schools in Florida, Illinois, Nebraska and Utah have joined with fund-raising companies, hoping to raise millions of dollars to provide new and desperately needed funding. While school-business partnerships are long standing in the education world, new twists continue to be developed as revenue bases cannot keep up with the demands placed on schools.[23]

The complexity of the issues only serves to underscore the need for effective school business leaders who understand the bigger picture and can argue for effective tax systems. Although the challenge to education is great, there are improvements that can be made to the fiscal support structure of education that would improve its equity without harming tax yield. While variations to such a system would be equal to the number of critics, the principles represent a starting point from which policymakers and educators could build a better system. Four such principles, with six specific recommendations for improvements, provide a basis for new directions in funding education.

The first principle is to recognize a powerful point made in the first edition of this book. Simply stated, it is unreasonable to expect any one tax to have all the characteristics believed appropriate for the tax system as a whole. Using this logic, it has long been argued that a variety of taxes make up the best tax system, since some people will escape any one tax that can be devised. While this principle requires some careful attention to avoid too much leniency in evaluating any given tax and further does not excuse using taxes that systematically violate the elements of a good tax, its advice is sensible.

The second principle argues that any local tax is inherently less fair than a tax more widely applied. The ramifications of this statement are extremely broad, as it can be hypothesized from this point that greater fairness will be obtained with fewer school districts, that statewide taxation is required to achieve uniformity, that any decision made at the local level results in wider variability than is true when decisions are made at some higher level, and so on. Obviously there is a proper role for local communities in education, and that role should be preserved and promoted. Mathematically, however, variations in funding are exacerbated as the number of taxing units increases—the dispositive proof of which are the many school funding lawsuits since the late 1960s. While there is a proper balance that must be struck between local, state, and federal governments in education, parochialism must yield to some greater extent because it is presently producing the negative behaviors described earlier.

The third principle argues that taxes performing poorly on elements of good tax systems should be minimized. The principle simply states that no tax is inherently useless or is so flawed as to be abandoned, but it does suggest that the most important functions of government ought to be funded by the most equitable tax available.

The fourth principle argues, absent radical redesign of the whole tax structure in the nation and including constitutional reallocation of federal and state powers, that the present system for funding schools could be improved in specific ways relating to the overworked property tax. These suggestions would ameliorate some of the concerns about the present tax system without dramatic upheaval. These issues also go to the heart of a close companion problem known as state aid distribution formula design.[24]

First would be to consider implementation of a statewide property tax, or at least implementation of a uniform property tax rate across a given state. Assuming no change in school district boundaries and assuming equitable property assessments, the effect of such a tax would be to place all districts on the same level of consideration for purposes of calculating state fiscal aid. While another set of issues such as permissible expenditure per pupil would have to be discussed, the property tax would take a giant step forward by bringing all residents in the state into more equal tax effort.

A second improvement would be to assure that property valuations across the state are accurate and equitable and removed from political influence. Various methods of equalizing valuations exist, ranging from statewide reappraisal using highly trained assessors to such devices as sales-to-assessment-ratio studies that can be used to produce adjusted assessed valuations for formula aid purposes. Removing property valuation from the political arena is by far the more difficult of the tasks, but considerable unfairness in property taxation will continue to exist until valuations are equalized nonpreferentially.

A third improvement to equitable school funding could be obtained through elimination of property tax exemptions and abatements. Hundreds of millions of dollars are lost annually in every state, either as a result of political pressuring of local officials or as a result of policymakers believing that economic development outweighs the lost revenue. In either case, tax rates on nonexempted property are higher as a result of tax abatement, even though better mechanisms such as industrial revenue bonds (IRBs) produce the same result but without loss of tax revenue.

A fourth improvement echoes historic critics of tax systems because they make an excellent point. Because it is acknowledged that total system redesign is unlikely and because it is admitted that no tax performs perfectly on every criterion of a good tax system, state and local policymakers should make the best effort possible to assure that tax systems are sufficiently broad to capture everyone. From a practical perspective, this would require greater use of the income tax and broader use of other wealth such as personal property. While some equity improvement would be incurred by relying less on real property, some loss of equity would also be experienced by taxing lower income groups on property not now subject to tax. Overall, however, less reliance on real property would benefit all groups, and significant effort should be put into effective use of circuit breakers, homestead exemptions, and exemptions for items criticized earlier for the negative effect on low income groups.

A fifth improvement is perhaps more radical, but movement in this direction could help alleviate many of the problems described in this chapter. Inasmuch as possible, movement toward greater utilization of income as the basis for all taxation should be encouraged. Applied to schools, reducing or eliminating the property tax would eliminate certain inequities and tax burdens on taxpayers. Most importantly, utilization of income for direct support of schools would revolutionize wealth-based opportunity, as paying for education would be truly based on ability to pay. Obviously, the income tax for schools would require a finely tuned state aid formula.

Finally, a sixth improvement would be to carefully scrutinize school aid distribution formulas. The knowledge for equitable aid formulas exists if politics can be minimized. An equitable aid formula combined with a tax base built on ability to pay is a powerful combination that offers hope for the future.

Summary

As noted in he first edition of this book, the present period presents challenges to school business administrators and other educational leaders, forcing them to obtain new knowledge and skills beyond those that sustained the profession in the past. Simply mastering the technicalities of school funding and budgeting is no longer enough. Understanding the broader economic, social, and political trends will be essential in the future for administrative planning, influencing policymaking, and managing the schools for the twenty-first century.

Endnotes

[1] D.C. Thompson, R.C. Wood, D.S. Honeyman, *Fiscal Leadership for Schools: Concepts and Practices,* (New York: Longman, 1994), 131.

[2] U.S. Constitution, Amend. XVI reads: "The Congress shall have power to lay and collect taxes on incomes, from whatever source derived, without apportionment among the several States, and without regard to any census or enumeration."

[3] For a more complete discussion of the concept of federal revenue, expenditure, school aid, and educational needs, see Chapters 1-3 in D.C. Thompson, R.C. Wood, and D.S. Honeyman, *Fiscal Leadership for Schools.*

[4] For a more complete discussion of tax base domination and abandonment, see Chapter 3 in D.C. Thompson, R.C. Wood, and D.S. Honeyman, *Fiscal Leadership for Schools.*

[5] R.C. Wood, "New Revenues For Financing Education at the Local Level," *The Impacts of Litigation and Legislation on Public School Finance:* Tenth Annual American Education Finance Ass'n Yearbook, Chap. Three, New York, Harper & Row, 1989, 59-75.

[6] For a fuller development and appreciation of the culture of tax fairness and some controlling concepts of tax philosophy, see Chapters 1-3 in D.C. Thompson, R.C. Wood, and D.S. Honeyman, *Fiscal Leadership for Schools.*

[7] For a fuller development of the concepts of tax equity in this context, see D.C. Thompson, R.C. Wood, and D.S. Honeyman, *Fiscal Leadership for Schools:*, 176-179. For extended development of value judgments in equity, see R. Berne and L. Stiefel, *The Measurement of Equity in School Finance: Conceptual, Methodological, and Empirical Dimensions,* (Baltimore: Johns Hopkins, 1984).

[8] Critics will be quick to point up another favorite argument which suggests that nonproperty owners do in fact pay these taxes by virtue of tax shifting. This argument will be acknowledged later in text when discussing tax incidence and tax impact.

[9] Even this comparison is subject to the same imperfect limitations described in text, in that entire neighborhoods may be assessed special taxes to pave access roads or lay main utility lines that serve a large area. In other words, some special assessments are laid broadly across taxpayers, while other special taxes conform more closely to the benefit principle.

[10] A technical exception to this is corporate income tax, which can be reasoned to be passed on to shareholders or owners. The logic is that corporations do not pay income tax; rather, people pay taxes. Neither can this tax be shifted because it will always circle back to individuals, e.g., if increased taxes were added into the prices charged for goods or services in the corporation, the increase would show up as increased income to shareholders or owners, triggering more taxes.

[11] See Chapter 3, *Principles of School Business Management,* R.C. Wood, (Reston, VA: Association of School Business Officials, 1986), 83.

[12] A mill is defined as 1/1000 of a dollar, so that 1 mill per $1,000 tax value= $1 tax yield. A few states express tax rates using cents per $100.

[13] State laws on property valuation are quite distinct once certain elemental principles are discussed. Likewise, a discussion of the impact on schools is highly state-specific. For instance, some states have state-wide property tax levies that are uniform in rate. Other states allow tax rates to float free, as in the example in text where every school district theoretically has a tax rate different from every other school district. Still other states have tax lids that cannot be exceeded. In a few states, school districts and other units of government are prevented from setting

budgets and tax rates in that they are fiscally dependent on higher levels of government which must approve subunits' budget and tax requests. The multiplexity is impossible to describe fully, but school districts are profoundly affected by how the property tax is constructed and administered.

[14] D.C. Thompson, R.C. Wood, and D.S. Honeyman, *Fiscal Leadership for Schools:* 185.

[15] In 1990, local property tax accounted for 97.4 percent of all local revenues to schools in the 41 states where there are at least some independent school districts (i.e. districts that can determine tax rates). In these states, there were only four states where local revenues was less than 98 percent comprised of property tax. These states were: Louisiana which was principally sales tax-dependent (55.1 percent) but closely followed by property tax (44.7 percent); Kentucky with 66 percent from property taxes and the remainder from income, utilities and miscellaneous taxes; Pennsylvania with 83 percent from property taxes and the balance from income and miscellaneous taxes; and South Dakota where property produced 95.6 percent with the balance coming from sales and miscellaneous taxes. Source: American Education Finance Association, *Public School Finance Programs of the United States and Canada 1990-91, Vol. 1,* ed. S. Gold (Albany: American Education Finance Association, State University of New York at Albany, 1992), 16.

[16] D.C. Thompson, R.C. Wood, and D.S. Honeyman, *Fiscal Leadership for Schools*, 190.

[17] Source: American Education Finance Association, *Public School Finance Programs of the United States and Canada 1990-91,* 11.

[18] Ibid..

[19] National Commission on Excellence in Education, *A Nation at Risk: The Imperative for Educational Reform* (Washington, DC: National Commission on Excellence in Education, 1983).

[20] Citing the Comprehensive Compliance Investigation report, *Education Week* (Aug. 3, 1994), 20.

[21] *Education Week* (Aug. 3, 1994), 5.

[22] Report issued by the American Legislative Exchange Council, *Education Week* (Sept. 28, 1994), 21.

[23] R.C. Wood, "New Revenues For Financing Education at the Local Level."

[24] For a fuller development of state aid formulas, see Chap. 4 in D.C. Thompson, R.C. Wood, and D.S. Honeyman, *Fiscal Leadership for Schools:*.

CHAPTER 4
Management Theories

Introduction

School systems today are large complex organizations. The school business administrator is responsible for ensuring that many critical functions work together smoothly and efficiently to support the educational process. An argument is often made that schools should be run more like businesses. Although there are substantial differences between governments and private businesses,[1] many of the functions performed by the school business office are similar or even identical to those performed by private businesses. As a result, there is much that can be learned from existing theories of management to help the school business administrator perform his or her tasks more effectively and efficiently. The purpose of this chapter is to relate these theories to the technical aspects of the school business administrator's position to improve the administration of the schools in general, and to help individuals in these positions better understand the organization in which they work.

This chapter describes historical and theoretical developments in administrative theory, and relates the information to the job performed by school business administrators. Four major topics are presented:

- The development of management theory.
- Theories of organizational structure.
- Theories of organizational culture.
- Organizational change.

The goal of this chapter is to make the student of educational administration aware of the literature in each of these areas, and provide a framework for organizing a school business office in a way that meets the needs of the students and other professionals employed by the district.

The first section deals with the development of management theory, describing four major stages of theoretical development. These stages include: scientific management, human relations, behavioral science, and systems theory. Each is described in detail herein.

An important component of any management system is its organizational structure, which is the focus of the second section. The emphasis of this section is on the formal relationships among individuals, starting with the bureaucratic model and continuing with participatory management, organic and mechanistic organizations, compliance theory, and strategy-structure typology. The third section examines more closely organizational culture, examining how cultures are created, maintained and changed.

Finally, the chapter closes with a discussion of organizational change, looking at the factors that cause pressure for change and the reasons school employees might resist changes in the organization.

Because of the myriad of complex issues that a school business administrator must deal with, and because of the implications his or her decisions will have on the school organization, it is critical that individuals in school business management positions have an understanding of the organizational issues that are dealt with through management theories.

The Development of Management Theory

The development of management theory can be divided into four major stages:

- Classical organizational theory.
- The human relations approach.
- The behavioral science approach.
- Systems theory.

Classical Organizational Theory

Classical organizational theory first emerged in the early years of the twentieth century and can be divided into two management perspectives: scientific management and administrative management. Scientific management focuses on managing the specific work and tasks of workers, while administrative management focuses on the structure of the overall organization.

Scientific Management

It is generally accepted that formal management theory first made an impact on organizational structure at the turn of the century. Before that time, organizations were generally much simpler and tended to be managed by an authority figure or leader. The movement away from the single leader management structure resulted from problems associated with managing large numbers of people and the growing complexity of society and industry. As our society entered the industrial age in the later half of the 1800s, organizations became larger and production processes more interdependent, requiring alternatives to traditional hierarchical structures. Scientific management was one of the first efforts to deal with these complexities.

Scientific management is based on the theories of Frederick W. Taylor, known as the "Father of Scientific Management."[2] Taylor was an engineer by training, and he thought that through careful scientific analysis of work tasks, the "one best way" to perform a task could be determined. Taylor's theory can be summarized in four principles:[3]

- **Scientific Job Analysis:** The use of observation, data gathering, and measurement to determine the "one best way" to perform a task or job.

- **Selection of Personnel:** Once the best way to do a job had been determined, the next step was to select and train individuals to perform that job. Prior to Taylor's writing, many workers chose their own work and often were expected to train themselves.

- **Management Cooperation:** It was Taylor's belief that management and workers should cooperate so that work would be completed in accordance with the scientific principles that had been developed.

- **Functional Supervision:** Taylor argued that there should be a division of work between managers and workers, with management responsible for planning, organizing, and making decisions, and workers simply performing their jobs.

Taylor's scientific management relied on research and experimentation to improve production. He would establish explicit guidelines for each position designed to maximize productivity of the individual assigned to that specific task.

Administrative Management

While scientific management focuses on individual jobs, administrative management theory focuses on the overall administration of an organization. The major contributors to this theory include Henri Fayol, Luther Gulick, and Max Weber.

Fayol was a French steel and coal-mining executive who is credited with defining a set of principles by which organizations should be managed.[4] Fayol argued that there were five basic functions of management: planning, organizing, commanding, coordinating, and controlling. In addition, he developed fourteen principles of management to guide the management of organizations. These are summarized in table 4.1.

One of Fayol's most important contributions was his realization that management is a continuous process, and not something that is established once and then forgotten. Lunenburg and Ornstein argue that Fayol's fourteen principles emphasize chain of command, allocation of authority, order, efficiency, equity, and stability.[5]

Table 4.1
Fayol's Fourteen Principles of Management

Component	Description
Division of work	Improved efficiency through reduction of waste, increased output, and simplification of job training.
Authority	The right to give orders and the power to extract obedience.
Discipline	Respect for the organization's rules and clear statements of agreements between the organization and its employees.
Unity of Command	Each employee should have only one superior from whom he or she receives orders to avoid breakdowns in authority and discipline.
Unity of Direction	Activities that are similar and aimed toward a singular goal should be grouped under one manager.
Subordination of Individual Interest	The interests of the organization as a whole should take precedence over the interests of individuals and small groups.
Remuneration	Compensation should be satisfactory and fair to employees and to the organization.
Centralization	Subordinates should be given enough authority to do a task successfully, but managers must retain final responsibility. The level of centralization depends on the individual situation.
Scalar Chain	Lines of authority should be clear and followed at all times. The scalar chain refers to the chain of command from the highest official to the lowest employee ranks.
Order	All resources, both human and material should be coordinated so that they are in the right place at the time they are needed.
Equity	Managers should consider equity and equality of treatment when dealing with employees.
Stability of Personnel	Fayol believed that a stable workforce was necessary to a successful organization and felt that managers should encourage long term commitments of employees to the organization.
Initiative	All employees should be encouraged to develop and carry out plans for improvement.
Esprit de corps	Teamwork, team spirit, and an sense of unity and togetherness among employees should be encouraged by managers.

Source: Adapted from Fred C. Lunenburg and Allan C. Ornstein, *Educational Administration: Concepts and Practices,* (Belmont, CA: Wadsworth Publishing Company, 1991), p. 7.

Another classical theorist, Luther Gulick identified seven functions of management which can be summarized as follows:[6]

- **Planning:** The development of an outline of things to be done and how they are to be accomplished. This action forecasts the future actions of the organization, and establishes the direction it will take.
- **Organizing:** The establishment of an authority structure through which work is organized, defined, and coordinated.
- **Staffing**: Selecting workers, training them for the jobs they are expected to perform, and maintaining favorable working conditions.
- **Directing**: Communicating with others, making decisions, implementing decisions, and evaluating employees.
- **Coordinating**: All activities necessary to keep the organization focused on its common goals.
- **Reporting**: Verification of progress toward goals through records, research, and inspection. The purpose of this step is to ensure things are done properly, and ensure that corrective action is taken if necessary.
- **Budgeting**: Fiscal planning, accounting, and control.

The last of the classical theorists was Max Weber. Weber was not a manager like Taylor and Fayol, but rather an intellectual who studied the growth of large organizations in Germany and predicted there would be a need for more formalized procedures for administrators if such organizations were to succeed. He called this structure "bureaucracy,"[7] which he identified as an ideal form of organization. Based on his analyses of the Catholic Church and the Prussian army, Weber promoted bureaucracy as the most efficient form of organization. Weber considered bureaucracy to be the humane way to manage complex organizations, and that it was a solution to the problems of personal subjugation, nepotism, cruelty, emotional vicissitudes, and subjective judgments which characterized management in the early 1900s.[8]

Classical management theory focused on efficiency of production and ignored individual psychological and social factors that impacted the work place, assuming financial incentives were adequate to motivate workers. The focus was on the task, not the individual or his or her working conditions. This gap was the major reason for the development of the second approach to management theory, the human relations approach.

The Human Relations Approach

The origin of the human relations approach to management was a series of studies conducted by Elton Mayo and his associates at the Hawthorne Plant of Western Electric near Chicago between 1927 and 1933.[9] Widely known as the Hawthorne Studies, this work has had a significant impact on administrative theory. During the course of his studies, Mayo conducted a number of experiments. In one, the lighting in one room was varied while the lighting in another room of female employees doing the same work remained constant. The productivity of each work group was then analyzed. Mayo found that regardless of the light level or other changes in rest periods and lengths of workdays and work weeks, the productivity in both the control group and the experimental group improved. The worse conditions got, the more productivity rose.

In a second experiment, researchers paid men working in the Bank Wiring Observation Room on a piecework incentive system. Since each individual was rewarded for improved productivity, the researchers expected productivity to improve over time. Instead, the researchers found that the group established an acceptable level of output and relied on considerable peer pressure to not exceed or under produce.

Since the results of these experiments were not what had originally been expected, Mayo and his team conducted interviews with over 20,000 employees who participated in the experiments over the six years of research. He concluded that increases in productivity were more an outgrowth of group dynamics and effective management than employer demands or physical factors. In the light test, the improvement in productivity was attributed to the fact that the groups participating in the test felt important because of that participation, hence productivity increased. This is often referred to as the "Hawthorne" effect in research conducted on employee productivity and describes situations where just being part of the experiment leads to productivity increases in the control group. In the incentive pay experiment, researchers concluded that informal work groups merged and established their own norms for appropriate behavior among individuals in the group.

From these studies, researchers concluded that human social factors such as morale and belongedness were important to improved productivity, and that effective management included such interpersonal skills as motivating, leading, participative decision making, and effective communications. Mayo's major contribution to management theory was establishing the importance of human behavior in management theory.

Other important contributors to the human relations approach include Kurt Lewin, Carl Rogers, and Jacob Moreno. Lewin emphasized field theory and group dynamics, concluding that democratic groups that allow for active participation of members are more productive in terms of both human satisfaction and achievement of group goals than are authoritarian groups.[10] Carl Rogers, one of Mayo's associates in the Hawthorne Studies, concluded that the best way to understand behavior is to consider the individual and his or her experiences.[11] Jacob Moreno's contribution to the human relations approach consisted of his view that people develop selective affinities for other people, and that groups composed of individuals with similar affinities for one another will perform better than groups composed of people who do not have such preferences.[12] The major assumptions of the human relations approach are summarized in figure 4.2.

The Behavioral Science Approach

A number of theorists considered both of the models presented to be incomplete in terms of understanding how employees participated in the work setting. The attempt to reconcile the differences between the two models led to the development of the behavioral science approach. The first author to take this approach was Chester Barnard.[13] His concept of the cooperative system is important to school business managers today as they implement a system, e.g., site based management systems.

Barnard's theory of cooperative systems is an attempt to integrate both the human relations and classical management principles into a single model. He suggests that to achieve financial success and cooperation, an executive must meet two conditions, effectiveness and efficiency. Effectiveness is defined as

Figure 4.1
Major Assumptions of the Human Relations Approach

1. Employees are motivated by social and psychological needs as well as by economic incentives.

2. These needs, including but not limited to recognition, belongingness, and security, are more important in determining worker morale and productivity than the physical conditions of the work environment.

3. A individual's perceptions, beliefs, motivation, cognition, responses to frustration, values, and similar factors may affect behavior in the work setting.

4. People in all types of organizations tend to develop informal social organizations that work along with the formal organization and can help or hinder management.

5. Informal social groups within the workplace crate and enforce their own norms and codes of behavior. Team effort, conflict between groups, social conformity, group loyalty, communication patterns and emergent leadership are important concepts for determining individual and group behavior.

6. Employees have higher morale and work harder under supportive management. the human relationistws believe that increased morale results in increased productivity.

7. Communication, power, influence, authority motivation and manipulation are all important relationships within an organization, especially between superior and subordinate. Effective communication channels should be developed between the various levels in the hierarchy. The human relationists emphasize democratic rather than authoritarian leadership.

Source: Lunenburg, Fred C. and Allan C. Ornstein, *Educational Administration: Concepts and Practices,* (Belmont, CA: Wadsworth Publishing Company, 1991). p. 10-11.

the degree to which the common purpose of the organization is achieved, while efficiency deals with the satisfaction of individual employees. Barnard's point is that an organization can only be successful when both the organization and the individuals who work for the organization share the same goals.

A number of other theorists have followed in this vein. E. Wright Bakke viewed organizations as having a fusion process where the individual attempts to use the organization to further his or her own goals and vice versa. Thus the organization and individual are fused together to accomplish the goals of each.[14]

Chris Argyris suggests that there is an inherent conflict between the individual and the organization, and that this conflict is the result of incompatibilities between the growth and development of the individual's personality and the repressive nature of the formal organization. Argyris believes that this conflict leads to adaptive behavior on the part of the individual such as apathy. Such an indifferent orientation to work would thus lead to the undermining of the organization's goals.[15]

Other contributions to this field of thought include Abraham Maslow's hierarchy of needs and Douglas McGregor's Theory X and Theory Y concepts of management. Both of these are described in more detail in the sections on organizational structure that follow. Another major theory in this area is that of Fredrick Herzberg who developed a two factor theory of motivation.[16] Herzberg's two factors are hygiene factors which cause or prevent job dissatisfaction, and motivation factors which cause job satisfaction. In Herzberg's view, only the later group can lead to effective productivity gains.

These theories deal with the development and structure of an organization, however, they only deal with individual components of that organization, i.e., the employees, the managers, and so on. Many organizations are very complex, and it is helpful to think of all of the functions that must be included in the operation of the organization when searching for a management theory. Systems theory provides one way in which this need to consider all aspects of the organization at once can be accommodated.

Systems Theory

It is often useful to think of organizations as a system of interrelated parts. Systems theory was developed to provide a structural framework for such analyses. As such, it suggests that organizations must be analyzed as a whole, taking into consideration these interrelationships.

A basic systems model includes inputs, a transformation process, and outputs. In addition, the system considers the environment in which the organization operates and includes feedback loops to provide information concerning the outputs or the process of the organization. This may lead to changes in both the transformation process and the future process outputs. Figure 4.3, adapted from Lunenburg and Ornstein shows how systems theory could be applied to school administration generally.

This model is useful for school business administrators because of the range of functions for which they are responsible and because they often have to ensure that these functions work together in a coordinated way to support the education program of a school district. For example, not only does the transportation system have to operate clean, safe buses, it has to ensure that they operate within the schedule constraints of the educational program. At the same time, the purchasing department needs to be able to provide schools with the supplies and materials each learning center needs on a timely basis while the maintenance and repair department ensures that the schools are well maintained. It is important that all of these "subsystems" operate efficiently so that the schools can focus on learning. Thus, a great deal of coordination is required to make sure that all areas work to support that focus, and that major repairs or regular services to facilities or equipment be done such that the program is not interrupted. Systems theory can help establish processes to met all of the diverse demands on the school business office and ensure that they work together smoothly.

Figure 4.2
A Systems Model of School Administration

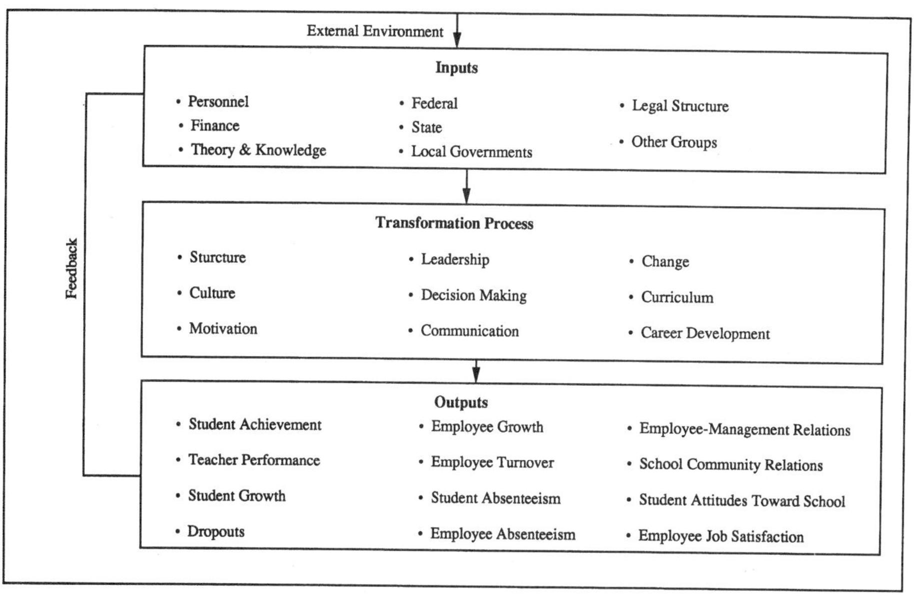

Source: Lunenburg and Ornstein, Educational Administration: Concepts and Practices, 1991, p.19

Summary

This section provides a summary of the development of management theories and thought in the twentieth century. Beginning with Taylor's concept of scientific management where the focus is on the production task, and moving to the human relations approach which focuses on the ways human differences and reactions can affect outputs and productivity, this section summarizes some of the most important literature in the field. In addition, brief summaries of the behavioral science approach, which seek to use both knowledge of production processes and motivation theory to improve individual productivity, and the systems model, which looks at an organization from a more global perspective, have been provided. Figure 4.4 offers a summary of the major theorists in each of these schools, and the time frame in which they were developed.

Theories of Organizational Structure

An important task facing any school business administrator is determining how to organize or structure the functions and staff of his or her office. Since it is helpful to have a set of basic concepts in hand before delving into the theories of organizational structure, the first part of this section provides a brief description of the basic concepts of organizational structure including such things as job specialization and departmentalization, chain of command, authority, and the span of management. Next, the bureaucratic model of organizations is presented in some detail, followed by a discussion of the participatory management model. Finally, some alternative organizational structures will be briefly considered.

Basic Concepts

This section will give the reader background on the basic concepts of organizational structure to facilitate understanding of the analysis of bureaucratic, participatory management and alternative models. These concepts aid in the understanding of how organizational structure is used to provide for vertical control and horizontal coordination within an organization.

Job Specialization and Departmentalization

To efficiently organize the work of an organization, it is generally divided into specialized tasks and often distinct units. In schools, examples of specialization include the use of a separate business office, often with a number of individual departments such as transportation, maintenance and operations, personnel, risk management, and so on. This allows for individuals to become experts at the tasks assigned to them and to provide others in the school district with consistent information, be those individuals the school board, the superintendent, school principals, teachers, parents, taxpayers, or patrons.

Figure 4.3
Summary of the Development of Management Thought

Period	Management Elements	Time Frame
Scientific Management	Fredrick Taylor Henri Fayol Max Weber	1900-1930
Human Relations	Elton Mayo Fritz Rothlisberger Kurt Lewin Carl Rogers Jacob Moreno	1930-1950
Behavioral Science	Chester Barnard E. Wright Bakke Chirs Argyris Abraham Maslow Douglas McGregor Fredrick Herzberg	1950-1970
Systems	Kenneth Boulding Jacob Getzels Egon Guba	1960-present

Chain of Command

The flow of authority and responsibility in an organization requires a chain of command. There are two important principles of a chain of command. The first is unity of command which implies that a subordinate is only accountable to one person. The second, the scalar principle, means that authority and responsibility should flow in a direct line vertically from top management to the lowest level. This will establish a hierarchical division of work within the organization.

Authority and Responsibility

Authority is the right to make decisions and direct the work of others. Administrators must have the authority to carry out the jobs for which they are responsible. At the same time, authority and responsibility must be linked so that individuals responsible for a certain task have the authority to carry it out as assigned.

In school districts, authority is vested in the Board of Education. The board delegates to the superintendent the authority necessary to run the schools, who in turn delegates narrower authority to other individuals, including the school business administrator.[17] Each lower level has narrower amounts of authority.

Centralization/Decentralization

When most of the authority is delegated to subordinates and decisions tend to be made at lower levels of an organization, the organization is generally following a philosophy of decentralization. On the other hand, if most of the authority is maintained by school administrators at the highest levels, the organization is more centralized. The important question facing all organizations is what level of centralization or decentralization best fits that organization's needs. The trend in recent years has been toward decentralization, particularly in education, where educators are carefully analyzing the value of site-based management in making schools better places for student learning. The advantages of decentralization are that it makes greater use of human resources, unburdens top-level administrators, ensures that decisions are made close to the problem by personnel with technical knowledge, and permits more rapid response to external changes.[18] Site-based management is discussed more thoroughly in chapter 23.

Line and Staff Authority

Staff and line authority offer another way to consider organizational structures. Line authority is the relationship between a superior and a subordinate, and refers to a direct line in the chain of command. Line authority follows the scalar principle previously detailed. For example, the superintendent has line authority over his or her subordinate assistant or deputy superintendents, including the chief business administrator. The chief business administrator would in turn have line authority over the various department heads who make up the business office of the school district. Depending on how the district chooses to organize its functions, such authority could include some or all of the following: personnel, accounting, risk management, maintenance and operations, budgeting, transportation, and others.

Staff authority is advisory in nature. Staff personnel do not have the authority to implement decisions. Rather they are responsible for the development and analysis of data for line personnel. Staff assistants and legal counsel are examples of personnel with staff authority in an organization.

Span of Control

Span of control refers to the number of subordinates reporting directly to a supervisor. While there is no agreed upon ideal number of subordinates for a span of control, there is a general view that the span can be larger at lower levels of an organization than at higher levels since subordinates at lower levels typically perform more routine activities and can be more effectively supervised. For example, in elementary schools, teachers often report directly to the principal, whereas in high schools, department heads may provide an additional level of supervision, due to both the increased complexity of the material taught, and the generally larger number of teachers in a high school.

As elementary schools become larger, there are assistant principals or team leaders who serve in this same function.

In a school business office, it makes little sense for the bus drivers or the payroll clerks to report directly to the chief business administrator. Rather, bus drivers are the responsibility of the transportation supervisor, and the payroll clerks of the accounting department. One individual can supervise a substantial number of bus drivers or payroll clerks since all are assigned the same basic task, driving school buses, or handling personnel payroll matters. However, their supervisors might report directly to the chief business administrator. He or she may not have as large a span of control due to the increased complexity arising from the number of different issues brought to that office, and the variation in the types of decisions that must be made.

These concepts are used to describe alternative models of organizational structure. The first is Weber's bureaucratic model.

The Bureaucratic Model

Although the term "bureaucracy" has a negative connotation today, it originally evolved from Max Weber's studies of organizations. Weber viewed the bureaucracy as the "ideal form of organizational structure." In Weber's view, the ideal bureaucracy has the following six characteristics:[19]

- **Hierarchy of Authority:** All jobs are arranged in a hierarchical fashion, with the lower offices under the control or direction of higher offices and a clear chain of command from the top of the organization to the bottom.

- **Division of Work:** The division of all work tasks into specialized jobs. Job holders are given the authority they need to fulfill their responsibilities.

- **Rules and Regulations:** Each task is performed according to a consistent set of rules to ensure that tasks are performed in a uniform manner regardless of who is in the position.

- **Uniform Procedures:** There would be uniform procedures to deal with all work-related situations so that individual preferences or prejudices would not influence work outcomes or personnel decisions.

- **Impersonality**: Impersonal attitudes are to be maintained toward subordinates to ensure that rational decisions are made and that such decisions are not based on favoritism or prejudices.

- **Competence**: Employment and promotions are based on job performance. Similarly, employees are protected from arbitrary dismissal. Weber thought both of these concepts would result in high levels of loyalty among employees.

These characteristics can be found in many large organizations in the United States, including most, if not all public school districts. Unfortunately, Weber's "ideal" model has not always been implemented as he envisioned, and often there are strong criticisms of the problems of bureaucracies. Critics argue that it is the bureaucracy that keeps our schools from functioning well and wastes valuable resources on administration rather than focusing on instruction and student performance. Although Weber's model is based on rational behavior, it reflects certain basic and fundamental flaws.

One of the major problems is that a high degree of specialization of work tasks may also lead to a reduction in the challenge of work. The result could be reduced job performance due to boredom, increased absenteeism, or job turnover.

Another problem with bureaucratic structures is that reliance on rules and regulations can lead to inefficiency and inertia if the rules become the end in and of themselves rather than the means to achieve the organization's goals.

A third problem has to do with the hierarchical structure advocated by Weber. In Weber's mind, communication would flow both up and down the organizational chain to improve coordination across the organization. In practice, communication has often had only a downward orientation, with subordinates withholding information from their superiors and exhibiting frustration with not being able to participate in the decision-making process.

Weber wanted promotions to be based on performance. However, in many complex jobs, like teaching, performance is difficult to assess and is often based on subjective criteria. As a result, many organizations tend to promote individuals on the basis of seniority and/or loyalty rather than competence and merit.

One final criticism of bureaucracies in recent years is their impersonal nature. Critics argue that bureaucracies are rigid, control oriented organizations, which emphasizes structures over people.

One of the strongest critics of the bureaucratic model is Warren Bennis, who, over twenty years ago, forecast that bureaucracy as an organizational structure would fail.[20] In his recent book, *Why Leaders Can't Lead,*[21] Bennis argues that organizations with entrenched bureaucracies that have a commitment to the status quo will undermine unwary leaders, creating an unconscious conspiracy that prevents leaders from taking charge and making changes. A number of other authors in recent years have also criticized bureaucratic organizational structures, calling for alternatives in organizations where nurturing, caring, trusting, and challenging structures help improve individual productivity.[22] There has been a similar movement in school districts, with a growing emphasis on site-based management and flatter organizational structures.[23]

These criticisms have led to the development of alternatives to minimize the deficiencies of bureaucratic models. A number of them are more fully described herein.

Participatory Management Models

Placing more emphasis on employee job satisfaction and overall morale, participatory management models are an extension of bureaucratic models that stress the importance of employee motivation. The concept is to build an organization that is designed to satisfy employee needs under the theory that this will, in turn, lead to improvements in employee productivity. The topics covered in this part will include the following:

- Theory X and Theory Y.
- Individual vs. Organization.
- System 4 Organization.
- Transformation Leadership.
- Value-Added Leadership.
- Site Based Management.

Theory X and Theory Y

Based on the work of Douglas McGregor[24] these theories reflect the concept that managers' actions flow from their assumptions about their employees. McGregor divided managers into two groups with contrasting sets of assumptions about subordinates, naming them Theory X and Theory Y.

Managers with Theory X beliefs view people as disliking work and attempting to avoid it. Therefore, Theory X managers believe they have to control, coerce, and carefully direct their subordinates to have the work completed. Moreover, McGregor argued that Theory X managers believe that the average person prefers to be directed and controlled by individuals in authority.

On the other hand, Theory Y managers believe that work is as natural as play or rest and that a worker's commitment to the organization's objectives is a function of the rewards he or she receives for achievement. Theory Y managers believe that people will accept and seek responsibility under the right conditions. McGregor argued that Theory Y is essential for the success of today's organizations.

Individual vs. Organization

Chris Argyris argues that rigid impersonal organizations hinder employees from using their full potential.[25] He suggests that the restraining effects of bureaucratic structures can be reduced by relaxing the rules and regulations and reducing the division of labor. He argues that greater delegation of authority and more participation in decision making will result in greater growth and development of human personalities and reduce the conflict between individuals and organizations. This, he argues, will lead to improved productivity.

System 4 Organizations

Rensis Likert has similarly proposed an organizational structure that focuses more heavily on individual differences and needs. Likert's System 4 Organization is based on team-oriented participant groups who develop high levels of trust and confidence in themselves and in their superior. In this model, information flows freely up and down the organizational chain with an emphasis on teamwork and high performance goals.[26]

The System 4 Organization relies on a manager's use of supportive relationships, group decision making, and high performance goals for the organization. In other words, if the organization were to be effective, all relationships between supervisors and subordinates must be perceived as enhancing both individuals' sense of value to the organization. Likert argues that to make an organization work under this structure, all administrators must be trained to build work groups and focus on high performance outcomes.

Transformation Leadership

Terrence Deal argues that organizations cannot be changed or transformed through changes in the organizational structure alone. He argues that attempts to change the structure overlook cultural values that underlie everyday behavior. These deeper patterns provide meaning and continuity, and to modify them requires transforming the organization, not simply changing the organizational structure. Examples are Lee Iacocca's transformation of Chrysler and AT&T's efforts to survive following the deregulation of the communications industry. Such changes are very difficult to manage however, and still remain the exception not the rule.[27]

Value Added Leadership

A number of writers, Amitai Etzioni among them, suggest that moral authority provides a basis for management.[28] He argues that what most people respond to is based on what they believe and how they feel. Important to them are the shared norms, values, and cultural symbols that emerge from groups with which they identify. He contends that these are more important motivators than basic, extrinsic needs and motives. The result of this is an organizational structure that treats purposes, values, and commitments as most important. Administrators who can develop this moral authority will be able to derive extraordinary commitment and performance from their subordinates.

Site-Based Management

Site-based management presents a dramatic change in the organizational structure of a school, and offers a number of important challenges to school business administrators. The major changes implied by site-based management are increased autonomy of school sites through relief from rules and regulations and the authority to make decisions at the school site with participation from teachers, parents, students and community members. In addition, increased fiscal autonomy at the school site has important implications for the way school business offices operate. More information on site based management is presented in chapter 23.

As this discussion shows, there are a number of competing management theories that suggest the rigid controls of bureaucratic structures are not the best way to organize complex organizations. Each of the alternatives presented suggests a stronger focus on individual employees and their needs. The general concept of each is that if the employees are satisfied with their jobs and properly challenged to work harder and to do the best they can, the performance of the entire organization will improve.

Alternative Organizational Structures

The bureaucratic and participatory management models described offer a framework for more complex models. Three specific models are described. They include compliance theory, mechanistic-organic organizations, and strategy-structure typology.

Compliance Theory

Amitai Etzioni has developed a theory of organizational structure based on the type of power used to direct the behavior and involvement of members of the organization. Known as compliance theory,[29] Etzioni identified three types of organizational power–coercive, utilitarian, and normative. Each is related to three types of involvement–alienative, calculative, and moral. Figure 4.5 provides a simplified model of the relationships among these three types of power and three types of involvement. Each of these is described.

- **Coercive Power:** The use of force and fear to control lower level participants in the organization exemplifies coercive power. Examples of organizations that rely on this power include prisons and the military.

- **Utilitarian Power:** This form of power relies on extrinsic rewards such as salary, merit pay, fringe benefits, working conditions, and job security to control lower-level members of the organization. Unions, farmers' co-ops, and many government agencies use this form of power in addition to business concerns.

- **Normative Power:** The use of intrinsic rewards is the key to normative power. Examples of such rewards include interesting work, identification with

Figure 4.4
Etzioni's Types of Compliance

Type of Involvement		Type of Power: Coercive	Utilitarian	Normative
	Alienative	X		
	Calculative		X	
	Moral			X

goals, and making a contribution to society. Management power is based on its ability to use symbolic rewards, esteem, and prestige.

Normative organizations include churches, political organizations, hospitals, and universities. Schools are most likely part of this category, although in many instances business functions may rely on utilitarian power. Examples include the transportation and maintenance departments.

The three types of involvement determine how well each of these types of power encourage or obtain employee participation in the organization. Etzioni's three types of involvement are:

- **Alienative Involvement:** This method relies on intense, negative actions resulting in the alienation of individuals from their organizations. Prisoners offer perhaps the best example of this type of involvement.

- **Calculative Involvement:** A negative or positive orientation of low intensity is the basis of calculative involvement. In a calculative environment, individuals interact with the organization for personal gain.

- **Moral Involvement:** This relates to a positive orientation of high intensity such as involvement with a church or synagogue or devotion to a political party. In this case, the rewards to the individual are at least partially intrinsic. This is a good model in educational systems because of the autonomy that exists for many school employees.

Mechanistic-Organic Organizations

Researchers have begun to find a paradox between bureaucratic and professional norms. That is often, occupants of hierarchical positions do not have the technical competence to make decisions involving professional knowledge. Jerald Hage has suggested a theory of organizations that offers a framework for determining what kind of organizational structure is best. He offers two structures, mechanistic (bureaucratic) and organic (professional).[30] In this theory, Hage identifies four variables constituting organizational means–complexity, centralization, formalization, and stratification; and four variables that constitute organizational ends–adaptiveness, production, efficiency, and job satisfaction. Based on these variables, Hage identifies seven organizational propositions that he says show the relationships among the eight variables. It is his thesis that maximization of one of the variables results in the minimization of another in predictable ways. Figure 4.6 shows how they are related.

No organization has a pure mechanistic or organic structure. The degree to which one or the other of these structures is used by an organization depends on its goals and characteristics. Clearly, schools would tend to be more organic given the efforts to professionalize teaching, the high complexity of the task, the lack of centralization given the authority of the teacher in a classroom, and such things as the need to be adaptive to individual students and the generally low production and efficiency of which schools are often accused. However, there is evidence that many teachers experience low job satisfaction due to the number of centralized rules and regulations from above. Models like site-based management may be a vehicle to further professionalize the teaching profession.

In a similar vein, many functions of the business office are best organized in an organic or professional manner, while others may be better organized more mechanistically. For some functions, there is most likely a middle ground. For example, school custodians do a number of fairly routine tasks, but often have considerable autonomy within their school. On the other hand, the book-

Figure 4.5
Characteristics of Mechanistic and Organic Organizational Forms

Mechanistic (Bureaucratic)	Organic (Professional)
Low complexity	High complexity
High centralization	Low centralization
High formalization	Low formalization
High stratification	Low stratification
Low adaptiveness	High adaptiveness
High production	Low production
High efficiency	Low efficiency
Low job satisfaction	High job satisfaction

keeping staff in the central office may be better organized bureaucratically. One of the complexities of the school business office is the range in professionalization of the tasks assigned to it. Finding the right mix across functions and individuals is a very difficult job and presents the school business administrator with a number of complex organizational decisions.

Strategy-Structure Typology

This theory of organizational structure suggests that an organization's strategy dictates the most appropriate organizational structure. Mintzberg offers five parts of an organization and suggests that the type of structure depends on the part of the structure being considered. His five key parts of an organization include the strategic apex, the operative core, the middle line, the technostructure, and the support staff.[31]

Depending on which part of the organization is being considered, Mintzberg suggests there are five coordinating mechanisms including: direct supervision, standardization of work processes, standardization of skills, standardization of output, and mutual adjustment. Each of these represents a method for keeping the organizational structure in balance and represents a different way to coordinate output depending on each unit's function within the organization.

Organizational Culture

The previous section on organizational structure focused on how organizations are structured and how power and authority flow across the organization. The student of educational administration must note that much of the success of an organization depends on the cooperation of employees. An important part of assuring that an organization's workforce is committed to the goals and mission of the organization is the organizational culture.

The culture of an organization according to Lunenburg and Ornstein is all the beliefs, feelings, behaviors, and symbols that characterize an organization. Specifically, the culture includes shared philosophies, ideologies, beliefs, feelings, assumptions, expectations, attitudes, norms, and values.[32] Lunenburg and Ornstein develop six characteristics of organizational culture as follows:

- **Observed Behavioral Regularities:** The use of common language, terminology, and rituals or ceremonies when organizational members interact.

- **Norms**: Behavior standards that evolve in work groups resulting in standards and yardsticks by which performance can be and generally is measured.

- **Dominant Values:** The sharing of major values across an organization.

- **Philosophy**: This relates to the policies that guide an organization's beliefs regarding employee and customer treatment. This category would include a school's mission statement.

- **Rules**: The guidelines for getting along in an organization and the things a new member of the organization needs to learn to become an accepted member of the group or organization.

- **Feelings**: The overall atmosphere based on the physical layout of the organization and the way members interact with clients or other outsiders.

Taken collectively, these characteristics can be used to define an organization's culture. In the following sections the development, the maintenance, the change of organizational cultures, and the effects of organizational cultures on school districts are discussed.

Development, Maintenance and Change of Organizational Cultures

This section discusses how organizational cultures develop, how they are maintained, and how they can be or are changed over time.

Creation of Organizational Culture

The development of a corporate culture is complex and difficult to describe. The discussion that follows is based heavily on Deal and Kennedy's book, *Corporate Cultures*.[33] They suggest that the creation of an organizational culture relies on heroes, rites and rituals, and communication networks. Each is considered below.

Heroes: Heroes, whether born or created, are an important part of any organization's culture. Examples of born heroes include such individuals as Henry Ford, founder of Ford Motor Company, and Mary Kay Ash, founder of Mary Kay Cosmetics. Created heroes are individuals who were made heroes by the organization to commemorate important events in the organization. Lee Iacocca of Chrysler is an example of such a situational hero because he turned the company around and brought it from near bankruptcy to profitable status. Other such heroes include Sam Walton of Wal-Mart and Green Bay Packer coach Vince Lombardi. Heroes provide role models and offer a symbol of the organization to others. They also set performance standards used to motivate participant achievement.

Rites and Rituals: Rites and rituals are those events, activities, and celebrations managed by a company or organization to recognize achievement. Awards such as the Teacher of the Year Award and National Merit Schools Program are examples of such rites and rituals. The purpose of these rites and rituals is to recognize individuals for their performance and encourage others in the organization to establish similarly high goals.

Communication Networks: Stories or myths of heroes are transmitted through communication networks within an organization. Deal and Kennedy identify a number of storytellers who interpret what is going on. Their interpretation of the information influences others. "Priests" are the individuals who guard the organization's values. They listen to others and provide alternative solutions to problems. "Whispers" are individuals who have the boss' ear. As a result, people who want to get things done often go to the whisperer in hopes that they can convince the boss of the wisdom of their own idea. "Gossips" carry the day to day activities of the organization through the communication network. A large part of their role is to maintain the status of the heroes of the organization. Finally, individuals who keep everyone informed concerning what is going on in the organization are known as "spies" in this framework.

Maintaining the Organization's Culture

To maintain an organization's culture, there are a number of mechanisms that organizations utilize. These mechanisms include the following:

- **Selection of Entry-Level Candidates:** By choosing new employees carefully, organizations can employ individuals who share the values of the organization.

- **Humility-Inducing Experiences**: These are experiences which cause employees to question prior beliefs and values, making them more receptive to the values of the culture of the organization they have just joined. Examples of these experiences include assigning more work than can reasonably be completed or assignment of work for which the individual is overqualified. The message given to newcomers is "You must pay your dues."

- **Job Mastery:** The purpose of this step is to focus on the development of an employee's technical knowledge. The organization assesses individual performance and assigns new responsibilities to people on the basis of their progress. Often these processes include a step-by-step procedure where promotional objectives are clearly laid out and performance levels required to reach each step enumerated to all employees.

- **Reward and Control Systems:** Organizations pay attention to measuring results and rewarding performance. Reward systems focus on aspects of the organization that are tied to success and the values of the organization's culture. Individuals who do not meet the established performance standards are often assigned to tasks that are innocuous or that have relatively little responsibility, pushing their careers "off track." On the other hand, those who succeed are promoted to new and more responsible positions over time.

- **Adherence to Values:** Over time, individual behavior begins to more closely match the values of the culture. This helps employees reconcile personal sacrifices caused by their belonging to the organization.

- **Reinforcing Folklore**: Throughout the socialization process, members of the organization are exposed to rites and rituals, stories myths, and heroes that portray and reinforce the culture.

- **Consistent Role Models:** Organizations rely on individuals who have served the organization well to serve as role models to newcomers. They are identified as employees symbolizing success, and others are encouraged to behave and perform in similar ways.

Changing Organizational Culture

Changing an organization's culture is a complex and lengthy process which requires a number of things to happen. Lunenburg and Ornstein describe a change cycle with the following parts:[34]

- **External Enabling Conditions**: Conditions that are supportive of cultural change in the organization. These include things that are in the external environment, but impact the organization. In schools, this could include a large enrollment increase or even a legal ruling that requires the district to act in new ways.

- **Internal Permitting Conditions:** There are four internal conditions that must exist if an organization is to succeed in changing its culture. These are:

    - A surplus of change resources.
    - System readiness.
    - Minimal coupling.
    - Change agent power and leadership.

- **Precipitating Pressures:** There are four factors the could precipitate a change in organizational culture. They are:

    - Atypical performance.
    - Pressure exerted by stakeholders.
    - Organizational growth or decrease in size, membership heterogeneity, or structural complexity.
    - Real or perceived crises associated with environmental uncertainty.

- **Triggering Events**: Culture change is often a response to one or more triggering events such as an environmental disaster, a managerial crisis, an external revolution such as mandated desegregation, or internal revolution such as a new administrative team.

- **Cultural Visioning**: The creation of a new vision of the organizational culture is an important step in changing cultures. Leaders attempt to ascertain what future conditions will be like and then create an image of the organization within that future.

- **Culture-Change Strategy:** Once the vision for a new culture is established, a strategy for achieving that culture is essential. This strategy would outline the general process for transforming the existing culture into the new one.

- **Culture Change Action Plans**: A series of plans must be developed to actually change and manage the development of the new culture. Members of the organization must be encouraged to accept the new cultural values, and a management strategy must be developed to have members to accept that change.[35]

- **Implementation of Interventions:** Interventions based on the way a particular organization works must be selected for each phase of the action plan developed in the previous step.

- **Reformation of Culture:** Once fully implemented, the intervention plan results in a new organizational culture.

While straightforward on paper, the process of changing an organization's culture is a complex and time consuming process that requires all members of the organization to accept the new cultural values. Inability to bring along members of the organization will lead to friction and unhappiness among members of the organization, further complicating the task of changing the culture, and making it more difficult to achieve the organization's mission and goals. For that reason, such changes should be contemplated very carefully, and if it is important to make such changes, they should be organized and planned to ensure full implementation across the organization.

Effects of Organizational Culture

The culture of an organization affects many parts of the administrative process. The impacts of culture are felt in motivation, leadership, decision making, communication, and change. Since administrators are evaluated on the results they achieve, an organization's culture is an important concept because it affects the results that are produced, and provides a framework for the evaluation of those results.

Excellence as an Organizational Goal

One of the major focuses of organizational culture is achievement of excellence in the organization's stated goals. While many have written concerning this topic, one of the most popular and widely read treatises on excellence and organizational success is Thomas Peters and Robert Waterman's pathbreaking book, *In Search of Excellence.*[36] Based on extensive research on

America's "best run companies," Peters and Waterman identified eight attributes that characterize excellent companies:

- **A Bias Toward Action**: The organization continually does and experiments. For example, in a school district, the administration might attempt to implement a new management by objectives program to guide its mission and measure its results.

- **Close to the Customer:** The company looks to the customer for direction in the development and design of new products. Listening to parents' concerns and needs is the school district equivalent of this important characteristic.

- **Autonomy and Entrepreneurship:** The company values and encourages risk taking and innovation. This is very difficult for many school districts. The risk of failure in an educational setting is quite unpopular, and it is possible that many of the individuals working in a public agency have chosen public service because they are risk averse. A teacher has relatively high job security, as do most administrators, and consequently, they are often willing to accept lower salaries in exchange for that security. The related result is that there are fewer individuals willing to risk failure in school systems than are found in the private sector, making such autonomy less common and less often taken advantage of when it is available.

- **Productivity Through People:** The belief in the organization's employees is shown through shared decision making and encouragement of new ideas. School districts that rely on site-based management programs offer greater confidence in employee ability, along with parent ability, to participate in educational decision making.

- **Hands-On, Value Driven Effort:** Core values are promoted and clarified to employees. In strong culture schools that emphasize high achievement levels of students and high performance and growth of faculty, hands-on value-driven efforts are taking place.

- **"Sticking to the Knitting":** This refers to sticking to the business the company knows how to run. In education, this implies that school districts stick to the education of students and not focus on all of the other problems of society. In recent years, society has thrust many additional responsibilities on the public schools, and this has led not only to the inability to meet all of the demands placed on the schools, but in many cases to the deterioration of the quality of schooling in general.

- **Simple Form, Lean Staff:** Similar to site-based management, this concept calls for small, central staffs with minimum levels of "bureaucracy" at all levels.

- **Simultaneous Loose-Tight Properties:** While this may seem a paradox at first glance, it implies that the company relies on tight maintenance of such things as cultural values, and is loose about autonomy and decentralized decision making. Schools can similarly provide tight, well reasoned sets of values, and leave the implementation of those values to the school site managers and staff.

Theory Z Management and Schools

In recent years, Japanese management techniques have received a great deal of attention in both the popular press and in management theory. In a path breaking comparison of the two management styles, William Ouchi compared high producing companies in the United States to those in Japan to see if there were any commonalties across the two cultures. He developed Theory Z to explain the success of these companies.[37]

An extension of McGregor's Theory X and Theory Y, Ouchi's Theory Z is concerned with the culture of the whole organization, rather than with individual supervisors. Ouchi argues that when applied to schools, Theory Z's features include trust, subtlety, and intimacy; shared control and decision making; training in planning, organizational processes, budgeting systems, and interpersonal skills; motivation through self-interest; rewards over the long run, and the importance of high quality education.[38]

Summary

This section has described the importance of culture on successful organizations. It has shown how many of the concepts of good culture can be applied to public schools and provides examples of how such cultural values can be applied at the school district and school level. However, these values and goals cannot be carried out without a good educational leader. The important traits of leadership are described in the next section of this chapter.

Organizational Change

Given the previous research on organizational theory, structure, and culture, it is important to recognize that organizations also experience change. There are many reasons for an organization to change. These include decisions to refocus the mission or goals of the organization and responses to external circumstances. Change in and of itself is not necessarily positive nor negative. However, school district administrators are often reluctant to initiate change or respond to situations demanding organizational changes. There are many reasons why school business administrators may want to change the way they operate. For example, dramatic changes in funding levels could lead to substantial changes in the organization of the school business office, as could governance changes like district decentralization. In addition, state policy mandates that require different methods of operation, or community pressure on the school system to do things differently could also put pressure on a school business office to make significant changes in its organizational structure.

During the last decade, there has been a change in the approach taken by most organizational theorists on how to deal with change. They have moved from an emphasis on innovation in organizations to development of frameworks for change based on the organization's strategy for human resource management and its relation with the external environment.[39] This section describes internal and external pressures for change in schools, looks at why many organizations and/or leaders experience resistance to change, and provides some approaches for making change happen in a school.

Pressures for Change

There are a number of places where pressure for change can occur in a school organization. These pressures include government intervention, societal values, technological change and the knowledge explosion, and the effect of administrative processes and the fulfillment of employee's needs.[40] Each is described in more detail herein.

Government Intervention

Government intervention in the operation of a school district comes in many forms. Since school districts are a creation of state government, state policymakers often exert considerable control over what school districts can and cannot do. This ability to dictate what districts must do can lead to changes in the organizational structure of a school district. For example, if a state changes its accounting and fiscal reporting system for school districts, the school business office of each district must comply with the new requirements, even if it leads to substantial changes in the way the accounting and budgeting offices operate. Similarly, a number of states have enacted legislation enabling schools to establish themselves as charter schools and break away from the home school district. Such actions on the part of a school site can have considerable impact on the district's fiscal offices as well. The district business office has to determine what its new revenue level will be (assuming the charter school is entitled to its own revenue). In addition, many charter schools may continue to rely on the former school districts for certain services such as major repairs, payroll services, and so on. However, since the charter school is newly independent, this may necessitate the school district establishing new or different systems to accommodate both the charter school and the public schools that remain part of the school district itself.

The federal government can also have a tremendous impact on the organization of local school districts and the fiscal operations. Two examples of the kind of impacts that the federal government can have are Chapter 1 and special education. Chapter 1, which provides categorical assistance to schools and districts with high incidence of poor children, distributes over $5 billion a year to schools across the nation. However, those funds presently come with a number of mandates, including the requirement that the funds "supplement, not supplant" existing district expenditures. School districts are subject to audits regarding the use of Chapter 1 funds and risk having to pay back funds that are misspent and losing future funding. As a result, most districts that receive Chapter 1 funds go to considerable lengths to make sure that they are using the funds in the manner intended, so as not to lose these moneys in the future nor lose access to other federal funding programs as well.

Another example of the influence of the federal government is in special education. Since the passage of Public Law 94-142 in 1974, school districts have been required to assure that children with special needs receive appropriate educational services. The federal government provides some financial reimbursement for these programs, but nearly 96 percent of the funds for special education come from state and local funds. Meeting the requirements of this mandate to education children with special needs, and of the more recent IDEA (the Individuals with Disabilities Education Act), has led to considerable change in the way public school districts and even school business offices are organized.

One last manner in which the federal and state governments have forced organizational change in public school districts is the reform movement of the 1980s. Following the publication of *A Nation at Risk*, many states enacted legislation calling for school districts to increase high school graduation requirements, use longer school years and longer days, increase teacher salaries, and enact other reforms. These have led to changes in the way school organizations operate. As this discussion shows, higher levels of government (the state and Federal governments) can play a very strong role in forcing organizational change in school districts.

Society's Values

Societal values influence school organizations in two ways. Herzberg's hygiene factors (salary, job security, working conditions, supervision, organizational policies, and status) are today an expected part of an employment package, and the absence of them, or even the requirement that some kinds of benefits be partially paid for, can lead to employee dissatisfaction. Thus, we have seen considerable changes in organizations to meet the needs of employees.

The second societal value impact is closely related to the government interventions discussed. In the 1960s and 1970s, the values of equity and efficiency dominated educational policy, leading to such programs as Chapter 1 and special education. In the 1980s under the Reagan administration, there was a general shift to a greater emphasis on excellence. Equity appears to be making a comeback, but the demand for excellence has not diminished. In each case, changing school organizational structure within school districts have been needed.

Technological Change and the Knowledge Explosion

The dramatic changes in technology, combined with the virtual explosion of new information require school organizations to work harder than ever to keep up with both the technological change and the vast storehouses of new information that becomes available. To do so often requires changes in organizational structures. One of the most obvious changes is the growth in research and development departments in many large school districts.

Process and People

Administrative process and the people working for an organization may also be causes for change within the organization. If there are breakdowns in communications, decision making, and leadership for example, then changes may be needed to put an organization back "on-line." Moreover, if people are not performing to their full capacity, changes in the organizational structure to help motivate them and improve performance may bear examination.

Resistance to Change

Unfortunately, change is not easy, and there is often considerable resistance to changes in the organization. These barriers include such things as:

- **Interference with Need Fulfillment**: When an individual feels that the changes taking place will either result in lower income, or reduced job satisfaction due to changes in work tasks, or has similar fears about change, there will be resistance to that change.

- **Fear of the Unknown**: Many people like stability and feel secure and invested in the status quo. Therefore, they will resist changes that lead to establishment of new work routines and job duties. They may be afraid of learning a new job or new tasks, and thus will resist any changes in organizational structure.

- **Collective Bargaining Agreements**: Most school district employees today work under the terms of a collective bargaining agreement with their employer school district. These contracts, which represent one of the most significant changes in educational structures in recent years, can also be substantial barriers to change. These agreements restrict the flexibility of the board of education to allocate resources when they provide for multi-year salary increases, and they often define behavior of administrators and teachers, inhibiting their ability to effect changes in structure if they are desired.

- **Limited Resources**: Often change requires new resources. Additional personnel or capital may be needed to effectively implement any organizational change. On the other hand, limited resources can also be an inducement to change because it is impossible to continue doing things as they have always been done in the past. One responsibility of a school business manager is to identify programs or projects that are no longer necessary and use the funds saved from elimination of those programs to implement changes. The district may want to maintain its investment in personnel by helping them gain the requisite skills for new tasks that are assigned as a result of organizational changes.

These pressures to resist change are often very strong, making it difficult to enact changes even if decision makers feel it is in the best interests of the organization. The following section offers some suggestions on how to successfully manage change processes.

Managing Organizational Change

Cornell and Herman argue that there are three basic characteristics of successful organizational change. They are:[41]

- **Establishment of Need**: All individuals who are impacted by the change must recognize the need for the change. This will provide the incentive for them to participate in the change effort.

- **Involvement of People**: In addition to shaping changes on the basis of the needs of the organization, it is important to secure the agreement and cooperation of the groups that are affected by the change. In a school district, this most often means the teachers and administrators, but if the changes being contemplated will change the way employees of the business department do their jobs, they should be included as well. For example, if the district implements new standards for campus appearance, it is important that the school custodians as well as central maintenance and repair personnel be consulted and participate in the planning for the implementation of the new program. This will secure greater levels of cooperation from the parties involved and hopefully make implementation of the change more effective.

- **Sustaining Commitment to Change**: Once implemented, the school district leadership must remain committed to the new structure if they were to expect the new process to succeed.

Overall Summary

This chapter has summarized a great deal of information about management theory and organizational structure and change. These concepts, most of which were developed for private sector organizations, are highly applicable to the operation of school districts generally, and to school business operations and activities in particular. Many of the functions of a school business office are similar to those performed by similar departments in private organizations. The most obvious are payroll, accounting, maintenance and repairs, and even transportation. The material presented is an attempt to offer the student of school business management a management framework and perspective for organizing the multitude of functions performed by a school business office.

Endnotes

[1] See, e.g., R. C. Wood and W. M. Worner, "Run It Like a Business," *American School and University*, Aug., 1981, 24, 26, 28.

[2] F. W. Taylor, *Principles of Scientific Management* (New York: Harper, 1911).

[3] *Ibid.*

[4] H. Fayol, *General and Industrial Administration*, (New York: Pitman, 1949).

[5] F. C. Lunenburg and A. C. Ornstein, *Educational Administration: Concepts and Practices*. (Belmont, CA: Wadsworth Pub. Co., 1991).

[6] L. Gulick and L. Urwick (eds.), *Papers on the Science of Administration* (New York: Columbia Univ. Press, 1937).

[7] M. Weber, *The Theory of Social and Economic Organizations* (translated by T. Parsons), (New York: Oxford Univ. Press, 1947).

[8] W. H. Knight, "Theories of Modern Management," in R. C. Wood, *Principles of School Business Management* (Reston Va: ASBO International, 1986).

[9] E. Mayo, *The Human Problems of an Industrial Civilization* (New York: Macmillan, 1933).

[10] K.Lewin, R. Lippitt and R. White, "Patterns of Aggressive Behavior in Experimentally Created 'Social Climates,'" *Journal of Social Psychology*, 10(1939), 271-299.

[11] C. R. Rogers, *Client Centered Therapy* (Boston: Houghton Mifflin, 1951).

[12] J. Moreno, *Who Shall Survive?* (New York: Beacon House, 1953).

[13] C. I. Barnard, *The Functions of the Executive*. (Cambridge, MA: Harvard Univ. Press, 1938).

[14] E. W. Bakke, *The Fusion Process* (New Haven, CT: Yale Univ. Press, 1955).

[15] C. Argyris, *Personality and Organization,* (New York: Harper and Row, 1957).

[16] F. Herzberg, *Work and the Nature of Man*. (Cleveland: World, 1966).

[17] In a few states, the chief fiscal officer is charged legally with dual responsibility with the school superintendent, i.e., he or she answers directly to the school board. While this is generally rare, it should be noted.

[18] See J. Hannaway, and M. Carnoy, *Decentralization and School Improvement: Can We Fulfill the Promise*, (San Francisco: Josey Bass) 1993.

Chapter 5
Strategic Planning

Introduction

As will be seen throughout this text, planning forms a critical and indispensable role in modern public school districts. While planning in some form has always been important to public education, its role in the modern school system has increased to enormous proportions by virtue of an uncertain and constantly changing environment that forces school districts to anticipate a highly unstable future to an extent never encountered by its forebears. Indeed, the many faces of administration include a wide variety of opportunities for leadership and frontline drama, and it is in the arena of planning that most leadership acts occur that permanently alter the fundamental nature of public education.

The task for school business administration has never been more compelling, since the modern district has come to recognize the interrelatedness of all areas of school operations and the interdependence of all system elements on a wide variety of resources, including money. School leaders of all types have, therefore, been forced into a continual planning mode in order to survive, making it essential that successful administrators understand the elements of planning to strategically position their districts for optimum benefit and service to constituencies.

Given such reality, this chapter is intended to introduce and explore in a broad context the role of planning in school districts. The chapter begins by laying an historical basis for planning through the development of formal study of the planning process, moving from very general considerations through the phases of planning theory up to the present. The second section becomes more specific to the context of planning in a governmental organization, such as schools, by outlining some concerns school districts face that may limit how aggressively they are able to plan. The third section moves into the broadest aspects of the actual planning activity by considering demographic, economic, and political considerations that frame much of the planning activity. This type of thinking links directly to the fourth section, which moves the planning activity inside the school by considering the scope of planning and issues relating to values surrounding the public schools.

The fifth and final section turns the discussion specifically to the task of strategic planning itself, defining and differentiating strategic planning from other types of planning activities and working through the elements of the strategic planning process. Although this chapter is not intended to be an exhaustive expose of strategic planning, it lays a foundation for informed and productive work on the part of administrators by defining a healthy respect for a vital element of one facet of the administrator's leadership role.

Planning Theory

The role of planning in the modern context of American education seems to present a confounding dilemma. On one hand, Americans have cherished individual freedom, initiative, and responsibility more than any other nation, while on the other hand there has been a widespread recognition of the need for long-term decisions and commitments on matters affecting a community of collective interests. The railing of one political candidate against "pointy-headed social planners who can't even park a bicycle straight" captures the invective commonly felt by planners who seek to bring public and individual interests together for a unified purpose.

Nevertheless, there has been a clear recognition that planning is an essential element of progress and social order, so much so that "getting the government off the backs of Americans" is no longer the most common sentiment throughout society. In fact, there is evidence to suggest that there is a growing demand for planning, as calls for accountability rise and as government itself takes a leading role in demanding effective and efficient use of public moneys. This includes long-range planning for the purpose of controlling and recording moneys spent and linking the development of an organization to neutral and powerfully important goals or shared values. In the modern context, the issues of accountability and goal-directedness are increasingly driving the engine of change in American education.

Rational-Technical

As noted in the first edition of this text, much of what has passed for planning in schools has its basic origins in other disciplines and has not stood as a separate literature base in education.[1] A rational-technical perspective has been at the root of much planning, however, in that such planning is closely related to managerial and organizational behavior models that are discussed later in chapter 8 under the topic of personnel administration. This view of organizations and people has argued that good planning can resemble a well-oiled machine in which outcomes are predictable based on manipulation of inputs such as labor and capital. This model has also pervaded planning, as rational-technical perspectives concerning planning have advocated establishing clear goals, making cause-and-effect relationships between means and ends, and explicitly predetermining monitoring measures and procedures. The planning method is thus ultimately rational in that it starts with overt and explicit statements of intent regarding desirable outcomes, and reasons deductively to link various means of realizing the desired end. The net sum of a rational-technical planning model is thus highly linear, i.e., identifying the end goal and reasoning backward to determine the efficient steps leading causally to the anticipated outcome.

For planners, the obvious advantage of the rational-technical view is its focus on achievement. The process is nearly impossible to implement, however, without having carefully refined at least one, and generally several, goals that serve to guide the planning process. In a technical sense, the conception is a model of efficiency, an engineered design that links part A and part B with process X and Y with highly predictable outcomes. Its organizational analog is a bureaucracy in true Weberian fashion, in that all rational-technical models assume that organizational goals provide the direction for product manufacturing and that there is a reasonable congruence between the large goals of the organization and the work of multiple groups and individuals within the organization.[2] By this standard, the measure of success can be found in clearly stated goals, specified ends-means relationships, and benchmarks of progress that force the organization to stay on-task in a world full of diversions and reasons to justify the failure to accomplish its aims.

In the technical-rational planning scheme, the underlying solution to a distracting world is control. Control implies the ability to change or overcome the conditions leading to failure, and control is comprised of acts that cause planning to be built around technical production processes. In order to carry out the process of coordinating the technical work in organizations, a rationalized bureaucratic conception of planning is necessary—a reality that introduces the planocrat.[3] The ultimate bureaucrat with organizational skills and adeptness for carrying out the fine details of an organizational plan, these people keep the process running according to plan where stability, production, and control are achieved through a standardized process that is deliberately static, systematic, and logical.

While highly appealing to production-based organizations, educators have had difficulty implementing a technical-rational perspective on planning in school organizations. One reason is that the most basic goals of public education are imprecise and abstract. Quality, excellence, and educating for maximum performance are tidy and clean goal statements that are to be coveted, but are open to multiple and often competing interpretations as to how best reach them. Worse, the full and precise interpretation of such goals is dependent on who is interpreting. While the goals are multifaceted, the difficulty stems from their conceptual complexity in terms of their relative priority. Some goals are means to other goals in which the interactive effects of organizational dynamics are enormously complex.

Although a hallmark of educators has been to attempt to simplify the goal statements, by the time goals are "translated" into objectives and means, the interpretation of others (particularly implementors) clouds the original meaning of the goals. In any case, priorities are usually left to implementors or others in the organization, thus diverting the responsibility to a technical or organizational analysis rather than to a political (value preference) choice.

In addition, there are technical difficulties that relate to the knowledge and information concerning the best, or even a very good, way of doing things. Often boards of education, administrators, and teachers lack specific knowledge concerning what the alternatives are, much less what will work best in the long view. Sometimes, the shortcomings can be remedied by obtaining additional information, at least to the point where a reasonable judgment can be made concerning what will work in a particular instance. But this requires a trial and error approach that fosters an acceptability of potential failure—a fatally flawed approach in a technical-rational model that presumes perfectibility of cause-and-effect relationships for prediction purposes. In peopled organizations, particularly those with highly abstract skills and tasks, perfect cause-effect knowledge is simply too difficult to obtain in predicting long-range specific outcomes. The model is well captured by a corporate manager who observed that even though the information system in the organization could track the "fall of every sparrow," everyone secretly knew the absurdity of projecting numbers out five years to decimal points.[4]

To complicate the model further, the approach requires additional rationalization about who in the organization is best suited to carry out the task, the right amount of time, money and personnel resources to be reallocated, and the specific measures for success. Under all but the most engineered circumstances, the task quickly becomes impossible from a management perspective—ironically, the original task was to make control easier by a highly intellectual technical-rational model of scientific management prediction.

As schools have belatedly discovered, the general intent of this approach to planning works well in theory, but not in practice, because the requirements of the model cannot be met in abstract knowledge organizations such as schools. The model is better suited to a technical and scientific view of the world in which the environment (laboratory) can be held constant, carefully controlled ends-means relationships can be created, and multiple demands can be met sequentially and within a much narrower definition than typically describes the work of education. The root flaw of the technical-rational planning model for peopled complex organizations, then, is that the world of broad goals, human subjects, and organizational dynamics cannot be made to fit the template of the rational conception.

Disjointed-Incremental

Limitations of the rational-technical planning model led to genuine criticism of the highly stylized and idealistic view of the world. In direct reaction to these problems, new models were conceptualized that sought to address the weaknesses and to provide a better vehicle for enhancing organizational goal achievement. Rather than starting with a conception of how the world should behave, theorists such as Chester Barnard and Charles Lindblom countered with a model that grew out of empirical research concerning how organizations really function. As a consequence, models of disjointed-incremental planning emerged from describing how goals are actually established and decisions are made and implemented.

While the work of these theorists was not written for the purpose of contributing to planning theory, the implications for planning are significant. Clearly relevant to planning theory was the contention that decisions are made under conditions of extreme uncertainty. The purpose of planning is to reduce uncertainty, and insofar as the arena of the plan can be adequately controlled, the purpose is met. If controls were not adequate, however, the purpose is either defeated or only partially satisfied. What Lindblom and others found was that there are multiple and competing influences on the organization, many of which cannot be adequately stabilized or held sufficiently steady to stay within the closely reasoned series of cause-effect relationships inherent to the rational-technical planning style.

For example, for schools such observation meant that influences outside the district, such as economic and political conditions or federal and state policy, cannot be adequately forecast and controlled to permit operating confidently from a rational style when engaging in long-range planning activities. Further, events inside the organization are equally complex in that there are many decision points involving a large number of actors, each of whom interprets the goals and policy of the organization from a completely individualistic perspective. Under such conditions, each decisionmaker tries to "make sense" of the operation for which the person is responsible, or, in a more Machiavellian light, seeks to turn the organization toward personal gain. Either way, the conclusion is that organizations are sufficiently complex so as not to be able to be controlled to the extent demanded by rational planning without smothering the organization in red tape.

These descriptions can be readily seen to apply to schools as complex organizations. Even at the district level, educators have seen the futility of attempting to impose a single template of sameness on all schools or even on "similar" classrooms. Quite aside from the question of professional autonomy, this recognition admits to the fact that it is impossible to force the enormous variability within even singular classrooms into a one-sized manageable model. The essential question under such variability becomes "Who has the best information at the operational level to make the kind of decisions needed to keep the organization moving forward?"

The uncertainty of events and the enormous complexity of organizations prevent top-level administrators from having the information or overall knowledge needed to make optimally correct long-range decisions. According to some theorists, then, administrators in complex organizations would be better served by a descriptive model that closely corresponds to the conditions under which organizations operate. In 1959, Lindblom coined the term disjointed incrementalism to describe the nature of decisions as being disjointed, not by deliberate design, but as a result of incomplete knowledge in any one place in the organization. While broad purposes are part of any organization and constitute the initial reason for organizational existence, planning in a real sense takes place at the operational level where decisions—as events, exceptions, and anomalies—occur. These theorists would then argue that it is through such decisions that the real policy of the organization is made.

Lindblom further described decision making as incrementalism, meaning that it seldom leads to radical, high risk changes, but rather marginal ones. Organizational decisionmakers add or subtract a little here and there, making small changes to accommodate environmental shifts. Incrementalism, as Lindblom saw it, is necessary because proposed means of control have faltered, as in zero-based budgeting and program planning and budgeting systems (PPBS) due to a demand for more operational knowledge than was possible to accumulate and interpret. By keeping closely in touch with the internal and external environments, the organization keeps itself adaptable to a constantly shifting environment.

For people seeking purer rationality in planning, disjointed incrementalism may seem to some to be a way of simply rationalizing the status quo. To others, it may seem like no planning at all. Planners have viewed disjointed incrementalism differently, however. Since theorists recognized that decisions are made within the broad context of organizational purposes, planners could feel, with some degree of assurance, that transformational decisions will not be made in the disjointed manner, thus leaving some room for deliberate and long-range planning. They also recognized that to be effective, planning would have to be attentive to the reality of the organization into which it is to be integrated.

Although highly appealing in comparison to the limitations of the rational-technical perspective on planning, other scholars nonetheless saw problems with the disjointed incremental approach. While they agreed that the technical-rational scheme was too rigid for practical purposes, they were skeptical of the broad flexibility of disjointed incrementalism, claiming that it was too difficult to differentiate "planning" from what it set out to replace.[5] Thus by the late 1960s, organizational planning was ahead of where it had been, although only slightly, in that theorists were searching hard for a middle ground that could be described as lying somewhere between a tightly structured rational-technical model and the loosely structured disjointed incremental position touted by those theorists who were awed by the complexity of peopled organizations.

Mixed Scanning

An alternate middle ground was proposed in 1967 by Etzioni,[6] a theorist discussed later in this text for his other useful contributions to theories of relationships between people and organizations. Etzioni took the position of contrasting rational-technical perspectives on planning with disjointed incrementalism, offering a critique of both that laid bare the assumptive flaws. The rational model was portrayed by Etzioni as making impossible demands in that it requires constant comprehensive vigilance of planners over the goals and operations of the organization—the time and human resources required to implement it are simply not available. But Etzioni was no less harsh in his criticism of incremental models, arguing that this approach was too attentive to short-term operations definable as generally no further into the future than the next year's budget. Etzioni also noted that the model allowed the organization to drift for years without considering either imaginative new directions or potentially disastrous environmental shifts on the horizon—all functions of the shortsightedness of the model due to its focus on unpredictability.

From this perspective of opposite extremes on a planning continuum, Etzioni proposed a mixed scanning model in which he offered the metaphor of a camera in which the broad view of the purposes of an organization might be thought of as a wide-angle lens. In contrast, a sharply focused close-up lens would symbolize immediate operations of the organization. In this model, each of the earlier models retained some beneficial utility to the organization; for example, the rational-comprehensive view was too demanding for normal operations, but it was essential to take a wide-angle view of the future, particularly when the environment could be described as turbulent. The wide-angle view derived its usefulness from the opportunity for creativity and risk, as well as for minor adjustments based on a long-range view—effectively, an organizational snapshot "from the air," so to speak, which could take place at regular intervals or when a major decision had to be made or when a crisis must be confronted.

In contrast, the close-up view would show top management the fine detail of the landscape or, in organizational terms, provide the necessary view for scrutiny of specific organizational operations. From Etzioni's perspective, such planning would evolve as "successive but limited opportunities for choice" presented to those closest to the work of the organization (disjointed), with each decision varying slightly from previous operations (incrementalism).

The mixed scanning model provided an inviting logical response to the previous planning models, although it too has been subjected to some criticism. A major criticism is that mixed scanning follows the tradition of rationality and is, therefore, subject to similar limitations. These limitations particularly apply to the demands for depth of understanding in all areas, the information requirements, and to some extent the cause-effect mechanical linkages that burden mixed scanning, even though the demand for order and absolute control from top management is alleviated. Further, mixed scanning has been criticized for not making clear what constitutes a "big" decision—that is, when is it appropriate to use the wide-angle lens as opposed to the sharp focus of the camera on the organization? In school terms, how will the administrator know when to make more than an incremental decision? In short, while the theory added conceptual depth to an otherwise polarized view of planning, practical limitations went unaddressed in an empirical context.

Organizational Development

As will be seen later in chapter 8 when discussing the evolution of organizational behavior, the 1960s and 1970s brought intense questioning of rationality as the model for anything associated with human behavior in organizations. In the case of a theory known as organizational development, the reaction focused on how to mobilize human resources, not only on achieving organizational goals, but also toward employee self-development, growth, and satisfaction. Organizational development was introduced to the planning literature as a human relations counterpart to rational theories of planning. According to organizational development, clean lines of logical thought are prone to overlook human motivation and organizational dynamics. The response was to focus on the interpersonal side of the organization with the intent of making it part of the long-range plan, as well as part of the planning process.

Although there is insufficient room in this text to fully explore each planning theory's conceptual roots, the criticisms of organizational development are important elements to the evolution of planning as presented in this chapter. The first criticism was that organizational development tends to focus on a single environmental influence to the exclusion of other important environmental elements that induce just as much instability, if not more. The best capturing of this criticism was found in a comment by Warren Bennis as he stated, "I have yet to see an organization development program that uses an intervention strategy other than an interpersonal one, and this is serious when one considers that the most important strategies of change in our society are political, legal, and technological."[7] From this view, many important elements of the environment are missed by a minute focus on particular aspects of planning, a view supported by another comment Bennis made when he described rational-systems as focusing on "structure without people . . ." while the opposite extreme focuses on " . . . people without organization."[8]

Bennis' comments also echo a second criticism of organizational development, it attempts to apply a single strategy to all organizations, the same criticism leveled earlier at rational-technical systems. For school purposes, organizational development suffers in this regard as it does not distinguish between professionally staffed organizations and assembly line units, or between organizations operating in turbulent and stable external environments—

conditions dramatically different for schools than for other types of organizations.

Organizational development was an attempt to represent a new integrated and multifaceted approach to planning and decision theory. However, it actually revealed itself as a remanifestation of rational-technical thought, keeping intact the characteristics of goal orientation and the analytical and objective mentality of cause-effect better suited to stable environments.

New Wave

Although the search for better planning schemes made progress over the course of several decades, planning theory entered the 1980s largely fixated on rational-technical models, in that searching for rationality is so inherent in human motivation and behavior. Evidence of recent models based in rationality is easily portrayed, as in the near cult-like movements typified by such popular programs as Management by Objectives, the *One Minute Manager*, the Effective Schools Movement, production function research, outcomes-based education, site-based management, teacher training models, the tremendous surge of professional staff development programs, and so forth. The advent of cheap microchip technology and the thrust to unseat the complacency of teaching through the introduction of research into laboratory-like clinical school settings is another example of the rational-technical approach.

Renewed research efforts in schools and other organizations to collect massive data banks and to manipulate these data in search of "solutions" are ample proof of the fascination with prediction and application of machine-like metaphors to education's basic tasks of teaching and learning. All such efforts can be described as attempts to control and manipulate an environment using static principles within the context of a dynamic set of variables. These attempts have been made because it is unacceptable to admit randomness or unmanageable footing in a nation and a profession preoccupied with controlling outcomes.[9]

Despite the fact that most organizations, including schools, seem determined to make rational and bureaucratic theory work against all odds and against the experience of such work environments, the 1980s brought a renewed attempt to move beyond planning a priori. Such theories have struck at the heart of blind application of scientific management to education, arguing that there has been little adaptation or consideration for the difference in purpose, level of skill, or type of motivation based on the kind of organization to which theory has historically been applied. New theories question the benefit of goal-oriented bureaucracies, arguing that schools have ignored what parents, teachers, and administrators have always known, i.e., that children cannot be counted upon to have uniform behavior, motivation, and expectations, and that it is not even reasonable to strip away from children their humanity and individuality in a frantic search for absolutely predictable order. Such theory has fundamentally propounded that, while there has been great benefit in the intellectual history of western civilization that has sought order in the universe, there has been a stigma attached to reality that is perhaps undeserved in that "order is good, while disorder is bad." In other words, "messy," "unpredictable," and "dissipative" are words that have been applied pejoratively to administrative styles, when in fact those words may better describe reality as it should be met when seeking to be a "good" administrator—in other words, rationality may not describe reality.

Although only recently developed, the early 1900s actually saw the first explosive break with this line of thinking. Physicists found that not only was more than one interpretation of the same phenomenon plausible, but also that the various interpretations may not make sense under the old view of harmony in nature.[10] For example, light could be both wave and particle at the same time, thus violating the idea that nature is explicit or definitive. Darwin's account of evolution provided another example, showing that progress, random behavior, failed mutations, and dead ends were simultaneously possible. Still another case in point is provided by thermodynamics, which as a science has provided extreme accuracy in predictability, while admitting to random and unpredictable events.[11] Likewise, the work of economists, demographers, and actuaries is all based in prediction. Yet none of these experts is able to account for the anomalies that establish broad generalizations but refuse to permit particulars when predicting such issues as economic conditions, population characteristics, or life and casualty expectancies. This phenomenon was well described in the first edition of this text when it described how everyone fits the general pattern, but no one has exactly 1.8 children.

The result of new wave theory was not necessarily to denigrate the importance of science or systematic observation in the scientific world, but rather to admit that not all worlds are identically structured. To some extent, rationality must form the basis for all worlds—there would be no reason to even think about planning if there were no hope or basis for prediction, order, and control. To relate Darwin's evolution theory to Etzioni's camera analogy, Darwin saw nature as having both a general direction of evolution with clear and broad patterns of development and, at the so-called microlevel where life is lived, confusion or "variance" in which the opportunity exists to take initiative, to create, to succeed, and to fail. Thus the broad contribution of new wave theory was to set forth a view of unpredictability as something not to be disparaged, i.e., that it should not be a point of consternation that schools do not have a one-to-one correspondence with the prediction of light particles or natural environments. Indeed, this may be nothing more than a restatement in a different context of the systems theory discussed later in chapter 8 where bureaucratic goose-stepping (the nomothetic dimension) must admit to the presence of the informal organization (the idiographic dimension) if the work of the organization is not to be dysfunctional.

The impact of new wave theory on educational planning was highly significant for its implications for practice. Stated succinctly, the impact was that theoreticians, researchers, and students of organizational behavior had finally confirmed what teachers and administrators have known from the outset—that schools, in order to be effective, must be both orderly and controlled, and highly adaptive, always ready to give way to idiosyncratic variation and difference.[12] The paradox of "both/and" is difficult to manage intellectually, administratively, and politically, but at least the admission makes it possible to no longer defend, for ideological purposes, that important educational goals and means can be vague, ambiguous, and subject to legitimate reinterpretation based on conditions and context.

A useful way of understanding organizational context in new wave ideological terms is to systematically observe how productive organizations actually function. From these observations have come two important points. First, people need to believe in what they are doing—it is the role of the leader to provide vision and to create an atmosphere of shared purpose among the employees through the leader's own actions and attitudes. As will be seen later when discussing strategic planning, the actions of leaders shape attitudes and beliefs about what is really important. While many specific and major operational decisions may be made from the center (top), they "evolve" as they are filtered and adapted by the makers (of products) and doers (of services) who are openly encouraged to make their own interpretations of how values apply and how to get the job done—indeed, these concepts form the basis of many current practices such as decentralization through site councils and relocating resources closer to the point of utilization and the point of greatest expertise.

This is not to say we should throw away all order—rather, connecting links remain the same for new wave organizations as for old traditional bureaucracies. The policies, legal mandates, contractual agreements and local connections of job specifications and reporting relationships discussed in the next section continue to define and limit the way organizations work. Yet these connections are reinforced by a "social glue" that comes from recognition of mutual dependency, of doing something important collaboratively, and a sense of excitement. Leaders of new wave organizations recognize their roles as motivators of people by creating the meanings by which people live. The essence of a fluid organization is defined by needs and values and characteristics, rather than an attempt to make the organization fit a uniform mold. Using new wave language, this is precisely the shortcoming of the rational model, which prefers analytical detachment over demonstrations of strong commitment. In other words, a technological environment breeds a technological response, rather than a human commitment.

Numerous theoretical perspectives based on new wave thinking have emerged. Although all such theories are too numerous to detail here, Weick's loose coupling theory is illustrative of new wave thinking.[13] Weick argued that organizations tend to have both loosely coupled and tightly coupled operations.[14] Under such a perspective, schools having tightly coupled operations are controlled from, and made accountable to, the center in such aspects as transportation, payroll, and building codes for construction purposes. Aspects of tight coupling, according to Weick, are characterized by rules, agreements on rules, inspection, and feedback to improve compliance; conversely, when one of these aspects is missing, the system is more loosely coupled until it arrives at the point of virtual autonomy—a condition familiar to most experienced school leaders. The natural extension of the perspective, then, is to accept both tightness and looseness as benefits, adjusting when needed, and maximizing the benefit from each aspect.

In contrast, traditional rational management theory has sought to tighten all couplings in the organization. However, attempts to control the highly idiosyncratic and specific activities of teaching and learning generally meet with minimal, if any, success. The corollary has been complaints about systems' lack of cohesiveness without understanding the basic nature of different types of organizations—again, a one-size-fits-all-approach that is ill-fitted to the nature of work in complex organizations. The evidence suggests that the unintended consequences of tightening control of instruction with conventional means is to produce an environment that in the long view is inimical to the very interests that control seeks to promote.

Conversely, however, new wave ideology does not promote lack of control, disjointed incrementalism, or even mixed scanning as the appropriate approach to leadership and planning. Rather, it is an approach that comes from outside the rationalistic tradition entirely by acknowledging ambiguity, paradox, messiness, individual difference, decentralization, and dissipative structures as normal and as descriptive of organizations.[15] Sergiovanni describes this condition by using the graphic terms of Mystics, Neats, and Scruffies to categorize how administrators seek to run their organizations.[16] These labels identify the theoretical stance or "mindscape" of the administrator in perceiving and constructing individual reality—a reality created individually and limited externally by forces that must count into how effectively planning can occur.

The role of planning is thus central to every organization, and the assumptions and beliefs underlying planning help form the basis (theory) from which planning will be pursued.

General Limitations on Planning

Planning in a Government Context

As seen in the foregoing section, theory of planning has a powerful impact on how planning actually occurs in any organization. From one very important perspective, theory defines practice by studying reality and then constructing a post hoc generalizable explanation. From another perspective, theory should dictate practice, in that the purpose of theory is in fact an attempt at prediction to avoid a black box practice syndrome. Certainly a part of constructing a theory of planning is to account for external forces that assist or limit planning. It would be nonsensical to attempt to fit one model onto a wide variety of organizational types.

Limitations regarding planning are particularly important to schools in that they are fundamentally engaged in planning within a government context. Unlike private organizations, restrictions and limitations apply that both force external environments on schools in ways that cannot be rejected and which drive how schools must plan in both substance and process. While such forces seem to be ever-increasing, the most omnipresent can be found in statutory law, administrative law, auditing and evaluation, case law, disputes, and categorical programs. The importance of these limitations on planning have grown since about 1965 as the role of the federal and state governments have increased to enormous proportions.

Statutory Law

The planning environment of an educational organization is heavily influenced by the presence and substance of statutory law. By definition, a law has broad applicability and is the starting point for a plethora of administrative laws that are promulgated through rules and regulations of the monitoring agency. For example, the explosion of state aid to education has resulted in massive rules codified in statute and regulation codes, granting badly needed resources but heavily restricting the use of such aid in order to ensure compliance with the original intent of the law. Similarly, development of state board of education rules and regulations on educational programs are often codified into law, and, in the instance of freestanding regulations, have the same force and effect of law in that state education agencies are themselves extensions of the legislature with powers for enforcing rules and regulations.

There are always three purposes for understanding the impact of statutory law on the planning process, all of which affect the planning process by limiting behavior at the grass roots level. The first purpose of reviewing statutory law is to determine minimum standards; perhaps to qualify for aid as in the instance just cited, to determine compliance with building requirements or transportation safety codes, or to guarantee integrity in the use of moneys or resources through vehicles such as audits and accounting procedures. All such standards are explicitly and exhaustively codified, placing severe restrictions on the latitude school district administrators and other leaders would have were it not for the existence of such standards.

The second purpose of reviewing statutory law is to determine generally the level of flexibility that a district has under the law by understanding the broad intent or spirit of the law in any given area. In contrast to legislatures, which are plenary bodies absent specific restriction, local units are less plenary in scope, although local powers are generally construed fairly liberally by virtue of extensions of the state.

The third purpose of reviewing statutory law is to enable the district to wisely and effectively become involved in the legislative process, influencing legislation when appropriate. For many years, local administrators treated federal and state legislation as an arena in which little control could be exercised and in which passiveness was the proper response. The increased volume of legislation affecting education, however, has forced a change in attitude, moving administrative duty into an arena that includes active policy involvement and advocating for education's benefit among power brokers at state and national levels. In essence, understanding the impact of statutory law on planning is now a prerequisite to planning itself, with the natural extension being active attempts to influence policy at the highest levels.

Administrative Law

As noted, the planning environment of an educational organization is also heavily influenced by the presence and substance of administrative law. The distinction of statutory law provided earlier noted that legislation is frequently interpreted by the courts. Additionally, there is a body of rules and regulations flowing from each statute with administrative law. Efforts have been made in recent years to assure that legislative intent is not understated or overstated by bureaucratic arms of the federal and state governments, but regulations remain a major source of concern and specific constraint to local planning activities. Aside from being involved in the legislative process, local districts need representation in the design of the regulations which, in turn, are often interpreted even more strenuously when disseminated as administrative procedures.

Beyond the role of sentinel, local districts should be advocating for flexibility, local option, and looser arrangements to permit school districts to operate in ways that best fit the local environment. Such a stance must be moderated by common sense, however, in that fire regulations, and other such safety measures, should have a universally applicable uniform code. The limitations of administrative law are often so restrictive, however, that a major goal for planners should be to provide sufficient freedom to conduct the planning activity as free as possible from outside constraint that serves no purpose other than to impose a "foreign" will on the district or to impede its planning through over application of bureaucratic self-perpetuation by successively higher levels of government.

Auditing and Evaluation

A further limitation to creative planning rests in federal and state auditing requirements and procedures that present special barriers to school business administrators, program managers, and school district administrators. The audits are fundamentally attempts to measure how the school is performing on certain aspects of educational functions. Although such audits serve a useful and legitimate purpose, they are also politically sensitive in ways that prevent creative planning and stifle efforts to take risks that might benefit the district. Further, such audits demand inordinate amounts of time through recordkeeping and reporting requirements that take away valuable time from planners who otherwise could be envisioning the future of the district. While these burdens have lessened somewhat since the numerous federal programs that were previously part of the Elementary and Secondary Education Act were collapsed into the Education Consolidation and Improvement Act in the 1980s, the burden is still significant and the advent of technology has only created new work for school districts as massive data bases are now within reach of higher government. The concern here is not for withholding information from other units of government—rather, the concern is for the fact that data efficiency creates demand for new data instead of streamlining existing functions in a manner that frees up time for more creative planning tasks.

Another area of concern in this arena that has a heavy impact on educational planning is the frequently overlooked relationship between the program audit and evaluation process, and the legislation and regulations. Program auditors and evaluators are highly visible actors in the policy development arena, for it is this function of higher government that may pass weighty judgment on how school districts operate educational programs. A quick review of how many such special program audits occur in the average district produces a stunning quantity of state and federal inquiries into the local district, an issue that is increasing rapidly as states take on massive reform of teacher and administrator certification, program review, and fiscal retrenchment policies. Under the conditions of dizzying social change in the United States, many school programs are successful by one definition that takes demography into account, while a single achievement standard may impose a different view.

As a result, audits and evaluations place local districts in a position of zealotry in order to ensure that audits and evaluations are both methodologically and substantively accurate and in accord with explicit and implicit legislative intent and district purposes. It is important that districts of similar demographic characteristics, goals, and size should be vigilant in keeping higher units of government honest in the intense scrutiny characterizing the current nature of evaluation.

Case Law and Disputes

The earlier review of statutory law can be extended logically by the simple statement that statutory law produces case law by virtue of disputes concerning the statute itself or how the statute is applied. While no mathematical formula exists by which to indicate the relationship between these two aspects of government, it is accurate that as more statutes are passed, there follows new case law arising from disputes through the courts. Conversely, the more that is litigated through the courts, the more likely legislatures will react to restrict or expand the courts' interpretations. Case law is a complex interaction that affects every aspect of schooling and certainly has a tremendous impact on district planning in that every change must be considered for liability reasons and for the expense of litigating the district's position.[17] The restrictive impact on planning is completely intuitive, as it must be recommended that an attorney's advice is required every time the district believes a planned program or project might precipitate a legal challenge.

Categorical Programs

Although the experienced administrator can recall many more events restricting freedom in local planning than are cited in this section of the textbook, a final area of limitation rests in the many restrictions on how districts receive funding and how such funding may be spent. The case of categorical fiscal aid provides a perfect illustration, as school districts operate many categorical programs, such as special education, vocational education, foodservice, special awards and grants, transportation, and a host of other restricted funds that must be strictly accounted for and audited for program compliance. Many of these topics are the basis for entire chapters in this book, such as foodservice aid and transportation. The restrictions placed on these funds are understandable, but they affect planning by effectively driving areas of the school's program that might not otherwise be a priority, since by definition categorical moneys are not available for any other purpose.

An added effect is that the program of a school can become fragmented by virtue of many different programs that are not well linked together, particularly when categorical programs are viewed locally as part of an agenda that is not well integrated into local needs and preferences. For planners, the task is to deal with fragmentation, multiple agencies, separate accounting bases, and multiple aims and objectives that in some cases are at odds with the aims and objectives of other legitimate programs. The ultimate task, of course, is to link individual projects to the district's immediate and long-range goals for the benefit of students, but the task is made more difficult by single-purpose legislation that was not designed comprehensively and holistically.

Social Context of Planning

As illustrated in other portions of this text and other excellent sources describing the social impact of education,[18] the concept of planning is driven entirely by the social context of modern education. The organizational context of a school is shaped by its internal and external environments, and the school's clientele are compelled to come and to bring their advantages and disadvantages with them. Indeed, the social context of a school forms the school's mission in that it must take its clients as they are and attempt to improve their lives for individual and social benefit for all future generations. As a result, schools' planning is driven fundamentally by demographics, economics, and politics.

Demographics

At the basest level, demographics impinge upon local planning in two important ways: the total size and the direction of change in student population, and the educational attainment level and socioeconomic characteristics of the adult population that tend to set students' expectations and abilities.

For reasons understood by every school administrator, a primary demographic concern centers on the size of a school district's student population. Size drives planning from every perspective relating to budget, program, staffing, space, and so forth. Large districts tend to have planning specialists, while in small districts planning is either added to an existing job description or is spread among several positions. The issue of size is of local, state, regional, and national proportions, and in recent years has even become international in some districts as immigration has swelled public school classrooms. Intelligent interpretation of population data is a complicated matter as many districts have discovered, having gone through the population booms of the 1950s and 1960s and the uneven enrollment trends following thereafter. These demographic issues may dictate when to sell buildings, when to lease, or whether to keep teachers through temporary short swings and how to find teachers to deal with such issues as bilingual instruction or special needs populations.

Of equal importance is the direction of change in communities per se. School districts have changed enormously in the last decade, although such change began in the 1960s and has been moving fitfully forward ever since. Districts have moved from rural to urban, from blue collar to white collar, from predominantly one ethnic make-up to another, from industrial prosperity to abandoned slums, with all the attendant problems arising from expectations relating to different needs in programs, tax base, and highly varying willingness of patrons to be taxed for educational purposes. The task is infinitely more complex for planners because the fundamental task is to plan long-range for something that is elusive on a daily basis. Generally, socioeconomic characteristics taken from highly targeted census tract data are among the best indicators about educational goals and planning, as it is possible to use socioeconomic status among adults as the basis for predicting the types of programmatic needs that will emerge in the foreseeable future.

Demographic planning has become a field of its own with a high degree of specialization. The volatility and mobility of populations in the United States is legendary, making it difficult for administrators to develop the necessary skills to stay abreast of accurate projections. Nonetheless, such data must be used for planning purposes, and school administrators can both enlist the aid of professional experts and make local use of demographic data that is tracked by other people such as brokers, real estate developers, and city and county planners who also have an interest in community affairs. While the task is complex, failure to observe this aspect of planning virtually assures that the district will not be prepared to face a future that will be very different from the one imagined just a few years earlier.

Economics and Politics

If demographics drive much of schools' planning, it is also accurate to say that economics and politics drive the remainder. In fact, it is often difficult to separate economics and politics from demographics, as it is true that most things in the world are connected in some fashion. For immediate purposes, however, economics and politics are harsh realities for school planners, in that it may be simply and powerfully stated as follows: the key to converting dreams and shared values is money. While reformers may become indignant by such brutal truth, it is no less true. As the famous French philosopher Voltaire said, "when it is a question of money, everybody is of the same religion." This simply serves to point up the fact that money is at the root of all momentum, an argument that says planning for money is a political task and one that must be successful if all the dreams and good intentions of a district were to be realized.

Factors in the earlier discussion of demographics suggest that local differences may directly affect school budgets. This has been painfully apparent over the last decades, particularly as litigation surrounding funding of education has become a national pastime as discussed next in chapter 6. The relationship between federal, state, and local dollars has been a fundamental economic and political issue in that local districts have historically provided the largest share of education's resources. While in recent years the proportion has shifted more to the states, the process has only become more political and economic in that decisions are further removed now from local control and placed in a competitive arena with other communities whose interests are also at stake. Similarly, the federal government has been notorious for mandating programs and providing seed money that is removed as federal politics cycle. Likewise, the states' economies have a tremendous impact on attitudes toward funding education, and at the local level the increase in state responsibility has tended to produce apathy toward education's financial needs since control is now further removed. These events can be catastrophic, as in California where post-Proposition 13 schools have been gutted financially and as other states have been war-torn over the court battles to equalize funding.

Federal categorical funding has presented a particular problem for education as it has been a function of three specific variables: the willingness of Congress to fund education generally; the specific laws governing the programs to be funded; and the amount of funds actually appropriated for a given program. This has been difficult for planners as they have sought to become involved in prefunding access points at which educators can make wishes and advice known and can contribute to a stabilized policy environment. The case of Chapter 1 funding provides a case in point, as local school involvement recently led to a shift from single-year to multi-year funding guarantees, in turn leading to greater financial stabilization. Such critical moments have abounded, as Congress in the 1990s has again debated the federal role in education just as it did under the Reagan administration in the 1980s when many previously categorical programs were turned into block grants to the states.

Yet, in most instances, state funding is more important in terms of dollars and flexibility. State funds are normally used for "general" purposes, while federal funds are typically targeted for earmarked programs. Despite the foibles of politics and economics, long-range planning is generally much more stable at the state level, with planners less interested in purposes for funds than for how many dollars will be available across a multi-year perspective. For planning purposes, such issues are serious as states generally do not commit multi-year dollars to schools, making accurate forecasting under best case, worst case, and moderate assumptions an important planning activity.

Clearly, planners face a major limitation due to the lack of multi-year funding. It is not impossible, however, for local districts to receive commitments from key legislators to a level of funding, assuming certain economic forecasts hold true. Similarly, districts can seek provision in the state aid formula for hold-harmless clauses preventing drastic changes in state aid from year to year. Each of these mechanisms is a planning tool, serving to stabilize an otherwise volatile environment that would make planning a totally inane exercise.[19]

Broad Planning Concepts

The perspective presented thus far has argued that planning is a critical element of organizational behavior that prevents total chaos in the system. Planning theory has not been complete in that too many variables affect the process to permit a production line planning operation in peopled organizations. Schools have particular planning problems in that they are limited by virtue of operating in a government context, in that many issues press upon them from statutory and administrative origins. School planners also are fundamentally driven by demographics, economics, and politics. Within this context, compulsory taxation resembles a voluntary charity in that resources forcibly taken from taxpayers still must be cajoled from them by virtue of the power of voters to elect new officials who more closely align with their personal views. Further, the impact of employee protection laws causes districts to plan within an even further constrained context. The result is that from the broad perspective, schools plan in an environment characterized by public school values and environmental values that shape the purpose and scope of the planning function. Figure 5.1 represents various elements of the planning process. Useful depictions of the social forces impinging on the educational system also appear in chapter 8.

Figure 5.1
The General Planning Process

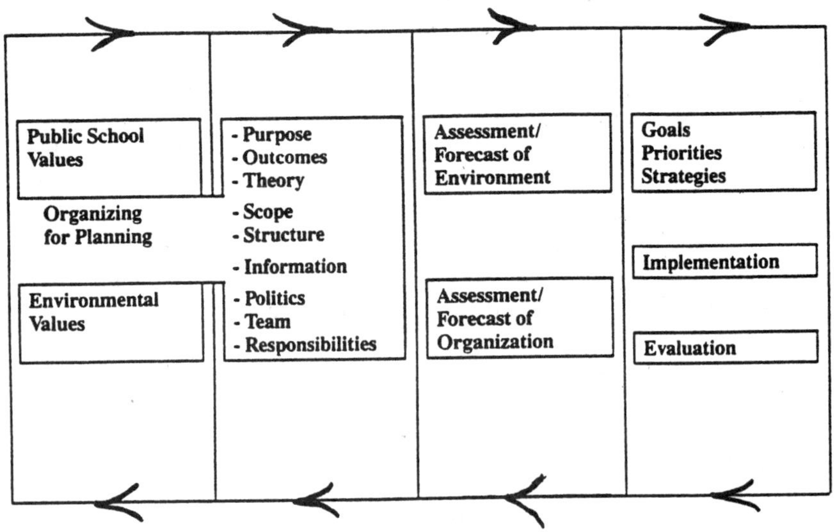

Public School Values

The issue of values shapes everything people do. Although it is now popular to speak of a value-free context, reality indicates that this is not possible, in that advocating for a value-free context actually proclaims virtuosity in tolerance—a value itself. Public schools are no different than any other institution in that values shape the environment in which planning occurs. Any experienced administrator will quickly say that planning that is contrary to local values is destined for absolute failure.

For reasons central to the theme of this chapter, a critical value in place in public schools is that the planning function should take place in a manner that involves nearly everyone, including district employees. The process is crucial, perhaps even more so than the product. Led by high level school district officers, the discussion process should be aimed at defining basic values in order to gather broad-based support for any eventual plans. These discussions are a genuine effort to set the general but fundamental direction implied by beliefs and values. This initial process is simultaneously loose and tight, open and closed. There is no mistake in purpose, but there must be enough openness and looseness to let ideas run the course. Compulsive and hurried activity at this point is the first clear message that the system really does not care.

An exciting vision of the future is an appropriate first step. A clearly articulated vision has the capacity to stir people to move with high energy, dedication, and performance. Without a value-sense, organizations tend to live in the past, with memory killing innovation. Yet a broad visionary statement is only the beginning, as care must be taken to ensure that the statement of values becomes a shared belief, a channel through which to create a set of specific achievement goals and strategies. Without an intent to act, vision is mere illusion.

Another danger is that the vision will become so broad that everything, including past practices, could perhaps flourish. If it means everything, it means nothing. Still another danger is that the vision that raises the moral and ethical issues will slip into self-righteousness, a self-defining absoluteness and, in the end, a self-defeating moral bankruptcy. The difference is in attitude, as the pursuit of excellence can free or enslave. As will be seen later in the section on the specifics of strategic planning, there are steps that can be taken to ensure productive envisioning of public school values. As seen in figure 5.1, the role of public school values is a fundamental shaping aspect of the entire planning process.

Environmental Values

As seen in the depiction in figure 5.1, a second fundamental element to effective planning relates to environmental values. The values chosen by the organization are important not only on its own terms, but also in how they integrate with the specific contextual environment. As earlier discussion illustrated, schools exist in a set of cultural values that must be dealt with effectively in order to produce harmony and momentum. Citizens, in the form of patrons and taxpayers, should be informed about what professional educators feel to be the important, base values, and educators, in turn, need to be well informed about the community values and beliefs through interaction and face-to-face contact.[20] A variety of techniques for such data-gathering will be given later, including surveys, questionnaires, interviews, and other means that can be used to help in a systematic, informal, and formal discussion of environmental values. School planners must attune to the historical and often deeply embedded values of the community and link these with events that portend trends for the future.

A second part of environmental values needs to be gleaned from the larger context as well. State and national trends have been shown in this chapter to impinge upon and limit local planning, and the environmental value context is an important aspect as well. School districts cannot hope to achieve harmony easily if local values run deeply counter to national and state values, a reality attested to daily. Severe conflict now exists in such arenas, and local planners cannot ignore either local or larger value judgments if planning is to take meaningful form. While balancing irresolubly competing elements appears risky, it is even riskier to ignore either side of the larger gestalt.

The most difficult task conceptually is to develop a framework within which all environmental values can be understood. Initially the framework should not be overly tight. Nevertheless, it is important to have a definite direction that includes four basic elements. The first element is a firm stance, i.e., the fundamental purpose and reason for being. The second element is that of eliciting community values in the form of deeply embedded, present manifestations, and emerging trends. The third element is eliciting state and national values and trends for later overlay and assimilation when possible. The fourth element relates to analyzing how these values interrelate and have implications and limitations for school operations. The process will focus the district planning effort on those concerns that are most important to the community and the role the school plays in it. As the environmental values influencing the community are uncovered and explored, broad paths of action consistent with these fundamental values will hopefully become more apparent.

Purpose and Scope of Planning

As seen, school values directly affect the purposes of planning and how the planning activities are organized. The fundamental purposes of planning in a school system are: *what* planning will be expected to accomplish; *who* will be involved in the planning; and *what information* will be most crucial. As a result, the following activities form the basis for a planning organization within schools, of which several are discussed, with the other elements imbedded in the later discussion of strategic planning:

- The purpose of planning must be established.
- The intended outcomes of planning must be clarified.
- A planning theory must be agreed upon.
- The scope of planning must be determined.

- A structure for planning must be identified.
- Information requirements must be reviewed.
- The politics of planning must be assessed.
- A planning team must be created.
- Responsibilities must be clarified.

The first organizational activity is to determine the purposes planning is to fulfill. This is the key to all that follows. Planning can serve the purpose of control, it can serve a symbolic purpose, and it can be used for pedagogical purposes. These are not necessarily either/or categories, as all can be used simultaneously, but one or the other is often predominant in school districts.

In the discussion of planning theory and again in considering organizational context, control has historically predominated planning purposes. There is broad agreement today that the rational-technical theory on which this purpose is based is too rigid for real-world organizations. Nevertheless, three important considerations emerge from this theory.

First, there are some areas in which tight structure and control are necessary, such as payroll, transportation, safety measures, and financial audit procedures. The key question is what areas, if routinized, make the operation more efficient for the purpose of educating children?

Second, there should be consideration for the fact that the organization can be both centralized and decentralized in that the conditions may dictate the organization's structure. Thus, the perspective on the organization is legitimate when people see things as loose or tight, depending on where the person is in relation to the organization. Routinely overlooked by classical rational theory, simultaneous loose-tight properties of organizations are now commonly recognized by theorists and managers as both extant and appropriate.

Despite the reality and newly emerging theories, however, conventional planning models still reflect control as the primary purpose of planning. For this reason alone, it is easy to follow a rational-technical model in spite of the weight of experience indicating that "tight" is not the one best way. Yet at the same time, the political importance of a school district's perception as "tight" should not be underrated in that the rational model conforms to one of the public's expectations and is, therefore, very important for symbolic reasons and, not incidentally, for funding. The questions that a planner must answer is how much control is the true primary purpose, and to what extent the role of control is causing dysfunction in the system.

The third consideration or purpose of planning is pedagogical—that is, providing information, perspectives, and alternatives to some or all other functions of the organization. In this way, planning can be viewed as closely relating to research. Both research and planning function as enhancements to the organization's ability to learn. If the structure and the function of planning follow this purpose, planning is seen as a service arm of the organization rather than as a control function. Assisting others in solving their own problems becomes more important than solving problems for them. The planner is a team member, rather than a technical authority. Beyond this function, the planner also provides a long-range perspective not only regarding purposes of the organization, but also assists in tracing the consequences of alternatives designed to resolve immediate problems. Thus, planning is both short-term and long-term in its informational role.

The purposes seen by the administrator as satisfying to relevant needs will, in fact, affect the organization and the nature of planning. In the abstract, the response is likely to be "all of the above." If such is the case, the job increases in complexity, for "purposes" are central to decisions relating to centralizing or decentralizing the planning function, loosening or tightening it, controlling or teaching, and so forth. Even if no decision is made initially about the purpose of planning, over time planning will tend to be characterized by one purpose or another. As Lindblom observed, numerous small decisions have a way of drifting into and becoming policy.

The second organizational activity is to assess the outcomes of planning by answering what is generally expected as an end result of the significant expenditure of time and money. Criteria for success should be stated for political reasons, as well as to provide a basis to justify the expenditure of resources. Outcomes relate to the issue of process versus product noted earlier. If products, such as planning documents, are the chief measure of success, the entire planning effort will tend to be directed toward document production, regardless of values or planning theory. If planning is viewed as a process, on the other hand, it is less easily measured. In such cases, it must be very clearly stated as an emphasis. Ideally, one should not focus on product or process, but rather on achieving a desired balance between the two. The process itself should be viewed as a product.

A third organizational activity relates to establishing a planning theory. In earlier sections of this chapter, various approaches to planning with the assumptions and implications were noted. Without a planning theory to guide discussion, the planning process can become a rambling directionless activity that produces more frustration than benefits. Society is undergoing a tremendous transition, and planning styles must be commensurate with a degree of comfort, as well as local organizational reality. Under any circumstances, however, one will have to choose, implicitly or explicitly, to adopt some variant or combination of the rational model, an incremental, mixed-scanning, organizational development approach or a new wave view of organizational planning. It is precisely from this reality that the last part of this chapter is devoted to a concept identified as strategic planning.

The fourth organizational activity in setting up for planning involves considering the scope of planning. There are primarily three alternatives, all of which have political and resource implications: a comprehensive planning approach, a key areas approach in which related areas such as student academic progress and staff development are the focus, or an approach that features specific items within these or other categories.

A comprehensive plan, irrespective of planning theory, offers an advantage in taking initiatives simultaneously in a number of related areas and thus creating a synergy of forces in which the whole is greater than its parts. Comprehensive planning also permits a wide range of trade-offs or compromises, both within the organization and with outside political forces. Quite commonly, for example, extra moneys are linked to accountability measures, but this can also be tied to everything from curricular redesign to methods for evaluating teachers. On the other hand, comprehensive planning requires enormous resources and energy and, perhaps above all, a mindset for reform. The board of education, the community, the superintendent, and other administrators, and the teachers must believe that such effort is necessary and will produce desirable results. A small nucleus team can be assembled to initially promote such a vision and assist in building commitment to change, with additional people brought on board as momentum gathers.

An alternative to comprehensive planning is to select several key areas as a planning focus. The areas ideally should be interrelated to take advantage of synergistic effects. Under a key areas approach, for example, building maintenance and staff development may be selected as priorities for political or practical reasons. The advantage of this approach is that the targets are broad enough to gain support and garner enthusiasm, yet do not appear too ambitious given available resources. In a hostile environment, the approach has the political advantage of protecting proposed programs from public scrutiny until the climate is appropriate. Proposals for reform are vulnerable to public criticism, particularly when interest groups have years to mount an attack.

A third alternative is to conduct planning with a highly limited focus, designed to show that planning does work, and that someone is taking action toward resolving a particularly visible problem. Focused planning on a limited area must demonstrate thoughtful analysis and preparation for future trends or events that are foreseeable. It has the advantage of demonstrating within a short time frame how a plan translates into action and resolves problems. If done well, it has the added advantage of demonstrating how the planning function contributes to organizational decision making. This helps to counteract criticism that planning is peripheral to operations or, worse, another layer of useless bureaucracy.

The second and third strategies require fewer resources than comprehensive planning and are particularly appropriate when there is a need to demonstrate the value of planning or to make rapid shifts in response to uncertain environments. Indeed, there is a strong argument not to undertake comprehensive planning when the risks are high by virtue of an unstable environment.

As will be seen later in the discussion of strategic planning, there are innumerable questions that must be asked to obtain the values describing the business of schools. For immediate introductory purposes, these questions may be characterized as driving the nature and scope of planning:

- Is the culture of the organization prepared for significant reform? If not, is it possible to create a desire for reform? If the response to both were no, is there any reason to consider comprehensive planning?

- What are the human resources needed for the chosen approach to planning? Are these available or can they be made available?

- Does the district have the fiscal capacity and the political will to carry through on planning, comprehensive or otherwise?

- Who is directly accountable for keeping planning efforts going?

- Is the proposed organizational placement of planning appropriate to the goals?

- Who will report to whom about what and in what form for planning? For operations?

- What are the approximate time lines in mind for accomplishing the various stages of planning? Of implementation?

- Who is responsible for assuring that the planning effort has a designated and realistic budget?

- Are evaluations and assessments of, for example, students, teachers or programs included in the responsibilities of the planning team?

- Who will be responsible for implementation of the operations? At what points do responsibilities shift upon implementation?

- What are the quality checkpoints to assure successful planning and implementation?

These and other questions form the basis of planning in a school district. As seen thus far in this chapter, different models exist for reaching goals that are appropriate under different types of circumstances. The complex nature of schools, however, begs for complex models that understand the dynamics of social organizations involved in work that is sometimes known for its vagueness and ambiguity, rather than a production model that is ill-fitting and forcefully overlaid. As a consequence, the chapter to this point has raised a series of issues and questions about appropriate design for planning and has been headed toward the next section, which is devoted to the concept of strategic planning as a specific and useful tool in shaping and guiding the difficult discussions involving governmental organizations operating in a heavily restricted and value-laden context.

Strategic Planning

The need for effective planning in school districts derives from a number of observations about education in the modern day. Demands and pressures placed on schools stemming from popularization of concepts such as decentralization, school-based management, and economic limits placed on ever-growing government have left an indelible mark on school organizations, with the concomitant requirement to adapt or die. New methods of working in schools have also forced a need for good planning, as technology and learning theory have moved at a rapid pace over the last few decades. The concept of a learning culture has also emerged as a product of the information revolution, requiring collaborative work in groups that has implications for educational redesign in the arenas of programming, simulations, computers, videos, and shared decision making. A corollary to such change has been that empowerment is really a one-way street in that once autonomy and professionalism are experienced, it is difficult or impossible to reverse. All these elements have worked jointly to result in an increased need for good planning, both by virtue of keeping up with change for the sake of currency and as a consequence of surviving in an increasingly competitive resource environment.

Strategic Planning Defined

A tool that has emerged to help meet the challenges described is known as strategic planning. Although the aspects of planning discussed earlier could all be seen as strategic planning by some definition, the concept as discussed at this point is a specific strategy that is seen as thoughtfully incorporating the strengths and weaknesses of earlier models with the intent of minimizing shortfalls and maximizing strengths.

Although there are similarities between long-range planning and strategic planning, there are important differences that distinguish one from the other. Indeed, during the early days of strategic planning in schools, there was perhaps little difference. During those times, strategic planning focused on the act of planning itself with some additional attention paid to doing some environmental scanning. Although the fundamental elements were present, the understanding of environment was so weak that scanning did not play a major part and the focus was merely old-style planning by a new name. The difference also lay in the fact that no technology was present to link planning to the larger community in a way that would reach the heights envisioned for this tool. In contrast, strategic planning should have been a vehicle for capturing specific organizational strategies for assessing and linking environments, probing and assimilating culture and values, and responding appropriately to limitations on planning inherent to the school's origins in the public context.

By definition, long-range planning seeks to anticipate and prepare for the future. Strategic planning, by definition, encompasses the same goal and moves on to form a set of pre-planned rational steps leading to that end while seeking to incorporate a holistic understanding of the forces of change. Strategic planning is a process that also has at its root the intent to change and transform the organization actively, as opposed to seeking to maintain equilibrium in the face of change. Similarly, strategic planning is a philosophy, a way of life, a problem-solving tool, and a community involvement device.

Specifically, strategic planning pursues a series of steps designed to understand external environments and their implications for the organization, develop a vision and mission for what the organization should stand for and be about, develop an action plan to reach those goals, and review progress toward goals and to adjust both the vision and the means as needed. Indeed, strategic planning differs from long-range planning in that the latter seeks adaptation, while the former constantly seeks change by trying to move the organization forward at a rate of speed equal to or greater than its surroundings. The key word is transforming, and the key descriptor for administrators is to be a transforming leader. In essence, strategic planning is a rational model, but it incorporates psychological dimensions to address the shortcomings of earlier rational-technical models that were fixated on prescriptive rigidity.

Differentiating between long-range and strategic planning, however, does not imply that one is useless while the other is omnipotent. In most cases, the difference in utility lies in the task at hand. Long-range planning is a useful tool for stable environments and for internal planning in areas where the environment is relatively closed. Likewise, long-range planning is well-suited to areas where data are readily available and where "mushy" contextual factors such as values and norms do not heavily weigh on results. Similarly, long-range planning is a useful tool when projections of hard numbers are the expected outcome, especially in terms of laying plans that are unlikely to be affected by a volatile gestalt.

In contrast, strategic planning is a useful tool when working with social systems where exchange among environments is a predominant feature, where vision and mission involve wide and perhaps divergent groups, and where intuition and creativity play an important part in imaging the future. Long-range planning is well-suited to budgetary planning, for example, while strategic planning is more likely to take in community and social issues, as well as overall vision for the district. Both planning types, however, are mutually inclusive of one another in that each adds to the observations of the other and may be sequentially linked to produce the district's overall strategy. For purposes of this chapter, however, strategic planning is discussed because it should precede the firmer plans that emerge from long-range planning activities.

Strategic Planning Phases

In most school districts, planning of any sort proceeds in phases after having been decided upon at any of several levels. As discussed throughout this text, the policy level is a very important aspect of school district operations in that board policy should be at the root of any district activity to give both legitimacy and impetus to the school district's operations. Because of its envisioning power, strategic planning is well-suited to adoption because it is at the board level where legitimacy is granted to the district's mission and goals, and it is at the board level where expectations for performance and outcomes are established. Strategic planning does not stop at the policy level, however, in that it is a useful tool for creating the necessary framework to guide all further activities, particularly in the twin areas of program development and program delivery.

These phases, roughly described, ask three important questions corresponding to the three types of planning that fall under the broad umbrella making up strategic planning. At the policy level of strategic planning, the question arises of where the district is generally headed. In this instance, the expected outcomes would normally relate to developing a mission statement, identifying goals, and reaching decision points about whether to continue the process and defining more specific areas of interest. As implied earlier, these elements should occur at top levels, although some dialogue should occur with stakeholder groups in order not to alienate the process before it has sufficient momentum.

At the program planning level, the question addresses the specific strategy for accomplishing the goals identified at the previous level. In this instance, the expected outcomes would include more detailed plans such as instructional goals, curriculum plans, resource identification, and establishment of budgets designed to make achievement possible. These elements naturally move to those people with appropriate expertise, such as specialists, building administrators, teachers, and so forth. At the program delivery level, the third question arises regarding the actual operationalization of all the foregoing elements by seeking the specific methodologies required to implement and carry out the goals and mission. Expected outcomes naturally entail carrying out the plan through the many types of work done in schools such as lesson plans, teaching, and evaluating student progress.

These phases are obviously sequential, although overlapping, and are all equally required for strategic planning to be either complete or effective. Although it is obviously possible to broadly outline strategic planning in a single statement, its implementation is a much longer process taking months to years from initiation to evaluation. The phases may be broadly described in five steps: creating a base for change through the process of environmental scanning, developing a strategic plan, developing an implementation plan, implementing and monitoring the plan, and renewing the plan.[21]

Environmental Scanning

Although some implication of concern for knowing the external and internal environments of schools was present in the earlier discussion of other planning models, strategic planning bases a significant amount of its justification on the process of environmental scanning. As its name infers, environmental scanning is a process that seeks to gather relevant information from its internal and external circumstances in the belief that these elements drive much of its fate. In many ways, environmental scanning is a recognition of the resource dependency model cited earlier, and it also captures the general limitations of planning that derive from schools' position in a governmental context. In the case of strategic planning, great attention is paid to analyzing data from entities known as stakeholders. Stakeholders include parents, students, staff, community, and the agencies that have a vested interest in the educational process. Stakeholders, then, may be described as the gestalt of the organization with care to exclude no relevant factor under the assumption that any neglected aspect may become a source of trouble.

As the name implies, external environmental scanning is used to collect information that can be analyzed to assess the future that may be imposed on the organization. The earlier discussion regarding the value of data on demographics, economics, and politics showed that these elements make up a major portion of external scanning activities. Data on a wide variety of items are usually accumulated for the purpose of pattern analysis and trend analysis within the headings of national, regional, state, and local concerns. Problem areas are then identified and used to project implications for schools. Recent examples might include assessing the economic, political, and educational impact of proposed changes in foodservice regulations at the federal level;[22] similarly, changes in the state's method of funding education would also trigger such scanning, as would increasing numbers of disadvantaged children in a district brought on by immigration or local economic problems. The types of data interesting to external scanning would include:

- Economic structure in terms of income, growth, business trends, and prospects for expansion.
- Employment factors such as labor force statistics and unemployment.
- Demographic data and trends by age and need for school privileges.
- Social data on health, welfare, housing, and government.
- Educational data with implications for schools, such as parental educational attainment, available resources, community support, and so forth.

The goal is to capture a comprehensive view of trends that are likely to affect schools and to describe those trends in ways that have common and graphic meaning to all participants, with conclusions of significance drawn—a process calling for the steps seen in figure 5.2, a process that can also be modified to conduct the internal scan discussed next.

Figure 5.2
Overview of Environmental Scanning

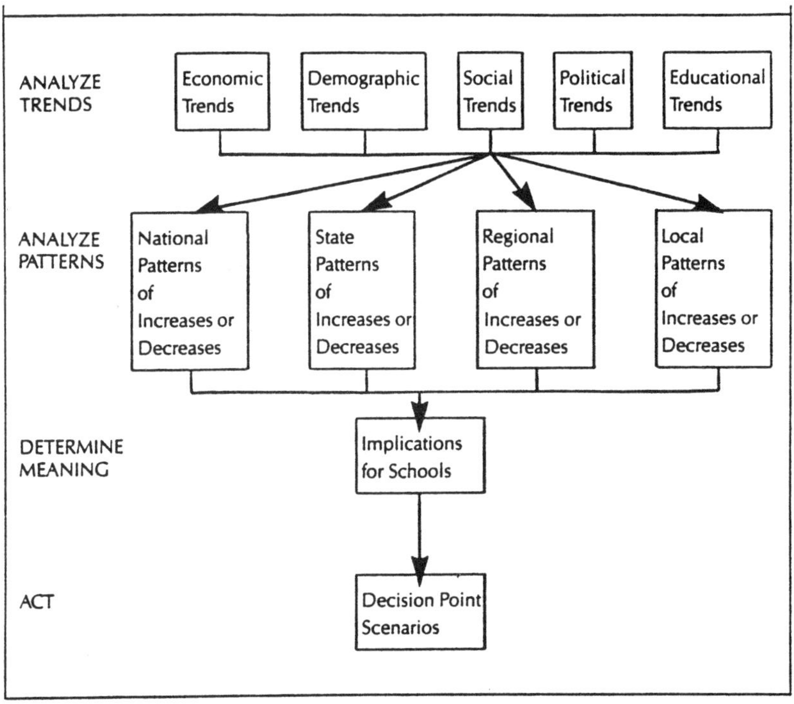

The environmental scanning process also takes a strong interest in looking inward to determine a wide variety of information vital to effective planning. During this aspect, the focus concerns where the organization is and what it is capable of doing. A process generally familiar to most educators through such activities as accreditation reviews, internal scanning seeks to identify strengths, weaknesses, and issues that confront the district on an immediate and long-term basis.

The process of internal scanning is often similar to how external scanning is conducted. Again, the economic, social, political, and other elements of the system become the focal point for questions that seek to gather information that will help build a profile of the system. Generally, a database is built that provides collection and disaggregation of data relating to all facets of the district's operations, including:

- Social and demographic trends such as enrollment projections, income of students, achievement data by various descriptors, staff characteristics, and student needs by various definitions.

- Organization of the district and individual schools, including direct instruction, management, and support services.

- Community data from an internal perspective relating to students in school and visible community support from within the school.

- Data on available resources, particularly from the perspective of money by the many ways school finance people can break down fiscal data.

Data from internal and external scanning provide a comprehensive overview of the organizational health of the school and its contextual environment. These data are then used in a variety of ways throughout the remainder of the strategic planning process, although initially a benefit is to communicate with stakeholders (broadly defined) to discuss how the school relates to and is shaped by the community. The process of using data from environmental scanning is fundamentally one of educating the district and its interested stakeholders in the conditions making up all environments affecting the school, with an added benefit of helping people see the logic behind why things are the way they are.

Time to absorb the results of environmental scanning is needed, as presentation of these data requires some thought as to how to best proceed and because different people need different amounts of time to realize the import of such data. A further benefit follows from adjusting these data based on input from the various stakeholder groups, as community presentations, internal school committee meetings, and administrator and board groups take time to digest and contemplate what these data mean to the district. The outcome of a carefully constructed environmental scan is to meaningfully raise the level of awareness of the district's potential and its limitations, making possible the next step of preparing a strategic plan.

Developing a Strategic Plan

As seen in the previous section, information is the key that drives strategic planning. While environmental scanning comprises the first phase of strategic planning, concerted attention to developing the plan comprises the second phase. Based on the information collected during the scan, a series of questions can be asked that is intended to shape what the school district should do next, both in terms of the legitimacy of its goals and in terms of what should be done to enhance the likelihood of goal attainment.

McCune provides a useful list of probable issues that make up the development of a strategic plan. The list cuts to the heart of the second phase of strategic planning, as these questions describe most of the events making up development of the strategic plan. Development of the plan is predicated on earnest involvement by stakeholders of all types in that collaboration and commitment to the eventual product is essential to successfully developing the strategic plan. The list of issues making up the majority of the second phase include:

Review of External Scanning:

- Where is the community?
- What are the national, state, regional, and local trends?
- What implications are these likely to have for schools in this community?
- What opportunities (needs) are in this community?
- What threatens the future well-being of schools in this community?
- What activities might be developed in the future?

Review of Internal Scanning:

- Where is this organization?
- What are the organizational strengths? (staff, curriculum, foodservice, business management, staff development, etc.).
- What are the critical issues facing the organization?
- What internal steps must be taken to strengthen the school?
- Where are the possible matches between community needs and the resources and strengths of this organization?

Developing the Vision:

- What is the preferred future?
- What should the organization be doing five years from now?
- What role should schools perform in the future?
- What should be the core purpose or mission of the schools?
- What things are most valued about the schools now?
- What words or phrases describe this district's mission and goals?

Developing Strategic Goals:

- How might the preferred future be achieved?
- What general or large tasks must be accomplished first?
- Over what period of time should these tasks be accomplished?
- What should have been done within five years?[23]

As McCune notes, the outcome of this activity should be a strategic plan draft. Of critical importance are four products flowing from such a draft: the *need* for such a document, the *beliefs* and *assumptions* beneath the plan, identification of the district's *mission*, and the district's strategic *goals*. As stated earlier, these elements provide a framework that can guide the district as it moves from the policy arena into program planning concerns that begin the process of effectuating the district's strategic plan.

Developing an Implementation Plan

The third phase of strategic planning focuses on developing a plan for implementing the accomplishments and needs identified in the earlier stages of the planning process. Unlike the broad policy statements making up the first two phases of strategic planning, the implementation phase specifies techniques for putting the plan into action. An important aspect of the implementation plan is developing alternative ways of reaching the district's goals and understanding that more than one way is possible. Likewise, there is a need to understand that multiple ways may be used by different parts of the organization to reach the same goals, as it is not required that a lock-step top-down model be adopted.

The extended nature of developing an implementation plan deserves comment from two perspectives. The first perspective is that the extended nature of implementation planning is too deep to describe here in detail. The second perspective follows immediately thereafter in that implementation is an extended process using expertise closest to the source of where the plan will eventually be carried out. This is what makes it both possible and desirable to engage in strategic planning. This means that much time and resources are required to build implementation plans and that teachers, administrators, and other stakeholders must be intimately involved in order to capture the expertise and ownership that will eventually dictate the plan's relative success. In effect, implementation demands a much earlier recognition that strategic planning is a consuming process not to be approached casually or to be underestimated.

Implementation and Monitoring

The weightiness of strategic planning is further reinforced when one considers that the process begins after scanning, developing the plan, and developing the implementation plan are completed. This, in effect, argues that implementation and monitoring take even more time because it is at this level that the plan is actually carried out. If strategic planning first occurred at the policy level and then proceeded to the program planning level during the implementation construction phase, it is through implementation and monitoring that action really is attached to the process through program delivery plans.

Implementation involves actually placing visions and programs into place. It is at this level that administrators and teachers become so involved that they live strategic planning on a daily basis. Everyone must be engaged and committed, and everyone must be competent and energetic to enable the district's vision to be operationalized. Most strategic planners recommend functioning oversight committees with separate responsibility for the district perspective, the building perspective, and the "trench warfare" perspective, with considerable coordination and collaboration. Monitoring is precisely what is implied, in that oversight must be combined with other strategies to assess and adjust elements of the plan, and perhaps the plan itself, as seen through actual experience.

Continuity and communication are key words making up implementation and monitoring. Although implementation and monitoring are easily described, neither is easily accomplished. It requires staff training and time and expensive commodities that are available only through the commitment of financial resources in a resource-finite environment. Importantly, the monitoring phase is the key to the final phase of renewal discussed next.

Renewing a Strategic Plan

Strategic planning is a process that is never complete. Like most good plans, things go awry or simply become outdated over time. Alternatively, good plans can always be improved—all these are reasons why a strategic plan should be in a constant state of revision and renewal. Revision implies change based on feedback, and renewal suggests reinvigoration, new vision, and enthusiasm. Indeed, to do less is to return to the status quo and to stagnate—a condition totally inconsonant with the initial aims of engaging in strategic planning. By definition, such planning is intended to strategically place the district on the cutting edge and to "strategize" the way into a promising future.

Inasmuch as strategic plans are meant to encompass extended periods of time, adjustment is required for all the reasons just stated. Much modern theory work now focuses on reflective practice, a term that is meant to remove schools from frenetic behavior and to place them in a state of operation best described as thoughtful, contemplative, and reasoned in their responses to issues. Strategic planning fits those descriptors very well, requiring that evaluation and revision make up an integral part of the ongoing strategic planning concept. McCune states these issues well, suggesting that renewal should be comprised of at least the following questions:

- What conditions have changed in the external environment since the plan was written?
- Do the original assumptions of the plan still apply?
- Do the mission and goals continue to express the current vision of schools and the things needed to reach that vision?
- What has been learned that should modify the strategic plan?[24]

These questions, among others, should drive renewal of the strategic plan on a continual basis, as good school administrators should not rest on their laurels in hopes that the environment will not change—a futile hope, for change is the only constant in the modern world.

Summary

As stated at the outset of this chapter, the task of planning forms a critical and indispensable role in modern public school districts. While planning in some form has always been important, its role has increased in response to an uncertain and constantly changing environment that now forces districts toward an unstable future. The many complex faces of administration, in fact, do include opportunities for leadership and frontline drama, and it is true that in the arena of planning that good planning offers the chance to invoke change for the betterment of society. School leaders are, therefore, vanguards of change or, alternatively, are guardians of a dead past. Successful administrators who understand the elements of good planning are strategically positioned to provide outstanding service to their constituencies who depend on them for leadership and progress in America's public schools.

Endnotes

[1] Portions of this chapter are based on the previous edition written by A.P. Johnston and K. Hood.

[2] However, see chapter 8 later in this book to consider further development of the limitations of such rational-technical efficiency in workgroups and informal organizational norms.

[3] "The New Breed of Strategic Planner," *Business Week* (Sept. 17, 1984).

[4] Ibid.

[5] A. Wildavsky, "If Planning is Everything, Maybe It's Nothing." *Policy Sciences* (4) (1973) 127-153.

[6] A. Etzioni, "Mixed Scanning: A "Third' Approach to Decision-Making." *Public Admin. Rev.* 27 (Dec. 1967) 385-392.

[7] W. Bennis, "Editorial," *Journal of Applied Behavioral Science* 4 (no.2, 1968) 228.

[8] W. Bennis, "Leadership Theory and Administrative Behavior." *Administrative Science Quarterly* 8 (1972) p266.

[9] See , discussion in chapter 8 about organizations operating in random environments.

[10] See, e.g., G. Zukav, *The Dancing Wu Li Masters.* (New York: Bantam Books, 1979).

[11] J. Campbell, *Grammatical Man.* (New York: Simon & Schuster, 1982).

[12] E. Misher, "Meaning in Context: Is There Any Other Kind?" *Harvard Educ. Rev.* 49 (February 1979), 1-19.

[13] See also one of the best attempts to put the theory "weedpatch" in perspective in a single source in J. Pfeffer, *Organizations and Organization Theory.* (Boston: Pitman , 1982) and many others.

[14] K. Weick, "Administering Education in Loosely Coupled Schools." *Phi Delta Kappan* (June 1982), 673-718; also Weick, "Educational Organizations as Loosely Coupled Systems, *Administrative Science Quart.,* 21 (1976) 1-19.

[15] J. March and J. Olsen, "Organizational Choice Under Ambiguity," in *Ambiguity and Choice in Organizations* (Bergen, Norway: Universtetsforiaget, 1979).

[16] T. J. Sergiovanni, *The Principalship: A Reflective Practice Perspective, 3rd ed.* (Boston: Allyn and Bacon , 1995) 30.

[17] See particularly chapters 6, 7, 8, and 19 in this text for legal concerns faced by school districts.

[18] See, for example, chapter 1 and related references in D. C. Thompson, R. C. Wood, and D. Honeyman, *Fiscal Leadership for Schools: Concepts and Practices.* (New York: Longman , 1994).

[19] Interestingly, see chapter 8 for a discussion of organizational theories, including a resource dependency perspective which argues that organizations are jerked around unmercifully by external environments over which they generally have no control.

[20] J. Dewey, *The Public and Its Problems.* (Chicago: The Swallow Press, 1927).

[21] S. McCune, *Guide to Strategic Planning for Educators.* (Alexandria, VA.: Association for Supervision and Curriculum Development, 1986) 38. This

discussion in text is based in part on McCune's detailing of strategic planning, 38-54.

[22] A discussion of this proposal in the 104th Congress of the United States appears in chapter 22.

[23] Adapted and modified from S. McCune, *Guide to Strategic Planning for Educators.* (Alexandria, VA: Association for Supervision and Curriculum Development, 1986) 48-49.

[24] S. McCune, *Guide to Strategic Planning for Educators.* (Alexandria, VA.: Association for Supervision and Curriculum Development, 1986) 54.

Chapter 6
Fiscal Equity Challenges

Introduction

As may be inferred from each of the chapters in this text, there has been a tremendous struggle over fiscal resources for public education. While some struggles have been born of radically different views on what schools should represent to the many different segments of society,[1] there is little question that perhaps the greatest struggle has been over how tax dollars are distributed to public schools. A great part of the struggle has arisen simply because there has never been enough money for public education, making distribution of available funds even more critical because of obvious scarcity. This struggle has been heightened by increased knowledge concerning the relationship between money and educational quality, and because of increased legal sophistication and greater willingness to aggressively pursue intensely confrontive solutions, such as court challenges to school funding formulas.

While it is true that correlations between money and educational achievement cannot be made, there is a great assumption that more money makes better education imminently more possible. While many aspects of this belief are open to challenge, there are at least six hypotheses that support an activist view of the importance of school finance. The first hypothesis is that schools distribute economic and social opportunity. Inasmuch as higher earnings can be shown to correlate to higher levels of education, this hypothesis is entirely reasonable. The second hypothesis is that opportunity depends in large measure on the quality of schools that children attend. Insofar as people treasure and seek out good schools for their children, this belief is nearly without challenge. The third hypothesis is that despite the lack of a strong statistical relationship between money and quality, school quality is heavily conditioned by purchased resources such as teachers. This view is readily confirmed in the research literature.

The fourth hypothesis is that, absent ability to purchase these inputs, schools will fail to live up to public expectations. This is automatically provable since schools cannot operate without resources, leaving only the question of how much is enough to create an effective school setting. The fifth hypothesis is that people who argue for the irrelevance of money in the educational achievement equation still prefer a larger share for their own schools—an observation needing no further explanation. The sixth hypothesis, then, is that until money is convincingly proved to have no effect on learning, a meaningful effect must be presumed from the behavior of wealthy people who create and fiercely protect good schools for their children.

Effective school business managers have a great need to be current in the concepts of equity, adequacy, and the many derivations on these concepts of fundamental fairness.[2] It should not be surprising that the vehicle for examining equity rests in a review of school finance litigation, since fairness in school funding has been the vigorous subject of lawsuits challenging the design of state aid distribution plans. Anyone familiar with modern school business management knows that money is the object of intense envy, and it is accurate that such envy has led to massive restructuring of state funding formulas, particularly via recent litigation efforts.

This type of litigation first began at the federal level. Because litigation has focused for the most part on the individual states, the chapter also considers the outcome of pioneering suits that were brought at the state level. Because there has been intense litigation since the first edition of this book, the chapter provides a summary of more recent cases and also identifies the funding lawsuits and issues currently before the courts.

Finally, the chapter synthesizes those broad principles that have been laid down by courts and provides discussion of the future role of litigation in an increasingly litigious educational environment. As the first edition of this book noted, only through an understanding of equity is one well positioned to be an effective fiscal manager in a public education arena that has become exceedingly complex and contentious.

Origins of School Funding Challenges[3]

For more than a hundred years, school finance has been of vital concern to both courts and policymakers. Although education finance as a discipline has only developed during the Twentieth Century, issues of taxation have been a central feature of school funding debates since the earliest days of the nation. Likewise, schools have long been the object of debate regarding equality of educational opportunity as it relates to discrimination, and it is no far leap of faith to connect the concept of discrimination to differences in amounts of school funding available to children. Thus, education finance litigation began long before the birth of the discipline known as education finance, particularly when the discipline is viewed in a sweeping manner that considers a whole series of cases such as *Brown v Board of Education*[4] for the financial implications. Although the full range of equality lawsuits underlying the strands leading to fullscale school finance litigation is too broad to examine within this chapter,[5] these threads should be recognized. The chapter restricts discussion to those questions of law that have emerged in the context of frontal attacks to school funding schemes.

The history of education finance litigation has developed in federal and state courts. Since courts do not initiate legal disputes and generally confine themselves to interpreting whether some law has been violated,[6] education finance litigation has been at once constitutional and statutory. At the federal level, litigation has focused on the United States Constitution in the context of interpreting the limits of federal responsibility toward education and on any right to education that can be interpreted from the Constitution.

At the state level, litigation has been focused on both the constitutional and statutory provisions of each state. At both levels, these questions have been difficult, as constitutional interpretation is swayed by the times and the individual temperament of courts, particularly when there is wide variation among states in the constitutional and statutory provisions for funding public education. Regardless of whether suit has been brought in federal or state court, litigants have had to be first concerned with questions of law in order to create a justiciable controversy.

In both federal and state cases, litigants have sought rulings that would determine the meaning and limits of equal opportunity and test the strength and expansiveness of constitutional and statutory language. Depending on the level at which a lawsuit is brought, these attacks have centered on three key claims: education as a fundamental right, the equal protection of the law, and the education articles of individual state constitutions. Each of these elements can be traced from the federal and state origins into modern education finance litigation strategy.

Federal Origins[7]

Although litigation involving state aid schemes is believed by many people to be a recent activity, legal struggles over money in schools have been a long standing reality. While this chapter does not seek to develop that lengthy history, it is important to note that the modern litigation context is based on both federal and state questions that have roots in particular strategies aimed at various features of federal and state law. The federal case predates all other claims, as aggrieved plaintiffs sought equality in fiscal resources in a logical attempt to capture a U.S. Supreme Court ruling. The logic of attacking in federal court was that if a favorable ruling were won, that ruling would influence the states to conform. At the same time, litigants also sought federal grounds on which states could be required to equalize fiscal resources in schools. The issues on which both federal and state litigation has been pursued have theories in common, as well as differences, that have contributed to their outcomes.

Historically, bringing a funding lawsuit at the federal level was a reasonable act. Concern for equality has been an important social issue in American society since the days when the colonial charters sought freedom from outside rule. The Declaration of Independence was a clear statement of equality, as the nation declared that " . . . all men are created equal, that they are endowed by their Creator with certain inalienable rights . . . . "[8] Aggressive concern for equality was evidenced in the Bill of Rights,[9] and anticentrist sentiments were jealously carried over into individual state constitutions.

The Emancipation Proclamation in 1863 and the Thirteenth Amendment in 1865 abolishing slavery[10] were also strong statements of American views on equality, but nowhere was the demand for fairness more evident than in the Fourteenth Amendment to the U.S. Constitution, which guaranteed federal equality under the law. The Fourteenth Amendment not only granted full rights of citizenship to all persons born or naturalized in the United States, but it also applied its provisions to the individual states. In powerful language, the amendment declared:

> No State shall make or enforce any law which shall abridge the privileges or immunities of citizens of the United States; nor shall any State deprive any person of life, liberty, or property, without due process of law; nor deny to any person within its jurisdiction the equal protection of the laws.[11]

It is difficult to imagine any words more powerfully directed at the concept of equality under law. These words were to prove useful in federal and state litigation in a sweeping variety of equality lawsuits. While no federal school finance lawsuits emerged for more than 100 years after adoption of the Fourteenth Amendment, foundations for these suits were laid by a series of broader cases with school finance implications. As noted elsewhere, the Fourteenth Amendment's equality provisions led to three litigation strands[12] that would have a great impact on school funding.

The first strand was a series of lawsuits relating to segregation, in which funding was a natural implication.[13] The second strand was a series of cases commonly called reapportionment decisions, establishing the principle of "one man, one vote."[14] The third strand arose from lawsuits known as the indigent defendants and administration of criminal justice cases, which established that defendants may not be denied the right of appeal due to inability to pay for a trial transcript because such denial is comparable to wealth discrimination.[15] Although at first glance these cases seem unrelated to school finance, these strands were actually destined to lay a framework for equal protection grievances relating to how education funds are distributed.

An entire philosophy of equality was thus laid in a series of legal challenges that were instrumental in inducing courts to ask difficult questions concerning the meaning of equality under the U.S. Constitution. The third clause of the Fourteenth Amendment was especially argued to this end, as it declared that no citizen may be unequal under the Constitution. Segregation was a natural basis for building an inequality argument, and these cases had deep relationships to the costs and structure of education.

The reapportionment decisions accelerated the equality debate, as all people were given federal protection and equal status under the law. The third strand, however, was critically important because it asked the difficult question of whether differential wealth could be an aspect of inequality under law. If wealth inequality were ruled unconstitutional, a new meaning to equality would declare unequal funding in schools to be a constitutional abridgment.

These three strands were actually an extension of judicial sympathy to the meaning of equality that had already resulted in other fundamental rights under the law. In addition to rights enumerated by the Constitution, the U.S. Supreme Court had ruled on several other rights which it found to be so fundamental that they could not be denied except by due process of law. Several such rights were found in the cases making up the three strands. For example, the Supreme Court had held for fundamental rights to interstate travel,[16] procreation,[17] voting,[18] and criminal appeal.[19] The importance of fundamental rights could not be overstated; they must be protected at all costs. The Fourteenth Amendment thus guaranteed that all persons, regardless of wealth or social status, would be assured equal protection against discrimination.

The effect of this third clause of the Fourteenth Amendment was to produce two litigation thrusts. The first thrust resulted from defining unequal treatment of suspect classes. Various conditions could lead to establishment of suspectness wherein "prejudice against discrete and insular minorities may be a special condition, which tends seriously to curtail the operation of those political processes ordinarily to be relied upon to protect minorities, and which may call for a correspondingly more searching judicial inquiry."[20]

Litigation was to create suspectness for race,[21] national origin,[22] and alienage,[23] and to heighten sensitivity toward other social groups. The first thrust thus laid its basis in illegitimate differential treatment on the basis of immutable differences wherein people were seen to be discriminated against for things they could not change, such as race. The second thrust resulted from denial of some fundamental right. This thrust thus laid its basis on existing constitutional provisions or, alternatively, as the Supreme Court created new fundamental rights based on arguments for essential freedoms.

Suspectness and fundamentality were crucial to equal protection litigation for one simple reason. If a court could be convinced that a fundamental right or suspect class were the object of wrongful treatment, it would examine the matter with exacting strict scrutiny. For example, race was obviously a suspect class because of immutability, and violating the constitutional rights of members of a suspect class would require strict scrutiny under antidiscrimination equal protection laws. Likewise, violation of a fundamental right would require the same scrutiny. Under these conditions, establishing new suspect classes and creating new fundamental rights became a driving force behind federal equal protection lawsuits.

These concepts were the strategy of the first order because the burden of proof would shift to the defendant to show a compelling interest in the offending statute—a nearly impossible task for defendants because there is always a better way to meet some duty, regardless of how carefully a law is structured. This test of strict scrutiny became the coveted judicial standard, as the usual standard of rational basis would require only some sensible reason related to a legitimate governmental purpose in order to allow a law to stand.

These concepts, derived from the broader litigation strands, were quickly applied to education, wherein reformers made a concerted effort to create suspect classes and fundamental rights for education. Although more complex than developed in this brief introduction, these strands illustrate why it was reasonable in an historical sense to first bring school finance lawsuits in federal courts. Irrespective of whether suits were based on race or fiscal resources, the goal of such lawsuits was to seek federal protection wherein perceived inequality would be alleged, with consonant claim of denial of equal educational opportunity.

Extended to its logical outcome, if education merited Fourteenth Amendment equal protection, new federal constitutional interpretation would be controlling on the individual states. The coveted language became strict scrutiny equal protection analysis, accompanied by an equally intense search for fundamental right to education so as to evoke the same searching scrutiny under which nearly any challenged law would likely fail.

The federal origins of school funding challenges were clearly reasonable. If the analysis underlying broader cases could be shifted to education's claims, the successful application would mean the creation of new constitutional protections in a fruitful and expansive arena of social reform justice. However, if neither a fundamental right to education nor a suspect class of poor could be established, claims would have to turn to individual states since the Tenth Amendment's silence on education would foreclose a further federal case. The risks and stakes were high, as funding laws nationwide would have to be rewritten to provide a high degree of fiscal equity if the challenge prevailed, while the window of opportunity would be greatly reduced if the case failed. As noted earlier, the search for equality was first successfully tested in the school desegregation arena, followed by federal school finance litigation seeking equalization of how school funds are distributed.

The Federal Case

Although federal racial equality litigation spanned many decades,[24] it was in the *Brown v Board of Education* [25] case in 1954 that equality of educational opportunity struck a new court response. In a landmark ruling, the U.S. Supreme Court invalidated the "separate but equal" provisions in states that provided for segregated schools. In a decision that overruled the social and economic heritage of American society, the Court held that separate but equal was inherently unequal, and that education was important to the health and well being of the nation. The now famous quotation of the Court was prophetic of the growing struggle over the distribution of education's benefits as it stated:

> ...[E]ducation is perhaps the most important function of state and local governments. Compulsory school attendance laws and the great expenditures for education demonstrate our recognition of the importance of education to our democratic society. It is required in the performance of our most basic public responsibilities, even service in the armed forces. It is the very foundation of good citizenship. Today it is a principal instrument in awakening the child to cultural values, in preparing him for later professional training, and in helping him to adjust normally to his environment. In these days, it is doubtful that any child may reasonably be expected to succeed in life if he is denied the opportunity of an education. Such an opportunity where the state has undertaken to provide it, is a right which must be made available to all on equal terms.[26]

With the ruling in *Brown,* equality and equal educational opportunity seemed to be the watchwords of the future. *Brown* clearly invoked the equal protection clause of the Fourteenth Amendment, yielding credence to the concept that education was shielded from irrational discrimination. *Brown* also spoke strongly to the value of education by calling it one of the most important functions of government, yielding great hope for the claim of a fundamental right to education.

Using strong language, *Brown* represented the best hopes of reformers as it seemingly upheld all the theories and claims advanced by reform advocates seeking a new social order in America's schools. Almost singlehandedly, the language in *Brown* opened a new era from which an entire genre of civil and education rights litigation would emerge to demand the attention of the nation's court system for many years into the future.

Fresh from the victory in *Brown*, reformers turned their attention to fiscal inequality, believing that similar analysis could apply to inequality in school funding. Reformers argued that if the Court's ruling in *Brown* were reasonably construed, it would be immediately recognized that unequal revenue is fundamental to many inequalities. Reformers knew, however, that a major obstacle lay before them because, while their logic seemed unshakable, there was no precedent for fiscal inequality claims that was comparable to the record of racial discrimination suits. The force was merely a moral plea, rather than a case to be made based on legal precedent. By the 1960s, plaintiffs had formulated their arguments and were prepared to launch a first attack on fiscal inequality in schools.

The first strand of equality seemed supportable by the *Brown* case because the Court had held that children must be given equal opportunity. The second strand of equal protection analysis also appeared viable. It was logical to draw an equal protection analogy to geographic discrimination, since it was easily demonstrable that revenue and educational opportunities varied greatly based on residence in school districts of differing wealth. The third strand also seemed sensible, as there was existing case law from which to argue that wealth may not frustrate equality under law. Of significant support to this theory was an argument that variability in school district wealth made a good case for wealth discrimination claims, which could lead to creation of a wealth-based suspect class.

To reformers, wealth as a suspect class was grounded in case law by the indigent cases, and it was only a matter of applying *Brown*-like logic on racial inequality to fiscal inequality. Plaintiffs undergirded this position by relying on the U.S. Supreme Court's ruling in *United States v Carolene*, as the Court noted that "[p]rejudice against discrete and insular minorities may be a special condition, which tends seriously to curtail the operation of those political processes ordinarily relied upon to protect minorities, and which may call for a correspondingly more searching judicial inquiry."[27] By this logic, wealth-disadvantaged children were a perfect case in point. Supported by this logic and further armed with arguments derived from other extant rulings from the voting rights and criminal appeals cases in which the Court had held that classifications based on wealth were to be strictly scrutinized,[28] plaintiffs built a strong case for high stakes constitutional litigation.

The first suit to actually appear was *Burruss v Wilkerson*,[29] which was filed in Virginia in 1968. Plaintiffs in *Burruss* based their claims on the Fourteenth Amendment, arguing that state aid in Virginia was not allocated to local school districts on the basis of educational need. The United States District Court rendered its decision in May 1969, stating that while "[t]he existence of such deficiencies and differences is forcefully put by plaintiffs' counsel...we do not believe they are creatures of discrimination by the State....[O]ur reexamination of the Act confirms that the cities and counties receive state funds under a uniform and consistent plan. We can only see to it that the outlays on one group are not invidiously greater or less than that of another...[N]o such arbitrariness is manifest here."[30] The court added that although "plaintiffs seek to obtain allocations of state funds among the cities and counties so that the pupils in each of them will enjoy the same educational opportunities...the courts have neither the knowledge, nor means, nor the power to tailor the public monies to fit the varying needs of these students throughout the state."[31]

A second federal case of *McInnis v Shapiro*[32] was actually the first case to be heard by the U.S. Supreme Court and was filed in Illinois in the same year as *Burruss*. Originally heard in United States District Court and later upheld by the U.S. Supreme Court,[33] *McInnis* was also a Fourteenth Amendment equal protection suit. It attempted to overturn the state funding plan on the basis that unequal expenditures linked to property values and local tax rates were an impermissible violation of equal protection. The court ruled for the state, however, using language that would become common in defining the federal posture.

While the court agreed that there were wide differences in expenditures per pupil based on wealth, the court stated its inability to redress these issues on the basis of what it perceived to be nebulous conditions. First, the court observed that differences in revenue were not by any inherent nature automatically invidious or arbitrary. Second, the court noted that the legislature's policy of allowing local districts to tax and spend at different rates was a reasonable goal of local control of schools, and that differences were softened by virtue of the state's common school fund, which provided a $400 minimum base amount per pupil.

Third, the court held that there was no constitutional requirement for absolutely equal expenditures under the Fourteenth Amendment's equal protection clause. Finally, the court stated that revenue distribution is a policy prerogative of the legislature when no constitutional claim is breached. This posture was upheld in a later critique of *McInnis* by the U.S. Supreme Court, when the Court stated that "the two standards (posed by plaintiffs) are mutually irreconcilable since educational needs may demand a much higher expenditure for certain children than for others."[34]

In contrast to the previous two cases, a third federal case was heard in 1972 in *Van Dusartz v Hatfield*[35] as a Minnesota federal district court agreed with plaintiff's contention that wealthier school districts in that state not only had greater revenue per child but also enjoyed a lower tax rate. The court found that the level of spending for publicly financed education in Minnesota was profoundly affected by the wealth of each school district. At the same time, the court observed that a system of public school financing that made spending per pupil a function of the school district's wealth was a violation of the federal equal protection guarantee.

Although plaintiffs in the *Brown* legacy had reasoned well in their view and had even managed to gain sympathy in *Van Dusartz* for equal protection analysis, the decisions in *Burruss* and *McInnis* did not support such logic. In fact, in both earlier instances federal courts had uniformly refused to intervene on three grounds. The first basis was a plain reading of the Fourteenth Amendment, noting no equal protection mandate for unequal revenue. The second basis was also important, as these courts bowed to legislative prerogative by invoking separation of powers in the absence of invidious discrimination. The third basis was a particularly important one that would be examined by the nation's highest court, as these courts complained of a lack of judicially manageable standards on which a favorable ruling could be based. Equality in federal court was stated negatively in that the absence of resources was not the same as the discriminatory denial of moneys. From a federal perspective, the federal claim was both legislatively proscribed and judicially unmanageable.

Reformers realized, however, that a definitive U.S. Supreme Court ruling was not yet in place, and a test case was carefully chosen. The vehicle was found in a case known as *San Antonio Independent School District v Rodriguez*[36] which had actually been filed in 1968 and in which the United States District Court for the Western District of Texas had upheld plaintiff's position that the state must exercise fiscal neutrality in financing public schools. This court's 1971 decision had also buoyed reformers' hopes as it accepted the argument that education is a fundamental interest to the state and should be subjected to strict scrutiny equal protection analysis.[37]

After carefully crafting the attack, the Texas case was taken on appeal by the U.S. Supreme Court. The approach argued that the Texas funding system violated the federal equal protection clause by discriminating against a class of poor, and that students were denied their right to an education. As such, plaintiffs argued wealth as a suspect class and fundamentality at the highest level in a desperate effort to sustain their most basic premises of equality. The Supreme Court, however, rejected plaintiff suspect class arguments as it saw plaintiffs to be students living in poor school districts, rather than being poor themselves.

Speaking for the majority, Justice Powell noted that wealth discrimination had always been confined to personal wealth, and that the class in *Rodriguez* was not one for which special protection has been provided; i.e., it was neither politically powerless, discrete, nor an insular minority.[38] The Court noted that individual income did not readily correlate with district wealth, and that even if the correlation had been strong, the Court's application of wealth suspectness had been limited to absolute deprivation rather than mere differences.[39] Since no student was deprived of an education, fiscal inequalities were of only relative difference and not deserving of new suspect identification.

The Court then turned to plaintiffs' fundamentality argument, again rejecting their position. Plaintiffs had argued that education was so predicatory to other fundamental rights that it established an important nexus to those other established rights.[40] According to this argument, education was so tied to the right to vote and to free speech, and these rights were dependent on an equal education. The Court rejected these arguments, however, seeing no more connection between education and these rights than it saw between housing, food, or other subsistence and the right to vote.[41]

The Court particularly noted a difference between hindering a child's education and the state aid scheme which, in its view, sought to improve available offerings.[42] While the Court pointedly questioned the efficacy of wide disparities among Texas school districts, it rejected strict scrutiny, stating that only a rational relationship was required to defend an aid plan absent invidious discrimination. A rational basis could be found in Texas' goal of promoting local control of schools, and the Court was clearly reluctant to intervene in such a sacred, complex, and political arena:

> Education, perhaps even more than welfare, presents a myriad of intractable economic, social, and even philosophical problems. The very complexity of the problems of financing and managing a statewide public school system suggests that there will be more than one constitutionally permissible method of solving them, and that, within the limits of rationality, the legislature's efforts to tackle the problems should be entitled to respect.[43]

Hopes for a sweeping federal decision fell with the ruling in *Rodriguez*. In apparent contrast to *Brown*, no fundamentality or suspect class or equal protection for funding was available, except in cases of total educational deprivation or in the established instances such as race discrimination. The Court had sanctioned legislative prerogative and claimed judicially unmanageable standards.

The ruling in *Rodriguez* had two predictable effects on school finance litigation. One effect was to lessen the importance of federal lawsuits, while the other effect was to increase efforts in state courts as an alternative path to the same end. Although the balance clearly shifted toward the state reform strategy, it would be incorrect to close the federal discussion without recognizing that a limited pursuit of federal educational justice has continued.

Subsequent Federal Litigation[44]

Although *Rodriguez* had a substantial chilling effect on new federal litigation, other cases have nonetheless been brought directly in federal court, in addition to continuing to raise federal issues in state level litigation. These subsequent cases have kept the federal question alive, particularly in light of the fact that the Supreme Court did not so decisively close the door on a federal claim as to prevent any possibility of a different ruling at some future time. These additional cases were important for defining a future federal role in education.[45]

The first subsequent federal case[46] of importance followed thirteen years after *Rodriguez,* as plaintiff districts in Mississippi argued violation of federal equal protection in revenue disparity based on Section 16 lands lost during the Civil War.[47] Although the state provided aid to offset losses, by 1981 state funds were only $.63 per pupil compared to $75.34 per pupil in districts whose lands had not been taken. Originally dismissed in federal district court, the Fifth Circuit Court of Appeals held in *Papasan v Allain*[48] that equal protection was not barred by the Eleventh Amendment, but also held that *Rodriguez* was the federal standard regarding fiscal disparity.

The U.S. Supreme Court affirmed the immunity decision,[49] but reversed on the equal protection claim and remanded the case for development. *Papasan* was thus notable for two important reasons. First, the complaint was so narrowly written as to limit its relevance to traditional federal claims, never raising the broader question of fundamentality. Second, a small window of federal interest in school funding was seen by virtue of remand to the lower court, as the Supreme Court noted that unreasonable government actions would raise significant constitutional questions.

A second important federal case arose in Texas a few years later as the U.S. Supreme Court ruled in *Plyler v Doe*[50] that refusal by a state to educate illegal aliens could stir court interest in constitutional pleas for equal protection. While the Court in *Plyler* stopped significantly short of declaring education a fundamental right, it did make significant overtures in that direction: it approved a higher level of scrutiny in cases of absolute educational deprivation, while using language that qualified its closed posture to fundamentality under certain conditions.

The majority in *Plyler* stated that while its ruling in *Rodriguez* was unaffected by the present case, it was concerned that education is more than a mere service and convenience, and that some quantity of education might deserve closer scrutiny for its inalienability. The Court's own words pointed eloquently to its hesitancy to completely disallow fundamentality as it stated:

> Education provides the basic tools by which individuals might lead economically productive lives to the benefit of us all. In sum, education has a fundamental role in maintaining the fabric of our society. We cannot ignore the significant social costs borne by our Nation when select groups are denied the means to absorb the values and skills on which our social order rests.[51]

The third and most recent federal case occurred in 1988 in *Kadrmas v Dickinson Public Schools.*[52] Plaintiffs in this North Dakota case had argued in trial that charging for bus service denied equal protection of the laws, since the plaintiff child was poor and could not afford to pay for transportation as readily as many other children. Although the U.S. Supreme Court found for the defendant state, its five to four decision was highly illustrative of the unsettled nature of federal education claims. The Court was particularly sharp in warning that *Rodriguez* should not be viewed as the final word on education in that there are reasonably many variances and exceptions that would interest the federal judiciary. The sizable minority opinion expressed this view exceptionally well, particularly in its last sentence, stating that:

> The Court therefore does not address the question whether a state constitutionally could deny a child access to a minimally adequate education. In prior cases this court explicitly has left open the question whether such a deprivation of access would violate a fundamental constitutional right. That question remains open today.[53]

Although *Rodriguez* has been considered by many to absolutely foreclose a successful federal case, the litigation record appears far less certain on the matter than downcast critics seem to believe. Federal courts are in fact sympathetic to the problem of judicially manageable standards, and they are generally disposed to uphold legislative prerogative. Likewise, the nation's highest court is reluctant to adopt a social activist posture that would result in a forthright declaration of education as a fundamental right. The Court also is not willing to create the suspect classifications longed for by education's reformers. But it is also clear the Court has a latent interest in education that emerges each time it revisits and qualifies earlier rulings, and this interest is being slowly developed by the urgent questions of the times. Until then, however, the case for reform has clearly turned to state courts.

State Origins

As stated earlier, education finance litigation is often thought to be of recent origin. This misperception has occurred largely due to two factors that have drawn national attention to education finance litigation as if it were entirely new. First, federal claims beginning in 1968 give the impression that education finance litigation was newly emergent. Second, state court reform following after *Rodriguez* has further lent weight to the notion of recency.[54] Because state finance litigation has had a powerful impact that has accumulated for many years, it is important to understand the state case in the post-*Rodriguez* light.

The State Case[55]

Development of the state case for equalizing fiscal resources in schools has paralleled the federal case in many ways. Lines of chronological demarcation became blurred as suits were brought simultaneously in state and federal courts in the early days of reform. For example, *Burruss, McInnis,* and *Rodriguez* were all filed in federal courts in the 1960s, but the California case of *Serrano v Priest*[56] had already been decided at the state supreme court level before the U.S. Supreme Court ruled in *Rodriguez* in 1973. Common legal claims likewise occurred, as state cases similar to *Serrano* grieved both federal and state constitutional claims. While success on the federal front has been extremely limited, state litigation has occurred more frequently and with different results in many instances.

Early State Cases

The first of the state fiscal equalization cases to gain national attention was the *Serrano v Priest*[57] decision in which the California Supreme Court was asked to rule on what was destined to become a classic model for state school finance litigation. Laying out a pattern that would be widely copied, this lawsuit claimed that the state aid plan for public schools violated federal and state constitutional requirements for equal protection. A central feature of these claims was reliance on concepts developed earlier in federal cases, wherein plaintiffs sought a favorable ruling on questions concerning a fundamental right to education, wealth as a suspect class, and equal protection under the federal and state constitutions.

To prove their case, plaintiffs charged that the state plan for funding education had permitted great disparities in the funding available to children and that these differences affected the quality of education. Plaintiffs also charged that taxpayers were required to pay higher tax rates, but that they received a poorer education in spite of greater tax effort. The combined result, plaintiffs claimed, was that these injustices jointly denied children equal protection, violated their fundamental right to education, and made the quality of education a function of residence that corresponded impermissibly to differences in local wealth.

In a sweeping victory for plaintiffs, the state supreme court ruled that the method of funding public elementary and secondary education in California was unconstitutional by violating the Fourteenth Amendment and the state constitution's equal protection clause because the state school finance scheme made the quality of education dependent on local school district wealth. This ruling ran completely counter to every observed trend in school finance litigation up to that point in time, giving new hope to reformers on the state front. The state's highest court was harsh in its condemnation of unequal educational opportunity, declaring:

> We have determined that this funding scheme invidiously discriminates against the poor because it makes the quality of a child's education a function of the wealth of his parents and neighbors. Recognizing as we must that the right to an education in our public schools is a fundamental interest which cannot be conditional on wealth, we can discern no compelling state purpose necessitating the present method of financing. We have concluded, therefore, that such a sys-

> tem cannot withstand constitutional challenge and must fall before the equal protection clause.[58]

The method of examination by the court was a great strategical learning activity for state level reformers. In examining the facts, the court noted that the fatal flaw was that state aid was too low to offset disparate wealth per pupil. The differences examined by the court were not even the greatest actual differences, which approached a 50:1 ratio statewide, but differences demonstrated before the court of 1:13 between Baldwin Park and Beverly Hills were nonetheless sufficient cause for alarm. The court was quite unimpressed with the fact that state aid actually exacerbated the problem, as aid was given to both rich and poor districts. The court observed under these conditions that aid became meaningless to poor districts, as differences were merely perpetuated by such a system.

Plaintiffs also were taught significant strategy as the court ruled on their legal claims. The court held that wealth could indeed be a suspect class and that education must not depend on local wealth, stating "[t]o allot more educational dollars to the children of one district than to those of another merely because of property is to make education dependent upon the location of private commercial and industrial establishments."[59] The court continued, saying "we reject defendants' underlying thesis that classification by wealth is constitutional so long as the wealth is that of the district, not the individual. We think that discrimination on the basis of district wealth is equally valid."[60]

Similarly, plaintiffs gathered new arguments for fundamentality, as the court held that education in an industrial state is indispensable, that education determines one's chances for success, and that education is a unique influence on the development of citizens. The court justified itself as it declared "[w]e think that from a larger perspective, education may have far greater social significance than a free transcript or a court-appointed lawyer."[61] So stating, the state supreme court held that strict scrutiny equal protection analysis was the proper judicial standard for such state claims.

The fact that *Serrano's* federal claim was later overturned by the 1973 ruling in *Rodriguez* did not detract from the impact of *Serrano* at the state reform level. *Serrano* provided a significant model for state litigation by its success on fundamentality and equal protection, and it clearly showed that state constitutions might be vulnerable in ways that were not available at the federal level. In addition, success on wealth suspectness was a victory for plaintiffs in that this argument had been among the weakest due to lack of precedent. The impact of *Serrano* was tremendous for these reasons. In addition, it provided a catalyst for litigation across the nation.

A *Serrano*-like decision followed shortly thereafter, when in 1972, the Michigan Supreme Court issued *Milliken v Green*.[62] Although many aspects of the case followed complimentarily after *Serrano*-type claims, particularly relating to equal protection, the victory was shortlived because the state supreme court experienced a change of judges, resulting in reversal upon review of the case. Disappointingly for reformers, the new court issued its different ruling[63] only one year later, vacating the prior court's decision on the basis that evidence did not prove that the equal protection of children in low wealth districts was violated.

Of particular importance to the court was the question of linkage between fiscal inputs and achievement outcomes, so that additional inputs would produce greater and more equal outcomes. The court refused to accept the logic that the production equation was so simply and causally defined, with the court explaining its logic in clear and concise language:

> Without disagreeing with either theory of defining equal opportunity—inputs or outputs—we cannot accept without criticism either of the further narrowed definitions offered by the parties. The reduction of the sum total output to the accomplishment of the pupils on a few achievement tests would be grossly unjust to both the educators and the pupils, for education must extend far beyond the limits of verbal facility or mathematical proficiency. With respect to the input received by the school, the level of taxable resources within a district is only one of the myriad inputs into an educational system.[64]

The case for reform again went badly for plaintiffs as, shortly after *Rodriguez*, the Arizona Supreme Court ruled for the defendant state in *Shofstall v Hollins*.[65] The state supreme court had been asked to review whether that state's school finance law violated the state equal protection clause and its "general and uniform" provision for schools that appeared in the state constitution. The court modestly interpreted general and uniform to mean that the state would provide a minimum school year, certify personnel, and establish requirements and standards for courses.

The Arizona court was clear in accepting *Rodriguez* as the standard and indicated that there was no constitutional issue except when there was absolute denial of educational services. Although the court held for a fundamental right to education and recognized that there were problems across the state, its view was clear that legislative redress was the proper solution to political problems that could not be solved by the judiciary.

Still a third instance of defeat for reformers occurred in 1973 when the Illinois Supreme Court issued a ruling in *Blase v Illinois*[66] denying a narrow reading of constitutional provisions, preferring instead a contextual analysis that recognized a broad legislative role. Plaintiffs had based their claims on strict interpretation of the state constitution's rather strong language on the role of education in the state, wherein the constitution required:

> A fundamental goal of the People of the State is the educational development of all persons to the limits of their capacities.
>
> The State shall provide for an efficient system of high quality public educational institutions and services.
>
> The State has the primary responsibility for financing the system of public education.[67]

Of particular importance to plaintiffs was the use in the state constitution of words like fundamental, efficient, and primary responsibility. Plaintiffs contended that the state should provide no less than 50 percent of all educational revenues, along with other claims for strict interpretation of equality as might be read into various constitutional provisions. The state supreme court rejected that argument, however, ruling that the constitutional language was meant only to express a goal or objective and was not meant to be a specific command to the legislature.

Yet another defeat for reformers' claims was experienced in a case in the state of Washington in 1974 in *Northshore School District v Kinnear.*[68] In this case, the petitioner offered the following four contentions of unconstitutionality: (1) that children are denied equal protection in violation of the Fourteenth Amendment and comparable state constitutional protections; (2) that taxpayers in poorer districts are denied equal protection because they pay higher tax rates than taxpayers in wealthier districts to raise the same revenue per pupil; (3) that the state had failed to obey a provision of the state constitution which read "[i]t is the paramount duty of the state to make ample provision for the education of all children;"[69] and (4) that the state had failed to provide a general and uniform system of public schools as mandated by the state constitution.

The state's highest court rejected the first claim because data reflected that assessed valuation had little, if anything, to do with the quality of education. The court also indicated that there was no evidence to support the claim that one district or another provided superior or inferior educational services. In considering the equal protection argument the court said "[t]hat it takes more millage to raise the same amount of dollars on low valued property than it does on high valued property is no more than a meaningless truism and can be answered with another truism that the lower the value of one's property the lower one's taxes, neither truism having anything to do with the equal protection clause of both constitutions so long as everybody in the taxing scheme pays the same rate."[70]

The court in *Northshore* was emphatic in pointing out other problems with plaintiffs' arguments. The equal expenditure argument was pointedly rejected, as the court noted that even if the state were one giant school district, expenditures per child would still vary due to infinite variations in geography, climate, terrain, social and economic conditions, transportation, special services, and local choices in curricula and so forth. Likewise, the strength of constitutional language regarding ample provision for education was regarded unfavorably, as the court noted that "constitutionally speaking, the duty or function is the same as any other major duty or function of state government."[71]

Plaintiffs' argument for the plain meaning of a general and uniform system was also rebuffed, as the court ruled that such a system is one which provides every child with free access to certain minimum and reasonably standardized educational programs, noting that differences in children's needs preclude absolute uniformity. A similar posture was observed in Idaho just one year later, as that state's supreme court wrote in *Thompson v Engleking*[72] that "the record does not demonstrate a failure by the legislature to comply with its mandate to establish a system of basic, thorough, and uniform education..."[73]

Although plaintiffs seemed to be losing challenges at the state level in the face of the gains made in *Serrano*, reform gained a significant victory as the supreme court in New Jersey ruled in 1975 in *Robinson v Cahill.*[74] A favorable decision by the superior court dating back to 1972 was taken on review by the state high court, wherein plaintiffs had charged that the state school finance plan violated federal and state equal protection laws and denied students' fundamental right to education because tax revenue varied with district wealth and was not adequately equalized by the state. According to plaintiffs, the state had abrogated its responsibility to education because the laws were not equal in effect on all citizens where equal tax effort did not produce equal tax yield, despite the fact that state aid provided approximately 28 percent of all district revenues.

The state supreme court ruling was notable for very important reasons. To reformers' dismay, the high court rejected fundamentality, citing an overlooked point made by the U.S. Supreme Court in *Rodriguez*. The U.S. Supreme Court had said that "every claim arising under the Equal Protection Clause has implications for the relationship between national and state power under our federal system...[i]t would be difficult to imagine a case having a greater potential impact on our federal system than the one now before us, in which we are urged to abrogate systems of financing public education presently in existence in virtually every State."[75]

The New Jersey court also noted that the U.S. Supreme Court had never cited *Brown* as a case involving the fundamental right concept, stating that *Brown* would point the opposite direction since it declared education to be a most important function of state and local government.[76] Further, the New Jersey court refused to find wealth a suspect class, noting that "if this is held to constitute classification according to 'wealth' and therefore 'suspect,' our political structure will be fundamentally changed."[77] Under these conditions, the court could find no basis for fundamentality or federal equal protection. But despite its rejection of reformers' core views, the court nonetheless ruled the system unconstitutional by invoking the education article of the state constitution which demanded a thorough and efficient education system—a requirement that was not met as a result of the lack of equalization in revenues which violated the state's equal protection clause. The court directed the legislature to define and develop a school finance program that would satisfy the thorough and efficient requirement under the state constitution.

After various attempts to devise a system that was thorough and efficient, the legislature was not able to provide sufficient funds to make notable improvements in the state's school system. In 1976, the New Jersey supreme court enjoined state and local officials from expending any funds for schools until appropriate legislative funding was achieved. Schools were closed for two weeks beginning July 1 of that year, as the legislature struggled to find enough money to mollify the court. The legislature was successful in enacting an income tax, which in turn persuaded the court to lift the injunction.

New success by reformers, however, did not immediately overwhelm adverse court rulings. Three additional cases were decided in Oregon, Kansas, and New York that indicated the reluctance of courts to sweepingly accept plaintiff arguments. In *Olsen v State of Oregon*[78] the same issues were heard regarding unequal revenues and general and uniform education systems. The state supreme court, however, held "[w]e hold that the Oregon system of school financing does not violate the Oregon Constitution. Our decision should not be interpreted to mean that we are of the opinion that the Oregon system...is politically and educationally desirable. Our only role is to pass on its constitutionality."[79]

The same result effectively occurred in Kansas in *Knowles v State Board of Education*,[80] as that state's supreme court dismissed a suit over the state's requirement to make "suitable" educational provision because the legislature had rendered the cause moot by enacting a different school finance law. Similarly, in reversing a trial court in New York that had declared the state finance scheme invalid in *Board of Education of Levittown v Nyquist*[81] after focusing on the problems of large metropolitan areas such as municipal overburden, the state supreme court ruled against fundamentality and stated that because average pupil expenditure exceeded all but two states, the system did not violate the state's education article.

Success returned to reformers, however, in a series of cases that followed. In a Connecticut case, that state's supreme court ruled for plaintiffs in *Horton v Meskill*.[82] The supreme court noted that there was a link between the financial resources of districts and the range of course offerings, number of teachers per special program, library resources, and pupil/teacher ratios among the state's school districts. In addition, the court expressed concern for taxpayer equity. The court also chose to view fundamentality more sympathetically under the state constitution, stating " . . . we must conclude that in Connecticut the right to education is so basic and fundamental that any infringement of that right must be scrutinized . . . the state system of financing public elementary and secondary education as it presently exists and operates cannot pass the test of 'strict scrutiny' as to its constitutionality."[83]

Interestingly, the supreme court in the state of Washington partially reversed itself in *Seattle v State*[84] by ruling that the state system was invalid as it applied to Seattle. In this case, the focus was on the fact that the state's school districts had the potential to become overreliant on voted special excess levys so that the state's foundation aid program was held to be inadequate.

A 1979 decision in *Pauley v Kelly*[85] suggested a new direction by illustrating the lengths to which courts might be willing to examine in becoming involved in a quasilegislative role. The West Virginia supreme court reversed a lower court's dismissal of the claim that the state system violated the thorough and efficient requirement of the state constitution, ruling that education was a fundamental right in that state, and remanding the case for development and retrial.

In 1982, the trial court issued a lengthy opinion overturning the system, following that decision with a supplemental opinion directing the state board of education and the state superintendent of schools to develop a master plan for improving education. The case was remarkable because it represented a highly activist bench which chose to examine in detail the specifics of a good education system, including in its order that it wanted specifications that would detail every facet of schooling such as curricula and extracurricular opportunities, even including appropriate square footage for programs as well as storage space and supplies.

Despite these rulings, another series of defendant victories emerged during the next few years. The case of *Danson v Casey*[86] in Pennsylvania in 1979 rejected plaintiff claims that the state funding scheme failed to provide a "normal" program of education to students in Philadelphia. The court concluded that no child was denied an adequate, minimum, or basic education, and cited the 1938 *Teachers' Tenure Act* cases[87] to illustrate that words such as "thorough and efficient" should not be narrowly or constrictively viewed. Rather, the court said such words were positive models and that narrow interpretation would unnecessarily restrict future legislatures in pursuit of their duties. It said "...everything directly related to maintenance of a 'thorough and efficient' system of public schools must at all times be subject to future legislative control...."[88]

A similar outcome was seen Colorado, as the state supreme court reversed a lower court ruling in *Lujan v Colorado State Board of Education*[89] by concluding that all the actions of the legislature were rational and that constitutional language merely required action by the legislature, rather than the burdensome requirements that would be involved in a fundamental right to education. Similar results were obtained in Ohio in 1979, as that state's supreme court overruled the trial court in *Board of Education of the City of Cincinnati v Walter*[90] because it found that no school district was totally starved for funds and no district lacked teachers, buildings or equipment. The court stated that while legislative prerogative is not absolute, wide discretionary powers are necessary for the legislature, making the issue of defining "thorough and efficient" very difficult when dealing with matters such as taxation and educational policy.

Still other decisions favoring plaintiffs or defendants emerged over the next several years. One of the more expansive state supreme court decisions regarding the constitutionality of a state school finance system occurred in Wyoming in *Washakie v Herschler*,[91] rendered in 1980. The court indicated that the poor school districts in the state showed a consistent pattern of less total revenue per student than rich districts, based on assessed valuation per pupil. The court accepted plaintiff arguments that the quality of educational services is related directly to availability of financial resources. The court cut to the core of the issue with a blunt statement. It said " . . . until equality of financing is achieved, there is no practicable method of achieving equality of quality."[92]

The Wyoming court reached this decision based on the fact that certain provisions of the state constitution were more demanding than the equivalent federal equal protection guarantees, and because education was of such compelling interest in that state so as to be a fundamental right. Unlike most courts, the Wyoming supreme court accepted the argument of wealth as a suspect class, stating " . . . the state has the burden of demonstrating a compelling interest . . . which is served by the challenged legislation and which cannot be satisfied by any other convenient legal structure. The state has failed to make any showing that other methods are not available or possible."[93]

A similar ruling was made in Arkansas in *Dupree v Alma School District*[94] as that state's high court held that the state public school finance system relied too heavily on local property taxes, with a resulting violation of equal protection for students in property-poor districts, particularly since the finance scheme was not rationally based on the needs of individual districts.

Defendant victories were also recorded in the states of Maryland, Georgia, and Michigan. Plaintiffs in Maryland had argued in *Somerset v Hornbeck*[95] that the education clause and the Maryland Declaration of Rights were violated by the state school finance plan. While the trial court agreed with plaintiffs on the basis of evidence showing a lack of thorough and efficient provision when wealthier districts were permitted virtually unlimited spending, the state supreme court overruled and held for defendants. The high court held that equal expenditures were not required to fulfill that state's mandate for a thorough and efficient system of public schools, and further held that no violation had occurred to either equal protection or a fundamental right to education.

A similar scenario occurred in Georgia, where the trial court held for plaintiffs on the grounds of fundamentality and strict scrutiny where the state finance plan gave nearly identical revenues to rich and poor districts alike. However, the state supreme court reversed in *McDaniel v Thomas,*[96] declaring that the aid plan bore a rational relationship to a legitimate state purpose since the state's aim was to provide funding for a basic education for all children in the state. The court stated that the Georgia Constitution did not require the state to equalize educational opportunities among districts.

Michigan successfully defended its funding scheme in *East Jackson Public Schools v State*[97] where the high court held that no fundamental right existed, that equal protection was not violated, and that state obligation to provide an educational system was not synonymous with an obligation to provide equal funding per pupil. Further, the data analysis had showed that the system was less disparate than it had been under *Milliken* in 1972, thereby severely weakening plaintiff claims and gaining no court sympathy in the face of time-honored legislative prerogative.

Later Reform Cases[98]

An additional series of reform cases followed after the early years of school finance litigation, moving the struggle for reform up to the present era. Courts in more recent times have continued to struggle with constitutional interpretation, infringement on separation of powers, and pressure to use the bench as an activist weapon for social reform agendas where legislatures are charged with failure to respond to social needs.

In 1987, plaintiffs contended in *Britt v North Carolina*[99] that the aid distribution scheme violated the general and uniform and equal protection clauses of the state constitution. The appellate court held that the framers of the state constitution had intended students to have equal access to public education regardless of race and to have a fundamental right to an education. The court further stated, however, that this did not mean that each student within the state was guaranteed an education equivalent to every other student in the state. Only one year later, the state of South Carolina held in *Richland County v Campbell*[100] that plaintiff arguments regarding vast disparities under the state aid formula were not convincing, in that the court believed the formula appropriately considered individual wealth of each school district and that the state constitution did not require the state to pay for the total cost of providing educational programs and services.

In contrast, the issue that began in Texas with *Rodriguez* was continued at the state level, with the state supreme court in 1988 overturning an appeal court's decision in *Kirby v Edgewood.*[101] Disagreeing with the trial court, the high court held that education in Texas was a fundamental right and that the state constitution required uniform delivery of educational opportunity. The court took note of the fact that evidence showed more than one-third of all districts in the state had insufficient money to offer adequate educational programs and ordered the legislature to create a remedy within a specific time period.

A different interpretation on a similar question, however, was issued only one year later in Wisconsin, when that state's supreme court ruled in *Kukor v Grover*[102] that uniformity of a minimal standard was sufficient to meet the constitutional requirement. The court held that education was a fundamental right, although it chose to apply a rational basis standard in that plaintiffs had failed to prove or allege a complete denial of educational opportunity.

Two recent cases in Kentucky and New Jersey have brought national focus to the nature and extent of the education reform agenda. One of the more touted reform suits in the later years of the state case was the 1989 Kentucky supreme court's decision in *Rose v Council for Better Education.*[103] Holding for the plaintiffs, this court was reminiscent of the behavior seen in West Virginia in *Pauley* in 1982, as the Kentucky court held that the system of common schools was not efficient. Holding for a fundamental right to education, the court found that this right was violated when the state's schools were significantly underfunded and inadequate in educational programming. Pervasive inequalities in the state's 177 districts caused the court to order a complete overhaul of the entire public education system, resulting in a massive infusion of new dollars and the total redesign of the education system in that state.

Similarly, the New Jersey case of *Abbott v Burke*[104] in 1990 created nationwide consternation as that state's finance system was again declared unconstitutional because the state aid formula did not meet the needs of poor urban districts and because it was not fiscally neutral, resulting in violation of the state's thorough and efficient requirement. Plaintiffs had narrowly drawn their complaint to show only the worst and best examples of education in New Jersey, and had technically asked only that the poorest district be made equal to the wealthiest district.

The *Abbott* methodology hinged on absolute differences, rather than on relative differences that had been common in other state litigation strategy and which provided a far less dramatic picture of the extremes of education in a state—generally an easy task in states with huge urban areas. The court agreed with the results, stating "[f]rom this record we find that certain poorer urban districts do not provide a thorough and efficient education to their students.... We find the constitutional failure clear, severe, and of long duration."[105]

Other recent cases have been decided by state supreme courts that also have contributed to the development of school finance litigation history. The Oregon Supreme Court issued a ruling in 1991 in *Coalition for Equitable School Funding v State*[106] after plaintiffs had argued that lack of uniformity among school districts conflicted with the state constitution. Plaintiffs had complained that the state should be required to fund all state mandates and should be further required to provide equal funding per pupil. The state supreme court rejected those contentions, however, saying that the state constitution and statutes were adequate by allowing local districts to utilize local property taxes to meet state standards.

The intense flurry of litigation that emerged in the 1990s has also recently begun to produce the expected state supreme court decisions. In addition to the Oregon case just discussed, high courts in Alabama, Arizona, Kansas, Minnesota, North Dakota, and Tennessee have issued rulings on plaintiff claims that were consonant with the issues first broached in the early years of school finance litigation. In a 1994 ruling that surprised no one and in which several defendants actually sought to join with plaintiffs in seeking reform, the high court in Alabama ruled for plaintiffs' claims of fundamentality and equal protection in one of the worst-case scenario instances that could be imagined. Plaintiffs in *Harper v Hunt*[107] demonstrated deplorable educational differences in that state's largely minority public school system.

A surprising reversal occurred, however, in the Arizona case of *Roosevelt v Bishop,*[108] where plaintiffs had brought a narrowly drawn case focusing on capital facility needs. Reversing the trial court's holding for defendants that had used *Shofstall* as a precedent, the Arizona supreme court ruled that education was a fundamental right and that a limited reading of the complaint did not do justice to the issue before the court. Consequently, that court decided that the entire education system should be scrutinized since all aspects of the system were interdependent.

In contrast, the Minnesota system of public school finance was upheld in *Skeen v Minnesota,*[109] as that state's supreme court perceptively held that while education was a fundamental right, only a rational basis should be applied to whether funding for the educational system was thorough and efficient.[110] Likewise, a plaintiff victory in Tennessee was reversed by that state's high court, as the justices issued a ruling upholding defendants in *Tennessee Small School Systems v McWherter.*[111] That ruling effectively made unimportant a fundamental right to education because, in holding for fundamentality, there was no concomitant finding of great importance sufficient to overturn the state funding scheme.

By similar logic, the state supreme court in North Dakota[112] reversed a trial court's holding for plaintiffs, while agreeing with plaintiffs on disparities in the educational system. The high court interpreted the state constitution to demand a fundamental right to education and found that the system did not provide the required obligation. However, it held that the trial court had severely overstepped its authority by demanding specific remedy of the legislature, leaving no choice but to reverse on appeal with stern warning to the legislature to redress inequality. Similarly, the state supreme court of Kansas held in *Unified School District v State*[113] that the system met all requirements of the constitution " . . . even though the law may be unwise, impolitic, or unjust . . . the remedy lies with the people through the political process."[114]

Active State Cases

The high-risk stakes and uncertainty in constitutional litigation, however, have not deterred plaintiffs from all quarters of the nation from aggressive new school finance lawsuits. In fact, it may be argued with some merit that a more definitive court trend might produce less activity because the outcome would be more predictable. The uncertainty and risk may be causing plaintiffs to approach courts more readily, since the chance to prevail appears nearly equal to the opposite outcome. Likewise, it has become more acceptable to challenge state funding schemes, and there is a certain political strategy to litigation in that inadequate revenues can be made out to be the fault of external forces beyond the district's control.

Obviously, some current litigation is a function of compliance monitoring, such as the repeated tribulations of school finance systems in states like Texas, New Jersey, and California. But beyond these instances, the 1990s has brought an intense wave of new filings seeking to force judicial review of how states distribute aid to school districts. Both the historic and present record stand as evident proof of the popularity of litigation as a solution to districts' economic woes. As of late 1994, plaintiffs had won twenty victories at the state supreme court level, although this number included compliance litigation as in New Jersey, beginning with *Robinson* in 1972, and continuing through four rounds of the *Abbott* case. By contrast, states had successfully defended the school finance distribution formulas in twenty-nine instances, obtaining rulings which declared the systems constitutional, or vacated or dismissed plaintiff claims.

Although the record would seem to slightly favor defendants, the popularity of litigation as a favored recourse is evident in the new suits that continually emerge, while active cases continue to wend their way to the highest level. The result is an enormous amount of litigation of record and litigation in process as plaintiffs initiate lawsuits on almost a daily basis, and in many instances withdraw them as well. Some litigation never moves beyond obscurity, as some suits involve only small issues that are quickly resolved at the lowest court level. Further, lawsuits may be dismissed, either by virtue of a court's ruling that a claim is nonmeritorious or through technical points of law relating to states' motions to dismiss.

Although modern school finance litigation is so fluid that a complete count of petitions before the courts can never be accurate at any given moment, assessment of current litigation in an overview of major known actions is useful because it monitors the nation's mood. At this writing, the slate of litigation was accelerating at a rapid pace. At the opening of 1995, litigation existed in more than half the states in which no supreme court decision had been rendered. In addition to these known states, rumors were raging that new filings would soon emerge, as one national advocacy group had threatened new filings in half the states. When these active struggles are added to compliance monitoring and appeals from trial courts, the climate of school finance in recent years has been one of great anxiety and tension.

Summary[115]

The importance of finance and school business management has been illustrated throughout this text, and its sometimes contentious nature has been graphically revealed in this chapter on finance litigation. There can be no question that inclination toward calm reasoning is overwhelmed when citizens and other policymakers are persuaded to believe by a litigious climate that they are disadvantaged monetarily in the process of providing a good education system.[116] The concept of equality is so deeply engrained in the American psyche that real, and imagined, inequalities take on a special stature requiring drastic action.

Of nearly equal proportion is the modern propensity to wage war using litigation in order to seize upon new fiscal resources that are commonly believed to make the final difference in social and economic mobility. The reality of a society that so covets resources and equality that it is willing to risk vast sums of money to play in a high stakes constitutional litigation game virtually assures that school business managers will directly experience or be touched by school finance litigation at some point in their careers.

An equal reality argues that school business leaders should be highly knowledgeable in this arena, since it falls to such professional people to secure the resources required to mount an assault on state funding schemes or to join with defendants as interested parties. As such, it is important to close this chapter with some effort to discern what has been learned from the last quarter century of intense litigation and to suggest what the future may hold in this volatile public policy arena.

The decisions and strategy first laid down in *Rodriguez*, *Serrano*, and *Robinson* have provided a legacy of fierce litigation in most of the fifty states. Since 1971, state aid plans have been held unconstitutional at the state supreme court level in the states of Arkansas,[117] Arizona,[118] California,[119] Connecticut,[120] Kentucky,[121] Montana,[122] New Jersey,[123] Texas,[124] Washington,[125] West Virginia,[126] and Wyoming.[127] These states embodied the dreams of social reformers, as each case held opportunity for divining the elements of successful constitutional assault. The generalizability of legal strategy has not been overwhelming, however, as state courts have reached different decisions when asked to interpret the unique provisions of each state's constitution and statutes.

While there has been winning litigation in many states, school finance formulas have also been upheld in other states where courts have refused to read constitutional provisions so narrowly as to embrace reformers' views on fundamentality and strict scrutiny equal protection. Because there has been no overriding common thesis adopted by courts in analyses, school aid plans have been upheld by state supreme courts in the states of Arizona,[128] Colorado,[129] Georgia,[130] Idaho,[131] Kansas,[132] Maine,[133] Maryland,[134] Michigan,[135] Minnesota,[136] Montana,[137] Nebraska,[138] New York,[139] North Carolina,[140] North Dakota,[141] Ohio,[142] Oklahoma,[143] Oregon,[144] Pennsylvania,[145] South Carolina,[146] Tennessee,[147] Washington,[148] and Wisconsin.[149]

Recently, the supreme court of Virginia ruled that no requirement for the equal distribution of state moneys existed under the state constitution.[150] The trial court in South Dakota rejected a challenge to the constitutionality of the distribution of state moneys in that the court ruled the plaintiffs failed to prove their case in light of the overwhelming defense of the state.[151]

Several such states have had more than one ruling, often with opposite outcomes for a variety of reasons. Likewise, federal rulings have unanimously turned plaintiffs away, as the few instances of federal pleas in Florida,[152] Illinois,[153] Louisiana,[154] Minnesota,[155] Mississippi,[156] North Dakota,[157] Texas,[158] and Virginia[159] have all had disastrous results when attempting to overturn a state system.

From these decisions, important observations have emerged. First, the supreme courts in ten states have declared education a fundamental right. Second, a sizable number of states have declared that education is not a fundamental right. Third, the theory that fundamentality would require strict scrutiny equal protection analysis has not been uniformly upheld, although the majority of courts holding for fundamentality have also used the more stringent standard. Fourth, failure to rule for fundamentality has not invalidated plaintiffs' claims in all instances, as there has been a legacy of rulings using "thorough and efficient" provisions to invalidate state aid plans, thereby achieving the same outcome under a lesser standard. Fifth, efforts to establish wealth as a suspect class have met with nearly total disaster, as only two states have accepted this standard that was so eloquently first rejected in *Robinson* for its implications for all other governmental areas. The results of these decisions are seen in matrix form in table 6.1.

The matrix indicates a wide spectrum of rulings, confirming the colorful language that describes school finance lawsuits as high stakes constitutional litigation. Of striking absence is the paucity of rulings for fundamentality. As stated very early in *Robinson,* the implications for society embodied in fundamentality are simply too dangerous for most courts in that all other government services could be immediately subject to the same claim. Second, fundamentality has suffered because courts are slow to construe new rights from state constitutions and for which federal precedent is adverse.

Table 6.1
Litigation Matrix*

STATE	NAME OF CASE	FILED FED CT.	STATE SUP. CT.	PLNF WON	PLNF LOST	RULED SUSPECT	RULED FUNDA.	FILED ON ED. ART	FILED ON EQ PROT
Alabama	*Harper*		x	x			x	x	x
Arizona	*Shofstall*		x		x		x		x
	Roosevelt		x	x			x	x	x
Arkansas	*Dupree*		x	x				x	
Calif.	*Serrano*		x	x		x	x		x
	Serrano II		x	x					
	Serrano III		x		x				
Colorado	*Lujan*		x		x			x	x
Conn.	*Horton*		x	x			x	x	x
	Horton II		x	x					
	Horton III		x		x				
Delaware									
Florida	*Askew*	x	x	(vacated)					
Georgia	*McDaniel*		x		x			x	x
Hawaii									
Idaho	*Thompson*		x		x	–		x	x
	Frazier		x		x			x	x
Illinois	*McInnis*	x			x				x
Indiana									
Iowa									
Kansas	*Knowles*		x		x			x	x
	USD 229		x		x				
Kentucky	*Rose*		x	x			x	x	x
La.	*Scarnato*	x			(dismissed)				x
Maine									
Maryland	*Hornbeck*		x		x			x	x
Mass.	*McDuffy*		x	x				x	x
Michigan	*Milliken*		x		x				x
Minn.	*VanDusartz*	x		x					x
	Skeen		x		x		x	x	x
Miss.	*Papasan*	x			x				x
Missouri									
Montana	*Helena*		x	x			x	x	x
	Woodahl		x		x			x	x
Nebraska	*Gould*		x		x			x	x
Nevada									
N. Hamp									
N. Jersey	*Robinson*		x	x				x	x
	Abbott		x	x				x	x
	Abbott II			x					
	Abbott III			x					
	Abbott iV			x					
N. Mex.									
N. York	*Levittown*		x		x			x	x
	Spano		x		x			x	x
N. Car.	*Britt*				x				
N. Dak.	*Kadrmas*	x			x				x
Ohio	*Walter*		x		x			x	x
Okla.	*Fair*		x		x			x	x
Oregon	*Olsen*		x		x			x	x
	Coalition		x		x			x	x
Penn.	*Danson*		x		x			x	x
R. Island									
S. Car.									
S. Dak.									
Tenn.	*Small Sch.*		x		x		x	x	x
Texas	*Rodriguez*	x			x			x	x
	Edgewood		x	x				x	x
Utah									
Vermont									
Virginia	*Burruss*	x			x			x	x
Wash.	*N.Shore*		x		x			x	x
	Seattle		x	x				x	
W. Va.	*Pauley*		x	x			x	x	x
Wis.	*Buse*		x	x			x	x	x
	Kukor		x		x		x	x	x
Wyo.	*Washakie*		x	x		x	x	x	x

* Includes only decisions rendered at the state supreme court level and federal court decisions. Does not include active or threatened litigation.

Third, federal and state equal protection are automatically included in any legal challenge, but it is apparent that claims for state equal protection have been the only successful vehicle—and one that has worked fairly well whenever a strong state education article can be invoked in a plain reading.

As the Texas supreme court noted in *Edgewood* in 1989 when refusing to rule for fundamentality, "[w]hether the legislature acts directly or enlists local government to help meet its obligation, the end product must still be what the constitution demands,"[160] i.e., an efficiency clause may be as compelling as fundamentality. The result of massive litigation taken collectively strongly suggests that plaintiff causes depend heavily on specific constitutional language.

For example, the likelihood of success in Mississippi seems much lower than in Wyoming or Washington; similarly, litigants in the many states with only moderate language can be even less certain about the outcome—in fact, many such states are among those represented by trial courts ruling for plaintiffs, only to be reversed at the highest level. As seen in table 6.2, states have tremendous variation in constitutional language, as well as significant commonalties, that can be linked at least phenomenologically to litigation outcomes.[161]

Table 6.2

State Education Articles

Alabama	The legislature shall establish, organize and maintain a liberal system of public schools throughout the state for the benefit of the children thereof between the ages of seven and 21 years. *Ala. Const. art. 14, sec. 256.*
Alaska	The legislature shall by general law establish and maintain a system of public schools open to all children of the State, and may provide for other public educational institutions. *Alaska Const. art. VII, sec. 1.*
Arkansas	The State shall ever maintain a general, suitable and efficient system of free schools whereby all persons in the State between the ages of six and twenty-one years may receive gratuitous instruction. *Ark. Const. art. 19, sec. 19.*
Arizona	The University and all other State educational institutions shall be open to students of both sexes, and the instruction furnished shall be as

nearly free as possible. The Legislature shall provide for a system of common schools by which a free school shall be established and maintained in every school district for at least six months each year, which school shall be open to all pupils between the ages of six and twenty-one years. *Ariz. Const. art. II, sec. 6.*

California A general diffusion of knowledge and intelligence being essential to the preservation of the rights and liberties of the people, the Legislature shall encourage by all suitable means the promotion of intellectual, scientific, moral, and agricultural improvement. *Cal. Const. art. 9, sec. 1.*

Colorado The general assembly shall, as soon as practicable, provide for the establishment and maintenance of a thorough and uniform system of free public schools throughout the state, wherein all resident of the state, between the ages of six and twenty-one years, may be educated gratuitously. *Colo. Const. art. IX, sec. 2.*

Connecticut There shall always be free public elementary and secondary schools in the state. *Conn. Const. art. 8, sec. 1.*

Delaware The General Assembly shall provide for the establishment and maintenance of a general and efficient system of free public schools and may require by law that every child, not physically or mentally disabled, shall attend the public school, unless educated by other means. *Del. Const. art. X, sec. 1.*

Florida Adequate provision shall be made by law for a uniform system of free public schools and for the establishment, maintenance and operation of institutions of higher learning and other public education programs that the needs of the people may require. *Fla. Const. art. IX, sec. 1.*

Georgia The provision of an adequate education for the citizens shall be a primary obligation of the State of Georgia, the expense of which shall be provided for by taxation. *Ga. Const. art. VIII, sec. 1.*

Hawaii The State shall provide for the establishment, support and control of a statewide system of public schools... *Hawaii Const. art. IX, sec. 2.*

Idaho The stability of a republic form of government depending mainly upon the intelligence of the people, it shall be the duty of the legislature of Idaho to establish and maintain a general, uniform and thorough system of public, free common schools. *Idaho Const. art. IX, sec. 1.*

Illinois A fundamental goal of the People of the State is the educational development of all persons to the limits of their capacities. The State shall provide for an efficient system of high quality public educational institutions and services... *Ill. Const. art. X, sec. 1.*

Indiana Knowledge and learning, generally diffused throughout a community, being essential to the preservation of a free government; it shall be the duty of the General Assembly to encourage, by all suitable means, moral, intellectual, scientific, and agricultural improvement; and to provide, by law, for a general uniform system of Common schools... *Ind. Const. art. 9, sec. 1.*

Iowa The Board of Education shall provide for the education of all the youths of the State, through a system of Common Schools... *Iowa Const. art. IX, sec. 3.*

Kansas The legislature shall provide for intellectual, educational, vocational and scientific improvement by establishing and maintaining public schools, educational institutions and related activities which may be organized and changed in such manner as may be provided by law. *Kan. Const. art. 6, sec. 1.*

Kentucky The General Assembly shall, by appropriate legislation, provide for an efficient system of common schools throughout the State. *Ky. Const. sec. 183.*

Louisiana The legislature shall provide for the education of the people of the state and shall establish and

	maintain a public education system. *La. Const. art. 8, sec. 1.*
Maine	A general diffusion of the advancement of education being essential to the preservation of the rights and liberties of the people; to promote this important object, the Legislature are authorized...to require, the several towns to make suitable provision, at their own expense, for the support and maintenance of public schools. *Me. Const. art. VIII, sec. 8.*
Maryland	The General Assembly, at its first session after the adoption of this Constitution, shall, by law, establish throughout the state a thorough and efficient system of free public schools. *Md. Const. art. VIII, sec. 1.*
Massachusetts	Wisdom, and knowledge, as well as virtue, diffused generally among the body of the people, being necessary for the preservation of their rights and liberties;...it shall be the duty of legislatures...to cherish the interests of literature and the sciences... *Mass. Const. ch. 5, sec. 2.*
Michigan	Religion, morality and knowledge being necessary to good government and the happiness of mankind, schools and the means of education shall forever be encouraged. *Mich. Const. art. VIII, sec. 1.*
Minnesota	The stability of a republican form of government depending mainly upon the intelligence of the people, it is the duty of the legislature to establish a general and uniform system of public schools. The legislature shall make such provisions by taxation or otherwise as will secure a thorough and efficient system of public schools throughout the state. *Minn. Const. art. XIII, sec. 1.*
Mississippi	The legislature may in its discretion provide for the maintenance and establishment of free public schools for all children between the ages of six and twenty-one years. *Miss. Const. art. 8, sec. 201.*

Missouri	A general diffusion of knowledge and intelligence being essential to the preservation of the rights and liberties of the people, the general assembly shall establish and maintain free public schools... *Mo. Const. art. X, sec. 1(3).*
Montana	It is the goal of the people to establish a system of education which will develop the full educational potential of each person. Equality of educational opportunity is guaranteed to each person of the state. *Mont. Const. art. X, sec. 1.*
Nebraska	The Legislature shall provide for the free instruction in the common schools of this state of all persons between the ages of five and twenty-one years. *Neb. Const. art. VII, sec. 1.*
Nevada	The legislature shall provide for a uniform system of common schools... *Nev. Const. art 11, sec. 2.*
New Hampshire	It shall be the duty of legislators and magistrates, in all future periods of this government, to cherish the interest of literature and the sciences. *N.H. Const. art. 83.*
New Jersey	The legislature shall provide for the maintenance and support of a thorough and efficient system of free public schools for the instruction of all the children in the state between the ages of five and eighteen years. *N.J. Const. art. 8, sec. 4.*
New Mexico	A uniform system of free public schools sufficient for the education of, and open to, all the children of school age in the state shall be established and maintained. *N.M. Const. art. XII, sec. 1.*
New York	The legislature shall provide for the maintenance and support of a system of free common schools, wherein all the children of the state may be educated. *N.Y. Const. art. 11, sec. 1.*
N. Carolina	The General Assembly shall provide by taxation and otherwise for a general and uniform system of free public schools...wherein equal opportunities shall be provided for all students. *N.C. Const. art. IX, sec. 2(1).*

N. Dakota	A high degree of intelligence, patriotism, integrity and morality on the part of every voter in a government by the people being necessary in order to insure the continuance of that government and the prosperity and happiness of the people, the legislative assembly shall make provision for the establishment and maintenance of a system of public schools....This legislative requirement shall be irrevocable without the consent of the United States and the people of North Dakota. *N.D. Const. art. VIII, sec. 1.*
Ohio	The general assembly shall make such provisions, by taxation, or otherwise, as, with the income arising from the school trust fund, will secure a thorough and efficient system of common schools throughout the state. *Ohio Const. art. VI, sec. 2.*
Oklahoma	The Legislature shall establish and maintain a system of free public schools wherein all children of the state may be educated. *Okla. Const. art. 13, sec. 1.*
Oregon	The Legislative Assembly shall provide by law for the establishment of a uniform and general system of common schools. *Or. Const. art. VIII, sec. 3.*
Pennsylvania	The General Assembly shall provide for the maintenance of a thorough and efficient system of public education to serve the needs of the Commonwealth. *Pa. Const. art. 3, sec. 14.*
Rhode Island	It shall be the duty of the general assembly to promote public schools and to adopt all means which they may deem necessary and proper to secure to the people the advantages and opportunities of education. *R.I. Const. art. 13, sec. 1.*
S. Carolina	The General Assembly shall provide for the maintenance and support of a system of free public schools open to all children of the state... *S.C. Const. art. XI, sec. 3.*
S. Dakota	The stability of a republican form of government depending on the morality and intelligence of the people, it shall be the duty of the legislature to establish and maintain a general and uniform

	system of public schools wherein tuition shall be without charge, and equally open to all... *S.D. Const. art. VIII, sec. 1.*
Tennessee	The General Assembly shall provide for the maintenance, support and eligibility standards of a system of free public schools. *Tenn. Const. art. 11, sec. 12.*
Texas	A general diffusion of knowledge being essential to the preservation of the liberties and rights of the people, it shall be the duty of the legislature of the state to establish and make suitable provision for the support and maintenance of an efficient system of public free schools. *Tex. Const. art. VIII, sec. 1.*
Utah	The Legislature shall provide for the establishment and maintenance of a uniform system of public schools... *Utah Const. art. X, sec. 1.*
Vermont	Laws for the encouragement of virtue and prevention of vice and immorality ought to be constantly kept in force, and duly executed; and a competent number of schools ought to be maintained in each town unless the general assembly permits other provisions for the convenient instruction of youth. *Vt. Const. ch. II, sec. 64.*
Virginia	The General Assembly shall provide for a system of free public elementary and secondary schools for all children throughout the Commonwealth and shall seek to ensure that an educational program of high quality is established and continually maintained. *Va. Const. art. VIII, sec. 1.*
Washington	The legislature shall provide for a general and uniform system of public schools. *Wash. Const. art. IX, sec. 2*; and It is the paramount duty of the state to make ample provision for the education of all children residing within its borders... *Wash. Const. art. IX, sec. 1.*
W. Virginia	The legislature shall provide, by general law, for a thorough and efficient system of free schools. *W. Va. Const. art. XII, sec. 1.*

Wisconsin	The legislature shall provide by law for the establishment of district schools, which shall be as nearly uniform as practicable... *Wis. Const. art. 10, sec. 3.*
Wyoming	The legislature shall provide for the establishment and maintenance of a complete and uniform system of public instruction... *Wyo. Const. art. 7, sec. 1.*

Data in tables 6.1 and 6.2 argue for several conclusions. First, it is clear that federal litigation is a minimalist exercise in futility. Unless the improbability occurs that the Supreme Court will reverse its traditional posture on fundamentality, wealth suspectness, or Fourteenth Amendment equal protection short of absolute deprivation of educational services, such litigation will be largely unfruitful. Wise reformers will simply have to wait patiently for changes in the Court itself, or engage in oblique attacks in other arenas where equal protection and suspect classes are already available, such as race and educational deprivation.[162]

Second, these data show that litigation in state courts will continue and even escalate into the foreseeable future. Plaintiffs' only success has been at this level, and the growing needs of children virtually guarantee that resources will be increasingly inadequate to meet their needs.[163] Third, expectations for litigation in state courts should be tempered with realism and defined with less zeal, as few states have the language needed to confidently predict application of strict scrutiny and even fewer courts will move to aggressively adopt or invent new standards that infringe on legislative prerogative. Fourth, a requirement of fully equal revenues or expenditures is not generally likely to become a common standard.

This posture is highly evident in compliance litigation, as in *Serrano III*[164] when the court was satisfied by a $100 range in per pupil variation. This was also echoed in *Horton II,*[165] as the court required only that disparities not be so great as to be unconstitutional. This same posture has appeared frequently as courts have heard testimony arguing a lack of basic wisdom in overscrutinizing equality to the point of ignoring larger problems or even blaming the wrong object for a state's economic and educational woes.[166]

A fifth conclusion observes that the absence of fundamentality will continue to drive litigation that turns on scrutinizing the education clause of individual state constitutions. *Robinson* proved that fundamentality and suspectness are not absolute prerequisites to plaintiff causes, and most winning litigation stands as further proof. A sixth observation notes that plaintiffs will continue to have only equal chances of winning a lawsuit, if for no other reason than the fact that different constitutions say different things. A confounding factor is clearly that the language in many state education articles is nearly empty, leaving the court to its own ability and inclination to earnestly determine intent of the framers of the constitution. Seventh, it may be expected that courts will continue to be reluctant to provide specific judicial prescription as a remedy, as appeal often rejects such activism and as courts are generally respectful of the separation of powers.

The powerful history of legislative inaction and resistance in the face of court mandates also provides an eighth conclusion, arguing that plaintiffs will not necessarily achieve their goals through litigation since a favorable decision does not guarantee either immediate or receptive legislative response. For example, response to *Edgewood* was a call for a constitutional amendment nullifying the court's decision, and reform in New Jersey is still incomplete more than twenty anguished years later. Current decisions are expected to take years to be carried out, as in Arizona, where no action is expected until at least three years beyond the supreme court's ruling.

As a result of court struggles to chart the way through difficult issues, and as a consequence of the enormous sums involved and the political machinations that fall out naturally in distributing state aid to schools, it is likely that reform move slowly for many years into the future. The net sum of a quarter-century of scorched earth litigation has proved only that the outcome of future lawsuits cannot be known. Too many variables can affect an issue that is ultimately value-laden, and the courts themselves seem uncertain of whether to lead a social agenda or to limit themselves to reading the law in historical context.

Good arguments can be constructed to support either position, and it is precisely that issue which polarizes the courts, society in general, and the legislatures. So much controversey pervades the discussion that "normal" behavior has come to be defined as advocacy for social and political agendas on multiple sides, rather than an honest searching for truth and justice apart from political preferences.[167] There has been great progress as a result of litigation—states have assumed greater shares, taxes have been better equalized, and expenditures have been increased. Thus, while equity has far to go, it is certain that less progress toward fairness and justice in school finance would have been made if litigation had not been a factor in school funding.

Although these conditions indicate only that a difficult path lies ahead, neither side should build a silent and defeated camp. Education remains a great and noble pursuit since life's opportunities are in large measure a product of education. Since children are the most evident expression of the future, it is more important than ever to seek mutually satisfactory solutions that are based on accepting political differences and recognizing that fiscal limitations are a legitimate factor in how education is structured within our nation.

Endnotes

[1] For an indepth treatment of the whole concept of how society has historically attached various values to education, see chapter two in D. C. Thompson, R. C. Wood, and D. S. Honeyman, *Fiscal Leadership for Schools: Concepts and Practices* (New York: Longman, 1994).

[2] For a discussion of adequacy and equity see, R.C. Wood, "Adequacy Issues in Recent Education Finance Litigation," *Developments in School Finance*, ed. W. J. Fowler, (Washington DC: National Center for Education Statistics, 1994) 27-38.

[3] This section borrows from chapter five in D. C. Thompson, R. C. Wood, and D. S. Honeyman, *Fiscal Leadership for Schools:* and from R. C. Wood and D. C. Thompson, *Educational Finance Law: Constitutional Challenges to State Aid Plans—An Analysis of Strategy*, (Topeka: National Organization on Legal Problems of Education, 1993).

[4] *Brown v Board of Educ. of Topeka,* 374 U.S. 483 (1954).

[5] For treatment of the concept of equality in the law with broad implications including school finance, see generally R. Salomone, *Equal Education Under Law* (New York: St. Martin's Press, 1986).

[6] An argument supporting a contrary view is available which views courts as conservative or activist in a "legislation from the bench" concept. This observation has found some courts more willing to be disposed to the idea that a court should do more than scrutinize the law for technical correctness. By this logic, some courts have become involved in the social impact of finance laws, while others have refused to impose remedies. These dispositions will be seen in text, particularly in differentiating the federal posture from that of state courts, and also in contrasting state courts against each other. The intent here, however, is to point out the role traditionally assigned to courts.

[7] These arguments generally, and sections in part, of this discussion have been developed elsewhere by the authors. Specifically see D. C. Thompson, R. C. Wood, and D. S. Honeyman, *Fiscal Leadership for Schools*; see also R.C. Wood and D. C. Thompson, *Educational Finance Law*; see also D. C. Thompson, "School Finance and the Courts: A Reanalysis of Progress," West's *Education Law Reporter*, 59:4 (1990), 945-961; see also D. C. Thompson, et. al. "Equal Protection Under the Law: Reanalysis and New Directions in School Finance Litigation," in *Spheres of Justice in Education*, ed. D. Verstegen and J. Ward. The Eleventh Annual Yearbook of the American Education Finance Association (New York: Harper Business, 1990).

[8] In Congress, July 4, 1776. A Declaration By the Representatives of the United States of America, In General Congress assembled.

[9] Ten Original Amendments: The Bill of Rights. In force December 15, 1791.

[10] The Thirteenth Amendment to the U.S. Constitution states, "Neither slavery nor involuntary servitude, except as a punishment for crime whereof the party shall have been duly convicted, shall exist within the United States or any place subject to their jurisdiction.

[11] Proposed by the 39th Congress, June 13, 1866 and declared ratified in a proclamation by the Secretary of State, July 28, 1868.

[12] P. Kurland, "Equal Educational Opportunity: The Limits of Constitutional Jurisprudence Undefined," *Univ. of Chicago Law Rev.*, 35 (1968), 583-600.

[13] The history of racial equality is too complex to describe here. In an educational context, it is obvious that *Brown v Board of Educ.*, 374 U.S. 483 (1954) is the most critical. However, this analysis is aided by recognition that a long history of equality preceded *Brown*, which in an educational context included, e.g., *Gong Lum v Rice* (275 U.S. 78 (1927); *Missouri ex. rel. Gaines v Canada*, 305 U.S. 337 (1938); *Freeman v County School Board*, 82 F.Supp. 167 (E.D. Va. 1948); *Butler v Wilemon*, 86 F.Supp. 397 (N.D. Tex. 1949); *Pitts v Board of Trustees*, 84 F.Supp. 975 (E.D. Ark. 1949); *McLaurin v Oklahoma State Regents*, 339 U.S. 637 (1950); *Sweatt v Painter*, 339 U.S. 629 (1950); *Carter v School Board*, 182 F.2d 531 (4th Cir. 1950); and *Davis v County School Board*, 103 F.Supp. 337 (E.D. Va. 1952).

[14] These cases are multiple and meaningful. E.g., see *Baker v Carr*, 369 U.S. 186 (1962) where accident of geography and arbitrary boundaries of governments may not be a basis for discrimination among otherwise equal citizens, in this instance forbidding the requirement that one must pay property taxes in order to vote; see also, *Gray v Sanders,* 372 U.S. 368 (1963); *Wesberry v Sanders*, 276 U.S. 1 (1964); *Reynolds v Sims*, 377 U.S. 533 (1964); *Davis v Mann*, 377 U.S. 678 91964).

[15] These cases are multiple and far-ranging. C.f. e.g., *Shapiro v Thompson,* 394 U.S. 618 (1969), where a law against bringing indigents into a state violates the right to interstate travel; see also *Lubin v Panish,* 415 U.S. 706 (1974) where a filing fee as prerequisite to the right to vote is wealth discrimination; see also *Harper v Virginia Board of Elections*, 383 U.S. 663 (1966) where a state poll tax is wealth discrimination; and *Griffin v Illinois,* 351 U.S. 12 (1956) where denying an indigent defendant a transcript of trial for appeal is wealth discrimination.

[16] See, e.g., *Shapiro v Thompson*, 394 U.S. 618 (1969).

[17] See, e.g., *Skinner v Oklahoma*, 316 U.S. 535 (1942).

[18] See, e.g., *Reynolds v Sims*, 377 U.S. 533 (1964).

[19] See, e.g., *Griffin v Illinois*, 351 U.S. 12 (1956).

[20] *United States v Carolene Products Co.*, 304 U.S. 144 at 153 (1938).

[21] See, e.g., *Loving v Virginia*, 388 U.S. 1 (1967).

[22] See, e.g., *Oyama v California*, 332 U.S. 633 (1948).

[23] See, e.g., *Graham v Richardson*, 403 U.S. 365 371 (1971).

[24] Litigation on this issue is longstanding. E.g., a Massachusetts court in 1850 addressed school segregation under Massachusetts' 1780 equality statute in 59 Mass. (5 Cush.) 198 (1850).

[25] *Brown*.

[26] *Brown*.

[27] *United States v Carolene*, 304 U.S. 144 at 153 (1938) n.4 (dictum).

[28] E.g., *Griffin v Illinois*, 351 U.S. 12 (1956); *Douglas v California*, 372 U.S. 353 (1963); and *Harper v Virginia Board. of Elections*, 383 U.S. 663 (1966). This logic was especially supported as the Court had stated in *Harper*, in ruling against a poll tax, that "[l]ines drawn on the basis of wealth or property, like those of race, are disfavored." (quoting at 668).

[29] *Burruss v Wilkerson*, 310 F.Supp. 572 *aff'd mem.*, 397 U.S. 44 (1970).

[30] *Burruss* at 574.

[31] *Id.*.

[32] *McInnis v Shapiro*, 293 F.Supp. 327 (N.D. Ill. 1968).

[33] *Aff'd sub nom. McInnis v Ogilvie*, 394 U.S. 322 (1969).

[34] *San Antonio Indep. School Dist. v Rodriguez*. 411 U.S. 1 (1973).

[35] 334 F.Supp. 870 (Minn. 1971).

[36] *San Antonio*.

[37] 337 F.Supp. 280 (W.D. Tex. 1971).

[38] *San Antonio* at 27-28.

[39] *San Antonio* at 20-22; noting *Griffin v Illinois*, 351 U.S. 12 (1956) on transcripts; *Douglas v California*, 372 U.S. 353 (1963) on hiring counsel; and others.

[40] It will be seen later that this claim was not entirely novel, as it had been successfully used in *Serrano v Priest,* 487 P.2d 1241 (1971) in California. It had not, however, been tried at the federal level even though the state court in *Serrano* invoked the Fourteenth Amendment. The overlapping federal/state chronology should be noted, as state litigation had already begun following the debacle of *McInnis* and *Burruss*, but prior to *Rodriguez*.

[41] *San Antonio* at 37.

[42] *San Antonio* at 39.

[43] *San Antonio* at 42.

[44] This section is based on Thompson, "School Finance and the Courts," 945-961

[45] Although only four federal cases are discussed here, there were actually five cases, as in 1987 plaintiffs in *Livingston v Louisiana State Board*, 830 F.2d 563 (5th Cir. 1987) claimed equal protection, arguing that the state aid plan was arbitrary and discriminated against districts with large homestead exemptions. This case is noted herein since like *Papasan*, it was narrowly drawn and did not claim inadequate education or a fundamental right. *Livingston* stood, however, as another case of federal escape from the basic issue. This analysis also considers only Supreme Court cases—thus perfect correlation will not occur, e.g., *Scarnato v Parker*, 415, F.Supp. 272 (1976) granting summary judgment and dismissing does not factor into this analysis.

[46] While the cases discussed in text are termed important, it should be recognized that others have attempted to raise federal claims but have been so unsuccessful as to be dismissed. Specifically, see e.g., *Scarnato v Parker,* 415 F.Supp. 272 (La. 1976) where a challenge to Louisiana's minimum foundation formula was dismissed in its entirety due to a federal court's view that the case was foreclosed by *Rodriguez.*

[47] *Papasan v Allain*, 478 U.S. 265 (1986); *Papasan v United States*, 756 F.2d 1087 (5th Cir. 1985).

[48] Amendment XI to the United States Constitution reads "The judicial power of the United States shall not be construed to extend to any suit in law or equity, commenced or prosecuted against one of the United States by citizens of another State, or by citizens or subjects of any foreign state."

[49] *Papasan v Allain*, 106 S.Ct. 2932 (1987).

[50] *Plyler v Doe*, 457 U.S. 202 (1982).

[51] *Plyler* at 221.

[52] *Kadrmas v Dickinson Public Schools,* 487 U.S. 450, 108 S.Ct. 2481 (1988). For a detailed discussion of these issues see, R. C. Wood, "*Kadrmas v. Dickinson Public Schools*: A Further Retreat From Equality of Educational Opportunity," *J. of Educ. Finance,* v. 15, n. 3, Winter, 1990, 429-436.

[53] *Kadrmas,* 108 S.Ct. at 2491.

[54] School finance litigation can be traced to the Nineteenth Century where complex roots of other issues impacting school finance can be understood. E.g., school finance-related litigation can be seen in *Stuart v School District of the Village of Kalamazoo*, 30 Mich. 69 (1874) where a court determined the power to lay and collect taxes for the support of secondary schools; similarly, see *Sawyer v Gilmore*, 109 Me. 169 (1912) in which a court enunciated the *Rodriguez*-like principle of legislative discretion in the manner and amount of tax distributions to schools;

similarly, *Morton Salt Co. v City of South Hutchinson*, 177 F.2d 889 (10th Cir. 1949) in which the court noted that no direct benefit need accrue to taxpayers if the taxes are uniform and for public purposes benefiting the entire public; and *Lewis v Mosley*, 204 So. 2d 197 (Fla. 1967) in which the court ruled that laws providing for taxation must be construed in favor of the taxpayer when questions of court discretion arise. These cases are not discussed since the focus of this discussion is on modern school finance policy.

[55] These arguments again have been developed elsewhere by the authors. Specifically, see Chapter 5 in D. C. Thompson, R. C. Wood, and D. S. Honeyman, *Fiscal Leadership for Schools*; see also R. C. Wood and D. C. Thompson, *Educational Finance Law:*; see also D. C. Thompson, "School Finance and the Courts,"; see also Thompson et.al., "Equal Protection Under the Law."

[56] *Serrano v Priest*, 487 P.2d 1241 (1971).

[57] *Serrano*.

[58] *Serrano* at 1244.

[59] *Serrano* at 1253.

[60] *Serrano* at 1252.

[61] *Serrano* at 1258.

[62] *Milliken v Green,* 203 N.W.2d 457 (Mich. 1972).

[63] *Milliken v Green, vac.,* 212 N.W.2d 711 (Mich. 1973).

[64] *Milliken* at 716.

[65] 110 Ariz. 88, 515 P.2d 590 (Ariz. 1973).

[66] 55 Ill. 2d 94, 302 N.E.2d 46 (Ill. 1973).

[67] Illinois Const., Art. X § 1 (1970).

[68] 530 P.2d 178 (Wash. 1974).

[69] 530 P.2d at 184.

[70] 530 P.2d at 191.

[71] 530 P.2d at 198.

[72] 357 P.2d 635 (Idaho 1975).

[73] 357 P.2d at 653.

[74] *Robinson v Cahill*, 287 A.2d 187 (N.J. Super. 1972), *aff'd as mod.*, 303 A.2d 273 (N.J. 1973).

[75] *San Antonio Indep. School Dist. v Rodriguez*, 93 S.Ct. at 1302 (1973).

[76] *Robinson v Cahill*, 303 A.2d 273 at 284 (N.J. 1973).

[77] *Robinson* at 283.

[78] 554 P.2d 139 (Or. 1976).

[79] 554 P.2d at 149.

[80] 547 P.2d 609 (Kan. 1976).

[81] 408 N.Y.S.2d 606 (N.Y. 1978); 529 N.E.2d 359 (1982), *appeal dismissed* 459 U.S. 1138, 103 S. Ct. 775 (1983).

[82] 376 A.2d 359 (Conn. 1977).

[83] 376 A.2d at 373-374.

[84] *Seattle School Dist. v State* 585 P.2d 71 (Wash. 1978).

[85] 255 S.E.2d 859 (West Va. 1979).

[86] 399 A.2d 360 (Pa. 1979).

[87] 399 A.2d at 366.

[88] 399 A.2d at 365-367.

[89] 649 P.2d 1005 (Colo. 1982).

[90] *Board of Educ. v Walter* 390 N.E.2d 813 (Ohio 1979).

[91] 606 P.2d 310 (1980).

[92] 606 P.2d at 344.

[93] 606 P.2d at 335.

[94] 651 S.W.2d 90 (Ark. 1983).

[95] 458 A.2d 758 (Md. 1983).

[96] 285 S.E.2d 156 (Ga. 1981).

[97] 348 N.W.2d 303 (Mich. 1984).

[98] This section is based on R.C. Wood and D.C. Thompson, *Educational Finance Law.*

[99] *Britt v North Carolina State Board of Educ.* 357 S.E.2d 432 (N.C. Ct. App, 1987).

[100] 364 S.E.2d 470 (S.C. 1988).

[101] 761 S.W.2d 859 (Tex. 1988).

[102] 436 N.W.2d 568 (Wis. 1989).

[103] 790 S.W.2d 186 (Ky. 1989).

[104] 575 A.2d 359 (N.J. 1990).

[105] *Abbott* at 408.

[106] 811 P.2d 116 (Or. 1991).

[107] *Harper v Hunt*, 624 So. 2d 107 (Alabama 1994). It should be noted that only representative state supreme court cases are discussed herein due to space restrictions. Other state supreme court cases have also emerged that have relative merit in the litigation scheme: e.g., *Idaho Schools For Equal Educ. Opportunity v. Evans,* 123 Idaho 573, 850 P.2d 724, (1993); *McDuffy v Secretary,* 415 Mass 545, 615 N.E.2d 516, Supreme Court of Massachusetts (1993); *Campaign for Fiscal Equity v. State,* 616 N.Y.S.2d 851 (Supp. 1994).

[108] *Roosevelt v Bishop*, 179 Ariz 233, 877 P.2d 806, (Ariz. 1994).

[109] *Skeen v State* 505 N.W.2d 299 (Minn. 1993).

[110] *Skeen* 505 N.W.2d at 315.

[111] *Tennessee Small School Systems v McWherter* 851 S.W.3d 139 (Tenn. 1993).

[112] *Bismarck Pub. Sch. Dist. v State*, 511 N.W.2d 247 (N.D. 1994).

[113] *Unified School District v State of Kansas,* __ P.2d __, 1994 WL 671552 KS.

[114] *Unified School District* at 3.

[115] Some discussion here is adapted from chapter five in D. C. Thompson, R. C. Wood, and D. S. Honeyman, *Fiscal Leadership for Schools.*

[116] See, in general, R.C. Wood, *School Finance Litigation in America,* General Session Presentation, National Organization on Legal Problems of Education, Annual Meeting, Phoenix, November, 1992.

[117] *Dupree v Alma School District*, 651 S.W.2d 90 (1983).

[118] *Roosevelt v Bishop.*

[119] *Serrano*, 487 P.2d 1241 (1971) ('*Serrano I*'); 557 P.2d 929 (1976) ('*Serrano II*').

[120] *Horton v Meskill*, 376 A.2d 359 (1977) ('*Horton I*'); 486 A.2d 1099 (1985) ('*Horton II*').

[121] *Rose v The Council for Better Education*, 790 S.W. 186 (Ky. 1989).

[122] *Helena School District v Montana*, 769 P.2d 684 (Mont. 1989).

[123] *Robinson,* 303 A.2d 273 (1973); *Abbott v Burke*, 119 N.J. 287 (1990).

[124] *Edgewood v Kirby*, 777 S.W.2d 391 (Tex. 1989).

[125] *Seattle v Washington*, 585 P.2d 71 (Wash. 1978).

[126] *Pauley v Kelly*, 255 S.E.2d 859 (W. Va. 1979).

[127] *Washakie County School Dist. v Herschler*, 606 P.2d 310 (Wyo. 1980).

[128] *Shofstall v Hollins*, 515 P.2d 590 (1973).

[129] *Lujan v State Board of Educ.*, 649 P.2d 1005 (Col. 1982).

[130] *Thomas v McDaniels*, 285 S.E.2d 156 (Ga. 1981).

[131] *Thompson v Engleking*, 537 P.2d 635 (Idaho 1975).

[132] *Knowles v State Board of Educ.,* 547 P.2d 609 (Kan. 1976) and *Unified School Dist. v State of Kansas.*

[133] *Sawyer v Gilmore*, 83 A. 673 (Me. 1912).

[134] *Hornbeck v Somerset County*, 458 A.2d 758 (Md. 1983).

[135] *Milliken v Green*, 212 N.W.2d 711 (Mich. 1973); *East Jackson v State*, 348 N.W.2d 303 (1984), *ev. den.*, 419 Mich. 943 (1984).

[136] *Skeen v State of Minnesota* , 505 N.W.2d 299 (Minn. 1993).

[137] *State ex rel. Woodahl v Straub*, 520 P.2d 776 (Mont. 1974).

[138] Citation omitted (Neb. 1994).

[139] *Board of Educ. v Nyquist*, 57 N.Y.2d 27 (1982), *app. dismissed*, 459 U.S. 1138; *see also Spano v Board of Educ.*, 328 N.Y.S.2d 229 (N.Y. 1972); *also Campaign For Fiscal Equity v. State..*

[140] *Britt v State Board of Educ,* 357 S.E.2d 432, *app. dismissed* 361 S.E.2d 71 (1987).

[141] *Bismarck Pub. Sch. Dist. v State*, 511 N.W.2d 247 (N.D. 1994).

142 *Board of Educ. v Walter*, 390 N.E.2d 813 (1979), *cert. den.*, 444 U.S. 1015 (1980).

[143] *Fair School Finance Council v State*, 746 P.2d 1135 (Okla. 1987).

[144] *Olsen v Oregon*, 554 P.2d 139 (Ore. 1979); *Coalition for Educ. Equity v Oregon*, 811 P.2d 116 (Ore. 1991).

[145] *Dansen v Casey*, 399 A.2d 360 (1979).

[146] *Richland County v Campbell*, 364 S.E.2d 470 (S.C. 1988).

[147] *Tennessee Small School Systems v McWherter* 851 S.W.3d 139 (Tenn. 1993).

[148] *Northshore v Kinnear,* 530 P.2d 178 (Wash. 1974).

[149] *Kukor v Grover*, 436 N.W.2d 568 (Wisc. 1989).

[150] *Scott v. Commonwealth.*

[151] *Bezdichek v. State*

[152] *Hargrave v Kirk,* 313 F.Supp. 994 (M.D. Fla. 1970), *vac. and remanded, sub. nom., Askew v Hargrave,* 401 U.S. 476 (1971).

[153] *McInnis v Shapiro*, 293 F.Supp. 327 (N.D. Ill. 1968). *Aff'd sub nom. McInnis v Ogilvie*, 394 U.S. 322 (1969).

[154] *Scarnato v Parker,* 415 F.Supp. 272 (M.D. La. 1976).

[155] *Van Dusartz v Hatfield,* 334. F.Supp. 870 (Minn. 1971).

[156] *Papasan v Allain*, 478 U.S. 265 (1986); *Papasan v United States*, 756 F.2d 1087 (5th Cir. 1985).

[157] *Kadrmas v Dickinson Public Schools.*

[158] *San Antonio Indep. School Dist. v Rodriguez.* 411 U.S. 1 (1973).

[159] *Burruss v Wilkerson*, 310 F.Supp. 572 *affirmed mem.*, 397 U.S. 44 (1970).

[160] *Edgewood v Kirby*, 777 S.W.2d 391 at 398 (1989).

[161] For further development on the importance of language in state constitutions, see e.g. M. McUsic, "The Use of Education Clauses in School Finance Reform Litigation," *28 Harvard J. on Legis.* 307 (1992).

[162] E.g., the authors make the case elsewhere that desegregation is an oblique form of school finance litigation, for which an entire history of litigation stands evident. Similarly, other cases present opportunities for plaintiffs to approach fiscal issues, such as disadvantaged education: e.g., a current case in the state of Washington in

which plaintiffs argue underfunding for bilingual programs in *Sang Van v Seattle School District v State of Washington,* King County Cause No. 92-2-28350-6.

[163] See D. C. Thompson, "Special Needs Students: A Generation at Risk" Chapter 5 in *Who Pays for Student Diversity? Population Changes and Educational Policy*, J.G. Ward (ed.). Yearbook of the American Education Finance Association (1991) 97-124; See also R. C. Wood and D. S. Honeyman, "Rapid Growth and Unfulfilled Expectations: Problems for School Finance in Florida," Chapter 8 in *Who Pays for Student Diversity? Population Changes and Educational Policy*, J.G. Ward (ed.). Yearbook of the American Education Finance Association (1991) 160-179; see generally this yearbook.

[164] *Serrano III*, 557 P.2d 929 (1976).

[165] *Horton II*, 486 A.2d 1099 (1985).

[166] See, e.g., D. C. Thompson and R. C. Wood, *Findings of Fact and Opinion on the Equity and Fiscal Neutrality of South Dakota's State Aid Formula to Public Schools: Expert Analysis on Behalf of the State in Bezdichek v State of South Dakota.* (Manhattan, KS Wood, Thompson & Associates, Inc. 1993). See also R.C. Wood and D.C. Thompson, *The Study of Resource Accessibility, Wealth Neutrality, and State Yield in Montana Rural Education Association v State: Expert Witnesses' Analysis for Defendants.* (Gainesville, Fl: Wood, Thompson & Associates, Inc. 1992); See also, R.C. Wood, D.C. Thompson; M.D. Miller, *Funding Public Education in Montana Based on the Concept of Cost of Living Indices in Montana Rural Education Association v State. Expert Witnesses' Analysis for Defendants.* (Gainesville, Fl: Wood, Thompson & Associates, Inc. 1992); See also R.C. Wood, D.C. Thompson, and M.D. Miller, *Findings of Fact and Expert Opinion on the Equity of Pennsylvania's State Aid Formula to Public Schools: Expert Analysis on Behalf of Defendant Intervenors in PARRS v State of Pennsylvania.* (Gainesville, Fl: Wood, Thompson & Associates, Inc. 1995).

[167] R.C. Wood and J. Maiden, "An Analysis of Resource Allocations Issues within Education Finance Litigation," in *Where Does the Money Go? Resource Allocation in Elementary and Secondary Schools*, Sixteenth Annual Yearbook of the American Education Finance Association, ed. L. Picus, (Newbury Park, CA: Corwin Press, 1995).

Chapter 7
Legal Liability

Introduction

Throughout the first six chapters of this text, there has been a great intentional focus on critical issues affecting school administrators. While the chapters have addressed different topics, an underlying theme has been the weighty responsibility for ethical behavior and wise professional decision making by those people entrusted with the nation's greatest single enterprise known as the education of children. This chapter continues the theme of tremendous responsibility on the part of administrators and other people actively connected to schools by considering legal liability, a topic that concerns most reasonable people by its scope and gravity.

Although some critics have very correctly bemoaned the increasing interest of courts in public schools and have castigated a litigious modern society that is frequently overzealous in sensitivity to both real and imagined personal affronts, the courts have made important contributions to civilized society and to the overall preservation and furtherance of human rights that would not have been available if courts were not intimately involved in education's operations. Yet, while the contributions through establishment and interpretation of the foundations of law have been great, the liability of school personnel has been at least commensurately increased, underscoring the importance of legally informed and thoughtful administrators and other school personnel. Thus, school professionals face a host of broadranging legal liabilities for their actions. Given such reality, this chapter is meant to introduce, review, and update general knowledge of legal liability for both prospective and practicing school business administrators. Although not exhaustive of the many legal concerns districts face, this chapter discusses many common issues that form a foundation for informed and careful behavior on the part of school personnel by granting proper respect for a very complex set of laws about equally complex human interactions.

Historical Perspectives

The relationship between the law and American public schools has been an old one indeed, stemming from long before the birth of this nation. Although every country's laws are based in its own philosophical and political traditions, these realities are in fact grounded in historical antecedents that led each nation into existence for its own peculiar reasons. In the United States, law is based on the English tradition, primarily due to the formative influence of England on the establishment and character of the original American colonies.

The lengthy historical development of education systems in the United States is chronicled in many sources.[1] While that history is too vast to develop here, it is important to note that there has long been a sustained struggle over how education should occur and who should be charged with its responsibility. Such questions give immediate rise to conflict between differing viewpoints, particularly whenever laws are enacted that strongly contravene the views of those who argued for a perspective apart from the prevailing view. Under these conditions, it can be argued that a reasonable pinpoint in history for conflict over education in the United States began during the colonial era, when the Massachusetts legislature enacted the Ye Old Deluder Satan Law in 1647 for the purpose of teaching citizens to read the Holy Scriptures in order to ward off evil influences. The law itself referenced the concept of a court, as it stated "[I]t is therefore ordered by this court and authority thereof, that every township in this jurisdiction, after the Lord hath increased them to the number of fifty householders, shall then forthwith appoint one within their town to teach all such children as shall resort to him to write and read, whose wages shall be paid either by the parents or masters of such children, or by the inhabitants in general..."[2]

While the concept of a court relationship to education was far different in 1647 than it is today, school laws were promulgated in the colonies in rapid succession, as New Hampshire seized on the benefit of education laws in 1680 and with Connecticut following in 1700. The important point, however, is the concept of a social responsibility for education, together with some requirement for creating and maintaining schools for a set of community purposes.

The idea of compelling any aspect of human behavior, however, can certainly promote conflict between citizens who quickly resort to legal battles to protect and further their interests. While perhaps most colonists agreed with formalized education, opportunities for strife were multiplied in at least equal proportion to the number of laws being enacted as state legislatures increasingly passed school laws throughout the first 200 years after founding of the colonies. By the middle of the Nineteenth Century, the seeds of dissent were deeply rooted and destined to grow, as in 1852 when Massachusetts enacted the nation's first compulsory education law. The many legal struggles from the broader concept of constitutional equality discussed in the last chapter were also carried forward into education, giving rise to a host of issues that would have a permanent impact on schools.[3]

While these cases are clearly too numerous to trace in detail, it is important to note that legal disputes concerning education have been so long-standing as to cause litigation to be seen as a "normal" means by which to resolve political and philosophical differences and causing the courts to repeatedly examine and reaffirm the benefits of education to individuals and society.[4] A critically important aspect of this struggle that holds since education is so vital to the interests of the nation and to individual welfare, the state can in some cases have a compelling claim on compliance that must be weighed against the individual's right to liberty.[5] Similarly, development of state constitutional provisions regarding education has provided fruitful cause for legal dispute, as citizens have complained vigorously about a broad spectrum of grievances represented by fierce battles over student and parental rights,[6] the right to appropriate education,[7] and the fiscal resource distribution struggles witnessed in the previous chapter under school finance litigation—all of which serve to underscore the astute observation that education is all about class, wealth, and power.[8]

Struggles over American education have been a philosophical and developmental groping through a morass of difficult moral and ethical questions. These struggles affect a wide variety of people because of the simultaneous implication of wrongful acts, or alternatively, failure to act correlatively on the part of those responsible for education. The sweeping nature of education as structured in the United States has also made for many people's justifiable concern for how to fulfill their duty to education. Because legal disputes have embraced a broad spectrum of charges ranging from constitutional duty to actionable issues of physical or other harm, responsible parties have had to worry about whether their duty has been met or whether some liability might emerge from an allegation. Most affected have been those with first-line responsibility for education, capturing states themselves as legal entities, state legislatures, school boards, district and building administrators, and instructional and support personnel in the nearly 16,000 school districts in the United States. While a significant body of litigation surrounds these individuals, the liability question must first consider whether there is in fact a legal basis for liability in schools.

Doctrine of Sovereign Immunity

Although complete and total sovereign immunity exists in a few instances, the doctrine is still powerful and instructive. Of all the concepts inherited from English law, none has been more important than sovereign immunity of government from suits seeking to establish liability of the state. Descended from medieval law, sovereign immunity held that, "the king can do no wrong" and denied any recourse for offensive sovereign acts against the people. This rule prevailed for centuries in England and found its way into American, appearing as early as 1812 in a decision that held that government was immune from liability. This law came from the fact that courts feared a proliferation of such suits, the lacked a treasury from which judgment could be paid, and had no legislation providing for such liability.[9] Such logic was clearly appealing to government and was incorporated into United States Supreme Court decisions, with the Court noting as early as 1869 that "every government has an inherent right to protect itself against suits . . .." and that "the principle is fundamental and applies to every sovereign power."[10]

Although immunity from tort liability was initially grounded in the concept of sovereignty, the American basis for denial of governmental liability has generally followed the reasoning of the *Nichols* decision in 1869, with additional arguments added over time. Some courts historically have held that the law provides no funds for payment of damages in liability suits. Another basis has argued that school funds have specific public benefit purposes, of which payment for judgments was never contemplated nor specifically authorized and may not be considered to benefit the public. Still another argument has been that an act violating the law cannot be taken to properly represent the district, from whose funds a judgment would have to be paid. A further basis has been that committing an illegal act is never within the authority of an individual or a legal entity so that the wrongful act exceeds any statutory powers. Yet another argument has been that government and its employees do not function as in a master and servant relationship, so that *respondeat superior* is not applicable.[11] Historically, these reasonings have been so compelling that courts have been reluctant to abandon this position except when there is specific cause or order to do so.

The strength of sovereign immunity for governmental liability was so strong that it continued as a virtually unassailable precept of law in the United States until the middle of the twentieth century when recognition of legitimate claims against government was first contemplated. The first steps toward abrogation of sovereign immunity were taken in 1946 as the federal government enacted the Federal Tort Claims Act[12] in recognition that some claims against the state should be validated. The Act generally abrogated federal immunity from tort liability and established the conditions for lawsuits and claims against the federal government, although it continued to preserve immunity in the categories of intentional torts and acts of omission falling within the discretionary functions or duties of federal agencies and employees.

At the state level, sovereign immunity was the universal rule, although some states began to adopt a federal-like posture following after the new Federal Tort Claims Act. A significant early case for abrogating sovereign immunity occurred in *Molitor*[13] as the state supreme court of Illinois traced the origins of immunity into the United States[14] and its subsequent adoption into Illinois state law via two cases known as *Waltham*[15] and *Kinnare*[16]. Immunity was applied to Illinois for towns and counties in *Waltham* and extended to school districts in *Kinnare* wherein it was held that a school board was not liable for the death of a laborer who fell from a rooftop, even though it was alleged that the school district had failed to provide proper safeguards against accident such as scaffolding.

Abrogation of this position was first ruled in *Molitor* in 1959 as the court held against the traditional immunity arguments in a case involving injury to a student when a school bus hit a culvert and exploded, allegedly due to the driver's negligence. The court examined each defense argument, rejecting the view that sovereignty was a justifiable basis, that diversion of funds to pay judgments was invalid as an argument against liability, and that depletion of the treasury was an unlikely event in the modern era. The court was sharp in its criticism of these defenses, citing other cases from other state jurisdictions to illustrate its findings. The Illinois court's posture was perhaps best summed up by its verbatim adoption of a New Mexico court's wording, as it noted "The whole doctrine of governmental immunity from liability for tort rests upon a rotten foundation. It is almost incredible that in this modern age of comparative sociological enlightenment, and in a republic, the medieval absolutism supposed to be implicit in the maxim, 'the King can do no wrong,' should exempt the various branches of government from liability for their torts . . . ."[17]

Exceptions to Sovereign Immunity

The unconscionability of absolute immunity of government for tortious acts has led to a variety of controlling state statutes for the concept. These controlling state statutes have generally been established in one of two important ways. The first way has been by enactment of state statutes abrogating sovereign immunity, while the second way has come about by doctrinal modifications by judicial requirement. In either instance, a rulemaking body has been dissatisfied with the result of absolutism in this arena. In the instance of judicial modifications, partial or complete abrogation has been by virtue of three particular issues relating to proprietary acts, nuisances, and Eleventh Amendment immunity.

Proprietary Acts

Some courts' have expressed unwillingness to accept absolute governmental immunity, believing instead that neither extreme of abrogation or strict adoption is the proper posture in modern society. Instead, these courts have examined the events that transpired, attempting to determine whether the activity in question was a governmental one or whether it could be classified as a proprietary act. A governmental function is defined as "an act in exercise of police power or in exercise of constitutional, legislative, administrative, or judicial powers conferred upon federal, state or local government for benefit of public . . ." and further, "the functions of a municipality which are essential to its existence, in the sense of serving public at large, and are to be distinguished from those which are private, which are not necessary to its existence, and which endure to advantage of its inhabitants."[18] In contrast, a proprietary function is one in which "acts [are] done . . . for general betterment and improvement of such . . .. The 'governmental functions' of a municipal corporation are those functions exercised as arm of state, and for public good generally, whereas 'proprietary functions' are those exercised for peculiar benefit and advantage of citizens."[19]

The difference in immunity by these definitional criteria rests in modification of viewing the concept of immunity, in that whether liability may exist or not is related to what role government was playing at the time the event transpired. Under this distinction, if it can be established that the act was part of the normal course of events in which government can be expected to engage, then it may be held, under controlling state statutes, that no liability exists. In contrast, a finding that the act was proprietary by such controlling state statutes could result in abrogation of immunity.

For education, the distinction between proprietary and governmental acts generally falls into a gray area separating the educational functions of schools from the host of other school sponsored events that are at times seen as peripheral to the central mission of schools. This gray area has been a contentious source of debate, with courts ruling on both sides of the issue. A prime example rests in athletic contests wherein these events exceed the normal limits of a school day, are offered on a voluntary participation basis and for which admission is often charged. If someone were injured, the liability question arises. Courts in some instances have held that such contests are governmental ones, while in other instances proprietary control is found so that liability is imposed.

An oft-cited case holding for liability of a school district is found in *Sawaya*[20] where two districts had rented a football stadium from a third district. Due to a bleacher railing defect, a spectator was injured after falling from a considerable height. The question before the court rested in whether the district that had rented out its stadium forsook its governmental role by entering into a proprietary act. The state supreme court of Arizona, in finding for plaintiff, held that the act exceeded the normal scope of governmental responsibility.

In contrast, however, a case only two years later in Michigan reached an opposite conclusion based on a differentiation of facts. The state supreme court of Michigan held in *Richards*[21] that no liability was present in a case where the facts were nearly the same. In that instance, the court distinguished the fact that the lessor district was one of the two teams playing in the contest, and further noted that the district did not intend to make a profit and was merely providing an educational activity of its own.[22] While courts have differed in language regarding whether an act is proprietary or governmental, or whether it is discretionary or ministerial,[23] the result has been the same with courts coming down on both sides of the issue.

The issues involved in determining whether districts have liability for such instances do not seem to rest on whether the immunity concept might be argued to have diminished over time, although it is clearly true that unquestioned sovereign immunity is now generally rejected in most instances.[24] Rather, whether there is liability is more contingent on how courts approach the facts of a case and the extent to which specific state laws control immunity. For example, historically, many courts have made findings that uphold the state, as in *Davis*[25] where a Michigan court found no liability resulting from a field trip on which a student was bitten by a farm animal. Likewise, no liability was found in *McManus*[26] when a student was burned by an open fire at a homecoming event. In similar fashion, no liability was present in *Hutt*[27] when a student was alleged to have been injured by unsafe physical education facilities.

The difference in whether liability exists or not is controlled in the federal instance by the Tort Claims Act of 1946, while in states it is dependent on the controlling state statutes. The chasm is illustrated by two courts reaching two different conclusions, as in Michigan and Texas. The Michigan court stated "when a governmental agency engages in mandated or authorized activities, it is immune from tort liability unless the activity is proprietary in nature."[28] In contrast, the Texas court said "since a school district is purely a governmental agency . . . it performs no proprietary functions that are separate from governmental functions."[29]

Nuisances

If some abrogation of immunity has been available by distinguishing between nuances of the nature of an act committed by government, a second and more obvious area where immunity often has been limited relates to the concept of a nuisance. Numerous excellent definitions of a nuisance have been developed by courts, several of which takes into account when defining various facets of nuisance:

> Nuisances are commonly classed as public, private, and mixed. A public nuisance is one which affects an indefinite number of persons, or all the residents of a particular locality, or all people coming within the extent of its range or operation, although the extent of the annoyance or damage inflicted upon individuals may be unequal. Maintaining a public nuisance is by act, or by failure to perform a legal duty, intentionally causing or permitting a condition to exist which injures or endangers the public health, safety or welfare....As distinguished from public nuisance, a private nuisance includes any wrongful act which destroys or deteriorates the property of an individual or interferes with their lawful use or enjoyment thereof, or any act which unlawfully hinders them in the enjoyment of a common or public right and causes them a special injury different from that sustained by the general public....It is evident that the same thing or act may constitute a public nuisance and at the same time a private nuisance. A mixed nuisance is of the last kind described."[30]

While defining a nuisance obviously takes many offensive acts into account, the area in which schools tend to be most concerned relates to liability for nuisances, which may result in actual physical injury to someone. In these instances, the claim by plaintiff invariably charges that the nuisance was foreseeable and that there was no attempt or insufficient effort to abate the nuisance. An often quoted longstanding case occurred in Michigan in 1899 in which the plaintiff in *Ferris*[31] claimed that snow had fallen from the roof of a school, damaging his property and causing him to slip on the ice. The court, in holding the district liable for maintaining a nuisance, said " . . . plaintiff had the right to the exclusive use and enjoyment of his property, and the defendant had no . . . right to erect a building in such a manner that the ice and snow would inevitably slide from the roof, and be precipitated upon the plaintiff's premises . . .."[32] Similarly the high court in Connecticut, while not discussing the facts of the case, ruled that a child who was pushed in such a manner as to cause injury from a hazardous condition creates a liability whereupon immunity does not provide protection in a "cause of action founded on a nuisance created by positive act"[33] of government. Such question is particularly important with a type of nuisance known as an attractive nuisance. The attractive nuisance doctrine states that a "person who has an instrumentality, agency, or condition upon his own premises, or who creates such condition upon the premises of another, or in a public place, which may reasonably be apprehended to be a source of danger to children, is under a duty to take such precautions as a reasonably prudent man would take to prevent injury to children of tender years whom he knows to be accustomed to resort there, or who may, by reason of something there which may be expected to attract them, come there to play."[34]

While cases pointing to liability for nuisances indeed exist, there is also case law that goes opposite to such findings. An earlier footnoted case of *Smith*[35] found against the plaintiff, holding that an open stairwell upon which an injury was sustained during a community meeting did not impose liability on the district beyond the duty to refrain from intentional injury. Similarly, another Kansas case in *Jones*[36] found for the defendant district wherein failure to maintain a safe place nonetheless held the district harmless from liability where children throwing wet towels on a slippery floor had caused injury to a student. Still a third case in Kansas found for the defendant in *Sly*,[37] as the court saw no nuisance even though the district had failed to supervise students on a playground whereupon an injury occurred.

More recently, a Michigan court held that the district had not created or maintained a nuisance when it piled snow on a playground, resulting in injury to a child. The court said that "to establish a claim of intentional nuisance against a governmental agency, a plaintiff must show that there is a condition which is a nuisance and that the agency intended to create that condition" and that "[w]e do not believe that the Legislature intended its immunity exception to include the temporary condition of snow piles."[38] In this case, as in many others, the issue of intent to create a nuisance played an important role.

While nuisance liability is somewhat unpredictable by depending on specific laws and specific jurisdictions and the unique facts of each case, the concept of nuisance represents one of the several instances in which at least partial abrogation of sovereign immunity has been provided in the modern context. Clearly the opportunity for liability is greater when considering the standard of care relating to the age of injured persons, and it is also clear that reasonable efforts to provide safeguards for such items as attractive nuisances should be utilized as a wise legal and moral investment on the part of school districts and administrators.

Eleventh Amendment Immunity

A special case regarding immunity relates to the third area in which the doctrine of sovereign immunity has been at least partially abrogated. The Eleventh Amendment was adopted into the United States Constitution in 1795 and provided that "[t]he judicial power of the United States shall not be construed to extend to any suit in law or equity, commenced or prosecuted against one of the United States by citizens of another State, or by citizens or subjects of any foreign state."[39] The Eleventh Amendment had been enacted following after two lawsuits in 1791 and 1792, in which claims against government for recovery were brought in the states of Maryland and Georgia. In the case of *Vanstophorst,*[40] foreign national residents had brought suit against the state of Maryland seeking to recover bad debts. In the other case, plaintiff residents in South Carolina were seeking a favorable judgment against the state of Georgia in *Chisholm.*[41] In both instances, reaction was swift as the uncontemplated possibility of federal court liability for claims by citizens and foreigners provided the impetus for passage of the Eleventh Amendment so as to deny judicial recourse to claims through federal courts under applicable constitutional protections.

The effect of the Eleventh Amendment was to institutionalize the concept of sovereign immunity by intentionally closing a potential window of opportunity for litigants that had been up until this time characterized by considerable vagueness. This foundation for claiming constitutional immunity was successful for a number of years, as the U.S. Supreme Court ruled on several occasions that the intent of the amendment was express, including barring the right of suit against a state by citizens, and including extension of the provisions of the Eleventh Amendment to apply to citizens attempting to bring suit in federal court against their own state.[42] School districts have also enjoyed Eleventh Amendment protection to a certain extent, provided that it could be shown that the state itself was in fact implicated in the lawsuit to such extent that liability would extend to the state. In some instances, this protection has been available under the test of whether the state treasury would be invaded to pay any adverse judgment.[43]

For the most part, however, Eleventh Amendment immunity has been controlled by four possible interventions that serve to limit its absolute application. The first intervention references action on the part of states in voluntarily waiving Eleventh Amendment immunity. The Supreme Court noted in *Atascadero*[44] that a state may choose to waive immunity, in which case suit in federal court is in fact permissible. This intervention limits such specific abrogation to tight interpretation, however, as courts have been cautious in granting such an opening by waiver, ruling in *Edelman*[45] that availability of immunity waiver will be sustained only when the state unquestionably permits and sets forth clearly its intent to allow suit against itself.

The second intervention relates to specific provision by the federal government when writing laws, as in the case of provisions in the Individuals with Disabilities Education Act[46] (IDEA) wherein Congress intentionally made specific provision to subject the act to Eleventh Amendment waiver, stating that states are not immune from federal suits for violations of IDEA and that all rights and remedies of equity usually available under law are specifically intended to apply to IDEA.[47]

The third intervention relates to precedent in observing specific laws, as in the instance of the U.S. Supreme Court's ruling that Eleventh Amendment immunity is unavailable under questions invoking Fourteenth Amendment equal protection provisions, in that Congress has granted itself power to write legislation invoking penalties for successful equal protection claims.

The fourth intervention is particularly important for schools, as it relates to how courts see government agencies in the relationships to the state. Under this scenario, Eleventh Amendment immunity availability may turn on whether the court sees the agency as a strict extension of the state. For schools, this question was found in a most important case in *Mt. Healthy*[48] wherein the Supreme Court held local school districts in a similar light to city and county governments, rather than in the context of traditional thinking that sees schools as an arm and extension of the state. Instead, the Court saw school districts more as political subdivisions of the state because of their taxing power and autonomy, a view consistent with the Court's other decisions in cases such as *Monell*[49] and *Pugh*.[50]

Statutory limitations on immunity have been an important source of penetrating government's liability, particularly in the context of federal enactment of the Eleventh Amendment that has continued the cloak of immunity while at the same time providing for specific exception when considering either federal statutes specifically waiving immunity or when considering the weighty matter of violation of other constitutional rights. The enactment of state statutes abrogating the doctrine of sovereign immunity has resulted in a variety of limitations on and general availability of immunity. Similarly, doctrinal modifications by courts in areas such as proprietary acts and nuisances have provided certain openings to aggrieved parties with concomitant limits placed on the immunity enjoyed by government, with one court stating the incredibility of absolute immunity in a modern context.[51] Like many aspects of law, state statutes provide the most immediate test of immunity availability, making it vitally important for administrators to know and understand the risks inherent to acts they may commit in the course of professional employment.

Tort Liability

Results of earlier discussions concerning governmental immunity has several implications for people who have responsibilities in school organizations. Of first importance is the realization that rejection of absolute immunity means that liability, if present, must be attributed to someone acting individually or collectively, since all acts are committed by people working within organizations or outside the parameters of the organization. As a result, determining that there are instances where liability will be found for educational organizations has the direct outcome that board members, administrators, and other personnel involved in the education process may be held responsible for acts or omissions that may carry some requirement for redress under the law. While various types of liability may arise for acts ranging from civil offense to criminal acts, the most common type of liability in schools arises under a legal concept known as *tort* law. By definition, a tort is a civil wrong not involving contracts and which does not qualify under criminal statute. By such definition, then, tortious acts encompass a sweeping scope of legal offenses that may be committed in violation of either statutory or constitutional requirements of law.

The basis of a tort is a noncriminal act that offends a law that was designed to protect some right of individuals. Since all organizations are peopled and consist of dispensers and receivers of the organization's mission (e.g., teaching and learning in the case of schools, the rights of people on both sides of an act must be preserved and protected). The opportunity for committing a tort is particularly great because constitutional and statutory rights are both broad and specific. For example, tort claims may arise from breaching contracts, defaming or impugning a person's reputation, denying or violating the civil rights of citizens, or from engaging in some behavior wherein a negligence claim is raised alleging a wrongful act or a wrongful failure to act in the presence of an affirmative duty.

Although the scope of possible claims is great, most sources group torts into three categories of intentional torts, torts involving negligence, and strict liability[52] torts. Two of these categories are particularly important to school personnel in educational organizations that may be held liable, and each is further important to those persons individually since they may face personal liability as well. For instance, depending on the particular state, abrogation of sovereign immunity and nonapplicability of immunity claims to school districts may mean a district is liable for wrongful acts such as proprietary or nuisance claims or serious infractions of the civil rights of pupils or teachers. In a related vein, there has been considerable consternation that educational malpractice claims might arise wherein courts might fix some liability if it were determined that the rights of individuals had been violated. By direct transference, then, board members and administrators and other personnel may be held personally liable in an action. Similarly, intentional torts and negligence claims may also be alleged against school personnel.

Intentional Torts

Although an intentional tort may seem oddly named in that no one generally sets out intending to break the law, an intentional tort arises from an act resulting in some type of harm to another person, regardless of whether harm was intended. An intentional tort is defined as a "wrong perpetrated by one who intends to do that which the law has declared wrong, as contrasted with negligence in which the tortfeasor fails to exercise that degree of care in doing what is otherwise permissible. It involves elements of intent or purpose and malice or ill will, but malice or ill will may be shown by indifference to safety of others, with knowledge of their danger, or failure to use ordinary care to avoid injury after acquiring such knowledge."[53]

An important element thus rests in whether it can be established that the behavior of an individual resulted in direct harm to another. For example, a person having no ill will toward some person but who commits an act leading to injury of that individual's rights or property or person may be held to have committed an intentional tort. Certainly in some instances wrongful intent or malice is present, as in cases involving trespass and false imprisonment or in knowingly denying someone's civil rights as in discrimination cases, but such instances are rather scarce in school settings. In most instances, claims for intentional torts in schools involve situations where it is alleged that some otherwise permissible act moved beyond the line of mere negligence to become motivated by a desire to do harm or in which an impermissible act was accompanied by a reasonable expectation of impending harm.

While relatively few school cases fit the typical description of an intentional tort, the most common claim may relate to issues of corporal punishment. In the absence of a state statute that forbids physical punishment in and of itself, or controlling state statutes that either specifically allow or not allow local school boards to forbid such actions, its application is an instance where school personnel are said to have crossed the line, whereupon assault and battery charges may be lodged against the person administering the punishment. The definition of these charges is important to the nature of an intentional tort, in that assault is defined as "any willful attempt or threat to inflict injury upon the person of another, when coupled with an apparent present ability so to do, and any intentional display of force such as would give the victim reason to fear or expect immediate bodily harm...An assault may be committed without actually touching, or striking, or doing bodily harm."[54] Likewise, battery is importantly involved in such cases and is definitionally distinguished from assault in that "the actual offer to use force to the injury of another person is assault; the actual use of it is battery, which always includes an assault."[55] The critical aspect is that an intentional tort via assault and battery can be filed in such cases, in that it is not necessary to show malice or the intent to harm another in order to have a valid claim.

As a practical matter in schools, however, courts have long struggled with allegations of assault and battery in cases involving corporal punishment. As a general rule, in the absence of controlling state statutes, courts have been unwilling to remove this power from school personnel and have approved the use of reasonable force to cause students to comply with behavior codes under the doctrine of *in loco parentis* which recognizes that schools have a respected limited parental right by virtue of the teacher's relationship to pupils. The often relied-upon U.S. Supreme Court case of *Ingraham*[56] said that corporal punishment did not violate the Eighth Amendment's prohibition of cruel and unusual punishment,[57] and this ruling has been the basis for many subsequent decisions upholding the right of schools to physically punish students. The record is replete, however, with numerous instances of assault and battery charges leveled against teachers and administrators who have sought to control students and have been charged with wrongful conduct under intentional tort claims.

The *Ingraham* case bears brief review here since it controls the way courts have viewed intentional tort claims involving school liability for corporal punishment. In a divided five to four vote, the U.S. Supreme Court upheld the physical punishment of two students in which one student was injured sufficiently to miss eleven days of school, allegedly due to the nature of the injuries. The punishment administered by school officials was sufficient to cause one student to suffer a hematoma. The parents' claim that the beating violated Eighth Amendment prohibition of cruel and unusual punishment was rejected by the Court, ruling that this Constitutional provision did not apply to schools since children need less protection than do criminals due to the open nature of school systems and community ability to police the school's activities. The Court further noted that parents' have recourse by filing charges of assault and battery within the court system.

Recent cases have also tended to uphold the same leniency toward schools' right to discipline students. In *Mathis,*[58] a whipping that produced bruises was brought to court two years later claiming post-traumatic stress disorder.

The appeal court was unsympathetic, however, noting that permissible corporal punishment can be expected to cause pain and a potential for bruising, and that the punishment was not sufficiently severe as to violate the law. Similarly, a case in which it was alleged that district policy was not followed when applying corporal punishment failed as a federal appeals court ruled in *Woodard*[59] that the punishment did not fit the definition of arbitrary or capricious or an inhumane or shocking abuse of official power.

The continued applicability of *Ingraham* as a standard was applied as a direct case in point in the Michigan case of *Willoughby,*[60] in which the appeal court held that a teacher's grabbing of a student and holding him in a "full nelson" did not constitute battery and unreasonable use of force. Many other such cases exist, as in *Thompson*[61] where a teacher kicked a student who had been repeatedly disobedient. Although the teacher's behavior was unusual, the court held that the blow was inconsequential and that it was more embarrassing than injurious. These and many other cases illustrate that corporal punishment is often decided in favor of teachers because there is a presumption of appropriate behavior on the part of the teacher and to which there is a heavy burden of proof on the complainant to show overwhelming impropriety in order to prevail in such claims.

On the other hand, courts have also been willing to overcome the heavy presumption of correctness on the part of the school when the facts support plaintiff claims. The longevity of such dispute is evident in cases on record, dating from the mid-Nineteenth Century wherein courts have severely condemned punishment found to be cruel, unconscionable, or even intemperate.[62] Courts have also agreed with plaintiffs under similar circumstances, as in *Frank*[63] where a teacher was found guilty of assault and battery for having broken a student's arm by shaking him against a hard object and then dropping him to the floor. Likewise, a teacher was guilty in *Sansone*[64] for having broken a student's collarbone as a result of being thrown against a blackboard.

In *Garcia,*[65] a federal court found school officials could not hide behind good faith immunity due to the facts of the case. In this instance, a third grade student was held upside down by her ankles while the principal hit her with a paddle, followed by five swats on the front of her leg, leaving a permanent scar. In a later incident, she was again subjected to corporal punishment wherein she was physically forced to bend over a chair while receiving additional blows. The court rejected the argument that legal standards for corporal punishment were not well established, noting that at some level of force such punishment invades the child's right to bodily security, holding that in this instance the beating was sufficiently brutal that officials should have known their acts were violative of constitutional guarantees. In still another recent case, a teacher's dismissal for excessive force was upheld as the court ruled in *Burton*[66] that there had been failure to obey district policy to consider a student's age, size, mental, and emotional condition when using physical discipline.

As well-illustrated by the cases briefly discussed here, intentional torts represent a difficult arena for courts. In the case of schools, there are often conflicting pressures to respect the prerogative of schools to engage in valid activities that are naturally aggravated by an immature clientele. As a general rule, only a relatively few cases involve intentional torts wherein harm is intentionally caused, fitting rather the definition given earlier wherein no genuine malice is present but which still conforms to the legal definition of an intentional tort. In the few instances, where such malice or intent to cause bodily harm is in fact present, an intentional tort may more accurately become a criminal act. A more frequent description of tort liability questions is found in the next section wherein negligence claims against school districts are considered.

Negligence

The most common tort action against teachers and administrators is a claim for negligence. As contrasted against the definition of an intentional tort, a tort of negligence is defined as conduct falling below a legally established standard of care, resulting in injury to another person. It is failure to exercise due care when subjecting another to a risk or danger that causes harm. As courts consider negligence claims, there are four specific elements leading to properly establishing negligence, all of which must be present if negligence is to be upheld. These elements are duty, breach of duty, proximate cause, and actual injury. Proofs to these elements are attempted by plaintiffs, while defendants offer several expected defenses.

The element of duty refers to the fact that before a person can be held liable for another's injury, it must be clearly proven that the actor had responsibility for, or duty toward, the injured person. A duty can arise from statutes, contracts, or even common sense. The courts have identified three areas of duty for schools and educators. The first area relates to adequate supervision. A second area relates to proper instruction. The third area relates to maintenance of a safe environment. A distinct duty must be found to hold someone liable, so that some duty or connection between the injured party and the person being sued must be clearly established. For example, the administrative duty to maintain safe equipment would include a duty to purchase and maintain appropriate and safe equipment in all areas where students may have access. Of these three areas, probably the most common negligence action involves a claim that a lack of adequate supervision resulted in injury to a student. Clearly, when students are in school there, is a duty to supervise them, but the claim also often appears when students are off school grounds or on school grounds before or after school hours.

Once a duty has been established or undertaken, it must be exercised reasonably so as to not breach the duty. This is one of the most commonly disputed elements in a negligence action. To be liable for the injury the complaining party must show that the duty was not carried out reasonably, and therefore the person breached the duty. This standard of reasonableness varies among the many different circumstances making up each case; what may be negligence in one case may be a reasonable exercise of one's duty in another case. This reasonable standard has been personified by the "reasonable person" test. This person is a fictitious individual who is prudent and uses ordinary care and skill under the same or similar circumstances.[67]

In general, courts assume that professional educators, by virtue of their professional education training and certification have superior skill and knowledge as compared to ordinary laypersons. This latter element is particularly important in work-related claims, as in education where educators are assumed to possess superior skills and understanding of the educational process so that the required conduct becomes that of a prudent professional under the same or similar circumstances. The result becomes that in determining whether a duty has been breached, it must be decided what the reasonable professional would do under the same or similar circumstances. If the actions of the actor fall short of this, the duty has been breached.

The third element in a negligence claim relates to whether the act actually provided cause for the injury. If the injury were for reasons other than the defendant's actions, there is no liability. Courts require that the negligence of the defendant be the substantial cause of the harm to the plaintiff, substantial enough to conclude that the act was indeed a direct cause of the injury. To make such a determination, there must be an unbroken chain of events between the act and the resulting injury, so that if negligence were not a substantial factor, no liability can be found. However, a defendant can sometimes be found liable for injury even if the most immediate cause were not a direct result of the action. Acts that occur between the negligence and the injury are known as intervening acts. If any intervening acts were foreseeable, the defendant's negligence may still be found to be the legal cause of the injury.

The fourth and final element in a negligence claim holds that the defendant is liable only if the alleged act caused an actual injury. Usually a physical injury to the plaintiff is required in order to receive an award for damages, so that absent actual injury, even nominal damages cannot be awarded as a token punishment to the offender.[68] The whole concept of damages applies to negligence acts, however, in that courts are able to make a wide range of reparations to the injured party. Damages are generally defined as "a pecuniary compensation or indemnity which may be recovered in the courts by any person who has suffered loss, detriment, or injury, whether to his person, property or rights, through the unlawful act or omission or negligence of another . . .. Damages may be compensatory or punitive according to whether they are awarded as the measure of actual loss suffered or as punishment for outrageous conduct and to deter future transgressions."[69] As a result, proving all the elements of negligence can be quite costly to a defendant, in that pursuant to state controlling statutes, a damage award may encompass compensation for physical damages, for such expenses as medical bills, and loss of prospective earnings, nonphysical injuries such as emotional distress and pain and suffering, and other damages such as exemplary and punitive damages depending on the nature and severity of the case.

There are several defenses that schools and administrators generally use in rebutting negligence claims. In any negligence action, a common line of defense, pursuant to state statute, is contributory or comparative negligence. Likewise, a defense claim may arise for assumption of risk.

Both comparative and contributory negligence involve some fault or negligence on the part of the injured person. To successfully use this defense, and where allowed by state statute, the defendant must show that the plaintiff in some way through personal negligence contributed to the injury. In determining whether the plaintiff was negligent, the same four elements of negligence are applied to the plaintiff's action regarding reasonable care and causation. The doctrine of contributory negligence seeks to determine whether the plaintiff's own actions contributed to the injury in question.

This principle was concisely stated in the *Restatement of Torts*,[70] wherein contributory negligence was defined as "[c]onduct on the part of the plaintiff, which falls below the standard to which he should conform for his own protection, and which is legally contributing cause cooperating with the negligence of the defendant in bringing about the plaintiff's harm."[71] If it were established in this manner that the plaintiff in fact contributed to the injury, the doctrine operates as a total bar to recovery, subject to some intervening variables such as a court's consideration of the standard of conduct expected due to the age and physical traits of the injured person (e.g., a child is incrementally less capable of contributory negligence at incrementally younger ages), and the mental abilities of retarded persons relieve their negligence to some extent. In older cases, however, courts prevented a plaintiff's recovery if any amount of negligence was shown to have caused the injuries, no matter how slight. This doctrine had drastic results when the injuries were great and the plaintiff's negligence was slight. Recognizing this in more modern cases, courts have generally barred recovery if the plaintiff's negligence has been a substantial factor in the injury sustained. Again, it must be stressed that these issues operate largely under controlling statutory guidance.

In response to the drastic results sometimes caused by the doctrine of contributory negligence, most states have converted to the doctrine of comparative negligence. In using comparative negligence the amount of negligence and causation of each party is compared to determine a degree of fault, so that recovery is based on the relative degree of negligence. For example, if a plaintiff's negligence resulted in only 10 percent of the injury, the defendant would be liable for only 90 percent of damages due. The specific provisions of comparative negligence vary among states, but the concept of comparative negligence is generally consistent.[72]

A third defense may arise by claiming assumption of risk on the part of the injured party. When raising this defense, it is claimed that the injured party's actions effectively relieved defendant from a prescribed level of care or duty to the injured party. The argument is based on the theory that the injured person voluntarily engaged in the act in question and knew by express or implied admission of the danger involved. This doctrine, also known as *volenti non fit injuria*, "means legally that a plaintiff may not recover for an injury to which he assents . . .. The requirements for the defense of *volenti non fit injuria* are that the plaintiff has knowledge of facts constituting a dangerous condition, he knows the condition is dangerous, he appreciates the nature or extent of the danger, and he voluntarily exposes himself to the danger."[73]

A fourth defense against negligence rests in the sovereign immunity concept discussed earlier in the chapter. Under this common law inheritance, it is argued that governmental units are free of liability under the theory that the doctrine is justifiable because it limits the flow of public moneys to private citizens in damage awards and that liability payments deplete governmental treasuries and detract from other legitimate functions of government. It was established earlier in this chapter that various instances exist in which immunity may be abrogated, either statutorily or by the judiciary. This has especially been true since many states have recognized the injustice that can be done by blind application of this concept, and a particular reason has been because there is now widespread availability of insurance covering tortious acts that can be purchased by governmental units. Under the latter concept, it is reasoned that both plaintiff and defendant are benefited by allowing tort claims since costs to the treasury in purchasing insurance are reasonable in comparison to direct treasury vulnerability for damages and because justice may be better served by not exempting government from liability for its wrongful acts. Limited abrogation of immunity has been the recent rule, as only Maryland and Mississippi continue to hold to total sovereign immunity.

In contrast, several states have created administrative agencies to hear claims, controlling liability limits through such mechanisms as shortened statutes of limitations and caps on allowable damages. Other states have abrogated immunity in those instances where it is statutorily permissible for governmental agencies to purchase insurance.[74] In other states, legislatures have abolished immunity and enacted "save harmless" statutes. These statutes are tantamount to liability insurance policies against personal liability for teachers, administrators, and other officials when acting within the scope of employment.

Recent cases involving negligence claims illustrate the viability and application of these concepts of negligence and accompanying defenses. While a broad scope of cases is available for discussion, the following cases illustrate the principal elements discussed in this chapter by involving questions of governmental immunity, duty of care, lack of supervision, assumption of risk, and intentional torts.

A large number of school cases appearing under negligence claims relates to injuries received while participating in certain kinds of classes or in extracurricular activities. Of particular concern to schools are vocational classes where it is well established that dangers must be minimized and safety devices should be in place. In *Velmer*,[75] a high school student lost a finger due to a shop class machine, which was not equipped with a manufacturer's approved guard. Dismissed at trial under immunity, the appeal argued that Michigan made exception to governmental immunity for unreasonably dangerous conditions in public buildings. The appeals court upheld the trial court since the machine was not bolted to the floor so as to make it a permanent fixture, whereupon a supreme court appeal reversed by holding that permanent attachment was not the only way the machine could become a permanent part of the premises. Similarly, a student argued in *Moore*[76] that the teacher's leaving the woodshop classroom during which the student was injured was a violation of his due process rights. The trial court dismissed after finding governmental immunity during the scope of employment, but that due process would be available if the student could show malicious, egregious, or callously indifferent behavior by the teacher. Since the student alleged only that the teacher left the room, the court held for the defendant. Still a third instance is found in *Grubbs*[77] as a student was injured while using a tablesaw that was not equipped with a safety guard. The student sued under 42 U.S.C. § 1983 stating that the act of allowing him to use the saw without a guard was an act of conscious indifference violating his civil rights. The court held, however, that the district was not liable because the student had not alleged that the unconstitutional act was caused by an official policy of the school and because he had no constitutionally protected right to a safe school environment.

A somewhat opposite result was obtained in *Higgins*[78] as a mentally disabled student was severely burned in a metal shop class. His suit claimed negligent supervision, whereupon as his mental age was only eleven while his chronological age was eighteen, raising the claim that the appropriate standard of care was that of a child. The trial court awarded $50,000 damages, but reduced the award to $15,000 since it also held the student to be 70 percent negligible. Although the court noted his mental age, it also noted a statute stating that, unless specifically provided for by law, all persons should be deemed to be of full age for all purposes at the age of eighteen.

A second area of concern for negligence claims relates to physical education classes. An Illinois case of *Ward*[79] contained facts wherein an elementary and a secondary playing field were adjacent, allowing a collision between a high school boy and a nine-year-old child and resulting in permanent injuries for the younger boy. Suit was brought alleging willful misconduct and negligence, whereupon the trial court held that there was no duty of care owed to the boy by the district. On appeal the court held that there was no anticipation of unreasonable risk to be found in the cojoinment of the playing fields. A similar result was reached in *Stevens*[80] when a fourteen-year-old boy with a prior injury was sent out to the football field and was standing on the sideline when he was re-injured by a student who ran into him. His suit alleged negligent supervision. Both the trial court and the supreme court held that the teacher had a reasonable supervision duty but could not be expected to supervise each individual student at every moment and that the student was of such age as to show reasonable care for himself.

Contributory negligence was also a factor in *Joliuette*[81] as a student slipped on a wet gym floor. The trial court held the girl 50 percent responsible for her injury as she had been warned not to go onto the floor, but the appeal court reduced her negligence to only 10 percent, stating that her age, intelligence, and experience, together with the board's position of superior actor in creating the hazard, warranted lowering her degree of fault. Similarly, in a Pennsylvania case of *McCloskey*[82] a high school student was rendered quadriplegic when he fell off the gymnastic rings during physical education class. His suit charged the board with failing to maintain safe conditions and failure to provide supervision. The trial court ruled for the district under the Pennsylvania Political Subdivision Tort Claims Act which limited liability to real property causation. On appeal, the evidence showed that the rings had been set at the wrong height and that proper care and control were lacking. The court reversed under the real property exception, sending the case back for further proceedings.

Still a third area of negligence concern relates to unsupervised accidents, both on and off school property. In the case of *Maldonado*[83] an eleven year old child died from strangulation after catching his bandanna on a cloakroom hook. His suit alleged lack of supervision for the twenty-minute period in which he remained undiscovered, charging deliberate indifference under a 42 U.S.C. § 1983 right to protection, particularly in light of the state compulsory attendance laws. The trial court held for the defendant teacher, whereupon the appeal court sustained the lower court, quoting the U.S. Supreme Court that had held in *DeShaney*[84] that the Constitution does not impose an affirmative duty on the state to protect except under limited circumstances, of which compulsory attendance did not qualify as a liberty restraint.

A similar judicial posture was seen in *Waltz*[85] where a student alleged an unsafe playground after tripping over an exposed tree root. On plaintiff's appeal, the court held that the district's duty was to use reasonable care to keep the grounds safe and to warn of hidden danger, but that such duty did not extend to ensuring the student's safety. Contributory negligence was also argued in *Byrd,*[86] as a high school student severed his leg while trying to stop the rotating drum in a clothes dryer. The student sued the school board charging lack of proper instruction and supervision. Although the trial court dismissed the case under contributory negligence defense, the appeal court accepted the argument that the board had violated state child labor laws by allowing him to work with the machine, stating that the student's negligence did not bar recovery.

Cases involving liability off school grounds also present a variety of results. In *Gilmore*,[87] a student who was late for school was injured at an intersection after the crossing guards had left their post. Originally dismissed, the appeal alleged that the school district had exhibited "conscious disregard and utter indifference" in failing to perform the duty of having a crossing guard present, arguing that the voluntary decision to use guards at any time created a duty. The court held, however, that a board has wide discretion in exercising its powers, and that liability only follows after intentional or reckless disregard. A similar Oregon case held in *Fielding*[88] that the plaintiff had not produced sufficient evidence showing lack of standard of care regarding signs, signals, and speed limits, that the school district was not obliged to adequate measures in that the law only stated that school districts may conduct safety patrols, and that districts were immune based on discretionary acts.

In another case, a student waiting for a school bus was shot by gang members whereupon the student sued the school in *Brownell*,[89] claiming inadequate supervision. A trial court awarded damages, but on appeal it was held that thc district's duty was to supervise students on school grounds and to enforce rules for their protection, but it was not liable since schools must be only reasonably supervised rather than impenetrable to gang violence.

Reaching somewhat different results, the Colorado Supreme Court overturned a dismissal in *Justus,*[90] holding that a student could recover damages for injuries received while riding a bicycle home from school. The appeal noted that the district's handbook and its voluntary assignment of patrols created a legal duty to prevent a first grader from riding his bicycle. The case was remanded for development on the question of the district's duty of care owed to the student. Likewise, a junior high student in *Joyce*[91] was hit by a car while crossing at an uncontrolled intersection. The intersection gave access to an open gate to school property, causing the court on appeal to remand the case to be tried on its merits since there was a question of whether the open gate resulted in the district encouraging students to take the shortcut, rather than the controlled crosswalk about 450 feet away.

A fourth area of grave liability rests in acts involving intentional conduct. In *Doe*[92] the federal appeals court held that compulsory attendance laws create a special relationship of constitutional proportion between officials and children. In this case, a teenage boy was sexually molested by a teacher, bringing suit that the district had a constitutional affirmative duty to protect his liberty rights. Noting the difference between the present case and the Supreme Court in *DeShaney,*[93] the court held that the high court's ruling was altered by a special relationship between government and individuals, wherein such a relationship enters when an individual is incapable of acting for himself and that compulsory attendance created such relationship. The court thus ruled in *Doe* that when a child is placed in school, it renders the child's guardian unable to act for the child and that the child is in the functional custody of school officials and must rely on them for protection.

Similarly, in another case also known as *Doe,*[94] a girl was sexually molested in a school restroom by an unknown attacker. She brought suit against her teacher, the principal, and the school board. The court dismissed charges against the teacher and principal but awarded damages by the board. On appeal, the court held that the board could be liable even though the teacher was not because the board had the duty of supervision independent of any duty falling to the teacher. In contrast, however, the court held in *W.L.O.*[95] that a kindergartner who was sexually assaulted by two men on the playground did not have a claim against the district for negligence and failure to provide a safe place because the evidence could not sustain the negligence charge and because the school did not have a duty to prevent criminal acts of a third person absent reason for the school to know that the act was about to take place.

The cases discussed in this section have provided a wide panorama of possible outcomes of allegations of negligence against school districts. Taken by individual case, it would appear that there is little consistency regarding school district liability among state and federal courts. That impression, however, does not do justice to the scope of court rulings, making it important to consider general conclusions from commonalties in these and other cases and from how the law is applied to negligence claims. While no amount of summary can erase the uncertainty found in confronting different situations and law, five important observations help meet this goal by providing a brief summary of how courts see negligence.

The first observation returns to the actual elements of a negligence claim. To make a valid case, the claim must demonstrate that there is an actual legal duty to protect others from a defined unreasonable risk of injury, that such duty in fact has been breached, the breach directly led to the injury, and the plaintiff suffered an actual loss. Under these weighty conditions, it is a heavy burden on the plaintiff to prove a case and it is equally difficult for a court to establish and find all the conditions of negligence. The difficulty is only increased by the fact that negligence may include inadvertence, carelessness, or failure to foresee harm. Obviously, under these conditions some significant overlap occurs between negligence and intentional torts.

The second observation relates to the concept of assumed risk. Most cases indicate that when someone is injured and the injured party could have avoided the injury through common sense, liability will not follow. In cases involving athletics, courts generally hold that students assume the risks inherent to participation and that liability and damages are not available, absent a showing of negligence by the school or its personnel.

However, courts generally find liability for districts or physical education teachers if injuries result from known hazards or from failure to properly instruct or supervise pupils. Likewise, schools are usually held to a standard of care for safe work environment, such as shop classes, wherein failure to use all safety equipment and to supervise and warn students of danger evoke findings for negligence. As a general overall rule extending beyond vocational arts and physical education classes, courts hold schools and agents liable for injuries during the course of regular school events that result from failure to provide a reasonably safe environment or from failure to warn of known hazards, and particularly from failure to provide supervision suitable to the activity and ages of participants.

The third observation relates to other supervised or unsupervised activities. Courts have generally held that school districts have a duty to use reasonable care in protecting against hazards on school property and have held school districts liable whenever no supervision is provided or where supervision is negligently performed. Similarly, school districts can be liable for maintaining or tolerating hazards on the school grounds, although the burden rests heavily on plaintiff and courts will closely scrutinize the nearby events to determine how well plaintiff's claims meet all necessary conditions of establishing negligence.

The fourth observation moves beyond mere negligence into acts of intentional or willful harm. Districts face grave problems in this arena and the outcome is never known until the verdict is rendered. In this area, courts have been harsh in dealing with intentional misconduct, and courts have also found districts liable for the intentional acts of third parties, holding that the district should have reasonably foreseen the opportunity for misconduct. Importantly, a finding of wanton or willful misconduct may defeat a sovereign immunity defense.

The fifth observation requires emphasizing the differences among individual states on the laws governing liability. Each state has specific statutes controlling liability, although the general legal principles common to all these claims carry over into the practice of law. States have differing laws on immunity so that establishing negligence in any two states may produce a different ruling on liability because one state may provide no immunity while the other state provides complete immunity under similar factual circumstances. It is precisely these differences that make sweeping summary impossible, leaving only that application of federal justice may be controlling depending on the claims raised. As a consequence, persons with responsibility in schools are well advised to know the general risks and to work closely with legal counsel in reducing and managing these specific risks.

Civil Rights

In addition to liability for claims relating to negligence and other civil torts discussed previously, tort claims may arise regarding civil rights violations under federal and state law. As a result, this section identifies issues of concern to districts and school personnel.

Recovery for violation of federal constitutional or statutory rights is possible by virtue of the Civil Rights Act of 1871.[96] This law, originally enacted to prevent discrimination against African-Americans following after the Civil War, has been the basis of a wider set of lawsuits seeking remedy for alleged wrongs and has been broadly construed and interpreted by courts to confer liability on school personnel and school districts for actions that go beyond racial discrimination, including acts resulting in the impairment of other federally protected rights. The basic concept of liability under this federal provision extends general liability to include individual and personal risk for public officials who violate federal statutory or constitutional rights of another individual, such as a teacher or student. This law, commonly known as Section 1983, provided a clear basis for prohibiting civil wrongs as it stated:

> Every person who, under color of any statute, ordinance, regulation, custom, or usage, of any State or Territory, subjects, or causes to be subjected, any citizen of the United States or other person within the jurisdiction thereof to the deprivation of any rights, privileges, or immunities secured by the Constitution and laws, shall be liable to the party injured in an action at law, set in equity, or other proper proceeding for redress.[97]

The word " person" in Section 1983 has been interpreted by the United States Supreme Court to include local governments or institutions,[98] including school districts.[99] Under these conditions, Eleventh Amendment immunity protecting states as governmental entities from suit in federal court is not available.[100] By this interpretation, a person who alleges damages by policies or practices in school districts can sue the individuals and institutions charged with a Section 1983 civil rights violation.

Civil rights violations under Section 1983 received its greatest clarification in *Wood*[101] which addressed school board members and their liability. The courts have generally held these persons immune when acting in good faith, but in *Wood* the court noted "a school board member is not immune from liability for damages under Section 1983 if he knew or reasonably should have known that the action he took within his sphere of official responsibility would violate the constitutional rights of the student affected, or if he took the action with the malicious intention to cause a deprivation of constitutional rights or other injury to the student."[102]

Wood thus enunciated a new standard of immunity known as qualified good faith immunity, wherein a board member can be held liable for acts violating constitutional rights under three particular conditions. The first condition finds liability when board members knowingly violate constitutional rights in that they are legally bound to make decisions consistent with current law. Importantly, the second condition finds liability when unknowingly violating constitutional rights in that ignorance is no excuse if the violator should have known the law. Of great significance to unknowing violation is the burden to know the law. The third condition finds liability when acting on the basis of malice. Any immunity that might otherwise be presumed or available is lost if it can be shown that the violator set out to intentionally harm someone or if actions originate in vindictiveness or spite.

The findings on Section 1983 have placed a significant burden on defendants in that the conditions under which immunity would still apply have been rather narrowly drawn. The burden of establishing good faith immunity clearly rests with the defendant in that the plaintiff does not have to prove that immunity does not apply.[103] Further, the requirement to know the law was reinforced by the Supreme Court when ruling in *Harlow*[104] as it stated "government officials performing discretionary functions generally are shielded from liability for civil damages insofar as their conduct does not violate clearly established statutory or constitutional rights of which a reasonable person would have known."[105] This has led to rulings where qualified immunity has been denied due to failure of accused persons to observe well established law, as in *Gavrilles*[106] where board action against a teacher, who had utilized a union grievance procedure, gave the appearance of retaliation and was interpreted to violate the teacher's constitutional rights. Similar results were obtained in *Littlejohn*[107] as liability for Sect. 1983 violation was found when a superintendent recommended against a teacher's contract renewal in part because of unsavory involvement in a divorce proceeding. Perhaps the best summary of Section 1983 liability was stated as:

> Indeed courts generally look at two basic elements in determining whether Section 1983 should be imposed upon governmental entities and their employees. With respect to the liability of the entity, courts will impose liability only if a state official acts in a "callously indifferent" manner toward the rights of individuals pursuant to governmental policy or custom. In essence, a state official must not only perform an outrageous action, but such actions must also be common within that particular governmental entity. Individual liability revolves around the "callous indifference" standard. If an employee acts with "callous indifference" toward the rights of an

individual, he or she may be subject to liability in his or her individual capacity.[108]

Section 1983 clearly provides that a litigant may bring suit under the Act for compensatory damages, as well as for equitable relief. Generally, actual damages and attorney fees are awarded to a prevailing party in a Section 1983 action. To receive any award, some actual damage must be shown. In *Carey*[109] the Supreme Court dealt with an issue of nominal actual damages. The case involved two students who had been suspended from school for drug use on school premises. The trial court ruled that the students were denied procedural due process. The court stated that plaintiffs were technically entitled to damages, yet it dismissed the claim as none were shown to exist. A federal appeals court reversed, however, stating that plaintiffs were entitled to substantial nonpunitive damages without proof of actual injury because the suspensions were irrelevant. The Supreme Court in final review ruled that, in absence of proof concerning actual damages, the students were entitled only to nominal damages, which the Court set at one dollar. However, court cases involving Section 1983 have consistently allowed for compensation of mental stress and emotional anguish whenever it can be shown that actual damages were incurred.

In addition to damages, attorney fees are normally awarded to the prevailing party in a Section 1983 action.[110] Because there is no set scale in these actions by which to determine such fees, the amount is left to the discretion of the court, which makes a reasonable award for time spent on the litigation. Generally, such factors as time, complexity, size of the award, and expertise of the attorney are taken into account in awarding fees. In *Hensley,*[111] the Supreme Court found that a court could determine reasonable attorney fees by multiplying the hours spent by a reasonable hourly rate. As a consequence, fees for Section 1983 civil rights cases vary considerably dependent on the unique circumstances. Similarly, there are disputes over what is considered work in such claims, particularly related to how much time an attorney may be expected to put into a case before it actually comes to trial. In *Webb,*[112] the Supreme Court held that Section 1983 generally does not permit a court to award fees to a prevailing party for costs incurred for legal services in an optional state administrative proceeding. In contrast, however, the Supreme Court held in *New York Gaslight*[113] that attorney fees for state administrative hearings under Title VII could be awarded under Section 1983. The difference between the two cases was in that the hearing in the latter case was mandatory and was part of the Section 1983 litigation, while in the first instance the Section 1983 action could have been initiated without going through the administrative process.

The outcome of Section 1983 liability has been to raise a high level of anxiety over liability. The fear has been real for schools and the people who work in education because there is much evidence of an increasingly litigious society that sees constitutional infraction in every real and imagined instance of slight. The cost has been significant, as litigants have won significant amounts of money in damage awards. It is a tremendous burden on districts and employees when civil rights violations are filed simply because of the knowledge that if the opposing party wins, the school district may be faced with its own attorney fees and opposing counsel fees, in addition to potential damages of various kinds ranging from actual to punitive. It is precisely such knowledge that has caused so many claims to be settled out of court because of the weight of establishing innocence in the face of serious constitutional rights.

Educational Malpractice

A related area of tort claims that has been of considerable concern to school personnel has been the area of educational malpractice. The source of concern has been a rising interest in the obligations of school districts to perform at some predetermined level of student achievement, coupled with the rising tide of litigation in the United States that sees every perceived injustice as an actionable cause. Since the 1970s, it has been feared that educational malpractice claims would succeed at some point in time, causing administrators and others to carefully avoid situations that might result in such a lawsuit. Although fear of such a ruling has been great, there have actually been only a few instances in which malpractice lawsuits have actually been filed.

The concept of educational malpractice is an attempt to apply tort law to educational outcomes. Under this concept, the idea was to identify whether a student had educational deficiencies resulting from inadequate instruction. From one perspective, malpractice could be seen as an intentional tort, while from another viewpoint it could be seen as a negligence claim. Neither claim has been successful as each has experienced considerable difficulty in garnering court sympathy for these claims. In the first instance, malpractice as an intentional tort has been very difficult to establish since it would require proving that the school system or its agent maliciously sought to deprive a child of educational achievement. While quite difficult to prove, however, at least one court recognized the basis for the claim as it said "where an individual engaged in the education process is shown to have willfully and maliciously injured a child . . . [S]uch outrageous conduct greatly outweighs any public policy considerations which would otherwise preclude liability."[114] Under such conditions, any presumed immunity would likely fall if the facts were to support an intentional tort claim. However, the likelihood of a school district intentionally harming a child's education appears to be unrealistic under normal expectations and circumstances.

The second instance of claiming negligence has been more common, however, as several courts have considered that question. The earliest educational malpractice lawsuits focused on the question of whether students have any right to a prescribed achievement level by virtue of the fact that they are compelled to attend school under compulsory attendance laws, while more recent cases have asked the courts to determine whether schools have been negligent in diagnosing and prescribing for identifiable educational needs.

The seminal case in educational malpractice occurred in *Peter W*[115] wherein a student received a high school diploma from the San Francisco public school system. Contrary to California statute requiring graduates to be able to read above the eighth grade level, the student was functionally illiterate in that his reading skill never surpassed the fifth grade. The student sued the district, claiming that it had been negligent in not teaching him the required skills and permitting him to graduate despite his failure. He also claimed that his skills had not been properly represented to his parents, who themselves testified that they were never told of his educational deficiencies. The lower court dismissed the action and was upheld by the appeals court, which discussed at length the "duty" requirement as it existed in negligence law, finding no duty on which to base an action. The court's ruling centered in part on the inability to find a workable standard of care for teaching against which the defendant's action could be judged:

> Unlike the activity of the highway or the marketplace, classroom methodology affords no readily acceptable standards of care, cause, or injury. The science of pedagogy itself is fraught with different and conflicting theories of how or what a child should be taught, and any layman might—and commonly does—have his own emphatic view on the subject.[116]

In addition, the court noted that the degree of certainty that the plaintiff had suffered an injury, the extent of the injury and establishment of a causal link between defendants' conduct and plaintiff's injuries were all highly problematic. The primary motive, apparently, for the court's reluctance in examining these problems was the public policy involved. The court found a public policy against allowing such suits because of the burdensome litigation that would be generated:

> Few of our institutions, if any, have aroused the controversies or incurred the public dissatisfaction, which have attended the operation of the public schools during the last few decades. Rightly or wrongly, but widely, they are charged with outright failure in the achievement of their educational objectives; according to some critics, they bear responsibility for many of the social and moral problems of our society at large. Their public plight in these respects is attested in the daily media, in bitter governing board elections, in wholesale rejections of school bond proposals, and in survey upon survey. To hold them to an actionable "duty of care," in the discharge of their academic functions, would expose them to the tort claims—real or imagined—of disaffected students and parents in

> countless numbers. They are already beset by social and financial problems, which have gone to major litigation, but for which no permanent solution has yet appeared. The ultimate consequences, in terms of public time and money, would burden them—and society—beyond calculation.[117]

A similar tone was taken by a New York court of appeals in a case brought by a learning disabled student in *Donohue*[118] who argued that he was inadequately prepared to deal with everyday life, including the inability to fill out simple job applications. In dismissing the $5 million negligence suit, the appeals court based its decision equally on public policy concerns. The court expressed concern that recognition of this cause of action would require the courts to make judgments on the validity of broad educational policy and might eventually require their hand in review of day-to-day implementation of educational policies. The court argued that this would contravene the judicial policy of not becoming involved in educational questions, calling on Supreme Court precedent to support its position of removing itself from educational policymaking arenas.[119] Additionally, the court noted that the public has administrative avenues available for review of complaints through government itself. The court concluded: "[R]ecognition in the courts of this cause of action would constitute blatant interference with the responsibility for the administration of the public school system lodged by constitution and statute in school administrative agencies."[120]

In subsequent action, the same New York court dismissed a case in which a student had gained a lower court award of $500,000 due to what the lower court had held to be unusual circumstances meriting award. The child's claim in *Hoffman*[121] argued that the district had known for more than ten years that he had been misdiagnosed and wrongly placed in special education when in fact his test scores were within normal limits. The lower court had held that an affirmative act of negligence was evident by having damaged the student's education by wrongly prescribing a course of instruction inadequate to his skill level. The appeals court, however, again ruled that the judiciary had no role in intervening in the educational policymaking arena and that the student had the same administrative recourse it noted earlier in *Donohue.* Unwillingness to confront the enormous problems presented by tort claims for malpractice has also been demonstrated by the various courts that have reviewed such cases.[122]

Very recent cases continue to illustrate the reluctance of courts to become enmeshed in questions of malpractice. Many of the issues of negligence discussed previously come together in three illustrative cases involving educational malpractice claims. In *Rubio*,[123] parents charged that the school was negligent in hiring and retaining a teacher who had provided illicit drugs to students and permitted them to use drugs during school hours. The parents claimed educational malpractice and violation of constitutional rights under 42 U.S.C. § 1983 and § 1988. The trial court dismissed the case and awarded attorney fees to the district. On appeal, the court held that the district could not be liable for negligence in hiring or retaining a teacher since negligent acts were not within any exception to the New Mexico Tort Claims Act. By the same logic, the court held that the district could not be guilty of malpractice since malpractice was itself a form of negligence. Further, since the district itself did not sanction the distribution of drugs, the distribution could not have occurred under state supervision so as to invoke negligence.

Similar technicalities caused a ruling for defendant in *Jones*.[124] Although illiterate, the student received a diploma in 1968 while playing basketball and went on to a junior college whereupon he alleged that his athletic talent had been exploited. A lower court finding for negligence was reversed by the Michigan Court of Appeals, holding that immunity protected the board because claims arising before 1970 were covered by immunity, including intentional torts provided they were within the scope of governmental function.

The third recent case occurred in 1992 in *Ross*[125] as a talented high school basketball star was recruited to a university. The university told him that he would receive support for his academic deficiencies, but his skills never rose to a high school level during four years at the school. The student sued the university for negligence and breach of contract based on educational malpractice and negligent admission. A federal district court dismissed the complaint for failure to state a claim upon which relief could be granted. On appeal, the court stated that malpractice claims have been a persistent problem, noting that eleven states have previously rejected such claims.[126]

The court noted the reasons for such rejection. First, the lack of a satisfactory standard of care by which to evaluate an educator stands as a significant barrier to establishing negligence. Second, such claims have inherent uncertainties about the cause and extent of damages brought about by malpractice. Third, the potential for a torrent of malpractice litigation continues to cause courts to shy away from affirming such negligence. Fourth, the threat of engaging courts in educational policy debates continues to stand as a forbidding spectra. The breach of contract claim was only slightly more successful, as the court allowed it to proceed because the university had been charged with failing to deliver certain services.

Although educational malpractice has been an area of legitimate and growing concern for school districts, courts have not yet enthusiastically entered the fray sometimes said to be the "legalization of education." The fears that began in the 1970s have been largely unrealized, although much good has come of such tension. It is obvious that administrators and others have been benefited in that they have been more careful to avoid situations leading to such a lawsuit. Similarly, extant litigation has helped establish the posture of courts, making it possible to ameliorate some of the unfounded fears of districts, which otherwise might behave in an overly cautious manner. Although there has been little window of opportunity for plaintiff success, it nonetheless remains an open issue, dependent upon unique factual circumstances, whereupon it is still possible that an educational malpractice claim will succeed at some future point in time.

Defamation and Student Records

Although the concept of liability extends into many areas of school operations, one final area of concern in this chapter for school districts and school personnel is that of defamation and student records. School districts and school personnel have a particular concern in this sensitive area since schools are highly people-intensive in that staff and pupils represent the main resource and the main business of schools. Schools by the nature of operations prepare, preserve, and transmit information about staff and students. In the case of staff, administrators and boards are legally charged with evaluating and improving teachers and dismissing incompetent or undesirable personnel. In the case of students, the mission of public education is to prepare students for transition into the world of work and to become productive citizens within society. In both instances, written records and oral communication are the key sources of information transmittal which, if improperly conducted, may create enormous liability for districts and personnel. As a result, it is important to consider how defamation and student records relate to liability.

Defamation

The concept of defamation lies at the root of problems involving both student and professional personnel in school settings. By definition, defamation is described at considerable length, indicating the seriousness of the charge:

> Defamation is that which tends to injure reputation; to diminish the esteem, respect, goodwill or confidence in which the plaintiff is held, or to excite adverse, derogatory or unpleasant feelings or opinions against him.....A communication is defamatory if it tends so to harm the reputation of another as to lower him in the estimation of the community or to deter third persons from associating or dealing with him. The meaning of a communication is that which the recipient correctly, or mistakenly, but reasonably, understands that it was intended to express.[127]

As may be seen from the strong language of the *Restatement of Torts*, the concept of defamation is both broad and inclusive. Defamation may take the form of libel in which harm is done to another person in a permanent written form, or it may take the form of slander in which the same harm is done orally. Defamation may occur either *per se* or *per quod*, meaning that defamation *per se* is words that require no proof of actual harm beyond their clear meaning, "in respect of words, those which by themselves, and as such, without reference to extrinsic proof, injure the reputation of the person to whom they are applied"[128] and as defamation *per quod* means "in respect of words, those which require an allegation of facts, aside from the words contained in the publication, by way of innuendo, to show wherein the words used libel the plaintiff."[129] Either form is an actionable tort for which individuals may be held personally liable, and for which the only defenses are those of privilege and good faith.. For schools, defamation thus takes on special meaning in that the whole nature and purpose of a school is the process of evaluating and making statements about students and teachers.

The defenses of privilege and good faith against defamation are important in that they form much of the basis for the right of schools to conduct their duties of preparing and disseminating information about pupils and teachers. The defense of privilege in school settings is not the same as for other public officials, as with justices and state officers who generally have absolute privilege so long as they are carrying out their official duties or by other persons participating in judicial or legislative proceedings. In contrast, generally, school personnel have a qualified privilege in they may be protected so long as they are carrying out a common duty assigned to such organizations and provided that they are acting in good faith[130] and "upon a proper occasion, from a proper motive, in a proper manner, and based upon reasonable or probable cause."[131]

The qualified privilege under these restrictions has further limits, however, in that even truthful statements without good intent or statements honestly believed to be true may be actionable, as in a case where a teacher had improperly described a student as " . . . ruined by tobacco and whiskey."[132] For a communication to be protected by these defenses, it must be "made in good faith, without malice, not voluntarily but in answer to an inquiry, and in the reasonable protection of his own interest or performance of a duty to society, that the defendant must honestly believe the communication to be true, [that] there must have been reasonable or probable grounds known to him for the suspicion, [and] that the communication, if made in answer to an inquiry, must not go further than to truly state the facts."[133] Defamation under such conditions thus extends to written and oral communication in schools. For teachers, the issue is largely related to slanderous statements about their fitness for teaching and written evaluations regarding their performance. For students, the issue is more often related to permanent achievement records or anecdotal records that may arise to affect their employability after leaving school.

Pupil Records

Although the issue of defamation of pupils has been longstanding,[134] two recent events have heightened the liability concern for schools. The first event has been an increased interest in challenging pupil privacy rights, accompanied by a series of court rulings sympathetic to student privacy claims. This phenomenon largely resulted from a Supreme Court ruling in *Griswold*[135] in which the court held that there must be a compelling interest for any governmental act that infringes on privacy rights, including the confidentiality of personal information contained in records. This ruling and accompanying litigation led Congress to create the second event, which occurred with passage of the Family Rights and Privacy Act (FERPA) in 1974,[136] better known as the Buckley Amendment. This act stipulated in detail the records that must be made available to parents and students and stated the rules by which school information may be considered public and confidential.

The Act set out seven basic rules of compliance, causing many schools to reorder how they conducted student information accumulation and dissemination. The first rule demanded that individual student records identifying any single student must not be released without written consent of the parent[137] or of the student once the child reaches the age of eighteen.[138] The second rule demanded that all pupil records must be open to the parents of children under age eighteen and to the pupils themselves once the age of eighteen is reached.[139] The third rule required that no school could refuse to release such information on the basis that information was not identifiable by individual child.[140] The fourth rule required that staff members permitted to examine student records must be named and that the names of these people be made known to the "owners" of the records.[141] The fifth rule required that anyone accessing student records must sign a permanent record showing access to the records.[142]

The sixth rule demanded that the contents of a file may be challenged and set forth procedures for a hearing to resolve the issue.[143] And the seventh rule required the school to set out what it considered to be public directory information and to identify the circumstances under which that information will be released.[144] In addition, the Act provided that students under age eighteen may gain access to records if the school chooses to permit it, or if the parents grant access to the child as an authorized third party.[145]

In addition to the Act's provisions protecting student privacy, other provisions were made regarding how records may be released without parental permission. Records may be released to school officials in a district to which the student intends to transfer after the parent has had a chance to inspect the records.[146] In *Klipa*,[147] the district was found not liable for invasion of privacy where it had inadvertently permitted a student's psychological and medical records to be transferred to the new school when the parents had requested that only academic records be sent. The court held that there was no liability because "the consent of the parents was not necessary in order to permit the transfer of all the child's records to the new school."[148] Further, all records may be released to school officials in the same district who have been determined to have a legitimate educational interest in the records.[149] Additionally, records can be given to student financial aid officers, but only to the extent necessary to determine eligibility,[150] and records can be released to accrediting agencies for the purpose of accreditation.[151] Records can also be released to various state and national educational agencies when enforcing federal laws,[152] and records can be released in compliance with a court order after the school has made a reasonable effort to notify the parent or student of the order prior to compliance.[153]

In *Rios*,[154] a class action involving hundreds of files resulted in subpoena of records by one of the parties. The court held that even when a number of records are subpoenaed, the school still had the duty to make efforts to contact parents and students, although notice by publication or mail would suffice. The court noted "it would seem sensible to require in the disclosure order that the recipients of the student records avoid revealing the litigation and destroy the data when it is no longer needed. But it is neither required or necessary that the defendants remove the names of the students from the records and substitute neutral identifying information."[155] Further, names and addresses may be considered public directory information,[156] although notification to such effect must occur and parents and students may request deletion from the list.[157] Finally, information that is not personally identifiable can be released,[158] and emergencies involving health or safety will qualify as well.[159] The provisions of FERPA extend to all information directly related to students, recorded in any form and maintained by the school with primary exception for notes made by a teacher in a teacher's log, which are not disclosed to others and physician's or psychologist's notes, which are used for treatment and are not disclosed to others.[160]

As noted earlier, provisions of the Act require schools to prepare annually a list of procedures and policies governing access to student records. The general rules on inspection of records include:

- A school must respond to written or oral request to inspect within a reasonable time, not to exceed 45 days.[161]

- Authorized persons are entitled to physically inspect all records regardless of their location. They may request access to all their records without having to specify particular records in which they are interested. A parent or eligible student may bring another along, although a written consent form may be required by the school.[162]

- As described earlier, a school must provide copies of the records on a parent's or student's request whenever records are transferred to another school, whenever information is released to a third party, or whenever denial of copies would effectively deny the right of access.[163] A school may charge a reasonable fee for copying but may not charge for the labor expended in searching for or retrieving records.[164]

- A school may condition receipt of transcripts on payment of all fees[165] and loans.[166]

In addition, rules on hearings for challenging records discussed earlier call for a hearing to be scheduled within a reasonable time with notice given to participants allowing sufficient time to afford an opportunity to prepare and must be conducted before an impartial hearing officer. Further, the parent or student has the right to be represented, to present evidence, and to receive a reasonably prompt decision.[167] If the hearing officer denies the request, the complainant may place a statement of explanation in the record. Finally, the remedy for violation of the Act is withdrawal of federal funds, as the courts have determined that no alternate private remedy exists.[168]

Recent cases have continued to hold schools and school personnel to close observance of privacy rights, to personal responsibility for actions, and to requirements of law for proving claims against defendants. In *McKee,*[169] a private school student was successful in an action seeking injunctive relief and damages because the school refused to grant a transcript due to failure to pay tuition. The court held that the student was entitled to recover the transcript on the theory of detrimental reliance. In another student records case, the court ruled in *Demers*[170] that, where a graduate student had sought data from police department internal affairs documents, nonpending complaint forms were government information subject to disclosure under the Minnesota Government Data Practices Act.

Similarly, a former student was successful in a higher education case in which a class directory was published containing data from a questionnaire in which the alumnus had made the statement that he had "come to terms with his homosexuality and the reality of AIDS."[171] The federal court ruled that the alumnus could proceed with defamation claim without alleging or proving special damages. Opposite findings, however, can be costly as in *Jungherr*[172] where a citizen was named as a defendant in a defamation suit brought by the board president because of statements made at a public meeting about the president's handling of board business. The citizen then brought countersuit claiming first amendment right to freedom of speech, and seeking more than $28,000 in attorney fees incurred in exonerating himself. The court denied the claim, however, holding that a citizen is free to exercise his first amendment right, but such right does not guarantee that he will never need to defend himself in court.

Likewise in *Sander,*[173] two board members failed in their suit when suing another board member for public statements he had made about their alleged malfeasance, fraud, and mismanagement of district funds. A jury awarded $1.25 million to each plaintiff in actual damages and $750,000 in punitive damages, but the state supreme court set the judgment aside saying that the dollar amount was arbitrary and capricious and so grossly excessive as to indicate passion, prejudice, or other improper motive. While different cases obviously reach different conclusions for diverse reasons, defamation claims are weighty matters deserving of careful behavior so as to not violate either constitutional or statutory protections.

Teacher Defamation

Although considerable space was devoted earlier to avoiding charges of defamation relating to students, most of that discussion was actually centered on student records because there have been relatively few cases involving traditional defamatory claims by and among students. The issue of teacher defamation, however, has been much more contentious largely due to teachers claiming that they have either been victims of slanderous statements about their fitness for teaching or victims of libelous written evaluations regarding their performance. Such claims are inevitable to schools, in that every state requires administrators to observe and evaluate teacher performance and because the nature of schools finds constant opportunity for potentially offensive comments by parents or by school administrators about teacher performance and teacher fitness.

As a general rule, courts are fairly slow to find defamation in schools except in clear factual instances not otherwise explainable. As noted in *Malia*,[174] qualified immunity is a common rule except in instances where it is shown that comments were motivated by personal malice or animosity or where the weight of the different requirements of defamation discussed earlier under privilege and good faith immunity is overwhelming. This reality is evident in commonly cited cases, such as *Santavicca*[175] in which a written reprimand to a coach by a superintendent was covered under absolute privilege and his comments to the press were protected by qualified privilege. Similarly, the court held in *Agins*[176] that statements made by a superintendent in the course of his official duties in connection with investigating complaints of wrongdoing were not actionable in defamation.

Likewise, the court held for the defendant district in *Hayes*[177] in a case where the superintendent was alleged to have said that a teacher was homosexually involved with one of the defendants. The court held that there was no evidence that such characterization would expose the plaintiff to public hatred or contempt, and importantly noting the Supreme Court's hostility toward expanding the scope of defamation *per se*. Similarly, the court held in *Anderson*[178] that the district could not be liable for newspaper articles that quoted statements by a district employee. The court held that the district was not liable under a Section 1983 claim for unconstitutional acts of employees under the doctrine of *respondeat superior*, that plaintiff failed to prove the acts were a result of official policy, and that both the district and employee were immune from torts under the state code.

Privilege reaches in both directions, however, as parent complaints about teachers generally receive the same consideration. For example, a court held in *Garrett*[179] that a list of grievances and allegations against a teacher presented to a board of education did not constitute defamation of character. Similarly, the court held in *Martin*[180] that channels for airing grievances were naturally and intentionally part of the educational process and should not be obstructed. In the same vein, the court held for the defendant in *Weissman*,[181] noting that a law requiring complaints to be in writing weighed heavily against a finding for libel by making it counterproductive for parents to complain if their acts would then lead quickly to liability for defamation. Likewise, the court in *O'Brien*[182] ruled for defendants as a group of twenty-seven teachers sued newspapers, reporters, and a news wire service for defamation and invasion of privacy resulting from a news report about a meeting of parents responding to the expulsion of a student for fighting with a teacher and to request investigation of possible sexual misconduct by teachers. In holding for defendants, the court found that the news article did not defame the teachers since no false statements were made.

While different cases reach different conclusions, defamation claims are thorny matters for school districts and educational personnel. In the case of student records, FERPA caused many districts to purge records and to refuse to enter any negative comments into such files. In fact, the Act did not intend that negative notations should never be made, rather the Act merely required that such information must be factual and related to the student's education in order to avoid defamatory liability. Similarly, overcaution has occurred in more general defamation issues. Although caution is always wise when dealing with a definition of defamation so broad as to encompass "that which tends to injure reputation, to diminish the esteem, respect, goodwill, or confidence in which the plaintiff is held, or to excite adverse, derogatory, or unpleasant feelings or opinions against him,"[183] the courts have recognized that it is possible to so severely restrict communication to the extent that fear of liability becomes an inordinate stricture on the conduct of normal and productive business.

In the case of teacher defamation, educational personnel in fact enjoy qualified privilege whereby written or spoken communication may not be tortious as long as such communication is made to appropriate individuals and with proper purposes. Likewise, the burden of proof is on the complainant in that the public nature of schools requires open lines of communication that sometimes stretch the limits of normal permissibility. Clearly the best advice to districts and educational personnel is to be aware of individual rights and consider those rights before acting and to review policies with current constitutional and statutory standards in mind. Similarly, the prudent district and the wise administrator will also seek to minimize exposure by ensuring heavily against tort liability, as discussed elsewhere in this text.

Summary

Although hostile litigation seeking to enforce claims of aggrieved persons has been a longstanding concern for schools, the latter half of the Twentieth Century has seen a tremendous increase in the amount of litigation and in the development of legal concepts regarding liability of schools and school personnel. As noted in the last edition of this text, the burning issues of the last decade focused on matters involving policies and the educational process including personnel management, negligence, civil rights actions, and interpretation and application of school rules. This has proven accurate, and it is equally correct to say that this trend has continued unabated throughout the first half of the 1990s and will undoubtedly maintain or even increase its pace as we approach a new millennium. While administrators and other school people can expect no relief from the vigilance of a society eager to take offense and to profit from a legal system that tolerates dissent and is in fact fundamentally adversarial in nature, these realities do not suggest that the parties should draw apart into nervous and defensive camps. As noted earlier, education remains a great and noble pursuit since life's opportunities are in large measure a product of the education received by children. Since children are the most evident expression of the future, it is more important than ever to learn to live with unpleasant realities while seeking to make improvement where needed.

The material in this chapter has not claimed to exhaustively review every area of law that imposes liability on school districts and educational personnel. The chapter has, however, focused on selected concerns and issues facing every person associated with education, and it has argued that exposure to liability has placed every educator and policymaker in a position that demands knowledge of vulnerability and knowledge of the rights and responsibilities of everyone involved. Under these circumstances, administrators will need to know the law and respond in appropriate fashion to minimize exposure to liability.

Endnotes

[1] *See, e.g.,* Chap. 2 in D.C. Thompson; R.C. Wood, and D.S. Honeyman, *Fiscal Leadership for Schools: Concepts and Practices.* New York: Longman (1994); *see also* G.S. Counts, *Education and American Civilization.* New York: Columbia University Teachers College (1952); *see also* L. Cremin, *American Education: The Colonial Experience 1607-1783.* New York: Harper & Row (1970); *see also* J. Spring, *The American School 1642-1990.* 2d New York: Longman (1990).

[2] Sect. 1, Mass. School Law of 1647.

[3] For treatment of the concept of equality in the law, *see generally* R. Salomone, *Equal Education Under Law* (New York: St. Martin's Press, 1986). *See also* a wide spectrum of cases cited in notes in Chapter 6 of this text relating to civil rights litigation, the broader equal protection and fundamental rights cases, and so forth.

[4] *See generally,* N. Edwards, *The Courts and the Public Schools.* The Univ. of Chicago Press (1955).

[5] *Meyer v State of Nebraska,* 262 U.S. 390, 43 S.Ct. 625, 67 L.Ed. 1042 (1923), holding that a law forbidding public school instruction in German in order to develop English skills and American political ideals violated the rights of teachers to teach, students to learn, and parents to control their children's educations.

[6] *See, e.g., Pierce v Society of Sisters,* 268 U.S. 510, 45 S.Ct. 571, 69 L.Ed. 1070 (1925) regarding the right to a private education; *Wisconsin v Yoder,* 406 U.S. 205, 92 S.Ct. 1526, 32 L.Ed.2d 15 (1972) regarding rights of Amish believers to choose contrary to social conventions about the value of modern education.

[7] *See* the many special education cases reflecting struggles over defining an appropriate education, *e.g., Pennsylvania Ass'n of Retarded Children v Commonwealth of Penn.,* 344 F.Supp. 1257 (E.D.Pa.1971) where by consent decree the state capitulated to demands for appropriate education; *Mills v Board of Educ. of the District of Columbia,* 348 F.Supp. 866 (D.D.C. 1972) where plaintiffs won a ruling (as opposed to consent decree in PARC) holding improper the absence of specialized educational program for handicapped children; *Lau v Nichols,* 414 U.S. 63, 94 S.Ct. 786, 39 L.Ed. 2d 1 (1974) ruling discrimination under Title VI of the Civil Rights Act of 1964 against language-deficient children where instruction in their native language was not equally available.

[8] Two very early struggles reflecting this view are found in *Commonwealth v Hartman,* 17 Pa. 118 (1851) in which a school board refused to obey a law requiring them to provide common schools for all children, rather than merely poor children; likewise, the famous case of *Stuart School Dist. of the Village of Kalamazoo,* 30 Mich. 69 (1874) in which the court held that a citizen's protest against local maintenance of a high school absent express statutory authorization did not exceed the board's scope of control. A sense of such "wars" over education can be found in D. Ravitch, *The Great School Wars: New York City, 1805-1973.* New York: Basic Books (1974).

[9] *Mower v The inhabitants of Leicester,* 9 Mass. 247 (1812).

[10] *Nichols v United States,* 74 U.S. 122 (1869).

[11] *Black's Law Dictionary,* 6th ed. (St. Paul, MN: West Pub. Co., 1990), 1179. This doctrine states "let the master answer," so that a master is liable for wrongful acts of servants. However, the doctrine applies only where injury occurs while the

servant is acting within the scope of legitimate authority, so that illegal acts do not fall to the employer.

[12] Federal Tort Claims Act (1946).

[13] *Molitor v Kaneland Comm. Unit Dist.,* 18 Ill.2d 11, 163 N.E.2d 89 (Ill. 1959).

[14] This court held that sovereign immunity was transported to this nation in 1788 in *Russell v Men of Devon,* 2 Term Rep. 671, 100 Eng. Rep. 359 (1788).

[15] *Town of Waltham v Kemper,* 55 Ill. 346 (1870).

[16] *Kennare v City of Chicago,* 171 Ill. 332, 49 N.E. 536 (1898).

[17] *Molitor,* 18 Ill.2d 11, 163 N.E.2d, quoting *Barker v City of Santa Fe,* 47 N.M. 85, 136 P.2d at 482.

[18] *Black's Law Dict.,* 626.

[19] *Black's Law Dictionary,* 1097-1098.

[20] *Sawaya v Tucson High Sch. Dist.,* 78 Ariz. 389, 281 P.2d 105 (1955).

[21] *Richards v School Dist. of City of Birmingham,* 348 Mich. 490, 83 N.W.2d 643 (1957).

[22] A similar decision favoring nonliability was reached in *Kellam v School Board of Norfolk,* 202 Va. 252, 117 S.E.2d 96 (1960).

[23] The distinction is more than semantic, as districts are also held to have both ministerial and discretionary duties. A ministerial duty is one in which administrative acts are performed, while a discretionary duty is a policymaking one (this concept appears elsewhere in this text, wherein the focus is on whether a board may delegate duties to employees). Courts, in holding on the liability question analogous to the one relating to proprietary/governmental acts, have sometimes ruled that districts are liable for ministerial acts but are immune from liability for discretionary acts; the logic, regardless of terminology, is the same. *See, e.g., Greider v Shawnee Mission Unif. Sch. Dist. 512,* 710 F.Supp. 296 (D. Kan. 1989) where discretionary immunity is not available if the district fails to meet its statutory responsibility; *see also, Smith v Board of Educ. of Caney School Dist.,* 204 Kan. 580, 464 P.2d 571 (1970) where an open stairwell upon which an adult injury was sustained during a community meeting did not impose liability on the district beyond the duty to refrain from intentional injury; *Heigl v Board of Educ. of Town of New Canaan,* 587 A.2d 423 (Conn. 1991) where an open campus fell properly within the definition of an immune discretionary activity; *see also, e.g., Koehn v Board of Educ. of City of Newton,* 193 Kan. 263, 293 P.2d 949 (1964); *Ross v Consumers Power Co.,* 420 Mich. 567, 363 N.W.2d 641 (1984); *Stout v Grand Prairie Ind. School Dist.,* 733 S.W.2d 290 (Tex. App. 1987).

[24] An exception to this statement is reserved for later discussion under abrogation by legislatures.

[25] *Davis v Homestead Farms, Inc.,* 359 N.W.2d 1 (Mich. Ct. App. 1984).

[26] *McManus v Anahuac Indep. Sch. Dist.,* 667 S.W.2d 275 (Tex. Ct. App. 1984).

[27] *Hutt v Etowah County Bd. of Educ.,* 454 So. 2d 973 (Ala. 1984).

[28] *Ross v Consumers Power Co.,* 420 Mich. 567, 363 N.W.2d 641 (1984).

[29] *Stout v Grand Prairie Ind. Sch. Dist.,* 733 S.W.2d (Tex. App. 1987) at 296.

[30] *Black's Law Dict.*, 961.

[31] *Ferris v Board of Educ. of Detroit,* 122 Mich. 315, 81 N.W. 98 (1899).

[32] *Ferris.*

[33] *Sestero v Town of Glastonbury,* 19 Conn. Sup. 156, 110 A.2d 629 (1954).

[34] *Black's Law Dict.* 119.

[35] *Smith v Board of Educ. of Caney Sch. Dist,* 204 Kan. 580, 464 P.2d 571 (1970).

[36] *Jones v Kansas City,* 176 Kan. 406, 271 P.2d 803 (1954).

[37] *Sly v Board of Educ. of Kansas City,* 213 Kan. 415, 516 P.2d 895 (1973).

[38] *Hendricks v Southfield Public Schools,* 178 Mich. App. 672, 444 N.W.2d 143 (1989).

[39] U.S. Const., Amend. XI. This amendment was proposed to the legislatures of the several states by the Third Congress on March 4, 1794, and was declared to have been ratified in a message from the President to Congress, dated Jan. 8, 1798.

[40] *Vanstophorst v Maryland,* 2. U.S. (2 Dall.) 401 (1791).

[41] *Chisholm v Georgia,* 2 U.S. (2 Dall.) 419 (1792).

[42] *Hans v Louisiana,* 134 U.S. 1, 10 S.Ct. 504 (1890); *Great Northern Life Ins. Co. v Read,* 322 U.S. 47, 64 S.Ct. 873 (1944).

[43] *Urbano v Board of Managers of New Jersey State Prison,* 415 F.2d 247 (3d Cir. 1969), *cert. denied,* 397 U.S. 948 (1970). Other lesser tests are also applied, including questions about statutory provision for immunity and the agency's legal configuration.

[44] *Atasadero State Hosp. v Sanlon,* 473 U.S. 234, 105 S.Ct. 3142 (1985).

[45] *Edelman v Jordan,* 415 U.S. 651, 94 S.Ct. 1347 (1974).

[46] Individuals with Disabilities Education Act (1990), formerly Education Handicapped Act (1976).

[47] Individuals with Disabilities Education Act, *Abrogation of State Sovereign Immunity,* Sect. 1403.

[48] *Mt. Healthy City Sch. Dist. Bd. of Educ. v Dole,* 429 U.S. 274, 97 S.Ct. 568 (1977).

[49] *Monell v Department of Social Services of City of New York,* 436 U.S. 658, 98 S.Ct. 2018 (1978).

[50] *Alabama v Pugh,* 438 U.S. 781, 98 S.Ct. 3057 (1978).

[51] *Molitor.*

[52] Although many sources mention strict liability, the topic is seldom developed beyond mere mention. *Black's* definition of strict liability states that it is "a concept applied by the courts in product liability cases in which a seller is liable for any and all defective or hazardous products which unduly threaten a consumer's personal safety . . .. The concept of strict liability in tort is founded on the premise that when manufacturer presents his goods to the public for sale, he represents they are suitable for their intended use" (*Black's Law Dictionary,* 6th ed. (St. Paul, MN: West Pub. Co., 1990), 1275). The concept is rarely applied to schools, being reserved almost exclusively for instances involving abnormal danger generally through ultrahazardous conditions. The scarcity of such claims in schools has been the reason why the concept receives little notice in education litigation literature and forms the basis for omitting a discussion in this text.

[53] *Black's Law Dict.,* 1335, 1435.

[54] *Black's Law Dict.,* 105.

[55] *Black's Law Dict.,* 139.

[56] *Ingraham v Wright,* 430 U.S. 651, 97 S.Ct. 1401, 51 L.Ed.2d 711 (1977).

[57] U.S. Const., Amend. VIII, reads: Excessive bail shall not be required, nor excessive fines imposed, nor cruel and unusual punishments inflicted.

[58] *Mathis v Berrien County Sch. Dist.,* 378 S.E.2d 505 (Ga. App. 1989).

[59] *Woodard v Los Fresnos Indep. Sch. Dist.,* 732 F.2d 1243 (5th Cir. 1984).

[60] *Willoughby v Lehrbass,* 388 N.W.2d 688 (Mich. App. 1986).

[61] *Thompson v Iberville Parish Sch. Bd.,* 372 So. 2d 642 (La. Ct. App. 1979), *writ denied,* 374 So. 2d 650 (La. 1979).

[62] *See, e.g., Gardner v State,* 4 Ind. 632 (1853); *Cooper v McJunkin,* 4 Ind. 290 (1853).

[63] *Frank v Orleans Parish Sch. Bd.,* 195 So. 2d 451 (La. Ct. App. 1967).

[64] *Sansone v Bechtel,* 429 A.2d 820 (Conn. 1980).

[65] *Garcia v Miera,* 817 F.2d 650 (10th Circ. 1987).

[66] *Burton v Kirby,* 775 S.W.2d 834 (Tex. App. Austin 1989).

[67] W.L. Prosser, *Law of Torts* (St. Paul, Minn: West Pub. Co., 1971), 124.

[68] *Black's* defines such damages as "Nominal damages are a trifling sum awarded to a plaintiff in an action, where there is no substantial loss or injury to be compensated, but still the law recognizes a technical invasion of his rights or a

breach of the defendant's duty, or in cases where, although there has been a real injury, the plaintiff's evidence entirely fails to show its amount." *Black's Law Dict.,* 351.

[69] *Black's Law Dict.,* 351-352.

[70] *Restatement of Torts, Second.*

[71] *Restatement,* § 463.

[72] W.L. Prosser, *Law of Torts* (St. Paul, Minn: West Pub. Co., 1971), 274.

[73] *Black's Law Dict.,* 113.

[74] The opposite case illustrates the concept. In instances where immunity has been the rule, there is an argument about whether insurance may be legally purchased because it is said to be an unauthorized use of public moneys to insure against a risk that does not exist since there is no risk when government is immune; see *Board of Educ. of County of Raleigh v Commercial Casualty Ins. Co.,* 182 S.E. 87 (W. Va. 1935). In contrast, other cases have reached different conclusions on whether availability of insurance abrogates immunity: e.g., *Lehman v Wansing,* 624 S.W.2d 1 (Mo. 1981) where immunity may still be available even though statutory authority to buy insurance exists; other states have ruled that buying insurance acts to voluntarily waive immunity, *e.g., Crowell v School Dist.,* 805 P.2d 522 (Mont. 1991); *School Bd. of Orange County v Coffey,* 524 So. 2d 1052 (Fla. Dist. Ct. App. 1988), *rev. den.,* 534 So. 2d 401 (Fla. 1988); *Dugger v Sprouse,* 364 S.E.2d 275 (Ga. 1988), and others.

[75] *Velmer v Baraga Area Schools,* 424 N.W.3d 770 (Mich. 1988).

[76] *Moore v Port Arthur Indep. School Dist.,* 751 F.Supp. 671 (E.D. Tex. 1990).

[77] *Grubbs v Aldine Indep. Sch. Dist.,* 709 F.Supp. 127 (S.D.Tex. 1989).

[78] *Higgins v East Valley Sch. Dist.,* 704 P.2d 630 (Wash.App. 1985).

[79] *Ward v Cmty. Unit School Dist. 220,* 572 N.E.2d 986) (Ill. App. 1st Dist. 1991).

[80] *Stevens v Chestine,* 561 So. 2d 1100 (Ala. 1990).

[81] *Joliuette v Iberia Parish Sch. Bd.,* 601 So. 2d 812 (La .App. 3d Circ. 1992).

[82] *McCloskey v Abington Sch. Dist.,* 515 A.2d 642 (Pa. Cmwlth. 1986).

[83] *Maldonado v Josey,* 975 F.2d 727 (10th Cir. 1992).

[84] *DeShaney v Winnebago County Dept. of Social Services,* 109 S.Ct. 998 (1989).

[85] *Waltz v Wake County Bd. of Educ.,* 409 S.E.2d 106 (N.C. App. 1991).

[86] *Byrd v Bossier Parish Sch. Bd.,* 543 So. 2d 35 (La. App. 1989).

[87] *Gilmore v City of Zion,* 605 N.E.2d 110 (Ill. App. 2d Dist. 1992).

[88] *Fielding v Heiderick,* 832 P.2d 1244 (Or. App. 1992).

[89] *Brownell v Los Angeles Unif. Sch. Dist.,* 5 Cal. Rptr. 2d 761 (Cal. App. 1992).

[90] *Jefferson County School Dist. v Justus,* 725 P.2d 767 (Colo. 1986).

[91] *Joyce v Simi Valley Unif. School Dist.,* 7 Cal. Rptr. 2d 783 (Cal. App. 2d Dist. 1992).

[92] *Doe v Taylor Indep. Sch. Dist.,* 975 F.2d. 137 (5th Cir. 1992).

[93] *DeShaney.*

[94] *Doe v City of New Orleans,* 577 So. 2d 1024 (La. App. 1991).

[95] *W.L.O. v Smith,* 585 So. 2d 22 (Ala. 1991).

[96] 42 U.S.C.A. 1983.

[97] 42 U.S.C.A. 1983.

[98] *Monell v New York City Dept. of Social Services,* 436 U.S. 658, 98 S.Ct. 2018 (1978) contravening *Monroe v Pape,* 365 U.S. 167 (1961).

[99] *Wood v Strickland,* 420 U.S. 308 (1975).

[100] *Mt. Healthy City Sch. Dist. Bd. of Educ. v Doyle,* 429 U.S. 274 (1977).

[101] *Wood v Strickland,* 420 U.S. 308 (1975).

[102] *Wood* at 322.

[103] *Gomez v Toledo,* 446 U.S. 635 (1980).

[104] *Harlow v Fitzgerald,* 457 U.S. 800 (1982).

[105] *Harlow* at 818.

[106] *Gavrilles v O'Connor,* 611 F.Supp. 210 (D. Mass. 1985).

[107] *Littlejohn v Rose,* 768 F.2d 765 (6th Cir. 1985), *cert. den.,* 475 U.S. 1045 (1986).

[108] J. Horner, "The Anatomy of a Constitutional Tort," 47 *Ed. Law Rep.* 1 (1988), 13.

[109] *Carey v Piphus,* 435 U.S. 247 (1978).

[110] 42 U.S.C.A. (1985).

[111] *Hensley v Eckerhart,* 461 U.S. 424 (1983).

[112] *Webb v County Bd. of Educ.,* 53 U.S.L.W. 4473 (1985).

[113] *New York Gaslight v Carey,* 447 U.S. 54 (1980).

[114] *Hunter v Board of Educ.,* 292 Md. 481, 439 A.2d 583 (1982).

[115] *Peter W. v San Francisco Unif. Sch. Dist.,* 60 C.A.3d 814, 131 Cal. Rptr. 854 (1976).

[116] *Peter W.,* 860.

[117] *Peter W.,* 861.

[118] *Donohue v Copiague Union Free Sch. Dist.,* 47 N.Y.2d 440, 418 N.Y.W.2d 375, 391 N.E.2d 1352 (1979).

[119] *See, e.g., Board.of Curators of Univ. of Missouri v Horowitz,* 435 U.S. 78, 98 S.Ct. 948 (1978).

[120] *Donohue* at 1354.

[121] *Hoffman v Board of Educ.*, 410 N.Y.S.2d 99 (N.Y. App. Div. 1978), *rev'd*, 424 N.Y.S.2d 376 (1979).

[122] *See, e.g., Aubrey v School Dist. of Philadelphia,* 437 A.2d 1306 (Pa. Commw. Ct. 1981); *D.S.W. v Fairbanks North Star Borough Sch. Dist.*, 628 P.2d 554 (Alaska 1981); *Doe v Board of Educ. of Montgomery County,* 453 A.2d 814 (Md. 1982); *Keech v Berkeley Unif. Sch. Dist.*, 210 Cal. Rptr. 7 (Cal. Ct. App. 1984); *Johnson v Clark,* 418 N.W.2d 466 (Mich. Ct. App. 1987); *Poe v Hamilton,* 565 N.E.2d 887 (Ohio Ct. App. 1990).

[123] *Rubio v Calsbad Mun. Sch. Dist.*, 744 P.2d 919 (N.M. App. 1987).

[124] *Jones v Williams,* 431 N.W.2d 419 (Mich. App. 1988).

[125] *Ross v Creighton Univ.*, 957 F.2d 410 (7th Cir. 1992).

[126] Alabama, Alaska, California, Florida, Idaho, Iowa, Kentucky, Maryland, New Jersey, New York, and Wisconsin.

[127] *Restatement, Second, Torts* §§ 559,563.

[128] *Black's Law Dict.*, 376.

[129] *Black's Law Dict.*, 376.

[130] *Restatement, Second, Torts* § 596.

[131] *Baskett v Crossfield,* 228 S.W. 673 (Ky. 1920).

[132] *Dawkins v Billingsley,* 172 P. 69 (1918).

[133] *Baskett v Crossfield,* 228 S.W. 673 (Ky. 1920).

[134] *See, e.g., Dawkins v Billingsley,* 172 P. 69 (1918).

[135] *Griswold v Connecticut,* 381 U.S. 479 (1965).

[136] 20 U.S.C.A. § 1232 g; 34 C.F.R. § 99 *et seq.*

[137] 20 U.S.C.A. § 1232 g(b) (1) and 34 C.F.R. § 99.30 (a)(3).

[138] 20 U.S.C.A. § 1232 g(d) (1) and 34 C.F.R. § 99.5.

[139] 20 U.S.C.A. § 1232 g(a) (1) (A) and 34 C.F.R. § 99.10.

[140] 34 C.F.R. § 99.6 (a)(2)(iv).

[141] 34 C.F.R. § 99.6 (a)(4).

[142] 34 C.F.R. § 99.6 (a)(5) and § 99.32.

[143] 30 U.S.C.A. § 1232g(a)(2) and 34 C.F.R. § 99.6 (a)(7).

[144] 30 U.S.C.A. § 1232g(a)(5)(B); 34 C.F.R. § 99.3; § 99.6(a)(6) and § 99.37.

[145] 34 C.F.R. § 99.4(c).

[146] *Page v Rotterdam,* 441 N.Y.S.2d 323 (1981).

[147] *Klipa v Board of Educ.*, 460 A.2d 601 (Md. App. 1983).

[148] *Klipa* at 608.

[149] 34 C.F.R. 99.31 (a)(1).

[150] 34 C.F.R. 99.31 (a)(4).

[151] 34 C.F.R. 99.31 (a)(7).

[152] 34 C.F.R. 99.31 (a)(3).

[153] 34 C.F.R. 99.31 (a)(9); *Mattie T. v Johnston,* 74 F.R.D. 498 (D.C. Miss. 1977).

[154] *Rios v Read,* 74 F.R.D. 589 (D.C.N.Y. 1977).

[155] *Rios* at 602.

[156] *Hathaway v Joint Sch. Dist.,* 110 Wis. 2d 254, 329 N.W.2d 217 (1982); *Kestenbaum v Michigan State Univ.,* 294 N.W.2d 228 (Mich. App. 1980).

[157] 34 C.F.R. 99.31 (a)(10).

[158] *See, e.g., Dryston v Board of Educ.,* 430 N.Y.S.2d 688 (1980) where access to standardized test scores was allowed after names were removed and scores were randomly reordered.

[159] 34 C.F.R. 99.3 and 99.37.

[160] 34 C.F.R. 99.3.

[161] 34 C.F.R. 99.11(a).

[162] 34 C.F.R. 99.11(a).

[163] 34 C.F.R. 99.11(b)(2).

[164] 34 C.F.R. 99.8(a)(b).

[165] *Spas v Wharton,* 106 Misc, 2d 180, 431 N.Y.S.2d 638 (1980).

[166] *Girardier v Webster Coll.,* 563 F.2d 1267 (8th Cir. 1977).

[167] 34 C.F.R. 99.21 (c)(d).

[168] 20 U.S.C.A. 1232g (a)(1)(A) 1976.

[169] *McKee v Southfield Sch.,* 613 So. 2d 659 (La .Ct .App. 1993).

[170] *Demers v City of Minneapolis,* 468 N.W.2d 71 (Minn. 1991).

[171] *Sleem v Yale Univ.,* 843 F.Supp. 57 (M.D.N.C. 1993).

[172] *Jungherr v San Francisco U.S.D. Bd. of Educ.,* 923 F.2d 743 (9th Cir. 1991).

[173] *Sander v Prince,* 403 S.E.2d 640 (S.C. 1991).

[174] *Malia v Monchak,* 543 A.2d 184 (Pa. Cmmwlth. Ct. 1988).

[175] *Santavicca v City of Yonkers,* 518 N.Y.S.2d 29 (N.Y.App. Div. 1987).

[176] *Agins v Darmstadter,* 544 N.Y.S.2d 635 (App. Div. 1989).

[177] *Hayes v Smith,* 832 P.2d 1022 (Colo. Ct. App. 1991).

[178] *Anderson v Blankenship,* 790 F.Supp. 695 (E.D. Tex. 1992).

[179] *Garrett v Olson,* 691 P.2d 123 (Or. Ct. App. 1984).

[180] *Martin v Kearney,* 124 Cal.Rptr.281 (Cal. Ct. App. 1975).

[181] *Weissman v Mogel,* 462 N.Y.S.2d 383 (N.Y. 1983).

[182] *O'Brien v Williamson Daily News,* 735 F.Supp. 218 (E.D. Ky. 1990).

[183] *Black's Law Dict.,* 375.

Chapter 8 Personnel

Introduction

Throughout the preceding chapters of this text, there has been a tremen dous emphasis placed on critical issues affecting public school administrators. Although the topics have varied widely, the overriding emphasis has been on the relationship of schools to the people who make up the school community in all its various aspects. Schools are first and fundamentally peopled organizations, with far-flung responsibilities that range from the tasks performed by the highest ranking administrators and board executives to a wide host of jobs carried out by professional and service personnel. This vast personnel panorama takes in people serving as district and building administrators, classroom teachers, professional support staff, permanent and itinerant teaching or support staff, paraprofessionals, and classified employees including custodians, maintenance workers, foodservice personnel, and clerical and other administrative classified staff.

The personnel function is an area of great interest to school business administrators, since the basic nature of schools causes most of the educational budget to be consumed by employee costs. Further, every problem and every opportunity faced by school systems originates with or relates to personnel, from a solution perspective or from the fact that no goal or objective can be carried out without involving the district's personnel.

The Personnel Function

As has been emphasized in previous chapters, the business of school districts and school business administration has come under recent attack in the United States. While the reasons for this criticism have been varied, one of the driving forces has been dissatisfaction with the high costs of education. These costs are driven in large part by salaries and benefits. As the costs of personnel have skyrocketed, concerns have been raised among citizens about whether education's benefits equal or exceed its expense to society. While such questions cannot be answered simply, they can at least in part be addressed by considering the scope of education's personnel function and the goals and objectives that underlie personnel operations in schools.

Scope of Expenditures

Although nearly every text addressing the concept of education funding has noted that education is big business, the size of education's fiscal demands and outlays are often understated by a fairly innocent public that perceives a great cost but fails to understand or appreciate the extent and uses of money in education.[1] These costs are consumed largely by people working in school organizations, and the sheer enormity of education's demands on human and fiscal resources is astounding. Although the public has become accustomed to massive numbers relating to the American economy and the American tax system, few people can appreciate the vastness of public finance.

Vast amounts of money are being spent in states and local communities for public elementary and secondary education. In 1994, over $160 billion were expended for public schools. Expenditures for schools have traditionally been funded on a current basis requiring that each year's revenue must equal or exceed expenditures. The burden of current expenditures increasing annually can be well illustrated using some historical data. Although recordkeeping was not emphasized in the earlier years of the nation, there are sufficient data to indicate steady growth in expenditures for the last hundred years.

In 1900, expenditures in the United States for K-12 education were $215 million, or 1 percent of the Gross National Product (GNP). A tremendous growth spurt had become evident only a decade later, as by 1910 expenditures had increased to $426 million (+98 percent), comprising 1.2 percent of the GNP. A sustained growth phenomenon occured shortly afterward as expenditures by the end of the next decade had reached a record $1.036 billion. Ten years later, expenditures had increased to $2.3 billion or 2.2 percent of the GNP by 1930.

Increasing educational expenditures were only slightly slowed by the worst economic devastation ever seen in the nation during this period, the Great Depression. Expenditures grew from 1930 to 1940, but at a slower pace following the Wall Street crash in 1929. By 1940, public education expenditures had risen by 1.3 percent over the decade, but the growth was nonetheless extraordinary since the GNP had dropped -3.4 percent during the same time. Spiraling education expenditures resumed with the economic boom brought about by World War II, so that from 1940 to 1950 spending jumped 148 percent to $5.8 billion, increasing its share of GNP to 2 percent. The "red scare" of the 1950s sparked by the Russian space program led by the launching of Sputnik caused a tremendous increase in educational funding, as the nation embarked on a national education agenda under the National Defense Education Act. Federal aid to education resulted in a tremendous emphasis on math and science education, foreshadowing decline of the industrial economy and sparking the beginning of the present technological and information age.

Educations expenditures have also grown rapidly as a consequence of tremendous social upheaval in the nation. Earthshaking change occurred from 1950 to 1970, as the nation moved from a rural to an urbanized political society. The new political nature of the nation was profoundly shaped by the famous *Brown v Board of Education*[2] decision in 1954, when the United States Supreme Court overturned racial segregation in the nation's schools and thereby caused restructuring of the entire education system, including a tremendous impact on how schools were configured and funded. Similarly, the nation underwent an unprecedented social revolution in the 1960s with the war on poverty. This movement gave birth to an entitlement state that focused on human rights and dignity, particularly for groups identified as disadvantaged. It also focused attention on education through programs such as the Elementary and Secondary Education Act of 1965.

As a result of changes during these twenty years, educational expenditures grew from $15.61 billion (+169 percent) by 1960 and another 160 percent by 1970 to $40.68 billion, reaching a record 4.7 percent of the GNP. Data illustrate, however, that social change has not been the only factor driving expenditures, as the same type of growth occured in the next two decades. As seen earlier in chapter 6, the years from 1970 to 1990 were dramatically affected by many types of reform, particularly as many states had their school funding systems overturned by courts on grounds of unconstitutionality.[3] Similarly, the high inflation years of the 1970s and the fiscal retrenchment of the 1980s had a significant impact on education's costs. By 1980, expenditures had grown another 136 percent to $95.96 billion, and still further growth was seen as it more than doubled to $198.9 billion in 1990. The pace had still not slackened by 1994, as expenditures reached an astonishing $160 billion, making education a truly vast American industry.

Scope of Service

The vastness of education's cost has also given voice to critics who have not been satisfied with nearly a hundred years of steady growth. Some critics point with consternation to the tremendous growth and complain about declining educational productivity, while others point with indignation and justification at the increasingly complex role of schools.[4] The productivity question has proved particularly vexing, yielding only weak responses to critics' charge that more is being spent for less with no promise for improvement. Response by education's defenders, however, has also been compelling, as data show both the cause for decline and the case militating against reversal. These arguments show tremendous growth in services, most of which are frighteningly unrelated to achievement issues. The reasons for change have been identified by demographers such as Hodgkinson, who has argued compellingly that the combination of age, births, and family status is the future.[5]

Other data on services also point to reasons for increased costs of education, while holding out little hope for improvement. Recent data have shown that much of education's expenditures are for children with learning problems. For example, some data indicate that 20 percent of children suffer from developmental problems, learning disabilities, or emotional problems.[6] Similarly, high costs are associated with non-English speaking students. Growth in special education programs has also taxed the fiscal limits of education by demanding vast resources to serve children. The Department of Education's Office of Special Education Programs (OSEP) has reported that special education populations have grown so rapidly that they now represent more than 12 percent of all public school enrollments, particularly since implementation in 1986 of Part H of the Early Childhood Handicapped Act under Public Law 99-457, which mandated programs for handicapped preschoolers. Obviously, these demographic data have clear implications for hiring and staffing issues.

Expectations for increased special education needs have grown recently, given new information on the student population. With Specific Learning Disabilities (SLD), Speech/Language (S/L), Mental Retardation (MR), and Seriously Emotionally Disturbed (SED) presently accounting for 94 percent of all disabilities, schools are facing an explosion in new special populations and the resultant personnel issues.

Scope of Personnel

If the scope of expenditures and services has been great, it has been by virtue of the fact that nearly all such expenditures are earmarked for the budget personnel, who represent the single greatest expenditure line item in public schools. In 1869, the average salary of instructional staff in public schools was $189 per year. By 1900, salaries had only risen to $325 annually. But by 1930, salaries had more than quadrupled to $1,420 annually, and the number of teachers was growing rapidly. By 1950, salaries had again more than doubled to $3,010, and by 1970 they had tripled to $8,840 per teacher. By 1990, salaries had again quadrupled to $33,041, and by 1994 salaries had broken all records, reaching a total of $102.7 billion.

These salaries were distributed to nearly three million public school teachers, with estimates on teacher supply and demand placing annual growth in the number of teachers required at more than 200,000 positions in the decade between 1990 and 2000.[7] At projected rates of increase in teaching staff, holding all salaries constant in 1994 dollars, another $23 billion would be needed by the year 2000 to fund only those new positions. The sum of current and new salaries would reach more than $125.3 billion, and would likely represent losses in real income to teachers. No careful economic factoring would have been accounted for since salaries had been held constant at 1994 levels. Given the enormous demographic change and growth in personnel needs, it is unlikely that any foreseeable increase in money to schools could result in a meaningful net increase in real educational resources.

A partial assessment of the scope of the personnel function in dollar terms can be seen in table 8.1, which shows only a portion of the instructional side of education's costs. Among other relevant data, table 8.1 shows the number of classroom teachers and instructional aides by individual states and the average teacher pay in those same states. These data indicate that in 1991 there were more than 2.4 million teachers and nearly 410,500 classroom aides. If all teachers in the United States had earned the national average that year, total salary expenditures would have reached nearly $83 billion, exclusive of benefits or other salaries of professional and classified personnel. In actuality, salaries ranged from a low average of $23,291 in South Dakota to a high average of $46,971 in Connecticut.

The high cost to states of instructional salaries can also be illustrated by multiplying the number of teachers by the average salary in any given state, as in California where the state's 224,000 teachers earned an average $40,192 for a total salary demand of more than $9 billion. Similarly, even South Dakota, with the nation's lowest average teacher salary, spent the significant sum of $206.5 million for instructional salaries. This takes on added meaning when it is considered that the state's small number of teachers and low salaries were supported by an equally small citizenry with concomitantly low per capita income and fewer than 132,000 pupils. Thus, the data on virtually any state must be viewed in the local economic context in order to determine the extent of tax effort in support of public schools.

Similar data on other personnel in education can be seen in table 8.2, which considers selected information on administrative and classified personnel over a period of more than two decades. Data indicate that between 1975 and 1992, average superintendent salaries increased from $30,388 to $83,342 for a gain of about 274 percent. Comparable data can be produced for other administrative ranks such as associate superintendents, business managers, and other line and staff supervisors. For example, staff services increased 290 percent, while business manager salaries increased an average 260 percent. Likewise, while these data do not provide numbers of staff by position description, data on classified personnel can be seen to have grown by significant amounts. Central office secretarial and clerical salaries more than tripled during the time period, growing on average in the case of secretaries from $7,318 in 1975 to $22,309 in 1992. Hourly employees also contributed to increased costs, as instructional aide salaries climbed 280 percent and bus driver salaries grew by 268 percent.

Table 8.1
Selected School Statistics
Elementary and Secondary Public Schools
Fall 1991

	Districts	Teachers	Enrollment	PTR	Teacher's Average Tchr Pay	Instructional Aides	Expend per Pupil
United States	15,173	2,431,622	42,000,343	17.3	$34,098	410,498	$5,245
Alabama	129	40,480	722,004	17.8	26,951	3,543	3,627
Alaska	56	7,118	118,680	16.7	44,718	1,626	8,330
Arizona	229	33,978	656,980	19.3	31,176	5,489	4,309
Arkansas	325	25,785	438,518	17.0	27,070	4,511	3,700
California	1,005	224,000	5,107,145	22.8	40,192	55,290	4,491
Colorado	176	33,093	593,030	17.9	33,072	4,441	5,064
Connecticut	166	34,383	481,050	14.0	46,971	5,547	7,602
Delaware	19	6,095	102,196	16.8	34,548	732	5,865
Dist. Col.	1	6,087	80,618	13.2	38,798	308	9,259
Florida	69	109,939	1,932,131	17.6	31,070	21,384	5,276
Georgia	183	63,816	1,177,569	18.5	29,509	17,534	4,466
Hawaii	1	9,451	174,747	18.5	34,528	1,049	5,166
Idaho	114	11,626	225,680	19.4	26,334	1,402	3,386
Illinois	942	110,143	1,848,166	16.8	36,461	15,607	5,520
Indiana	296	54,546	956,994	17.5	34,809	11,912	4,930
Iowa	435	31,395	491,363	15.7	29,202	4,536	4,679
Kansas	304	29,324	445,390	15.2	30,731	3,150	4,874
Kentucky	176	37,571	646,024	17.2	30,870	7,652	4,354
Louisiana	66	46,118	765,589	16.6	25,948	8,433	4,146
Maine	283	15,416	216,400	14.0	30,097	3,386	5,458
Maryland	24	43,616	736,238	16.9	38,728	6,231	6,566
Massachusetts	352	55,963	846,155	15.1	37,256	9,009	6,366
Michigan	561	82,967	1,591,120	19.2	41,149	12,969	5,883
Minnesota	434	44,903	773,571	17.2	34,451	7,888	5,239
Mississippi	150	28,111	504,127	17.9	24,367	9,788	3,187
Missouri	543	52,306	827,404	15.8	28,895	5,009	4,754
Montana	539	9,883	155,779	15.8	27,590	1,373	5,204
Nebraska	777	19,069	279,552	14.7	27,231	3,065	5,038
Nevada	17	11,409	211,810	18.6	33,857		4,653
New Hamp.	174	11,464	177,138	15.5	33,170	2,253	5,672
New Jersey	608	80,515	1,109,796	13.8	41,027	10,000	8,645
New Mexico	88	17,498	308,667	17.6	26,389	4,080	3,895
New York	717	171,914	2,643,993	15.4	43,335	24,110	8,565
North Car.	133	65,326	1,097,598	16.8	28,791	19,407	4,488
North Dakota	276	7,733	118,376	15.3	24,495	1,135	4,199
Ohio	613	103,372	1,783,767	17.3	33,243	9,229	5,245
Oklahoma	593	37,650	588,263	15.6	25,339	5,746	3,791
Oregon	296	26,745	498,614	18.6	34,100	4,746	5,683
Pennsylvania	503	100,475	1,692,797	16.8	38,715	11,316	6,541
Rhode Island	37	9,709	142,144	14.6	36,417	1,217	6,343
South Car.	95	37,115	627,470	16.9	28,068	6,597	4,351
South Dakota	189	8,868	131,576	14.8	23,291	1,205	3,965
Tennessee	140	43,062	833,651	19.4	28,621	7,963	3,782
Texas	1,051	219,192	3,464,371	15.8	29,041	33,976	4,438
Utah	40	18,305	456,430	24.9	26,339	3,812	2,960
Vermont	279	7,031	97,137	13.8	33,646	2,052	6,738
Virginia	141	64,537	1,016,204	15.7	31,657	10,283	4,836
Washington	296	42,931	869,327	20.2	34,823	6,797	5,000
West Virginia	55	20,997	320,249	15.3	27,366	2,748	4,911
Wisconsin	428	52,028	814,671	15.7	35,227	7,741	5,871
Wyoming	49	6,564	102,074	15.6	30,425	1,221	5,723

Source: National Center for Education Statistics (1992).

Table 8.2
Average Salaries and Wages in Schools by Selected Positions—1975-1992

POSITION	1975	1980	1985	1990	1992
Annual Salary					
Central office administrators:					
Superintendent	30,388	39,344	56,954	75,425	83,342
Deputy superintendent	30,074	37,440	52,877	69,623	76,796
Assistant superintendent	26,460	33,452	48,003	62,698	69,315
Administration for:					
Finance & business	21,850	27,147	40,344	52,354	57,036
Instructional services	22,608	29,790	43,452	56,359	62,102
Public relations	21,470	24,021	35,287	44,926	50,625
Personnel services	21,470	29,623	44,182	56,344	62,269
Subject area supervisors	18,601	23,974	34,422	45,929	50,580
School building administrators:					
Principals:					
Elementary	19,061	25,165	36,452	48,431	53,856
Junior high/middle	21,136	27,625	39,650	52,163	57,504
Senior high	22,894	29,207	42,094	55,722	61,768
Assistant principals:					
Elementary	15,968	20,708	30,496	40,916	45,558
Junior high/middle	17,868	23,507	33,793	44,570	48,956
Senior high	18,939	24,816	23,587	31,278	34,565
Auxillary professional personnel:					
Counselors	14,479	18,847	27,593	35,979	39,563
Librarians	12,546	16,764	24,981	33,469	37,227
School nurses	10,673	13,788	19,944	26,090	28,721
Secretarial/clerical personnel:					
Central office:					
Secretaries/stenographers	7,318	10,331	15,343	20,238	22,309
Accounting/payroll clerks	7,588	10,479	15,421	20,088	22,215
Clerk-typists	6,089	8,359	12,481	16,125	17,646
School building level:					
Secretaries/stenographers	6,046	8,348	12,504	16,184	17,784
Library clerks	5,052	6,778	9,911	12,151	13,347
Hourly Wage Rate					
Other support personnel:					
Teacher aides:					
Instructional	2.91	4.06	5.89	7.43	8.15
Noninstructional	2.81	3.89	5.60	7.08	7.70
Custodians	3.54	4.88	6.90	8.54	9.35
Cafeteria workers	2.61	3.78	5.42	6.77	7.39
Bus drivers	3.75	5.21	7.27	9.21	10.04

Source:Educational Research Service, Arlington, VA.

The scope of personnel has broad reaching dimensions that powerfully impact levels of expenditure and provision of school services. Services have greatly expanded in recent history, and expenditures have soared in an attempt to keep pace with rapid changes in education. Change has been a product of demographic alteration of the American landscape and a response to changing perceptions of equality and opportunity as interpreted by legislative mandate and judicial decree. As discussed, the basic nature of schools causes the majority of educational budgets to be consumed by employee costs, and it is equally true that every problem and every opportunity in school systems has personnel issues at its roots and as a fundamental part of any solution. Table 8.3 reflects the fiscal implications of one selected state in light of American demographic changes and growth.

Table 8.3
Selected School District Fiscal Data in One State: Changes From 1972-1994

Dollars in Thousands

Category	1972	1994	Increase 1972-94
General State Aid	$104,562	$1,365,719	$1,261,157
Gen Aid as % of GF	25.5%	66.9%	41.4%
Other state aids			
Teacher Retirement	14,937	53,859	38,922
Special Education	4,183	149,026	144,843
Food Assistance	0	2,350	2,350
Adult Basic Education	0	295	295
Bilingual Education	0	(moved to Gen Fund)	n/a
Motorcycle Safety	0	57	57
Inservice Education	0	2,475	2,475
Asbestos Aid	0	0	0
Sexuality-AIDS Educ.	0	0	0
Building Based Educ.	0	1,485	1,485
Parent Education	0	2,475	2,475
Driver Education	0	1,364	1,364
K-12 Enrollment	474,747	439,873	(-34,875)
Certified Staff	29,542	36,780	+7,238

Source: Kansas State Department of Education, *School Districts--Selected Data 1972-1994* (Topeka: Kansas State Department of Education, 1994).

Goals and Objectives

The scope of the personnel function underscores the point that a school district must be firmly anchored by a set of goals and objectives that express a clear understanding of the importance of personnel. Many purposes for schools have been formulated during the American experience, with periodic policy statements issued by major national organizations. The calls for school reform that many people associate with *A Nation at Risk*[8] actually have occurred with frequency throughout history, particularly in the current century.

For example, the Commission on Reorganization of Secondary Education in 1918 published Seven Cardinal Principles in which it set forth the purpose of schools as preparing students for the world of work. In 1938, the Educational Policies Commission promoted four goals for schools, among which economic efficiency was a central tenet. In 1951, the Ten Imperative Needs of Youth issued by the National Association of Secondary School Principals touted vocational education, and in 1955 the White House Conference on Education stated fourteen goals for education, among which work habits was a prominent theme. Such calls have been frequent and have depended on current issues of the day to set their agendas, as in *A Nation at Risk* when the National Commission on Excellence in Education declared that the United States must reform schools "[i]f only to keep and improve on the slim competitive edge we still retain in world markets."[9]

Under the tense and dramatic conditions that so often surround American schools, clear goals and objectives for the personnel function are absolute requirements. Accordingly, the personnel function can be viewed as having both general goals and specific objectives designed to assist in carrying out the district mission. General goals of the personnel function center on at least six main activities, all of which are inclusive of the scope of the personnel function and are intended to coordinate satisfactory delivery of the educational product. General goals encompass determining staffing needs, recruiting and retaining competent people, assisting in development of individual competencies, assuring that employees are assigned and utilized efficiently according to the full extent of their skills, increasing and improving employee satisfaction, and establishing clearly defined work expectations along with competent evaluations.

The operational objectives of the personnel function are thus inclusive of a wide variety of specific tasks which may be seen to encompass the following ten major responsibilities:

- Projecting both long range and short term personnel needs.
- Preparing policies to direct an overall master scheme to implement and develop the human resources of the organization.
- Building a specific plan to recruit the quantity and quality of persons suited to the program the organization is seeking to deliver to its clients.
- Selecting only those personnel meeting organizational requirements as identified in the previous planning phases.
- Developing plans for induction of new personnel and carrying out plans to help all organizational placeholders in acculturating and assimilating those new individuals.

- Developing and administering an information system that tracks and assists both the district and its personnel in organizational and individual needs.

- Developing plans to appraise current personnel.

- Planning and carrying out reward systems for those personnel who contribute positively to organizational success.

- Developing vertical and horizontal feedback systems to provide for a free exchange of ideas and information at all levels in the system.

- Developing linkages between the various functions of the district in order to coordinate and integrate district activities to facilitate organizational goals.

While these tasks are forbidding in scope and complexity, the alternative is highly unflattering and can unfortunately be seen in many school systems that have not taken the time to carefully or accurately develop personnel systems built around these issues. Evidence of a lack of proper personnel planning can be found in schools where failure to achieve a coordinated outcome is exhibited by common dysfunctionalities, including:

- Inability to meet high student achievement standards because the system has failed to establish policies that recruit and retain competent instructional and support staff.

- Inability to design and implement outstanding academic programs because the system lacks a personnel policy capable of attracting and keeping competent personnel who know how to properly conceptualize the school's educational program.

- Inability to coordinate the total system and to align the district with desired educational outcomes because administrators have not clearly conceptualized and established a personnel policy that is able to seamlessly integrate the various functions of the system.

- Inability to deal with organizational and individual frustrations because the system does not recognize and adjust to the realities of people working in organizations.

- Inability of the system to avoid undue stress related to conditions in the external environment such as fiscal shortfalls and oversupply of staff during periods of instability because the system has not learned to adjust to extraneous conditions.

- Inability to continually improve the system because proper evaluation and development programs have not been developed.

In most instances, such failure is either a function of insufficient skill and knowledge of the elements of good personnel systems, or alternatively, failure on the part of administrators and other district executives to understand and cope with organizational behavior. The personnel function thus involves leadership for improvement and management for enhancing improvement—areas which require active administrator involvement in understanding the behavior of people in organizations and skills in the process of recruiting, selecting, orienting, compensating, evaluating, retaining, and even dismissing employees.

People and Organizations

It is a fact that organizations are controlled by people, and that people are controlled by organizations. In truth. the purpose of an organization is to shepherd and control people in directions that are consonant with organizational ends, and it is equally accurate to say that organizational ends are conceived and carried out by people who exercise organizational control. Recognizing the dual nature of behavior on the part of people in organizations is a critical step in understanding that organizations and the people in them have lives of their own.

Interest in the study of human behavior has been longstanding for school administrators and others involved in aiding organizations seeking to achieve their goals. Whether the administrator chooses to perceive the world as a collection of individuals independently pursuing their own aims or to recognize organizations as purposeful and somewhat independent creatures, the intended result of such interest is to understand and predict the consequences of any act undertaken as part of the educational enterprise. To suggest prediction, however, is to automatically invoke theory—an observation that must be followed by the understanding that administration is both theory and practice. In fact, the definition of theory forces administrators to become theorists, assuming they intend to act rationally.

Although numerous definitions of theory have been offered, Kerlinger gives a particularly useful definition, stating that "a theory is a set of interrelated constructs, definitions, and propositions that present a systematic view of phenomena by specifying relationships among variables, with the purpose of explaining and predicting phenomena."[10] By such definition, then, theory should define practice in that the ability to predict the outcome of some practice originates in sound theory, thereby precluding any possibility of rational administrative practice without a good working knowledge of theories of individual and organizational behavior. Likewise, a theory is only as good as its predictive ability, wherein practice may be said to validate theory when predictive validity is high or, alternatively, disprove theory if its hypotheses do not withstand the test of practice.

Classical Organizational Theory

Although theory pertaining to organizations has been of longstanding concern, the beginning of classical organizational theory is often marked with Frederick Taylor. Taylor's engineering background led him to observe that work in organizations could be improved by studying the tasks to be performed and to program people to do those tasks more efficiently. Taylor's major contribution to the theory of organizations lay in his view of people as machines, leading him to formulate a theory of scientific management in which he viewed organizational improvement as the basis for increasing productivity.

In 1911, Taylor published a monumental work entitled *Principles of Scientific Management.*[11] His theory of scientific management called for several firm principles which he believed would increase the productivity of any organization. Scientific management called for engineering the job in an ordered fashion that would include five principal conditions leading to greater productivity. First, Taylor argued that a large daily task should be assigned to the worker. A full day would be required to finish the job, and which a clear beginning and end would be evident to everyone, including the worker. Second, he called for standard conditions in the workplace. Workers would all function under the same conditions using the same tools, making it easy to judge workers' results individually and in comparison to one another. Third, Taylor called for high pay for success in the assigned task. Fourth and contrastingly, he called for loss in the case of failure. And fifth, Taylor introduced the concept of expertise in organizations wherein tasks would be made difficult enough to develop skills and to later remove those from the organization who were unable to develop the skills needed for success.

Taylor's engineering interests also led him to engage in time and motion studies, prescribing in close detail how workers should perform their tasks in order to make them more efficient. His time and motion studies were so detailed as to require that tasks be structured to minimize arm and body movements and to prescribe locations of tools and machines to result in smooth and continuous motions of short reaches. The result of Taylor's work was a factory production model that can be observed in many aspects of schools today, including how teachers teach and administrators lead.

The scientific management school received an additional boost with the work of Henri Fayol. Fayol argued that the task of administration was comprised of planning, organizing, commanding, coordinating, and controlling. To Fayol, planning was the systematic study of anticipating the future and arranging events in the organization to achieve the desired end. In the same vein, Gulick formulated an acronym learned by many school administrators in their educational preparation programs, in which he held that successful administration is the result of seven functions abbreviated as POSDCoRB: Planning, Organizing, Staffing, Directing, Coordinating, Reporting, and Budgeting. All these studies held to a rational view of the world in which order and success could be created by structuring the work of organizations around specific tasks for which people were trained in machinelike behaviors.

Theory of scientific management contributed greatly to American organizations, including schools. A look at most business or industrial organizations will reveal many of the characteristics of scientific management, and schools themselves are often significantly characterized by these same traits. Among the most evident of such tenets are division of labor and specialization of tasks, and span of control. For example, schools divide labor into the fields of instruction and support services and supervision. Schools are hierarchically structured into job categories such as food service, transportation, administration, and departments of languages, mathematics, science, and so forth under supervisors who are responsible for only a limited number of personnel doing similar jobs.

Span of control is so prevalent that as the number of people under a supervisor increases, the manager is generally given assistants to help supervise the tasks of the organization. The influence of scientific management is so pervasive that it only requires looking at a school's organizational chart to understand the presence of hierarchy, division of labor, and so forth to identify that virtually every organization is shaped like a pyramid under authority and task relationships. In fact, the greatest proof of scientific management in schools and other organizations is the prevalence of the term "chain of command." The contribution of scientific management theory to organizations cannot be understated, as most organizations conform to some extent to the scientific elements of time and motion, division of labor and specialization, standardization of tasks, unity of command, span of control, and strong commitment to the concept of formal organization as the tool to productivity.

Human Relations Theory

Building organizations around a machine concept was not without critics, however. The human relations approach to organization was formed largely in response to negative perceptions of treating people as machines. Under the initial leadership of Mary Parker Follett, a theory of people in organizations was developed in which the goal was to encourage strong relationships among the people who work in organizations.[12] Follett's work focused in significant part on the problems flowing from conflict on the job, and it was her thesis that minimizing such problems would aid in organizational aims. Although her work began the human relations theory, it received its greatest impetus in the famous Western Electric Company experiments at its Hawthorne plant in Chicago from which the familiar phrase "the Hawthorne effect" was coined. These studies were briefly addressed within chapter 5 of this text.

Since the Hawthorne study asked more questions than it answered, additional research was conducted. Mayo and Roethlisberger[13] embarked on a study of questions concerning how worker production is influenced. Among the questions were the effect of rest periods, shorter and longer work days, and employee attitudes about work and organizations. The study, conducted on assembly workers, tested these questions by various group designs manipulating work conditions such as changing physical conditions, permitting workers to socialize freely, and so forth. The results were also perplexing, in that worker productivity generally continued to rise regardless of changes in work conditions. The final phase of the study looked at the informal social structure and revealed new information about how work is performed.

In this experiment, it was found that worker behavior did not actually conform to the organization's expectations of a scientific model—rather, the group operated with a strong sense of informal social structure that included group norms, values, and attitudes. On further inquiry, the researchers found evidence of behavior affecting performance—for example, shirking produced censure from other employees as did rate-busting by producing more work than the group wished to see occur. Censure involved various behaviors, including ostracism and sarcasm to transmit group displeasure.

The human relations approach to organizational productivity yielded important observations about the nature of work and how to enhance the goals of the organization. Unlike scientific management, the focus on individuals showed that extrinsic incentive might be less powerful than group pressure. The Hawthorne studies showed, among other things, that workers respond as individuals and that their reference group is not superordinates, but rather peers, and that production is more heavily influenced by the informal organization than by physical limits. Thus, the human relations approach contributed powerfully to the task of organizational administration by illustrating the fact that organizations are peopled and that motivation and satisfaction of employees are critical facets of productivity.

Behavioral Science Theory

Because scientific management considered only the technological dimensions of work and because the human relations approach was often accused of warmth and fuzziness to the exclusion of formal organization, the behavioral science approach to work was formulated, primarily under the direction of Chester Barnard in the 1930s. Barnard was himself an executive who was interested in productivity issues, laying down many of the precepts of the movement in *Functions of the Executive*[14] in 1938. Expanded significantly by Herbert Simon in *Administrative Behavior*,[15] the behavioral approach focused on balancing the views of the formal and the informal organization and hypothesizing that each may play an important part in the life of work.

In fact, Simon argued for the concept of organizational equilibrium as a principal element of motivation for work, and he saw these issues as an exchange system where harmony between the formal and informal organization would produce greater work efficiency. Simon argued further that people stay in organizations and contribute to productivity because they perceive the rewards to be greater than their contribution, thus satisfying a basic tenet of any market system of exchange.

Although less firmly research based than the rigid time and motion studies of scientific management and the informal group studies of the human relations approach, major contributions regarding work in organizations were made by the behavioral science approach. Profound observations were contained in the desire of behavioral science to meld the formal and informal organization into a dynamic scheme of exchange. Further, Simon's observation that administration can be defined as a process of rational decision making in that dynamic context represented a major step in unlocking the puzzle of productivity. Simon's work also offered the observation that it may be impossible to optimize work, and that it may be more realistic to understand that no problem is perfectly resoluble, and that it may be acceptable to merely satisfy. The weaknesses of the earlier models were captured succinctly by Simon as he argued "no formal organization will operate effectively without an accompanying informal."[16]

A Systems Theory

The tremendous contributions of earlier theorists to the field of administration led to additional thinking concerning organizations in a manner that can be described as a systems theory. A system conjures up images of something joined together with all its pieces connected as integral elements in a natural and appropriate state. Most administrators have been exposed to systems theory, such as the work of Talcott Parsons or the model in figure 8.1 proposed by Getzels and Guba,[17] wherein organizations are constructed of, and respond to, both internal and external environments. According to the model, schools are bounded social systems made up of their parts summed into a whole. The model must accept input from the external environment over which it does not exert absolute control, and the input impacts on the internal environment to create responses that eventually lead to a feedback loop for organizational adjustment. These actions and reactions are functions of the interaction of the internal environment, which itself is composed of elements analogous to earlier theorists' beliefs about an organization as a formal entity and its human side or informal organization.

As can be seen in the model, the internal environment offers the best opportunity for administrative leaders to apply formal and informal organizational characteristics to problem resolution. For example, social systems theory accepts that systems are made up of interdependent parts, that systems are

Figure 8.1
Systems Model

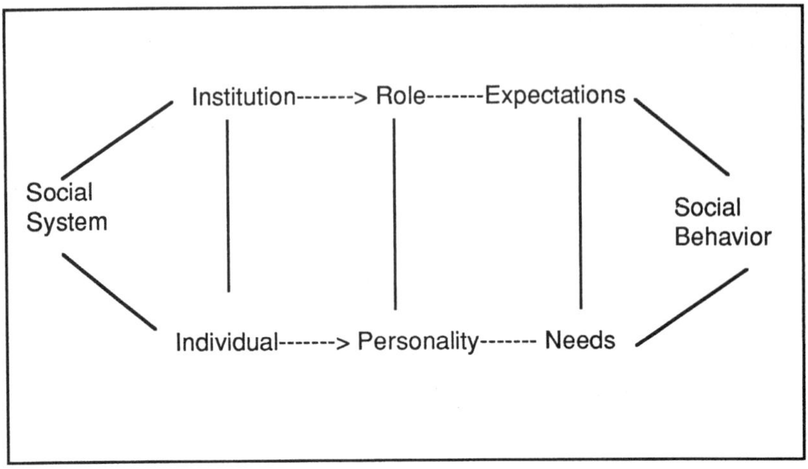

both peopled and structural, that systems are influenced by being open to both internal and external forces, and that the roles played by placeholders in systems represent an opportunity for directing the success of the organization. The model is operationalized by the formula B= f(R x P) in which Behavior (B) in and of organizations may be analogized as a function *(f)* of Role (R) as defined by formal expectations and Personality (P).

The model fundamentally recognizes two parallel elements of the nomothetic dimension and the idiographic dimension. The nomothetic dimension makes up the formal organizational structure wherein it is seen linearly as the institution which defines the role which establishes the institutional expectations of the role incumbent. Since organizations are peopled, the idiographic dimension is simultaneously linear. The individual defines the personality, which establishes the needs of the individual, which seek to be met. Simultaneous interactions among the individual elements of both dimensions seek homeostasis and equilibrium, whereupon social behavior is produced which enters a feedback loop to regenerate and adjust the cycle. The task, of course, is alignment of both dimensions so that the system is balanced or, alternatively, engaged in balancing behaviors.

Further refinement of the model has followed as researchers have studied organizations. Many different conceptualizations and additions to general systems theory have been offered in an attempt to explain and predict behavior and to maximize organizational functioning. Etzioni, for example, developed a compliance theory in which he argued that people respond to organizations by producing behavior consonant with the kind of power relationships that exist in the organization. Etzioni's typology of power, as seen in figure 8.2, is generally familiar to administrators. He sees power exerted by organizations in pursuit of their goals as being either coercive, remunerative, or normative. He then conceives of responses to applied power as falling into categories of either alienation, calculation, or commitment.

Figure 8.2
Etzioni's Compliance Types

Power \ Involvement	*Alienation*	*Calculation*	*Commitment*
Coercive	Coercive		
Remunerative		Utilitarian	
Normative			Normative

According to Etzioni, prediction follows upon application of types of power. If coercive power is applied, the response will usually be alienation. Etzioni suggests that prisons and mental hospitals best fit this description, but there is significant research showing that schools often use coercive power to achieve the desired ends. If remunerative power is used, the natural response is calculative in which the "client" will calculate the rewards for compliance and the sanctions for noncompliance (e.g., the industrial scientific management model discussed earlier). If normative power is applied, the response tends toward commitment. Interestingly, at times schools exhibit each of these patterns. The task for school administrators, of course, is to align the goals and structure of the institution with the desired behavioral responses by the people within the organization.

While many other conceptualizations of systems have been offered, it can be seen that systems theory offers a tremendous insight into the functioning of productive organizations. The interdependence of the system with all its elements is a key to understanding and structuring the organization to meet the demands of both the nomothetic and idiographic dimensions and to engage in helpful exchange between the external and internal environments.

Post-Rational Organizational Theory

Development of organizational theory has also gone beyond the traditional assumptions and beliefs about how the world operates by moving into an arena that can be defined as post-rational theory. By definition, post-rational organizational thought may be said to question the very foundations on which life is perceived to be ordered by arguing that the appearance of reality and reason may in fact be something other than what is seemingly sensible. Pfeffer captured the essence of organizational theory development when he noted that "the domain of organization theory is coming to resemble more of a weed patch than a well-tended garden."[18] Although the topic is too complex to explore fully in a text of this nature, a sense of development in this post-scientific age provides a final assessment of organizational functioning before moving into more specific issues of personnel administration.

Sergiovanni has noted that traditional organizational theory has been bound by specific limits that have slowed understanding of how to best design organizations to meet both institutional and individual needs.[19] Sergiovanni argues that traditional management theory is best equipped to deal with only certain situations characterized by narrow conditions in which prediction can be expected. He suggests that traditional management theory will only work under linear conditions in which situations of practice can be tightly structured and connected without causing unanticipated harmful effects. The essence of such conditions is described as those situations in which the need exists to bring about a routine level of competence and performance wherein a high degree of linearity of events and outcomes can be expected. Sergiovanni describes linear conditions using ten descriptors:

- Stable, predictable environments.
- Tight management connections.
- Loose cultural connections.
- Discrete goals.
- Structured tasks.
- Single solutions.
- Easily measured outcomes.
- Sure operating procedures.

- Determinate consequences of action.
- Clear lines of authority.

Sergiovanni argues that simplicity, clarity, order, and predictability are the key elements of these ten conditions, so that accurate prediction of the consequences of any action can be made.

In contrast, he argues that nonlinear conditions are also present in organizational settings, making it difficult to apply linearity to every situation in a rational manner. Sergiovanni describes nonlinear conditions using a different set of ten descriptors that contrast markedly with assumptions about management in organizations:

- Dynamic environments.
- Loose management connections.
- Tight cultural connections.
- Multiple and competing goals.
- Unstructured tasks.
- Competing solutions.
- Difficult-to-measure outcomes.
- Unsure operating procedures.
- Indeterminate consequences of action.
- Unclear and competing lines of authority.

He further argues that the majority of human interactions that take place in organizations, including schools, can be described as nonlinear, with the peculiar result that "one cannot predict the conditions...until they are experienced."[20] Observations of such nonlinearity in organizations have led to theories that seek to take into account a "post-rational" organizational perspective which is more tolerant of irrationality than was true in all the earlier theories discussed in this chapter.

Pfeffer's attempt to weed the theory patch provided some insight into rational and post-rational perspectives as he argued that all theories fall into one of three perspectives.[21] The first perspective can be identified as those theories in which behavior is seen as purposive, boundedly or intentionally rational, and prospective or goal directed. Under this view, theories follow the assumed natural order of the world in which cause and effect are paramount; i.e., administrators are taught to predict a certain consequence of a given action. Pfeffer notes that this view is idealized and venerated under the Judeo-

Christian heritage and philosophical basis for American life and culture and is fixated on reason as the basis for action.

A critical aspect of Pfeffer's thinking in this regard is that he sees rational theories as based in choice among competing alternatives, so that rationality is prospective--reasoning through possible consequences precedes behavior. Pfeffer argues that rational theories attempt to explain how organizations act intentionally and rationally to achieve some goal, such as efficiency or higher performance. For example, goalsetting theory argues that people's behavior is so governed by the goals they seek to attain that the results of setting goals can maximize organizational functioning. Likewise, needs theory and task design theory are rationally based in that they see people evaluating situations in the context of their individual needs and work values and then reacting to situations accordingly. Pfeffer notes that even Marxist approaches to organizations presume thoughtful and foresightful behavior by the owners of capital.

People are so conditioned to rationality that they may question whether there are any other alternatives. Pfeffer, however, argues that a second perspective is available which sees behavior from an external constraint or situational control perspective. As Pfeffer states, "situationism, or the external constraint perspective, tends to ignore individual or organizational factors, such as personality, preferences, goals, and information processing, or at least to see them as less important than the primary impact of the external stimulus or effect."[22] Under these conditions, behavior of the organization and its people is seen as responding to the external environment so that action is a result of constraints, contingencies, and demands confronting the organization. In contrast to totally rational theories, the external constraint perspective may have little to do with choice so that rationality is retrospective, i.e., making sense of events after they have occurred.

Under these conditions, motivation is a very different matter, because survival and remodeling values to fit the mold become important in an operant-conditioning style that seeks to avoid punishment. Although several theories fit this perspective, the most easily understood is the resource-dependence perspective in which organizations and people are jerked around helplessly by the environment because they are dependent on external environmental elements such as funding for survival—a reality that results in goal displacement and adoption of values that may not be congruent with individual preferences.

Pfeffer advances a third perspective on organizational theories which he terms the almost random/emergent process view of action. Blending elements of the two previous perspectives, the random view denies either an internally directed or an externally determined rationality of behavior. The essence of such theory is that organizations are a context in which people, problems, and solutions are joined, with the results determined by an emerging process that is limited by constraints on the process. Under these conditions, rationality cannot guide such a fluid action because rationality, goals, and preferences emerge from the action, rather than guiding the action prospectively. Again, rationality is retrospective since knowledge of external constraints and forces is insufficient to predict action and is complicated by internal differing goals and objectives. Under these conditions, motivation is a collective process of group decision making and ownership, restricted by any constraints from the external environment.

While several theories fit this model, the familiar model of schools as loosely coupled organizations provides an analogy to explain this view. As Weick has argued, schools are connected to each other in such a way that one influences the other but that connections are confounded by other connections, and that any influence is rarely strong or direct. For example, universities offer a prime case wherein they are said to be so loosely coupled that they are held together by plumbing and parking lots. Under a random perspective, rationality is an emerging process wherein prediction is not foreseeable as the group creates its own context.

Pfeffer's work sought to bring order to the organizational theory weedpatch by boxing his view of the dominant perspectives into a typology in which he characterized the principal theories of organizations into groups of similar fundamental assumptions. As seen in figure 8.3, he grouped major theories by perspectives on action and by unit of analysis. Perspectives on action are identified by rationality, external constraint, and random/emergent assumptions. Level of analysis considers whether the focus of theory prefers the individual or the total organization. While a full explanation is not possible

Figure 8.3
Pfeffer's Organizational Theory Typology

		PERSPECTIVES ON ACTION		
		Purposive, Intentional, Goal Directed, Rational	Externally Constrained and Controlled	Emergent, Almost-Random, Dependent on Process and Social Construction
LEVEL OF ANALYSIS	Individuals, Coalitions, or Subunits	Expectancy theory Goal setting Needs theories and job design Political theories	Operant conditioning Social learning theory Socialization Role theories Social context effects and groups Retrospective rationality Social information processing	Ethnomethodology Cognitive theories of organizations Language in organizations Affect-based processes
	Total Organization	Structural contingency theory Market failures/ transaction costs Marxist or class perspectives	Population ecology Resource dependence	Organizations as paradigms Decision process and administrative theories Institutionalization theory

in this short space, it can be seen that there are differing views on why organizations and the people in them behave as they do—a difference so fundamental as to question the very nature of reality itself.

Many other theories of behavior by organizations and individuals exist. For example, Maslow's hierarchy of needs, Herzberg's two-factor theory, the conflict theory of decision making, the garbage can model, Fiedler's contingency theory, and many other models familiar to educators all seek to explain and predict as best possible the complex behaviors exhibited by organizations and the people in them. The array is at once confusing and fascinating, but there should be no doubt that great value has come of questioning the assumptions that underlie organizational functioning because such is the very process of administration—indeed, to fail to question is to never improve and to suggest that the practice of administration is no better than mysticism or sorcery. For administrators, organizational theory results in one profound question that should drive a constant search to improve the organization: "Are public schools, then, rigid bureaucracies that need to be loosened or organizational anarchies that need to be tightened?"[23] Such a question is not musing, as the answer determines the relationship between administrators as organizational leaders and other personnel who are identifiable as both role incumbents and as individuals.

Personnel Planning

General Planning

The survey of personnel concerns thus far in this chapter leave no doubt that planning represents a key element of any organization, and particularly schools. Since the purpose of schools is to prepare the next generation for productive lives in an increasingly complex world, it follows that personnel planning is growing increasingly complex as well, because schools cannot carry out the mission without the careful work of highly competent people in a completely labor-intensive industry. Castetter cuts to the heart of personnel issues confronting school systems in the modern age as he lists twelve difficult problems that must be addressed through the personnel function. Specifically, he demands that schools must learn:

- How to create and infuse the school system with a vision of how the personnel function can best cope with and respond to an uncertain future environment.

- How to attract, retain, and develop qualified, competent system members.

- How to design the personnel selection process, and how to minimize position-position holder mismatches.

- How to develop a compensation structure that is equitable, competitive, and performance-oriented.
- How to integrate individual and organizational objectives.
- How to distribute power, authority, and responsibility.
- How to shape and maintain a supportive performance culture.
- How to develop performance appraisal systems that are focused on teaching-learning outcomes.
- How to utilize and enhance the skills of all system members.
- How to utilize modern information systems more effectively in managing the personnel function.
- How to rid the system of . . . careless or indifferent persons.
- How to achieve union-system cooperation so that the parties become allies rather than adversaries...and devise alternative solutions that are nonconfrontational and will prevent losses in pupil learning time and inconvenience to the public.[24]

Based on his list it is apparent that most, if not all, of the organizational theory questions raised in the last section are integral to the personnel function in schools.

The importance of planning presumes the rationality and deliberate goals of guiding organizations and people to desirable ends. Planning is a method of making sense of an environment where chaos would rule if it were not for deliberate attempts at intervention and redirection. Schools plan for personnel administration by considering a multitude of different questions, most of which can be identified within the areas of system planning, program planning, pupil planning, and staff planning, with other consideration for information and external environment planning. By systematically planning for input and control in these important areas, schools are able to build a strategic plan that moves from the general to the specific, and which addresses the components that make up a peopled organization which has as its sole purpose to serve clientele effectively and efficiently.

System Planning

Most people thinking of school personnel would tend to think first about the professional teaching staff. While such tendency would be understandable in that the act of teaching is at the core of the school, that view would be incomplete for several reasons. First, the personnel function takes in a much larger group by including in both certified staff (i.e., teachers and other professional employees) and classified personnel (i.e., custodians, school bus drivers, teacher aides, cafeteria workers, and other noncertified staff), all of whom perform valued roles in the district. Second, the existence of such people presupposes a high degree of extended planning about the system itself, including system structure in terms of organizational aims and purposes, the district's hierarchical shape, the development of assumptions and projections about the future, and the preparation of master plans, policies, programs, and budgets that will operationalize the goals of the system.

For example, a philosophy of strong central leadership in a district will produce one set of goals and objectives that are supported by a distinct pattern of authority lines for setting out educational programs and budget behaviors. In contrast, a philosophy of decentralized, team-centered decision making will produce a very different organizational chart, and procedures and outcomes may also be markedly different as a result. In either case, philosophies and action channels must be created, and in all cases certain data regarding program offerings and personnel are required for moving toward organizational goal attainment.

Development of a system philosophy is likely to be an arduous and lengthy task. Much literature exists on developing mission statements and goal setting where the focus is on a sense of direction and clarity of organizational purposes. The premise is that the system will be without a clear path to any desirable ends until it has articulated what it is about and until all personnel understand and internalize the aims of the organization. The progression moves from a broad mission statement to general goal statements about how the institution intends to meet its objectives. The district may decide that its mission statement embraces the motto that "All Children Can Learn" and that its goals are to assure that all personnel individually accept this belief so that every child will demonstrate positive measurable annual growth according to a set of standards.

Figure 8.4 illustrates one district's mission statement and accompanying goals, identifying that the focus of all district efforts will be on excellence in the areas of curriculum, instruction, instructional leadership, school environment, and staff development. These initial planning steps are the prerequisite for developing the master plans that are needed to operationalize and carry out the organization's mission and goals. Master plans must then be developed to direct how goal statements will be translated into programmatic and operational terms and how goal attainment will be evaluated.

Establishing responsibility for carrying out the mission and goals is usually accomplished by assigning line and staff positions to oversee and carry it out. Organizational charts are familiar items in nearly every school district, although in actuality too few people in the trenches either see or fully value them. Constructing an organizational chart goes back to district philosophies such as decentralized decision making, wherein power and authority are distributed across various staff positions according to core values concerning

Figure 8.4
Sample Mission and Goals Statement

-----------------------*PURSUIT OF THE VISION*-----------------

The district 's new mission statement provides a clear focus for the efforts of the district and its staff. In order to pursue that vision of excellence in our schools, a number of actions are planned or are underway. Those are and will be based on 1) internal and external research in relation to effective schools, (2) site-based management, and (3) student learning outcomes.

ACTIONS/PLANS FOR EXCELLENCE

Curriculum

- *More Focused Curriculum Planning*
- *Better Curriculum/Test Alignment*
- *Learning Strategies Program*
- *Outcome-Based Education*

Instruction

- *Clinical Teaching*
- *T.E.S.A. Training*
- *Cooperative Learning*
- *Classroom Management*
- *Mastery Learning*
- *Time on Task Study*

Instructional Leadership

- *Instructional Leadership Study*
- *Training in Teacher Evaluation*
- *Training in Clinical Supervision*
- *More Systematic Teacher*
- *Beginning Teacher Program*
- *Management Effectiveness*
- *Improvement of Administrator Evaluation System*

School Environment

- *School Climate Study*
- *Site-Based Management*
- *School Improvement Teams*
- *Peer Coaching*
- *Assessment*

Staff Development

- *Training to support implementation of new practices/programs*
- *Peer coaching and support groups during implementation*
- *Training conducted in manner which assures successful implementation*

how the district should operate. In the typical school district, the organizational chart may resemble a pyramid, with ultimate authority resting in the board of education and descending through line positions to the superintendent, associate or deputy superintendents, division directors, area specialists, and so forth until reaching the classroom. If the district were large enough to justify the additional costs, staff positions usually attach to this hierarchy by reporting to line positions in an advisory capacity.

The sample organizational chart in figure 8.5 illustrates one district's philosophy of providing a reporting hierarchy which groups relationships according to proximity of service delivery to the school's clients. Collaborative relationships may also be seen in figure 8.5, as there are ample opportunities for horizontal and vertical two-way exchange among the system elements, although the system is clearly hierarchically structured in a fairly traditional manner.

A district may also refine its structure by creating extended organizational charts. Figure 8.6 makes this point, illustrating the fact that a district may benefit from multiple organizational charts addressing each broad function. In figure 8.6, where the business office has constructed an additional organizational chart which thoroughly defines the lines separating the broad functions of central administration and building level units in carrying out business office operations.Regardless of how personnel operations are structured,

Figure 8.5
Sample Organization Chart

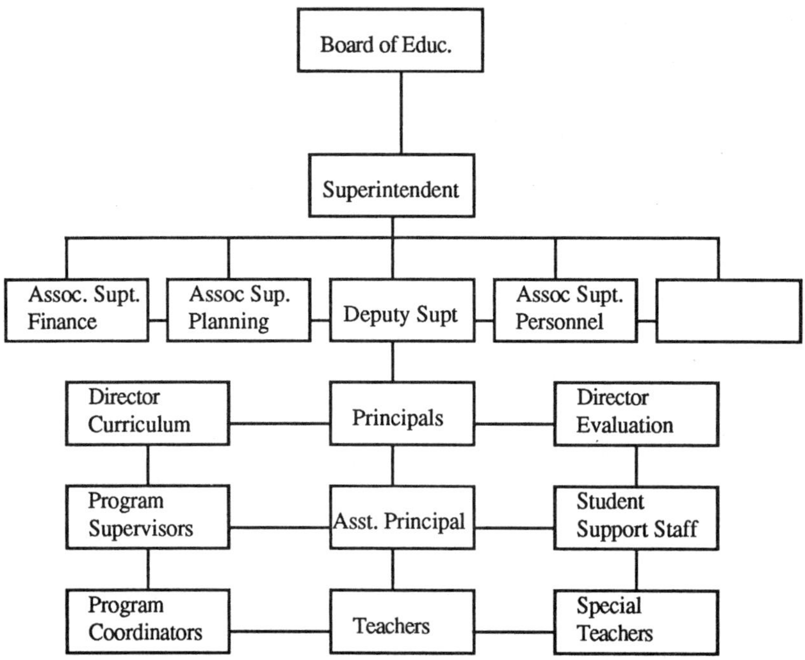

all areas of the district should be linked together under one umbrella. The umbrella should be provided in a board of education policy manual that contains written policies governing all aspects of personnel administration. Written policies assure uniform communication and, if properly prepared, assure a reflective and interdependent relationship between the goals of the school

Figure 8.6
Sample Business Office Organizational Chart

WHO IS ACCOUNTABLE FOR...	Learning Center	Joint	District Education Center
Who is responsible for supporting accountability programs?		X	
Who is responsible for keeping abreast of new learning and instructional methods?		X	
Who is responsible for providing and maintaining an effective communication system?	X		
Who is responsible for informing the public of school programs and needs?	X		X
Who is responsible for disseminating results of the educational system to the public?	X		X
Who is responsible for disseminating revised and updated policy statements to the public?			X
Who is responsible for insuring clarity, continuity, and relevance of student outcomes?		X	
Who is responsible for insuring that the curriculum planned is actually taught and tested?	X		
Who is responsible for insuring that student outcomes data (achievement tests, grades, etc.) are used by teachers to improve student learning?	X		
Who is responsible for monitoring and improving teacher instructional behavior?	X		

district and expectations of personnel in carrying out those goals. Policies must be clearly stated so as not to create ambiguity for either the district or the staff member.

A benefit of policy development is that it provides a blueprint for administrative action that goes beyond the organizational chart, which in turn has a greater opportunity for consistent administrative behavior that will enhance organizational morale and productivity at all levels. The critical nature of the

WHO IS ACCOUNTABLE FOR...	Learning Center	Joint	District Education Center
Who sets the amount of monies that will be available to the learning center?		X	
Who makes the decision on how mcney allocated to learning centers will be expended?	X		
If original allocations are not sufficient to meet the needs of the learning center, who will make the decision to allocate additional funds?			X
Who can make budget line adjustments during the fiscal year?	X		
Who is responsible for making decisions concerning Professional Staff Development?		X	
Who does budgetary accounting for site-hased management?			X
Who generates purchase orders for site-based expenditures?	X		
Who makes final approval on purchase orders?	X		X
Who is responsible for site-based budget line item over-expenditures?	X		
Who is responsible for total district budgets?			X
Who is responsible for new staff employment?		X	
Who is responsible for the final "checking in" of supplies at the learning centers?	X		

personnel function has led most districts to centralize this operation in a single office, placing one highly qualified person with line authority over personnel operations. Although titles differ according to custom and district size, an assistant superintendent for personnel, a director of human resource development, or even the superintendent of schools may have first line responsibility for the personnel function.

In many large districts the responsibility is often further subdivided—for example, the chief fiscal officer may also serve as the personnel director for support staff, while the director of human relations might serve as personnel officer for certified staff. In such districts certain aspects are likely to be centralized between the two offices, while others remain clearly separated.

	Learning Center	Joint	District Education Center
WHO IS ACCOUNTABLE FOR...			
Who is responsible for bidding supplies and equiprnent in compliance with district policies and state statutes?		X	
Who is responsible for the Food Service operations?			X
Who is responsible for assigning of transportation?			X
Who is responsible for assignment of maintenance projects and personnel?			X
Who is responsible for learning center custodians?		X	
Who is responsible for learning center secretaries?	X		
Who is responsible for learning center teacher assistants?	X		
Who is responsible for itinerant teachers?	X		
Who is responsible for the District Copy Center?			X
Who is responsible for what is being copied in the Copy Center as requested by teachers?	X		
Who is responsible for the volume and flow of materials to be copied in the Copy Center?	X		
Who is responsible for planning the annual painting schedule for buildings, rooms, etc.?	X		
Who is responsible for the districts site-based management plan?	X		

The purpose of a centralized office is to ensure smooth coordination of all vital aspects of the school district because the system will be in disarray if personnel issues are not effectively addressed. However, it should be recognized that while a central point person for personnel is required, every administrator is a personnel officer in that administration is the act of leading and coordinating people and the conditions surrounding teaching and learning.

The presence of a well-articulated mission, a set of meaningful goal statements, a clear organizational chart, and a well-designed personnel office are the products of another set of planning activities that are both antecedent to

WHO IS ACCOUNTABLE FOR?

Who is responsible for maintaining an accurate assets inventory at the building level?	X		
Who is responsible for care of property at the building level?	X		
Who is responsible for insuring that policies are implemented at the learning center level?	X		X
Who is responsible for evaluating, revising, and updating policies?	X		X
Who is responsible for adherence to Board of Education policies at the building level?	X		
Who is responsible for cooperation of teachers with other groups in the accountability program?		X	
Who is responsible for positive learning center climate?	X		
Who is responsible for establishing fair employment practices for staff and personnel?			X
Who is responsible for providing instructional leadership to school personnel?		X	
Who is responsible for insuring maximum product-ivity of personnel?		X	
Who is responsible for providing a suitable climate for learning and teaching?		X	
Who is responsible for keeping regular school attendance?	X		

Source: Adapted from Great Bend Public Schools.

and concurrent with the existence of a system exhibiting these qualities. The district's mission and goals depend on forethought about how the educational program should be structured, and on planning for the number and kind of pupils who will come to school. These considerations in turn have a tremendous impact on the area of staff planning in terms of numbers and types of staff needed—all questions obviously precursing how the organizational chart will finally look.

Similarly the other areas mentioned earlier relating to information planning and the external environment help shape the mission and goals of the school, as well as the other interdependent elements being discussed. As a result, moving from the broad to the specific in terms of beginning with a mission and progressing to operationalizing instruction and support services provides the key to recognizing that the personnel function is one of simultaneous leadership and management at all levels. Implications for leadership are obvious in that the district's goals and objectives cannot be fully met except through proper control and management of both instructional and noninstructional staff.

Pupil Planning

Although the various planning elements are difficult to isolate as discrete sequential activities because of their interdependence, pupil planning follows as the next step beyond conceptualizing the system in broad generalizations about mission and goals. Pupil planning is perhaps the single most critical element, since personnel and programs are logically built around the nature and number of students that the district and individual schools will serve. As a consequence, the district's skill in collecting and managing demographic information drives its success in all other areas of personnel administration.

While many areas of pupil planning are required to effectively set up a school system, the heart of the effort rests in gathering information about enrollments. Failure to properly plan for enrollments has a powerful negative impact on system continuity. In the case of overestimating enrollment, damage is likely to be suffered from reduced state aid, reductions in staff, inefficient class sizes, and closing school facilities that cannot be justified on the basis of low enrollment. Conversely, underestimating true enrollment has potentially damaging effects such as understaffing, inadequate facilities, overcrowding in classrooms, and the severe system stress that can result from too hastily hiring new staff and building new facilities to meet growth demands. It must be remembered above all else that in most states every item of the budget is driven by student enrollment. The critical importance of good demographic planning is unarguable in that student enrollment is the prime factor in the amount of revenues and expenditures a district receives and consequently the programs that can be provided.

Pupil planning through enrollment estimation is a data-intensive task requiring the collection of a variety of information. At the very least, pupil planning requires gathering data on local and statewide birth rates, the present school population, community population by age and family status, family mobility, nonpublic school enrollments, community and school transportation systems, board policies concerning attendance boundaries and open enrollments, zoning regulations, business and industry's current and future plans, as well as historical economic stability of the community, and the results of community planning by other government or commercial organizations that affect the local school district. Each type of information is valuable to projecting enrollments and student needs.

Data on local and statewide birth rates and information on family status are needed to provide historical perspective on population stability and growth. Public and nonpublic school enrollments are also vital, as these data must be joined with birthrates and migration data to give a baseline that will be used in the enrollment projection formulas discussed later. Similarly, information on economic stability and the future planning activities of other governments and local businesses gives a better view of the community's future in terms of population growth, stability, or decline. Other data such as transportation system and attendance areas yield useful information on future costs to the district, as well as movement of school populations among the various attendance centers in the district. Of paramount importance is a single goal: determining how many children will come to each school at what grade level.

While there are several ways to estimate enrollment, it is critical to understand that no system is perfectly accurate, because there is often no way to predict how a free people will choose to structure their lives. Large or rapidly expanding urban or suburban districts that have purchased powerful computer mapping and population analysis software that plots the number of children by age and grade in each house on every street, including developing areas not yet built, are vexed by sudden inmigrations and outmigrations of families due to human factors and economic shifts. The same problem can be applied to medium and small districts where the loss of each student has a proportionately greater impact on budget-program-personnel stability than is true in larger systems where changes are softened by sheer numbers of pupils.

Enrollment predictions underenrollment projection techniques are remarkably accurate, however, because all such techniques are based on two important features. First, despite some inherent uncertainty, there is empirical evidence that basic conditions experienced in the past will continue into the future. Second, any inherent uncertainty can be mitigated by constant monitoring, reevaluation, and updating of assumptions underlying the projection model. Trend analysis and the cohort survival technique are the most common methods used estimating enrollment, each with its own strengths and weaknesses.

Trend analysis is a highly sophisticated statistical technique that uses a mathematical equation known as least-squares regression. The equation predicts future enrollment based on the enrollment from a series of prior years. Fundamentally, trend analysis predicts future enrollments based on how far previous enrollments deviated from a straight line. The formula for the regression line (a straight line) is expressed as $Y = Mx + b$ where Y is the estimated enrollment, M is a coefficient calculated during the regression operation, x is each future year for the prediction, and b is a constant showing the relationship between enrollment and year at some starting date. A future year is entered into the equation and the result is a predicted enrollment.

Figure 8.7 gives the result of a sample regression equation. The regression formula yields a line of best fit, with the projected enrollment for next year calculated to be 1,400 pupils. Trend analysis for these data indicate a substantially increasing enrollment trend, with year one enrollment at 1,000 students, increasing to about 1,300 students by year five. By substituting other years for the x variable, additional future years can be projected as well.

In most instances, trend analysis is more useful in larger districts. Smaller districts are more at risk using trend analysis simply because the averages on which the line of best fit is based are too broad; the loss of each student in small districts represents an incrementally larger portion of total enrollment.

Figure 8.7
Trend Line

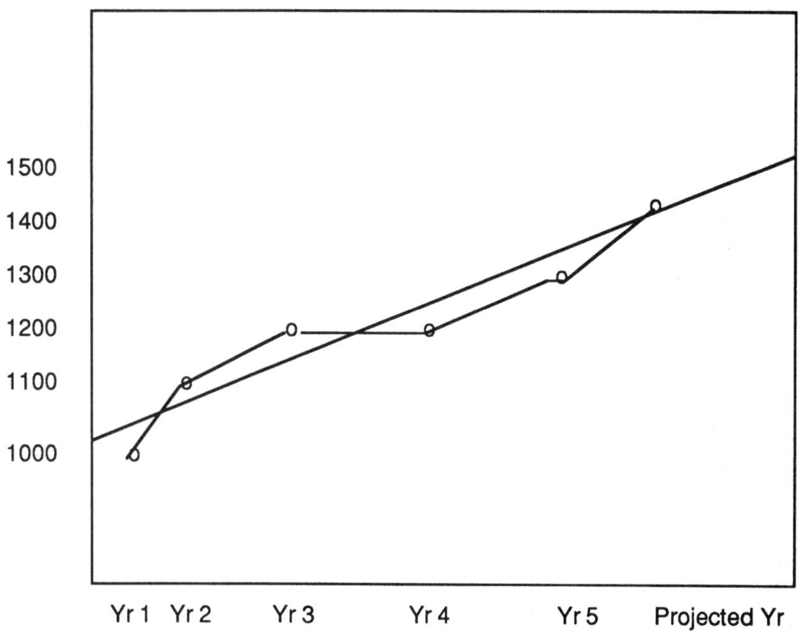

The averages in a line of best fit are better cushioned in larger districts because each student is proportionally a smaller part of the budget. Additionally, the complexity of calculation of trend analysis requires sophisticated computer work and considerable training in interpretation. As a result, many school districts prefer the cohort survival technique because it is more intuitive and more sensitive to changes in enrollment units.

The cohort survival technique is an enrollment projection method that groups students by grade level upon entry in school and tracks them through each year that they stay in the school district. This is not to say that individual students are tracked identifiably—rather, cohort survival sees entering kindergarten students, for example, in year one and calculates how many enter first grade in year two, second grade in year three, and so forth until exiting the system. Cohort survival is thereby a method evaluating the longitudinal relationship of the numbers of students passing from one grade to the next in the following year. Such a system more closely accounts for grade failures, dropouts, and inmigration and outmigration on a grade-by-grade basis.

Cohort survival requires enrollment data from the prior year and the current year for each grade in the district. The model finds a percentage change by grade level and defines survival as the percentage of the previous year's grade that moves forward in the system. For example, the middle part of figure 8.8 shows that past enrollments exiting first grade into second grade averaged only 76 percent of the prior year's first grade enrollment, indicating that migration or other factors such as grade failure resulted in a 24 percent average loss between first and second grades. Single year data, however, are not accurate enough for planning purposes, making it necessary to find the multi-year average shown in the bottom part of figure 8.8 where the actual current first grade enrollment can be multiplied by .76 to project only 19 students in second grade next year, compared to 25 students in first grade in the present year.

This process uses historic five-year data repeated for each grade, and should only be extended forward five years into the future, with enrollments found for individual grades and district totals. A data hole, however, relates to kindergarten students in that such data are obviously not obtainable from school attendance records. The value of earlier discussion about birth rates, family status, and so forth relate to exactly this problem.

Estimating kindergarten enrollment requires looking at past enrollment trends, reviewing birth records for the last five years, obtaining preschool program data, estimating migratory patterns, conducting kindergarten roundup, and so forth. Obviously, enrollment projections form the basis for determining staffing needs, making staff assignments, purchasing supplies and equipment, projecting facility utilization and new building programs, and many other revenue and expenditure problems. Because of its sensitivity to incremental

Figure 8.8
Cohort Survival Technique

Part I.
Historic Enrollments

Grade	Year 1	Year 2	Year 3	Year 4	Year 5	Year 6	Current Year	Avg
K	28.0	30.0	34.0	26.0	27.0	15.0	22.0	26.0
One	17.0	25.0	32.0	32.0	34.0	23.0	25.0	26.9
Two	25.0	17.0	15.0	22.0	21.0	26.0	20.0	20.9
Three	17.0	25.0	16.0	11.0	16.0	20.0	21.0	18.0
Four	12.0	20.0	24.0	22.0	9.0	16.0	22.0	17.9
Five	16.0	14.0	21.0	21.0	16.0	12.0	16.0	16.6
Six	13.0	14.0	16.0	19.0	17.0	18.0	12.0	15.6
Seven	17.0	15.0	19.0	15.0	13.0	23.0	16.0	16.9
Eight	17.0	20.0	16.0	13.0	15.0	16.0	17.0	16.3
Nine	18.0	19.0	19.0	16.0	15.0	20.0	15.0	17.4
Ten	11.0	18.0	17.0	20.0	10.0	19.0	15.0	15.7
Eleven	13.0	15.0	18.0	16.0	23.0	13.0	17.0	16.4
Twelve	19.0	13.0	11.0	17.0	14.0	18.0	11.0	14.7
TOTAL	223.0	245.0	258.0	250.0	230.0	239.0	229.0	

Part II.
Survival Ratio

Grade	Year 1	Year 2	Year 3	Year 4	Year 5	Year 6	6 Yr Avg
One	0.89	1.07	0.94	1.31	0.85	1.67	1.12
Two	1.00	0.60	0.69	0.66	0.76	0.87	0.76
Three	1.00	0.94	0.73	0.73	0.95	0.81	0.86
Four	1.18	0.96	1.38	0.82	1.00	1.10	1.07
Five	1.17	1.05	0.88	0.73	1.33	1.00	1.03
Six	0.88	1.14	0.90	0.81	1.13	1.00	0.98
Seven	1.15	1.36	0.94	0.68	1.35	0.89	1.06
Eight	1.18	1.07	0.68	1.00	1.23	0.74	0.98
Nine	1.12	0.95	1.00	1.15	1.33	0.94	1.08
Ten	1.00	0.89	1.05	0.63	1.27	0.75	0.93
Eleven	1.36	1.00	0.94	1.15	1.30	0.89	1.11
Twelve	1.00	0.73	0.94	0.88	0.78	0.85	0.86

Part III.
Enrollment Projection

Grade	6 Yr Avg	Actual	Year 1	Year 2	Year 3	Year 4	Year 5
0→K	0.187	22.0	24.0	21.0	18.0	19.0	19.0
One	1.12	25.0	24.7	26.9	23.5	20.2	21.3
Two	0.76	20.0	19.1	18.8	20.5	18.0	15.4
Three	0.86	21.0	17.2	16.4	16.2	17.7	15.5
Four	1.07	22.0	22.5	18.4	17.6	17.4	18.9
Five	1.03	16.0	22.6	23.1	18.9	18.0	17.8
Six	0.98	12.0	15.6	22.0	22.5	18.5	17.6
Seven	1.06	16.0	12.7	16.6	23.4	23.9	19.6
Eight	0.98	17.0	15.7	12.5	16.3	23.0	23.5
Nine	1.08	15.0	18.4	17.0	13.6	17.6	24.9
Ten	0.93	15.0	14.0	17.1	15.9	12.6	16.4
Eleven	1.11	17.0	16.6	15.5	19.0	17.6	14.0
Twelve	0.86	11.0	14.7	14.4	13.4	16.4	15.2
TOTAL		229.0	237.8	239.8	238.8	239.8	239.1

Source: Adapted and modified from David C. Thompson, R. Craig Wood, David S. Honeyman. FISCAL LEADERSHIP FOR SCHOOLS: CONCEPTS AND PRACTICES. New York: Longman (1994) p331

changes in population, cohort survival is usually the best choice among various techniques in estimating enrollment for most istricts.

The importance of accurate pupil planning cannot be overstated because enrollments drive budgets and state aid, and because other data collected in the process feed directly into the numbers of staff that will be needed and will dictate in large part the educational programs to be provided to pupils.

Program Planning

When enrollments are known, program planning can be undertaken with the concurrent recognition that programming overlaps with other areas related to developing mission and organizational goals. It is only reasonable to expect that the educational program in schools should reflect the system's mission and goals, and it is likewise reasonable to expect that the program should be driven in large part by information about the children coming to school. In other words, the educational program should not only reflect the broad educational mission generally and the mission of the individual district—it should also reflect the particular needs of the clientele who will receive the system's benefits.

Castetter has argued that a wide spectrum of issues should help drive human resource planning assumptions. He poses a series of thirteen general points, many of which can be transferred directly to the program planning activity with tight relevance regarding the tasks to be carried out. Castetter raises important questions relating to program planning as he asks:

- What persons in what age groups should be educated at public expense? What pre- and postcompulsory age groups should be included in the educational program?

- What purposes are schools expected to achieve? What should be the range and nature of educational experiences provided by the school system to achieve its purpose?

- How should the educational program be structured?

- What are the most effective ways of organizing instructional groups?

- What methods, materials, and services should be used to make educational experiences meaningful to learners?

- What should be the size and composition of the professional instructional staff? The administrative staff? The noncertified staff? To what extent should the staffing of each attendance unit be decentralized?

- What special services should be provided in each of the schoo attendance units? In the central administration?

- What systemwide services to teachers—for example, continuing education, supervision, professional library, and curricular and instructional aids—should be provided?

- What central administrative services—for example, pupil personnel, staff personnel, logistics, plant, research, planning, external relations, and coordination—are necessary?

- What should be the size and deployment of the classified staff—for example, custodial, maintenance, transportation, food, clerical, and security services?

- What positions should be retained in the future organization structure? Added? Dropped? Modified?

- To what extent should existing personnel comprise the future organization structure? New personnel?

- How can the future structure be organized to satisfy member needs more effectively?[25]

These and other important questions frame the process and product of the program planning function of personnel administration. The areas listed must be worked through in each district, with the results flowing from the district's mission and goal statements to shape its personnel and program structures. The organizational chart seen earlier in figure 8.5 is defined by these processes in that personnel and services should be aligned to facilitate delivery of the program. Every issue, including facilities and budgets, is driven by decisions about who will occupy what roles, how the roles will be carried out, and where those roles will be located. These type questions make up the program planning aspect of the personnel function, driving decisions about program focus, program content, program depth and breadth, pupil grouping, and program instructional methods and control.[26] Program planning obviously flows from pupil planning activities which in turn flow from system planning, making a successful educational enterprise dependent upon thoughtful preparation.

Staff Planning

Although a high degree of exchange in fact occurs between all the issues just discussed making it impossible to say that staff planning must follow sequentially after all the other issues have been resolved, it is reasonable to say that over time a school district will be more likely to plan staff around the system's mission, goals, pupils, and programs than to plan all the foregoing around the staff, per se. That is to say, modern school systems rarely become involved in startup operations where such sequential behavior could occur—rather, staff are generally hired into and participate in the various planning activities as an ongoing district activity. The point is that staff planning is at least a concurrent activity to all other planning in the district.

From the perspective of the administrator, staff planning can be said to comprise the major headings of recruitment and selection, induction and compensation, evaluation and development, and other personnel problems and issues.[27] These benchmarks of the personnel function describe how the system seeks to find the right people for the job and to align both the system and the individual for organizational productivity.

Recruitment and Selection

The modern context of education requires that school districts act aggressively to recruit and select staff. Many reasons are apparent for this aggressiveness. Schools are increasingly faced with fierce competition for employees from other sectors of the economy, particularly since many noneducation occupations pay significantly higher salaries than public schools can afford to offer. Schools must also pursue good candidates simply to expand the selection opportunity to the largest possible pool to identify the best available talent. Likewise, the modern context of the world finds some propensity on the part of unqualified and undesirable candidates to attempt to enter the education profession, making it necessary for legal and moral reasons to closely screen candidates for suitability.

The significant numbers of new teachers leaving the profession after only a few years also increases the demand for a continual supply of candidates, and the concept of employee dropouts implies that schools must do a better job of recruiting and selecting staff in order to avoid hiring people who can be predicted to fail. Such reality also implies that schools must do a better job of nurturing talented staff who leave because they often believe they are not provided with adequate support. In sum, recruitment and selection offer the best opportunity for success in all other operations of the school district because any action directed toward bolstering inadequate skills among existing staff is inherently remedial.

Recruitment

Recruitment naturally precedes selection and is comprised of a series of activities that may be designed to meet the unique needs of the district. As a general rule, recruitment activities begin with development of a recruitment policy speaking to the aims of the district through the personnel function. Such policy is common to all school districts and speaks to critical matters including prohibiting discrimination, position qualifications, how positions will be filled from within the system or from outside candidate pools, and statements about promotions and other reward structures. Issues such as any probationary period, procedures for gaining tenure, restrictions on outside employment, background inquiries prior to employment, layoff or seniority rules, and so forth should all be written into the recruitment policy in order to inform candidates of the nature of the district being considered. The policy also should speak to how actual selection will occur.

For example, many districts have moved to schoolbased personnel policies where some combination of administrators and current staff collaborate in hiring new people to fill empty positions. Other districts have much more centralized recruitment and selection processes. While some of these areas overlap during the selection phase, such information should be made known in advance to candidates at the recruiting stage in order to be clear about employment opportunities, and to ward off any detectable mismatch.

The recruitment policy also should establish responsibilities for current administrators charged with carrying out the recruiting task. It is particularly important to avoid confusion and conflict about authority relationships in the personnel function. At a minimum the policy should provide for four central activities. The policy should state that the role of administration is to recommend candidates, while only the board makes the final selection. The personnel office should also be charged with initiating formal assessment of immediate and longrange personnel needs and developing position descriptions for existing and prospective jobs to both inform candidates and to provide consistency of employee qualifications across the system. The policy should also require administrators to engage in formal study of the success of the recruiting function in order to improve its effectiveness and efficiency. Further, the policy should direct the administration to develop extended plans for communicating the needs of the district to various constituents such as universities and other candidate pools including notification of vacancies, position guides, application forms, and publicity literature such as brochures and district fact sheets.

The sum of a recruitment policy should provide for central planning, coordination of information, establishment of employment needs and standards, preparation of a plan detailing how recruitment will occur, systematization of candidate information, and development of candidate appraisal systems that will carry desirable employee candidates to the selection stage. The result of the recruitment process should provide satisfactory answers to the following questions: What are the outcomes on behalf of the school system that the recruiting process seeks to provide? In what ways is the recruiting process successful, and in what respects has it failed? How successful has the process been in recruiting qualified candidates as defined by quality performance and longevity? What has been the record of recruits who have joined the school system in terms of tenure, promotion, termination, or other departure?

Selection

If a successful recruitment plan is in place, it follows that the next step is selection, to secure the best candidate for a vacancy. Regardless of whether a school district is very large or very small, selection is vital because schools are only as good as the staff chosen to carry out their purposes. Tremendous amounts of time and money are at stake in the selection process, from the perspective of waste if the selection process fails to produce good people, and from the perspective that there are great legal dangers that can arise if particularly unsuitable candidates slip past the selection screen, because schools are charged with a high standard of care for young children.

As in all other personnel areas, an important aspect of the selection process begins with development of a policy speaking to the major elements of the process. As with other policies, the purpose of a selection policy is to set out the district's commitment to choosing quality personnel and to broadly establish the manner in which selection will occur. Like the recruitment policy, the selection policy should inform all interested persons of the board's intent to choose only the most qualified candidates and should speak to how the district views vacancies—e.g., whether positions will be first opened internally and expanded later, whether it is the district's intent to promote internally or to go outside for newly vacated positions, and other issues such as nepotism and equal opportunity. The policy should set forth the district's procedure for selection, such as the use of position descriptions to weed out unqualified persons and the steps in the process such background checks, multiple interviews, skill demonstrations, and so forth. In order to avoid legal problems, however, care must be taken to ensure that any requirements placed on candidates are related to the vacancy. As a general rule, districts may inquire when appropriate about educational qualifications, licensure, experience, specialized skills, relevant personal characteristics, and quality of background—such as prior work history and so forth. Figure 8.9 provides a sample position description for the superintendent of schools which flows from the board's selection policy, and figure 8.10 provides a sample position description for a classified staff member. In both instances, the major elements discussed above are related to the actual requirements of the job and are appropriately discriminating for purposes of quality control.

In the field of education, there is little problem in establishing strong criteria for the educational qualifications of candidates. All unclassified personnel are expected to be academically strong, and this is reinforced by state certification requirements demanding high credentials for persons at all levels of the organization. For both unclassified and classified personnel, the nature of schools also permits stringent background checks to determine prior work history, evidence of unsuitability, and claims for experience. Acceptance or rejection, however, should be based on the data collected from the various standardized tools used in the selection process.

Figure 8.9
Superintendent Position Description

Position Title Superintendent of Schools
Reports to Board of Educatioin
Supervises all Certificated Staff and Classified Staff--delegates as appropriate.

Basic Function:
Administers the programs of the District according to the policies of the Board of Education. Provides leadership to both board and all personnel; manages and directs all activities concerned with the district. s

Performance Responsibilities:
* Demonstrates leadership through values, belief's, skills, and personal characteristics that inspire people to accomplish the district mission.
* Ensures effective instruction to accomplish district goals based on the tenet that all students can learn.
* Monitors, assesses, and supervises the implementation of the educational program of the district.
* Ensures evaluation of the performance of certified and classified staff members, and offers guidance, support, and assistance, as needed.
* Interprets, implements, and maintains the integrity of board policies, state school laws, and administrative practices and procedures.
* Develops a program of public relations to further community understanding and support of the district's educational program.
* Administers the district's budget and recommends annual budget requirements.
* Identifies/applies effective strategies for dealing with political issues and forces that impinge upon the district's operation.
* Engages in a program of continuing professional development; participates as a member of local, state, and national professional groups.
* Ensures orientation of newly assigned staff members and ensures their familiarization with district policies/procedures, teaching materials and school facilities, and assists in their development, as appropriate.
* Possesses skills in the use of conflict resolution methods, various decision-making techniques, and the process of consensus building.
* Requirements: Minimum of Doctorate degree in education or school administration: possession of a valid certificate. Five years teaching/administration experience. Salary commensurate with experience. Broad in-depth knowledge of educational programs: ability to plan, organize, supervise, and evaluate; ability to resolve conflicts; ability to communicate effectively in both speaking and writing; ability to work with professional and lay publics; knowledge and appreciation for the ethnic and cultural diversity of the district's population.

Regardless of how the selection process is structured, it should at a minimum, provide answers to the following questions: Do data on the candidate clearly indicate educational preparation, work experience, and personal qualifications, or are there vague areas that need further investigation? Are all areas of information supplied by the candidate accurate? Does the candidate possess the professional qualifications required in the job description? Does the candidate show a training and/or work record demonstrating emotional maturity and stability? Does the candidate show personal characteristics sought by the district such as proper motivation, cooperativeness, enthusiasm, and willingness to learn and contribute to the system? Does the candidate show promise of fitting both the culture of the organization and the community at large? Does the candidate stand out from the pool of possible choices? Finally, are there any persistent doubts or reasons suggesting that the candidate might not be the best match? While good answers to all these questions do not ensure a long and productive tenure, a rigorous selection process greatly enhances opportunity for success. As the old saying goes, "if you don't want to fire 'em, don't hire 'em if there is any doubt."

Figure 8.10
Sample Classified Position Description
SECRETARY TO THE ASST. SUPERINTENDENT
OF OPERATIONS

A-11 Non-Certified Personnel. Record hours and absences for non-certified employees.

C-1 Work Orders (computer). Receive and process all Supply Requests for office supplies.

C-2 Supply Requests (computer). Receive and process all Supply Request for office supplies

C-4 Orders (computer). Process carpentry, local, and Co-op bids.

C-6 Budget (computer). Assist with compilation/revision of district budget

C-12 Catalogs. Maintain file of merchandise catalogs.

C-13 A-V. Equipment. Check out major A-V equipment as requested.

C-16 Student Transportation. Organize and carry out the procedure of parents' signing for mileage reimbursement for student transportation (spring).

D-4 Music Activity Fund. Handle clerical duties which includes some ordering, writing checks, and preparing monthly reports.

D-9 Building Use. Record approved activities. Supply notification of each activity to the appropriate principal as well as to the Director of Buildings. Submit monthly reports to principals. Bill the organizations for assessed fees.

E-1 Typing. Do general typing

E-3 Receptionist. Serve as back-up lobby and telephone receptionist.

E-5 Assistant Clerk. Serve as Board of Education secretary. This includes preparation and distribution of meeting notifications, agenda, and minutes for regularly scheduled meetings, special meetings, and committee meetings; handling BOE communications; and making meeting/meal arrangements.

E-6 District Booklets. Maintain/update the following booklets: *Certified Employee Employment Policy; Classified Personnel*

Handbook; Board of Education Policy Book; and Personnel Directory.

E-7 Teacher History (computer). Maintain/update the "Teacher History" information.

E-9 Fuel Bids. Send request for quotations on fuels to local vendors, receive and compile bids, and notify of results.

E-10 Speech/Debate. Handle financial transactions for the speech/ debate coach of the High School, including writing checks, recording receipts, and handling banking duties.

E-11 Workers' Compensation. Accept and forward to the local insurance agent all Workers" Compensation claims.

E-12 Insurance. Maintain a file of current insurance polices.

Source: Adapted and modified from Great Bend Public Schools.

Induction and Compensation

Many school systems have made the fatal error of believing that the personnel function ends when a new employee enters the system. This is not to say that new hires are abandoned at the schoolhouse door, but it is to say that coordinated efforts to integrate new people have lacked follow through in many cases. While it is likely such schools would acknowledge that induction and compensation are important, systems neglecting these areas do not have a full understanding of the need for continued care and nurturing of their people. In fact, there is an argument to be made that induction and compensation policies continue throughout the life of an employment relationship, particularly in light of earlier discussion in this chapter about the need to nurture both nomothetic and idiographic dimensions of the organization. At the very least, however, new staff must be aided in becoming acculturated to the system to maximize the opportunity for an enduring relationship. Of greatest importance in this regard are the areas of induction of new employees and compensation policies for all staff.

Induction

When a newly hired person enters the organization, there is a great deal of ambiguity and uncertainty. In virtually all instances, the person exits one culture to enter another culture where "the way we do things around here" is inevitably different in some respects. Data suggest that people are required to change some behaviors by forming a new association, that behaviors change in relation to the newly joined group, and that there is a period of influence that occurs during the early stages of new employment that results in either a positive or negative organizational response.

Proper induction should address a host of different issues, including developing mutual respect between all employees and the system, nurturing a sense of belongingness, and building a sense of confidence on the part of the new hire. Induction also provides an opportunity to make an initial judgment concerning the appropriateness of the new hire's assignment, to verify preliminary assessments of the candidate's skills, to assist the new person with general accommodation, and to spot and improve areas of weakness. In other words, the induction phase begins a never ening process of evaluation and development.

As with all other areas of the personnel function, constructing a valid induction process involves thinking through and developing a policy that addresses the goals the organization wishes to accomplish with new people. Board policy should be provided which is based on assisting in the transition from new to permanent in order not to waste the district's time and money on people who leave the system prematurely. At the least, a policy should be in place indicating that the district is committed to establishing effective relationships and to diminishing the anxiety that can accompany learning a new setting. The policy also should describe the major elements of how the district inducts new hires. Naturally, the ultimate objective is to ensure development of a positive, productive attitude and level of performance that can be described as competent and that is marked by independence and initiative.

The scope of the induction process should be as broad as possible, giving significant time to working with the new person. An earlier statement in the chapter about every administrator serving in a personnel role is particularly applicable to the induction phase. Significant effort should be given to helping the person learn how the work environment is structured and the tasks expected of someone in that position, and in getting to know the other people in the system. Useful activities to assist in these objectives include formal orientation sessions with administrators and other personnel, informal assignment of new hires to experienced staffers who will provide a warm and patient attitude, a mentoring program that teams new and current staff, and invitations to join in work groups and social events that help the newcomer learn the organization's work and its informal culture.

Significant amounts of information should flow to the newcomer from the outset to avoid "learning by stumbling," which risks poor attitude development and early association with the inevitable cynic who exists in every organization. Notwithstanding, sensitivity to work overload of new employees is important since they must cope with a new environment, a new set of performance tasks, and a thorough induction process—all of which leave little time for calm reflection.

Castetter lists some important principles of induction promoted by the American Management Association (AMA) that reflect considerable wisdom when inducting new people into the system. These principles, shortened and augmented by other original text here, argue that:

- *"Early impressions are lasting."* Little explanation of this concept is needed.

- *"The pre-arrival period is part of orientation."* Considerable literature on morale curves exists, wherein it can be seen that new people experience ups and downs that begin immediately upon signing a contract, and that the opportunity to positively influence the new hire begins long before the first day at work arrives.

- *"Day One is crucial."* Nearly everyone has memories of their first day, which can be an unsettling experience. Induction should plan to provide close assistance during this period.

- *"New people have an interest in the total organization."* Earlier text discussion of people in organizations supports this point, arguing for concerted effort at helping people see, understand, and support the whole organization. The common expression, "whoever is not for me stands against me" suggests great wisdom in moving quickly to bring the new hire into the organization.

- *"Teach the basics first."* This concept strikes to the heart of many mistakes made by systems, and requires a great sensitivity to information overload or lack of information. Good advice is found in the expression "all things in good time," suggesting that more information given to new people is better than less information, but it also suggests that not all information is needed at once.

- *"Give the new person major responsibility for his or her own orientation."* This concept does not argue for turning people loose to sink or swim, but rather notes that the learning curve will accelerate if people can learn by doing under supervision. The best analogy is showing, rather than telling.

- *"Understand that community, social, and family adjustment is a critical factor."* This principle fits closely with earlier discussion in this chapter about idiographic dimensions of organizations and points out that individuals also come from other social environments that include people not directly in the work group; i.e., family. The AMA suggests that the views of the spouse are a critical factor in determining whether an employee stays or leaves the organization.

- *"Remember that it's the individual that counts."* The critical flaw of organizations focusing too severely on only the organization's needs is brought to attention by this principle.

- *"Make the supervisor the key."* For schools, the earlier concept of every administrator as a personnel manager is particularly relevant for many reasons, among which are the value of the organizational chart and the early window of opportunity to positively influence the person regarding organizational goals.

- *"Emphasize solid orientation as a 'must' for productivity improvement."* As stated earlier, productivity is the goal of the organization and cannot be left to haphazard development—i.e., productivity remains elusive and incomplete if people are not acculturated into system goals and objectives.

The importance of induction cannot be overstated—indeed, it is as important as initial selection because attitudes are virtually imprinted on people new to the system from the earliest moment. Induction provides the opportunity to ward off great problems.

Compensation

Of equal importance to proper personnel operations in a school district is the area of employee compensation. Compensation offers the best opportunity to understand the interplay between the various functions of a district, in that the personnel and business operations of the district each have a deep and abiding interest in the actions of the other. For clear reasons, the business office is vitally concerned about the activities of the personnel office since it falls to the fiscal function of the district to responsibly provide for the compensation of all district employees.

Conversely, the personnel office is interested in the fiscal aspects of the district since it stands without question that no amount of careful personnel planning and recruiting can offset the damaging effect on selection and retention of employees that occurs when the district is not competitive financially with other districts and organizations in the marketplace. Under these conditions, strong planning and thoughtful compensation policies are required attributes of well organized school systems.

Of critical importance to any compensation plan is a policy speaking to the district's goals and objectives for treating its employees fairly and respectfully. Castetter notes the value of strong and effective compensation planning as he lists thirteen goals, around which the formal organization should build compensation policies. Alignment between these goals and the earlier chapter discussion of exchange theory and social systems theory is remarkable, underscoring the need for conceptual theory bases in personnel administration:

- Attracting and retaining competent career personnel.
- Motivating personnel to optimum performance.
- Creating incentives to growth in individual competence.

- Getting maximum return in service for the economic investment made in the compensation plan.
- Developing confidence of personnel in the intent of the organization to build equity and objectivity into the compensation plan.
- Making the plan internally consistent and externally competitive.
- Relating compensation levels to importance and difficulty of positions.
- Making salaries commensurate with the kinds of personnel the organization requires.
- Establishing a compensation structure conducive to giving economic, social, and psychological satisfaction to personnel.
- Minimizing union and individual grievances.
- Exercising careful control over salaries, a budgetary item that generally encompasses four-fifths of the expenditure plan.
- Developing plans to ensure continuity of funds needed for an effective salary, wage, and benefits program.
- Minimizing personnel turnover.[29]

The stark realities of these thirteen points demonstrate the value of careful planning for compensation of employees as it is apparent that the personnel function is severely damaged if adequate and fair compensation is lacking. Aggressive recruiting and discriminating selection are hurt by low or inequitable pay structures and may even be impossible in a competitive marketplace. While scientific management theories have largely been discredited for their overemphasis on greed, it remains a valid point that compensation may be a significant dissatisfier that contributes to high turnover and lack of career development and organizational continuity.

Serious credibility questions also arise in districts where motivational speeches about community and mission are tied to quite another budget message that fundamentally devalues the work of individuals. Only trouble pursues from parsing salaries in a sense of false economy—trouble that manifests itself in poor working relationships, confrontive and hostile negotiations, low worker production and related performance, and so forth. Of paramount importance is a recognition that compensation is more than economic—it is at once monetary, ethical, and political, and should be treated with care from both fiscal and moral responsibility.

Development of compensation policy begins at the board level and is felt throughout the system. Policy should be developed which takes into account a variety of issues relating to several different realities. One such reality is that competitive wages must be paid because of the deleterious effects listed earlier. Another reality is that community perceptions have a powerful impact on how much a district can realistically afford to pay staff. Still another reality is that funding is always external to the district and this has a constraining effect on the district's abilities. Yet another inevitable reality is that personnel are never overpaid in their own perception, making inevitable some degree of dissatisfaction with working conditions.

These and other very real limitations frustrate and restrict the compensation system and must be balanced against each other. Board policy should admit to these problems and assure certain efforts to address the concerns. At the least, board policy should commit the district to attracting and retaining competent personnel, provide salaries that are differentiated by level of responsibility and the worth of different kinds of work, guarantee comparable pay for comparable work, and be designed to secure the broadest possible community support base. Further, policy should demonstrate that the district values all classes of employees and that the system does not focus on any one group—e.g., teachers—to the diminishment of other administrative or classified staff. This focus should be paramount regardless of the strengths of various professional associations that may exist within the school district.

Personnel Problems and Issues

Although it is impossible to identify every personnel problem that can arise, and it is equally impossible to discuss all such issues in a single text, three particular problems facing school districts merit some development in closing this chapter on personnel. The first area involves grievance administration. The second area involves difficult decisions and processes relating to reduction in force. The third and final area relates to the complex considerations involving dismissals for other reasons relating to cause. The discussion regarding grievance procedures is presented in the collective bargaining discussion found in the next chapter. However, dependent upon state law, it is important to note that grievance procedures can be found either in controlling state statutes or in collective bargaining agreements.

Reduction in Force

Another area of concern for the personnel office involves reductions in force (RIF). Reductions in force occur when districts have more personnel than can be justified for reasons such as declining enrollment. In these instances, it is critical for the finance and personnel functions to work jointly to assess the problem, to minimize the damage caused by either excess numbers of employees or having to reduce staff, and to justify the impact of any forthcoming drastic action. Chapter 10 of this text covers this topic in detail.

Other Dismissals[30]

In addition to RIF, other dismissals occur in educational organizations. Although it is commonly believed that tenure provides a safe haven against dismissal, there is abundant evidence that both tenured and untenured teachers are dismissed. There is further ample proof that local boards have all necessary power required to evaluate and dismiss public school employees, regardless of longevity or statutory tenure rights. These powers are by virtue of state legislatures' granting of implied and delegated authority to maintain and operate the public schools.[31]

The basis for dismissal of employees for reasons other than RIF is well founded in the literature and the law. Dismissal is conceptually based on several considerations. First, there is an unseverable right of an employer to expect and receive satisfactory performance of contractual obligations. Second, the organization has the right to conduct its business in an efficient and effective manner and to dismiss employees who do not assist the organization in meeting those goals. Third, the system has the authority to dismiss any person who refuses to accept instruction intended to improve performance. Provided that the system follows the law regarding certain procedures designed to protect employee rights, the power of the board is unlimited. Of paramount importance, therefore, is board policy addressing the rights and limits of the organization in termination matters and the rights of employees against abuses of power.

Even though boards have broad termination powers, dismissal is not to be taken lightly. While powers are broad, the limitations of law are particularly important in education, since tenure laws exist and since a First Amendment free speech right is more easily invoked than in noneducation fields. The result is that in the majority of states, significant differences exist between dismissal for nontenured and tenured employees. By definition, tenured employees are granted a permanent or continuing contract and are thus entitled to due process in the dismissal act. In contrast, nontenured teachers are usually provided only a term contract under whatever employment and dismissal procedures are granted to them under contract law.[32]

This does not mean, however, that nontenured staff have no protections, because no one can be denied constitutional rights regardless of status of employment.[33] If allegations are made that constitutional rights such as freedom of speech have been violated, nontenured employees of any classification may have a basis for demanding due process. States may also provide other rights to nontenured people, making it necessary for administrators to seek out state-specific statutes, the steps in termination proceedings, and the negotiated terms of the actual employment contract. Dismissal is an extremely complex and time consuming process, involving the areas of coordination, substantiation, consultation, evaluation, admonition, documentation, communication, charges, notification, hearing, appeal, and decision.[34]

Although it is in fact much easier to dismiss nontenured staff, there are many instances of successful dismissal of tenured faculty.[35] The procedure, however, is more difficult because state statutes usually provide narrowly drawn authority for terminating tenured employees. The important conclusion is that no dismissal can occur except for those reasons. Examples of reasons for dismissal include incompetence, immorality, insubordination, incapacity, neglect of duty, unprofessional conduct, and conviction of a felony. As discussed in detail within chapter 10, fiscal exigency is also a valid reason for tenured dismissal, provided that law is followed regarding seniority and so forth. When any of these reasons can be justifiably applied, employers have full right of termination. Penalty for failure to prove serious charges or to follow the law, however, may result severe liability for employee compensation and restitution.

Dismissal for incompetence depends on the statutory definition of incompetence as set out in the individual states. However, courts have often allowed broad construction of incompetence wherein it is usually defined as the "lack of ability, legal qualification, or fitness to discharge the required duty."[36] A difficult aspect for administrators, however, is that proof of incompetence is measured against other individuals performing the same or similar duties. Under this condition, the administrator is immediately confronted with having to prove that the dismissed person is the worst case in order to avoid other charges that are sure to follow. Incompetence must therefore be strictly construed and proved by the district to include one or more of the following:

- Lack of a proper teaching certificate.
- Lack of knowledge of subject matter.
- Lack of ability to establish reasonable discipline in class.
- Deficiency in teaching methodology.
- Emotional instability that demonstrates an inability to effectively teach students.[37]

Both nontenured and tenured staff may be terminated for any of these reasons. Tenured dismissal, however, requires great care to maintain documentation substantiating any charges because investigative due process hearings will certainly follow. As seen earlier, the elements of coordination, substantiation, consultation, evaluation, admonition, documentation, communication, charges, notification, hearing, appeal, and decision must be carefully observed.

For example, all states require remediation before dismissal, setting forth requirements that remediation must be carefully documented.[38] Each step requires careful data collection, as in remediation where an effort must be made to help the employee succeed in the specific job, define the activities and responsibilities of all parties, and demonstrate good faith on the part of the district to save the contractual interest of the employee. Documentation of failure of the employee to respond to remediation and other steps must be exhaustive, because the first defense is a challenge that the terminated employee is not the worst case.

Immorality also may be a basis for dismissal. In the modern world, immorality is not confined to sexual impropriety. If dismissal for immorality were challenged in court, immorality is usually based on community standards in which the test becomes whether the act is "inimical to public welfare."[39] Immorality can be quite broad, taking in corruption or indecency in that standards of immorality may include "[a] course of conduct as offends the morals of the community and is a bad example to the youth whose ideals a teacher is supposed to foster and to elevate."[40]

Although immorality is a broad area, it is one where districts are very open to liability for misapplication. For example, one of the areas historically of concern to administrators is unwed pregnant teachers. Generally, courts have not supported dismissal for immorality in such instances because it is hard to show that unwed pregnancy damages students or that the teacher's respect in the community is adversely affected.[41]

Such a claim may not be supported because in 1978 Congress passed the Pregnancy Discrimination Act[42] amending Title VII in terms of the definition of sex. As a result, administrators should be careful of how immorality is defined at the local level,[43] and further caution relating to pregnancy in general is prudent due to the pervasive influence of federal legislation such as Title VII of the Civil Rights Act of 1964, numerous laws promulgated under the Equal Employment Opportunity Commission, Title IX of the Education Amendments of 1972, the Civil Rights Act of 1991, and others. Under several of these laws, liability includes compensatory and punitive damages.

Insubordination, however, represents a less thorny issue when dismissing tenured or untenured staff, as willful disregard or refusal to obey reasonable directives constitutes insubordination.[44] To uphold a claim of insubordination, the courts will demand proof, particularly in regard to tenured staff because of property rights in teacher contracts. A hasty charge of insubordination can be unwise because insubordination generally will not be upheld when:

- The alleged misconduct cannot be proved.
- Existence of a pertinent school rule or a superior's order is not proved.
- The motive for violating the rule or order is admirable.
- The rule or order is unreasonable.
- The rule or order is invalid or outside the authority of the maker.
- Enforcement of the rule or order reveals bias or discrimination.
- Enforcement of the rule or order violates constitutional rights.[45]

If one of these reasons were to be shown to have led to dismissal, courts will not support the district, because directives must be reasonable and rational. Insubordination and willful neglect must be proved, rather than assumed.[46]

Finally, employees may be dismissed for neglect of duty. Neglect of duty may be quite complicated and may overlap with other claims of more general incompetence. Neglect of duty may be broadly construed, including failure to follow curricular guidelines, inadequate discipline, failure to follow teaching plans, and other errors or omissions.[47] Neglect of duty is different from insubordination, however, in that an insubordinate employee chooses to disregard directives in a defiant manner. Generally, dismissal for either neglect of duty or insubordination must be backed up by prior notice to the employee and clear existing policy or directive serving to prove that the employee should have known differently.[48] Finally, other dismissals for criminal acts such as felonies generally will be sustained due to the effect on performance and the standing of the offending employee in the community.[49] For example, many state statutes hold conviction of a felony to be proof of "unfitness to teach."[50]

Personnel problems are a serious matter, whether from the position that poor employee performance and relations is a drain on productivity or from the perspective that liability follows when districts and administrators fail to meet the requirements of law. Liability for unfair dismissal is extremely serious because improper dismissal damages the whole personnel structure and subjects the district to liability for back pay and damages if the complainant prevails. The modern environment places a heavy burden on boards and administrators to avoid unneeded confrontation—at the same time, however, personnel problems must be addressed promptly with aplomb while meeting all procedural[51] and substantive due process[52] requirements of law.

It must be remembered that determination of whether due process has been provided is a function of legal examination of the external mechanics of any proceedings and scrutiny of the total process for its objectivity. Administrators and boards must be careful to deal objectively with the factual situation and to follow the appropriate due process steps because the duty on the district is weighty, as the United States Supreme Court has ruled that employment rights in contractual disputes are "broad and majestic."[53]

Summary

The grave responsibility of the personnel function was well stated earlier when it was said that coordination, substantiation, consultation, evaluation, admonition, documentation, communication, charges, notification, hearing, appeal, and decision make up the elements of grievance and dismissal. Indeed, it could be argued that these elements fairly describe a major part of the entire personnel function, both from its negative tasks and from the course of carrying out the educational mission of schools. But to believe that these "administrative management" tasks are the sum of the personnel function is to inflict deep harm on the role of administration.

To these sometimes unpleasant tasks must be added a tremendous moral responsibility for addressing people in organizations, as well as a sense of respect for the enormous scope of the personnel function in schools. The course of this chapter has demonstrated that the personnel function is broad, complex, and people-oriented. The overriding task is to align organizational and individual needs as to best serve the school's youthful clientele. Under these conditions and purposes, the word "responsibility" should more appropriately be replaced with "opportunity." Indeed, the personnel function in education represents a great opportunity for improving the nation's schools—a task and burden that administrators should be glad to accept.

Endnotes

[1] This section is based in part on D.C. Thompson, R.C. Wood, and D. Honeyman, *Fiscal Leadership for Schools: Concepts and Practices*. New York: Longman (1994).

[2] 347 U.S. 483, 74 S.Ct. 686 (1954). A fuller discussion of the impact of this landmark U.S. Supreme Court case appears in Chapter 5.

[3] Chapter 6 discusses implications of court rulings on federal and state funding for schools.

[4] For further development of the economics of education, productivity issues, and descriptions of the scope of expenditures and educational services, *see* Chapter 1 in D.C. Thompson, R.C. Wood, and D. Honeyman, *Fiscal Leadership for Schools*.

[5] H. Hodgkinson, *All One System: Demographics of Education, Kindergarten Through Graduate School* (Washington, DC: Institute for Educational Leadership, 1985).

[6] N. Zill, *Developmental, Learning, and Emotional Problems: Health of Our Nation's Children* (Washington, DC: Child Trends, Inc., 1990).

[7] Drawn from National Center for Education Statistics, *Digest of Education Statistics 1991* (Washington, D.C: National Center for Education Statistics, 1991); National Center for Education Statistics, *The Condition of Education 1990* (Washington, D.C: National Center for Education Statistics, 1990); and National Education Association, *Rankings of the States* (Washington, D.C: National Education Association, 1991).

[8] National Commission on Excellence in Education, *A Nation at Risk: The Imperative for Educational Reform* (Washington, D.C: National Commission on Excellence in Education, 1983). See also the following reports: Commission on

Reorganization of Secondary Education, *Cardinal Principles of Secondary Education*, Washington, D.C., Bureau of Education (1918); Educational Policies Comm., *Education for All American Youth in American Democracy*, Washington, D.C., The Commission (1938); National Association of Secondary School Principals, *Planning for American Youth*, Washington, D.C., The Ass'n (1951); Commission on Life Adjustment Educ., *Vitalizing Secondary Education*, Washington, D.C., U.S. Office of Education Bulletin No. 3 (1951) Second Comm. on Life Adjustment Educ. and Youth, *A Look Ahead in Secondary Education*, Washington, D.C., U.S. Office of Educ. Bulletin No. 4 (1954); Comm. for the White House Conf. on Educ., *A Report to the President*, Washington, D.C., U.S. Government Printing Office (1956); *A Nation at Risk*, National Comm. on Excellence in Educ. (1983); "Report of the Task Force on Federal Elementary and Secondary Education Policy," Twentieth Century Fund (1983); "Academic Preparation for College: What Students Need to Know and Be Able To Do," The College Board (1983); "America's Competitive Challenge: The Need for a National Response," Business-Higher Education Forum (1983); "Action for Excellence: A Comprehensive Plan to Improve Our Nation's Schools," Task Force on Education for Economic Growth of the Education Commission of the States (1983); "Educating Americans for the 21st Century," National Science Board's Comm. on Precollege Education in Mathematics, Science, and Technology (1983); *High School: A Report on Secondary Education in America*, Carnegie Foundation for the Advancement of Teaching (1983); *A Place Called School: Prospects for the Future*, J. Goodlad (1984); *Horace's Compromise: The Dilemma of the American High School*, T. Sizer (1984); "High Schools and the Changing Workplace: The Employer's View," National Academy of Sciences (1984); "Beyond the Commission Reports: The Coming Crisis in Teaching," RAND Corporation (1984); "Changed Lives: The Effects of the Perry Preschool Program on Youths Through Age 19," High/Scope Educational Research Foundation (1984); "The Unfinished Agenda: The Role of Vocational Education in the High School," National Comm. on Secondary Vocational Education (1984); "Make Something Happen: Hispanics and Urban High School Reform," National Comm. on Secondary Schooling for Hispanics (1984); "Barriers to Excellence: Our Children at Risk," National Coalition of Advocates for Students (1985); "The Shopping Mall High School," A. Powell et al., (1985); "A Call for Change in Teacher Education," National Comm. for Excellence in Teacher Education (1985); "Becoming a Nation of Readers," National Academy of Educ. (1985); "The Governors' Report on U.S. Education: 1991," National Governors' Ass'n (1985); "Reconnecting Youth: The Next Stage of Reform," Business Advisory Comm. of Educ. Comm. of the States (1985);

"Investing in Our Children," Committee for Economic Development (1985); *The Last Little Citadel*, R. Hampel (1986); "Tomorrow's Teachers," Holmes Group (1986); "A Nation Prepared: Teachers for the 21st Century," Carnegie Task Force on Teaching as a Profession (1986); "First Lessons," U.S. Educ. Dept. (1986); "Time for Results," National Governors' Ass'n (1986); "School Boards: Strengthening Grass Roots Leadership," Inst. for Educational Leadership (1986); "Drop-outs in America: Enough is Known for Action," A. Hahn et al., Inst. for Educational Leadership (1987); "Public and Private High Schools: The Impact of Communities," J. Coleman and T. Hoffer (1987); "Bringing Down the Barriers," National Governors' Ass'n (1987); "Children in Need," Committee for Economic Development (1987); "James Madison High School: A Curriculum for American Students," W. Bennett (1987); "New Voices: Immigrant Students in U.S. Public Schools," National Coalition of Advocates for Students (1988).

[9] *Nation at Risk,* 5.

[10] F.N. Kerlinger, *Foundations of Behavioral Research,* 2d. New York: Holt, Rinehart & Winston (1986), 9.

[11] F. Taylor. *Principles of Scientific Management.* New York: Harper (1947).

[12] M.P. Follett, *Dynamic Administration: The Collected Papers of Mary Parker Follett,* ed. H. Metcalf and L. Urwick. New York: Harper (1940).

[13] F.J. Roethlisberger and W.J. Dickinson, *Management and the Worker.* Cambridge: Harvard Univ. Press (1939), 28.

[14] C.I. Barnard, *Functions of the Executive.* Cambridge: Harvard Univ. Press (1938).

[15] H. Simon, *Administrative Behavior.* New York: MacMillan (1947).

[16] Simon, *Administrative Behavior.* 148.

[17] J.W. Getzels and E.G. Guba, "Social Behavior and the Administrative Process," *The School Review,* 5 (1957), 429.

[18] J. Pfeffer, *Organizations and Organization Theory.* Boston: Pitman (1982).

[19] Discussion of Sergiovanni's argument is adapted from J. Thomas, *The Principalship: A Reflective Practice Perspective.* 3d. Boston: Allyn and Bacon (1995) 41-43.

[20] Sergiovanni, *The Principalship,* 42.

[21] Discussion of Pfeffer's typology of organizational theories is adapted from J. Pfeffer, *Organizations and Organization Theory.* Boston: Pitman (1982) 8-12.

[22] Pfeffer, *Organizations and Organization Theory,* 8.

[23] W. Hoy and C. Miskel, *Educational Administration: Theory, Research, Practice.* 4d. New York: McGraw-Hill (1991) 139.

[24] W.B. Castetter, *The Personnel Function in Educational Administration.* 5th ed. New York: Macmillan (1992) p3.

[25] Castetter, *The Personnel Function in Educational Administration.*, 39.

[26] Castetter, *The Personnel Function in Educational Administration.* 40.

[27] Although evaluation and development are critically important aspects of personnel administration, they are not as central to school business management issues and are usually assigned to the personnel office. In the interest of space, these issues are not discussed in the chapter.

[28] Material in quotations is taken from Castetter, *The Personnel Function in Educational Administration.*, 200.

[29] *The Personnel Function in Educational Administration.*, 360.

[30] This discussion is based in part on Chapter 8 in D.C. Thompson, R.C. Wood, and D. Honeyman, *Fiscal Leadership for Schools*, 425-430.

[31] *Graham v. Special School Dist.,* 462 N.W.2d 78 (Minn. Ct. App. 1990).

[32] *Clark v. Mann,* 562 F.2d 1104 (8th Cir. 1977).

[33] *Harris v. Florence City Board of Educ.,* 568 So. 2d 827 (Ala. Civ. App. 1990).

[34] Castetter, *The Personnel Function in Educational Administration,* 349.

[35] *School Dist. v. Gautier,* 13 Okla 194, 73 P. 954 (Okla. 1903).

[36] *Black's Law Dictionary*, 5d (St. Paul, MN: West, 1985), 688.

[37] D.C. Thompson, R.C. Wood, and D. Honeyman, *Fiscal Leadership for Schools,* 427.

[38] *Brown v. Wood County Board of Educ.*, 400 S.E.2d 213 (W.Va. 1990); and *Selby v. North Calloway Board of Educ.,* 777 S.W.2d 275 (Mo. Ct. App. 1989). The reader should be aware that certain offenses may be deemed irremediable so as not to require a board of education to engage in remediation.

[39] *Black's Law Dict.,* 885.

[40] *Horosko v. Mt Pleasant Twp. School Dist.,* 335 Pa. 369, 6 A.2d 866 (1939), *cert. denied,* 308 U.S. 553 (1939).

[41] *See generally, Acanfora v. Board of Educ. of Montgomery County,* 491 F.2d 498 (4th Cir. 1974); *New Mexico State Board of Educ. v. Stoudt,* 571 P.2d 1186 (N.M. 1977); *Avery v. Homewood City Board of Educ.,* 674 F.2d 337 Dist. Ct. App. (5th Cir. 1982), *cert. denied,* 461 U.S. 943 (1983); *Andrews v. Drew Municipal Separate School Dist.,* 507 F.2d 611 (5th Cir. 1975), *cert. granted,* 423 U.S. 820 (1975), *cert. dismissed,* 425 U.S. 559 (1976).

[42] This was done primarily as a response to *General Electric Co. v. Gilbert,* 429 US 125, 97 S. Ct 401 (1976).

[43] 42 U.S.C. §2000e(k) states, "Because of or on the basis of pregnancy, childbirth, or related medical conditions; and women affected by pregnancy, childbirth, or related medical conditions shall be treated the same for all employment-related purposes, including the receipt of benefits under fringe benefit programs, as other persons not so affected but similar in their ability or inability to work . . .."

[44] *School Dist. Pinal County v. Sup. Court,* 102 Ariz. 478, 433 P.2d 28 (1967). *See also, Alderstein v. Board of Educ of New York City,* 485 N.Y.S.2d 1 (N.Y. 1984).

[45] D.C. Thompson, R.C. Wood, and D. Honeyman, *Fiscal Leadership for Schools,* 428.

[46] *Cowdery v. Philadelphia Board of Educ.,* 531 A.2d 1186 (Pa. Commw. 1987): *Mechley v. Kanawha County Board of Educ.,* 383 S.E.2d 839 (W.Va. 1989); *Johnson v. School Board of Dade City,* 578 So. 2d 387 (Fla Dist. Ct. App. 1991).

[47] *See, e.g., Harvey v. Jefferson County School Dist No. R-1,* 710 P.2d 1103 (Colo. 1985).

[48] *Cowdery,* 531 A.2d 1186; *Board of Educ. City of Gilmer v. Chaddock*, 398 S.E.2d 120 (W.Va. 1990).

[49] For the definition of a felony, see *Black's Law Dict.*, 555.

[50] *See, e.g., Bertrand v. New Mexico State Board of Educ.,* 544 P.2d 1176 (N.M. 1975); *Skripchuk v. Austin,* 379 A.2d 1142 (Del. Super. Ct. 1977).

[51] Procedural due process ensures that parties whose rights are to be affected are entitled to be heard, and in order that they may enjoy that right, they must be notified

[52] Substantive due process is broadly defined as a constitutional guarantee that no person shall be arbitrarily deprived of his life, liberty or property and shall be protected from arbitrary and unreasonable action.

[53] *Board of Regents v Roth,* 408 U.S. 564 (1972).

Chapter 9
Collective Negotiations

Introduction

It is critical to note at the outset of this chapter that negotiations in the public schools of the nation are a function of controlling state statutes. Negotiations have been brought about by force of law and social change, as employees have joined together to speak with a collective voice for collective and individual benefit. Although in many states only teacher groups are permitted to negotiate, other states have statutes under which other support staffs also negotiate. In practical terms, however, teacher negotiations have a powerful impact on administrative and classified salary structures as it proves difficult and politically risky to deviate too substantially from raises received by one group since other groups watch the negotiations process hawkishly.

The financial aspect of collective bargaining can be intensive and complex. Because so much is at stake, both financially and in terms of harmony, it is important that policy and practice should seek to develop significant trust and respect. Many districts hold informal discussions on a frequent basis during the year regarding mutual concerns, including compensation practices. These discussions should always be with the appropriate association leaders so as not to commit a potential unfair labor practice. It should be a major goal throughout each school year and even in multiple-year contracts to establish clear communications with all bargaining groups in order to facilitate the compensation/negotiations process.

The negotiations process differs in each state because of different laws. Nonetheless, the steps are similar as well as the general and overall arbitration process that may be present within a given state. At whatever time of year discussions begin, the fiscal office is usually called upon to generate financial data, to make fiscal projections concerning forthcoming revenue, and to cost out salary proposals, before and during the negotiations, from both sides of the table. Short-and long-term costs must be estimated for all proposals before a board of education can agree to any plan.

The fiscal office must be able to provide cost analyses for each proposal and must include direct costs of salary and fringe benefits, and estimate cash flow, costs of employee time, changes in personnel placement on salary schedules, and other items related to negotiations. Each issue must be analyzed for both present and future costs, and an assessment of the long-range impact on district financial health must be made. The process calls for data to be collected and verified in order to reach a "satisficing" solution. No solution can be reached until both sides have sound data on cash flow projections, historical financial data, and data showing how the district compares to other districts of similar financial status in the state and region.[1]

Careful preparation and presentation always enhance the potential for a win-win situation at the bargaining table, as well as favorable results in fact finding and arbitration proceedings. The goal of both sides is to discuss and agree on the facts needed to determine a fair and just salary package. When agreement is reached, the various applicable salary schedules must be adjusted. For example, for certified teaching staff the most direct impact is adjustment to the single salary schedule, both in dollars and in any changes to the structure of the salary schedule itself. For noncertified staff, the impact occurs in similar application to a salary schedule or in adjusting each individual's salary. Fairness must, therefore, also extend to fringe benefits. Obviously, the personnel office must be knowledgeable about fiscal affairs and must be prepared to work closely with the division of financial services to coordinate any commitments and agreements that are obtained.

During negotiations, both sides come prepared with items they seek to negotiate. Preparation on the part of the school district usually centers on several types of data that anticipate employee demands and that also reflect changes desired by the school district. Employee organizations also have high interest in the same items, although their interest generally seeks to advance or defend the status quo. Generally, the items both sides bring to the table include:

- *Strengths and weaknesses* of the current schedule and the entire contract.

- *The current salary schedule*, including the number and actual costs of each cell of the salary matrix over the life of the contract and a projection into the near future.

- *Basic data* regarding minimum, maximum, and actual average cost per employee over the life of the contract.

- *Comparative salaries* in competing districts and industries.

- *Living standards* of the local community.

- *Personnel turnover*, including recaptured funds from pending retirements.

- *Movement on the salary schedule* due to advanced training, as well as vertical scale movements.

- *New programmatic needs* and curtailments.

- *Future revenues* and expenditures, including tax levy histories and state aid projections.

- *Opinions* on which portions of the contract have worked well or have not worked well in the past.
- *Opinions* on which contract sections have created difficult situations for management or employees.
- *Proposals* on which contract sections should be modified or dropped.
- *Comments or complaints* that have been raised by employees during the life of the contract.
- *Statements* about whether there has been a pattern of grievances over parts of the contract.
- *Indications* of new issues that have been raised by employees that may prove to be problematic.[2]

The chief fiscal officer is a vital and necessary link in the process of collective bargaining in public education. Collective bargaining exists for professional and/or noncertified employees in some form in the vast majority of states. Again, it is critical to note that the collective bargaining process is highly state-specific. There is no constitutional right of employees to engage in collective bargaining actions. Specifically, absent of state statutory directive, school districts are not under a legal or constitutional mandate to acknowledge, accept, or bargain with employee organizations.[3] Additionally, there is no federal right to bargain by public education employees. The National Labor Relations Act (NLRA) specifically exempts public elementary and secondary employees from federal bargaining rights. The enforcement agency of the NLRA is the National Labor Relations Board (NLRB). The NLRB rulings are not applicable to public elementary and secondary education. However, notwithstanding this nonapplicability, in the absence of state-specific rules and/or guidelines, or statutes, most state courts tend to find the logic of the NLRB rulings persuasive.

Thus, this chapter offers an introductory overview of collective bargaining for elementary and secondary education. Such an overview is limited by the nature of the different views, philosophical positions, and the politics of each state. Keep in mind, therefore, that this overview must be tempered with the highly specific state statutory rules and regulations.

Collective Bargaining in Public Education

Collective bargaining in public education did not reach any significant level until the mid 1960s and 1970s. Only within the last twenty-five years has such a policy been found in most states. A few states statutorily prohibit collective bargaining with public education employees. Thus, again it must be fully understood that collective bargaining presents a wide array of procedures, aspects, and outcomes within this overall process.

The collective bargaining model was largely adopted from the private sector and has traditionally, and currently, tended to deal with the issues of wages, salaries, and working conditions. Within the collective process, it should be noted that in recent years, the collective bargaining process has also taken in many of the discussions regarding various educational reforms. Site-based management, as well as teacher empowerment, have emerged in many school districts as one of the most bargained issues within the overall contract.

During the 1990s, much needed educational reform has occurred in the nation. With rare exception, school districts may not bargain away legal responsibilities as assigned by state legislatures. Of course, most states do allow some procedure by which a school board may request an exemption from a given set of rules or regulations. Nonetheless, school boards, while entering into collaborative projects, may not generally give up sole managerial prerogatives. States that allow bargaining on educational policies still retain the sole authority of local school boards in what is generally thought to be the core area of educational policies. On the other hand, the broad area of curriculum could be argued to be a condition of employment in a given jurisdiction.[4]

When curriculum is bargained a series of public policy arguments may be forthcoming. Such interpretations by applicable court ruling as to whether such issues are consistent with public policies are largely unsettled and not directive for generalization. On the other hand, most school districts presently set curricular practice with a broad array of teacher and administrator committees, which generally allows constructive development of such issues away from the bargaining table. In the vast majority of communities in America, it would not be politically wise or feasible to attempt to diminish the role of parents and the public from serious curricular issues. Such an approach would be unproductive and counter to the role of the public in such decisions for the support of the public schools and the education of students in a given community.

In recent years, many school districts have emerged from a traditional and adversarial style of collective bargaining to a more collegial style of mutual goals and responsibilities. The traditional model that exists in many school districts is primarily concerned with the authority of public school administrators and balancing the overall school district budget. These items will always remain of paramount importance to those charged with the legal responsibility of conducting public schools and to those who must answer to the taxpayers within a given school district.

Collective Bargaining Issues

The issues that may be placed before collective bargaining teams are varied and widespread even within the same states. Generally, state statutes allow for bargaining on wages, working conditions, and related issues. The crucial issues revolve around the state judicial interpretations and the operationalizing of such terms.

Importantly, issues at the bargaining table are traditionally defined as to what rights are management rights and which ones do not qualify as such. In the vast majority of states, management rights fall outside of wages, benefits, and working conditions. The role of the collective bargaining unit is primarily to provide better wages and benefits, better working conditions, and improved job security. Management rights are traditionally concerned with such issues as: determine the hiring and firing of employees, assignment of the staff, and those issues that tend to maintain an efficient and orderly work place that they believe enhances efficiencies for the taxpayer as well as fostering student learning. Generally, the teachers' organizations are concerned with higher salaries and fringe benefits, smaller classes, and job security. Of recent origin is the push for teachers' organizations to have an active role in school district policies.

Critical to this discussion is whether state courts have largely adopted a broad or narrow scope of the bargaining concept. That is to say that those states that have adopted a broad scope of bargaining have embraced lenient interpretations of the terms wages, hours, and conditions of employment.[5] For example, Black has described Michigan as being an example of a state with a broad interpretation of bargainable rights, i.e., teachers' unions may bargain such issues as curriculum and class schedules, size of classes, selection of textbooks, materials and supplies are functions of a condition of employment.[6] On the other hand, as Black has explained, a narrow interpretation in Montana as being, "The negotiations and bargaining for agreement shall not include matters of curriculum, policy of operation, selection of teachers and other personnel, or physical plant of schools or other school facilities."[7]

No Strike Clause

Generally, the legal ability of public school teachers to strike is severely curtailed and only exists in a very narrow vein in a handful of states. The vast majority of states have clear statutory guidance that does not permit a teachers' union to engage in a strike or a work stoppage. Notwithstanding these statutory guidelines, most local contracts have similar statements prohibiting such work stoppages. Such local language clearly assists the local school board in sanctioning those employees who are in violation of these statutory guidelines.

Recent Innovations

With the advent of educational accountability, educational improvement, site-based management, teacher empowerment, and school improvement there have grown additional public expectations for increased cooperation between management and labor. Collaborative efforts have focused on such issues as class size and the nature of classroom assignments. Many school districts have essentially defused many issues that now are decided by committees of teachers and administrators who design curriculum and who make recommendations to either the building principal or parent-teacher site advisory councils in individual schools. Thus, these issues are removed from the collective bargaining process as well as varying from school to school throughout the school district.

Win-Win and Other Strategies

In recent years several strategies have emerged in the public sector bargaining arena. Generally referred to as "win-win" these strategies are often called "collaborative bargaining," or "principled negotiations." Generally, they involve teacher involvement, a sense of mutual trust and cooperation, and a wide range of acceptable bargainable topics. Collaborative bargaining is said to help union leaders and management to isolate divisive issues, work toward avoiding future misunderstandings, and to develop an appreciation for the other party's interest.[8] In recent times, even the teachers' union has advocated that the bargaining process observe the following steps:

- Separate the people form the problem.
- Focus on interests, not positions.
- Invent options for mutual gains.
- Insist on using objective criteria.[9]

Regardless of the recent changes in collective negotiations, it is a means by which an employer interacts with a large group of employees.[10] Another way is for the employer to contract individually with each employee. Negotiation is a technique applicable to all phases of an employment relationship. The legal aspects of collective negotiations in public education are the driving forces concerning many items that present themselves at the bargaining table and during the administration of the contract.

When true collective negotiations occur, law has developed regarding the nature of an appropriate bargaining unit. There are restrictions on the topics that are properly subject to collective negotiation when employment contracts are being designed. Since contract negotiation is rarely trouble-free, when breakdowns in the negotiating process occur, the law has evolved several processes for resolving conflicts.

The legal impact of collective negotiations on the traditional "management" rights of school boards is significant. In this second edition of this text, it must be stated that the issues remain largely the same with the addition of greater site-based issues as well as solutions as discussed. The realm of public education remains faced with changing enrollment patterns, severe budget cutbacks, and calls for reform of the educational process.

Constitutional Considerations

When a governing body is the employer, constitutional issues automatically arise. The federal and state constitutions define the rights of an individual interacting with government. These rights continue to exist even when the individual is a teacher (or other school employee), the interaction is through an employment relationship, and the government involved is simply a local school board. All laws or regulations involving public employees must harmonize with constitutional protections.

When school employees protest a local ordinance or regulation of a board of education and the conflict is litigated, the offending requirement is evaluated against the appropriate constitution. Most laws and regulations are upheld because they will meet the minimum standard of being rationally related to promoting a legitimate governmental interest. However, when a public employer tries to interfere with a fundamental right of its employees or draws distinctions among them using illegal discriminatory grounds, the courts are quick to uphold the individual employee against the employer. Such laws or regulations are upheld only if they reasonably promote overriding governmental interests. Restrictions concerning the joining of labor unions are generally unconstitutional, violating the employee's freedom of association. Statutory restrictions on public employee union membership have been upheld only when they are limited to managerial or supervisory personnel. However, the courts have not been eager to find that the right to join labor organizations imposes a constitutional duty on the government to bargain collectively. The duty to bargain collectively is imposed by statute, rather than by any constitution. In most states where the legislature has imposed this duty, the laws have been found constitutional.

The Duty to Bargain

When state law requires collective bargaining, the employer has the duty to deal with the public employee organization. Both sides must bargain, but neither must agree. All that the law imposes is a duty to bargain "in good faith." Dependent upon state statutory requirements, this duty is ongoing and has different meanings at various stages of the negotiating process.[11] Generally, at the very beginning of contract discussions, it means that neither side can insist that a particular topic must be discussed first. Good faith bargaining requires both the employer and the employee organization to present demonstrably reasonable proposals. When requested, the facts and projections used to develop a proposal must be made known so that informed negotiation can occur. Finally, the requirement of good faith bargaining continues after an agreement is signed; unilateral changes in topics subject to collective negotiation will be struck down by the courts.

Public Labor Relations Boards

With the advent of public employee's labor organizations has come the formation of public labor relations boards by the state legislatures. These administrative bodies serve by resolving contract disputes and breakdowns in negotiation, evaluating the terms of dispute settlement, and occasionally imposing penalties on either employers or employee organizations found in violation of fair labor practices. Board authority varies from state to state, being dependent on the specific statutory grants of power. Public employee relations boards, given a broad grant of jurisdiction, hold hearings, make findings, and issue orders as charged by state statute. Generally, this behavior is not an unconstitutional usurpation of the power of the courts. However, in states where statutes covering broad board authority are narrow and specific, judges will often restrict boards' actions to the letter of the law. The law recognizes that an administrative agency can exercise only those powers explicitly granted to it by the legislature.[12]

The Legislature Veto Problem

In the private sector, contracts developed through collective bargaining go into effect upon ratification by union members. The employer who has consented to such a contract is bound by its terms. When the employer is a local school board, however, another layer of approval is required before an agreement can be finalized. Any terms of the agreement that require expenditures of government funds may be in jeopardy even when the parties to the contract, the employer and the employee organization, have agreed. If a statute so provides, the contract is still subject to approval by an appropriate fiscal body before it becomes binding. While the procedures vary, this is specifically applicable in many states that have fiscally dependent school districts. Many states specifically require that the financial terms are subject to legislative approval.

Typically, a legislative veto of financial terms of a negotiated agreement occurs in one of two ways. Either the contract is negotiated and takes effect after the legislature has passed the current year's appropriations, or the legislature fails to appropriate sufficient funds to cover increases in a multi-year contract. Generally, whenever the financial provisions of a negotiated agreement are vetoed, only the remaining noneconomic terms may be enforced.

Collective Bargaining vs. "Meet and Confer"

Collective negotiations impose a duty to make proposals and counterproposals in good faith. There is an unstated expectation that all items agreed on will be written down, signed by both parties, and used to regulate the employment relationship. Statutes requiring collective negotiations are usually modeled from the National Labor Relations Act (NLRA), which governs employment relations in the private sector.

In many states it is much different where the only duty is for the employer to meet and confer with the employees. In these states, the laws applicable to labor relations of public employees differ significantly from those affecting the private sector. Most noticeable is the retention of a higher degree of managerial discretion in the public employer. Conversely, public employee organizations have far fewer rights in states mandating "meet and confer." The employer's only duties are to meet at reasonable times and to discuss proposals submitted by the employee's representative. The employer need not act on any of these proposals. Generally, in such instances, there is no written statutory expectation that an agreement will be negotiated and nothing need be reduced to writing.

In a few states the local school board must join in collective negotiations on certain issues. These are typically related to wages, hours, and other conditions of employment, the so-called mandatory topics of bargaining that will be discussed in greater detail later in this chapter. The legislatures in these states have imposed an additional duty to meet and confer on other topics. This second group of topics generally involves managerial choices and policy decisions that less directly affect teachers and other school district employees.

Appropriate Bargaining Units

When a school district has only a few employees, there is no inherent difficulty in bargaining collectively at one time with all district employees who have similar interests. In fact, the general pattern across the country is that bargaining units are district-wide in scope. However, when a school district has a larger number of employees, the dynamics of the workplace tend to more closely resemble that of privately employed organized labor. Not surprisingly, a number of states have statutes specifying the acceptable composition of bargaining units of public employees. In addition, when two rival teachers' associations present differing philosophies on the issue, the dominance of one over the other in a school district can influence the composition of the bargaining unit.[13]

While particular state laws differ, typically units for certified teachers exclude custodial and clerical workers. The basis for this is found in private employment labor law where bargaining units are structured to contain only those employees sharing a "community of interest" in similar job requirements, hours, and work environments. The issues become complex when it is necessary to decide what other employees may be included in a unit with classroom teachers. In most jurisdictions, specialty teachers and permanent substitutes are included in the unit. The determination is based on the amount of time spent in the classroom. Using the same standard, per-diem substitutes generally are excluded from the teachers' bargaining unit. Problematic areas always tend to center around the inclusion of coaches, teachers' aides, school district coordinators, department heads, laboratory personnel, and so on.

The most frequent factor for excluding certain personnel is administrative or managerial function. In some states, building principals have bargaining rights. Generally, when building principals have such rights they must form separate units. It is on the issue of including supervisory personnel that the two major teachers' associations have historically differed.[14] The National Education Association (NEA) is profession-oriented and tends to view all professional employees as equal parts of the education team. The NEA generally disfavors separating classroom teachers from supervisors when bargaining collectively with local school boards. The American Federation of Teachers (AFT) has historically taken the more traditional labor view found in private employment. Focusing on the teacher as worker, the AFT generally regards supervisors as part of the management that teachers bargain against.

Appropriate Subjects for Bargaining

There are three categories of topics when bargaining in the public sector: mandatory subjects, permissive subjects, and unlawful subjects.

Mandatory Subjects

As the name indicates, in jurisdictions that require collective bargaining, mandatory subjects are those on which the employer is required to bargain in good faith until an impasse is reached. It has been said that the public sector places a great reliance on statutes to determine the exact scope of bargaining.[15] Generally, statutes mandating subjects to be collectively negotiated are modeled on the Taft-Hartley Act's triad of "wages, hours, and other terms and conditions of employment." In many states the legislature has closely followed this language.

Not surprisingly, much of the conflict between school boards and employees revolves around interpretations of what constitutes "other conditions of employment." In many states, the law provides a laundry list of mandatory subjects. Employers may not be required to bargain at all on unlisted topics. While the law in each state differs, in at least some jurisdictions mandatory topics in elementary and secondary education include teachers' work loads, daily schedules, assignment of teachers to supervise extracurricular activities, grading responsibility, the inclusion of a "just cause" provision for teacher reprimands, contractual benefits other than salary, grievance procedures related to contractual benefits other than salary, grievance procedures to be followed in contract disputes, class size, prep periods, school calendars, and layoff decisions affecting staff or eliminating departments on remaining teachers' terms and conditions of employment. At the same time, other jurisdictions have decided over the years that boards of education do not have to bargain on class size, budget decisions, school calendars, daily schedules, the reorganization of departments, selection of extracurricular activities, number of steps in the salary schedule, textbooks, the use of paraprofessionals or general education policy.

Permissive Subjects

Permissive subjects form the second group of topics that are bargained collectively. Discussion of any permissive topic lies within the discretion of the employer. Permissive topics resemble the interactions between employer and employee in meet and confer states. However, a stable labor relations policy encourages the public employer to discuss a broad range of permissive subjects. Some states have statutory or case law articulating permissive subjects. In general, the lines between mandatory and permissive subjects are not clearly drawn.[16]

Unlawful Subjects

In some states certain topics have been declared to lie beyond the scope of collective bargaining. When a subject is covered by law or constitution, it cannot be negotiated. Thus, in some states, agriculture[17] and certain moral virtues[18] must be taught; in other instances, textbook selection and curriculum must lie with the board of education; and in still others the employees' organization can have no say regarding retirement benefits, termination of probationary teachers, or any issue of managerial or educational policy.

Breakdowns in the Negotiating Process

Impasse

When a board of education and the employee's organization becomes deadlocked following good-faith collective negotiations of mandatory subjects, they are said to have reached an impasse.[19] Generally, the mere refusal of one or both sides to meet with the other or discuss any mandatory topic, is technically not an impasse. Rather, the refusal constitutes an unfair labor practice that can have dire consequences for the recalcitrant party. Utilizing the federal law as illustration, "'impasse' means that point in the negotiation of a labor agreement at which the parties are unable to reach full agreement, notwithstanding their having made earnest efforts to reach agreement by direct negotiation . . . ."[20]

In the vast majority of jurisdictions, the two sides must continue jointly to seek a solution. Once an impasse has been reached, the negotiators have available a variety of techniques to resolve the problem. Frequently a specific series of procedures already have been detailed in a contract between the employer and the employees. When these procedures fail, or when no prior arrangements have been made, statutory control dictates the procedures in most states as how to resolve the dilemma.

Mediation

The most common first step in impasse resolution is the use of mediation. Mediation is distinguished by its informality. An individual or panel having the confidence of both parties will seek to reduce differences, and perhaps propose compromise solutions. When the parties are unable to agree on a mediator, the public employee relations board of many states can provide one. Alternatively, the parties can agree to use the offices of any of several national organizations. These include the U.S. Federal Mediation and Conciliation Service and the American Arbitration Association. Occasionally state law will specify the use of such a group when the parties are unable to agree on a mediator. However, pursuant to state statutory control and labor board directives, mediation is an informal process and any person respected by both sides can be selected.

Mediation is generally a voluntary process. It requires two sides that want to reach a settlement but find themselves unable to do so without the help of a third party. The process works best if the mediator is called in before the positions of the parties become rigid.[21] However, because mediation is informal and there are no sanctions available against a noncooperative party, it cannot resolve all impasse situations.

Fact-Finding

Fact-finding is specified as the second step for impasse resolution in the laws of many states. In a few states, no mediation occurs prior to fact-finding. Fact-finding is significantly different from mediation. A formal procedure, it involves the designation of an individual or a panel empowered to conduct investigations. Fact-finders conduct public hearings to which the requirements of due process attach: adequate notice to all interested parties and the opportunity to be heard.

After deliberation and consideration of all evidence collected, the fact-finder(s) issues a written report. This summarizes the evidence and contains recommendations for settling the impasse. Typically the report is first issued privately to the parties. In the law of several states it is made public unless the parties resolve their differences within a statutorily mandated period. While the recommended solutions may not be binding in some states, the threat of publication of an unfavorable report is sometimes enough to prompt stubborn school boards and employees' organizations to change their bargaining positions.

Fact-finding differs in the public and private sectors. In private industry, a fact-finding report is usually sought only after a work stoppage has been called. But when the employer is the government and employees are prohibited by law from striking, a fact-finder's report acts as a strike substitute.[22] It is not a procedure that always resolves the impasse. The success of fact-finding seems to depend on several factors: (1) whether the procedure is truly voluntary, or is mandated by statute; (2) the experience and negotiating abilities of the fact-finder and the representatives of the parties; and (3) the source of funds that cover the costly procedures. When the parties bear the cost, the results are better than when the state government pays.[23]

Arbitration

Mediation and fact-finding both depend on the potential of a neutral third party to restore a rational tone to negotiations. Generally, no sanctions compel the parties to cooperate except, of course, the pressure of public opinion. This is not always enough. In some instances, it is necessary to resort to a procedure that forces an end to the impasse. Most often state legislatures that specify a step beyond fact-finding in the controlling statutes choose arbitration. Arbitrators unilaterally determine compromises to resolve an impasse. Arbitration can be used either immediately after mediation fails or as the sole procedure mandated by the legislature.

In the private sector, arbitration is ordinarily distinguishable from mediation and fact-finding by its binding nature. Traditionally, the decision of the arbitrator is enforceable by the courts. However, the reality of the public employment sphere has altered this. In certain states arbitration is statutorily identified as nonbinding. In nearly all states, the decision of the arbitrator is binding only to the extent that the governing body appropriates funds to implement it. Since the threat of legislative veto injects an element of doubt into the finality of public employment contracts, some states have found it advisable to indicate that the board of education may agree to be bound by an arbitrator's decision.

Other Solutions

Not all jurisdictions have opted to permit voluntary or statutorily mandated binding arbitration to end impasses in collective bargaining. Certain states have chosen more creative solutions. In a few states the final decision is left with the "appropriate legislative body," generally the one with financial control of the contract. Decisions are to be made with the public's best interest in mind. Thus, in these instances, the local board of education may not agree to a new contract that it does not approve. Thus, it is possible that the terms of the previous contract continue in force until the impasse is broken.

A few states use a process sometimes called "finality of decisions." Final resolution of disputes left after all other steps are ended is given to the school board, one of the recalcitrant parties. At first glance, this seems remarkably unfair to the employees. Effectively, the school board could refuse to resolve any disputed contract terms, knowing that ultimately it will dictate them. There are two protections against this occurring. First, such an attitude violates the duty to bargain in good faith and is an unfair labor practice. Second, the school board is an elected body and must remain aware of public sentiments. The declared advantages of finality of decisions relate to the fact that school boards are elected "governments."[24] Binding arbitration removes decision making from the people elected to determine educational policy. Finality permits these officials to control all issues that come up while running the schools.

Grievance Procedures

Generally, all states or school districts have some form of a grievance procedure. In many states, the grievance procedure exists in individual contracts as determined by the local contract. In some states the grievance procedure is set by state statute and is thus outside of local agreements. Grievance procedures vary widely as to the issues that may be grieved, the manner of the grievance, and who may actually grieve what issues. For example, some contracts only allow individuals who are directly harmed to file a grievance. Other contracts allow the union to file a grievance on behalf of employees who may be affected by some school board action or policy. Generally, courts are sensitive to the issue of whether an individual or union has followed the grievance procedure before filing a suit against the board. In other words, the grievance procedure offers an administrative remedy that must generally be fully resolved before adjudication may be successful.

The modern context of the world suggests an increasing propensity toward grievance proceedings that may eventually involve litigation. As seen earlier in chapter 7, legal liability now represents a significant area of concern for school administrators because the level of scrutiny to which employees and others hold school organizations resembles at least a hostile environment, if not one of near frenzy. As a result, grievance administration from the personnel perspective is an important issue fundamentally relating back to theories of organizational functioning. As with all other personnel issues, grievance administration begins with establishment of proper policies covering the areas of concern. It is no surprise that efficient policy speeds settlement of grievances, and it is further apparent that the absence of policy represents a problematic issue.

It is intuitive to suggest that a good grievance policy is designed to serve several purposes relating to good employee relations. First, a grievance policy is intended to provide a formal channel through which employees may air disagreements about terms and conditions of work. A second corollary of providing greater employee security through a grievance is intended to assure employees that there is enough justice in the system to prevent punishment for legitimate disagreement if the finding goes contrary to the employee's position. A third purpose is to reduce the number of grievances that might otherwise arise if organizational superiors were not cognizant of the watchdog effect of a readily available grievance procedure. A fourth purpose is exploratory and preventive, in that a good grievance policy should be attuned to anticipating areas of possible grievance and simultaneously remembering and correcting conditions leading to grievances that have actually arisen. As such, a grievance policy actually serves both management and labor by a system of checks and balances that keeps operations more honest and efficient.

Grievance policies should contain several essential elements that are either required by state statute or by sound management practices. The first element involves a statement of purposes, laying out the system's desire to provide communication channels for airing of grievances. Purposes should include assurances that the goal of the grievance procedure is to provide fairness and equity and that all parties should do their best to resolve the matter at the lowest possible administrative level. Purposes should also include statements about the ability of the grievance process to improve the school system's environment and that the school board intends to comply with all requirements of law and the spirit of grievance. Likewise, the policy may make statements about the school board's desire to listen to employee comments concerning improving individual and system effectiveness and its desire to work for mutually acceptable solutions. A variety of language and promises can be used to meet these goals, such as a board of education promise to not retaliate for filing legitimate grievances by guaranteeing that records of grievance matters and individual personnel files will be kept separately in order to assure open communication channels.

The second element of a grievance policy makes up most of its actual text by setting out the procedures by which grievances may be made and how complaints will be handled. As a general rule, many items fall within this arena. The policy usually sets out time limits for filing grievances so that employees do not choose to wait inordinate amounts of time before filing. The purpose is to address problems judiciously while the events can still be reconstructed and while the players are still in the system.

The policy also usually prescribes forms to be used in making complaints, generally requiring considerable detail in order to gather the greatest amount of information and facts to be studied. For example, the policy may require all grievances to be filed in writing within thirty days of the alleged violation, that all parties to the grievance must be fully named, and that full graphic descriptions of events and witnesses must be provided. Similarly, the policy should require timely responses on behalf of the system and the accused person, such as the requirement that any grievance presented to the administration must be answered in writing within five working days, whereupon the balance of the proceedings must be explained along with timelines and responsible parties for each subsequent step.

Although states prescribe grievance procedures in various ways, grievance procedures generally fall into four identifiable steps. The first step involves initiation of a grievance proceeding. At this level, the policy sets forth how to file a grievance, who should be notified, and the responses required at that level. The earlier example of requiring all grievances to be filed in writing within thirty days of offense fits a step one description. Similarly, the grievant is usually expected to file the complaint in writing, specifying the exact offense, citing names and witnesses and—if contractual—the guarantees or clauses that were violated.

The first step also requires response under the principle of settlement at the lowest administrative level, such as requiring a face-to-face meeting of the administrator and grievant within five days of receipt of the notice of complaint. Generally the grievant is entitled to legal or other representation such as a union official at all stages of the complaint, including the initial meeting. The first step also usually requires the administrator to reach a decision within a specified time period and to provide notification of a decision to the grievant and the administrator's superiors. If the issue is resolved at this level, no further proceedings are required.

A second step is taken if both parties are not satisfied by step one. The grievance policy should set out appeal procedures wherein the complaint moves up the organizational chart to the next appropriate level. Generally the next level is identified as the immediate supervisor of the administrator rendering the initial decision, such as the superintendent of schools when the grievant has charged the principal with a contract violation. The process at each level is similar, requiring the superintendent, in this example, to reach a written decision within ten days of a scheduled hearing that provides for face-to-face discussion and representation.

Still a third step involves a board hearing of the complaint occurs if there is not a satisfactory resolution following step two. Effort to reach resolution at each step is the goal—however, an alternative to affirmative resolution via "no contest" may also apply by failure on the part of the employee to file timely at each next step.

It is difficult to generalize about grievance policies beyond this point because state statutes differ significantly when the grievance cannot be settled in-house. In some states, the board's decision is final unless constitutional allegations are involved. In such cases, and in other instances where tort violations are alleged, a fourth step of litigation may follow, which takes resolution out of the school district's authority and places it in a court of proper jurisdiction. In other states with arbitration laws, additional steps may occur that expand the grievance to include outside negotiators who pursue a path of mediation, fact-finding, and binding arbitration using Federal Mediation and Conciliation Service guidelines. In still other states, milder versions of arbitration apply that attempt to reach settlement. In most cases, considerable cost is involved for representation, mediation, court fees, transcripts of proceedings, and so forth—costs generally borne by each side of the dispute. Additionally, there may be considerable psychic costs that produces significant tensions.

Within a given grievance procedure many issues may arise and present unique problems for one of the parties. From time to time, one of the parties misrepresents a key element on which, to their detriment, the other party relies. When this occurs, the procedural issue called "estoppel" is triggered. The party that made the misrepresentation cannot proceed further on the false set of facts. When the employee is the party guilty of misrepresentation, the grievance will not be processed. Estoppel does not change the contract or any rights created in the employment relationship. It does preclude the continued assertion of those rights.[25]

Another issue in grievance resolution is res judicata. Res judicata literally means that the subject has been settled in a prior judgment. Since the conflict has already been decided on its merits, it cannot be relitigated by the same parties.

The reader should note that when a grievance is being settled, the rules of evidence are far more relaxed than they are in a court of law. The primary concern is that both sides be treated fairly. In fact, there are no formal rules of evidence. An arbitrator is bound only by the constitutional restraints of due process.

Dispute Resolution Techniques

It is important to note that grievance arbitration, sometimes called "interests arbitration," is not an impasse procedure. Grievance arbitration is a procedure agreed to during collective negotiations of the contract. It occurs only when an employment contract is in force. Arbitration is provided by a neutral individual or panel. When binding arbitration is used, the arbitrator's decision is final and will be enforced by the courts, if necessary. The decision will not be overturned by the courts, unless it was made by an arbitrator acting outside his authority, there is strong evidence of an arbitrator's bad faith or bias, or the proceedings were inherently unfair to one of the parties.

An interesting question is the extent to which grievance procedures in an employment contract either supersede or are themselves negated by procedures contained in a statute. This is very much an open issue in many jurisdictions. States differ significantly on this issue.

Collective Bargaining and Management Rights

In all jurisdictions requiring school boards to negotiate collectively with employees, there has been an end to unilateral dictation of contract terms by management. Employee organizations naturally want to control, as much as possible, all conditions affecting their members. Local school boards, however, are elected bodies retaining full responsibility to the community. Notwithstanding school site-based councils, and other school-based issues, It is not appropriate, and often illegal, for school boards to relinquish control concerning overall educational policy.

The duty to negotiate in good faith does not equal a surrender of the statutory mandate to supervise the operation of a district's schools and the subjects taught. However, at least one study shows that mandatory collective bargaining laws do affect the perceived abilities of school board members to fulfill the duties of office.[26] Not surprisingly, state statutes requiring binding arbitration to end impasses are seen as intruding the most into traditional management rights. According to the U.S. Supreme Court, the "managerial decisions which lie at the core of entrepreneurial control" should not be restricted by collective bargaining.[27]

Elected officials are public trustees and their status as an employer does not alter this fact. The collectively negotiated agreement should spell out management's rights just as it does those of the employees. The following "model" language has been proposed:

> The Board of Education, on its own behalf, hereby retains and reserves unto itself without limitation, all powers, rights, authority, duties, and responsibilities conferred upon and vested in it by applicable law, rules and regulations to establish the framework of school policies and projects including, but without limitation because of enumeration, the right:
>
> 1. To the executive management and administrative control of the school system and its properties, programs, and facilities and the activities of its employees.
>
> 2. To employ and reemploy all personnel and, subject to the provisions of law or State Department of Public Instruction regulations,

determine their qualifications and conditions of employment, of their dismissal or demotion, their promotion, and their work assignment.

3. To establish and supervise the program of instruction and to make the necessary assignments for all programs of an extracurricular nature that, in the opinion of the Board, benefit students.

4. To determine means and methods of instruction, selection of textbooks and other teaching materials, the use of teaching aids, class schedules, hours of instruction, length of school year, and terms and conditions of employment.[28]

School boards may also find it advisable to include a clause banning strikes in negotiated contracts. The employment contract spells out the rights and responsibilities of both sides, and should contain no implied terms. This is the case even when state law prohibits strikes by education employees. Few laws have been effective in deterring teacher strikes, even when fines are imposed.[29]

Trends

As school boards seek to retain traditional management prerogatives and at the same time engage in creative reform solutions to the myriad of educational problems, their employees are pressuring to broaden the scope of collective negotiations. There has been a gradual erosion of the subject matter lying within administrative discretion. In 1983, the American Bar Association prepared a list of some of the issues most likely to affect labor relations in education. These issues are presented as to note that they have, with rare exception, proven to be almost timeless as well as continuing to illustrate many of the issues facing public education for the next decade.

Agency Shop Question

Agency shop, or "fair share," fees are collected from all employees regardless of union membership. An agency shop fee is money equal to union dues and fees. In a fair share system, nonunion members are assessed a pro rata share of collective bargaining costs. The fees must be paid, as a condition of employment. The issue can be summed up as having a "free ride," versus having to purchase the right to work.

In 1977, the Supreme Court declared that these fees were not unconditional.[30] The only requirement is that a rebate be available to nonunion employees who object to politically or ideologically motivated expenditures. A state of flux continues to exist after many years over religious exemptions, and interpretations of "political or ideological purposes." For example, in one state the fees can be used in lobbying related to legislation affecting employer-employee relationships, as well as toward campaigns for or against ballot propositions affecting employee interests. However, in another state court it was found that some of these uses could be opposed to an employee's individual ideologies, even when their status as employees might be improved.

It should be noted that although such fees are not unconstitutional, they may not be valid under a particular state's laws. Where these arrangements are valid, a secondary issue is the appropriate sanction when an employee refuses to pay. Firing may violate state tenure laws.

The Duty of Fair Representation

Employee organizations must fairly represent their members against the employer. When a refusal to process a grievance is made in good faith, without arbitrariness or inexcusable neglect, the union has not breached its duty to the employee. Suppose, however, that an employee has initially suffered damages because of the employer. If the union fails to properly pursue the grievance to the final stage in a timely manner and the delay causes the employee to suffer more damages, the employer may be able recover this increment from the union.

Impasse Arbitration

Even as more jurisdictions alter the statutes to allow for arbitration as the last step of impasse resolution, public employers in general are coming to regard strikes as the lesser of two evils.[31] Arbitration takes the final decision away from officials elected to decide such matters.

Merit Pay

The recent past has seen various methodologies for experimenting with merit pay, and other means, to reward exceptional teachers. While curriculum content is still very much a matter of policy set by local boards of education, merit pay clearly falls into the category of mandatory topics for negotiation. Traditional teacher organizations tend to be very reluctant to support this concept. Perhaps this is because merit pay would tend to create competition and friction among union members, while unions strive to minimize such conflict.[32]

"Sunshine" Laws

A number of states have passed legislation requiring all government decision making to be made in open, public meetings. States vary widely as to whether this requirement extends to negotiating sessions. In some jurisdictions, open negotiating sessions may be optional. When that is the case, neither side can unilaterally decide to hold negotiations in public.[33]

Collective Bargaining Within Site-Based Management

The issue of site-based management within the overall collective bargaining structure is multifaceted. Each school site must successfully operate within the overall collective bargaining contract as well as presenting unique site-based polices and practices. Each site-based council must focus its efforts on the immediate goals in order to determine the financial parameters of the budget they will work with. Many districts have essentially adopted a "living" contract as well as the "working" contract that is outside the confines of the "living" document. The working contract covers items related to monetary rewards.

Briefly, a living contract includes all items not concerned with wages, benefits, and working conditions, i.e., all issues that require money including the calendar and professional development. Thus, the living contract is one that either party can bring to discussions and resolution at any time during the contract. During the contract, bargaining issues are typically highly contained and separate from the other concerns. In many states, school districts, or individual schools, can request waivers from any rule or regulation. If approved by the appropriate agencies, often times through a state or local waiver procedure, such districts or schools may operate outside the confines of imposed polices by local school boards or the state.

Experience has taught local schools to begin with short-range needs and to work within the master contract and the existing school board polices. Site advisory councils along with administrators and teachers must focus on realistic issues and solutions and identify the issues that all parties wish to resolve. Such issues, that both parties will tend to agree are mutually beneficial, include items that will improve the quality of teaching and student learning.

Overall, the improvement of the operation of the schools is paramount. Both the teachers and management must constantly follow up and evaluate in order to monitor the change and the accountability of the changes. If the parities can focus on what everyone wants, public education and society will benefit.

The negotiating parties at the school site must understand the following in terms of being able to reach a successful site team within the overall master agreement of the school district. These items include:

- Understand the resources available to fit within the discussion of finance issues.
- Need to understand the level of support from the local school board.
- Both the living and working contract exist symbiotically.

In states that allow for individual districts or schools to request waivers, i.e., deviations at the site level from the existing master contract or state rules/regulations, successful petitions reflect the following overall format:

- State the issues/procedures to be changed.
- Cite the current authority to be waived.
- State the rationale behind the waiver being requested.
- List the expected educational improvements.
- Detail the methods to be used to evaluate the changes proposed in the waiver.
- Obtain recommendations to both parties' district-level review team.[34]

In these situations, individual schools may request waivers in order to minimize allegations of unfair labor practices. Many changes that may be requested are deviations from the master contract agreement. For example, a high school principal, with the support of the teachers' union in the high school might request that a waiver to the existing salary schedule be granted. The rationale would be to recruit a high dollar bilingual science teacher for this inner city school. The evaluation would center on the improvement in science class performance measurers. All requests for waivers should be well conceived, realistic, and in the best interests of students.[35]

The relationship among these issues is ongoing. That is, the living contract and the working contract exist simultaneously while the implementation of the site is ongoing, some actions of which are based on waivers. Accountability is interwoven in order that both contracts may be adjusted. Figure 9.1 illustrates the relationship.

In site-based-management districts with collective bargaining contracts, the relationship is complex. All the individuals must know their roles in order to have the most effective organization. Site leaders must have effective directions, data, and an understanding of the implications of any change in the individual schools. Decisions, at the site, must be reached by consensus. Communication channels must be opened between the negotiations team and the sites within the school district.[36]

School-site management within a collective bargaining contract must allow for waivers, changes, and an acceptance of different philosophical constructs within the school district. There are certain cautions to observe in these arrangements:

It is not:

- A process whereby the school board abrogates its responsibilities.

Figure 9.1
The Site-Based Bargaining Cycle (SBBC)

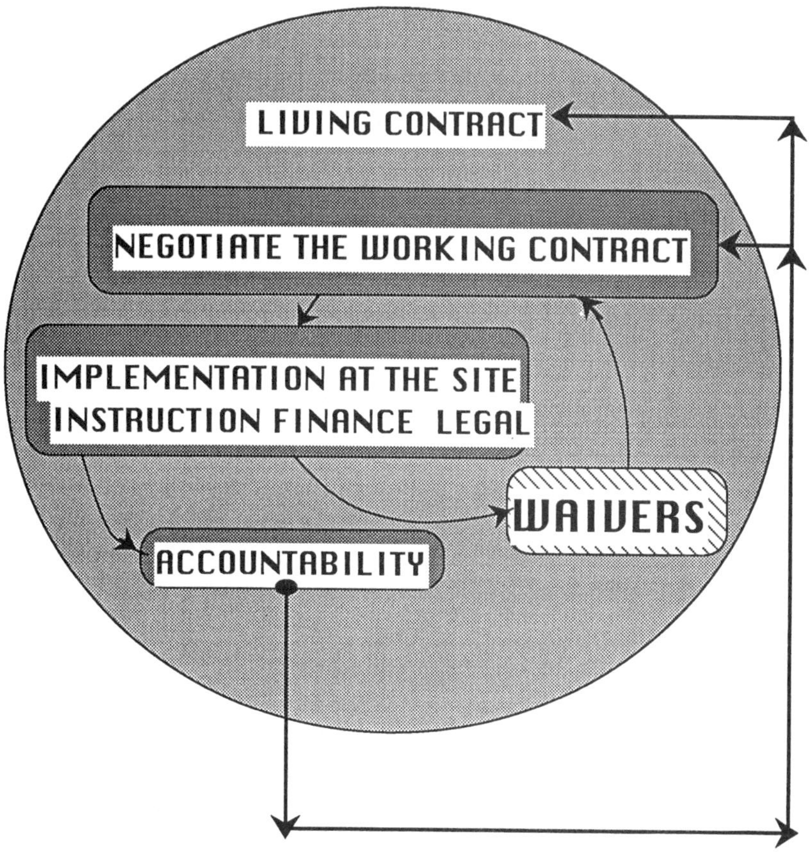

- A process whereby the superintendent relinquishes power.
- A method whereby everyone's ideas are accommodated.

It is:

- Time and labor intensive.
- A process with wide-ranging implications.

Roles of School Business Administrators

While the school business administrator has traditionally played the role of providing information within the negotiations process, this role is expanding to include serving as a change agent charged with merging those complex aspects of school business management with academic administration. This change is increasing the demands on the school business administrator to include the ability to foresee and communicate financial trends as well as the day-to-day operations of the district. These data also include costs and revenue while matching these to the district's objectives thereby ensuring cost effectiveness and efficiency within the district's overall operation.

This individual is recognized as a member of the strategic management team of the school district. The use of strategic management principles is paramount to the success of the business administrator in the collective negotiations process. The four aspects that make up strategic management provide a proactive stance on the part of the school district's management, and this stance is needed if school districts are to survive in the uncertain economic and changing atmosphere of education.[37]

This strategic management process calls for all management personnel to move from the typical reactive stance of the not-for-profit, governmental agency where events control the organization. The agency must move to a more creative, entrepreneurial stance that demands forecasting the future, being aware of trends in the total environment and planning for the long-term success of the district, both in academic and financial terms. This includes the foresight of planning to account for the overall effect that the district policies have on personnel through the collective negotiations process.

Initially, the school business administrator is concerned with: gathering information regarding the status and needs of the employees; trends in compensation; ranges and types of fringe benefits being offered by comparable school districts; analysis of the district's financial environment; and the overall goals of compensation as established by the school district management team. The concern of the district is to match the needs of the employees with the needs of the district to maintain and motivate the employees via a sound compensation program.

Second, the school business administrator deals with the positions of the various groups both inside and outside of the organization, all of whom influence management decisions. The school business administrator is concerned with the general tenor of the taxpayers and their willingness to support additional tax burdens. A final analysis of the ability of the district to pay is vital to the negotiations process. Further, the school business administrator is concerned with possible changes in the legislature and federal funding opportunities that yield revenue for the district.

Next the school business administrator is concerned with the financial aspects of the district. The resource requirements, funding possibilities, long-term sources of revenue and short-term debt funding are a concern. The accounting background and financial analysis skills of the business office become linked with the emphasis on the generation of various alternative programs for compensation and the funding of those alternative programs.

Lastly, the school business administrator is concerned with the control mechanism and the establishment of the means to oversee the entire financial operation of the school district in order to maintain optimum utility of the program for all aspects of the financial and personnel areas. This measure of control is needed as the long-range program planned by the school business administrator will permit incremental adjustments to accommodate the changes in circumstances of employees and the entire work force, and the objectives of the district.

Interwoven in this process are the strategies for negotiations. In the role of designer and developer of the financial information pertinent to strategies for negotiations, the school business administrator is concerned with each item in the negotiations package, particularly the financial issues. Once these items are recognized as part of the negotiations process, the administrator plans the bargaining objectives that might realistically be expected as the final outcome, as well as the range of possibilities. By measuring the impact of the various negotiating positions, the school business administrator permits analysis of the overall package to effect a solution to the various positions and to evaluate success of the negotiations process.

Change of Focus

Negotiations will not be the same in the next fifteen years as they have been in the past fifteen years. This is due in part to the maturity of the participants in the negotiations process and the tenure of the statutes that provide a structure for negotiations between teachers' groups and school boards. Although early experiences provided traumatic conditions and adversarial relationships for both sides of the negotiations table, after many years participants have had the opportunity to become familiar with the laws and processes and grow in maturity, reasonableness, and responsibility.

The future negotiating environment is further influenced by economic conditions. Where the initial thrust of negotiations between teachers and school boards was for winning the issues of wages, hours, and fringe benefits in an atmosphere of plenty, concerns for the future will center on job security while continuing to push for salary parity with the private sector. It now appears that recession and, perhaps, economic recovery will be the norm for the negotiations environment, amidst the changing enrollments, loss of revenue, and cries for quality and cost effectiveness in education. Other future issues are likely to include teacher evaluations, alternative pay plans, attempts to meet the external demands for quality within education, and affirmative action plans as they relate to layoffs.

From the settlements in the private sector, the use of concession bargaining may well be used as a tactic to assuage the general population and maintain the status of the district, both financially and academically. However, regaining the items granted via hard negotiating strategies by the associations is not an easy task, especially when associations view their role as protector of the work force. This will necessitate both parties to adopt a new stance of collegiality to work for effective and efficient solutions of overall problems facing school districts.

In the area of public sector negotiations, one issue needing to be addressed is whether there really is, as suggested, a commonality between the private and public sectors in the area of negotiations. While many states have adopted the system and strategies used in negotiations in the private sector in the area of education, there remains the question of whether the two sectors are really that similar. Support for this commonality includes the fact that little evidence has been found to suggest any real difference between public and private sector negotiations when it comes to the issues of allocation of resources and policy control.[38] This lack of difference is due to the nature of collective negotiations and the fact that both sectors are subject to the same external pressures for product quality and demands for change from the general public.

Another similarity concerns the internally imposed constraints that come from the organizational structure and the members of the organization. In particular is the effect of the budgetary cycle on the negotiating process giving rise to the "wait and see what the budget picture brings before completing negotiations" attitude or the "hold the line once the picture becomes clear" attitude. In both sectors, the economic conditions are set to a great extent by external forces, and the role of the business administrator to plan for these conditions is increasingly more important.

A second study reinforcing the commonality concerned an investigation of the budget-making procedure and the effects of fiscal adversity on the collective negotiations process.[39] These findings indicated that in tight economic conditions, ability to pay was the paramount issue. Ability to pay includes the entire budget and not just salary and personnel costs. Applying these findings to school districts, it is recognized that in tight economic conditions: (1) management postures that there is no money for increases; (2) collective negotiations finish before, or about the same time, as the budget-making process; (3) tight economic conditions dampen the collective bargaining process and create a negative tone; (4) money settlements are more modest; and (5) most settlements require additional funding.[40] These findings point to the need for a different posture with regard to collective negotiations—one where all parties work together in a collegial atmosphere to overcome the adversities of the environment and resolve the issues facing them.

This is where the school business administrator's role as a change agent becomes important in providing for quality education, cost effectiveness, and efficiency within the academic structure. With these environmental changes, the role of the school business administrator will resemble that of an accounting professional in the private sector who analyzes and evaluates both labor and management's proposals, gathers and prepares information for the various impasse procedures, and establishes monitoring procedures for contract administration.[41]

The remainder of this section will deal with four aspects of collective negotiations that are of primary concern for the school business administrator—compensation management, salary administration, fringe benefits, and the use of impasse procedures for resolving conflicts within the negotiations process.

Compensation Management

As indicated, one of the responsibilities of the school business administrator is to design, develop, and implement the financial strategies that will enable the school district to remain solvent. Nowhere is this role more important than in the collective negotiations process that concerns itself with an estimated 70 percent of the budgetary costs of the district. Although many labor agreements are for a one-year period, the school business administrator is concerned with the trends and long-range effects of collective negotiations and attendant costs to the district for the next five to ten years. Further, the concern is for the entire compensation package for the employees and its relationship to the overall revenue of the school district.

Compensation management is concerned with the entire range of compensation granted by the school board to the employee. Compensation includes the employee's salary with the traditional fringe benefits package, and other benefits, both of a cost and a "non-cost" nature. The school business administrator must select a mix of benefits that fit the needs of the individual employee and at the same time meet the educational, financial, and personnel objectives of the district. Recently, many firms within the private sector have moved to flexible, cafeteria compensation plans that permit individual employees to select their benefits or their total compensation package, based on individual needs. Such programs need to be examined for potential use within school districts.

Since the needs of employees and districts are unique, only guidelines can be offered concerning the planning and development of compensation packages. The following guidelines, derived from the private sector, have been adapted for school purposes:

• Clearly define the benefits currently offered by the organization. This includes a detailed listing of all benefits, both cost and "non-cost" benefits, and the reduction of complex benefits (such as hospitalization) to simple and single labels.

• Define the current level of costs for each of the benefits. This aspect needs to be detailed and as clear as possible. One method might be to reduce all costs of each benefit to a per-pupil per-day figure. This not only simplifies the process, but reduces the expenditures to an understandable level for the general public. This figure is computed by dividing the total dollar cost of the benefit by the number of employees receiving the benefit, then dividing that figure by the total number of pupils and dividing the resulting figure by the total number of days the district is in operation, including teacher days.

• Return to the planning principles of strategic management to determine the realistic options for near-term and long-term improvement, or demand, for each of the benefits. For example, based on an aging workforce with children beyond the "normal" orthodontic stage, a benefit for dental care including orthodontics, would reduce in value as a benefit, and hence be a cost to the district, while additional funding of retirement annuities might be a realistic rising cost.

• Determine the cost of each of the near-term and long-term benefits. This calls for gathering information from providers and potential providers about various benefits, and analyzing results of surveys regarding benefits being planned by competitive, comparable school districts within the region. The school business administrator is concerned with the need to provide those benefits necessary to maintain the workforce within the district.

• Determine the use of each of the benefits with the employee workforce. Gathering information about employees' needs, goals, and desires, facilitates near-term and long-term planning for changes in compensation packages and links those changes to the overall negotiating strategies of the district.[42]

The process outlined is completed on an annual basis with planning for the long-term program and its effects. Such an annual reexamination of the compensation program permits a realistic meeting of the employee needs, as well as the incremental shifts needed to maintain an effective compensation plan while meeting the long-term goals and objectives of the management of the school district.

The integration of compensation management with the objectives of collective negotiations and the principles of strategic management is an important facet of the changing role of the school business administrator. Tied to the analyses and planning process is the growing awareness of the need to merge compensation and benefits so that needs of both the employee and the employer are met. Additionally, there is the need to merge the reward system of the district with the management process in order for the district to fulfill its educational objectives.

There is a growing shift to provide after-tax and after-inflation compensation that allows for improvement in the employee's standard of living, while maintaining the costs to the district at a reasonable level. Further, as the nature of the workforce changes, and thus the needs of the employees change, management needs to integrate these two facets of compensation—salary and benefits—to meet the needs of the caliber of employees needed within the district. This calls for a constant reexamination of the overall compensation plan, as well as the reward structure and alternative compensation plans to meeting those needs.

Issues that appear to be a part of the future of compensation management for education include: compensation linked to national inflation via various cost-of-living adjustments (COLA); influences of state and federal legislation making it difficult to pay differentially due to charges of discrimination; and greater use of computerization in the personnel function, with emphasis on techniques for analysis of cost effectiveness and utility of the total compensation package. The school business administrator must be concerned with productivity, academic excellence, cost effectiveness, as well as the needs of employees. In the collective negotiations process, the school business administrator must develop and implement strategies that take into account these different emphases.

Salary Administration

Directly linked to the overall compensation program of the school district is the issue of salary administration. Although this area has traditionally been the domain of the personnel administrator, the school business administrator has an integral role in salary administration beyond the mere issuance of paychecks. As the consultant to the management team, the school business administrator is aware of the issues, problems, and current practices within this area to provide maximum use of the revenue of the district to meet the district's and the employees' needs.

The area of salary administration has changed drastically and will change even more with the advent of merit pay plans and other alternatives to the traditional step salary structures. Part of this change is due to the demand for improved quality and reform in education and the demand for assurance of quality of the teaching staff. Tied to this demand for quality is the lack of personnel in certain key areas and an overflow in other areas, all of which provide school district management with the opportunity to develop new salary structures to reward excellence.

There are additional issues affecting the planning and foresight needed by the school business administrator to prepare adequately for collective negotiations. Internal issues include the matter of equity—the fairness between job performance and job appraisal—that directly affect the alternative and merit pay plans that districts, due to increasing public pressure, are considering. With these new plans come problems that must be anticipated and planned for by school administrators. These include: the subjectivity of the appraiser; the possibility of favoritism; the value structure, which sees younger staff doing the job better; and the secrecy of the performance appraisal process that permits poor communication and the sense of inequity, This can lead to high turnover among personnel and other morale problems.

External issues involved in salary administration include competitiveness with the surrounding market, both in public schools and in the private sector. This is especially true regarding people trained in math, science, and computer science who can command higher salaries and better benefits outside of education. Nonprofessional staff is also involved in this issue, particularly when the private sector in the local area compensates workers at a higher rate and offers more prestige than the public schools have traditionally offered.

Related Issues

Other issues involved with collective negotiations and salary administration are the "just cause" clauses within the dismissal portion of the negotiated agreement.[43] What does the management do with employees who are protected by the just cause clause from dismissal but are not performing at a level that warrants the automatic step salary schedule increase? Although not solely a problem for the school business administrator, it affects the administration and equity of the salary program.

Finally, there is the issue of "comparable worth," which has recently emerged as a major social issue. While school districts have traditionally paid staff (at least for the professional teaching portion of the contract) on a single step salary schedule, the areas of extracurricular activities are now under closer scrutiny with respect to comparable pay. This affects all aspects of the salary administration program, as well as the overall objectives of the district. Aware of these conditions, the school business administrator provides information about the worth of each job based on a market analysis of local and regional competition.

Effective salary administration calls for limiting the use of quantitative decision-making tools and considering external forces affecting the school district and its employees. Proper cost-effective determination and administration of a salary program should be based on a series of checkpoints against which the school district, and the school business administrator, can evaluate the program.

Market-indexed compensation program calls for the blending of two schemes that include the use of quantitative decision-making tools and procedures. The following steps might be used to determine the district's salary program.[44]

1. Management personnel both at the building level and district level prepare complete and detailed descriptions of each position within the district. These descriptions include clearly delineated tasks, responsibilities, preparation needs, and expectations of the position within the overall accomplishment of the district's educational goals and objectives.

2. Management personnel then rank each of the positions based on the importance of the position in accomplishing the goals of the district. This may well mean that certain teaching positions are ranked higher than other positions due to the overall influence of this position on the total education of the student and the accomplishment of the district's objectives. These ranking are reviewed by all management personnel and ranked on a continuous basis until consensus is reached as to the relative rank of each position within the district.

3. Following the ranking of each position, weights or factor values are given to each position. This gives each position a weight that is relative to each other position within the district and is used in combination with the position rank to provide for total position value for compensation purposes.

4. The position value is then converted to either salary grades (not unlike a step salary schedule) or to salary ranges for each position. In either event, the position value is keyed to the market index for salary for similarly prepared and experienced personnel, both within the public education sphere and the private sector.

5. To convert the system to a salary range, which appears to be most typical in using this configuration, the final position value becomes the midpoint of the range from an entry level salary for the position, through semi-interquartile ranges (twenty-fifth and seventy-fifth) and the premium level to be paid an individual for that position. The determination of the range of salaries for each position is based on educational preparation and experience (either needed or available) on the part of individuals to successfully fulfill the position's requirements.

The essence of this salary determination system is linked to the goals and objectives of the district and provides an appropriate compensation for individuals based on the value of the position in the district and their performance. The use of this program also permits more effective hiring practices by allowing management to offer candidates a range of salary based on their potential worth to the district rather than a salary based on the traditional step salary structure. Further, this program permits annual review of the value and place of each position in the organization and keeps each position competitive with other districts and sectors in the economy.

The salary determination procedure described is a radical departure from the traditional step salary schedule. Yet, given the economic environment, the demands for quality in education and accountability of educational personnel to meet these standards, and the need to attract and maintain better qualified and difficult-to-retain personnel in such areas as math, science, and special education, such a procedure may well be considered by the management of school districts.

The Step Salary Schedule

More familiar to school management than the previously described salary program is the step salary schedule. It is based on years of experience and level of preparation. This form of salary determination does not need a detailed review. The step salary schedule system is greatly enhanced by typical spreadsheet analysis, which can be used not only to determine the salary schedule but also the total costs of the salary/fringe benefits package for the district. The use of various spreadsheet programs can simplify and speed the operation of both design and development of the salary schedule and the determination of the total cost of the salary/fringe benefits package.

These data are vital to compensation policies, since without data the costs are unavailable and the ability to fund contracts is largely unknown. Thus, the fiscal stability and condition of the district may be at great risk through a lack of information. Since many negotiations and contract resolutions occur during the course of state legislative sessions, data must be continually updated based on the latest and best information prior to signing a labor agreement.

While terms and conditions make up a large part of negotiations, the most critical element involves establishing salaries. As stated earlier, it is at salary adjustment time when the personnel and fiscal functions are most closely joined because satisfactory contract agreement prevents staff attrition damage to the personnel function. When employees at all levels are unhappy, children suffer as well because low morale affects performance in subtle ways. As a result, the fiscal and personnel offices in the district must work together to understand and address the most common problem areas in salary negotiations.

For certified staff, discussions about salary usually focus on altering the physical structure of the single salary schedule and on increasing the amount of the base teacher salary. Figure 9.2 shows a very basic salary schedule for a school district. This schedule offers a good illustration of the changes that will be proposed through negotiations in the vast majority of districts. Figure 9.2 thus provides a basis for examining the general elements of the negotiations process.

Figure 9.2
Sample Salary Schedule

EXPERIENC	BS +0	BS + 15	BS + 30	MS	MS + 15	MS + 30	DOCTORATI
1	100	102	104	106	108	110	113
2	102	104	106	108	110	113	115
3	104	106	108	110	113	115	117
4	10	108	110	113	115	117	120
5	108	110	113	115	117	120	122
6	110	113	115	117	120	122	124
7	113	115	117	120	122	124	127
8	115	117	120	122	124	127	129
9		120	122	124	127	129	132
10		122	124	127	129	132	135
11			127	129	132	135	137
12			129	132	135	137	140
13			132	135	137	140	143
14				137	140	143	146
15				140	143	146	149
16					146	149	152
17					149	152	155
18						155	158
18						158	161
20							164
1	25000	25500	26010	26530	27061	27602	28154
2	25500	26010	26530	27061	27602	28154	28717
3	26010	26530	27061	27602	28154	28717	29291
4	26530	27061	27602	28154	28717	29291	29877
5	27061	27602	28154	28717	29291	29877	30475
6	27602	28154	28717	29291	29877	30475	31084
7	28154	28717	29291	29877	30475	31084	31706
8	28717	29291	29877	30475	31084	31706	32340
9		29877	30475	31084	31706	32340	32987
10		30475	31084	31706	32340	32987	33647
11			31706	32340	32987	33647	34320
12			32340	32987	33647	34320	35006
13			32987	33647	34320	35006	35706
14				34320	35006	35706	36420
15				35006	35706	36420	37149
16					36420	37149	37892
17					37149	37892	38649
18						38649	39422
19						39422	40211
20							41015

Although negotiations differ according to local custom and state law, it is probable that three changes will be proposed to the salary schedule in figure 9.2. First, the union will ask for an increase in the base salary. Each side should realize that the effect is multiplicative because a change in the base changes all other steps in the schedule. This is because salary schedules are usually indexed in some manner. Second, the union will propose increasing the number of columns on the right hand side, probably to count additional hours not locked into a degree or to create a new column for doctorate degrees. Third, the union will propose adding steps at the bottom of columns to increase pay for longevity among those who are frozen in step.

The board's proposals, however, will try to negotiate new restrictive language on nonsalary items such as leave or duties, while trying to avoid changing to the salary schedule structure. Almost everyone recognizes that board efforts to negotiate any other items—for example, negotiating a salary decrease—is usually nonproductive and could even be construed, in some states, as an unfair labor practice. In sum, the board's position is fiscal damage control that must be balanced against the need to create fair salary structures that align with district policies on attracting and retaining quality staff members.

The changes proposed by teachers are likely to succeed to some extent simply because increases are customary and because they seem highly reasonable expectations. These changes should not be made lightly, however, due to the impact on both the personnel and finance functions of the district. The effect of increasing the base is illustrated in figure 9.2, which is built on a 2 percent step index. If the base salary of $25,000 were raised by 4 percent, it will increase by an innocuous $1,000 to a beginning salary of $26,000. Again, the effect must be remembered for its multiplicative impact because increasing the base causes a ripple effect throughout the schedule, as other steps will take the same increase and benefit by simultaneous vertical movement.

For example, a teacher already on step 1 will get the 4 percent base increase plus 2 percent for longevity in the district so that for just one teacher, the cost of increasing the base is actually 6 percent. If that teacher moved across one column, the increase would become 8 percent, double the base increase. In many districts the effect is even larger since increments between steps often increase more rapidly as the number of years of experience in the district increases. Great care must be taken to cost out the effect of increasing base salaries because every district has tens, if not hundreds, of staff members qualifying for both base and step increases. Proposals to increase base salaries are often far more costly than a simple dollar or percentage increase in base salary might suggest.

The cost of the second proposal to increase the number of columns on the schedule cannot be found using these data in figure 9.2 because decisions would need to be made concerning the new column's structure. The basic nature of new columns can be estimated, however, from figure 9.2. Even though only teachers with credits beyond the last column would be able to move horizontally to a new column, the impact in any district would be significant because there are always sizable numbers of teachers who could move to the new column. The effect is even greater than counting how many people would qualify because any teacher moving horizontally is already experienced. Adding another column would permit all such people to move to the right and downward—again illustrating that the effect is cumulative.

In figure 9.2, if the new column were represented by the doctorate degree, applying the same structure would result in a teacher with nineteen years of experience and a new doctorate degree receiving a salary increase of $2,445 over the prior year (assuming a $100 base increase, a 2 percent vertical step move, and a 2 percent horizontal new column move) for an increase of 6.08 percent, a move for which many other senior teachers might well qualify. Obviously, salaries at the top end of the schedule restrain any inclination to liberally increase the base salary amount and likewise augur against proposals to add steps to existing columns that would move teachers who have bottomed out on experience. In fact, the effect of unfreezing teachers through step and column changes is highly multiplicative.

Development of these data, as seen in figure 9.2, has many benefits in the negotiation process. One important benefit is the ability to automate the salary schedule in a spreadsheet that can be used at the bargaining table. The impact of data at the table has an added benefit in that both sides are confronted with stark reality, this may allow a compromise to be reached more quickly. Under this scenario, both sides have a better opportunity of reaching a workable and realistic solution.

As noted earlier, neither the personnel nor the finance function should take negotiations lightly because the budget is spread across the entire school system. One important effect is that a decision to improve the teaching salary schedule will affect other operations by reducing the amount of available moneys that might have been used for other purposes such as facilities and educational programs. The offsetting reality is that failure to improve salaries also damages district health through low morale.

Very importantly, teacher contract settlement has a secondary effect on the rest of the system because an increase in teacher salaries will be followed by increases for other staff in administrative and classified positions. The same multiplicative reality discussed earlier becomes even more significant when an increase is costed out vertically and horizontally for teachers, administrators, cooks, bus drivers, maintenance workers, and so forth.

Yet, despite best efforts at nurturing organizational and individual needs, negotiations may reach an impasse. As discussed, in most states, districts must engage in fact-finding followed by binding arbitration or unilateral contracts. Regardless of the process, the results are usually disastrous because the organization becomes highly dysfunctional in the sense that the needs of the formal organization and the individuals remain unmet. Further disaster follows in that the finance and personnel functions cannot issue contracts or set firm budgets until negotiations are settled.

In the few states where staff may strike, complete work stoppage damages everyone involved, including children. Situations are not much better in states where public employee strikes are prohibited because work is grudging and relationships severely strained. Finally, it must be remembered that negotiations are repeated to some extent with each employee group. Although easier said than done, the formal organization's needs are best served when the needs of people are met. Administration, of course, has the difficult task of carrying out that mandate using public tax dollars.

Regardless of which system is used, the school business administrator plays an integral part in the salary determination process. While academic management is concerned with the job descriptions, the matching of personnel to position and the communication of these decisions to the workforce, the school business administrator is concerned with the preparation of information and design of an equitable and competitive salary system that reflects the demands of the school district. To accomplish this the school business administrator gathers information regarding the compensation paid employees in competitive and comparable school districts and in the local and regional private sectors. Such information will include: base salaries, salary ranges, benefits, bonuses, merit pay/alternative pay option and the matching of position descriptions to assure compatibility of compensation programs.

The key to effective salary administration is the matching of the needs of the employees to the needs of the district. The needs of both of these groups are changing and will continue to change on a yearly basis. Therefore, careful attention to the development and administration of a sound salary program is essential.

Fringe Benefits

While costs due to salary increases continue to rise at roughly the rate of inflation, a second area of compensation, fringe benefits, is rising at a phenomenal rate. During the past decade, the total dollar costs and the percent of the payroll have increased during the past decade. Although it is expected that the percent of payroll will level off, the dollar benefit costs will continue to rise. This is linked to the security needs of the workforce. Specific costs for fringe benefits may be reflective of vast social and economic issues. These vast social and economic issues are clearly illustrated with the rising cost of health care. Public school districts that have agreed to relatively large percentages of support for medical care plans have seen the cost of such plans rise beyond any previous projections.

The overall dramatic rise in the cost and provision of fringe benefits is due to many factors. These include the need to reduce the taxable income and provide greater security for the workforce. The work of Paine and Seltz indicates that in the recent past there has been an increase in fringe benefits while the net pay in real dollars has only increased modestly.[45]. The dramatic increase of taxable income coupled with a decrease in buying power make fringe benefits extremely attractive.

The need to recoup lost income due to tax increases and inflation has led to the belief among many that fringe benefits can no longer be considered "extra." Rather, they are now viewed as a necessary ingredient of a compensation package, something almost akin to a right. Integral to most fringe benefit packages, provided as a kind of protection against the vagaries of life, are the following insurances: life, health, unemployment, disability, and much more.

Another reason for the rise of fringe benefits is the increased demand by employee organizations during the bargaining process as they attempt to protect employee benefits and security while achieving a more workable balance among the salary/benefit/taxable income mix. The overall effect of these points in negotiations is to increase the benefits paid all employees of the district.

Still another influence on the size and nature of fringe benefits is the extent to which management decides to provide a real-wage increase each year. The school business administrator, saddled with a budget that reflects rising total costs and the possibility of decreased revenue, must examine the variety and cost of fringe benefits and determine the mixture that is needed to meet the demands of the associations and still fulfill management's objectives.

What can the school business administrator do to plan for appropriate inclusion of fringe benefits within the total compensation package and at the same time contain the costs of these benefits? Recognizing that benefits, as part of the compensation package, are important for the attraction, motivation, and retention of good employees, the right mix of base pay, incentives, benefits, and other intangible rewards must be planned, designed, and implemented. Objectives must be established that permit maintenance of a competitive position, adequacy of benefits to assure basic needs, and flexibility within the benefit plan structure. The benefits must also be linked to projected rising costs and the funding available for meeting the costs with plans for providing benefits under different cost containment procedures.

Through surveys, the school business administrator can determine when employees plan to retire, the proper benefits-to-taxation mixture employees desire, as well as the types, status, age, and needs of the employees. The latter—types (male/female), status (single, married, divorced), age and needs of employees—is critical to developing a successful mixture of a benefits/salary package. This information should be updated yearly. How best to provide these benefits is discussed in chapter 19 of this text.

New Benefits

Several benefits are becoming increasingly popular during collective negotiations. The first is the use of flextime within the educational program of the district. Although not a benefit in the traditional sense, part-time employment may be considered a benefit by those individuals who seek to be actively working, but have other responsibilities and interests that make full-time work difficult or unattractive. A flextime plan can work to benefit both the educational program and the employee, since with part-time employees, salary and benefits (if any) are prorated or eliminated, thus providing a savings to the district.

Outplacement, and its corollary, retraining, are also becoming important benefit-type concepts. With the increasing decline in enrollments and the need for districts to contain and reduce costs, it is realistic that reduction-in-force, or layoffs, will occur. Retraining and outplacement assistance gives the employee a certain degree of security, at the same time that they demonstrate an active concern for the employee by the district and underscore the importance of the individuals to the organization. Although a cost to the organization, the morale benefits and good will that are created work to strengthen the organization. What is more, the retraining of personnel who might otherwise be laid off can actually reduce costs to the district if the employee finds satisfactory employment outside the district.

The area of fringe benefits is not only important from the standpoint of employee morale but from the standpoint of cost containment/reduction as well. Fringe benefit costs are rising and will continue to rise at a rate faster than wage increases. This causes the school business administrator to examine all options and plan carefully for the proper balance among the wage/benefit mix so as to meet both the needs of the employees and the objectives of the district.

Resolving Negotiation Impasses

When the negotiations process meets an impasse, how can both the associations and management have a modicum of control while attempting good faith negotiations and protecting the rights of both parties? To break an impasse, a balance must be struck in which the rights and needs of both parties in the negotiations process are taken seriously.

If, as discussed earlier, a commonly used means for resolving impasses in the private sector, arbitration, becomes the general trend in the public sector, then conflict resolution in the negotiations process between teachers associations and school districts will be reduced to a manageable level.

A contributing cause to impasses is the uncertainty of costs and benefits that can be accrued from continued negotiations. There are benefits and costs in compulsory arbitration regardless of the specific techniques employed.[46] Benefits range from the protection of the public interest through the prevention of work stoppages or strikes, to the protection of employees' interests by permitting bargaining from equal strength, to the use of a third party to resolve the dispute in a manner suitable to both parties.

Costs range from the restriction placed on a third party who is not responsible, through the uncertainty of costs and benefits of continued negotiations, to the "chilling" effect of arbitration on the negotiations process by virtue of the uncertainty of the outcome and the restriction to make concessions. The latter usually arises in the use of conventional interest arbitration, where the arbitrator splits the difference between the two parties. It is often suggested that there are additional means to avoid the "chilling" effect and the apparent loss of control by the policy-making body. Some have suggested that the use of final-offer arbitration (either issue by issue or total package) reduces the uncertainty about the costs and benefits as well as encourages reasonable offers and willingness to resolve disputes.

With these concerns in mind, what is the role of compulsory arbitration within education? Legally there are provisions and guidelines for the arbitrator to use concerning: the authority of the public sector body, the issues at hand, the interest and welfare of the public at large, the comparison of hours, wages, and terms and conditions of employment of the organization with those in other segments of the economy, overall compensation, changes in the organization depending on the outcomes of arbitration, and any other factors which might be considered in the "normal" resolution of impasse dealing with the issues of hours, wages, and terms and conditions of employment.[47] It appears that the final impasse resolution procedures available for participants within the private sector—advisory arbitration (fact-finding), conventional (split the difference) arbitration, final offer (issue by issue) or final offer (total package)—are viewed as effective ways to move both parties to want reasonable positions prior to the institution of arbitration, without damaging the overall bargaining process.

For example, the use of final-offer, total-package arbitration can be implemented successfully within the sphere of public education for the resolution of impasse disputes. It is the opinion of some that the process used in final-offer arbitration encourages both sides in the impasse dispute to make more concessions than they would otherwise, thereby bringing parties closer prior to arbitration, while increasing the level of commitment to the bargaining process following negotiations and the impasse resolution process. With final-offer arbitration, both sides are generally more satisfied with the process than they would be with more conventional methods that provide for splitting the award between the stances taken by either side.[48]

Careful consideration should be given to the role of the school business administrator as implementor of the process for negotiations. He or she must be careful to avoid taking a narrow view of the potential outcomes and seek to develop realistic positions for management in light of existing and predicted employee demands and the financial status of the district. By using quantitative data, the school business administrator is more likely to achieve a realistic view of the situation, a view of the larger picture and the ability of the management personnel to meet the financial and academic objectives of the district while meeting the needs of the employees.

Studies in which negotiators were trained in final-offer (total package) arbitration techniques and provided information and training in developing a more realistic perspective of the negotiations and arbitration situation, yielded more concessions on the part of the negotiators prior to arbitration. This was a posture of bargaining that resembled problem-solving rather than the traditional adversarial relationship.[49] The latter is particularly relevant, since the conventional arbitration process does not result in a feeling of "ownership." Knowing that the arbitrator will split the difference when reaching a settlement, negotiators on both sides tend to maintain the initial and distant positions. Although not always the best method for resolving impasses, final-offer arbitration does reduce conflict and produces lower costs and greater benefits in the long run.

In preparing for arbitration, particularly final-offer arbitration, the school business administrator must be concerned with at least four matters. The first area of concern is the final status of the unresolved issues. The school business administrator must determine the exact points of disagreement and the final offers that both sides have produced on these issues. A clear delineation of these points is essential for accurate and effective arbitration. This includes adopting a realistic perspective toward the issues and not fixating on the optimal, least-cost solutions that would be beyond the realm of possibility for the arbitrator to select.

Second, the school business administrator is concerned with the history of concessions made by both parties, the support and clearly demonstrated need for the item, and the degree of strength proposed by the parties. How important is the item and how important is the level at which the demand for the item is set? Consideration must be given to the level of support of the taxpaying public and practices of local and regional competition when formulating a position with respect to an item under dispute.

Third, the school business administrator is concerned with the evidence that will be used to establish the realistic nature of the management team's position. This is the most important function of the school business administrator in the arbitration process. In carrying out this function, the school business administrator is responsible for ensuring that the districts used in preparing comparable data are in fact comparable. This can be accomplished by examining size, location (rural, urban, suburban), state aid ratio, amount received in federal funds, the tax burden and effort, the number of staff, pupil/teacher ratios, the populations income levels (both mean income levels and per capita income levels), buying power of the general public, projected revenue and expenses (again a realistic projection), per pupil expenditures, and similar types of information. To obtain a realistic cross section, ten to twenty districts from the immediate geographic region should be identified initially and compared. Then a similar cross section from across the state should be selected and compared to those deemed really comparable from the local region.

Fourth, once the comparable districts have been located, the school business administrator needs to compile and produce information on the current status and proposed offering of each of the unresolved items. For instance, if salary were in dispute, a table of starting and ending salaries by level would be appropriate for each of the comparable districts including the difference (either plus or minus) between the comparable district and the district the school business administrator serves. Such tables of information should be prepared for each of the unresolved issues including "non-cost" items as well. These tables should also include information about the local and regional private sector as a means to bolster the position of the district by demonstrating comparability of rewards and compensation and the value of the positions under dispute.

The tables of comparison serve two purposes. First, they provide the management team with information regarding the district's standing comparable to similar districts. This enables management to adopt a realistic posture on the unresolved issue. This is particularly important if the district were out of alignment with comparable districts, thereby facilitating a change in the final offer prior to arbitration, if such a change in position were warranted. Second, assuming that the position of the district were realistic, the clarity of these tables will assist the arbitrator to ascertain management's position in rendering the final decision. By demonstrating the relative and realistic position of the district to comparable districts, management's final position on the issues is strengthened.

Although the control of the hearing process is in the hands of the arbitrator and the spokesperson for the team, the school business administrator should expect figures and points of comparison to be challenged by the other side.Providing that the tables and other evidence are well prepared, let the evidence speak for itself. When appropriate also, the school business administrator should challenge any perceived inaccuracies in the opposing side's presentation and ask for clarification of positions and evidence presented to ensure that the management team understands fully the implications of what is being presented. The focus of attention of the team members should be on the arbitrator, who will render the final decision.

It should be noted that challenges by either side should not be taken personally. Many arbitration hearings have been "lost" when the challenges have been taken as personal affronts and the expected reaction has been elicited. A spirit of mutual cooperation and civility must be maintained.

Although arbitration may not be the best means to resolve disputes and solve problems between the management and the employees of the district, it does provide for solutions without resorting to more drastic measures. Also, the use of final-offer arbitration does permit more realistic concessions and bargaining positions between the affected parties. This spirit of mutual cooperation and collegiality can work to foster better relations between management and employees while maintaining the balance of power and responsibilities between the two parties.

Summary

Regardless of the issues presented within this chapter it must be remembered that providing the best education possible for the next generation should be the ultimate goal of all parties. Effective schools are able to keep this as the mission regardless of the inherent differences that may exist between and among the parties.

This chapter has dealt with, in large part, the role of the school business administrator within the collective negotiations process. While remaining a provider of information, the role of this administrator has expanded to include that of designer, coordinator, consultant, and change agent in both the management of the district and in the collective negotiations process. In considering the various aspects of the negotiations process, from compensation management through salary administration, the mixture of fringe benefits to salary, and the place of final arbitration within impasse resolution, the role of the school business administrator calls for creativity and flexibility aimed at assisting the management team in meeting its goals and objectives, facilitating an understanding of the issues, and incorporating the needs of the employees into the negotiations process.

Endnotes

[1] *See generally*, J.L. Grogan, "The Salary and Fringe Benefit Package," *School Business Affairs,* 45:1 (1979), 22-23, 36.

[2] D.C. Thompson, R.C. Wood, and D. Honeyman, *Fiscal Leadership for Schools: Concepts and Practices.* New York: Longman (1994) 415.

[3] E.g., Florida Statutes (F.S. § 447.03) state, "Employees shall have the right to self-organization, to form, join, or assist labor unions or labor organizations or to refrain from such activity, to bargain collectively through representatives of their own choosing, and to engage in concerted activities, for the purpose of collective bargaining or other mutual aid or protection."

[4] Interestingly, one school superintendent has observed that "a teacher's working conditions are a student's learning conditions," B.M. Nygren, "Union-Administrator Collaboration that Works," paper presented before the annual meeting of the Amer. Ass'n of School Admin., Las Vegas, 1988. 4.

[5] See, S. Black, "Is Curriculum on Your Bargaining Table?" *American School Board J.*, April, 1992, 30-32.

[6] Black, "Is Curriculum on Your Bargaining Table?"

[7] Black, "Is Curriculum on Your Bargaining Table?"

[8] See, I. Tyler-Wood, M. Smith, and C.L. Barker, "Adversary Into Ally," *American School Board J.*, July, 1990, 26-28.

[9] .L. Hiraoka, "The New Interest in Bargaining," *NEA Today*, Oct., 1994, 12

[10] Portions of this chapter are based on the previous edition written by T. Jones and G. Baldwin.

[11] Freeman, "The Negotiating Impasse in Labor Relations for Teachers," 5 *Vermont Law Rev.* 39 (1980), 42.

[12] B. Werne, *The Law and Practice of Public Labor Rel.*, (1974), 225-228.

[13] Werne, 400.

[14] Werne, 400.

[15] H.T. Edwards, "The Emerging Duty to Bargain in the Public Sector", 71 *Michigan Law Rev.* 885, 909.

[16] Werne, 247.

[17] Oklahoma.

[18] In North Dakota honesty, temperance, civics, and the work ethic.

[19] Werne, 173.

[20] 5 CER 2470.2.

[21] Werne, 173.

[22] Werne, 175.

[23] Werne, 181.

[24] See Freeman, 43-45.

[25] Werne, 384

[26] Smit, Frank, and Rosemier, "The Impact of Mandatory Collective Bargaining Laws on a School Board's Ability to Govern," 4 *Gov't Union Rev.* (Spring 1983).

[27] *Fibreboard Paper Products Corp. v. NLRB*, 379 U.S. (1964).

[28] Wisconsin Ass'n of School Boards, Language for Use by Wisconsin School Boards in Writing Labor Agreements 8.

[29] Waren, 413.

[30] *Abood v. Detroit Bd. of Educ.*, 431 U.S. 209 (1977).

[31] Jascourt, "Looming Legal Issues in Labor Relations in Education," *J. of Law and Educ.*, (July 1983), 446.

[32] Lieberman, "Educational Reform and Teacher Bargaining," 4 *Gov't Union Rev.* 59, 62 (Summer 1983).

[33] *See, e.g., Talbot v. Concord Union Sch. Dist.*, 323 A. 2d 912 (New Hampshire, 1984).

[34] R.C. Wood and D. Honeyman, *School Site Based Management*, Association of School Business Officials National Workshop, Snow Bird, Utah, Dec., 1991.

[35] Wood and Honeyman, *School Site Based Management.*

[36] Wood and Honeyman, *School Site Based Management.*

[37] See e.g., A.J. Rowe, R.O. Mason, and K.E. Dickel, *Strategic Management and Business Policy: A Methodological Approach*, (Reading, MA: Addison-Wesley) 1980; H.I. Ansoff, Strategic Management, (New York: John Wiley & Sons) 1979

[38] See, C.R. Perry, "Teacher Bargaining: The Experience in Nine Systems," *Industrial and Labor Relations Rev.*, 33 (Oct., 1979), 3-17.

[39] M. Derber and M. Wagner, "Public Sector Bargaining and Budgetmaking Under Fiscal Adversity," *Industrial and Labor Rel. Rev.*, 33 (Oct., 1979), 18-25.

[40] Derber and Wagner, 23.

[41] J.R. Palmer, *The Use of Accounting Information in Labor Negotiations*, (New York: National Ass'n of Accountants) 1977.

[42] P. Kienast et. al., "The Modern Way to Redesign Compensation Packages," *Personal Administrator*, 28 (June, 1983), 127-131.

[43] P.R. Lewis, "Legal Trends," *Labor Law J.*, 28 (Aug., 1983), 15-16.

[44] D.E. Britten, "Rational Salary Administration," *Personnel J.*, 62 (Oct., 1983), 832-835; T. Turnasella, "Market Indexed Compensation," *Personal Administrator*, 29 (May, 1984), 27-39

[45] T.A. Paine and C.A. Seltz, "Development of Employee Benefits in the U.S.," *Financial Executive*, 44 (July, 1976), 54-61.

[46] See e.g.,: S.B. Rynecki, "Can Compulsory Arbitration Work in Educational Collective Bargaining: A Second Look: A Management View," *J. of Law and Education*, 9 (Jan., 1980),93-101; C.N. Lentz, "Can Compulsory Arbitration Work in Educational Collective Bargaining: A Second Look: The Teacher Organization Perspective," *J. of Law and Educ.*, 9 (Jan., 1980), 85-101; D.S. Chauhan, "The Political and Legal Issues of Binding Arbitration," *Monthly Labor Rev.*, 102 (Sept., 1979), 35-41.

[47] See e.g., F. A. Starke and W. W. Notz, "Pre and Post Intervention Effects of Conventional Versus Final Offer Arbitration," *Academy of Management J.*, 24 (Dec., 1981), 832-838; D. W. Grigsby and W. J. Bigoness, "Effects of Mediation and Alternative Forms of Arbitration on Bargaining: A Laboratory Study," *J. of Applied Psychology,* 67 (Oct., 1982), 549-554; G.A. Olson, "Does 'Final Offer' Allow the Bargaining that Conventional Arbitration Chills," *Monthly Labor Rev.*, 102 (May, 1979), 38-39.

[48] A. S. DeNisi and J. B. Dworkin, "Final Offer Arbitration and the Naive Negotiator," *Industrial and Labor Relations Rev.*, 35 (Oct., 1981), 78-81; M.A. Bazerrnan and N.A. Neale, "Improving Negotiation Effectiveness Under Final Offer Arbitration: The Role of Selection and Training," *J. of Applied Psychology*, 67 (Oct., 1982), 543-548.

[49] Bazerrnam and Neale.

Chapter 10
Reduction-In-Force

Introduction

Reduction-in-force (RIF) is an extremely difficult problem in school districts throughout the nation because it proposes reducing spending for instructional or operations budgets. RIF is a most difficult concept in schools because it runs counter to the culture of education. This is especially true when it involves instructional personnel since RIF can be defined as suspension of a tenured or nontenured classroom teacher for reasons unrelated to performance.[1] Conversely, if classroom performance led to enrollment declines, under most controlling state statutes, dismissal or nonrenewal may still be treated as a RIF.[2] The latter issue is more complex than the first, but generally the reason for engaging in RIF is fiscal insolvency. It must be understood that when financial insolvency is claimed by a board of education, counter claims by the appropriate unions will challenge the validity of the board's fiscal data.[3] The local board must be able to defend its actions with hard data.[4] In this arena, the need for clear and precise data is critical and demands cooperation between the personnel and finance divisions of a school district.[5] When these data are clear, local boards of education will be supported by the court.[6]

Planning for RIF

Planning for RIF is a necessary act even in school districts that presently do not face difficult financial issues. In the absence of specific controlling state statutory guidance, all school districts should have RIF policies in place so that an orderly process will be available in the event that RIF is ever required. When state statutory guidance is in place, school districts should have a policy that reflects congruence with statute as well as local decision making for these future events. As with all other types of policies discussed in this chapter, certain critical elements are essential. RIF policies should assure employees that RIF is a drastic last resort measure and all possible steps will be taken to preclude actual RIFing (through measures such as early retirements and not filling voluntary vacancies) and that all laws involving due process and property rights to contracts will be carefully observed.

Similarly, the actual procedures for RIF should be laid out in detail by prescribing in advance the conditions under which RIF will be initiated, notification of RIF, seniority and bumping status, and recall procedures during layoffs, employment of new personnel, and so forth. These latter procedural issues must be detailed because state statutes and applicable local collective bargaining agreements must be known and followed in order to avoid liability for unfair labor practices. The school district incurs liability if the rights of individuals were violated—for example, failure to properly follow RIF that results in the wrong people being laid off will result in liability. Likewise, if the school district's fortunes improve enough to recall some laid-off staff, failure to follow reinstatement procedures will also result in liability. The opportunity for error relating to teaching staff is perhaps greatest, since teachers represent the largest proportion of the school's workforce and because of contractual rights under the law.[7] From this perspective, a clear and correct RIF policy is vital to organizational functioning.[8]

School administrators should keep in mind that courts have extended some rights to nontenured teachers in these instances,[9] although other courts have ruled that nontenured teachers do not possess seniority rights.[10] Likewise, knowing the individual state law is critical because in many states teachers may bump those teachers who possess less seniority within certification areas.[11] Again, it must be understood that these issues are often dependent on state statutes and local collective bargaining agreements.[12] A universal rule, however, is that once RIF is declared, the school district must complete the task and not create new positions that violate the true intent of RIF.[13] The net sum is that administrators must be familiar with the collective bargaining agreement, as well as appropriate statutory requirements for such drastic fiscal actions and be backed up by thoughtful policies addressing reduction in force.

In specific sections of the nation, as well as specific sections of certain states, school business administrators continue to face problems of declining enrollments. The problems will be compounded by increasing reluctance of the public to continue funding education at current levels. Despite a drop in student numbers, the aggregate costs for education have not declined. Ten years ago in the first edition of this text, it was stated that while the aspiring school business administrator will find that much has been written on costs and the economies of scale related to the growth of public education, much less is available that describes planning activities for fiscal retrenchment. This statement still holds true now.

The role functions of the school business administrator are classically illustrated by figure 10.1.

Figure 10.1 outlines the necessary skills for the practitioner facing a complex and continuing problem such as declining enrollment. In dealing with this, the school business administrator must have a strong academic knowledge of public school finance as well as law. With this technical basis and knowledge, the school business administrator is able to develop the task-oriented skills to solve the problems of declining enrollment for a local school district. While identifying and clarifying the task, the school business administrator is also conceptualizing various alternatives. The last step, that of syn-

Figure 10.1
School Business Administrator Role Functions

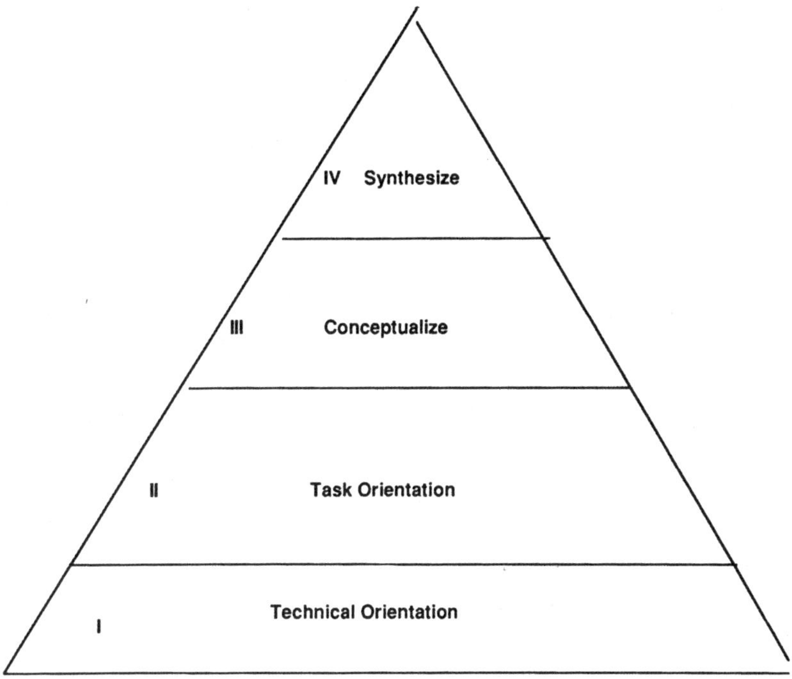

thesis, is the most difficult. The synthesis level involves blending technical knowledge, bringing that knowledge to the task, and conceptualizing the fiscal and legal options that lead to a solution. The synthesis stage brings these components together as practical solutions for a board of education and superintendent to consider.

The role of the school business administrator is that of a technician, resource person, and decisionmaker. For example, as a technician, the school business administrator has calculated the effects of declining enrollment on state aid and the potential compensating options such as increasing local property taxes. As a resource person, the school business administrator may serve as a guide to the superintendent and board of education in developing procedures for reducing staff. As an analyst, the school business administrator presents alternatives and recommendations for the superintendent and the board to consider.

The school business administrator exerts tremendous influence on the decision-making process by deciding what plans are presented as well as recommending policy for consideration. Decision-making responsibility is exercised also when deciding what not to recommend to the school board. While there is no doubt that the board of education has the final decision-making authority, the impact of those who identify options and make recommendations as a part of the decision-making process is significant.

Financial Exigencies

Financial exigencies continue to be an ever-increasing reality for many public school districts.[14] In many school districts, as the cost of education has risen dramatically, so has the decline of pupils, forcing the necessity of the dismissal of teaching staff. School districts have little, if any, flexibility when faced with fewer students and have been forced to engage in RIF. Specifically, RIF refers to layoff, leave, furlough, nonrenewal, or termination, and is the suspension of teachers in a manner unrelated to performance. Within guidelines, boards of education have authority over the hiring and firing of employees.[15] The school board is a governmental agency clothed in authority delegated by the state legislature. Education is considered to be a state function, in partnership with local educational agencies, as reflected in the Tenth Amendment of the United States Constitution. The various state legislatures have generally delegated much of the governance of schools to local school boards, and in this manner, provided for education by discharging a state function.

Because a state legislature can create and also destroy, school board functions within the prescribed legislative and judicial limits of its authority. The degree of flexibility that school boards may exercise is dependent on the role that each state legislature has delegated to the local education agencies under its jurisdiction. Generally, school boards possess no inherent powers and thus must operate within state legislated parameters. Accordingly, school districts often find themselves in the difficult position of discharging a state-determined function while they seek to provide quality educational programs under severe budgetary constraints.

An additional factor affecting a district's ability to achieve its goals and meet obligations is the presence and relative strength of the unions representing classroom teachers. Many of the historical and current concerns of the teachers' associations have been focused toward protecting as many positions as possible, while at the same time maximizing the economic reward for all those protected. Never is this dichotomy more evident than in situations of declining enrollments when the union seeks two mutually exclusive objectives, such as step and pay raises for everyone and job security for all who are covered by the contract. Specifically, a senior school business official has written:

> In the school system that will be facing a decrease in student enrollment . . . or an increase in technological teaching innovations such as computer or satellite teaching, thus reducing the number of em-

> ployees, the unions will increase demands in the area of job security. System building, and class/program seniority will come to the table as factors to be dealt with when determinations are being made concerning reductions in the number of personnel employed. Super seniority for union officers, stewards, and representatives that will allow them to be the last employees removed from the system as long as a job exists . . . regardless of their seniority, will definitely be demanded by all unions, especially those representing blue-collar employees.[16]

From a management point of view, such a position is difficult to comprehend when faced with the reality of budgetary constraints and priorities.

The fiscal exigencies within a school district are generally not a major concern of the teachers' association. Observers of collective bargaining in the public sector should note that during financial insolvency, the teachers' union should be vitally concerned. In the past, concern for financial ability of school districts in the long run was outweighed by a preoccupation with the short-term objectives. In the face of enrollment declines, teachers' associations have typically recommended raising taxes, short-term borrowing, cutting certain nonteaching areas (e.g., maintenance and custodial services) before reducing the numbers covered by the contract. Recent years appear to have brought a more realistic and long-range view from many of the teachers' associations.

Job Security

The strongest means by which a teachers' association may ensure security in the face of financial exigencies is through a collective bargaining agreement that contains a job security clause. This clause may be worded in such a manner as to severely restrict, if not eliminate, a board of education's decision-making options in this area. Job security clauses may present acute problems for a school district facing declining enrollments, and may be compounded as states generally do not permit deficit spending by school districts.

Two court cases from the public sector in New York serve to typify judicial logic regarding job security clauses. In *Burke v. Bowen,*[17] the New York State Court of Appeals examined a job security clause that stated:

> All tours shall consist of a minimum of six (6) Fire Fighters. . . It is agreed by and between the parties that the best interest of public safety as regards full protection with the minimum complement of thirty-four (34) active Fire Fighters, shall be acceptable for the term of this agreement . . . . Notwithstanding any provision herein, it is agreed that upon a reevaluation as to minimum complement, that in

> no event shall the presently agreed upon minimum be readjusted downward. The specific intent of this provision is to issue public safety standards as well as minimum job protection for the Fire Fighters.[18]

The court ruled that job security clauses, although not mandatory, were permissible if granted over a limited period of time. Nevertheless, the court stated that since the city faced extreme financial pressure, it did not have to reinstate the workers. The court did state, however, that the workers might have cause to litigate damages due to breach of contract. An examination of *Burke* reveals the obvious ramifications and legal issues involved in the litigation of RIF policies.

In a similar case in New York, *Board of Education v. Yonkers Federation of Teachers* the contract clause in question read:

> During the life of this contract no person in this bargaining unit shall be terminated due to budgetary reasons or abolition of programs but only for unsatisfactory job performance as provided for under the Tenure Law.[19]

In examining the issue, the New York Court of Appeals considered the following factors:

> (The clauses) were of relatively brief duration. . . . nor were they clauses negotiated in a time of financial emergency between parties of one equal bargaining power. Most important, the job security clause in this instance is explicit in its protection of the teachers from abolition of their positions due to budgetary exigencies. . .(citations omitted).[20]

Although the court acknowledged that not all job security clauses were valid or enforceable depending upon the circumstances, the court stated:

> A job security provision ensures that, at least for the duration of the agreement, the employee need not fear being put out of a job. Such absence of fear may be critical to the maintenance and efficiency of public employment just as the fear of inability to meet its debts may destroy the credit of the municipality. A job security clause is useless if the public employer is free to disregard it when it is needed.[21]

In *Burke* and *Yonkers* the courts ruled that depending on the circumstances, a board of education may have to honor the collective bargaining agreement in terms of job security. Although a state may outline via statute the RIF procedure, the teachers' association may, dependent upon the jurisdiction, still be allowed to bargain the definition of seniority by way of defining and expanding the seniority rights of teachers.[22] For example, in the state of Washington, a court has ruled that RIF is not a mandatory item for discussion at the bargaining table, as a contract settlement was not a prerequisite for dismissal of employees based on budgetary consideration.[23]

Generally, job security is still a permissible subject in the bargaining process. Under certain factual situations, school districts in some states may be unable to renege on a contract based on the *Burke* and *Yonkers* logic. Obviously, school districts outside New York are not bound by such decisions, although they may face similar interpretations.

Job security is generally a high priority of most teacher associations. If job security for all cannot be achieved, the union normally aspires to a straight "last hired-first fired" clause, hence preserving jobs for those with greatest seniority. It is common practice in many communities for the teachers' association to oppose RIF action by engaging in some combination of activities including the attendance at all public meetings in order to seek and display any evidence of preconceived ideas by members of the board of education, monitoring all budgetary documents that might reflect a board of education's unwillingness to seek additional funding, presenting and publicizing various alternatives concerning the dismissal of teachers, publicizing RIF as a ploy to close more neighborhood schools, gathering all financial data relevant to administrative positions for purpose of comparative analysis, and monitoring the collective bargaining contract for possible interpretive or factual violations.[24]

Most states have clearly delegated the authority for teacher termination to local boards of education.[25] Although there are many causes that result in RIFs, the majority fall in three areas: financial, enrollment, and programmatic reasons. Generally, most state tenure laws are not intended to guarantee continuous lifetime employment.[26]

Financial Insolvency

Court cases involving RIF, based on financial insolvency by school districts generally center on attempts to determine if such a situation does, in fact, exist. Generally, the focus of such cases is not on the board's authority, but rather the financial facts from which the board reached its conclusions.[27] Challenges to RIFs based on breach of contract may be successful under certain factual and statutory conditions, regardless of data supportive to a RIF action.[28] In the event that an unanticipated action such as a court order, prevents contract fulfillment, RIF might be successful without a successful breach of contract suit.[29]

It is important to recognize that each state may have a different interpretation of what constitutes a financially insolvent or potentially insolvent situation. For example, the term "insufficient funds" is used in the Missouri statute. In *Frimel v. Humphrey*[30] the court ruled that RIF was permissible in view of the erosion of expected sources or revenue."[31] In California, the statute permits RIF on the basis of "lack of funds." However, the court did not equate bankruptcy with a "lack of funds" as the district in question had a cash reserve.[32] The potential usage of bankruptcy protections for abandoning a valid contract is outside the confines of this discussion.[33]

When financial exigency is claimed by the school board, the school board must be able to refute any bad faith allegations by being prepared to substantiate the actions with empirical data. Several cases typify different situations and different outcomes. In *Unified School District v. Epperson*,[34] two teachers were dismissed due to alleged "budgetary cuts, necessitated by a decrease in state aid."[35] On appeal, the court determined that the teachers' due process had been violated. Testimony indicated that their selection for termination might have been associated with their union activity. In a Washington state case, a dismissed teacher successfully challenged the dismissal in court as the board failed to show sufficient documentation to eliminate the position.[36]

On the other hand, the courts have reached different conclusions in those cases where the financial conditions were indeed bonafide and warranted major budgetary adjustments. In Connecticut, the court upheld a teacher termination when the Board of Selectmen failed to pass the budget of the board of education.[37] In Pennsylvania, a charge brought by a dismissed teacher alleging the board had insufficient evidence to warrant RIF was rejected by the Pennsylvania Commonwealth Court.[38]

These cases serve to underscore the need for administrators to be prepared with empirical evidence of a clear, concise, factual, and financial nature that supports the board of education's position in the event of adjudication. The preponderance of court decisions reflects that school boards which act in good faith, and whose decisions are based on substantial empirical evidence, experienced minimal judicial interference.

Enrollment Declines

RIF, based on declines in enrollment, share similarities with those carried out because of financial exigencies. An important distinction, however, is that a given district may have an increase in overall enrollment but experience a decline within a given academic area or program. For example, an enrollment decline may result from a junior high to middle school reorganization or a shift from an open space to a traditional style of instruction. Again, judicial examination seems to center on the factual circumstances and the quality of empirical evidence presented. In *Philadelphia v. School District*[39] the court considered a decline of 661 students over six years as "substantial" under Pennsylvania statute.

A case from 1929 illustrates a specific declining academic area juxtaposed with an overall pupil increase. In that case, a teacher of a millinery class was laid off when changing women's fashion resulted in a reduced demand for the class.[40] The court agreed that the layoff due to specific enrollment decline was proper.[41]

In a similar case, a Pennsylvania court supported a board of education's decision to abolish the kindergarten program and its attendant jobs. At the time kindergarten was not required under state statute. The court stated:

> Many new fields of experimentation have been launched by our school authorities. Some have been unsuccessful in accomplishing their purpose, and if an employee's position is upheld, no school board would risk the establishment of any new department, which in the event of failure must be continued or its idle teachers paid.[42]

However, at least one state court has ruled that simply because the position was abolished, the employee had certain rights. In *Dugas v. Ascension Parish School Board*[43] the court stated that the board of education had to place the teacher in a position of equal standing and, if possible, pay the teacher's salary. Boards may not unilaterally make reassignments in the absence of well defined grounds.[44]

Authority of the Board of Education

The authority of a board of education to engage in RIF generally rests in the implied powers given to boards of education by the state legislature. As discussed previously, boards of education are creations of the state legislature, which generally have been granted broad and sweeping powers varying from state to state. These powers derive from specific legislative enactments or interpretations thereof, and in forty-nine states,[45] by rules and regulations promulgated by the state board of education.[46]

Understandably, a local board of education must have reasonable flexibility in dealing with declining enrollments, as well as a myriad of other problems. In doing so, the board of education may not violate any state procedures, regulations, statutes, federal laws, or function in a manner that is in conflict with the state's interest.[47] In nearly all states, the local board of education's authority to engage in a RIF is sanctioned by state statute.[48] In *Weisman v. Board*[49] the Colorado Supreme Court spoke to the issue of school board authority:

> The power of the board of education to dismiss and discipline teachers is not merely punitive in nature and is not intended to permit the exercise of personal moral judgments by board members; rather it exists and finds its justification in the state's legitimate interest in protecting the school community from harm and its exercise can

> only be justified upon showing that such harm has occurred or is likely to occur.[50]

School boards may, in some instances, engage in RIF in the absence of clearly defined statutory procedures and reasons. Selected Iowa cases illustrate judicial interpretation of certain statutory phrases. In *Hartman,*[51] the Iowa court ruled that "any good cause" did not include RIF layoffs due to declining enrollments, but rather only to personal faults of teachers. Later in *Briggs v. Board of Directors*,[52] the court upheld RIF deeming financial considerations to constitute "just cause." In a case taken from higher education, *Johnson v. Board of Directors*,[53] the federal district court upheld the termination of tenured faculty members under the statutory language of "just cause." The court stated that this was permissible as long as the Fourteenth Amendment and other individual rights were not violated. The courts elaborated by saying that if the termination were not due to impermissible reasons (such as race), and was not arbitrary, capricious, or unreasonable, it was legally defensible.[54]

The work behaviors of those employees who remain in the school district are affected. This behavior can be either positive or negative dependent upon how the school district handles the RIF actions. In fact, research clearly indicates that the degree of humaneness may be quite significant in terms of reducing the negative affects of RIF procedures in most organizations.[55]

School districts in the throes of declining enrollment generally wish to explore alternatives to RIF. If a steady pupil decline is inevitable, however, temporary solutions will not solve the problems of declining enrollment.[56] Alternatives described in popular educational literature include the following items:

Adult Education

The public school district broadens its appeal and programs to reach a larger clientele and creates new jobs in the process. Dependent on several conditions, this option, however, may not address long-range problems. Opponents of this concept question the savings that accrue to the taxpayer, saying it is not the purpose of government agencies to create work for individuals or create an ever-expanding nontaxpaying agency. In many states such students would not generate fiscal aid in terms of FTE generations for state aid calculations. It should not be suggested these plans are doomed, but merely that certain logistical and practical problems must be resolved before implementation.

Annual Pay Raises

Proponents suggest either doing away with or severely curtailing step raises. Depending on the bargaining climate, this may not be as practical as some would suggest. Case law clearly suggests that step raises pertaining to previously signed contracts must be honored. Whether step raises should or must be calculated in upcoming contracts is a lengthy subject outside the confines of this chapter. Arbitration panels have ruled in different manners in various states on this question.[57]

Attrition

The normal rate of leaves, resignations, and retirements should be accounted for and used whenever possible. This practice is common among school districts throughout the nation.[58]

Busing

Many school districts in certain sections of the nation have irregular pupil population trends (e.g., enrollments are increasing in some schools and declining in others). In areas where this occurs, and if legally and logistically possible, certain realistic, and perhaps limited, exchanges of students and teachers should be explored. With the recent court rulings regarding the unitary status of school districts the necessity of court-ordered busing plans appears to be declining.

Closing Facilities

One of the largest financial savings can be realized by reducing the number of facilities in use. This topic is explored in greater detail in chapter 18 of this text.

Early Retirement

In states where it is legally permissible for boards to participate in such plans, and under certain circumstances, some plans may produce limited savings. Generally, early retirement programs do not lead to the kind of cost savings claimed by the popular press and the teachers' unions.[59]

Helping Teachers Find New Jobs

This plan is used by many large urban school districts. When unemployment compensation benefits are accounted for, these plans may become cost effective.

Increasing Class Size/Increasing Work Loads

Many articles in the popular literature suggest that, if the class sizes/work loads were increased, many financial problems will be resolved. While research is unclear as to the optimum size of classes for learning, legal, political, and practical (facility) limitations restrict the extent sizes may be increased. It is not effective to increase class size if such a move results in disciplinary problems and increased vandalism. Generally, despite the inherent costs, most educators would agree that smaller class sizes are cost effective given the corresponding positive attributes of such plans.

Job Sharing

Two teachers filling one position on a prorated basis of salary and benefits is growing in popularity in some communities. There is, however, little research that examines how pupils learn under these arrangements.

Leave

Some school districts in the last several years have developed open and liberal unpaid leave policies. The rationale for such policies is that it encourages teachers to seek alternate work for one or more years. Some districts report that the majority of teachers who select leave find economic reward elsewhere and do not return. The practice consequently encourages attrition. As the need arises, some districts hire less experienced and therefore less costly teachers on a temporary contract. With careful design, the retiring teacher by virtue of his or her absence, has not progressed as quickly on the salary matrix because the step increases have not been awarded.

Multi-Age Classrooms

Through this technique, smaller classes are combined using modified instructional techniques.

Pre-Primary

As with the discussion of adult education, the school district expands its clientele and reassigns its teaching personnel. With more preschool children coming from homes where both parents work, the need for such programs in many communities will continue to grow. From an educational point of view, there is clear evidence that such programs are beneficial and cost effective. Such plans may not, however, solve the basic financial problems of the school district. In many states such students may not generate fiscal aid for state aid distribution purposes.

Reducing the Number of Part-Time Personnel

Many school districts reduce part-time positions before full-time positions. Although this seems to be wise, it should never be assumed that part-time employees are without legal and contractual guarantees.[60] The fact that part-time employees may not have been mentioned in the collective bargaining agreement does not mean necessarily they are not covered by such agreements in many states. Controlling state statutes, or lack thereof, will dictate if such plans are cost effective.

Short-Term Contracts

While in many districts short-term contracts are utilized wisely, they may be viewed as a violation of the procedural safeguards and the intent of many state teacher tenure acts. Those districts that have attempted such across-the-board action generally have not experienced a great deal of success.

Once RIF has been determined to be a necessity, the school board must deal with the many questions regarding the actual positions to be eliminated. Great care must be taken that the board of education does not act in an arbitrary, capricious, or unreasonable manner in making its decisions. Illustration of this principle is found in *Collins v. Wolson*[61] in which the court stated that the school board may:

> Employ any device or standard to implement the reduction, such as arbitrarily drawing lots or otherwise leaving the decision to chance, so long as the actual reason for the particular separation was not retribution against the instructor's constitutionally protected conduct.[62]

Minnesota courts[63] for example, have allowed administrators and school boards to have flexibility in deciding what positions to eliminate. The school officials may decide what positions to eliminate within categories as well as determine the nature of any new positions under alternative arrangements.

Notwithstanding this discussion, the Ohio Supreme Court has ruled that RIF may not occur unless there is "an actual continuing decline in enrollment." When a school board RIFs classroom teachers, it must do so not as a subterfuge for escaping the teacher tenure law.[64]

An alleged wrongful discharge under the concept of RIF is generally adjudicated in which the employee may be reinstated if the plaintiff prevails. In such instances, school boards generally argue that back pay and related benefits should be judged as to the actual moneys lost minus those moneys accounting for other employment the wrongfully discharged teacher may have received.[65]

Order of Dismissal

Decisions concerning which employees to dismiss will depend on statutory guidelines, collective bargaining agreements, and state court decisions. Minnesota, New York, and California,[66] for example, require by statute that the nontenured teacher be laid off first if the tenured teacher is certified and qualified to take over the nontenured teacher's position. This concept may be extended to other professionals such as teacher aids.[67] Unless state statute requires a given procedure or an applicable court case sets forth case precedent, it is to the school board's interests not to base RIF on a standard "last hired-first fired" procedure. A court in Nebraska[68] has upheld the right of a school board to use extracurricular activities as a basis for dismissing teachers when all other factual circumstances are similar. In Indiana, the courts[69] have clearly ruled that a tenured teacher may not be dismissed over a nontenured teacher if both are certified for the same position. In some instances, building and/or central office administrators may be allowed to bump nontenured teachers.[70]

Consolidation/Desegregation

RIF often occurs when school districts are consolidated, which normally results in the elimination of superfluous positions. In *Hensley v. State Board of Education*[71] the Supreme Court of New Mexico, examining the question of whether a classroom teacher loses seniority by virtue of the district's merging with another district, the court said that a teacher loses no rights when two districts merge to create a new district since the consolidated district is legally a continuation of the previously non-consolidated district. On the other hand, in *Beckett v. Roderick,*[72] the Maine Supreme Court ruled that a newly consolidated district was not obligated to hire several recently dismissed teachers. A crucial basis for the court's decision was that the enabling legislation for consolidation made no mention of personnel transfers. The court saw no inherent obligation on the part of the local school board to hire the previously dismissed teachers. It was, therefore, a matter to be left to the local school board's judgment.

When students are transferred across school district lines for court-ordered desegregation purposes, another school district may be instigating RIF procedures while another school district may be hiring additional classroom teachers. The impact of such applicable court orders is unclear as to the obligation of the receiving school district. In the absence of a definitive ruling by an appropriate court, school districts generally are not bound by legal obligation to hire the neighboring school district's classroom teachers.

Historically, during desegregation a great deal of consolidation has occurred by the merging of racially identifiable schools and school districts. In some instances, there has been evidence to suggest that minority administrators were dismissed as a result of consolidation. When RIFs involve minority dismissals,[73] the courts have generally looked to the "standard" found in *Singleton v. Jackson.*[74] *Singleton* stated that any dismissal or demotion of an employee must be on the basis of:

> . . . objective and reasonable nondiscriminatory standards from all the staff of the school district. . . . Prior to such reduction the school board will develop. . . nominal objective criteria. . . in dismissal or demotion. . . . Demotion as used above includes a reassignment 1) under which the staff member receives less pay or has less responsibility than under the assignment he held previously, 2) which requires a lesser degree of skill than did the assignment he held previously, 3) under which the staff member is asked to teach a subject or

> grade other than one for which he is certified or for which he has had substantial experience other than one for which he is certified or for which he has had substantial experience within a reasonably current period.[75]

Since the *Singleton* standard was first set forth, the interpretation of its applicability has been narrowed considerably. In the 1975 case of *Pickens v. Okolona*[76] the court found the *Singleton* standard not "relevant" to the overall reduction of the number of teachers.[77] In *Lee v. Chambers*,[78] the Fifth Circuit Court of Appeals ruled that *Singleton* was not applicable since the RIF was due to a loss of Title I funds and not desegregation. In *Lee* the court was more concerned with the rationale rather than the impact, in a clear departure from *Singleton*. In *Harkless v. Sweeny Independent District*[79] the *Singleton* standard was not applied as the evidence reflected that desegregation activities were not the result of a court order and had been implemented before *Singleton*. In *Barnes v. Jones County School District,*[80] the Fifth Circuit Court remanded the case to the federal district court to examine if the dismissal of teachers was racially motivated. However, in *Barnes* the Fifth Circuit Court stated that *Singleton* was not applicable because, although the number of minority teachers had been reduced, there was no overall RIF. The Fifth Circuit Court interpreting *Singleton* in *Wright v. Houston Independent School District*[81] stated that it must be clearly shown that desegregation was the primary cause of the RIF. In *Peques v. Morehead Parish School Board,*[82] the Fifth Circuit Court ruled that a "mere lapse of time" is a sufficient rationale for dismissing a claim under the *Singleton* standard.

Although *Singleton* has been narrowly interpreted by the courts, its basic thrust remains valid. Inasmuch as racial discrimination creates a "suspect class," greater judicial scrutiny will be present in adjudicated matters involving RIF and its impact on minority employees. Once again, the necessity of having empirical data as reflected in the objective teacher evaluations and sound school board policies is illustrated.[83]

Bumping

As a general rule, state courts have upheld the policy of dismissing nontenured teachers before tenured teachers in a RIF procedure when all the factual circumstances are the same. When a tenured teacher is released and he or she is certified in other areas currently held by a nontenured teacher, the tenured teacher must be restored and the nontenured teacher released. Note that this discussion assumes all things are generally equal with the absence of a statute or contract reflecting anything different and where the factual circumstances are without challenge.[84] In *Swisher v. Darden,*[85] the New Mexico Supreme Court ruled that the school board had to prove in its dismissal procedure that there were no other vacancies for which the dismissed teacher was legally qualified to teach. This standard was narrowed in *Fort Summer v. Parsons*[86] and *Penasco v. Lucew*[87] in which the New Mexico courts stated that *Swisher* was valid but only if the curriculum were not adversely affected by the procedure in question.

In summary, the vast majority of courts have upheld the concept of the tenured teacher "bumping" the nontenured teacher. These decisions reflect the logic and intent of the various state tenure laws. On the other hand, there are a significant number of exceptions, which either do not hold to this view or severely restrict such an interpretation. State court decisions in Connecticut,[88] Florida,[89] Oklahoma,[90] Pennsylvania,[91] and Maine[92] have issued highly specific, factually based decisions concerning across-the-board bumping even within certified categories.

When tenured teachers are dismissed, boards must often determine the position to be eliminated. Two cases serve to illustrate widely different views. In *Williams v. Board of Education of Lamar*,[93] the Alabama Supreme Court stated that a board of education may dismiss any tenured teacher and that seniority is not a criteria for RIF of tenured teachers. The Pennsylvania Supreme Court, on the other hand, in *Welsko v. Foster Township School District*[94] stated that seniority must be a criteria for selection of those being dismissed. Courts in New York[95] California[96] and Illinois[97] have also reached varying conclusions. Questions are often raised as to the determination of teaching experience in relationship to previous assignments involving certified and noncertified teaching experience.[98] States may have different definitions of the term seniority and thus lead to differing views.[99]

Often questions are raised concerning seniority rights of nontenured teachers. One view holds that all employees acquire seniority regardless of tenure. The opposing view embraces the concept that nontenured teachers have, by virtue of their status, little if any rights or privileges that normally accrue to tenured teachers. The seniority question is one of determining if considerations extend to and among nontenured teachers. A New York court[100] has ruled that nontenured teachers have seniority rights among nontenured teachers.[101] The opposite view was accepted by the Washington Supreme Court[102] where the court ruled that seniority rights do not necessarily extend to nontenured teachers. Courts has ruled that a nontenured teacher has no right to the district's RIF policy when the teacher is dismissed for reasons of staff reduction.[103] Within these overall broad concepts, in the absence of constitutional infringements, school boards are allowed discretion, in the absence of controlling state statutes or collective bargaining agreements.[104]

Full-Time to Part-Time

Often boards of education, due to financial conditions, wish to reduce certain positions to part-time positions. Some states require a set procedure for reducing full-time into part-time positions. In an Illinois case, *Caviness v. Board of Education*[105], the court ruled that the state statute concerning RIF was applicable where an employee was being reduced to a part-time position from a full-time position. In at least one instance, a tenured teacher who had agreed to a part-time status position could then bump a less qualified person for a full-time position.[106] Although relatively few statutes and court cases[107] offer specific guidance in this area, boards have found that great care should be exercised in this aspect of RIF.

Qualified versus Certified

Boards of education are often placed in an awkward position regarding conflicts in the status of "qualified" and "certified" personnel. Without a negotiated agreement or statute the terms are often interchanged to represent the same concept. Generally, certified is defined as "legally qualified." However, the Illinois Supreme Court[108] has ruled that "legally qualified" means more than the term "certified." If a school board were to make an argument that "qualified" reflects years of successful experience in teaching within a given area, or possessing certain teaching skills unique to the assignment and so on, then the board may have greater flexibility.[109] Several states have statutes that only apply to those who qualify for a given position. Thus, noncertified teachers are not protected in being placed in a position for which they are not certified.[110]

Absence of Policy

The failure of a board of education to have a RIF policy or agreement, in the absence of a state statute, is not entirely negative. Constitutionally protected freedoms only protect individuals from Constitutional infringements. They do not themselves require a policy to be in place before dismissal. Obviously, however, a board of education must comply with all applicable state statutes. It also should be noted that in the absence of a written policy, a court may interpret a course of conduct to constitute a policy which, in turn, affects the working conditions of all employees[111] under certain state statutes.

As a broad generalization and in the absence of written policies or collective bargaining agreements, administrative positions do not carry specific rights regarding RIF to other administrative positions[112]. Generally, such positions may revert to the teaching staff dependent upon the availability of positions and certification requirements.[113] In fact, it has been specifically ruled in at least one case that a building principal's right to equal protection was not violated when the principal failed to achieve rights given to the teaching staff regarding the issues of seniority within a reduction-in-force policy.[114]

In some instances, teachers who are RIFed may retain employment rights despite the fact they may be teaching in another school district.[115] This issue is highly state specific and in the absence of controlling state statute must be judged by an appropriate state court.

Summary

A review of the litigation surrounding RIF claims is revealing. First, the widely accepted practice of "last hired-first fired" as practiced by countless school districts need not be a legal necessity. Second, in the absence of any infringement on the constitutionally protected rights of citizens, federal courts have been reluctant to interfere with the discharge of classroom teachers by local school boards.[116]

Third, the courts are heavily persuaded by empirical data supporting demographic and financial claims by boards of education. With these data, challenges based on constitutional infringements are weakened considerably in the course of litigation.

As a generalization, where boards of education have operated within the state procedures and followed due process guidelines, the state courts have not substituted their judgment for that of the local school board. The evidence clearly indicates that although RIF is a difficult task and subject to litigation, local boards of education in the face of financial exigencies may reduce the teaching staffs through a variety of acceptable methods.[117] As illustration, one court has stated that the system is valid as long as the system "had some reasonable relevance to the teacher's qualities and qualifications."[118]

The school business administrator should keep the following in mind when assisting a local board of education with RIF planning:

- Appropriate state statutes should be followed in all procedures.
- Adequate legal counsel familiar with applicable public school law should be secured at an early point.
- All financial and demographic projections for RIF decisions would be as accurate as possible to withstand any potential challenge.
- Great care should be exercised not to violate any constitutionally protected rights or affirmative action guidelines throughout the decision-making process, and
- The school business administrator must stay abreast of all recent developments concerning RIFs to protect the board of education.

There are several excellent RIF models that a school business administrator may wish to consider. Policies will vary in each state dependent on the appropriate statutes and court cases. Thus, the policies contained within this chapter should never be adopted without careful analysis by school business administrator and legal counsel. Each board of education will have different needs that require policy adjustments within the appropriate legal context. Selected samples are contained at the end of this chapter.

Appendix
Sample Contract Clause

Lay Off Clause

1. In the event of a reduction of teaching staff, the parties recognize the principle of seniority in the layoff or recall of staff within areas of certification and qualification.

2 . For purposes of this section, the following definitions will apply:

a. "Certified" shall mean holding a current certificate in a subject matter area issued by the certification authority.

b. "Seniority" shall mean continuous service as a teacher commencing with the first day in the school district.

c. "Qualified" or "Qualifications" shall mean a teacher who is fully certified in the subject matter category as defined herein, and who is currently teaching or has successfully taught such subject matter category with the past (x) years in the school district. A teacher shall appear on each subject matter seniority list for which he or she is qualified as defined herein.

3. In the event of reduction in staff, teachers shall be laid off in inverse order of seniority within area of qualification.

4. In the event of recall of staff, teachers shall be recalled in order of seniority within area of qualification.

a. A teacher shall have the right to recall for a period of (x) years from the last duty day in the school district.

b. If a position becomes available for a qualified teacher on layoff, the school district shall mail written notice to such teacher at the last address on file as provided by the teacher and contained in school district records, who shall have (x) calendar days from the date of such notice to respond in writing accepting or declining the offer of recall and an additional (x) calendar days to report for duty, if the teacher accepts recall. Failure to provide written notice of acceptance or reemployment, or to report with the time period provided herein, shall constitute forfeiture of the right to recall and the teacher's right to recall shall be terminated.

5. In the event of equal seniority, the order of layoff and recall shall be within the discretion of the school district.

6. A teacher on layoff pursuant to this section shall retain such amount of experience credit for pay purposes and leave accrual and other accrued benefits, if any which he or she had accrued at the

time he or she went on leave for use upon recall. Additional experience credit for pay purposes or other benefits shall accrue for the period of time that an employee is on layoff under this section.

7. For purposes of seniority standing, a teacher's seniority date will be unaffected by layoff pursuant to this section, or any board-approved leave pursuant to this agreement.

Adapted from the School District of River Falls, Wisconsin.

Sample Board Policy

Reduction in Professional Staff Work Force

I. Definition of Terms

A. Service. The period of time from the effective date or an educational staff member's signed employment contract, commencing the most recent term of continuing service in the (named) School District, including sabbatical leave(s) and authorized leave(s) of absence. If two or more employees have the same length of service, they shall be ranked by date of contract offer that begins the most recent period of continuous service and, if necessary, date of receipt of most recent application resulting in employment, and finally, by lot.

B. Endorsement Areas

1. Subject Area. The teaching endorsement on the educational staff member's (Training) Certificate. Comprehensive endorsements will be divided into their constituent endorsements for the purposes of determining subject area, provided, however, that the individual endorsements comprising the history and social science endorsements will be aggregated into a single subject area indicated as history/social science. A list of subject areas is attached.

2. Grade Area. The teaching endorsement on the educational staff member's (Training) Certificate relating to elementary education. Elementary teaching endorsements will be grouped as follows: Kindergarten (K), grades 1-6.

C. Classification Areas. The position title currently held by an administrator or supervisor in the district as well as other miscellaneous positions.

D. Active Assignments. The endorsement area(s) in which the teacher is actively teaching (or if not actively teaching, the most recent teaching assignment) at such time as it becomes necessary to implement the destaffing or reduction-in-force procedure.

E. Layoff. A condition creating certain recall rights, attendant to termination resulting from displacement through the operation of this procedure.

II. Destaffing

A. Teachers

1. Whenever the board determines that there is an excess of teacher(s) in an endorsement area in a particular school or workplace for such reasons as reduction in school population or the closing of a facility, the teacher(s) with least service who is actively teaching in the endorsement area will be designated for destaffing by receiving a written notice to that effect. Consideration for transfer will be given to volunteers in the endorsement area in the affected school before this provision is implemented. Notice of transfer request status will be given to teachers concerned as soon as possible but no later than (x date).

2. Teachers designated for destaffing will be ranked in order of service and by endorsement area. Said teachers will be assigned, in order of service, to available vacancies in their active assignments. If no vacancies are available in the teacher's active assignment, he or she will be assigned to an available vacancy in an endorsement area in which the teacher is certified. Said assignment will be made in order of service, provided, however, that assignment of vacancies outside the teacher's active assignment will not be made in circumstances where there are teachers to be destaffed in the endorsement area.

3. If a teacher refuses a position in an endorsement area in which he or she was actively teaching at the time of destaffing, the teacher will be terminated and will not be entitled to recall rights hereunder. If a teacher refuses an assignment outside his or her active assignment, the teacher will be entitled to utilize the reduction-in-force procedure described below.

If no vacancies are available in a teacher's active assignment and if the teacher is either ineligible for or elects not to accept vacancies outside his or her active assignment, said teacher will be entitled to exercise prerogatives set out in the reduction-in-force procedure.

B. Personnel in Classification and Miscellaneous Areas

1. Whenever the board determines that there is an excess number of persons in a classification area, including miscellaneous posi-

tions, at a particular facility for such reasons as reduction in school population or the closing of a facility, the person(s) in the classification area with least service at the facility will be designated for destaffing. Consideration for transfer will be given to volunteers in the classification area of the affected facility before this provision is implemented. Notice of transfer request status will be given to the persons concerned as soon as possible but no later than (x date)

2. Personnel designated for destaffing will be ranked in order of service and by classification area. Said personnel will be assigned, in order of service, to available vacancies in their classification area. If no vacancies are available in the person's classification, he or she will be entitled to utilize the reduction-in-force procedure here under.

3 . If a person designated for destaffing refuses an available position in his or her classification area, said person will be terminated and will not be entitled to recall rights hereunder.

III. Reduction-in-Force

A. Whenever the board determines that there is an excess number of teachers in an endorsement area in the (named) school district, or when destaff teachers or school system personnel in classification areas are entitled to exercise reduction-in-force rights in the teacher endorsement areas, for such reasons as substantial reduction in student population or the closing of a facility, the following procedure will apply:

1. Systemwide service lists prepared by endorsement area and based on length of service in this school district will be established, a copy shall be placed in each school. These lists will include all persons presently employed by the (district) (including persons on authorized leave of absence) who have service as educational staff members. Endorsements attained subsequent to the date the service lists are prepared will not be included on the said list until all persons in said list are recalled or forfeit recall.

2. Persons eligible for inclusion on the service lists described in subparagraph 1 will be ranked in order of service and may, in their order of service, displace the junior actively employed teachers in the displaced teachers' active assignments provided that the person exercising displacement rights has greater service than the teacher to be displaced. Where there is more than one person entitled to displacement rights in an endorsement area, said person will be assigned to an equivalent number of positions held by teachers with less service in the endorsement area.

This provision is not intended to permit a person with less service to displace an active teacher with more service. A person refusing an assignment within said person's active assignment will be terminated and will not be entitled to recall rights hereunder.

3. In the event a person is not eligible to displace a teacher in his or her active assignment, the following procedure shall apply:

a. A master list of displaced personnel will be prepared in order of service. The list will include the endorsement areas of each person.

b. The person with the greatest service on the list will be assigned to the position held by the teacher with the least service in the displacing person's endorsement areas.

c. The procedure set out in subparagraph b above will be followed subsequently for the remaining displaced personnel on the master list.

d. In no event will a person displace a teacher having greater service.

4. A person displaced from a teaching position by operation of this procedure will be eligible to displace teachers with less service in the displaced teacher's endorsement area.

5. A person who elects not to exercise displacement rights in his or her active assignment will be terminated.

6. A person who elects not to exercise displacement rights in his or her endorsement area (other than his or her active assignment) will be laid off.

B. The reduction-in-force procedure contained in subsection 3a above will apply to classification area positions with the following modifications:

1. Personnel who are eligible for displacement prerogatives by operation of the destaffing procedure or by virtue of having been designated as excess will be entitled to displace the junior person in their classification area providing the displacing person has more service than the person to be displaced.

2. If a person within a classification area lacks sufficient service to displace another person within his or her classification area, that person may exercise reduction-in-force prerogatives in the teacher classification as set forth in subsection 3a. For purposes of exercising displacement rights within a teacher's endorsement area, the classification area employee's active assignment will be deemed to

be the endorsement area in which the employee most recently taught prior to assuming a classification are a position.

3. A person may only exercise displacement prerogatives with the classification area in which he or she is employed at the time he or she becomes eligible to exercise such rights, provided that intermediate and high school principals will be deemed eligible to displace intermediate and high school assistant principals according to the procedure set out in subsection IIIA, and elementary school principals will be entitled to displace elementary school principals; except as provided above, a person's displacement prerogatives are limited to his or her classification area, and thereafter, to the teaching endorsement area.

4. Classification area personnel who elect not to exercise displacement rights within their classification area shall be terminated and will not be entitled to recall rights hereunder. Classification area personnel who elect not to exercise displacement prerogatives in teacher endorsement areas and principals who elect not to displace assistant principals will be laid off and will be entitled to recall rights hereunder.

IV. Layoff and Recall

A. Educational staff members on layoff are terminated; however, upon recall, they will receive all previously accumulated service credit.

B. Teachers laid off as a result of this policy shall be ranked in order of service and by endorsement. When openings occur, the teacher on layoff with the greatest service in the endorsement area will be entitled to the position. If the opening is in the teacher's active assignment, the teacher must take the opening or lose all recall rights under this provision. If the opening is in the teacher's endorsement area but not in his or her active assignment, the teacher may elect whether or not to take the opening. If the teacher elects not to take the opening, his or her name will be struck from the recall list in that endorsement area, but he or she who retain recall rights in his or her other endorsement areas. Teachers who remain on layoff for up to three years from date of layoff will enjoy no further right of recall.

C. Classification area personnel who are laid off as a result of this policy will be ranked in order of service and by classification area. Classification area personnel who exercise displacement rights in teaching endorsement areas, and principals who exercise displacement rights in the assistant principal classification area, will be eligible for recall on the same basis as classification area personnel who are laid off. For purposes of this provision, they will be deemed to be "people on layoff status." When openings occur, the person of layoff status with the greatest service in the endorsement area will be entitled to the position. If the person elects not to accept the opening, he or she will forfeit entitlement of recall under this procedure.

D. When an educational staff member is to be recalled, he or she will be notified by registered mail (return receipt requested) at his or her last known address on record with the board. If an education staff member so notified of recall does not accept in writing the offer of reemployment within ten work days of receipt of notice, all right of recall under this provision will be forfeited. If the notice of recall is undeliverable, the educational staff member will forfeit the recall rights of this provision.

V. Part-Time Teaching Positions

As an alternative to layoff, an educational staff member may elect to accept an available part-time teaching or teacher aide position in the school district. For the purposes of this provision, an available part-time teacher or teacher aide position is defined as an opening created by the termination of an incumbent part-time teacher or teacher aide or the expansion of the available part-time teacher or teacher positions. Educational staff members eligible and certified for said positions will be notified by the Board and will be offered available positions in their order of service. An educational staff member electing to accept a position as a part-time teacher will retain the recall rights set out above. An educational staff member foregoing such alternative employment will retain his or her recall rights.

VI. Miscellaneous

Notwithstanding any other provisions of this policy dealing with displacement, educational staff members whose services are necessary to ensure efficient operation may be retained irrespective of length service and will not be subject to displacement provided they are within the following assignments, limited to one individual for each activity in a school and provided the incumbents presently have and are to be reassigned the same responsibilities the following year:

High school and secondary school sponsors and coaches who fall under Categories I and II of the salary supplements regulation, the band director, the choral director, and the chairperson of the major departments, intermediate school newspaper, yearbook and SCA sponsors, and chairperson of the major departments.

A maximum of (x) educational staff members in addition to the above may be detained irrespective of length of service and will not be subjected to displacement if they are necessary to maintain program demands.

During a period of reduction-in-force, the board shall limit the hiring of educational staff members to those positions that cannot be filled by qualified current educational staff members, including those returning from leaves of absence, as well as those employees being considered for reduction-in-force provided that this provision will not operate to permit an educational staff member to claim a vacancy in a higher classification.

VII. Notice

Appropriate, certified employee organizations shall be notified in writing at least 30 days in advance of any pending layoff.

Adapted from Fairfax County School Board, Fairfax, Virginia.

Endnotes

[1] See R.C. Wood, "Financial Exigencies and the Dismissal of Public School Teachers: A Legal Perspective," *Government Union Review,* 7:3 (1982), 262-276; and R.C. Wood, "Chapter Nine of the Federal Bankruptcy Act: A Fiscal Alternative for School Districts," *J. of Education Finance*, (Summer 1985), 56-68.

[2] In *re Gaetjens,* 485 A.2d 1057 (N.H. 1984).

[3] See, e.g., *Frimel v. Humphrey,* 555 S.W.2d 350 (Mo. 1977); *California School Empl. Ass'n v. Pasadena Unif. School Dist.,* 139 Cal. Rptr. 633 (Ct. App. 1977).

[4] *Gross v. Board of Educ. of Elmsford* , 577 N.Y.S.2d 200 (Ct. App. 1991).

[5] See, *e.g., Unified School Dist. v. Epperson,* 583 F.2d 1118 (10th Cir. 1978); *Bradfute v. Renton School Dist.*, 577 P.2d 147 (Wash. 1978); *Cigardski v. Lake Lehman School Dist.*, 407 A.2d 4609 (Pa. 1979); *Baston v. Ricci,* 391 A.2d 161, (1978); *Freiberg v. Board of Educ of Big Bay De Noc*, 283 N.W.2d 775 (Mich. 1979).

[6] See e.g., *Britten v. South Bend Comm. School Corp.*, 819 F.2d 766 (7th Cir. 1987), *cert den.* 108 S.Ct. 288 (1987).

[7] See other chapters in this text regarding rights of teachers.

[8] *Heruth v. Independent School Dist.* 434 N.W.2d 470 (Minn. Ct. App. 1989).

[9] *Alexander v. Delano Joint Union High School Dist.,* 188 Cal. Rptr. 705 (Ct. App. 1983); *Bohmann v. Board of Educ. of West Clermont,* 443 N.E.2d 176 (Ohio 1983); *Carmody v. Board of Directors,* 453 A.2d 965 (Pa. 1982).

[10] See, e.g., *Lezette v. Board of Educ.,* 319 N.E.2d 189 (N.Y. 1974); *Hill v. Dayton School Dist.,* 532 P.2d 1154 (Wash. 1975).

[11] See e.g., *McKeesport Area School Dist. v. Cicogna,* 558 A.2d 116 (Pa. Commw Ct. 1989).

[12] *Fort Sumner Mun. School Board v. Parsons,* 405 P.2d 366 (N.M. 1971); *Penasco Indep. School Dist No. 4 v. Lucero,* 526 P.2d 825 (N.M. 1974); *Karback v. Board of Educ. of Comm. Unit School Dist.,* 428 N.E.2d 1126 (Ill. 1981).

[13] See, *O'Hair v. Board of Educ,* 805 P.2d 40 (Kan. Ct. App. 1990).

[14] See, R.C. Wood, "Financial Exigencies and the Dismissal of Public School Teachers: A Legal Perspective," *Government Union Rev.*, III (Fall, 1982), 49-66.

[15] *See, in general, Childers v. Independent School Dist.*, 842 P.2d 355 (Okla. Ct. App. 1992).

[16] K.W. Nickoles, "Future Shock: What's Coming to the Bargaining Table," *School Business Affairs*, Feb., 1990, at 38, see also, J.G. Ward, "Reductions in Force and School Business Administration," *J. of Education Finance, Fall* 1984, 172-195.

[17] 353 N.E. 2d 567 (N.Y. 1976).

[18] Id. at 568.

[19] *Board of Educ. of Yonkers City School Dist. v. Yonkers Fed. of Teachers*, 386 N.Y.S. 2d 657 (N.Y. App. 1976).

[20] Id. at 568.

[21] Id.

[22] *Board. of Educ. v. Lakeland Fed'n of Teachers*, 381 N.Y.S. 2d 5115 (N.Y. 1976).

[23] *Spokane Educ. Ass'n v. Barnes*, 517 P.2d 1362 (Washington 1974).

[24] S.B. Neill and J. Curtis, *Staff Dismissal: Problems and Solutions* (Arlington, Va: American Ass'n of School Admin.) 80.

[25] For a discussion of the termination of classroom teaches for cause see R.C. Wood and G. Baldwin, "Dismissal Process for Classroom Teachers," *The Principal's Legal Handbook,* (Topeka, KS: NOLPE), Chap. 21, 183-192, 1993.

[26] See e.g., *State ex rel. Bing v. Board of Educ.*, 7 N.W. 2d 544 (Minn. 1942); *Foesch v. Independent School Dist.*, 223 N.W. 2d 371 (Minn. 1974); and *Milne v. School Comm. of Manchester,* 410 N.E. 2d 1216 (Mass. 1980).

[27] See e.g., *Bing,* 7 N.W. at 556.

[28] See e.g., *Bruinsma v. Wyoming Public Schools*, 197 N.W. 2d 95 (Mich., 1972), *Clune v. School Dist.*, 166 N.W. 11 (Wis. 1918); *Hall v. School Dist.*, 24 Mo. App. 213 (1887), *School Town of Milford v. Ziegler*, 27 N.E. 303 (Ind. 1890).

[29] See e.g., *Kuhl v. School Dist.*, 51 N.W. 2d 746 (Neb. 1952); *Ashby v. School Township,* 98 N.W. 2d 848 (Iowa 1959); *Neilson v. Lincoln Parish*, 125 So. 458 (La. 1929); *Morey v. Power*, 78 Tenn. 219, (1882).

[30] 555 S.W. 2d 350, (Mo. 1977).

[31] Id. at 352.

[32] *California School Empl, Ass'n v. Pasadena Unif. School Dist.*, 139 Cal. Rep. 633 (1977).

[33] See, R.C. Wood, "Chapter Nine of the Federal Bankruptcy Act: A Fiscal Alternative for School Districts," *J. of Educ. Finance,* v. 11, no. 1, 1985, 56-68.

[34] *Unified School Dist. v. Epperson,* 583 F.2d 1118 (10th Cir. 1978).

[35] Id.

[36] *Bradfute v. Renton School Dist.*, 557 P.2d 147 (Wash. 1978); *See also Cigarski v. Lake Lehman School Dist.,* 407 A.2d 460 (Pa. 1979).

[37] *Baston v. Ricci*, 391, A.2d 161 (Ct. 1978).

[38] *Penzenstadler v. Avonworth School Dist.*, 403 A.2d 621 (Pa. 1979). *See also Freiberg v. Board. of Educ. of Big Bay De Noc School Dist.* 283 N.W. 2d 775 (Mich. 1979).

[39] *Phillippi v. School Dist. of Springfield Township*, 367 A.2d 1133 (Pa. 1977). *See also Andresky v. Allegheny School Dist.*, 437 A. 2d 1075 (Pa. 1981).

[40] *Funston v. Dist. School Bd.* 278 P. 1075 (Ore. 1929).

[41] See also, *Piquard v. Board of Educ*, 610 N.W.2d 757 (Ill. App. Ct. 1993).

[42] *Ehret v. Kulpmont Borough School Dist.*, 5 A. 2d 188 at 191 (Pa. 1939).

[43] 81 So. 2d 817 (La. 1955).

[44] See, *Petition of Dallman*, 470 N.W.2d 148 (Minn. Ct. App. 1991).

[45] Wisconsin does not have a state board of education.

[46] See K.D. Moran, "Governance," *The Yearbook of School Law*, 1979, ed. P.K. Peile (Topeka KS: NOLPE, 1979) 8-9.

[47] See e.g., in re Miller, 17 *Educ. Dep't Report*, N.Y. Comm'r Dec. No 9600 (1978), and *Fair Lawn Educ Ass'n v. Fair Lawn Bd of Educ.* 401, A. 2d 681 (N.J. 1979), for a discussion of local interests versus state interests.

[48] *E.g., Ala. Code* §16-24-8; *Conn. Gen. Stat.* §10-151; *Me. Rev Stat. Ann.* §20-161(5); *Minn. Stat.* §125.12(6A-GG); *Neb. Rev. Stat* §79-1254; *N.C. Gen. Stat.* §115-142(e)(1)(1). *See also, King v. Berkeley Unified School Dist.*, 152 Cal.Rptr. 782 (1979); *in re Fallick,* 18 *Educ. Dep't Reporter.* N.Y. Comm'r Dec. No. 9979 (1979); *Shelton v. Lauderdale County Bd. of Educ.*, 380 So. 2d 835 (Ala., 1980).

[49] *Weisman v. Board. of Educ. of Jefferson City School Dist.*, 547 P.2d 1267 (Colo. 1978).

[50] Id.

[51] *Hartman v. Merged Area VI Comm. College*, 270 N.W. 2d 822 (Iowa 1978).

[52] *Briggs v. Board. of Directors*, 282 N.W. 2d 740 (Iowa 1979).

[53] 377 F.Supp. 227 (Wis. 1974).

[54] Id. at 239.

[55] See. J. Brockner, S. Grover, T. Reed, R. DeWitt, and M O"Malley, "Survivors' Reactions to Layoffs: We Get By With A Little Help From Our Friends," *Administrative Science Q.*, 32, 1987, 526-541.

[56] See, J.K. Oehm, "Enhancing Productivity: A Structured Approach to Downsizing," *School Business Affairs*, Sept. 1991, 22-24.

[57] See e.g., *Fredrich v. Independent School. Dist.* 465 N.W.2d 692 (Minn. Ct. App. 1991).

[58] In general see, N.O. Bolick, "School Budget Blues," *American School Board J.,*Sept. 1991, 34-36.

[59] For a complete discussion of early retirement cost effectiveness and legal considerations see R.C. Wood, "The Early Retirement Concept and A Fiscal Assessment Model for Public School Districts," *J. of Education Finance,* 7 (Winter 1982) 262-276: see also, H.R. Brown and J.T. Repa, "Do Early Outs Work Out?" *School Business Affairs,* Aug. 1993, 12-16; S.L. Jacobson and S. Kennedy, "Holding Off Layoffs, Deferred Salary Leave Programs, *School Business Affairs,* vol. 58, no. 8, 12-15; H.B. Polansky, "Early Retirement Incentive Plans," *School Business Affairs*, Dec. 1990. 26-29.

[60] See e.g., *Gross v. Board of Educ.*, 571 N.Y.S.2d 200 (N.Y. 1991); *see also, Sherek v. Independent School Dist.,* 464 N.W.2d 322 (Mich. Ct. App. 1990).

[61] See e.g., 298 N.W. 2d 371; J*ordahl v. Indep School Dist No. 129,* 225 N.W. 2d 224 (Minn. 1974); and *Hendrickson v. Indep. School Dist.*, 228 N.W. 2d 126 (Minn. 1975).

[62] Id., at 1103-04.

[63] *Foesch,* 223 N.W. 2d 371; *Jordahl,* 225 N.W. 2d 224 (Minn. 1974); and *Hendrickson,* 228 N.W. 2d 126 (Minn. 1975).

[64] See, *Phillips v. South Range Local School Dist.*, 543 N.E.2d 492 (Ohio 1989), see also, *Stone County School Bd. v. McMaster, 573 So. 2d 753, (Miss. 1990).; Dennery v. Passaic Valley Reg. High School, 597 A.2d 559 (N.J. Super. Ct. App. Div., 1991).* See also, *Phillips v. Trustees, 867 P.2d 1104 (Mont. 1994).*

[65] See, *Gross v. Board of Educ.*, 571 N.Y.S.2d 200 (N.Y. 1991). It should also be noted that any early retirement plan cannot be a subterfuge to discriminate against older teachers. The Age Discrimination in Employment Act (U.S.C. §621-634) does allow for bona fide occupational qualifications if they are reasonably necessary for the normal operation of a particular business. Under ADEA, RIF plans are allowed, see, *EEOC v. Chrysler*, 733 F.2d 1189, (6th. Cir. 1984).

[66] *Minn. Stat. Ann.* § 125.12; *N.Y. Educ. Law* §2588; *Cal Educ. Code* §444955.

[67] See, *Anderson v. Corland City School Dist.*, 554 N.Y.S.2d 769 (NY Sup Ct. 1990).

[68] *Dykeman v. Board. of Educ.,* 8th Dist of Cedar City, Nebraska No. 6688.

[69] See *Barnes v. Mendenhall*, 183 N.E. 556 (Ind. 1932); *Watson v. Burnett*, 23 N.E. 2d 420 (Ind. 1939); See also, *Rappold v. Board. of Educ. Cleveland Hills,* 447 N.Y.S. 2d 126 (N.Y. 1982); *O'Hair v. Board of Educ.*, 805 P.2d 40 (Kan Ct. App. 1990).

[70] See, *Holmes v. Board of Trustees*, 792, P.2d 10 (Mont. 1990).

[71] 376 P.2d 968 (N.M. 1962).

[72] 252 A.2d 427 (Me. 1969).

[73] See also, R.S. Davies and G.J. Stenmoe, "Disparate Impact Analysis in Subjective Employment Decisions: Watson v. Fort Worth Bank,: 52 *Educ. Law Rep.*, 9 (May 11, 1989).

[74] *Singleton v. Jackson Mun. School Dist.*, 419 F.2d 1211 (5th Cir. 1970), *cert denied* 390 U.S. 966 (1977).

[75] Id., at 1218.

[76] *Pickens v. Okolona Mun. Separate School Dist.*, 419 F.2d 358 (5th Cir. 1976).

[77] Id.

[78] *Lee v. Chambers County Bd. of Educ.*, 533 F.2d 1353 (5th Cir. 1977), *cert. denied,* 434 U.S. 966 (1977).

[79] 388 F.Supp. 738 (S.D. Texas, 1975), *rev.,* 554 F.2d 1353 (5th Cir. 1977), *cert. denied*, 434 U.S. 966 (1977).

[80] 544 F.2d 804 (5th Cir. 1977).

[81] 569 F.2d 1383 (5th Cir. 1978).

[82] Fifth Cir. Court, No. 78-3757, Dec. 18, 1980.

[83] For the use of teacher evaluations in RIF decisions *see, Mackey v. Newell-Province Comm. School Dist.,* 483 N.W.2d 5 (Iowa Ct. App. 1992).

[84] See e.g., *Barnes,* 183 N.E. 556; *Seidel v. Board. of Educ.*, 168 A. 297 (N.J. 1933); *Flechtner v. Board. of Educ.*, 174 A. 529 (N.M. 1955); *Watson,* 23 N.E. 2d 420; *Pickens v. Keasler*, 82 So.2d 197 (Ala. 1957); *Hankeson v. Board. of Educ.*, 146 N.E. 2d 194 (Ill. 1957); *State ex. rel. Marolt v. Independent School Dist.*, 217 N.W.2d 212 (Minn. 1974); *Contra, Garovoy,* 424 A.2d 297; *but c.f., Aken v. Board of Control of Lake County Area Voc. Center*, 604 N.E.2d 524 (Ill. App. Ct. 1992). An example of one such plan that appears in the literature is, R.D. Carfield, and J.K. Walter, "Teacher Evaluation and RIF-Can There Be Peaceful Coexistence?" *NASSP Bulletin*, Nov. 1984, 48-53. An example from Canada can be found in, S.L. Jacobson and S. Kennedy, "Deferred Salary Leaves in Education: A Canadian

Alternative to Reductions in the Teaching Work Force," *Educational Evaluation and Policy Analysis*, Spring 1992, 83-87.

[85] 287 P.2d 73 (N.M. 1955).

[86] *Fort Summer Mun. School Bd. v. Parsons,* 405 P.2d 366 (N.M. 1971).

[87] *Penasco Indep. School Dist. v. Lucew*, 526 P.2d 825 (N.M. 1974).

[88] *Garovoy,* 424 A.2d 297; *McKee v. Board of Educ. of Watertown*, 627 A.2d 951 (Conn. Ct. App. 1993), for administrator bumping *see, Trotta v. Board of Educ. of Plymouth, 628 A.2d 1343.* (Conn. Ct. App. 1993).

[89] *Costell v. Governing Board of Lee County Special Educ. Ass'n,* 623 N.E.2d 966 (Ill Ct. 1993).

[90] *Babb v. Independent School Dist.*, 829 P.2d 973 (Okla. 1992).

[91] *Pennsylvania Labor Rel. Bd. v. Greater Johnstown Area*, 410 A.2d 1290.

[92] *Board of Dir. v. Teachers Ass'n* 392 A.2d 1097 (Me. 1978).

[93] 82 So. 2d 549 (Ala. 1959). See also *Paradise v. School Admin. Dist* (Maine Sup. Jud. Ct. 1982); *O'Hair v. Board of Educ.*, 805 P.2d 405 (Kan. Ct. App. 1990).

[94] 119 A.2d 43 (Pa. 1956).

[95] See *Silver v. Board. of Educ.*, 362 N.Y.S. 2d 337 (1975).

[96] See *Karback v. Board. of Educ.*, 114 Cal.Rptr. 84 (Ct. App. 1974).

[97] *Higgins v. Board. of Educ.*, 428 N.E.2d 1126 (Ill. 1981).

[98] See *Arcurio v. Greater Johnstown School Dist.*, 582 A.2d 402 (Pa. Commw. Ct. 1990); *Wild v. Board of Educ.*, 560 N.Y.S.2d 563 (App. Div. 1990); *Greene v. Pateros School Dist.,* 799 P.2d 276 (Wash. Ct. App. 1990).

[99] See *Fredich,* 465 N.W.2d 692 (Minn. Ct. App. 1991).

[100] *Lezette v. Board. of Educ.*, 319 N.E.2d 189 (N.Y. 1974).

[101] See also *Frederich* 465 N.W.2d 692.

[102] *Hill v. Dayton School Dist.* 532 P.2d 1154 (Wash. 1975), see also, *Hampson v. Board of Educ.*, 576 N.E.2d 54 (Ill. App. Ct. 1991).

[103] *Jackson v. Randolph County*, 565 So. 2d 641 (Ala. Civ. App. 1990); *Kuta v. Joint Dist.*, 799 P.2d 379 (Colo. 1990).

[104] See *Cole v. Board of Educ.*, 562 N.Y.S.2d 209 (App. Div. 1990). Notwithstanding this issue, see *Theriault v. Board of Educ.*, 626 A.2d 781 (Conn. Ct. App. 1993), wherein Connecticut statute requires a different procedure.

[105] 375 N.E.2d 157 (Ill., 1978). See also *Dinersteein v. Board. of Educ.*, 409 N.Y.S. 2d 22 (App. Div. 1978); and *Leonard v. Board. of Educ.*, 373 N.E.2d 477 (Ill. 1978).

[106] *Deem v. Board of Educ.*, 558 N.E.2d 291 (Ill. App. Ct. 1990).

[107] See e.g., *Mroczek v. Board. of Educ.*, 400 N.E.2d 1362 (Ohio C.P. 1979); *Delagores v. Board. of Educ.,* 410 A.2d 461 (Conn. Super. Ct. 1979); *Nagy v. Belle*

Vernon Area School Dist., 412 A.2d 172 (Pa. Commonwealth Ct. 1980); *Krug v. Independent School Dist.*, 293 N.W. 2d 26 (Minn. 1980); *Jenkintown*, 441 A.2d 816 (Pa. 1982).

[108] *Lenard v. Board. of Educ.*, 384 N.E.2d 1321 (Ill. 1979).

[109] See e.g., *Johnson*, 377 F.Supp, 227; Hill, 119 A.2d 43.

[110] See, e.g., *Stewart v. Fort Wayne*, 564 N.E.2d 274 (Ind. 1990).

[111] See, e.g., *Brady v. Board. of Trustees*, 242 N.W. 2d 616 (Neb. 1976); *Hill*, 119 A.2d 43; *Schnabel v. Alcester School Dist*, 295 N.W.2d 340 (S.D. 1980).

[112] The issues of RIF for school administrators is outside the confines of this chapter. Notwithstanding, *see e.g., Curry v. Dawson County Board of Educ.*, 442 S.E.2d 919 (Ga Ct. App. 1994).

[113] See, *Board of Educ. v. Bowers*, 396 S.E.2d 166 (W.Va. 1990).

[114] *Murphy v. Western Line School Dist.*, 832 F.Supp. 178 (N.D. Miss. 1993).

[115] *Hampton v. Glendale Union High School Dist.*, 837 P.2d 1166 (Ariz. Ct. App. 1992).

[116] See e.g., *Kramedas v. Board. of Educ.*, 523 F.Supp. 1268 (Del. 1981).

[117] For an illustration of a school district that failed in this burden, see, *State ex rel. Bush V. Spurlock*, 588 N.E.2d 840 (Ohio 1992).

[118] *Kent v. Sawyer Public School Dist.*, 484 N.W.2d at 289, (N.D. 1992).

Chapter 11 Accounting

Introduction

Although this text has made the case that public school administrators are faced with many important challenges that must be addressed with great skill and aplomb, there are few areas of administrative leadership that so immediately impact on integrity and confidence as does the accounting function in school districts. Although good administrative leaders are first educators before they are financial experts, it is important to observe that high-quality education will not occur without administrators who are superbly skilled in all aspects of fiscal management. While this book has made the important case that school budgets are fiscal expressions of the educational philosophy of a district, it must be observed that the budget and all educational planning are without practical merit until proper accounting procedures are in place. The accounting function provides the means by which to control the entire educational enterprise.

As stated elsewhere, the tasks of accounting, reporting, and auditing are the tools by which school district budgets are controlled.[1] Despite the fact that educators are seldom trained in accounting, administrators must nonetheless be able to track revenues and expenditures in order to uphold the public trust and to use accounting information to make fiscal decisions, thus enabling the educational enterprise to function properly. Given the central role of accounting in school business administration, this chapter identifies the role of accounting in schools, offers a broad view of the accounting base of school districts, and examines the structure of funds and account groups in order to help administrators understand the nature of accounting.[2] Proficient skills in accounting are needed in order to interpret the school district's financial position to faculty and staff and to school boards and voters. Accounting tells school business administrators what resources are available, how they become financial resources of the district, how resources are converted into usable assets, how use of resources is planned, and how to actually use resources in carrying out the educational mission of schools.

Accounting Functions in School Systems

Various accounting functions are used in accounting for school funds. Among them are financial management, internal auditing, the annual outside independent audit, and audits conducted by governmental agencies. These functions are based on a set of purposes for an accounting system that provides a point of entry into discussion of the more technical aspects of accounting functions in school systems.

Purposes of the Accounting Function[3]

As will be seen later in chapter 12, the purpose of a budget is to state the educational goals of a district in fiscal terms. A supporting element of budgeting is the accounting system that tracks and reports budgetary activity. To reach that goal, the accounting function has specific purposes of its own that assist in operationalizing the educational goals of the district in financial terms. Based on the foregoing discussion, the first purpose of an accounting system is to provide a vehicle for accumulating, categorizing, reporting, and controlling the fiscal transactions of a school district. Accumulating transactions requires a method of collecting fiscal data on the district's financial activities in a central archive. Categorizing transactions implies that the fiscal activities of a district have commonalities and differences, and it requires segregating fiscal transactions into similar information groups. Reporting transactions first requires accumulation and categorization, all of which combine under a reporting method in order to understand the results. Schools are continually faced with finite resources that are seemingly never adequate to meet all the needs of children. The reality of finite moneys underscores the importance of fiscal control.

The second purpose of the accounting function is to create mechanisms that assist districts in estimating how well they are achieving the stated goals. Although most educators tend to believe that goal attainment refers to standardized testing and other indicators of student achievement, it is nonetheless true that accounting is an educational achievement tool because no amount of good planning can overcome a basic lack of fiscal resources or the wasteful use of existing funds. Since all resources in a school district, including teacher expertise, are purchased commodities, the accounting function can assist in gauging goal attainment by tracking the revenue and expenditure activity of a school district, which in turn yields a useful assessment of the district's financial condition. For example, the accounting function is used to monitor changes in the fund balances; if only 50 percent of the amount budgeted for a fixed expense is remaining by the second month of the school year, the accounting function will flag a serious problem. In contrast, expending all textbook funds at the start of the fiscal year may not send the same message. The accounting function can also help judge progress in academic achievement by linking fiscal information to student achievement data.[4] A major purpose of the accounting function therefore relates to fiscal and policy statements at various levels about expenditures and educational programs.

The third purpose of accounting is to report to state legislatures that school districts are meeting the constitutional and statutory educational responsibilities. As noted earlier in chapter 6, school districts are extensions of the state wherein districts must show that they are satisfying the state's constitutional obligation for the education of children. Development of state aid has only increased this responsibility and has resulted in enormous amounts of fiscal information that must be reported to state agencies in order for the state to track school revenue and expenditure patterns and to provide a thoughtful basis for sending equitable state aid to school districts. The accounting function is the vehicle by which such reporting and decision making occur. Uniform data, therefore, are required so that the state can accumulate comparable data on all school districts. Additionally, states use fiscal data from the accounting function to meet obligations for federal reporting wherein eligibility for federal grants depends on good data.

The fourth purpose of accounting is to provide the historic data needed in annual budget preparations. As seen next in chapter 12, school districts budget by assigning money to line items in the district budget so that the district's educational plan can be carried out. The accounting function contributes to budget building by providing historic data, and it also creates the current data that will become the historic data for the next budget cycle. Without the accounting function, no budgeting data method would be available if districts were not required to budget to accountable funds and individual lines within those funds. There would be no way of determining where moneys were located and how moneys were spent. In fact, educational planning would be an aimless exercise if budgeting and accounting structures were not in place. Just as budgeting requires an accounting structure to enable the educational plan, accounting requires a budget from which to report the fiscal information that describes the educational plan against which progress should be measured.

The fifth purpose of accounting is to provide proper handling and use of the public trust inherent to tax supported institutions. There is no greater trust than the responsibility that falls to school administrators because the public treasury and the lives of children depend on honest and competent management of fiscal resources. Many naive or careless accounting practices in schools can be observed, as experienced administrators can remember instances of loose or shoddy handling of activity funds, lunch moneys, expense accounts, and wrongful diversion of supplies from the proper use. The accounting function provides the structure required for proper handling of the public trust and provides the tripwires by which errors and wrongful acts are identified and corrected. The importance of preserving public confidence cannot be overstated, and it is precisely the accounting function that aids this purpose.

The purpose of accounting is thus planning and stewardship. Accounting is the means for structuring and organizing the budget, and the budget is the object around which the accounting function is structured. Accounting is also the vehicle for monitoring school district fiscal health, and educational goal attainment can also be estimated through the accounting function because accounting permits school district administrators to methodically monitor the flow of resources on a constant basis. The role of accounting is summarized as:

- Providing a complete record of financial transactions;
- Summarizing financial transactions in financial reports required for effective and efficient administration;
- Providing financial information used in budget preparation, adoption, and execution;
- Providing financial controls or safeguards;
- Providing a basis for the governing board to place administrative responsibility and to minimize waste, carelessness, inefficiency, and possible fraud;
- Providing clear and concise financial reports to the public as a basis for judging past, present, and future operation; and
- Providing an historical record that can be studied and analyzed for the purpose of aiding patrons, the board, and administrators in keeping pace with changes in education.[5]

Financial Management

As seen in the introduction of this chapter, financial management is the process of dealing with problems related to receiving the maximum revenue available to school districts while safeguarding all financial resources and controlling expenditures and financial commitments throughout the fiscal period. Thus, business administrators are the employees of the district who are primarily responsible for operating the school system within its financial resources and limitations as directed by policy of the board of education.

To illustrate the kinds of financial statements that the accounting function is capable of producing, illustrations have been chosen from some of the comprehensive annual financial reports (CAFR) that earned the 1994 *Certificate of Excellence in Financial Reporting* awarded by the Association of School Business Officials International (ASBO). It is beyond the scope of this chapter, however, to discuss all underlying public school accounting principles and financial reporting procedures in the detail required to actually perform the tasks. Rather, the goal is to inform or to enhance the competencies of administrators regarding the accounting function.

Internal Auditing

The concept of auditing is a familiar term to administrators, but the goals are not commonly understood. Various kinds of audits are performed in all school districts across the nation and generally fall into three kinds of audits, each designed for a specific purpose. At the local level, the *internal audit* refers to the accounting activities that are performed on a daily basis. Conducted by employees of the school district, this function addresses safeguarding the school district's financial resources at the most basic level. The internal auditor, as an employee of the district, and members of the internal audit staff monitor the operation of the accounting system on a regular and ongoing basis. Sometimes referred to as continuous auditing, the internal auditor reports to the board of education or the chief school officer and recommends ways by which the district can operate an effective system of internal accounting and financial controls capable of producing timely, meaningful, and reliable interim financial reports. The system of internal control and the work of the internal auditor are also primarily responsible for detecting and minimizing controllable losses and detecting fraud and embezzlement by employees of the school district.

All school systems engage in some amount of internal auditing. For example, financial reports provided to board members before each meeting are a type of internal audit because district employees had to engage in auditing to produce those reports. Producing a monthly budget status report requires maintaining electronic journals and ledgers for which income and expense statements were produced. Similarly, financial records maintained for state reporting requirements are based on internal auditing. Likewise, internal auditing occurs through the system of checks and balances used by most districts wherein multiple layers of approval are needed to expend funds. An encumbrance system that requires determining that the proper accounts and cash balances are in order before making a financial commitment on behalf of the district is a form of internal audit. Although internal auditing is not sufficient to guard against all error, internal auditing is an important element of good fiscal management.

The Annual Independent Audit

A second kind of audit is the *annual independent audit.* This is a function that relies heavily on the effectiveness of the school district's system of internal accounting and financial controls. The annual audit is conducted by one or more outside independent accountants who are qualified by laws of the appropriate state to examine the financial statements of the district in accordance with generally accepted auditing standards. These accountants are qualified by education, experience, practice, and professional examination to express, qualify, or deny an opinion on the district's comprehensive annual financial statements. They state whether in their opinion the annual financial statements are presented in accordance with Generally Accepted Accounting Principles (GAAP) applied on a basis consistent with the prior year.[6]

As a general rule, school districts preparing for an audit anticipate a general comprehensive audit of all financial records in the district. General comprehensive audits are designed to review all fiscal records in the district in order to test for sound accounting procedures and good fiscal management. In a comprehensive audit, the school district may expect auditors to be present for several days and to request copies of all financial records and supporting materials. These items are the tools used by auditors to perform the various tests that will permit objective and unbiased examination of the fiscal integrity and accuracy of the district. A general, comprehensive audit will result in a formal written report by the independent auditors. The report is meant to provide a permanent record of the audit results, to identify problems that must be corrected, and to make recommendations for improving the system. The report contains summaries of revenue and expenditures and compares cash balances against encumbrances to determine whether statutory requirements are observed. The report also gives results of examination of the financial records of all budgetary accounts, wherein each individual fund operated by the district is reviewed and reported separately.

The table of contents of the CAFR winning ASBO's *Certificate of Excellence* award is shown in figure 11.1 and illustrates the nature of reporting and auditing. As seen in figure 11.1, the comprehensive audit provides statements and reports on the district's various financial records. If no items of particular concern are noted, the audit report provides the financial statements listed in the table of contents, accompanied by a letter noting any exceptions to good practice. An audit report noting no deviations from good accounting practices is known as an *unqualified opinion* because it does not contain any *audit exceptions*. If concerns are raised, the letter is known as a *qualified opinion* because it sets out audit exceptions stating reservations about accounting practices. Generally, boards of education are expected to reflect receipt of the audit report in board minutes and to take action to correct deficiencies in order to comply with state law and the nature of public trust.

The comprehensive audit is meant to assure full compliance with the law and to improve financial and accounting operations. Board members, administrators, and the public are entitled to absolute conformance with the law, and administrators in particular need to be certain that auditing occurs. Being subjected to examination by independent accountants who are not employees of the school district or other governmental agency tends to safeguard the integrity of the school district's financial administrators. The audit report is a major report used by the board of education and the public to determine that the fiscal resources of the school district have in fact been used for the legal and intended purposes. Creditors also rely on the audit report as a basis for extension of credit such as in the case of bond ratings and establishing interest rates and creditworthiness for other long-term debt.

Audits Conducted by Governmental Agencies

A third kind of audit that occurs in school districts is known as a *governmental agency audit*. Usually these audits arise from a state interest in the district's resources. A general perspective on such audits can be stated by noting that each governmental agency providing revenue to school districts through distribution of taxes and commodities may require its own audit staff to examine whether the fiscal resources of the district are restricted in scope to those funds receiving revenue from taxes and commodities from that unit of government. These audits serve an entirely different purpose from general audits in that they are designed to test compliance with statutes and regulations involving distribution of state or federal moneys. Differences among states cause government audits to vary considerably, but in general these audits seek to determine whether the state's financial interests in education are protected. Governmental audits are thus built on auditing the various funds to which the agency supplies aid directly or to which intermediate units of government act as channels for federal aid to local districts. In the case of federal funds, states are interested in proper records of applications, expenditure reports, and transmittal documents. In the case of state aid funds, states are interested in all documents related to state aid claims.

A different kind of governmental audit may also arise. School districts are sometimes subject to *special audits*. The purpose of a special audit usually relates to error or fraud in the financial operation of a school district. Such audits are not a regular event, but they may be brought to the attention of officials through a variety of ways. The annual comprehensive audit may flag conditions leading to a special audit, and internal auditing can also alert a district to the need for a special audit. The possibilities leading to special audits are unlimited, but as a general rule the accounting function in school districts centers on normal internal and external comprehensive auditing.

Unique Characteristics of Accounting for Public School Systems

Public school accounting is actually a unique field of accounting. Although the uniqueness form the basis of how this chapter discusses accounting in school districts, there are nonetheless similarities with other fields of accounting. For example, the theory of debits and credits is used in virtually every accounting field, and textbooks used in university introductory accounting courses traditionally present debits and credits as applied to profit enterprises, i.e., proprietorships, partnerships, and corporations. The nature of nonprofit organizations, however, has forced recognition that several principles of accounting do not provide a good fit. As a result, basic accounting theory has been expanded to create a governmental accounting structure, a financial structure that has become generally accepted as the most appropriate basis of ac-

Figure 11.1
Table of Contents of a Comprehensive Annual Financial Report, Year Ending June 30, 1994

INTRODUCTORY SECTION

Letter of Transmittal ..... vii-xi
Board Members and Superintendent xii
Organizational Chart xiii
Certificate of Achievement for Excellence in Financial Reporting – GFOA xiv
Certificate of Excellence in Financial Reporting – ASBO xv

FINANCIAL SECTION

REPORT OF INDEPENDENT CERTIFIED PUBLIC ACCOUNTANTS 3

GENERAL PURPOSE FINANCIAL STATEMENTS
Combined Balance Sheet – All Fund Types and Account Groups 7
Combined Statement of Revenues, Expenditures and Changes in Fund Balances – All Governmental Fund Types and Expendable Trust Fund 8
Combined Statement of Revenues, Expenditures and Changes in Fund Balances – Budget and Actual (Non-GAAP Budgetary Basis) – All Governmental Fund Types 9
Combined Statement of Revenues, Expenses and Changes in Retained Earnings – All Proprietary Fund Types 10
Combined Statement of Cash Flows – All Proprietary Fund Types 11
Notes to General Purpose Financial Statements 13-19

COMBINING AND INDIVIDUAL FUND AND ACCOUNT GROUP STATEMENTS AND SCHEDULES
General Fund
Balance Sheet 24
Schedule of Revenues, Expenditures and Changes in Fund Balance 25-27
Schedule of Revenues, Expenditures, Encumbrances and Changes in Fund Balance – Budget and Actual (Non-GAAP Budgetary Basis) 28-29
Enterprise Funds
Combining Balance Sheet 32
Combining Statement of Revenues, Expenses and Changes in Retained Earnings 33
Combining Statement of Cash Flows 34
Schedule of Revenues, Expenses and Changes in Retained Earnings – Budget and Actual (GAAP Basis) 35

STATISTICAL SECTION

counting for tax revenue received by federal, state, and local units of government. This structure has been further modified by ASBO to make it a useful basis of financial and managerial accounting and reporting for school systems.

ASBO has, in effect, modified the governmental accounting structure to focus on a fund structure that meets the needs of school systems by eliminating special assessment funds, a fund concept widely used by other local and state units of government, but that does not apply to education. Instead of referring to a certain group of funds as *governmental-type* funds, ASBO recommends that this group of funds be referred to as *budgetary funds*. In effect, the theory of debits and credits has been expanded to apply not only to asset, liability, capital, income, and expense accounts as used by profit entities, but also to include budgetary accounts such as estimated revenue, expenditure, encumbrance, and appropriation accounts—all of which better reflect the non-profit tax structure of how schools receive and spend public moneys. A second example of useful changes to basic accounting principles relates to the account structure used for the general fixed assets account group, the general long-term debt account group, and agency funds. These three areas have been restructured to be significantly different from the account structure of other funds. The terminology for capital accounts used by business entities has been changed to *fund equity* or *fund balance*, and titles for income and expense accounts are changed to *revenue* and *expenditure* accounts in some funds. Additional accounts, frequently referred to as *budgetary* accounts, are used in budgetary funds to manage financial resources and to determine what administrative actions are needed to carry out the budget adopted by the board of education for the fiscal period.

Budgetary Accounts and Budgetary Funds

Budgetary accounts used for management purposes by school systems serve a useful purpose by comparing:

- *The budget's estimated revenue with revenue actually realized to date.* The difference between estimated revenue and actual revenue is the amount to which estimated revenue has not been realized, or the amount that actual revenue has exceeded the budgeted revenue estimate to date. When this comparative procedure is used, the variance between amounts budgeted and the amounts actually realized causes management to ask what additional administrative actions are needed if the district were to receive all its expected revenue from these sources.

- *Expenditures and encumbrances with legal limits established by appropriations.* When expenditures or encumbrances exceed the legal limit, one or more unauthorized transactions have taken place that can result in an operating deficit or a fund deficit. This comparison, when frequently made during the fiscal period, assists management in avoiding unauthorized expenditures and financial commitments, thereby creating either operating or fund deficits.

Encumbrances are contractual commitments in the form of purchase orders for goods or services that have not been filled. When the order is filled, it becomes an expenditure. In budgetary funds it is necessary to record encumbrances to prevent issuance of co¡ntractual commitments that could result in expenditures in excess of legal limits. An interim statement comparing budget (estimated) revenue with actual revenue to date and comparing expenditures and encumbrances with appropriations (legal limits) is presented in figure 11.2.

Other Terminology Changes

As noted, in school accounting some funds use the term *expense account* as it is used by a business entity, but in other funds different terms are used. For example, in a budgetary fund the term may be broadened and called an *expenditure account*. In funds of this type not only are entries made to record usual current expenses of the fund, also to record the fund's current expenses to acquire general fixed assets, current expenses caused by maturity of long-term debt, and current interest expense incurred on long-term debt. Transactions of these kinds are identified as expenditures and are recorded in the expenditure account.

Similarly, the fixed assets acquired by expenditure of budgetary funds are neither recorded as fixed assets of budgetary funds nor long-term liabilities of the school district recorded as liabilities in budgetary funds. Instead these entries are recorded in the general fixed assets account group and in the general long-term debt account group. Interest on long-term debt paid by a budgetary fund, however, is recorded as an expenditure.

Accounting procedures used to close expense and income accounts at the end of the fiscal period by a business entity are also used to close the revenue, expenditure and budgetary accounts of school systems, thereby making these accounting procedures consistent with accounting methods used by other types of financial and business entities.

The Fund Structure: A School System's Financial Structure

All organizations handling money need a uniform system by which to keep records of transactions. A business organized for the purpose of profit usually has only one group of self-balancing accounts known as its *general ledger* with adequate supporting *subsidiary ledgers* containing necessary subsidiary ledger accounts. Accounting for school systems, however, uses a more complex financial structure identified as a *fund* structure. By this accounting

Figure 11.2
General Fund Interim Statements—Actual and Appropriations

General Fund Interim Statements Comparing Budget Revenue with Actual to Date and Appropriations with Expenditures and Encumbrances to Date (For __________ month period ending* __________)

	See Schedule Number	Budgeted Estimated	Actual to Date		Unrealized Balance
Revenue:					
From local sources	2	$1,658,192	$1,647,533	Under	$ 10,659
From state sources, direct	2	1,763,835	1,773,346	Over	(9,511)
From state sources, redistribution of federal funds	2	35,595	26,509	Under	9,086
From intermediate sources, redistribution of federal funds	2	1,275	—	Under	1,275
From federal sources	2	2,874	9,200	Over	(6,326)
From other school districts in the state	2	23,453	25,171	Over	(1,718)
Total revenue	2	$3,485,224	$3,481,759	Under	$ 3,465
		Appropriations	**Expenditures**	**Encumbrances**	**Unencumbered Balance**
Expenditures:					
Instruction—elementary	3	$1,782,132	$1,467,843	$156,824	$157,465
Instruction—secondary	3	817,006	783,984	31,098	1,924
Instruction—special education	3	98,471	98,471	—	—
Instruction—summer school	3	23,000	—	23,000	—
Instruction—Title I	3	23,601	23,400	—	201
Instruction—federal grant	3	6,118	6,118	—	—
Administration	3	134,604	126,898	7,706	—
Transportation	3	58,486	51,742	5,000	1,744
Operation of plant	3	332,756	328,642	3,675	439
Maintenance of plant	3	36,038	34,750	985	303
Fixed charges	3	106,430	99,850	6,400	180
Capital outlay	3	26,415	26,415	—	—
Debt service	3	28,120	28,120	—	—
	See Schedule Number	**Appropriations**	**Expenditures**	**Encumbrances**	**Unencumbered Balance**
Tuition to other school districts in the state	3	2,047	2,047	—	—
Total expenditures		$3,475,224	$3,078,280	$234,688	$162,256
Excess revenues over expenditures		$ 10,000			

*Table is taken from *Financial and Managerial Accounting For Elementary and Secondary School Systems* by S. B. Tidwell, Reston, Va.: Association of School Business Officials, 1984.

convention, the financial structure of the organization is composed of a number of *funds* and *account groups*, each having its own general ledger with adequate supporting subsidiary ledgers. Each fund and account group used by the school system is designed to serve certain but limited purposes. The fund structure recommended by ASBO for use by school districts reflects those limited and specific purposes:

Budgetary funds

- General fund.
- Special revenue funds.
- Capital project funds.
- Debt service funds.

Account groups

- General fixed assets account group.
- General long-term debt account group.

Proprietary type funds

- Enterprise funds.
- Internal services funds.

Fiduciary type funds

- Trust funds, including both expendable and nonexpendable trust funds.
- Agency funds.

Each fund and account group listed has its own separate self-balancing group of accounts with underlying books of record that are maintained by some combination of handwritten and electronic record keeping.

Purpose of Each Fund and Accounting Group

The overarching fund structure provides the vehicle by which all moneys received into school districts can be deposited and expended by original revenue intent. All funds and account groups used by a school system must be created and abolished by recorded actions of the board of education. Because each fund and account group has a specific and restricted purpose, it is important to know the role of each. Brief descriptions of the purposes of the funds and account groups recommended by ASBO for use in school districts are as follows:.

Budgetary Funds

General Fund

The general fund is the most important of the funds and account groups. It is created at the inception of the school district and continues throughout the life of the school system. As implied by its name, its purpose is to provide the basic educational program encompassing a wide range of revenues and expenditures for general purpose use.

Special Education Funds

More appropriately called special revenue funds, these funds are useful when a temporary source of revenue is available to a school district to provide a special kind of educational program that is beyond the scope of the basic education or foundation program offered by the general fund, or for a program where the cost is beyond the financial resources available in the general fund. When revenue from a temporary source ceases to be available for a special revenue fund, the fund is abolished by action of the board of education.

When temporary sources of revenue become relatively certain to be provided in future years, the "special" kind of educational program may be declared by the board of education to be part of the district's basic education or foundation program, and the need for a separate special revenue fund may cease. The accounts needed to continue to identify the cost of this added basic education program may become a part of the school district's general fund. In most states, special education funds for handicapped children should not be confused with special revenue funds because traditional special education moneys are set out separately due to federal and state mandates.

Capital Projects Funds

Capital projects funds become useful to a school district when large amounts of moneys are available from federal, state, and/or local revenue sources for to acquire specific long-lived, general fixed assets such as land, buildings, machinery, classroom equipment, and so forth. A capital projects fund may be created for a project when all sources of the fund's revenue are restricted to only that project. A group of general fixed assets may also be acquired by a capital projects fund when all sources of the fund's revenue are restricted to that group of general fixed assets.

Expenditure accounts required to record the school district's acquisition of general fixed assets are closed at the end of the life of the capital projects fund. The general fixed assets acquired are not recorded in the asset accounts of capital projects funds, but instead are recorded in the general fixed assets account group where they continue to be maintained as long as they are owned by the district.

When all assets of a capital projects fund have been spent for the purposes specified by the fund's intent and when all fund liabilities have been paid and the general fixed assets have been acquired, any remaining cash or other current assets may be transferred by appropriate action of the board of education to another fund in the district. Assets may also be returned, if required, to the sources providing the resources in the capital projects fund, or they may be disposed in accordance with provisions specified by each of the sources at the time they were provided.

Debt Service Funds

A debt service fund is used to accumulate the assets necessary to pay off a specified general long-term debt of the school system when that debt matures and becomes due. A separate debt service fund should be created for each bond issue or long-term debt that has been authorized and outstanding.

When issuance of long-term bonds is authorized, an asset accumulation schedule, frequently referred to as an amortization schedule, is prepared showing the amounts of revenue that must be provided each year to pay off the debt when it matures. When the debt is authorized by the board of education, the accumulation schedule becomes the authorization or basis for the budgetary entry.

The debt service fund's budgetary entry each year records the amount of revenue that is required from property taxes or other taxable source and the amount of revenue required from interest on the debt service fund's investment if the tax burden were to be spread equitably and if the principal amount of bonds were to be accumulated each year over the life of the outstanding obligation.

The debt service fund's equity is restricted to retirement of the principal amount of the debt and, as such, is appropriated each year. The account title used to identify it as having been appropriated by action of the board is *reserve for debt retirement*. The balance in the reserve for debt retirement should correspond at the end of each year with the amount required by the accumulation schedule.

At the end of each year, a variance may exist between the amount of revenue that should have been provided and the amount of revenue actually provided. As a result, a *retained earnings* account is used to show the amount by which the debt service fund's equity exceeded or fell short of the accumulation schedule's requirements as shown in the reserve for debt retirement account. If earnings exceeded requirements, a credit balance will appear in the retained earnings account. If earnings fell short of the fund's requirements as shown by the balance of the reserve for debt retirement account, a debit balance will appear in the retained earnings account. This debit balance shows a need for administrative and/or board action in order to avoid default when the district's outstanding bonds mature.

Account Groups

General Fixed Assets Account Group

Tremendous amounts of money are invested by school districts in general fixed assets. These are long-lived assets such as land, building, machinery, and equipment that are acquired by the school system either through the general fund, special revenue funds, capital projects funds, or by transfer from proprietary type funds from some source outside the district. Because general fixed assets are expensive to replace, the general fixed assets account group is maintained to provide safeguards tending to minimize and disclose losses if they do occur, regardless of cause.

Any sale or disposition of general fixed assets should be authorized by action of the board of education or, as permitted under state statute, its designated representative. Proceeds from the authorized sale of any general fixed asset can then be traced to the fund that was designated by the board to receive it. The result of such careful accounting tends to minimize the possibility of unauthorized use or disposition of proceeds from any sale of valuable assets.

General fixed assets are intended to be used anywhere in the school system in the educational process, as distinguished from fixed assets used in the operation of an enterprise fund, an internal service fund, or a trust fund. The nature of each fixed asset determines the kind of internal controls required to assure taxpayers that the assets are being used for the original intended purpose for which the assets were acquired. Auditors reconcile expenditures from the general fund, special revenue funds, and capital projects funds for general fixed assets, as well as transfers of fixed assets from proprietary type and fiduciary type funds with the general ledger accounts and the subsidiary ledger accounts of the general fixed assets account group. Without the use of the general fixed assets account group maintained on a basis consistent with generally accepted accounting principles during the fiscal period, large financial losses could occur that would otherwise be prevented. Accounts that record depreciation of general fixed assets may also be used in this account group.

General Long-Term Debt Account Group

This account group is used when general long-term debt instruments are issued, usually in the form of long-term bonds or notes payable by amortization or on demand. This group of accounts focuses attention on the kind of long-term debt that is outstanding and on the amount, such as ten-year, 8 percent serial bonds issued by the district. This account group also shows the amount that must be provided by the district in future years to meet its payments and the amounts already paid in previous years for retirement of the principal of the debt. For management purposes, this account group can also show the total amount of interest payable in future years and the amount needed each year to pay interest. This procedure minimizes the possibility of overlooking the amount required for payment of interest and assures that each year's budget includes provisions for interest payments. No other accounts are maintained in this account group, as it has no assets, fund equity, revenue, or expenditures.

Proprietary Type Funds

Enterprise Funds

Accounting for enterprise funds parallels accounting for a business entity, in that enterprise funds are created to produce goods and services for the school system that may also be sold to the public. This fund structure is generally used for school food services, student bookstores, vocational arts programs with project receipts such as automotive repair, athletic programs, and theater presentations. The purpose of such a fund rests in recognizing that, unlike other revenues, income resembling a commercial enterprise operation forms the basis of receipts and disbursements.

If a product were bought and sold in its original form, this fund will need a structure that includes asset accounts, including accounts to record its own fixed assets and related accumulated depreciation, accounts to record current and long-term liabilities, capital, retained earnings, income, cost of goods sold, and general and administrative expenses. Other income and expense accounts record such things as interest income and interest expense where applicable.

If the enterprise fund purchases raw materials, places them in a production process such as manufacturing in vocational arts programs, adds labor costs and other production costs and sells the products to the public, either a job order cost accounting structure or a process cost accounting structure can be effectively used.

Internal Service Funds

An internal service fund recognizes still another type of revenue and expenditure in school districts. When applicable, an internal service fund is created to account for the cost of producing either goods or services that are meant to be used exclusively by the school system. This concept is useful to school districts that have central operations such as purchasing, warehousing and distribution, computer services, garage and transportation services, maintenance, printing services, and so on, in the sense of choosing to stay "in-house" as opposed to buying a service or commodity externally. This fund structure makes it possible for the school system to determine whether it is to its advantage to produce its own goods or services or to buy them from outside contractors. The accounting structure for internal service funds is comparable to the accounting structure needed for enterprise funds. The major difference in these two types of funds is that the goods or services produced by an internal service fund are exclusively used with the district itself.

Fiduciary Type Funds

Many districts also control moneys that has been given to the district for purposes relating to the nature of gifts and bequests. Generally such moneys are referred to as *fiduciary* or *trust* funds. Moneys held by a school district in trust have been placed with the board of education as trustee under the provision of a trust agreement. As a trustee, the board has duties and responsibilities not only as custodian for the assets held in trust, but also for management of the trust fund's assets, liabilities, principal, and income and expenses.

Trust funds fall into either of two types. One is a nonexpendable trust fund, and the other is an expendable trust. A trust fund may be created by the board from its own resources, but most often a trust fund is created by a gift to the district that is to be used for purposes specified by the trust agreement. Use of a separate fund for each nonexpendable trust fund and a separate fund for recording earnings and expenses of each expendable trust fund tends to minimize the possibility of depleting the trust principal of a nonexpendable trust fund. Each trust fund created should be authorized by specific action of the board of education.

Nonexpendable Trust Funds

A nonexpendable trust fund, by its nature, requires the board of education to act as administrator of the fund and to take whatever action is necessary to keep the principal of the fund intact. As such, only earnings on the principal are meant to be accessible for expenditure purposes. As trust administrator, the board is required to decide when to buy and sell investments of the type specified in the trust agreement.

A nonexpendable trust fund has asset accounts for cash and for each of its investments and liability accounts when necessary. The fund equity account is identified as the nonexpendable trust fund principal. Theoretically, the principal amount will remain intact. A nonexpendable trust fund therefore does not have revenue and expense accounts that could increase or deplete the trust fund principal. If, however, a loss is sustained in the assets of a nonexpendable trust fund, thus causing a reduction in principal, the trust agreement may specify that such losses be eliminated by transfer from the expendable trust fund. When a trust fund is created, a provision should be made in the trust agreement that gives the board of education some method of disposing of the nonexpendable trust fund principal in the event that operation of the fund becomes impossible.

Expendable Trust Funds

In contrast, an expendable trust fund may be created as a result of a transfer from any other fund of the school district that is authorized by the board or by earnings from investments held in the nonexpendable trust fund. An expendable trust fund has asset, liability, a retained earnings account, and revenue and expense accounts. The retained earnings account reflects the amount of net earnings that are pending disposition according to the terms of the expendable trust fund agreement.

Agency Funds

An agency fund is another type of fiduciary fund that is used for an entirely different purpose. The assets of an agency fund are held by the district as an agent rather than as an administrator or manager. Thus, the board of education has only custodial powers and duties rather than administrative and management responsibility as with accounting for trust funds. Therefore, an agency fund should have only asset and liability accounts, i.e., it has no equity accounts. The agency fund is useful when a school district operates a central payroll fund and a central treasury fund or when money is collected from a class, person or group, and is held for another person, group or organization.

The Central Payroll Fund as an Agency Fund

A central payroll fund, operated as an agency fund, is a device used to simplify the account structure of all funds of a school district by reducing the number of accounts used to record payroll expenses and related assets and liabilities. Absent such convention, numerous payroll accounts would be required—with a central payroll fund, however, only one payroll account is required, although records of transfers of amounts from specific funds are kept in order to be able to audit the appropriateness of expenditures from the various individual funds. As a result of a central payroll fund, overall financial reporting can be simplified to some extent. To do this, several accounts must be set up in the central payroll agency fund. These accounts are:

- Cash
    - Due from the general fund
    - Due from (each) special revenue fund
    - Due from (each) capital projects fund
    - Due from (each) enterprise fund
    - Due from (each) internal service fund
    - Due from (each) expendable trust fund
- Salaries payable

Federal income tax withheld and payable

Social security tax withheld and payable

State income tax withheld and payable

City income tax withheld and payable (where needed)

Employee retirement deductions payable

Group hospitalization insurance premiums withheld and payable

Union dues withheld and payable

Savings bond deductions withheld and payable

Other liability accounts (where needed)

Salary and wages expenses

Employer's social security tax expense

Employer's state unemployment tax expense

Employer's federal unemployment tax expense

Employer's retirement contributions expense

Workers' compensation insurance expense

Employer's group hospitalization insurance expense

Employer's group life insurance expense

The staff of the central payroll department, based on salary and wage authorizations by administrative officers of each fund employing people, makes all salary and wage computations using time cards or attendance reports submitted by each employing fund. In addition, the staff of the central payroll department computes the amounts to be withheld from salaries and wages, as well as the employer's payroll tax expenses for each fund. This centralizes the need for a trained staff position in payroll administration. The central payroll department maintains records regarding individual earnings, sick and annual leave, and all other payroll documents. It issues payroll checks to all employees and pays all payroll liabilities when due. Instead of each employing fund having to maintain accounts for payroll liabilities and employer expenses, this system reduces the number of accounts in the general ledger and financial reports. For example, the number of accounts needed in the general fund to record payroll could be reduced to the following through the use of a central payroll fund:

- Salaries and wages (classified adequately for management purposes).

- Employment expenses (classified adequately for management purposes).
- Due to central payroll fund.

Responsibility for determining the qualifications, administering the functions of hiring, dismissing and promoting personnel, reporting employee attendance, granting annual leave or sick leave, and raising moneys necessary to pay for personnel services remains the administrative responsibility of each budgetary, proprietary, and trust fund that uses the services of the central payroll department.

The Central Treasury Fund as an Agency Fund

The same advantage of economy in accounting structure can be gained from establishing a central treasury fund. Many advantages can result from centralizing the school district's treasury function, i.e., the function of collecting, temporarily investing, and disbursing all cash.

- The cash position of the school district can be determined at any time for the district as a unit, rather than by individual funds.
- Only one person operating with adequate internal financial controls is needed to manage the temporary investment of available cash that would otherwise be idle. This procedure tends to assure the maximum interest revenue that can be earned by the district through temporary investments of idle cash.
- All disbursements are made over the signature of the district treasurer, giving assurance that each disbursement is supported by an effective voucher system wherein each payment is supported by adequate and competent business documentation.
- Income earned on temporary cash investments in the central treasury can be distributed to each fund according to the average cash balance maintained by each fund in the central treasury. This procedure relieves each fund of the need to hire personnel who otherwise would be needed to manage cash and temporary cash investments.

Setting up a central treasury fund also requires establishing several accounts to transact the treasury function. Accounts needed in a central treasury fund operated as an agency fund are:

- Cash in bank.
- Due from (individual) fund.
- Temporary investments.
- Interest receivable.

- Due to (individual) fund.
- Interest earned distributed to school district funds.

The account needed by each fund using the services of the central treasury is *Due from Central Treasury*, in lieu of a cash account. Since the central treasury serves a banking function for each fund, each fund should prepare a bank reconciliation at the end of each month to confirm the accuracy of the fund's claim against the central treasury and to detect any errors that may have occurred.

Books of Record

The process of accounting is fundamentally one of keeping accurate and detailed records of fiscal transactions in all aspects of a business or nonprofit organization. Records require permanency, and the vehicle to this end is referred to as *books of record*. Books of record, including electronic records, for each of the various funds and account groups include a general journal, additional specialized journals where useful such as a cash receipts journal, a cash payments journal or check register, a voucher register, and a general ledger with necessary subsidiary ledgers. Each of these books of record serves to inform the accounting system and to reconstruct problems if errors in fact occur.

The journal, frequently referred to as the *general journal*, is a book of original entry in which each and every financial transaction of a fund or account group is first recorded. This analysis and recording procedure is referred to as *journalizing*. Journalizing records each transaction and shows how the transaction affected one or more of the accounts. These records include spreadsheets as well as electronic printouts.

Financial transactions of a similar kind, however, tend to repeat throughout the course of ordinary business. Therefore, the general journal may be relieved of some repetitive types of transactions by creating *special journals*. For example, the general journal could be relieved of recording all cash receipts by using a *cash receipts journal* that would record each cash receipt. Likewise, a *cash payments journal* could record all cash payments, and a *voucher register* could record all purchase of goods and services. Special journals are exactly what the name implies, in that the circumstances of the organization can help dictate which special journals would be useful. The use of special journals makes it possible not only to relieve the general journal of the need to record these repetitive types of transactions, thereby simplifying the recording routine, but also its use encourages a division of labor.

The General Ledger and Subsidiary Ledgers

Accounts that form the trial balance of a fund or account group are referred to as its *general ledger*. A trial balance is prepared at the end of an accounting period to verify that all general ledger accounts are in balance, i.e., that total debits are exactly equal to the total credits. If debits and credits do not balance, one or more errors have occurred and must be corrected. Maintaining the accounts of each fund and account group in accordance with the accounting equation is one of the most effective methods of internal accounting control available to school systems. A trial balance is prepared at the end of a day, a week, a month, or a fiscal period prior to undertaking the task of preparing a financial statement for a fund or account group. An example of a trial balance is presented in figure 11.3.

Individual general ledger account titles identify a certain asset, liability, fund equity, or revenue or expenditure account, or the title may identify a kind of a group of similar accounts by title, such as accounts receivable. If the latter is the case, the account in the general ledger is referred to as a *controlling* account. Because each account receivable may be a receivable from a different source or person or business entity, a supporting group of accounts is maintained separately to identify each source, person or firm, the amount receivable from each, and so on. The accounts supporting a controlling account in the general ledger comprise a *subsidiary ledger*.

Logically, the total of all accounts in a subsidiary ledger must equal the balance of the controlling account in the general ledger (that requires the use of a subsidiary ledger). When a trial balance is prepared, a schedule of each subsidiary ledger is also prepared to verify that the subsidiary ledger is in agreement with the controlling account in the general ledger. If it were not in

Figure 11.3
General Fund Trial Balance

(Date)		
Cash	$ 4,500	
Taxes receivable	22,000	
Temporary investment	3,000	
Loan to other fund - library	2,000	
Loan from bank		$ 1,000
Fund equity		10,000
Revenue from tax levy		30,000
Revenue from interest and penalties		500
Salaries expenditures	10,000	
Total	$41,500	$41,500

*Table is taken from *Financial and Managerial Accounting For Elementary and Secondary School Systems* by S. B. Tidwell, Reston, Va.: Association of School Business Officials, 1984.

agreement, one or more errors have occurred that must be corrected. A controlling account, its subsidiary ledger, and a schedule of subsidiary ledger account balances is presented in figure 11.4.

The Posting Process

All the various journals discussed thus far must be linked together. Information recorded in the general journal or in a special journal of a fund or account group is transferred to individual accounts in the general ledger and to related subsidiary ledgers of the fund or account group. This transferring process is referred to as *posting.*

If each financial transaction were recorded and posted when it occurs, the accounts of a fund or account group are "up to date." Financial statements prepared from current information therefore serve as a basis for effective and timely management actions as the school district moves through the fiscal period by knowing its exact asset and liability position.

Basis of Accounting

Depending on the kind of organization, there are several useful ways to record financial transactions based on how assets and liabilities are viewed. Timing the recording and reporting of transactions that involves recognition of revenue or income and expenditures or expenses is the factor involved by use of the term *basis of accounting.* Since schools are quite different from business organizations, the accounting base needs to reflect how schools relate to revenues and expenditures. Two bases of accounting are used by school dis-

Figure 11.4
General Ledger and Subsidiary Ledgers

The General Ledger

Taxes Receivable, 19-0			Taxes Receivable, 19-1		
CR23	280		CR25	175	

NOTE: ***the controlling accounts in the general ledger show the total of the amounts shown in detail in the subsidiary ledger accounts below.***

Subsidiary Taxes Receivable Ledger

Albert Bella			Abraham Adam		
CR23	150		CR25	100	

Douglas Daye			Charles Covenart		
CR23	130		CR25	75	

City of A School System
Schedule of Accounts Receivable (Date)

Albert Bella	$150
Douglas Daye	130
Total, General Ledger Controlling Account	$280

***Table is taken from *Financial and Managerial Accounting For Elementary and Secondary School Systems* by S.B. Tidwell, Reston, VA: Association of School Business Officials, 1984.**

tricts, *accrual* basis and *modified accrual* basis. The accrual basis is used in proprietary type funds and recognizes income in the accounting period in which it is earned or in which it becomes measurable, and expenses in the fiscal period in which they are incurred, if measurable. The modified accrual basis is used by the budgetary funds and expendable trust funds. According to this basis, revenues are recorded when they are susceptible to accrual, i.e., both measurable and available, and recognized in the accounting period in which the liability is incurred. "Available" refers to collectibility within the current fiscal period or soon enough thereafter to be used to pay current liabilities. Some revenue, however, may be more accurately reported as "deferred liabilities," and some expenditures may be more appropriately allocated to future fiscal periods by being reported as "prepaid assets." Decisions about whether a sum of money is deferred revenue, revenue, an expenditure, or a prepaid asset become important when moneys are material in amount. Results of these decisions are recorded as *adjusting entries* at the end of the accounting period. They affect the financial position and results of operation of a fund.

Documentary Evidence of Financial Transactions

If accounting were to serve its purpose effectively, every financial transaction must be analyzed and recorded as it affects the accounts of one or more of the funds and account groups used by the school district. It is therefore essential that each financial transaction be supported by a complete set of business papers that provide the school district with adequate and competent evidence that the transaction did, in fact, take place. The business papers must be retained for examination by the internal auditor, outside independent accountants during the annual comprehensive audit, governmental agency auditors, the board of education and, if necessary, the courts in case a dispute should ever arise.

Public School Fund Accounting Equations

To provide effective budgetary accounting control and current financial information for management purposes, a school district's accounting equation for a budgetary fund is expressed as follows:

Assets = Liabilities + Fund Equity + (Revenue - Estimated Revenue) - ([Expenditures + Encumbrances] - Appropriations)

The accounting equation for each type of fund and account group is different as detailed next. From the trial balance of each fund and account group, interim and annual financial statements are prepared through the use of accounting equations to express the true equity position of funds and accounts.

For the general fixed assets account group:

Assets = Equities in General Fixed Assets

The term *equities in general fixed assets* is preferable instead of a more widely used term *investments in general fixed assets* because the term "equities" requires presentation on the credit side of an account or financial statement in all other cases. The term "investment" implies that it is an asset that requires presentation on the debit side of an account or financial statement. Because equities in general fixed assets appear on the credit side, confusion can be eliminated by this change in terminology.

For the general long-term debt account group:

Resources = General Long-term Debt

The term *resources* is used instead of the budgetary account titles "amount to be provided" and "amount provided."

Assets = Liabilities + (Capital + Retained Earnings) + (Income - Expense)

This is also the equation used by business enterprises operated for profit.

There are three different equations for fiduciary type funds:

For nonexpendable trust funds:

Assets = Liabilities + Principal

For expendable trust funds:

Assets = Liabilities + Retained Earnings + (Income - Expenses)

For agency funds:

Assets = Liabilities

These equations are used in producing the financial statements of a district in order to inform interested stakeholders of the performance of the district's financial affairs. Obviously, the books of record and accounting procedures described are prerequisite to application of equations and to producing financial statements of any type.

Interim Financial Statements for Each Fund and Account Group

The individual funds and account groups are vitally important to the proper financial management of a school district. Members of the board of education need to know the financial position of each fund and account group and the results of operation at the end of each month in order to make effective and timely policy decisions. Each administrative officer having financial responsibility in the district also needs to know the financial position of each fund or account group at least once each month for effective management. In

Figure 11.5
General Fund Balance Sheet

	LOCAL CURRENT EXPENSE	STATE PUBLIC SCHOOL	FEDERAL GRANTS	CAPITAL OUTLAY	TOTALS 1994	TOTALS 1993
ASSETS						
CASH AND CASH EQUIVALENTS	$43,605,926		$ 160,636	$4,755,784	$48,522,346	$49,342,116
RECEIVABLES:						
Mecklenburg County	6,119				6,119	
State of North Carolina		$13,188,615			13,188,615	12,390,598
U. S. Government Agencies			1,807,688		1,807,688	1,087,275
Accrued Interest and Other	165,197			77,317	242,514	255,807
DUE FROM OTHER FUNDS						194,926
INVENTORIES	1,251,083				1,251,083	2,438,043
TOTAL	$45,028,325	$13,188,615	$1,968,324	$4,833,101	$65,018,365	$65,708,765
LIABILITIES AND FUND BALANCE						
ACCOUNTS PAYABLE	$ 3,113,652		$ 163,099	$ 286,834	$ 3,563,585	$ 3,921,134
ACCRUED SALARIES, WAGES AND BENEFITS:						
Teachers' Summer Supplement	20,937,457				20,937,457	17,815,363
Other	6,965,182	$12,271,261	989,524		20,225,967	17,486,682
DEFERRED REVENUE	1,011,844		162,272	169,448	1,343,564	3,893,681
LIABILITY FOR COMPENSATED ABSENCES	306,929	917,354	21,336		1,245,619	1,495,097
Total Liabilities	32,335,064	13,188,615	1,336,231	456,282	47,316,192	44,611,957
FUND BALANCE:						
Reserved by State Statute	11,049,658		632,093	4,369,690	16,051,441	17,948,681
Unreserved:						
Designated for Subsequent Year's Expenditures	1,600,000				1,600,000	2,300,000
Undesignated	43,603			7,129	50,732	848,127
Total Fund Balance	12,693,261		632,093	4,376,819	17,702,173	21,096,808
TOTAL	$45,028,325	$13,188,615	$1,968,324	$4,833,101	$65,018,365	$65,708,765

Source: Courtesy Charlotte-Mecklenburg Public Schools, CAFR. Recipient of ASBO *Certificate of Excellence Award* (1994).

Figure 11.6
Schedule of Revenues, Expenditures, and Changes in Fund Balance

	LOCAL CURRENT EXPENSE						TOTALS	
	COUNTY APPROPRIATION AND OTHER LOCAL SOURCES	STATE PROGRAMS	TOTAL	STATE PUBLIC SCHOOL	FEDERAL GRANTS	CAPITAL OUTLAY	1994	1993
REVENUES:								
State of North Carolina:								
State Public School				$240,183,590			$240,183,590	$228,328,199
Other:								
Vocational				12,330,539			12,330,539	11,632,543
Donated Textbooks		$2,309,748	$ 2,309,748				2,309,748	1,816,525
Other		77,992	77,992				77,992	211,580
Total State		2,387,740	2,387,740	252,514,129			254,901,869	241,988,847
Mecklenburg County	$117,000,000		117,000,000			$6,327,566	123,327,566	116,175,161
U.S. Government Agencies:								
Chapter I--Education of Children of Low Income Families					$ 6,969,177		6,969,177	7,876,684
Chapter II--Guidance and Counseling					708,450		708,450	860,751
Title VI--Education of Handicapped Chilldren					4,094,899		4,094,899	3,251,907
Vocational					989,319		989,319	1,002,065
Other					5,124,128		5,124,128	2,161,544
Total U.S. Government					17,885,973		17,885,973	15,152,951
Other:								
Sales of Surplus Properties						89,703	89,703	137,839
Insurance Proceeds						134,485	134,485	169,421
Inter-Fund Charges	2,098,717		2,098,717				2,098,717	1,929,658
Miscellaneous	4,373,354		4,373,354			109,899	4,483,253	3,956,981
Total Other	6,472,071		6,472,071			334,087	6,806,158	6,193,899
Total Revenues	123,472,071	2,387,740	125,859,811	252,514,129	17,885,973	6,661,653	402,921,566	379,510,858

	LOCAL CURRENT EXPENSE							
	COUNTY APPROPRIATION AND OTHER LOCAL SOURCES	STATE PROGRAMS	TOTAL	STATE PUBLIC SCHOOL	FEDERAL GRANTS	CAPITAL OUTLAY	TOTALS 1994	TOTALS 1993
CURRENT OPERATING EXPENDITURES:								
Instructional Programs:								
Regular	$39,444,494	$2,327,397	$41,771,891	$122,188,507	$ 2,155,037		$166,115,435	$150,342,383
Special	4,700,367	55,229	4,755,596	17,269,540	8,122,368		30,147,504	27,592,726
Vocational	1,359,020		1,359,020	8,851,843	826,735		11,037,598	9,852,044
Co-Curricular	1,731,207		1,731,207				1,731,207	1,735,551
Total	47,235,088	2,382,626	49,617,714	148,309,890	11,104,140		209,031,744	189,522,704
Supporting Services:								
Pupil Support:								
Director of Pupil Support	143,919		143,919				143,919	130,944
Attendance and Social Work	394,789		394,789	1,014,010			1,408,799	1,209,685
Guidance	984,644		984,644	4,232,295	407		5,217,346	5,145,648
Health	155,386		155,386	52,375	181,906		389,667	434,493
Psychological	796,619		796,619	3,143,337	143,750		4,083,706	4,033,635
Speech Pathology and Audiology	460,476		460,476	2,633,691	193,577		3,287,744	3,233,344
Vocational	76,639		76,639	846,166	175,991		1,098,796	1,269,952
Other	764,422		764,422	259,072	797,910		1,821,404	1,589,410
Total	3,776,894		3,776,894	12,180,946	1,493,541		17,451,381	17,047,111
Instructional Staff Support:								
Improvement of Instructional Services	3,276,781		3,276,781	2,725,018	899,121		6,900,920	6,003,766
Educational Media Services	2,708,877		2,708,877	4,745,485	307,480		7,761,842	7,565,664
Total	5,985,658		5,985,658	7,470,503	1,206,601		14,662,762	13,569,430
General Administrative Support:								
Board of Education	1,100,257		1,100,257				1,100,257	667,986
Executive Administration	1,154,295		1,154,295	334,790	494,347		1,983,432	1,629,477
Total	2,254,552		2,254,552	334,790	494,347		3,083,689	2,297,463
School Administrative Support:								
Office of the Principal	7,538,485		7,538,485	15,310,999	1,624		22,851,108	21,806,176
Business and Operating Support:								
Director of Business Support	825,723		825,723				825,723	2,179,597
Fiscal Services	1,422,388		1,422,388	260,611	533,815		2,216,814	2,018,750
Operation of Plant	15,369,501		15,369,501	4,303,359	3,546		19,676,406	18,036,105
Security	1,750,546		1,750,546				1,750,546	566,346
Transportation	4,942,004		4,942,004	11,533,938	224,865		16,700,807	14,874,370
Internal Services	1,930,218		1,930,218				1,930,218	1,968,231
Maintenance of Plant	11,922,941		11,922,941	5,424	45,065		11,973,430	12,536,424
Total	38,163,321		38,163,321	16,103,332	807,291		55,073,944	52,179,823

CHARLOTTE-MECKLENBURG BOARD OF EDUCATION
GENERAL FUND
SCHEDULE OF REVENUES, EXPENDITURES AND CHANGES IN FUND BALANCE
FOR THE YEAR ENDED JUNE 30, 1994 (Concluded)

	LOCAL CURRENT EXPENSE						TOTALS	
	COUNTY APPROPRIATION AND OTHER LOCAL SOURCES	STATE PROGRAMS	TOTAL	STATE PUBLIC SCHOOL	FEDERAL GRANTS	CAPITAL OUTLAY	1994	1993
Central Support:								
Director of Central Support	$ 45,309		$ 45,309				$ 45,309	$ 62,685
Planning, Research, Development and Evaluation	1,225,419	$ 158	1,225,577		$ 62,098		1,287,675	1,179,099
Information	404,817		404,817		13,685		418,502	420,488
Staff Services	2,041,083		2,041,083				2,041,083	2,030,277
Data Processing Services	3,179,662	4,956	3,184,618				3,184,618	3,194,639
Other	41,968		41,968				41,968	40,752
Total	6,938,258	5,114	6,943,372		75,783		7,019,155	6,927,940
Employee Benefits	16,573,006		16,573,006	$ 51,945,875	2,197,542		70,716,423	67,922,235
Community Services	503,105		503,105	32,837			535,942	606,204
Total Current Operating Expenditures	128,968,367	2,387,740	131,356,107	251,689,172	17,380,869		400,426,148	371,879,086
CAPITAL OUTLAY EXPENDITURES:								
Building and Site Improvements						$1,529,302	1,529,302	3,182,854
Furniture and Equipment						3,535,794	3,535,794	3,619,430
Total Capital Outlay						5,065,096	5,065,096	6,802,284
Total Expenditures	128,968,367	2,387,740	131,356,107	251,689,172	17,380,869	5,065,096	405,491,244	378,681,370
EXCESS (DEFICIENCY) OF REVENUES OVER EXPENDITURES	(5,496,296)	-0-	(5,496,296)	824,957	505,104	1,596,557	(2,569,678)	829,488
OTHER FINANCING USES:								
Operating Transfer to Enterprise Funds				(824,957)			(824,957)	(808,227)
EXCESS (DEFICIENCY) OF REVENUES OVER EXPENDITURES AND OTHER FINANCING USES	(5,496,296)	-0-	(5,496,296)	-0-	505,104	1,596,557	(3,394,635)	21,261
FUND BALANCE--Beginning of Year	18,189,557	-0-	18,189,557	-0-	126,989	2,780,262	21,096,808	21,075,547
FUND BALANCE--End of Year	$ 12,693,261	$ -0-	$ 12,693,261	$ -0-	$ 632,093	$4,376,819	$ 17,702,173	$ 21,096,808

Source: Courtesy Charlotte-Mecklenburg Public Schools, CAFR. Recipient of ASBO *Certificate of Excellence Award* (1994).

Figure 11.7
Schedule of Revenues, Expenditures, Encumbrances, and Changes in Fund Balance*Budget and Actual

	LOCAL CURRENT EXPENSE		STATE PUBLIC SCHOOL	
	BUDGET	ACTUAL	BUDGET	ACTUAL
REVENUES:				
State of North Carolina	$ 2,430,821	$ 2,387,740	$253,254,957	$251,255,211
Mecklenburg County	117,000,000	117,000,000		
U.S. Government Agencies				
Sales of Surplus Properties				
Other	7,257,707	6,472,071		
Total Revenues	126,688,528	125,859,811	253,254,957	251,255,211
EXPENDITURES AND ENCUMBRANCES:				
Current Operating Expenditures and Encumbrances:				
Instructional Programs:				
Regular	40,927,323	40,598,604	123,219,109	122,833,326
Special	4,895,285	4,835,217	17,907,838	17,270,853
Vocational	1,359,818	1,359,020	8,851,843	8,851,843
Co-Curricular	1,745,078	1,744,995		
	48,927,504	48,537,836	149,978,790	148,956,022
Pupil Support	3,907,297	3,778,502	12,235,382	12,180,946
Instructional Staff Support	6,057,020	6,034,412	7,630,913	7,394,083
General Administrative Support	2,262,211	2,253,677	335,289	334,790
School Administrative Support	7,538,779	6,980,462	15,341,166	15,312,100
Business and Operating Support	37,717,078	37,561,266	16,204,867	16,047,036
Central Support	6,650,263	6,376,030		
Employee Benefits	16,736,072	16,672,342	50,670,756	50,172,440
Community Services	505,545	505,541	32,837	32,837
Total Current Operating Expenditures and Encumbrances	130,301,769	128,700,068	252,430,000	250,430,254
Capital Outlay Expenditures and Encumbrances:				
Building and Site Improvements				
Furniture and Equipment				
Total Capital Outlay Expenditures and Encumbrances				
Total Expenditures and Encumbrances	130,301,769	128,700,066	252,430,000	250,430,254
EXCESS (DEFICIENCY) OF REVENUES OVER EXPENDITURES AND ENCUMBRANCES	(3,613,241)	(2,840,257)	824,957	824,957
OTHER FINANCING SOURCES (USES):				
Fund Balance Appropriated	3,613,241			
Operating Transfer to Enterprise Funds			(824,957)	(824,957)
Total Other Financing Sources (Uses)	3,613,241	-0-	(824,957)	(824,957)
DEFICIENCY OF REVENUES OVER EXPENDITURES, ENCUMBRANCES AND OTHER FINANCING SOURCES (USES)	-0-	(2,840,257)	-0-	-0-
FUND BALANCE--Beginning of Year	-0-	5,326,861	-0-	-0-
FUND BALANCE--End of Year	$ -0-	$ 2,486,604	$ -0-	$ -0-

CHARLOTTE-MECKLENBURG BOARD OF EDUCATION
GENERAL FUND
SCHEDULE OF REVENUES, EXPENDITURES, ENCUMBRANCES AND
CHANGES IN FUND BALANCE--BUDGET AND ACTUAL (NON-GAAP BUDGETARY BASIS)
FOR THE YEAR ENDED JUNE 30, 1994 (Concluded)

	FEDERAL GRANTS		CAPITAL OUTLAY		TOTALS	
	BUDGET	ACTUAL	BUDGET	ACTUAL	BUDGET	ACTUAL
REVENUES:						
State of North Carolina					$255,685,778	$253,642,951
Mecklenburg County			$6,330,066	$6,327,566	123,330,066	123,327,566
U.S. Government Agencies	$17,889,000	$17,885,973			17,889,000	17,885,973
Sales of Surplus Properties			100,000	89,703	100,000	89,703
Other			239,611	244,384	7,497,318	6,716,455
Total Revenues	17,889,000	17,885,973	6,669,677	6,661,653	404,502,162	401,662,648
EXPENDITURES AND ENCUMBRANCES:						
Current Operating Expenditures and Encumbrances:						
Instructional Programs:						
Regular	2,670,000	2,668,424			166,816,432	166,100,354
Special	8,119,700	8,118,598			30,922,823	30,224,668
Vocational	826,800	826,735			11,038,461	11,037,598
Co-Curricular					1,745,078	1,744,995
	11,616,500	11,613,757			210,522,794	209,107,615
Pupil Support	1,496,800	1,496,796			17,639,479	17,456,244
Instructional Staff Support	1,195,500	1,195,307			14,883,433	14,623,802
General Administrative Support	498,500	498,458			3,096,000	3,086,925
School Administrative Support	1,650	1,624			22,881,595	22,294,186
Business and Operating Support	807,300	807,291			54,729,245	54,415,593
Central Support	75,200	75,198			6,725,463	6,451,228
Employee Benefits	2,197,550	2,197,542			69,604,378	69,042,324
Community Services					538,382	538,378
Total Current Operating Expenditures and Encumbrances	17,889,000	17,885,973			400,620,769	397,016,295
Capital Outlay Expenditures and Encumbrances:						
Building and Site Improvements			1,442,533	1,442,533	1,442,533	1,442,533
Furniture and Equipment			5,841,559	5,840,755	5,841,559	5,840,755
Total Capital Outlay Expenditures and Encumbrances			7,284,092	7,283,288	7,284,092	7,283,288
Total Expenditures and Encumbrances	17,889,000	17,885,973	7,284,092	7,283,288	407,904,861	404,299,583
EXCESS (DEFICIENCY) OF REVENUES OVER EXPENDITURES AND ENCUMBRANCES	-0-	-0-	(614,415)	(621,635)	(3,402,699)	(2,636,935)
OTHER FINANCING SOURCES (USES):						
Fund Balance Appropriated			614,415	-0-	4,227,656	-0-
Operating Transfer to Enterprise Funds					(824,957)	(824,957)
Total Other Financing Sources (Uses)			614,415	-0-	3,402,699	(824,957)
DEFICIENCY OF REVENUES OVER EXPENDITURES, ENCUMBRANCES AND OTHER FINANCING SOURCES (USES)	-0-	-0-	-0-	(621,635)	-0-	(3,461,892)
FUND BALANCE--Beginning of Year	-0-	-0-	-0-	706,081	-0-	6,032,942
FUND BALANCE--End of Year	$ -0-	$ -0-	$ -0-	$ 84,446	$ -0-	$ 2,571,060

Source: Courtesy Charlotte-Mecklenburg Public Schools, CAFR. Recipient of ASBO *Certificate of Excellence Award* (1994).

sum, financial statements form the basis for wise administration in that such documents guide future decisions for both prudent and legal reasons.

A complete set of financial statements should be generated each month for review by board members and appropriate administrative staff. These interim statements will not only help the board make effective and timely decisions, but also help them better understand the fiscal structure of the district.

Budgetary Funds

Interim financial reports for each of the budgetary funds (general fund, special education fund, capital projects fund, and each debt service fund) may be reported using a comparative balance sheet that compares balances of

Figure 11.8
Statement of Changes in Financial Position

Resources provided:		
From revenues:		
Local sources	$ 767,995.79	
Intermediate sources	16,219.94	
State sources	849,904.04	
Federal sources	6,091.10	
Gifts and bequests	608.03	$1,640,818.90
From reduction in assets:		
Reduction in accounts receivable	$ 131,099.24	
Reduction in prepaid insurance	576.59	131,675.83
From increasing liabilities:		
Increase in accounts payable	$ 14,437.26	
Increase in salaries payable	33,367.80	
Increase in withholding payable	6,624.80	
Increase in due to other funds	2,221.90	
Increase in reserves and deferred revenues	4,759.56	61,411.32
Total resources provided		$1,833,906.05
Application of school financial resources:		
To expenditures of school district:		
Instruction	$1,097,646.48	
Administration	67,067.82	
Transportation of pupils	61,612.82	
Operation of plant	191,461.59	
Maintenance of plant	19,941.08	
Fixed charges	12,743.97	
Capital outlay	23,058.05	
Cafeteria funds deficit	9,609.97	
Transfers to other school districts (tuition)	117,054.89	$1,600,196.67
To increase assets:		
Increase in cash	$ 64,314.60	
Increase in taxes receivable	15,372.24	
Increase in inventory of supplies	984.02	80,670.86
To reduction of liabilities:		
Reduction of current and short term loans	$ 134,829.02	
Reduction of contracts for equipment purchases	18,209.50	153,038.25
Total school resources applied		$1,833,906.05

*Table is taken from *Financial and Managerial Accounting For Elementary and Secondary School Systems* by S. B. Tidwell, Reston, Va.: Association of School Business Officials, 1984.

accounts at the balance sheet date with account balances at the end of the prior year. An example is presented in figure 11.5.

A statement comparing actual revenue to date with actual revenue in the prior year is presented in figure 11.6. A similar statement showing actual revenue, expenditures and encumbrances to budgeted amounts provides a different view of the same type of statement as seen in figure 11.7. A further statement also could be prepared that would compare expenditures and encumbrances (financial commitments) to date with appropriations for the current year (an alternative presentation may combine this statement and the previous statement into a statement comparing actual and budgeted revenue and expenditure accounts as seen previously.

A statement of changes in financial position is also generally given. Figure 11.8 presents a statement of changes in financial position of a general fund when prepared to show gross financial resources provided and total school resources applied.

Account Groups

Interim reports may also be issued for the general fixed assets account group. Generally, these interim reports are:

- A statement of general fixed assets. (See figure 11.9.)
- A statement of changes in general fixed assets. (Figure 11.10 presents this statement showing changes by source.)
- When applicable, a statement of general long-term debt or a statement of changes in general long-term debt (not shown).

Proprietary Funds

Interim reports may also be issued for the enterprise funds and internal services funds when such funds are operated by a school district. These interim financial statements include:

- A balance sheet comparing current balances with balances at the end of the prior year. (See figure 11.11 showing a combining balance sheet for all enterprise funds). An alternative presentation would provide separate balance sheets for each enterprise fund.
- A statement of income showing gross income less sales discounts and sales returns and allowances, the cost of goods or services sold, gross profit, less selling and general expenses, and net income and a statement of changes in retained earnings. An alternative presentation may include both the statement of income and the statement of changes in retained earnings. (See figure 11.12.)

Figure 11.9
Schedule of General Fixed Assets by Function and Source

BY FUNCTION	LAND	BUILDINGS	FURNITURE & EQUIPMENT	VEHICLES	LIBRARY BOOKS	CONSTRUCTION IN PROGRESS	TOTALS 1994	TOTALS 1993
Instructional:								
Elementary Schools	$11,001,741	$152,102,613	$31,236,099	$ 1,200	$4,473,864	$17,462,712	$216,278,229	$190,154,101
Middle Schools	5,828,300	66,818,798	15,370,826	2,502	1,727,456	1,214,615	90,962,497	81,858,641
Senior High Schools	3,386,207	64,602,465	17,833,216	39,763	1,635,897	221,624	87,719,172	87,449,989
Special Schools	641,980	2,347,785	1,211,720		70,000	112,829	4,384,314	3,775,776
Administrative:								
General	2,324,387	14,657,798	7,306,708	69,584		919,963	25,278,440	24,553,387
Transportation	1,738,512	1,697,387	524,755	23,048,836	—	—	27,009,490	25,974,473
TOTAL GENERAL FIXED ASSETS BY FUNCTION	$24,921,127	$302,226,846	$73,483,324	$23,161,885	$7,907,217	$19,931,743	$451,632,142	$413,766,367
BY SOURCE								
Prior to 1976 (Not Determinable)							$134,764,587	$136,273,523
General Fund							71,258,501	67,220,929
Capital Projects Fund							243,393,024	208,240,995
Expendable Trust Fund							1,790,354	1,677,049
Donations							425,676	353,871
TOTAL INVESTMENT IN GENERAL FIXED ASSETS							$451,632,142	$413,766,367

Source: Courtesy Charlotte-Mecklenburg Public Schools, CAFR. Recipient of ASBO *Certificate of Excellence Award* (1994).

Figure 11.10
Schedule of Changes in General Fixed Assets by Source

	LAND	BUILDINGS	FURNITURE & EQUIPMENT	VEHICLES	LIBRARY BOOKS	CONSTRUCTION IN PROGRESS	TOTAL
GENERAL FIXED ASSETS--June 30, 1993	$22,904,115	$265,871,696	$67,584,040	$22,402,868	$7,640,040	$ 27,363,608	$413,766,367
ADDITIONS:							
Expenditures from Capital Projects	2,017,012	12,971,723	5,386,755		309,394	15,782,881	36,467,765
Expenditures from General Fund			7,411,059	963,921	46,122	168,681	8,589,783
Expenditures from Expendable Trust Fund			101,671	11,634			113,305
Donations			108,073				108,073
Transfer of Construction in Progress		23,383,427				(23,383,427)	-0-
Total Additions	2,017,012	36,355,150	13,007,558	975,555	355,516	(7,431,865)	45,278,926
RETIREMENTS			(7,108,274)	(216,538)	(88,339)		(7,413,151)
GENERAL FIXED ASSETS--June 30, 1994	$24,921,127	$302,226,846	$73,483,324	$23,161,885	$7,907,217	$ 19,931,743	$451,632,142

Source: Courtesy Charlotte-Mecklenburg Public Schools, CAFR. Recipient of ASBO *Certificate of Excellence Award* (1994).

Figure 11.11
Enterprise Funds Combining Balance Sheet

	AFTER SCHOOL ENRICHMENT PROGRAM	CHILD NUTRITION PROGRAM	TOTALS 1994	TOTALS 1993
ASSETS				
CASH AND CASH EQUIVALENTS	$1,290,051	$2,055,909	$3,345,960	$3,715,613
RECEIVABLES:				
U.S. Government Agencies		1,420,560	1,420,560	543,216
Accrued Interest and Other	25,711	35,576	61,287	64,284
INVENTORIES		820,105	820,105	609,567
FIXED ASSETS (Net of Accumulated Depreciation)	43,012	4,150,995	4,194,007	3,131,008
TOTAL	$1,358,774	$8,483,145	$9,841,919	$8,063,688
LIABILITIES AND FUND EQUITY				
ACCOUNTS PAYABLE		$ 82,170	$ 82,170	$ 97,979
ACCRUED SALARIES, WAGES AND BENEFITS		185,858	185,858	201,502
DUE TO GENERAL FUND				194,926
DEFERRED REVENUE	$ 218,227		218,227	217,682
LIABILITY FOR COMPENSATED ABSENCES	44,581	123,768	168,349	174,175
Total Liabilities	262,808	391,796	654,604	886,264
FUND EQUITY:				
Contributed Capital:				
Federal Grants		133,691	133,691	133,691
Less Amortization		(133,691)	(133,691)	(133,691)
		-0-	-0-	-0-
Local Sources		4,810,710	4,810,710	3,855,240
Total Contributed Capital		4,810,710	4,810,710	3,855,240
Retained Earnings:				
Reserved by State Statute	75,376	1,616,526	1,691,902	492,828
Unreserved	1,020,590	1,664,113	2,684,703	2,829,356
Total Retained Earnings	1,095,966	3,280,639	4,376,605	3,322,184
Total Fund Equity	1,095,966	8,091,349	9,187,315	7,177,424
TOTAL	$1,358,774	$8,483,145	$9,841,919	$8,063,688

Source: Courtesy Charlotte-Mecklenburg Public Schools, CAFR. Recipient of ASBO *Certificate of Excellence Award* (1994).

- Other statements as desired, such as a combining statement of cash flows (figure 11.13) and a schedule of revenues, expenses and changes in retained earnings comparing actual and budgeted amounts (figure 11.14).

Fiduciary and Agency Funds

The same type statements may be issued for fiduciary and agency funds. An interim financial statement for each nonexpendable trust fund consists of:

- A balance sheet comparing the current balances with those at the end of the prior year.

For each expendable trust fund, the following interim financial statements are used:

- A balance sheet comparing current balances with those at the end of the prior year.
- A statement of income and expense.
- A statement of changes in retained earnings.

Figure 11.12
Combining Statement of Revenues, Expenses and Changes in Retained Earnings

	AFTER SCHOOL ENRICHMENT PROGRAM	CHILD NUTRITION PROGRAM	TOTALS 1994	TOTALS 1993
OPERATING REVENUES:				
Food Sales		$ 11,432,287	$ 11,432,287	$10,815,910
Participant Fees	$4,686,569		4,686,569	3,804,805
Total Operating Revenues	4,686,569	11,432,287	16,118,856	14,620,715
OPERATING EXPENSES:				
Food Costs:				
Purchased Food	338,337	7,567,459	7,905,796	7,486,675
Donated Commodities		1,326,838	1,326,838	1,100,003
Salaries	2,999,447	8,758,938	11,758,385	10,754,578
Employee Benefits	484,778	2,160,993	2,645,771	2,534,752
Materials and Supplies	208,353	742,206	950,559	771,898
Utilities		172,393	172,393	178,938
Depreciation	13,535	523,402	536,937	441,779
Contracted Services	73,122	122,638	195,760	185,869
Equipment Rental		60,161	60,161	25,230
Indirect Cost	200,658	1,619,557	1,820,215	1,771,217
Other	116,530	149,413	265,943	223,427
Total Operating Expenses	4,434,760	23,203,998	27,638,758	25,474,366
OPERATING INCOME (LOSS)	251,809	(11,771,711)	(11,519,902)	(10,853,651)
NON-OPERATING REVENUES:				
U. S. Government Subsidy		10,201,686	10,201,686	9,382,466
U. S. Government Commodities		1,326,838	1,326,838	1,100,003
Interest Income	77,130	139,462	216,592	217,801
Other		4,250	4,250	
Total Non-Operating Revenues	77,130	11,672,236	11,749,366	10,700,270
INCOME (LOSS) BEFORE OPERATING TRANSFER	328,939	(99,475)	229,464	(153,381)
OPERATING TRANSFER FROM GENERAL FUND		824,957	824,957	808,227
NET INCOME	328,939	725,482	1,054,421	654,846
DEPRECIATION OF FIXED ASSETS ACQUIRED WITH FEDERAL GRANT FUNDS				1,170
INCREASE IN RETAINED EARNINGS	328,939	725,482	1,054,421	656,016
RETAINED EARNINGS--Beginning of Year	767,027	2,555,157	3,322,184	2,666,168
RETAINED EARNINGS--End of Year	$1,095,966	$ 3,280,639	$ 4,376,605	$ 3,322,184

Source: Courtesy Charlotte-Mecklenburg Public Schools, CAFR. Recipient of ASBO *Certificate of Excellence Award* (1994).

- A statement of changes in financial position.

For each agency fund they include:

- A statement of assets and liabilities.
- A schedule of cash receipts and disbursements.

Figure 11.15 presents a simplified form that can be used for balance sheets for both expendable trust funds and agency funds. Other forms of the same statement may include assets, liabilities, and fund balances. Figure 11.16 presents a schedule of revenues, expenditures and changes in fund balance for an expendable trust fund, and figure 11.17 presents a more detailed version of the same statement broken out by receiving schools.

Year-End Financial Statements for Funds and Account Groups

In addition to interim reports, a series of final statements are made at the end of the fiscal year. These documents include a balance sheet, a statement of revenue, expenditures and fund equity, and a statement of changes in financial position. These are prepared for the general fund and are the financial statements for the first of the budgetary funds.

For the remaining budgetary funds, combined financial statements are prepared. Combined financial statements also are prepared for all special revenue funds, all capital projects funds, and all debt service funds. Combined financial statements include a combined balance sheet, a combined statement of revenue, expenditures and fund equity, and a combined statement of changes in financial position.

Combined financial statements for each of these results in a set of columns that provides a total for each of the accounts included in the financial statements. This gives what the totals might be if there were only one general ledger for all special revenue funds, one general ledger for all capital projects funds, and one general ledger for all debt service funds. Examples of the "combining technique" for all proprietary and governmental fund types are

Figure 11.13
Combining Statement of Cash Flows: Enterprise Funds

	AFTER SCHOOL ENRICHMENT PROGRAM	CHILD NUTRITION PROGRAM	TOTALS 1994	TOTALS 1993
CASH FLOWS FROM OPERATING ACTIVITIES:				
Cash Received from Participants	$ 4,693,223	$ 11,422,161	$ 16,115,384	$ 14,669,764
Cash Paid to Employees	(2,993,951)	(8,785,904)	(11,779,855)	(10,839,170)
Cash Paid to Suppliers	(1,421,479)	(12,980,139)	(14,401,618)	(13,203,581)
Net Cash Provided by (Used in) Operating Activities	277,793	(10,343,882)	(10,066,089)	(9,372,987)
CASH FLOWS FROM NON-CAPITAL FINANCING ACTIVITIES:				
U.S. Government Subsidy		9,324,342	9,324,342	9,265,016
Operating Transfer from General Fund		824,957	824,957	808,227
Net Cash Provided by Non-Capital Financing Activities		10,149,299	10,149,299	10,073,243
CASH FLOWS FROM CAPITAL AND RELATED FINANCING ACTIVITIES:				
Proceeds from Sales of Equipment		4,483	4,483	
Acquisition of Equipment	(24,302)	(656,417)	(680,719)	(549,310)
Net Cash Used in Capital and Related Financing Activities	(24,302)	(651,934)	(676,236)	(549,310)
CASH FLOWS FROM INVESTING ACTIVITIES:				
Interest on Investments	78,294	145,079	223,373	213,430
Net Cash Provided by Investing Activities	78,294	145,079	223,373	213,430
NET INCREASE (DECREASE) IN CASH AND CASH EQUIVALENTS	331,785	(701,438)	(369,653)	364,376
CASH AND CASH EQUIVALENTS--Beginning of Year	958,266	2,757,347	3,715,613	3,351,237
CASH AND CASH EQUIVALENTS--End of Year	$ 1,290,051	$ 2,055,909	$ 3,345,960	$ 3,715,613
RECONCILIATION OF OPERATING INCOME (LOSS) TO NET CASH PROVIDED BY (USED IN) OPERATING ACTIVITIES:				
Operating Income (Loss)	$ 251,809	$(11,771,711)	$(11,519,902)	$(10,853,651)
Adjustments to Reconcile Operating Income (Loss) to Net Cash Provided by (Used in) Operating Activities:				
Depreciation	13,535	523,402	536,937	441,779
Loss on Disposals	299	35,954	36,253	25,838
Donated Commodities		1,326,838	1,326,838	1,100,003
Change in Assets and Liabilities:				
Accounts Receivable	6,109	(10,126)	(4,017)	(2,181)
Inventories		(210,538)	(210,538)	5,445
Accounts Payable		(15,809)	(15,809)	36,004
Accrued Salaries, Wages & Benefits		(15,644)	(15,644)	3,541
Due to Other Funds		(194,926)	(194,926)	(164,621)
Deferred Revenue	545		545	51,230
Compensated Absences	5,496	(11,322)	(5,826)	(16,374)
Total Adjustments	25,984	1,427,829	1,453,813	1,480,664
Net Cash Provided by (Used in) Operating Activities	$ 277,793	$(10,343,882)	$(10,066,089)	$ (9,372,987)
NON-CASH CAPITAL AND RELATED FINANCING ACTIVITIES:				
Contribution of Equipment from Other Funds		$ 955,469	$ 955,469	$ 499,431

Source: Courtesy Charlotte-Mecklenburg Public Schools, CAFR. Recipient of ASBO *Certificate of Excellence Award* (1994).

Figure 11.14
Schedule of Revenue, Expenses and Changes in Retained Earnings—Enterprise Funds

	AFTER SCHOOL ENRICHMENT PROGRAM		CHILD NUTRITION PROGRAM	
	BUDGET	ACTUAL	BUDGET	ACTUAL
OPERATING REVENUES:				
Food Sales			$ 11,360,246	$ 11,432,287
Participant Fees	$4,894,927	$4,686,569		
OPERATING EXPENSES:				
Food Costs	397,956	338,337	8,896,153	8,894,297
Salaries and Benefits	3,821,680	3,484,225	10,926,661	10,919,931
Other Operating Expenses	721,756	598,663	2,919,568	2,866,368
Depreciation	13,535	13,535	523,402	523,402
Total Operating Expenses	4,954,927	4,434,760	23,265,784	23,203,998
OPERATING INCOME (LOSS)	(60,000)	251,809	(11,905,538)	(11,771,711)
NON-OPERATING REVENUES:				
U. S. Government Agencies			10,906,260	11,528,524
Interest Income	60,000	77,130	162,321	139,462
Other			12,000	4,250
Total Non-Operating Revenues	60,000	77,130	11,080,581	11,672,236
INCOME (LOSS) BEFORE OPERATING TRANSFER	-0-	328,939	(824,957)	(99,475)
Operating Transfer From General Fund			824,957	824,957
NET INCOME	-0-	328,939	-0-	725,482
RETAINED EARNINGS--Beginning of Year	-0-	767,027	-0-	2,555,157
RETAINED EARNINGS--End of Year	$ -0-	$1,095,966	$ -0-	$ 3,280,639

Source: Courtesy Charlotte-Mecklenburg Public Schools, CAFR. Recipient of ASBO *Certificate of Excellence Award* (1994).

Figure 11.15
Expendable Trust Fund Balance Sheet

	Elementary Schools	Middle Schools
ASSETS		
CASH AND CASH EQUIVALENTS	$376,370	$223,925
FUND BALANCE		
RESERVED BY STATE STATUTE	$ 42,155	$ 27,822
UNRESERVED, UNDESIGNATED	334,215	196,103
	$376,370	$223,925

Source: Courtesy Charlotte-Mecklenburg Public Schools, CAFR. Recipient of ASBO Certificate of Excellence Award (1994).

given in figure 11.18 and figure 11.19 so that at the end of the fiscal year, a combined balance sheet, statement of revenue, expenses, and changes in retained earnings, and a statement of changes in financial position are presented for all enterprise funds and for all internal services funds. Similarly, a combined balance sheet, statement of revenue, expenses, and changes in retained earnings and a statement of changes in financial position are prepared for all nonexpendable trust funds, for all expendable trust funds, and agency funds. Each combined financial statement presents a totals column for each account for memorandum purposes only. This emphasizes that it is the individual funds' financial statements that provide the basis for effective and efficient financial control and management.

Combined Financial Statements

The totals columns on combined financial statements of budgetary funds are presented in columnar form with additional columns presenting the financial statements of the general fixed assets account group and the financial statements of the general long-term debt account group. These comprise the combined financial statements for the school district. They are:

- Combined Balance Sheet—All Fund Types and Accounts (see figure 11.20).
- Combined Statement of Revenue, Expenditures and Changes in Fund Balances—All Budgetary Funds and Expendable Trust Funds (see figure 11.21).
- Combined Statement of Revenue, Expenses and Changes in Retained Earnings—All Proprietary Type Funds, (i.e., enterprise funds, internal service funds and expendable trust funds). See figure 11.22.

Special Schools	Senior High Schools	TOTALS 1994	1993
$17,591	$629,230	$1,247,116	$1,221,764
$ 366	$170,470	$ 240,813	$ 270,541
17,225	458,760	1,006,303	951,223
$17,591	$629,230	$1,247,116	$1,221,764

Figure 11.16
Schedule of Revenues, Expenditures, and Changes in Fund Balance*Expendable Trust Fund

	Elementary Schools	Middle Schools	Special Schools	Senior High Schools	TOTALS 1994	TOTALS 1993
REVENUES:						
Fund Raising Activities (Net of Costs)	$308,437	$139,191	$ 4,611	$ 79,071	$ 531,310	$ 539,877
Athletics		116,350		678,241	794,591	779,788
Clubs and Organizations	99,469	535,635	7,068	1,626,025	2,268,197	1,966,952
Contributions and Other	415,780	192,931	19,842	236,110	864,663	714,589
Total Revenues	823,686	984,107	31,521	2,619,447	4,458,761	4,001,206
EXPENDITURES:						
Regular Instructional Programs	105,292	38,462	97	31,703	175,554	184,606
Co-Curricular Instructional Programs:						
Athletics		129,361		693,063	822,424	791,169
Clubs and Organizations	89,743	541,171	8,148	1,612,270	2,251,332	1,959,995
School Administrative Support	657,372	272,658	20,434	233,635	1,184,099	1,024,699
Total Expenditures	852,407	981,652	28,679	2,570,671	4,433,409	3,960,469
EXCESS (DEFICIENCY) OF REVENUES OVER EXPENDITURES	(28,721)	2,455	2,842	48,776	25,352	40,737
FUND BALANCE--Beginning of Year	405,091	221,470	14,749	580,454	1,221,764	1,181,027
FUND BALANCE--End of Year	$376,370	$223,925	$17,591	$ 629,230	$1,247,116	$1,221,764

Source: Courtesy Charlotte-Mecklenburg Public Schools, CAFR. Recipient of ASBO *Certificate of Excellence Award* (1994).

Figure 11.17
Detailed Schedule of Revenue, Expenditures and Changes in Fund Balance—Expendable Trust Fund

	REVENUES	EXPENDITURES	EXCESS (DEFICIENCY) OF REVENUES OVER EXPENDITURES	FUND BALANCE JUNE 30, 1993	FUND BALANCE JUNE 30, 1994	COMPONENTS OF FUND BALANCE	
						RESERVED	UNDESIGNATED
ELEMENTARY SCHOOLS:							
Albemarle Road	$ 21,522	$ 16,622	$ 4,900	$ 6,741	$ 11,641	$ 132	$ 11,509
Allenbrook	5,254	5,221	33	1,667	1,700		1,700
Ashley Park	12,142	10,491	1,651	2,273	3,924	93	3,831
Bain	11,624	13,365	(1,741)	5,795	4,054	1,260	2,794
Barringer	11,303	19,083	(7,780)	23,752	15,972	302	15,670
Berryhill	11,048	10,617	431	4,064	4,495	1,143	3,352
Beverly Wood	13,940	15,111	(1,171)	6,236	5,065	81	4,984
Billingsville	12,247	11,165	1,082	2,635	3,717	8	3,709
Briarwood	11,950	14,097	(2,147)	9,626	7,479	118	7,361
Bruns Avenue	9,300	12,286	(2,986)	8,894	5,908	3,756	2,152
Chantilly	1,576	1,586	(10)	949	939		939
Clear Creek	7,815	6,214	1,601	4,218	5,819	386	5,433
Collinswood	2,842	2,843	(1)	714	713	304	409
Cornelius	9,553	11,049	(1,496)	4,095	2,599	742	1,857
Cotswold	14,231	13,645	586	8,249	8,835	719	8,116
Crown Point	12,647	9,386	3,261	1,104	4,365	121	4,244
Davidson	13,307	13,854	(547)	3,835	3,288	114	3,174
David Cox Road	8,837	10,772	(1,935)	2,961	1,026		1,026
Derita	25,836	25,566	270	5,502	5,772	2,373	3,399
Devonshire	6,800	8,642	(1,842)	4,835	2,993	1,295	1,698
Dilworth	6,342	6,662	(320)	8,625	8,305	120	8,185
Druid Hills	3,708	5,691	(1,983)	2,644	661	76	585
Eastover	10,842	14,726	(3,884)	8,577	4,693	574	4,119
Elizabeth	16,557	11,162	5,395	8,095	13,490	1,030	12,460
First Ward	8,493	7,483	1,010	2,088	3,098	842	2,256
J.H. Gunn	8,748	11,523	(2,775)	2,946	171	171	
Hickory Grove	7,951	11,377	(3,426)	9,434	6,008	1,529	4,479
Hidden Valley	10,302	8,799	1,503	2,225	3,728	42	3,686
Highland	3,191	5,304	(2,113)	2,261	148		148
Hornets Nest	6,016	3,897	2,119	3,739	5,858		5,858
Huntersville	7,881	8,144	(263)	2,627	2,364		2,364
Huntingtowne Farms	4,812	7,313	(2,501)	4,433	1,932	578	1,354
Idlewild	18,252	17,481	771	1,185	1,956	671	1,285
Irwin Avenue	23,448	19,497	3,951	8,072	12,023	5,783	6,240
Amay James	3,539	3,542	(3)	2,212	2,209	20	2,189
Lake Wylie	16,424	16,443	(19)	7,160	7,141	228	6,913
Lansdowne	43,721	35,425	8,296	8,513	16,809	1,432	15,377
Lebanon Road	6,425	7,122	(697)	4,482	3,785	115	3,670
Lincoln Heights	6,175	5,843	332	1,276	1,608	346	1,262

CHARLOTTE-MECKLENBURG BOARD OF EDUCATION
EXPENDABLE TRUST FUND
DETAILED SCHEDULE OF REVENUES, EXPENDITURES AND CHANGES IN FUND BALANCE
FOR THE YEAR ENDED JUNE 30, 1994 (Continued)

	REVENUES	EXPENDITURES	EXCESS (DEFICIENCY) OF REVENUES OVER EXPENDITURES	FUND BALANCE JUNE 30, 1993	FUND BALANCE JUNE 30, 1994	COMPONENTS OF FUND BALANCE RESERVED	UNDESIGNATED
ELEMENTARY SCHOOLS (continued):							
Long Creek	$ 14,596	$ 14,845	$ (249)	$ 3,279	$ 3,030	$ 512	$ 2,518
Mallard Creek	28,063	33,502	(5,439)	14,594	9,155	1,940	7,215
Matthews	20,003	21,956	(1,953)	6,753	4,800	331	4,469
McAlpine	15,395	16,831	(1,436)	3,611	2,175	406	1,769
McKee Road	19,762	18,628	1,134	6,236	7,370		7,370
Merry Oaks	5,605	5,192	413	4,143	4,556	123	4,433
Montclaire	7,104	8,795	(1,691)	4,861	3,170	145	3,025
Myers Park	11,594	13,094	(1,500)	8,095	6,595	192	6,403
Nations Ford	9,624	16,291	(6,667)	15,522	8,855	2,522	6,333
Newell	13,849	10,522	3,327	8,812	12,139	924	11,215
Oakdale	4,857	5,024	(167)	3,146	2,979		2,979
Oakhurst	12,349	12,233	116	9,647	9,763	1,531	8,232
Oaklawn	4,817	7,240	(2,423)	3,144	721	46	675
Olde Providence	21,111	27,083	(5,972)	8,444	2,472	15	2,457
Park Road	3,326	4,835	(1,509)	3,545	2,036	192	1,844
Paw Creek	4,037	3,897	140	3,194	3,334		3,334
Pawtuckett	9,038	6,342	2,696	2,281	4,977		4,977
Pineville	7,815	9,980	(2,165)	5,654	3,489	166	3,323
Pinewood	5,797	5,887	(90)	3,612	3,522		3,522
Piney Grove	10,054	11,682	(1,628)	5,908	4,280	155	4,125
Rama Road	9,252	9,117	135	7,628	7,763	135	7,628
Reedy Creek	3,497	5,965	(2,468)	2,957	489	110	379
Reid Park (opening 8/22/94)	3,000	70	2,930		2,930		2,930
Sedgefield	9,046	10,160	(1,114)	2,302	1,188		1,188
Selwyn	6,160	6,001	159	3,143	3,302	281	3,021
Shamrock Gardens	2,396	2,713	(317)	2,899	2,582	23	2,559
Sharon	7,033	10,853	(3,820)	7,447	3,627	430	3,197
Smithfield	19,648	14,287	5,361	3,000	8,361	105	8,256
Starmount	12,228	12,405	(177)	2,797	2,620	98	2,522
Statesville Road	10,574	6,498	4,076	7,190	11,266	312	10,954
Steele Creek	7,258	14,227	(6,969)	10,558	3,589		3,589
Sterling	6,462	7,312	(850)	6,163	5,313	1,810	3,503
Thomasboro	5,597	6,580	(983)	4,573	3,590	120	3,470
Tryon Hills	5,010	4,595	415	1,611	2,026	67	1,959
Tuckaseegee	5,988	6,206	(218)	4,624	4,406	173	4,233
University Meadows	10,997	10,174	823	1,805	2,628	437	2,191
Villa Heights	3,785	4,216	(431)	1,038	607	89	518
Westerly Hills	8,337	7,159	1,178	3,821	4,999	2,261	2,738
Windsor Park	8,931	9,469	(538)	4,284	3,746		3,746
Winterfield	13,090	11,492	1,598	4,031	5,629		5,629
Total Elementary Schools	823,686	852,407	(28,721)	405,091	376,370	42,155	334,215

CHARLOTTE-MECKLENBURG BOARD OF EDUCATION
EXPENDABLE TRUST FUND
DETAILED SCHEDULE OF REVENUES, EXPENDITURES AND CHANGES IN FUND BALANCE
FOR THE YEAR ENDED JUNE 30, 1994 (Concluded)

	REVENUES	EXPENDITURES	EXCESS (DEFICIENCY) OF REVENUES OVER EXPENDITURES	FUND BALANCE JUNE 30, 1993	FUND BALANCE JUNE 30, 1994	COMPONENTS OF FUND BALANCE RESERVED	UNDESIGNATED
MIDDLE SCHOOLS:							
Albemarle Road	$ 52,250	$ 51,855	$ 395	$ 5,999	$ 6,394	$ 763	$ 5,631
J. McKnitt Alexander	72,329	78,306	(5,977)	30,875	24,898	2,807	22,091
Carmel	22,066	22,991	(925)	6,180	5,255	267	4,988
Cochrane	53,080	49,705	3,375	3,530	6,905	47	6,858
Coulwood	41,141	45,678	(4,537)	10,336	5,799		5,799
Marie G. Davis	10,730	9,774	956	570	1,526	140	1,386
Eastway	14,093	27,363	(13,270)	23,134	9,864	739	9,125
Alexander Graham	46,750	48,618	(1,868)	5,231	3,363	3,363	
Hawthorne	19,640	20,579	(939)	10,493	9,554	2,577	6,977
Robert F. Kennedy	44,284	48,271	(3,987)	14,594	10,607	1,453	9,154
McClintock	47,924	43,899	4,025	15,260	19,285	531	18,754
Northeast	63,523	68,898	(5,375)	9,796	4,421	127	4,294
Northwest	14,594	14,396	198	6,039	6,237	596	5,641
Piedmont	34,447	32,386	2,061	12,651	14,712	1,612	13,100
Quail Hollow	49,109	42,811	6,298	15,924	22,222	2,245	19,977
Randolph	46,652	39,645	7,007	3,884	10,891	1,228	9,663
Ranson	82,453	84,730	(2,277)	6,693	4,416	2,752	1,664
Sedgefield	13,516	9,516	4,000	573	4,573	1,614	2,959
J. Mason Smith	49,795	45,448	4,347	16,933	21,280	16	21,264
South Charlotte	85,193	76,534	8,659	8,766	17,425	3,342	14,083
Herbert Spaugh	27,403	27,989	(586)	5,946	5,360	745	4,615
John T. Williams	45,210	45,235	(25)	3,598	3,573	519	3,054
Wilson	47,925	47,025	900	4,465	5,365	339	5,026
Total Middle Schools	984,107	981,652	2,455	221,470	223,925	27,822	196,103
SPECIAL SCHOOLS:							
Dolly Tate Teen-Age Parent Services	2,973	2,456	517	4,799	5,316	23	5,293
Midway High School	4,462	5,254	(792)	3,361	2,569	343	2,226
Metro School	17,831	13,069	4,762	2,997	7,759		7,759
Plaza Road Preschool	6,255	7,900	(1,645)	3,592	1,947		1,947
Total Special Schools	31,521	28,679	2,842	14,749	17,591	366	17,225
HIGH SCHOOLS:							
East Mecklenburg	294,289	281,585	12,704	60,413	73,117	33,884	39,233
Garinger	231,076	198,098	32,978	36,622	69,600	30,360	39,240
Harry P. Harding	157,261	144,477	12,784	67,854	80,638	1,099	79,539
Independence	236,029	230,033	5,996	42,359	48,355	2,290	46,065
Myers Park	259,419	279,892	(20,473)	75,723	55,250	8,179	47,071
North Mecklenburg	262,543	269,283	(6,740)	41,289	34,549	7,556	26,993
Olympic	171,714	175,652	(3,938)	43,351	39,413	39,413	
Providence	343,103	318,051	25,052	51,807	76,859	9,887	66,972
South Mecklenburg	251,774	258,316	(6,542)	92,832	86,290	24,493	61,797
West Charlotte	239,132	240,630	(1,498)	21,070	19,572	12,617	6,955
West Mecklenburg	173,107	174,654	(1,547)	47,134	45,587	692	44,895
Total High Schools	2,619,447	2,570,671	48,776	580,454	629,230	170,470	458,760
TOTAL	$4,458,761	$4,433,409	$25,352	$1,221,764	$1,247,116	$240,813	$1,006,303

Source: Courtesy Charlotte-Mecklenburg Public Schools, CAFR. Recipient of ASBO *Certificate of Excellence Award* (1994).

Summary

The educational accounting procedures in this chapter are not only useful financial management tools for the complete accounting cycle, but they are also a vital part of effective school management. Accounting functions serve essentially three public constituencies: the general public (taxpayers), the school board (policymakers), and school managers (superintendents, business administrators, and so on). The comprehensive procedures described and illustrated in this chapter are therefore designed to demonstrate to the public that the highest standards of accounting and reporting are in place and that ultimately the public treasury is being managed efficiently and effectively to secure high quality education. Particularly for the school board, good accounting procedures provide vital information concerning the structure of the school district's finances, as well as timely information critical to good policy decisions. For school leaders, these procedures provide detailed data to help them understand the particular workings and needs of the district—a critical element in systematic planning for the future.

Figure 11.18
Combined Statement of Revenue, Expenses and Changes in Retained Earnings—All Proprietary Fund Types

	ENTERPRISE FUNDS	INTERNAL SERVICE FUNDS	TOTAL
OPERATING REVENUES:			
Food Sales	$ 11,432,287		$ 11,432,287
Participant Fees	4,686,569		4,686,569
Charges for Services		$17,110,797	17,110,797
Insurance Proceeds		61,200	61,200
Total Operating Revenues	16,118,856	17,171,997	33,290,853
OPERATING EXPENSES:			
Food Costs:			
Purchased Food	7,905,796		7,905,796
Donated Commodities	1,326,838		1,326,838
Salaries	11,758,385	7,612,247	19,370,632
Employee Benefits	2,645,771	1,971,723	4,617,494
Materials and Supplies	950,559	3,885,890	4,836,449
Utilities	172,393	176,329	348,722
Depreciation	536,937	1,089,773	1,626,710
Contracted Services	195,760	2,398,555	2,594,315
Equipment Rental	60,161	334,046	394,207
Insurance Premiums		751,645	751,645
Claim Payments		1,056,985	1,056,985
Indirect Cost	1,820,215		1,820,215
Other	265,943	100,569	366,512
Total Operating Expenses	27,638,758	19,377,762	47,016,520
OPERATING LOSS	(11,519,902)	(2,205,765)	(13,725,667)
NON-OPERATING REVENUES (EXPENSES):			
U.S. Government Subsidy	10,201,686		10,201,686
U.S. Government Commodities	1,326,838		1,326,838
Interest Income	216,592	771,852	988,444
Interest Expense		(2,923)	(2,923)
Gain on Sale of Fixed Assets		62,400	62,400
Other	4,250		4,250
Total Non-Operating Revenues	11,749,366	831,329	12,580,695
INCOME (LOSS) BEFORE OPERATING TRANSFER	229,464	(1,374,436)	(1,144,972)
OPERATING TRANSFER FROM GENERAL FUND	824,957		824,957
NET INCOME (LOSS)	1,054,421	(1,374,436)	(320,015)
RETAINED EARNINGS--Beginning of Year	3,322,184	1,811,234	5,133,418
RETAINED EARNINGS--End of Year	$ 4,376,605	$ 436,798	$ 4,813,403

Source: Courtesy Charlotte-Mecklenburg Public Schools, CAFR. Recipient of ASBO *Certificate of Excellence Award* (1994).

Figure 11.19

Combined Statement of Revenue, Expenditures, Encumbrances and Changes in Fund Balances—Budget and Actual—All Government Fund Types

	GENERAL FUND		CAPITAL PROJECTS FUND		TOTALS (MEMORANDUM ONLY)	
	BUDGET	ACTUAL	BUDGET	ACTUAL	BUDGET	ACTUAL
REVENUES:						
State of North Carolina	$255,685,778	$253,642,951	$ 1,850,000	$ 1,841,788	$257,535,778	$255,484,739
Mecklenburg County	123,330,066	123,327,566	31,200,000	31,140,665	154,530,066	154,468,231
U.S. Government Agencies	17,889,000	17,885,973			17,889,000	17,885,973
Other	7,597,318	6,806,158	7,050,000	7,041,339	14,647,318	13,847,497
Total Revenues	404,502,162	401,662,648	40,100,000	40,023,792	444,602,162	441,686,440
EXPENDITURES AND ENCUMBRANCES:						
Current Operating Expenditures and Encumbrances:						
Instructional Programs:						
Regular	166,816,432	166,100,354			166,816,432	166,100,354
Special	30,922,823	30,224,668			30,922,823	30,224,668
Vocational	11,038,461	11,037,598			11,038,461	11,037,598
Co-Curricular	1,745,078	1,744,995			1,745,078	1,744,995
	210,522,794	209,107,615			210,522,794	209,107,615
Pupil Support	17,639,479	17,456,244			17,639,479	17,456,244
Instructional Staff Support	14,883,433	14,623,802			14,883,433	14,623,802
General Administrative Support	3,096,000	3,086,925			3,096,000	3,086,925
School Administrative Support	22,881,595	22,294,186			22,881,595	22,294,186
Business and Operating Support	54,729,245	54,415,593			54,729,245	54,415,593
Central Support	6,725,463	6,451,228			6,725,463	6,451,228
Employee Benefits	69,604,378	69,042,324			69,604,378	69,042,324
Community Services	538,382	538,378			538,382	538,378
Total Current Operating Expenditures and Encumbrances	400,620,769	397,016,295			400,620,769	397,016,295
Capital Outlay Expenditures and Encumbrances:						
Buildings and Site Improvements	1,442,533	1,442,533	29,700,000	29,656,973	31,142,533	31,099,506
Furniture and Equipment	5,841,559	5,840,755	10,400,000	10,366,819	16,241,559	16,207,574
Total Capital Outlay Expenditures and Encumbrances	7,284,092	7,283,288	40,100,000	40,023,792	47,384,092	47,307,080
Total Expenditures and Encumbrances	407,904,861	404,299,583	40,100,000	40,023,792	448,004,861	444,323,376
DEFICIENCY OF REVENUES OVER EXPENDITURES AND ENCUMBRANCES	(3,402,699)	(2,636,935)	-0-	-0-	(3,402,699)	(2,636,935)
OTHER FINANCING SOURCES (USES):						
Fund Balance Appropriated	4,227,656	-0-			4,227,656	-0-
Operating Transfers to Enterprise Funds	(824,957)	(824,957)			(824,957)	(824,957)
Total Other Financing Sources (Uses)	3,402,699	(824,957)			3,402,699	(824,957)
DEFICIENCY OF REVENUES OVER EXPENDITURES, ENCUMBRANCES AND OTHER FINANCING SOURCES (USES)	-0-	(3,461,892)	-0-	-0-	-0-	(3,461,892)
FUND BALANCES--Beginning of Year	-0-	6,032,942	-0-	-0-	-0-	6,032,942
FUND BALANCES--End of Year	$ -0-	$ 2,571,050	$ -0-	$ -0-	$ -0-	$ 2,571,050

Source: Courtesy Charlotte-Mecklenburg Public Schools, CAFR. Recipient of ASBO *Certificate of Excellence Award* (1994).

Figure 11.20
Combined Balance Sheet—All Fund Types and Account Groups

	GOVERNMENTAL FUND TYPES		PROPRIETARY FUND TYPES		FIDUCIARY FUND TYPE	ACCOUNT GROUPS		TOTALS (MEMORANDUM ONLY)	
	GENERAL	CAPITAL PROJECTS	ENTERPRISE	INTERNAL SERVICE	EXPENDABLE TRUST	GENERAL FIXED ASSETS	GENERAL LONG-TERM DEBT	1994	1993
ASSETS									
CASH AND CASH EQUIVALENTS	$48,522,346	$ 2,325,241	$3,345,960	$14,055,399	$1,247,116			$ 69,496,062	$ 65,712,879
RECEIVABLES:									
Mecklenburg County	6,119	26,738,537						26,744,656	30,304,585
State of North Carolina	13,188,615							13,188,615	12,390,598
U.S. Government Agencies	1,807,688		1,420,560					3,228,248	1,630,491
Accrued Interest and Other	242,514		61,287	75,777				379,578	417,850
DUE FROM OTHER FUNDS									194,926
INVENTORIES	1,251,083		820,105	649,132				2,720,320	3,618,995
FIXED ASSETS			4,194,007	3,656,960		$451,632,142		459,483,109	421,216,876
AMOUNT TO BE PROVIDED FOR COMPENSATED ABSENCES							$16,397,923	16,397,923	14,429,627
TOTAL	$65,018,365	$29,063,778	$9,841,919	$18,437,268	$1,247,116	$451,632,142	$16,397,923	$591,638,511	$549,916,827
LIABILITIES, EQUITY AND OTHER CREDITS									
ACCOUNTS PAYABLE	$ 3,563,585	$ 3,612,702	$ 82,170	$ 113,485				$ 7,371,942	$ 9,110,452
CLAIMS PAYABLE				2,659,905				2,659,905	
ACCRUED SALARIES, WAGES AND BENEFITS:									
Teachers' Summer Supplement	20,937,457							20,937,457	17,815,363
Other	20,225,967		185,858	66,116				20,477,941	17,752,658
DUE TO OTHER FUNDS									194,926
DEFERRED REVENUE	1,343,564	2,325,241	218,227	1,069,163				4,956,195	5,694,586
OBLIGATIONS UNDER CAPITAL LEASES				14,687				14,687	47,800
LIABILITY FOR COMPENSATED ABSENCES	1,245,619		168,349	464,486			$16,397,923	18,276,377	16,560,201
Total Liabilities	47,316,192	5,937,943	654,604	4,387,842			16,397,923	74,694,504	67,175,986
EQUITY AND OTHER CREDITS:									
Contributed Capital:									
Local Sources			4,810,710	13,612,628				18,423,338	16,139,274
Investment in General Fixed Assets						$451,632,142		451,632,142	413,766,367
Retained Earnings:									
Reserved by State Statute			1,691,902					1,691,902	492,828
Unreserved			2,684,703	436,798				3,121,501	4,640,590
Fund Balance:									
Reserved by State Statute	16,051,441	23,125,835			$ 240,813			39,418,089	43,602,432
Unreserved:									
Designated for Subsequent Year's Expenditures	1,600,000							1,600,000	2,300,000
Undesignated	50,732				1,006,303			1,057,035	1,799,350
Total Equity and Other Credits	17,702,173	23,125,835	9,187,315	14,049,426	1,247,116	451,632,142		516,944,007	482,740,841
TOTAL	$65,018,365	$29,063,778	$9,841,919	$18,437,268	$1,247,116	$451,632,142	$16,397,923	$591,638,511	$549,916,827

Source: Courtesy Charlotte-Mecklenburg Public Schools, CAFR. Recipient of ASBO *Certificate of Excellence Award* (1994).

Figure 11.21
Combined Statement of Revenue, Expenditures, and Changes in Fund Balances—All Governmental Fund Types and Expendable Trust Fund

	GOVERNMENTAL FUND TYPES		FIDUCIARY FUND TYPE	TOTALS (MEMORANDUM ONLY)	
	GENERAL	CAPITAL PROJECTS	EXPENDABLE TRUST	1994	1993
REVENUES:					
State of North Carolina	$254,901,869	$ 1,841,788		$256,743,657	$244,352,544
Mecklenburg County	123,327,566	31,140,665		154,468,231	163,424,271
U.S. Government Agencies	17,885,973			17,885,973	15,152,951
Other	6,806,158	7,041,339	$4,458,761	18,306,258	10,195,105
Total Revenues	402,921,566	40,023,792	4,458,761	447,404,119	433,124,871
EXPENDITURES:					
Current Operating Expenditures:					
Instructional Programs:					
Regular	166,115,435		175,554	166,290,989	150,526,989
Special	30,147,504			30,147,504	27,592,726
Vocational	11,037,598			11,037,598	9,852,044
Co-Curricular	1,731,207		3,073,756	4,804,963	4,486,715
	209,031,744		3,249,310	212,281,054	192,458,474
Pupil Support	17,451,381			17,451,381	17,047,111
Instructional Staff Support	14,662,762			14,662,762	13,569,430
General Administrative Support	3,083,689			3,083,689	2,297,463
School Administrative Support	22,851,108		1,184,099	24,035,207	22,830,875
Business and Operating Support	55,073,944			55,073,944	52,179,823
Central Support	7,019,155			7,019,155	6,927,940
Employee Benefits	70,716,423			70,716,423	67,922,235
Community Services	535,942			535,942	606,204
Total Current Operating Expenditures	400,426,148		4,433,409	404,859,557	375,839,555
Capital Outlay Expenditures:					
Building and Site Improvements	1,529,302	32,545,146		34,074,448	39,006,957
Furniture and Equipment	3,535,794	9,736,021		13,271,815	11,904,827
Total Capital Outlay Expenditures	5,065,096	42,281,167		47,346,263	50,911,784
Total Expenditures	405,491,244	42,281,167	4,433,409	452,205,820	426,751,339
EXCESS (DEFICIENCY) OF REVENUES OVER EXPENDITURES	(2,569,678)	(2,257,375)	25,352	(4,801,701)	6,373,532
OTHER FINANCING USES:					
Operating Transfer to Enterprise Funds	(824,957)			(824,957)	(808,227)
EXCESS (DEFICIENCY) OF REVENUES OVER EXPENDITURES AND OTHER FINANCING USES	(3,394,635)	(2,257,375)	25,352	(5,626,658)	5,565,305
FUND BALANCES--Beginning of Year	21,096,808	25,383,210	1,221,764	47,701,782	42,136,477
FUND BALANCES--End of Year	$ 17,702,173	$23,125,835	$1,247,116	$ 42,075,124	$ 47,701,782

Source: Courtesy Charlotte-Mecklenburg Public Schools, CAFR. Recipient of ASBO *Certificate of Excellence Award* (1994).

Figure 11.22
Combined Statement of Revenue, Expenses and Changes in Retained Earnings— All Proprietary Fund Types

	ENTERPRISE FUNDS	INTERNAL SERVICE FUNDS	TOTAL
OPERATING REVENUES:			
Food Sales	$ 11,432,287		$ 11,432,287
Participant Fees	4,686,569		4,686,569
Charges for Services		$17,110,797	17,110,797
Insurance Proceeds		61,200	61,200
Total Operating Revenues	16,118,856	17,171,997	33,290,853
OPERATING EXPENSES:			
Food Costs:			
Purchased Food	7,905,796		7,905,796
Donated Commodities	1,326,838		1,326,838
Salaries	11,758,385	7,612,247	19,370,632
Employee Benefits	2,645,771	1,971,723	4,617,494
Materials and Supplies	950,559	3,885,890	4,836,449
Utilities	172,393	176,329	348,722
Depreciation	536,937	1,089,773	1,626,710
Contracted Services	195,760	2,398,555	2,594,315
Equipment Rental	60,161	334,046	394,207
Insurance Premiums		751,645	751,645
Claim Payments		1,056,985	1,056,985
Indirect Cost	1,820,215		1,820,215
Other	265,943	100,569	366,512
Total Operating Expenses	27,638,758	19,377,762	47,016,520
OPERATING LOSS	(11,519,902)	(2,205,765)	(13,725,667)
NON-OPERATING REVENUES (EXPENSES):			
U.S. Government Subsidy	10,201,686		10,201,686
U.S. Government Commodities	1,326,838		1,326,838
Interest Income	216,592	771,852	988,444
Interest Expense		(2,923)	(2,923)
Gain on Sale of Fixed Assets		62,400	62,400
Other	4,250		4,250
Total Non-Operating Revenues	11,749,366	831,329	12,580,695
INCOME (LOSS) BEFORE OPERATING TRANSFER	229,464	(1,374,436)	(1,144,972)
OPERATING TRANSFER FROM GENERAL FUND	824,957		824,957
NET INCOME (LOSS)	1,054,421	(1,374,436)	(320,015)
RETAINED EARNINGS--Beginning of Year	3,322,184	1,811,234	5,133,418
RETAINED EARNINGS--End of Year	$ 4,376,605	$ 436,798	$ 4,813,403

Source: Courtesy Charlotte-Mecklenburg Public Schools, CAFR. Recipient of ASBO *Certificate of Excellence Award* (1994).

The effect of good accounting procedures is broad. In recent years, financial analysts have increasingly relied on the statement of changes in financial position as a benchmark of financial health of a prospective borrowing school district. As a result, the American Institute of Certified Public Accountants has encouraged its members to show either a change in cash balance, the cash balance, the cash balance and temporary investments and working capital, or the changes in working capital. Perhaps the most telling justification for the complex accounting structure in schools rests in recognition that a statement of gross revenue does not represent the total picture of resources available to a school system. Significant financial resources are created each time an asset of the school system is converted, exchanged, or sold.

Using comprehensive accounting procedures enables the business administrator or chief financial officer to make decisions that take maximum advantage of the total resources available in the given cash flow environment. When the statement of changes in financial position is prepared on the basis of total financial resources provided and applied for each fund producing income and/or revenue, it graphically illustrates the financial magnitude of a school system at the local, state, and federal levels. It proves to management, the board of education, and the public at large that school business is indeed big business.

Endnotes

[1]D. C. Thompson, R. C. Wood, and D. S. Honeyman, *Fiscal Leadership for Schools: Concepts and Practices*. New York: Longman (1994), 346.

[2]Discussion in this chapter is rather technical and presumes prior exposure to accounting systems. For a more introductory approach to accounting, see chapter 7 in D. C. Thompson, R. C. Wood, and D. S. Honeyman, *Fiscal Leadership for Schools*, 345-400.

[3]This section is based on D. C. Thompson, R. C. Wood, and D. S. Honeyman. *Fiscal Leadership for Schools,* 347-349.

[4]Some discussion of this concept appears on chapter 15 of *Management Information Systems*. For an extended discussion, see chapter 13 in D. C. Thompson, R. C. Wood, and D.S . Honeyman, *Fiscal Leadership for Schools, 597-635.*

[5]D. C. Thompson, R. C. Wood, and D. S. Honeyman, *Fiscal Leadership for Schools,* 348-349.

[6]Some states use a GAAP waiver to exempt school districts from this requirement. That is not to say, however, that poor accounting practices are permitted—rather, states may have controlling state statutory guidance regarding proper reporting procedures.

Chapter 12 School District Budgeting

Introduction

School districts face growing pressure to provide an increased variety of educational services to a diverse student population within the context of limited resources. There is little reason to expect this trend to change in the near term given current concerns over economic growth, new spending requirements, increased federal and state mandates, deteriorating school facilities, and continued demand for improved educational quality. The effectiveness of school district responses to these pressures will depend to a considerable extent on the quality of their budget practices. Despite this critical place in the management and operation of educational organizations, it can be argued that surprisingly little is known about the factors that constitute successful budgeting and budget practice.[1]

Development of a budget requires detailed plans for the programs and services a school district will operate for the coming fiscal year as well as an estimate of the resources necessary to accomplish the goals of each program and service. Once the resources are identified, the costs of each must be determined. At the same time, the revenue that will be used to support the district's operations must be estimated, and expenditure plans adjusted to meet available funding levels.

Because a budget represents the district's plan for implementing its policies and goals, it must be approved by the board of education.[2] Additionally, a district's budget serves as a management tool to direct the fiscal operations of the district. The budget process creates a document that both serves to operationalize school district policies and ensure that scarce public funds are efficiently and legally allocated to achieve those policies.

This chapter provides a detailed discussion of school district budgeting processes, focusing on both the underpinnings of budgetary theory and the practical application of that theory to school district management and control.

Budget Defined

Most laypersons think of a school district's budget as a large collection of seemingly incomprehensible numbers, prepared by an obscure office somewhere in the central administration. Yet, these data tell, in surprising detail, a great deal about the school district, its educational philosophy, and its management style. The budget also provides a great deal of information on the priorities of the school district and offers a glimpse as to how those priorities will be carried out. Given the number of important functions attributed to a budget, it is not surprising that defining exactly what a budget is and what it is supposed to do, is a difficult undertaking.

Budgeting has been defined as "translating financial resources into human purposes." Thus, as a result of limited financial resources, some way must be found to "apportion available funds among competing people and purposes." Moreover, since government budgets must take resources from some citizens and distribute them to others, one outcome of budget processes is conflict. It is argued that given the "infinite variety of human desires, the budget of a government can never be just one thing—it must be many."[3] Thus, the budget serves at least six different functions including:[4]

- *A record of the past:* It reflects bargains and compromises that have been made in the past, reflects priorities of earlier administrations, provides a record of victories and defeats in earlier budget negotiations, and by inference indicates what has been eliminated in the past.

- *A statement about the future*: It links proposed expenditures with desired future events. As such, it represents a plan to determine future events through current actions.

- *Predictions of future actions*: Budgets specify connections between words and numbers in a budget document and future human behavior. Whether the behavior intended by the makers of the budget occur, is not a given, but rather a matter of observation.

- *A mechanism for allocating resources:* Funds are almost always more scarce than desires. As a result, a budget is a mechanism for allocating resources across competing demands, and if the intent were to reach certain objectives at the lowest possible cost, then a budget can also serve as an instrument for pursuing efficiency.

- *A form of power:* Budgets are mechanisms through which individuals or subunits bargain over conflicting goals, make side-payments, and attempt to motivate each other to achieve their own specific goals.

- *Signals:* Budgets serve as signals of the preferences of others, and provide a way for each group in an organization to communicate their priorities and desires.

A governmental budget may be examined and explained in a variety of ways. It can be suggested that a budget is first a political instrument to allocate scarce public resources among the social and economic needs of a jurisdiction. Second, it can be suggested that a budget is a management and administrative instrument that specifies ways and means of providing public services. Third, it can be argued that a budget is an economic instrument that can direct a nation's or a state's economic growth and can be used to redistribute income. Finally, it can be argued that a budget is an accounting tool used to hold government officials responsible for the expenditure of public funds.[5]

A variety of definitions of a school district budget exist such as the "translation of prioritized educational needs into a financial plan which is interpreted to the public in such a way that when formally adopted it expresses the kind of educational program the community is willing to support, financially and morally, for a one-year period."[6] It can be argued that a school district budget is "basically an instrument of educational planning and, incidentally, an instrument of control. It reflects the organizational pattern by categorizing the elements of a total plan into sectional, campus, and departmental components, allowing costs to be more easily estimated."[7] Further, it has been suggested that the "document resulting from the budgeting process, the budget, is often the most comprehensive description of what the school organization, be it a public or private school, intends to do."[8]

Hartman distinguishes between a budget and the process of budgeting. He defines the budget as "[a] document which specifies the planned expenditures and anticipated revenues of a school district in a given fiscal year, along with other data and information relating the fiscal elements to the educational philosophy, programs and needs of the district."[9] He suggests that there are three separate elements to a budget:

- A description of the educational program.
- An estimate of expenditures needed to carry out the program.
- An estimate of the revenue available to pay for those expenditures.

Furthermore, it can be argued that the process of developing a budget and the actual act of budgeting, is a cyclical activity that includes the activities of developing, approving, and implementing the district's budget. Thus, budgeting can be defined as "the sequence of activities involved in planning the district's educational program, estimating the needed expenditures and revenue to implement these programs, gaining the necessary approval, and using the budget to assist in managing the district's operations."[10]

The budget document is an important part of the operation of any school district. It not only describes the resources that will be used to conduct the district's goals and mission, but it summarizes the expenditures necessary for each of those resources.

Development of a budget is a complex process. Not only is it essential to determine what is needed and how much it will cost, but once that has been done, it is often necessary to make adjustments to the total if revenue is not sufficient to meet desired expenditure levels. When this happens, conflict occurs, and difficult decisions must be made. Where legally permissible, the school district may raise taxes or determine in some combination what programs or projects to reduce or eliminate. How this process plays out has a great deal to do with the final picture one receives of a school district.

Before describing how school district budgets are developed and used as a management tool, the next sections provide an historical description of budgets and budgetary processes in the United States, and offer a discussion of different budgetary methodologies that have been used in different settings.

An Historical Perspective on Budgets and Budgeting

Budgeting as we think of it today is a relatively new phenomenon. Prior to this century, budgets were often no more than the compilation of a series of appropriations passed by a legislative body. Rather than serving as a planning document, budgets simply summed up what was spent in each year. Presently, the most advanced budget techniques and methods can often be found within federal agencies. Moreover, most budget innovations have begun or initially been implemented at the federal level.

Federal Budget Policy: The Early Years

In 1784, the fourth act of Congress established the Treasury Department and directed the Secretary of the Treasury to prepare a system to estimate the revenue and expenditures of the federal government. However, the Constitution specifies that the authority to spend federal funds, the "power of the purse," belongs to the Congress. Moreover, the first secretary of the treasury, Alexander Hamilton, was so adept at putting the nation on a secure financial footing that his actions have been described as "primarily an exercise in getting rid of 'surplus funds' accumulated by tariff revenues or finding ways to fund major land purchases."[11]

The outcome of these factors was that for all practical purposes, Congress took responsibility for all federal budgeting. Congress received spending requests from each of the executive departments, considered the requests and appropriated funds. Often referred to as "particularism," the system resulted in detailed line-item appropriations of often insignificant amounts of funds. In addition, Congressional appropriations often contained direct instructions regarding the name of the individual to be hired for a specific position. With the exception of the President's veto power, the chief executive was largely removed from these procedures in terms of budgetary authority. Although the financial panic of 1837 and efforts to finance the Civil War led some to argue that a more sophisticated budgeting system was necessary, little changed in the way federal government funds were appropriated during the first 130 years of the nation's history.

One strategy used by presidents to have Congress appropriate funds for desired functions and purposes was the strategy of "coercive deficiencies." For example, President Jefferson's Louisiana Purchase of $15 million was 7.5 times the level of appropriations for that year. Perhaps the most famous use of this strategy was when President Theodore Roosevelt sent a Navy fleet halfway around the world and told Congress if it wanted the fleet to return home, it had to appropriate the funds for more fuel. Despite these attacks by presidents, for the most part, Congress was able to balance the budget. Deficits were rare, and many economists believe the federal budget deficit of zero in 1835 was one of the causes of the 1837-43 depression, the longest of that century.

Federal Budgets Following the Civil War

During the Civil War, President Lincoln was able to have Congress provide more "lump-sum" appropriations, in part because the level of spending made particularism more difficult to maintain, and, in part, because he demanded greater flexibility in determining how moneys should be spent on the war effort. Following the Civil War, the House Ways and Means and the Senate Finance Committees gave up their expenditure powers to appropriations committees in each House, but continued to maintain the revenue raising powers. Thus, for the first time, the power to raise and spend federal funds was split among different committees in both houses of Congress.

This spawned a 100-year debate over the budgetary roles of Congressional committees. With one committee responsible for revenue and one for appropriations, there was no real budget in the sense of ensuring that revenue and expenditures were balanced. In fact, this was not corrected until the Congressional Budget Reform Act was passed in 1974, and a third committee, the Budget Committee was created. The first Congressional budget was prepared in 1976.

The Movement for Budgetary Reform

Around the turn of the century, there were four factors that led to reform of the federal budget process. The first was the high level of governmental spending for the War of 1898 and World War I. Second, the change to an industrial economy limited the effectiveness of tariffs as a source of revenue. Third, the "good government reforms" of that time, both at the federal level and in New York, led to greater focus on governmental accountability and control. Finally, President Theodore Roosevelt was something of a proliferate spender, forcing Congress to appropriate funds for programs it might not otherwise have chosen to fund.

Federal budget reform became an issue with the publication of the Taft Commission report in 1912. Titled, *The Need for a National Budget*, this document argued for the establishment of a national budgeting system to serve the executive branch. Among the recommendations of the Taft Commission were the following:[12]

The president should submit an annual budget to Congress, which should contain:

- A budget message describing the significance of the budget and its proposals.
- A summary financial statement.
- A summary of expenditures.
- A comparison of estimated expenditures with actual expenditures in previous years.
- Estimates of revenue.
- A summary of changes in law that needed to be enacted to conduct public business with greater efficiency.

The secretary of the treasury should submit to Congress detailed reports on public revenue and expenditures.

The heads of all federal departments should be required to submit detailed expenditure reports to the secretary of the treasury.

The president should be required to keep financial accounts to keep track of all revenue and expenditures.

The president should propose appropriation bills to the Congress to conduct the business of the government.

The publication of the Taft Commission report highlighted a recurring theme of governmental budgeting, the conflict between the executive and the legislative branches over who, or what agency, would control the budget process. Because the Taft Commission Report suggested placing more power in the hands of the president, at the expense of Congress, its proposals were quickly rejected by the legislative branch. This conflict continued unabated, and when Congress passed legislation to establish a better budget system after World War I, President Wilson vetoed it because, in his view, it did not provide enough executive control.

In 1921, Congress did pass the Budget and Accounting Act, which was signed into law by President Harding.[13] It established a Bureau of the Budget within the Treasury Department, placing budget responsibilities with the executive branch, and created the General Accounting Office, which remained responsible to Congress. In 1939, the Bureau of the Budget became a direct staff office of the president, and has since been renamed the Office of Management and Budget (OMB). Today the OMB is one of the most powerful agencies within the federal government.

Federal Budgeting in the Twentieth Century:

The Stages of Reform

Following the passage of the Budget and Accounting Act, the federal budget process focused primarily on accountability and control. Budgeting was characterized by limited line-item budgets that were basically systematic, accounting-oriented ways to record public expenditures against various classification categories. These categories included salaries, travel, supplies, and equipment, all of which are familiar to individuals responsible for developing budgets today.[14] This has been referred to as the period of "Control Budget," one of the first of three important stages of budgetary reform to take place in the twentieth century.[15]

Concerned about the overemphasis on budget mechanics, V.O. Key wrote a path-breaking article in 1940 titled "The Lack of a Budgetary Theory," in which he posed a crucial question, "On what basis shall it be decided to allocate X dollars to activity A instead of activity B?"[16] Answering this question is still at the heart of public budget theory today.

Schick has suggested that while Key's question was critical, events of the 1940s and 1950s led governmental budgeting in a different direction, focusing instead on the use of the budget for management control. During this period, the budget processes focused mostly on the measurement of programs and outcomes, as opposed to the cost of the programs. Greater attention was devoted to assuring that resources were obtained and used effectively and efficiently to accomplish the organization's goals and objectives.[17]

The third stage of reform required using the budget as a planning document, where the process of deciding on objectives of the organization and the resources needed to attain those objectives becomes the focus of the budgetary process.[18] Reforms such as program budgeting and zero-based budgeting represent efforts to use budgets as more of a planning document.

As the next section of this chapter shows, the application of recent budgetary theories to school systems has met with limited success. Before discussing these reforms and the application to public school districts, a discussion of classical budget techniques is presented.

Approaches to Budgeting

Key's piercing question about what basis should be used to decide how resources should be allocated to different activities is at the heart of budget theory today. This section describes a number of alternative approaches that have been postulated for making these difficult decisions. Each has its strengths and weaknesses for governmental agencies in general, and for school districts specifically. The section begins with a discussion of economic and political approaches to budgeting, describes the classical components of a public budget, and assesses the valuc of two alternative approaches to budgeting that have been tried in public schools: program budgeting and zero-based budgeting.

An economic answer to Key's question would assume that rational decision making would lead to spending decisions that attempted to obtain maximum returns. Under economic theory, this occurs when the last dollar spent for each activity yields the same relative return, in other words, when the marginal rate of return for each service or activity provided by the agency was the same. In a market situation, where there is perfect information and values can be determined by price, it is theoretically possible to achieve such an optimal distribution of funds. Unfortunately, the goals and objectives of public services, particularly education, are not always easy to specify. If school leaders have difficulty reaching agreement on the desired outcomes of schooling (or at least on the relative importance of alternative outcome measures), then it is impossible to establish a rational measure of the best use of resources for educational expenditures.

The problem is further complicated by the fact that budget decisions have to be based not only on today's needs, but on forecasts of what the needs will be tomorrow, next year, and in the next decade. Thus, one of the problems is not so much the methodology for creating a budget, but the lack of firm numbers to use within whatever framework is chosen. Another issue is deciding who will obtain what from the budget, and who will pay for it. As a result of these uncertainties, and the need to reach compromises over who pays and who gains from governmental actions, it can be argued that budgets are very political, and should be thought of as political documents.[19]

Because of the many competing demands for governmental resources, developing a budget to accurately describe everything that is to be done is an immensely complex job. In classical budgeting, there is general agreement as to the size and scope of the budget and the overall distribution of expenditures. The conflicts are limited to the margins, in part to help deal with the complexity of budgeting. Public officials deal with this complexity in a variety of ways.

Wildavsky's Aids to Calculation

According to Wildavsky, to deal with the complexity of budgets, public officials rely on the following strategies.[20]

Consensus: Wildavsky argues that it is essential to have broad agreement on the general direction of public policy so that officials are not swamped with difficult choices. This does not mean that new directions cannot be introduced nor old policies ended. It simply suggests that on a regular basis most overall policy goals must be agreed upon.

In school districts, this is not as difficult as it may seem. For example, there is no doubt broad agreement in most, if not all school districts, that to provide educational services to the children residing within the district's boundaries, the school district must hire teachers to provide instruction and purchase textbooks to supplement and support that instruction. Along with those components, it is essential to provide administrative support to the teachers, along with adequate classrooms and other instructional facilities. Evaluation activities must be undertaken to ascertain how the students and the teachers are performing, and resources must be devoted to ensuring that the children are able to come to school each day. Few would disagree with these overall objectives. More likely, there would be discussion concerning the relative amount of resources to be devoted to each of these individual activities.

Historical: One way to deal with a problem of huge magnitude is to make rough guesses about costs and let experience accumulate. Overtime, the estimates of how much resources are needed for certain functions can be refined. By providing decisionmakers with historical data on budget decisions, it is possible to determine whether enough money was spent to achieve a certain goal, and if future appropriations should be increased or decreased.

Fragmented: Building a budget for an entire school district is a complex and difficult undertaking. Moreover, individuals in a central business office often have limited knowledge of the needs of individual schools or departments. As a result, budgets are made in fragments. District departments and individual schools develop budgets based on specific needs. These requests are channeled into a number of different places leading eventually to a central compilation of the entire budget document. It is likely that revenue and expenditures will not balance, and each department or school will be asked to revise its anticipated expenditure needs. Establishing priorities for increases or decreases in budget authority is, therefore, also an important component of the budget process.

Incremental: The largest single factor determining next year's budget is the current year's budget. In fact, most public budgeting is done by estimating the cost of the resources for the current year in the next year. For example, if salaries were expected to increase by 5 percent, and all other items will increase in cost by 7 percent, the budget for next year is first estimated by making those adjustments to the existing budget. The costs of programmatic changes are then estimated, and the budget adjusted accordingly.

Wildavsky coined the term "incrementalism" to describe this process. He argued that most governmental budgets are very incremental, and that most changes can only be made at the margin. Given that the technology of education has been exceedingly stable for many years, i.e., one teacher standing in front of a group of children in a classroom that is part of a larger school building, it is not surprising to find that most school system budgets are put together in a very incremental manner.

Simplification: Often referred to as "straining at gnats," policy makers frequently rely on a review of a sample of actions as indices of how other matters are handled. Rather than attempt to deal with all of the complexities of a large school district, board members may focus on a small administrative function. If that function were to be handled properly and efficiently, they may take it as an indicator that administrators can be trusted to operate larger functions equally well. This process can work very well if it is used as a testing device and if there is a reasonable carryover between the competencies needed for the simple function and the more complex one.

Satisficing: This term was originally coined by Herbert Simon to describe the way many individuals approach complex decisions. While it would be ideal to consider all possible options and rationally evaluate each before choosing the best alternative, it is often time consuming and expensive to do so. Consequently, individuals look at a subset of options, selecting one that meets most of their needs. Wildavsky suggests that government decisionmakers do the same thing in reaching decisions concerning complex budget issues. They do not actually refer to their actions as satisficing, but rather use terms like "get by," "come out all right," "avoid trouble," or "avoid the worst." Wildavsky points out that since budgeting is an annual function, this is a good method to correct glaring weaknesses when they occur.

Non-Programmatic: What this means is that the question of the desirability of a program usually only comes up once, or at most infrequently. Most budgeting relies on making marginal adjustments to existing programs. The ultimate desirability of most programs is only discussed occasionally.

Repetitive: Any particular problem is likely to be dealt with over and over again since budgeting occurs on an annual basis. As a result, most budget activity occurs in an environment where the decisionmakers realize that most problems do not have to be solved once and for all. Matters that are not considered in one year may be revisited in another. In many cases, problems may not be solved, but in Wildavsky's words, "worn down" by repeated attacks. He suggests that "problem succession," not problem solving, best describes what happens.[21]

Sequential: Many decisions made in one year are allowed to stand for many years, without review. Rather than deal with all problems at once, budget issues are often considered sequentially. If a problem is not resolved one year, it may be a topic of discussion the next.

What this discussion shows is that despite the myriad complexities that development of a budget seems to represent, policymaking bodies, be they Congress, a state legislature, or a school board, have a number of ways to simplify matters and deal with them in a way that is not overwhelming. The result of these actions is that most budgets show very little change from year to year. There is a great deal of inertia in a governmental budget. In times of growth, the goal is to add new programs, while in times of retrenchment, there is a great deal of effort devoted to maintaining existing programs and services.

Given these observations concerning the budget process, it is not surprising that most school district budgets are remarkably similar. There are, of course, differences based on state accounting and reporting requirements, and based on district size and management style, but in general, if school business administrators are doing a good job of preparing the budget document, it is possible to identify a basic set of components that constitute that budget.

Features of Classical Budgets

Although each school district's budget is, of necessity different, the basic components of the budget are the same. There are six features that are common to any good budget. They are as follows:

- *Unity*: The budget document should describe all of the programs and services of the school district. The budget document should include detailed revenue and expenditure forecasts for the general fund as well as for all other budgetary funds used by the school district, including capital funds.

- *Regularity*: Budgets must be prepared on a regular basis. In the United States, virtually all government agencies, including school districts, prepare annual budgets. Some states use biennial budgets, but the norm is the twelve-month or annual budget. More importantly, the budget should reflect the same twelve-month period each year. Most school districts use a fiscal year that runs from July 1 of one year to June 30 of the following year. The determination of a fiscal year is generally a state requirement, and several states rely on a calendar year, and at least one, Texas, requires school districts to use a fiscal year beginning on September 1 and ending on August 31.

- *Clarity:* A well-designed budget document makes it clear how the district's revenue is collected and spent. It will identify all sources of revenue, and indicate not only how much is spent, but in general, what those funds will be used to purchase. Particularly important are personnel counts by function or program, and separate breakdowns of expenditures for supplies and materials, travel, and other goods and services. The budget should be displayed in a manner that is understandable to the average citizen and provide aids to its review, such as descriptions of the budget process and its contents, and a comprehensive table of contents.

- *Balance*: The budget should be balanced. This means that revenue should equal expenditures plus any amount budgeted for contingencies. If a district's end-of-year actual revenue exceeds expenditures, the difference can be added, where legally permissible, to the district's financial reserves. In years where revenue is tight, districts may choose to draw down these reserves rather than make larger reductions in the upcoming year's operating budget. If this is done, the budget document should make it clear that that year's expenditures are expected to exceed that year's revenue and indicate what plans the district has to correct that imbalance before all of the reserves are exhausted.

- *Publicity*: School district budgets are public documents that describe how the district plans to use the tax revenue it collects to provide educational services to the children attending school in the district. As such, it is the right of every citizen in the district to review and comment on that budget. The budget development process (described in detail later) should provide ample opportunity for comment by the public, and all budget hearings should be open to the public. Moreover, the school district should make efforts to publicize the budget document's availability and the date and times of budget hearings. Generally, the number and time of these public hearings are proscribed by state statute. Many school districts make copies of the budgets available to the public through local libraries and at the school district offices.

- *Operational adequacy*: The expenditures detailed in the budget should be adequate to provide the services required to meet the district's mission and goals.

Given that the budget document must contain these six components, there are a number of approaches for using these components. The most common, even today, is the line-item budget. Other budget approaches that have been suggested, and are in some instances used in school districts, include program budgeting and zero-based budgeting. Each is described at length herein.

Line-Item Budgets

A line-item budget is a technique where individual lines describe allocations for various objects of expenditure, such as personnel salaries, textbooks, supplies, contractual services, and capital outlay. The individual line items are the focus of analysis, authorization, and control. This approach to budgeting was used almost exclusively through the 1960s. For educational agencies, it was particularly useful as it focused on identifiable items and provided educational leaders (both managers and school board members) with information that was straightforward and easy to understand. Moreover, it provides an excellent method to control expenditures as it is possible to compare actual expenditures with budgeted expenditures for any and all line items. The level of detail in a line-item budget is limited only by the desire for detail expressed by the school district's managers.

In standard line-item budgeting, attention is focused on what is purchased, rather than the purpose of those expenditures. This makes it difficult to use a line-item budget for planning or management purposes, particularly in today's complex school district environments.[22] Because similar items, such as teacher salaries, appear under the same line item, it is impossible to ascertain how much is spent for teachers in different programs, at different schools, or different types of pupils.

If one wanted to assess a district's progress toward meeting the National Goals for Education, it would be difficult to ascertain if its expenditures focused on areas deemed important by the drafters of those goals. For example, Goal Four states that "[b]y the year 2000, U.S. students will be first in the world in science and mathematics achievement." Using a line-item budget format, it is impossible to determine how much a district is spending for teachers in math and science compared to other subjects, and it would be impossible to determine whether teaching resources had been shifted to math and science since the publication of the national goals. This is because all teacher salaries in the district would be combined into one total.

Obviously, these problems can be easily overcome by increasing the number of line items in the district's budget. For example, separate line items for teacher salaries could be included by school, by grade level, or by subject matter. As Hartman has stated, "All of the other budgeting methods use a line-item approach as a basic element in the organization and development of their data."[23] There are alternatives to the line-item budget. Two that have received the most attention are program budgeting and zero-based budgeting.

Program Budgeting

Program budgeting, also known as Planning Programming and Budgeting System (PPBS) or the Planning, Programming Budgeting and Evaluation System (PPBES), is an approach to budgeting developed from a special study of resource allocations in the United States Department of Defense conducted by the RAND Corporation in the 1950s. The basic model for program budgeting was developed by economists David Novick and Charles Hitch, and was designed to link planning to budgeting. Robert McNamara introduced the concept to the Department of Defense when he became Secretary of Defense in 1961.

President Johnson was so impressed with program budgeting and its success in the Defense Department that he ordered the system be applied to all federal agencies in August 1965. This action was the first step in the budget reform movement that swept the nation and resulted in attempts to implement forms of program budgeting in many local governments across the United States.[24]

Many federal programs of the mid-1960s required school districts to implement program budgeting in order to receive federal funds.[25] Dade County, Florida, was the first district to implement program budgeting concepts in the nation. This provided a stimulus for national dissemination, and a number of other large school districts developed their own versions of program budgeting by the early 1970s.

Despite this interest, program budgeting did not live up to expectations. By June 1971, the United States Office of Management and Budget quietly pronounced program budgeting dead at the federal level, and in 1970, program budgeting systems in New York and California were abandoned as well.[26] California provides one of the best examples of program budgeting's short life. In 1966, the state legislature established a citizen's commission to facilitate installation of program budgeting in all of the state's school districts. Yet by 1972, the state gave up and abandoned program budgeting at a cost of several million dollars.[27]

The Elements of Program Budgeting

There are seven major steps to program budgeting.

- The school district identifies and defines its mission, goals, and objectives, and clarifies its desired outcomes.
- Alternative approaches or programs for achieving these outcomes are specified.
- The alternative programs are translated into fiscal and non-fiscal requirements. This includes planned expenditures and proposed revenue sources for a program. Ideally, this is done for a multi-year period, and not for a single budget cycle.
- Each alternative approach is analyzed to determine its cost effectiveness.
- With this information, it is possible to select the best combination of programs and establish the optimum course of action for the district.

- Once implemented, each program is reviewed and an assessment of the degree to which desired outcomes were achieved is made.
- Finally, these evaluations feed back into the system at the beginning of a new cycle.

As this discussion points out, program budgeting is a very top-down decision-making strategy. At the state and federal levels, additional specialized staff were employed to implement program budgeting systems. Many larger school districts were able to hire additional budget personnel as well to implement the components of program budgeting. However, many smaller districts that attempted to establish program budgeting systems elected to rely on existing staff to put the system in place.

One of the reasons program budgeting may have failed is the tremendous amount of work required to make it work well. In school systems that were able to employ new budget analysts, there was often conflict between the program budget staff and the other budget and accounting staff. In systems where the existing staff was given responsibility for program budgeting, they often found this added to their job descriptions, and frequently, even when they had the best of intentions, they simply did not have the time nor the resources to implement program budgeting methods adequately. In other instances, the conflict between program budgeting and the old ways of doing things scuttled the implementation of the new model.

Strengths of PPBS

Despite its lack of widespread acceptance in public schools, program budgeting offers a number of improvements over traditional line-item budgets. Primary among them is it forces school districts implementing a program budgeting system to improve planning capabilities. The multi-year facet of program budgeting becomes very important. Even in districts where program budgeting has been abandoned, there often remains a long range or strategic planning function. Developing the capability to predict future student enrollments and the consequent spending requirements is an important capability, regardless of the budgeting system in use.

Another strength of program budgeting is its focus on accountability through is stress on performance and the achievement of objectives. Program budgeting also helps decisionmakers consider the full range of available options, requiring administrators to evaluate several options before selecting the one most appropriate to his or her district or school.

Proponents of program budgeting also point out that the outcome orientation of the system requires the establishment of clear goals and objectives, and then forces planners to determine what resources are necessary to achieve those goals. When properly implemented, spending decisions are made on the basis of what is needed to achieve the district's goals, and not on some other basis such as "that's how we have always done it in the past." Programs are linked to outcomes, and the specification of activities and resources into a program structure allows inputs to be tracked to outcomes.

Weaknesses of Program Budgeting

As mentioned, one of the major problems with program budgeting was its cost. Full implementation of a program budget requires substantial personnel time, which is expensive. Schick estimated that it took five years to fully implement a program budget system, and that often regular staff had to prepare all the required documentation for a program budget along with their regular responsibilities. The time requirements for program budgeting were substantial. For example, if seven different alternatives were identified, each had to be costed out and analyzed before a decision as to how to proceed could be made. This is potentially seven times as much work as it takes to prepare a conventional line-item budget. As budget deadlines approached, program analyses were often set aside in favor of regular budget submission requirements.[28] Moreover, in many agencies, new staff hired for program budgeting programs were viewed with hostility by the existing budget staff. This limited their effectiveness and the ability of program budgeting proponents to make headway into the decision-making process of the agency.

Another potential weakness of program budgeting is its bias in favor of centralized planning. As pointed out, budgeting is political, characterized by negotiation and compromise. The central bias of program budgeting changes the power alignment within a school district and can result in friction or tension that is counterproductive. It has been argued that this centralization was instrumental in program budgeting's failure in California. Placing decision-making power in the hands of analysts at the central office was in direct conflict with the public's desire for more diffused power in school district governance.[29]

Another problem faced by educational organizations attempting to implement program budgeting was the difficulty they had developing clear and quantifiable goals. It is difficult to reach full agreement on the goals of our educational system, much less obtain complete agreement on the specific outcomes that are expected of each child. Moreover, even if such agreement could be reached, knowledge of the specific resources needed to achieve each of those goals is not available. Because program budgeting's strength revolved around the development of such goals, and evaluation of an agency's progress toward meeting those goals, it performed poorly in education settings were the necessary specificity could not be obtained.

Program budgeting is very complex. It has been argued that this complexity is the reason it was not widely adopted by governmental agencies and why it ultimately failed.[30] In many instances, the planning and program emphasis of program budgeting did not transfer very well from the Department of Defense to other government agencies. Similarly, many state and local government agencies did not have talented analysts available to implement the system. Thus, the implementation of a system originally developed for federal defense programs has yielded mixed results.

Zero-Based Budgeting

Zero-based budgeting (ZBB) was the most popular budgeting reform of the 1970s. The downward cycle of the economy in the early part of that decade, which resulted in declining resources for governments along with increasing demands for services, led government officials to look for ways to maintain rational decision-making methods in times of uncertainty.

ZBB, as we know it today, was first described by Peter Pyhrr in the *Harvard Business Review* in 1970.[31] Georgia Governor Jimmy Carter asked Pyhrr to help him implement the system for Georgia's state government in 1971. ZBB spread from Georgia to Texas and New Jersey, and developed a following in a number of cities across the United States. When Carter became president of the United States, he mandated that all federal agencies use ZBB as well.

ZBB has made major inroads in school districts as well. At its annual meeting in October, 1977, the Association of School Business Officials International passed the following resolution:

> The Association of School Business Officials believes that zero-based budgeting is an effective and rather easily understood management tool for planning and controlling the expenditure of funds. It is a concept that those who are responsible for providing funds for education can also understand and respect. Zero-based budgeting provides the school business official with a sound basis for explaining and substantiating budget requests. The Association believes that it is a tool that should be considered by all school systems in planning annual budgets.[32]

Unlike program budgeting, ZBB is still used in various forms by some government agencies.[33] ZBB is not a specific system, but a process that must be adapted to the individual needs of each organization. The generic form of ZBB has five basic steps.[34]

- *Identification of decision units:* A decision unit is any program that consumes resources. Decision units are the lowest level entities for which budgets are prepared. Decision units are usually based on existing organizational structure or traditional budget units. Each decision unit must contain, at a minimum, objectives, activities, resources, and costs.

- *Analysis of each decision package*: In this step, managers are asked to reassess and justify the operation of each program. Questions that are routinely asked and answered during this process include:

1, Does the decision package support and contribute to the goals of the district?

2. What is the impact on the district if the decision package is eliminated?

3. Is there a way to accomplish the decision packages goals and objectives in a more cost effective manner?

- *Decision packages are ranked*: All of the decision packages are ranked within the organization. ZBB itself does not specify how those packages should be ranked, nor does it specify a method for doing so. That is left up to each organizational unit. The final ranking is made by the person at the top of organization.

- *Acceptance of packages and allocation of funds:* In Pyhrr's model, once the decision packages have been ranked and the costs of each specified, they are funded until available funding runs out. Thus, the higher a package's priority, the greater likelihood of its being funded. In reality, most school district superintendents use the rankings to establish priorities for how much of the requested funding each program will receive. Given limited resources, those with the highest priority are more likely to receive all of the funding requested than are those lower on the list.

- *Budget preparation:* Once the list of budget packages has been selected and target funding levels determined, the final budget document is prepared and submitted for approval of the school board and any other required actions as specified in state law.

The term "zero-based" comes from the practice of requiring all programs to be subjected to this ranking process annually. In other words, for each budget cycle, all programs start with a base of zero, and are funded only if they remain important to the achievement of the goals and objectives of the school district.

ZBB is very much a bottom-up process, with each package's outcomes specified by the manager responsible for that package. Funding of each package is based on the relative importance of that package compared to all others in the school district.

Because ZBB was developed for business and assumes that the low priority packages are expendable, it has limited application to school systems where various programs and services must be maintained regardless of their priority level. Consequently, governments have modified the system so that rather than accept or reject a decision package, often the rankings serve to establish priorities for the amount of new funding to be allocated or the share of reduced funding that must be absorbed by each program. It can be argued that no government system warrants designation as using zero-based budgeting in its pure form.[35] Most educational institutions abandoned annual zero-based reviews of decision packages and modified or eliminated the ranking process by 1978.[36]

Strengths of ZBB

ZBB offers a number of advantages over traditional line-item budgeting. To begin with, it provides better information concerning the activities of each department within an organization. Moreover, unlike program budgeting, the ZBB budget process does not attempt to alter the program structure used by organizations. It requires participatory management in decision making, and allows all parties to review the objectives and goals of other parts of the organization. Finally, by requiring each department to build its program from zero each year, it provides an excellent way to make judgments about the true value of that program.

Weaknesses of ZBB

ZBB however, is not without its weaknesses. By incorporating performance measures in the formation of decision packages, it forces managers to establish a preference for effectiveness, efficiency, extensiveness, or equity as they try to rank decision packages. This makes the ranking process difficult. Also, it is hard to rank packages with diverse goals.

The use of a cutoff point for packages is difficult in public agencies. As a result when funds become tight, even in ZBB systems, each department is asked to make marginal reductions, rather than eliminate the lowest ranked programs. Moreover, similar to program budgeting, the implementation of a ZBB system requires a great deal of time on the part of agency staff, limiting their ability to perform other important functions.

This section has described the strengths and weaknesses of a variety of approaches to public sector budgeting. It identified six components of a good budget and described how government agencies develop budgets. It also discussed in detail two major budgeting reforms that were introduced with considerable anticipation for improved governmental budgeting, but have not been as successful as initially hoped. In the next section of this chapter, a detailed discussion of the school district budget development process is provided.

Steps in the Budget Process

Generally, it has long been accepted that the starting point for the budget process should be the school district's educational plan. The budget translates this plan into dollars.[37] Building the budget is a long-term process that requires the participation of many individuals within the school district. Hartman identifies five steps in the budget process. They include the development of budget guidelines, preparation of the budget document, making modifications to the original budget, obtaining approval of the budget, and finally, administering or managing the budget.

Development of Budget Guidelines

Management of the budget process is typically the responsibility of the chief business officer in a school district. The task of assembling a budget that accurately represents a school district's educational plan, i.e., balances revenue and expenditures, and is completed and approved on time, is a complex and difficult task that requires the cooperation of not only the staff in the school district's business office, but all of the managers and principals in the school district. Creation of a set of annual budget guidelines helps ensure that all the participants understand their responsibilities. Among the items that are often included in a set of budget guidelines are the following.[38]

- A message from the superintendent describing the fiscal context within which the budget must be developed. This will include a discussion of available funding, and how much, if any, additional resources can be expected for the next fiscal year. Moreover, it should lay out the superintendent's priorities for the next year, and relay to staff any important concerns of the board of education.

- A budget calendar that provides a time line for each step of the budget process, making it clear when each must be completed. Table 12.1 is a sample budget calendar from a school district.

- Guidelines for staff and community participation at the district and school level.

- Any budget forms that must be filled out by school and departmental staff.

- Information on the accounting structure to be used in preparing the budget.

- Other information participants will need to ensure that they are able to prepare their portion of the budget on time, and provide the district's central office with an adequate number of copies.

The budget guidelines should be distributed to the schools and staff early in the budget process to ensure that everyone has the time they need to collect necessary data and prepare their preliminary budgets. If this work were to be done in a consistent and orderly manner, then preparation of the final budget documents will be simplified.

Budget Preparation

Actual preparation of the school district budget includes three basic steps, estimation of expenditures, estimation of revenue, and a comparison of the two. If the budget does not balance, then proposed expenditures and/or revenue must be revised until the two are brought into balance. Before expenditures can be estimated, it is essential to have a projection of student enrollment for at least the next school year. Since the largest single component of expenditures will be for personnel compensation and benefits, it is also necessary to estimate personnel needs as part of the budget process. In the sections that follow, each of these issues is discussed in detail.

Enrollment Forecasting

One of the most important components of the budget process is estimating student enrollment for the budget year. The number of students determines many of the financial estimates that make up the final budget. For example, the number of teachers, textbooks, and classrooms needed to implement the district's educational plan are dependent on the number of students enrolled. In addition, most of the state and federal revenue a district receives is based on the number of pupils in the district. Therefore, knowing how many students will be enrolled is critical to the development of an accurate budget.

Essentially an art, forecasting can be effectively used in combination with other techniques to reduce the elements of risk and uncertainty in traditional management decisions, such as policy formulation and program planning. Enrollment forecasting is generally thought of as the projection of specified future student enrollments within a certain time frame. Typically, those charged with forecasting enrollments must do so on the basis of incomplete information. The accuracy of a forecast depends on the intellectual and material resources available and the validity of the methods used.

The amount of effort devoted to forecasting should be a function of the "stakes." That is, how much error can one afford. In theory, the marginal investment in the forecasting effort should equal the marginal value of any improved accuracy in the forecast. Enrollment forecasts improve to the extent that the following steps are carried out:

- Accurate and reliable data are obtained.
- Data are prepared properly.

Table 12.1
Sample Budget Calendar

August

- Business office begins gathering student, personnel, program, cost, and economic data.

September

- Superintendent and other district administrators hold planning/orientation meeting(s) with school board and any other groups directly involved in the budget process.
- Superintendent, district administrators, and principals holds input sessions for district patrons.
- Superintendent and business office develop proposed budget calendar.

October

- School board approves budget calendar.
- Superintendent and senior district administrators (including business manager) develop cost and program guidelines for the upcoming year.
- Superintendent and personnel office plan and conduct collective bargaining negotiations, if necessary.
- Research and /or business offices make enrollment projections for upcoming school year and distribute them to the schools and departments.

November

- Business office develops and distributes budgeting forms, instructions, guidelines, and other materials for school and district staff.
- Business office holds training sessions or workshops for all administrators and clerical staff involved in budget development.

December

- Schools and district level departments prepare expenditure requests.
- business office develops revenue estimates by source of revenue.

- The basic assumptions of the model are stated clearly.
- A consistent method of calculation and analysis is used.
- Multiple forecasting methods are used, and sensitivity analyses conducted.
- Sound judgment and intuition are used to review the forecasts.

Assumptions are crucial to this process. The kinds and numbers of assumptions made in projecting populations depend on three factors:

- *The level of detail desired by the school district:* If the district is only trying to measure aggregate change, for example, then only a measure of the overall projected change in the population is warranted. If more detailed data by age group or grade level is desired, other assumptions

Table 12.1 (cont.)
K-8 School District Student Enrollment History
1988-89 Through 1992-93

January

- Business office consolidates expenditure estimates to develop the overall district totals.
- Superintendent, business office, and other district administrators review expenditure requests and adjust if necessary to balance with available revenues.

February

- Superintendent and business office prepare complete budget message, budget document, and appropriate supplementary information for presentation to school board.

March

- Superintendent presents the completed budget document to the school board for their review and approval.
- School board holds hearings on the proposed budget of the school district.

April

- Business office updates revenue estimates as new information becomes available from the tax assessor, state legislature, state department of education, and federal government agencies.

May

- School district holds budget elections(s), if necessary.

June

- School board adopts the budget for the upcoming school year, makes the appropriations of moneys for the school district by major fund and budget category, and declares and certifies the property tax levy required to balance the budget.

July

- Business office closes out the accounts for the year just completed, enters the appropriations for the new year into the books of account, and opens the new accounts for the new year.

August

- Business office distributes the approved budget amounts for each cost center to the administrator responsible for its operation.

Source: W. Hartman, School District Budgeting. Englewood Cliffs, NJ: Prentice-Hall, Inc. p. 37

must be made. These assumptions would include projections of changes in development patterns across the school district, as well as analyses of demographic trends, particularly birth rates and household mobility.

- *The span of time between the present and the last projection:* If the projection is over a relatively short period of time, many important variables can be held constant. A three-year projection can assume that the birth rate, for example, will show the same trend as recent years. However, if you are projecting changes over twenty-five years, whatever is happening to the birth rate is very important and certain additional assumptions have to be made. If the policy dictates the need for estimating school enrollment for the next year or two, most long-term demographic variables can be held constant.

- *The degree of accuracy desired:* District needs may dictate that the enrollment estimates are to be as accurate as possible, but it is important to determine what level of accuracy is feasible within existing resources. Moreover, there is no point in spending the equivalent of ten times the district's revenue per pupil if it were only to improve the accuracy of the enrollment projection by three or four pupils.

There are six basic assumptions that must be made to accurately project student enrollments into the future. For the purposes of most school districts, the typical time frame is from three to five years. The basic assumptions of the model are:

- *Mortality:* There will be no major change in the death rate for any group of the population.

- *Fertility*: There will not be a dramatic change in fertility, or that the rate of change will be a constant. Census data show that there are dramatic differences in the birth rate by ethnicity.

- *Local economic conditions*: Trends in land use, e.g., industrial and commercial development, will affect enrollments. Therefore, it is important to establish baseline assumptions about the development that will take place in the future, and how that will affect student enrollments. There are a number of sources for this information. Among the most useful are the city and county planning offices, the major utility companies, i.e., water, power, and telephone, and transportation planning agencies. These organizations need accurate estimates of service demands for the future, and are generally willing to share their knowledge with local school officials. As part of each year's enrollment projection process, someone from the business or research office of the school district meets with representatives of these organizations to see if they have any new information that will improve the accuracy of the district's enrollment forecast.

- *District boundaries*: If changes in the school district's boundaries are anticipated, either through district consolidation or annexation, the impact of these changes on student enrollments should be included in any projections that are made.

- *Administrative policy:* Policies on attendance, promotion, and pupil retention and acceleration will not change significantly from one year to the next, or they will change at a predictable rate. A new attendance program that increases student attendance could improve enrollment, which might have an effect on revenue in a state that bases its financial aid on average daily attendance rather than a count of enrollment or membership.

- *Acts of God*: Enrollment projections should assume that no major catastrophe will occur.

Projection Methodologies

There is no single or best method for population projection. In general there are four types of demographic projection models, and several nondemographic methods that can be used to project student enrollments

Demographic Methods

- *Comparative projection:* The comparative or analogy methodology of population forecasting determines the population projections for the study area based on the growth pattern of an older area with similar social, economic, and political characteristics. The usefulness of this method for school districts is rather limited.

- *Projections by extrapolation:* There are two types of projections by extrapolations. Graphic extrapolation consists of plotting the population of past census years against time and sketching a line which, in the judgment of the forecaster, appears to fit past data. This line is then extended into the future to obtain population estimates for future years. Graphic extrapolation techniques are relatively quick and simple, but tend to be inaccurate for all but very short time periods. The modeling capabilities of today's computers make it possible to estimate both straight line and curvilinear lines to fit these data. Thus, this use of state-of-the-art software improves the accuracy of the enrollment projections.

- *Ratio and correlation methods:* One method uses the ratio of the population of the study area to the population of the larger area in which it resides, and then applies that ratio to projections of the larger area. A second method compares the total population to a related variable such as employment or school enrollments. This ratio is applied to estimates of this variable to determine future population. These methods only work if the relationship between the study area and the larger area or related variable can be shown to be historically accurate.

- *Component analysis*: There are two principal variations of this approach. One compares births, deaths, and in-and out-migration. These are combined to estimate future population. This works well for estimating the population of a geographical area, but is not as useful in projecting individual school enrollments. A second method is Cohort Survival, which forecasts future population by separate age, sex, and sometimes racial

Exhibit 12.1
Estimating School Enrollments With the Cohort Survival Method

The cohort survival method is one of the most accurate and commonly used methods for estimating future school enrollments. Its accuracy increases as the size of the population increases, meaning that is more accurate in larger enrollment school districts, and it provides better estimates of total school district enrollments that it does for individual schools. However, it is a very powerful and useful tool for estimating enrollments up to five years in the future. Moreover, microcomputers and spreadsheet programs commonly available in school offices today, make estimation of student enrollments using the cohort survival method a very simple task.

Because the cohort survival method relies on historical data to estimate future trends, the first step in the process is to obtain an accurate enrollment history of the school district by grade level. Table 12.2 displays a summary of grade by grade enrollments for a K-8 school district for the five year period beginning in 1988-89 and ending in 1992-92

Table 12.2
K-8 School District Student Enrollment History 1988-89 through 1992-93

Grade	1988-89	1989-90	1990-91	1991-92	1992-93
Grade K	270	394	481	601	786
Grade 1	285	345	526	600	787
Grade 2	247	345	403	541	686
Grade 3	222	299	404	466	610
Grade 4	199	233	326	441	552
Grade 5	219	232	304	355	529
Grade 6	164	223	264	342	425
Grade 7	149	198	262	316	431
Grade 8	127	168	212	279	361
TOTAL	1882	2437	3182	3941	5167

Once an accurate enrollment history has been created, the "survival" rate across cohorts can be calculated. The survival rate is the change in the number of students in a cohort from year to year, expressed as a percent. In the sample district shown in table 12.2, there is a great deal of annual growth in the number of students in each grade, indicating that the district is located in an area experiencing considerable population growth. The survival rate for a cohort from one year to the next is calculated using the formula:

$$Srxy = \frac{\text{Number of students in grade x in year Y}}{\text{Number of students in grade x-1 in year y-1}}$$

Using this formula, a cohort survival rate can be calculated for each pair of grade levels and each pair of years displayed in table 12.2. These survival rates are shown in table 12.3.

Once all of the individual survival rates have been determined, the average survival rate for each grade pair can be calculated. The calculations are shown in the last column of table 12.3. These average survival rates are then used to estimate the annual student enrollment by grade level for each year into the future using the following formula:

> Number of Students in grade x+1 and year Y+1 = (Number of students in grade x and year y * Average Survival Rate for grade x)

The estimates derived from this calculation are displayed in table 12.4. This method works for all grades except kindergarten. Kindergarten estimates can be calculated by seeking information on the number of births in the area served by the school district. Historical data will indicate what percentage of the births in one year enrolled in district schools five years later. This percentage can be applied to the number of births five years prior to each of the years for which enrollments are projected. If this method were utilzed to make estimates more than five years into the future, then estimates of the number of births will also have to be developed.

Table 12.4 shows that the district's enrollments are expected to increase dramatically in the next four years. This is because of the tremendous growth the district experienced in the five previous years. If there is some reason to suspect that future growth will not be as dramatic as the recent past, e.g., no more room for new housing developments, or a downturn in the

Table 12.3
Average Survival Rates Cohort Survival Method

Grade	1988-89 to 1989-90 (%)	1898-90 to 1990-91 (%)	1990-91to 1991-92 (%)	1991-92 to 1992-93 (%)	Average Survival Rate (%)
Grade K to 1	127.78	133.50	124.74	130.95	129.24
Grade1 to 2	121.05	116.81	102.85	114.33	113.76
Grade2 to 3	121.05	117.10	115.63	112.75	116.64
Grade 3 to 4	104.95	109.03	109.16	118.45	110.40
Grade 4 to 5	116.58	130.47	108.90	119.95	118.98
Grade 5 to 6	101.83	113.79	112.50	119.72	111.96
Grade 6 to 7	120.73	117.49	119.70	126.02	120.99
Grade 7 to 8	112.75	107.07	106.49	114.24	110.14

economy, then the individual responsible for estimating school enrollments should consider lowering the estimates accordingly.

The example presented gave equal weight to each of the years for which historical data were available. In some instance, it may be desirable to weight recent years more heavily. This can be accomplished in calculating the average survival rate by changing the formula in that column of the spreadsheet. In fact, it is possible to conduct a simple sensitivity analysis using different algorithms to determine the average survival rate, and then

Table 12.4
Future Enrollment Estimates Using the Cohort Survival Method

Grade	1993-94	1994-95	1995-96	1996-97
Grade K	870	788	760	750
Grade 1	1016	1124	1018	982
Grade 2	895	1156	1279	1159
Grade 3	800	1044	1348	1492
Grade 4	673	883	1153	1488
Grade 5	657	801	1051	1372
Grade 6	592	735	897	1177
Grade 7	514	717	890	1085
Grade 8	475	566	789	980
TOTAL	6493	7815	9185	10484

> compare the results. If large differences in projected enrollments emerge from this techniques, then the individual responsible for the projections will want to review all of the assumptions carefully to see if it can be determined what is causing these differences. Then, using local judgment, hopefully after discussions with others responsible for similar types of projections in the area, a most likely scenario should be selected.
>
> The cohort survival method described Is. commonly utilized to project student enrollments. As with other projections methods, its accuracy deteriorates the further in the future one attempts to make projections. However, given the relative ease which such calculations can be made using spreadsheet programs, this enrollment projection exercise should be repeat at least once a year.

group. For each cohort, birth, death, and migration components are determined. This is believed to be the most reliable method for projecting population. Exhibit 12.1 offers an example of how a cohort survival model can be used to project student enrollments in a school district.

Non-Demographic Methods

- *Build out rates:* This method is used to project population or enrollments by examining current and future housing developments in an area and multiplying by an assumed household size. This is most accurate in the short term.
- *Ultimate population analysis:* Used to determine the saturation population at full development of an area. An upper limit supposedly exists for any geographic area with fixed boundaries.

Estimating Personnel Needs

People are the single largest resource used by school districts. Accurate estimates of the number of personnel required each year are, therefore, essential to the development of an accurate budget. The number and type of personnel needed depends on how the district chooses to allocate staff positions. The most typical approach is to allocate personnel on the basis of staffing ratios.

For teachers, the most common metric for allocating personnel to school sites is the pupil/teacher ratio. Using this ratio, it is possible to determine how many teachers should be assigned to each school. Determination of these ratios is an important consideration. Many states have established legal guidelines for class sizes. For example, in Texas, class size in all K-4 classrooms must not exceed twenty-two students. Obviously, the number of students in each of these grades will have a considerable impact on the number of teachers each district is required to employ.

One of the highest priority policy goals of the education community has consistently been the reduction of class size as a way to improve student achievement. Although the research evidence to support the positive effect of class size reductions on student outcomes suggests certain limitations, there is almost universal agreement that reductions in class size are important if we are to improve our nation's schools. What the research does seem to indicate is that substantial reductions in class size are needed to have an impact on student performance. Research suggests that to be truly effective, a class size of approximately fifteen students per teacher or less is needed.[39]

On a national level, the average pupil/teacher ratio for K-12 public schools in the United States in 1991 was 17.2, very close to the 15 students per teacher emphasized in most research. Moreover, these data suggest that this ratio has declined consistently since 1955 when it stood at 26.9 pupils per teacher. In fact, except for an increase of 0.1 pupils-per-teacher between 1961 and 1962, the average pupil-teacher ratio across the United States has declined in every year since 1955.[40]

If a state does not have a class size policy, staffing ratios are typically set either by the school board as a matter of policy, or as part of the district's collective bargaining agreement with the professional staff. In addition, different ratios can be set for different programs. For example, many districts establish lower staffing ratios for high schools than for elementary schools. Special programs, such as those for students with disabilities, often use substantially lower staff ratios than do regular education programs.

Other school and school district personnel can also be allocated through staff ratios. Ratios based on the number of students are common, as are ratios based on other factors such as the number of teachers, total school enrollment, or the square footage of the school. Some of the more common personnel categories and possible allocation methods are discussed herein.

Certificated Staff

Teachers

As indicated, the number of teachers a district needs can be estimated through the use of pupil/teacher ratios. Allocation rules can be as simple as establishing of one ratio across the entire district, leaving individual teacher assignments up to site administrators and/or site-based management committees. Conversely, districts can use very complex allocation models, providing teachers at varying ratios depending on the program. In more complex models, there may be separate ratios by type of school (i.e., elementary, middle, or high school), by grade level, by subject matter (in secondary schools), by student needs (i.e., lower ratios for special education, compensatory education, and programs for students who are not proficient in English).

The advantage to simple allocation rules is that, in addition to making it easier to calculate teacher allocations to school sites, they provide more flexibility to the school-level decisionmakers. Some schools may elect to use all possible certificated staff in instructional programs to minimize class size, while others may choose to have slightly larger classes and use one or more of the certificated personnel as counselors. More detailed pupil/teacher ratio models offer school advisory councils less flexibility to manage personnel, but make district-wide planning for staff easier. It also ensures that previously agreed upon class size decisions are maintained.

Other Certificated, Non-Administrative Staff

Virtually all schools today have a number of certificated staff assigned to the site who do not have a specific classroom responsibility. These individuals fill a number of important functions in a school. They may serve as counselors, instructional leaders, mentors, student advisors, nurses, librarians, and so on. Determining how many of these positions are required can also be based on staff ratios. The ratios may be determined by state guidelines, accrediting standards, on the basis of the number of students, or on some other rational basis such as the number of teachers at a school, or the number of children meeting some specific requirement such as eligibility for Chapter 1 programs.

In many instances, the standard ratios used by a school district to assign these certificated staff to school may result in partial allocations, or in the worst case, a school may not have enough students to qualify for one or more of the positions. In some instances, such as school nurses and counselors, it may be valuable to have individuals with these qualifications available all day, regardless of the number of students they serve. Consequently, many allocation rules provide that each school receive a minimum of one such professional. For example, if a district policy were to allocate one school nurse for each 1,000 students, and to allocate at least one nurse to each school, then if it has a number of schools with fewer than 1,000 students, it will have to hire more nurses.

For example, suppose the district under consideration has 2,000 students in four schools of approximately 500 students each. At the district level, it appears that the district needs only two nurses, one for each 1,000 students. However, the policy as previously outlined also states that there will be at least one nurse at each school, implying that a total of four school nurses must be hired. Obviously, the district could also assign each nurse to two schools for half of each day. The final determination as to how to allocate such personnel is usually up to the school board, with advice from the district administration, and is based on such factors as the importance of a full-time nurse compared to other functions, and the costs of two additional personnel. For personnel with specialized, non-classroom assignments, decisions regarding how many to hire can be affected by both the total number of pupils used in the staffing ratio and the relative size of the schools in the district.

Site Administrators

Like the other certificated staff positions, school site administrators can be assigned to schools on the basis of student enrollment or the number of teachers who report to the school site. In a large district that provides special education services for children with severe disabilities at special school sites, it may be important to consider the number of teachers rather than the number of students in the staffing ratio for site administrators given the typically small pupil/teacher ratios found in special education programs for these children. In addition, all schools need to have a principal. Consequently, even in a school with as few as 200 students, a school district will have to provide a principal. Assuming that standard rules are established for determining when assistant principals are assigned to a school, it is clear that there may be substantial variations in the number of students and/or teachers per site administrator.

Central Office Administrators

Staffing ratios for central administrators are more difficult to establish. Some districts establish an overall pupil/administrator ratio for the central office staff and strive to keep within the bounds established. If necessary, they assign individuals to multiple functions. In other school districts, certain administrative positions are filled regardless of any pupil/teacher ratio.

Classified Staff

In addition to the certificated staff, school districts employ large numbers of classified personnel. These individuals are responsible for keeping school facilities clean and well repaired, maintaining the grounds and buildings, driving school buses, doing clerical work, and many other functions. Classified personnel are typically assigned to district cost centers on the basis of some staffing ratio. These ratios can be based on the number of students, the number of teachers at the site, or some other relevant factor. For example, it might make sense to assign school site office personnel on the basis of the number of students and/or teachers at the school. However, custodians and grounds keepers may best be assigned on the basis of a school facility's square footage, or the number of classrooms at the site. Similarly, bus drivers are needed based on the number of pupils bused each day.

Total Staff Estimates

Once the number of positions to be filled has been estimated, they are reported in terms of full-time equivalents (FTEs). FTEs provide a measure of the number of full-time employees needed. While many individuals in a school district may only work part of a day, for budget purposes, the number of FTEs is calculated. Thus, for example, if a district were to hire two individuals to work as librarians for half a day each, it has two employees, but only one FTE employee. Thus, a district's budget may report that school A has 4.5 FTE custodians. If a typical work day is considered to be 8 hours, 4.5 FTE translates into 36 hours a day of custodial time (i.e., 8 hours times 4.5 FTE). This does not necessarily mean that the school employees four full time custodians and one half time custodian. It is just as likely that the school has one full-time head custodian, and seven individuals working half time in that position. The exact distribution of personnel would be determined by the school and the district.

Budget documents should report the number of FTE personnel employed by program. Since personnel costs (salaries and benefits) often represent as much as 85 or 90 percent of a school district's budget, accurate estimation of personnel requirements is essential to budget development.

Estimation of Expenditures

Once the number of employees has been determined, it is much easier to estimate the expenditures necessary for carrying out the district's educational plan. It has been suggested that there are three main steps in estimating expenditures. They are very general, "but provide a structure for understanding the purpose and context of the specific expenditure estimations."[41]

- Identify the specific programs and/or functions around which the budget is to be constructed. The framework for this is the district's accounting system.[42]
- Ascertain what resources are needed to operationalize the tasks essential to each of the programs or functions identified in the previous step. These resources include the objects of expenditure needed to implement each program and include such items as personnel, benefits, books, and supplies, and so on. Determination of what resources are needed can be accomplished in a number of ways. District policies, standards, and formulas will help determine the type and amount of resources that are needed for each program. In addition, historical allocations offer an excellent guide to what resources will be needed to implement different programs in the future.
- Estimate the costs of the resources needed to implement each program or function. This step requires placing a cost on each of the resources identified in the previous steps, and summing up the total costs for each

program by school, district-level department, and finally for the district as a whole.[43]

Each cost center is responsible for making its own budget estimates using a standard model based on the district's needs. The basic unit of accounting for a budget are objects of expenditure. Objects represent actual items that can be purchased. Object-oriented budget systems can be very basic, or very specific. Table 12.5 displays the object classifications used by one state. In this example, each object has a four-digit accounting code. The table only displays detail at the first two digits. The third digit is used to distinguish between instructional and non-instructional expenditures, and between certificated and classified personnel in the benefit classifications.

The problem with budgets that only provide object-level data is that they do not give the reviewer any sense of the purpose for which the resources are being used. For example, an object-level budget might contain a line item summarizing total certificated salaries for teachers (e.g., the 1100 category in table 12.5). This would give the reviewer a concept of how much is spent

Table 12.5
California Object of Expenditure Classifications

Code	Classification
1000	Certificated Personnel Salaries
1100	Teachers' Salaries
1200	School Administrators' Salaries
1300	Supervisors' Salaries
1400	Librarians' Salaries
1500	Guidance, Welfare, and Attendance Services
1600	Physical and Mental Health Salaries
1700	Superintendents' Salaries
1800	Administrators' Salaries
1900	Other Certificated Salaries
2000	Classified Personnel Salaries
2100	Instructional Aides' Salaries
2200	Administrators' Salaries
2300	Clerical and Other Office Salaries
2400	Maintenance and Operations Salaries
2500	Food Services Salaries
2600	Transportation Salaries
2900	Other Classified Salaries
3000	Employee Benefits
3100	State Teachers' Retirement System
3200	Public Employees' Retirement System
3300	Social Security (OASDI and Medicare)
3400	Health and Welfare Benefits
3500	State Unemployment Insurance
3600	Workers' Compensation Insurance
3900	Other Benefits

Code	Description
4000	Books and Supplies
4100	Textbooks
4200	Books other than Textbooks
4300	Instructional Materials and Supplies
4500	Other Supplies
4600	Pupil Transportation Supplies
4700	Food Service Supplies
5000	Services and Other Operating Expenditures
5100	Personal Services of Instructional consultants, Lecturers, and Others
5200	Travel and Conferences
5300	Dues and Memberships
5400	Insurance
5500	Utilities and Housekeeping Services
5600	Rentals, Leases and Repairs
5700	Direct Costs of Interprogram/Interfund Services
5800	Other Services and Operating Expenditures
6000	Capital Outlay
6100	Sites and Improvement of Sites
6200	Buildings and Improvement of Buildings
6300	Books and Media for New School Libraries or Major Expansion
6400	Equipment
6500	Equipment Replacement
7000	Other Outgo
7100	Tuition
7200	Other Transfers Out
7300	Direct Support/Indirect Costs
7600	Interfund Transfers and other uses

Source: *California School Accounting Manual: 1988 Edition.* Sacramento, CA: California State Department of Education, 1988. 4-12 to 4-19.

on teacher salaries by the school district, but no sense of how those teachers are allocated among elementary, middle, or senior high schools or of the amount paid to substitute versus regular teachers. Moreover, a line item for teacher salaries at an individual school provides little information as to how those teacher resources are used by educational program. What is the distribution of teachers by subject matter at the high school level? What proportion of the teachers at an elementary school teach special education? Of those teachers, how many are resource room teachers and how many have self-contained classrooms?

To answer these questions, budgets can also be aggregated by function or program. Functions describe general areas of expenditure such as instruction, administration, operations and maintenance, pupil transportation, instructional support, and so on. Functional definitions vary across state accounting systems. However, the federal government has attempted to standardize functional definitions. Table 12.6 summarizes the current functional definitions as used by the federal government.

Table 12.6
School District Function Classifications as Defined by the Federal Government in Handbook II, Revised

Code	Function
1000	Instruction
2000	Support Services
2100	Students
2200	Instructional Staff
2300	General Administration
2400	School Administration
2500	Business
2600	Operation and Maintenance of Plant Services
2700	Student Transportation Services
2800	Central
2900	Other Support Services
3000	Operation of Non-Instructional Services
3100	Food Services Operations
3200	Other Enterprise Operations
3300	Community Support Operations
4000	Facilities Acquisition and Construction Services
4100	Site Acquisition Services
4200	Site Improvement Services
4300	Architecture and Engineering Services
4400	Educational Specifications Development Services
4500	Building Acquisition and Construction Services
4600	Building Improvements Services
4900	Other Facilities Acquisition and Construction Services
5000	Other Uses
5100	Debt Service
5200	Fund Transfers

Source: National Center for Education Statistics. *Financial Accounting for Local and State School Systems*. Washington D.C.: United States Department of Education, June 1980. 26-29.

Expenditures can also be classified by programs. The more detailed the programmatic distinctions, the more complex the budget process becomes, and the bigger the budget document. Clearly, some compromises need to be made in defining programs to be included. Florida has developed a very detailed program structure for school district budgeting and accounting. Table 12.7 displays the forty-four programs used in that state's system.

The advantage to a program budget is that is gives managers, as well as school board members and the general public, a better picture of what their funds are actually buying. One of the major problems that emerges in developing program budgets is that many services provided at a school are hard to assign to a single program. For example, custodial services benefit all of the programs at a school. If the custodial operation were treated as a separate function, there is a great deal of information available concerning what it costs to provide those services at the school. However, no information is provided regarding the level of those resources devoted to each instructional program. Since some programs, e.g., science laboratories, physical education, industrial arts, art, and home economics classes require more space and materials, it is likely that the custodians spend more time and resources cleaning in classrooms and areas devoted to those activities.

Table 12.7
Florida Program Classifications

Code	Program Description
Basic Programs	
101	K-3 Basic Education
102	4-8 Basic Education
103	9-12 Basic Education
Special Programs for At-Risk Students	
120	Dropout Prevention
131	Intensive English/ESOL K - 3
132	Intensive English/ESOL 4 - 8
133	Intensive English/ESOL 9 - 12
Special Programs for Exceptional Children	
210	Educable Mentally Handicapped
202	Trainable Mentally Handicapped
203	Physically Handicapped
204	Physical and Occupational Therapy (Part Time)
205	Speech, Language and Hearing (Part Time)
206	Speech, Language and Hearing
207	Visually Handicapped (Part Time)
208	Visually Handicapped
209	Emotionally Handicapped (Part Time)
210	Emotionally Handicapped
211	Specific Learning Disability (Part Time)
212	Specific Learning Disability
213	Gifted (Part Time)
214	Hospital and Home-bed (Part Time)
215	Profoundly Handicapped

Table 12.7 (cont.)

Code	Program Description
K-12 Vocational Programs	
301	Agriculture
302	Office
303	Distributive
304	Diversified
305	Health
306	Public Service
307	Home Economics
308	Technical Trade and Industry
309	Exploratory
316	Vocational Education Mainstream
Adult Job Preparatory Vocational Programs	
331	Agriculture
332	Office
333	Distributive
334	Diversified
335	Health
336	Public Service
337	Home Economics
338	Technical Trade and Industry
Adult Supplemental Vocational Programs	
361	Agriculture
362	Office
364	Health
365	Public Service
366	Home Economics
367	Technical Trade and Industry
Adult General Education Programs	
401	Adult Basic Skills
402	Adult Secondary Education
403	Lifelong Learning
416	Adult Handicapped

Source: State of Florida, Dept. of Educ., Tallahassee, FL, Florida Education Finance Program.

Typically, a defined algorithm is used to distribute the costs of these support functions across programs. For example, maintenance and custodial services might be allocated to instructional programs on the basis of classroom square footage, while administration is allocated on the basis of FTE teachers. These formulas help distribute support service costs across instructional programs, but are often confusing to school officials, and provide little information concerning expenditures for these services for comparison within a district and among districts. Therefore, data are maintained on the costs of each individual support program to understand the components of total program cost.

Estimation of Revenue

Revenue is the money the district receives from local property taxes, state aid, and federal aid. These funds are used to pay for the expenditures necessary for the operation of the school district. In almost all of the fifty states, revenues and expenditures must balance. If, after estimating all revenue and expenditures, expenditures exceed revenues, then expenditures must be reduced, revenue increased, or some combination of the two actions must take place pursuant to state statutes.

Proper revenue estimation is critical to implementing the district's educational plan. Without an accurate estimate of revenue, it is impossible to plan expenditures for the next fiscal year. Unfortunately, predicting revenue is not a simple task. Revenue projections are based on estimates of local property tax collections, assumptions regarding the level of state aid to the district, and the amount of federal assistance for which the school district will qualify.

Estimation of property taxes requires knowledge of the total assessed valuation of the school district, as well as, in most jurisdictions, an estimate of the number of taxpayers who will be delinquent in the payment of their taxes. In addition, a solid understanding of the property tax law in the applicable state is essential. In some states, school districts are granted authority to levy a property tax rate. In those cases, that rate is applied to the value of assessed valuation within the school district, and total revenue can be estimated (taking into account applicable delinquencies).

However, in other states, a school district is granted authority to collect a fixed sum of property tax revenue. In many instances a small increase in the taxation authority is allowed automatically, but larger increases are generally subject to voter approval. In states where property tax authority is based on a levy, rather than a tax rate, if property values were to go up substantially from one year to the next, the district does not benefit as its property tax rate will decline accordingly. Thus, a thorough understanding of the applicable state's property tax laws and how changes in economic factors will affect school district tax collections is essential to estimation of revenue.

State revenue is generally based on an aid formula designed to direct more moneys to school districts with low per-pupil assessed valuations. In addition, many states operate categorical programs to provide additional assistance for pupil transportation, special education, and a variety of other educational programs. Generally, the amount of moneys available to each school district is contingent upon the total state appropriation for public elementary and secondary education, something which is often not known until near the beginning of the next fiscal year when the state legislature finally completes its budget deliberations. While the state department of education may be able to provide relatively accurate estimates of a school district's state aid entitlement, the final allocation is often not determined until well into the school year.[44]

Presently, federal funds are distributed almost entirely through categorical programs, e.g., Chapter 1, which provides assistance for compensatory education for children from low income families. Although these programs represent a relatively small portion of school district revenue, they often have a considerable impact on local resource allocation decisions as they frequently come with matching requirements, or have certain fiscal performance requirements attached to acceptance of the federal dollars.

Sound school district financial management dictates that school districts budget funds for unexpected contingencies. If these funds are not used, or if spending during the current fiscal year falls below expectations for some reason while revenue remains stable, there may be a significant ending fund balance that can be included in the sources of funds available, pursuant to state guidelines, for the next year's expenditures. School district reserves, where statutorily applicable, are critical to the district should revenue fall below expectations during the course of the year, or if there are some unanticipated expenditure needs.

Budget Modification

If expenditure estimates were to exceed revenue projections, the district's administrators must make adjustments in one or both sides of the equation. Typically, it is easier to reduce expenditures than it is to increase revenue. Most state and federal programs have fixed revenue levels, and it is not likely that a school district will be able to seek additional funding from these sources over the short term. Local property taxes offer somewhat more hope, depending on the tax statutes in the particular state. In many states it is possible to seek voter approval for higher property taxes. However, recent tax limitation measures in many states have curtailed this option. Moreover, there appears to be less and less willingness on the part of local voters to approve tax increases for any purpose. Consequently, it may be necessary to reduce planned expenditures to balance the budget.

Reductions in expenditures often mean limited compensation for employees, or actually eliminating positions. Small reductions in projected expenditures may be possible without hurting the number of staff positions, but larger ones will most likely result in wage freezes, hiring freezes, or the elimination of positions. Most school districts attempt to make personnel reductions as far away from the classroom as possible, choosing to engage in reduction-in-force of custodians, central office administrators, clerical personnel, and other nonteaching staff.

If the budget reductions were to become severe, it may be necessary to eliminate teaching positions. For example, in the early 1990s, many school districts in California were forced to both increase class size (the logical outgrowth of eliminating teaching positions) and even reduce the wages of all school district employees, including teachers. These cuts were the result of years of reduced state revenue forcing the state to limit the amount of moneys available to schools. Because California's Proposition 13 places very strict limits on property taxes, school funding is largely a state responsibility. When state revenue actually declined from one year to the next, all state programs, including K-12, education suffered dramatically.

Although reductions in school budgets seem to be the norm in public education, the story is not as bleak as it appears. In fact, real spending for public elementary and secondary education has increased substantially in the last thirty years. Table 12.8 show that per pupil expenditures for education across the United States, after being adjusted for inflation, increased by over 200 percent between 1959-60 and 1989-90. While it is clear from the table that schools in some states have fared better than schools in other states, the table makes it clear that over time, schools have had additional revenue. If this trend continues over a period of time, then school business administrators will be forced to spend more time finding ways to use those resources to improve student achievement.

Budget Approval

The last step in the budget process is approval of the final budget. This step requires that the district superintendent approve of all revenue and expenditure estimates contained in the budget and that the budget be submitted to the board of education for review and approval. Pursuant to state statutes and guidelines, the school board will hold a number of public hearings on the budget before approving it, and a number of modifications may be suggested by the school board before final approval is granted. In many states, a school district's budget must also be approved by the state department of education, or an intermediate educational agency. In many states, which have fiscally dependent school districts, the budget must be accepted by another local governmental agency such as the county commissioner or city council. These approvals are essential to receipt of state funds in most of those states, and are generally designed to ensure that the district has a balanced budget that contains an accurate estimate of the revenue and expenditures for the forthcoming fiscal year.

As this discussion shows, development of a budget that accurately reflects a district's revenue and expenditures is a complex process that involves many individuals and a great deal of planning and compromise. As school districts begin moving toward more site-based decision making and management, the complexities of the budget process will be magnified. Not so much at the

Table 12.8
Change in Real Expenditure Per Pupil in Average Daily Attendance in Public Elementary and Secondary Schools By State: 1959-60 to 1989-90 (Constant 1989-90 Dollars)

State	Current Exp. Per ADA 1959-60 ($)	Current Exp. Per ADA 1969-70 ($)	Percent Change 1959-60 to 1969-70 (%)	Current Exp. Per ADA 1979-80 ($)	Percent Change 1969-70 to 1979-80 (%)	Current Exp. Per ADA 1989-90 ($)	Percent Change 1979-80 to 1989-90 (%)	Percent Change 1959-60 to 1989-90 (%)
United States	1,621	2,743	69.22	3,345	21.95	4,960	48.28	205.98
Alabama	1,042	1,828	75.43	2,281	24.78	3,327	45.86	219.29
Alaska	2,361	3,773	59.81	5,978	58.44	8,374	40.08	254.68
Arizona	1,744	2,421	38.82	2,980	23.09	4,057	36.14	132.63
Arkansas	973	1,908	96.09	2,190	14.78	3,485	59.13	258.17
California	1,832	2,915	59.12	3,351	14.96	4,391	31.04	139.68
Colorado	1,712	2,480	44.86	3,154	27.18	4,720	49.65	175.70
Connecticut	1,884	3,197	69.69	3,816	19.36	7,604	99.27	303.61
Delaware	1,969	3,025	53.63	3,711	22.68	5,696	53.49	189.28
D.C.	1,863	3,423	83.74	4,361	27.40	8,904	104.17	377.94
Florida	1,373	2,461	79.24	3,198	29.95	4,997	56.25	263.95
Georgia	1,095	1,976	80.46	2,586	30.87	4,187	61.91	282.37
Hawaii	1,403	2,825	101.35	3,378	19.58	4,448	31.68	217.03
Idaho	1,252	2,028	61.98	2,491	22.83	3,078	23.56	145.85
Illinois	1,895	3,057	61.32	3,717	21.59	5,118	37.69	170.08
Indiana	1,593	2,447	53.61	2,731	11.61	4,549	66.57	185.56
Iowa	1,589	2,837	78.54	3,086	8.78	4,453	44.30	180.24
Kansas	1,503	2,592	72.46	3,114	20.14	4,752	52.60	216.17
Kentucky	1,007	1,833	82.03	2,218	21.00	3,675	65.69	264.95
Louisiana	1,607	2,178	35.53	2,771	27.23	3,855	39.12	139.89
Maine	1,222	2,328	90.51	2,717	16.71	5,373	97.75	339.69
Maryland	1,697	3,087	81.91	3,838	24.33	6,196	61.44	265.11
Massachusetts	1,767	2,888	63.44	3,630	25.69	6,237	71.82	252.97
Michigan	1,794	3,038	69.34	3,737	23.01	5,546	48.41	209.14
Minnesota	1,837	3,037	65.32	3,785	24.63	4,971	31.33	170.60
Mississippi	890	1,684	89.21	2,149	27.61	3,096	44.07	247.87
Missouri	1,486	2,382	60.30	2,816	18.22	4,507	60.05	203.30
Montana	1,775	2,628	48.06	3,313	26.07	4,736	42.95	166.82
Nebraska	1,456	2,475	69.99	3,115	25.86	4,842	55.44	232.55
Nevada	1,860	2,586	39.03	2,913	12.65	4,117	41.33	121.34
New Hampshire	1,501	2,430	61.89	2,893	19.05	5,304	83.34	253.36
New Jersey	1,675	3,416	103.94	4,371	27.96	7,991	82.82	377.07
New Mexico	1,567	2,376	51.63	2,731	14.94	3,518	28.82	124.51
New York	2,427	4,460	83.77	5,659	26.88	8,062	42.46	232.18
North Carolina	1,025	2,058	100.78	2,677	30.08	4,268	59.43	316.39
North Dakota	1,585	2,318	46.25	2,724	17.52	4,189	53.78	164.29
Ohio	1,577	2,454	55.61	2,862	16.63	5,136	79.45	225.68
Oklahoma	1,346	2,032	50.97	2,517	23.87	3,512	39.53	160.92
Oregon	1,937	3,108	60.45	3,826	23.10	5,521	44.30	185.03
Pennsylvania	1,769	2,964	67.55	3,712	25.24	6,061	63.28	242.62
Rhode Island	1,786	2,996	67.75	3,933	31.28	6,249	58.89	249.89
South Carolina	951	2,059	116.51	2,531	22.92	4,088	61.52	329.86
South Dakota	1,499	2,319	54.70	2,626	13.24	3,732	42.12	148.97
Tennessee	1,029	1,903	84.94	2,429	27.64	3,664	50.84	256.07
Texas	1,436	2,098	46.10	2,607	24.26	4,150	59.19	189.00
Utah	1,393	2,105	51.11	2,510	19.24	2,730	8.76	95.98
Vermont	1,486	2,713	82.57	3,348	23.41	6,227	85.99	319.04
Virginia	1,185	2,379	100.76	2,941	23.62	4,612	56.82	289.20
Washington	1,817	3,077	69.35	3,419	11.11	4,681	36.91	157.62
West Virginia	1,117	2,252	101.61	2,567	13.99	4,359	69.81	290.24
Wisconsin	1,785	2,967	66.22	3,454	16.41	5,524	59.93	209.47
Wyoming	1,946	2,877	47.84	3,496	21.52	5,577	59.53	186.59

Source: NCES,. *Digest Of Education Statistics: 1992*. Washington D.C.: U.S. Government Printing Office.

district level, where there will be more summing of school-site budgets, but at the school-site level where decisions regarding educational priorities will have to be made and budgets reflecting those decisions developed. Site-based management may change the role of the central office from one of preparing the budget to one of providing technical assistance to local schools as they prepare their own site budgets.

Administering the Budget

The administering of the school district budget is a continual process involving the examination of the school district revenue and expenditure data. The school business office always monitors the status of the revenue and expenditure accounts on a timely basis. Often, this timely basis is done every week in order to project potential fiscal difficulties. In this manner, the school finance office is able to engage in a preventive and strategic management style regarding expenditures.

Conclusion: The Future of School District Budgeting

Despite the complexity of the task, each of the nation's nearly 16,000 school districts develops a budget each year. In addition, all other local government agencies prepare similar budget documents on an annual basis. At a recent budget symposium sponsored by the Government Finance Officers of America (GFOA), the following problems and issues with the preparation and implementation of local budgets were identified:

- The lack of a long-term perspective.
- Failure to integrate operating and capital budgets.
- Unrealistic revenue forecasts.
- Inadequate consideration of service delivery alternatives.
- Lack of performance measures to assess results.
- Political pressures to maintain or increase service levels.[45]

These problems can be found in school district budgets as well. Fortunately, GFOA has also developed considerable consensus on the characteristics of good budgeting practice. Among the practices that seem to help resolve these problems are the following:

- Integration of long-term planning with the budget process.
- Establishment of measures for continuous monitoring of performance.

- Obtaining citizen and other stakeholder involvement early in the budget process.
- Effective communication of goals, assumptions, and tradeoffs to citizens and other stakeholders.
- Provision of incentives and a sense of empowerment to managers and other government employees to implement effective practices and achieve goals.
- Incorporation of realistic revenue and expenditure forecasts.

Each of these concepts could prove to be helpful for local school districts in improving the budget process in the future, and should be given careful consideration by local school officials charged with responsibility for developing school district budgets.

Footnotes

[1] "State of the Art." *Government Finance Review,* Aug. 1993, 36-37.

[2] It should be noted that in many states the local educational agency's budget must be approved/adopted by another governmental agency in addition to the local school board, e.g., the voters at an annual meeting, local city and/or county officials, or some other agency.

[3] A. Wildavsky, *Budgeting: A Comparative Theory of Budgetary Processes.* Revised Edit. New Brunswick, NJ: Transaction Pub, 1986, 7.

[4] A. Wildavsky, *Budgeting,* 7-9.

[5] A.C. Hyde and J.M. Shafritz, eds .*Government Budgeting: Theory, Process, Politics,* Oak Park, IL: Moore Pub. Co., Inc. 1978, 1.

[6] T.L. Drake and W.H. Roe, *School Business Management: Supporting Instructional Effectiveness,* Boston, MA: Allyn and Bacon, 1994, 70.

[7] I.C. Candoli, W.G. Hack and J.R. Ray, *School Business Administration: A Planning Approach,* 4d. Boston, MA: Allyn and Bacon, 1992, 111.

[8] G.C.Hentschke, *School Business Administration: A Comparative Perspective,* Berkeley, CA: McCutchan Pub. Cor., 1986, 178.

[9] W.T. Hartman, *School District Budgeting,* Englewood Cliffs, NJ: Prentice Hall, 1988, 2.

[10] W.T. Hartman, *School District Budgeting,* 3.

[11] A.C. Hyde and J.M. Shafritz, eds. *Government Budgeting: Theory, Process, Politics,* 2.

[12] Taft Commission, "The Need for a National Budget." in A.C. Hyde and J.M. Shafritz, eds. *Government Budgeting: Theory, Process, Politics.* 4-11.

[13] 42 Stat. 18, 1921. Reprinted in A.C. Hyde and J.M. Shafritz. eds. *Government Budgeting: Theory, Process, Politics.* 12-19.

[14] A.C. Hyde and J.M. Shafritz, eds. *Government Budgeting: Theory, Process, Politics,* 2.

[15] A. Schick, "The Road to PPB: The Stages of Budget Reform." *Public Administration Rev*. vol. 26, no. 4 Dec. 1966, 243-258.

[16] V.O. Key, "The Lack of a Budgetary Theory," *American Political Science Rev.* vol. 34 (Dec., 1940), 1137-1140.

[17] A. Schick, "The Road to PPB: The Stages of Budget Reform." *Public Admin. Rev.,* 246.

[18] A. Schick, "The Road to PPB:" 247-8.

[19] See e.g., A. Wildavsky. *The Politics of the Budgetary Process.* 4d. Boston, MA: Little, Brown and Co. 1984.

[20] This section is derived from A. Wildavsky, *The New Politics of the Budgetary Process*. Glenview, IL: Scott, Foresman and Co., 1988, chap. 3.

[21] A. Wildavsky, *The New Politics of the Budgetary Process*, 81.

[22] W. Hartman, *School District Budgeting*, 26-27.

[23] W. Hartman, *School District Budgeting*, 27.

[24] R.D. Lee and R.W. Johnson, *Public Budgeting Systems*, Baltimore, MD: University Park Press, 1977. 94.

[25] J. Brackett, J. Chambers and T. Parrish, *The Legacy of Rational Budgeting Models in Education and a Proposal for the Future.* Stanford, CA: Inst. for Research on Educational Finance and Governance, Rep. #83-A21, Aug. 1983.

[26] R.D. Lee and R.W. Johnson, *Public Budgeting Systems*, 102.

[27] M. Kirst, "The Rise and Fall of PPBS in California." *Phi Delta Kappan,* April 1975, 536-38.

[28] A. Schick., Budget Innovation in the States. Washington D.C.: The Brookings Inst., 1971.

[29] M. Kirst, "The Rise and Fall of PPBS in California."

[30] A. Wildavsky, *Budgeting: A Comparative Theory of Budgetary Processes,* 180-200.

[31] P.Pyhrr, "Zero-Based Budgeting," *Harvard Business Review*, vol. 48, no. 4 (Nov./Dec., 1970), 111-121.

[32] S.W. Bliss, *Zero-Based Budgeting: A Management Tool for School Districts*, Reston, VA: ASBO, 1978, 3.

[33] J. Brackett, J. Chambers, and T. Parrish, *The Leg acy of Rational Budgeting Models in Education and a Proposal for the Future,* 26.

[34] These steps are derived from S.W. Bliss, *Zero-Based Budgeting: A Management Tool for School Districts*, 7-8.

[35] A.Schick, "The Road From ZBB," *Public Administration Review,* vol. 38, no. 2. March/April, 1978, 177-180.

[36] G. Hentschke, "Evaluating ZBB in Light of Earlier Budget Reforms" *J. of Education Finance.* vol.4, no. 1, Fall, 1978, 234-247.

[37] I.C. Candoli, W.G. Hack, and J.R. Ray, *School Business Administration: A Planning Approach,* 136.

[38] This list is adapted from W. Hartman, *School District Budgeting*. 8-9.

[39] See, A. Odden. "Class Size and Student Achievement: Research-Based Policy Alternatives," *Educational Evaluation and Policy Analysis.* vol. 12, no. 2, 1990, 213-227; R. Slavin. "Achievement Effects of Substantial Reductions in Class Size," in R. Slavin (ed.), *School and Classroom Organization,* Hillsdale, NJ: Erlbaum, 1989, 247-257; M.L. Smith & G.V. Glass, "Meta-Analysis of Research on Class Size and Its Relationship to Attitudes and Instruction," *American Educational Research J.,* vol. 17, no. 4, 1980, 419-433.

[40] National Center for Education Statistics, *Digest of Education Statistics,* NCES, Washington, D.C., U.S. Government Printing Office, 1992, 79.

[41] W. Hartman, *School District Budgeting,* 96.

[42] See Chapter 11 for a detailed discussion of school district accounting systems.

[43] W. Hartman, *School District Budgeting,* 96.

[44] For more information on state school finance formulas and the impact on local school districts, see D.C. Thompson, R.C. Wood, and D.S. Honeyman, Chap 4 in "Fiscal Leadership for Schools: Concepts and Practices, New York, NY: Longman, 1993, and A.R. Odden and L.O. Picus. *School Finance: A Policy Perspective.* NY: New York, McGraw-Hill, 1992. Chapters 7 and 8.

[45] "State of the Art," *Government Finance Rev.,* Aug. 1993, 36-37.

CHAPTER 13

Financial Management and Cash Flow

Introduction

Chapter 12 described the process school districts use to estimate the amount of money they will need each year to finance the educational activities they provide.[1] While a district's budget offers an overall concept of spending plans, it does not provide detail on the management of actual cash funds used for school district purchases. Moreover, as this chapter will reflect, revenues do not always accrue to the school district in a time frame that is related to expenditure needs. At times, the school district will have an excess of revenue compared to its spending needs, and at other times it may need to spend more than it is receiving. How a school district invests its excess cash, when available, and how it covers periods where expenditure demands exceed revenue streams are the topics of this chapter.

Cash management has been defined as "the process of managing the moneys of a school district to ensure maximum cash availability and maximum yield on investments."[2] Virtually every cash management action of a school district is governed by applicable state statutes. States have incorporated the concept of the "prudent person rule" very similar to trust activities of the private sector.[3] Traditionally, there are four specific activities associated with cash management:

- The conversion of accounts receivable to cash receipts.
- The conversion of accounts payable to cash disbursements.
- The rate at which cash disbursements clear the bank.
- Maximizing the utilization of cash.[4]

This chapter discusses the goals of cash management, defining how school business officials can maximize the amount of money available to their school district while ensuring all expenditure commitments are fully and promptly met. In addition, this chapter provides information on the management of cash budgets and instruments that can be used to either invest surplus school district funds or borrow when short-term spending demands exceed current cash resources.

It should be noted by the student of school business management that this chapter addresses cash management for school districts that are fiscally independent of other governmental agencies. In many states, e.g., Connecticut and Virginia, other governmental agencies control the cash management on a city or county basis based on classification schema of school districts. Generally, in fiscally dependent school districts, the school business administrator presents claims or warrants to the other governmental agency for payment. The town or county manages the investments along similar lines as discussed in this chapter. The school district, however, is usually not involved in such activities.

Cash Management Goals

It must be clearly understood that the investment of school district funds is clearly a fiduciary function, i.e., the conservation of principal is the utmost responsibility for the school business administrator.[5] While there are a variety of goals for cash management, there are three major financial goals that must be considered by school business officials. They are:

- *Availability:* Ensuring cash availability (liquidity) to meet the daily needs of the school district, and to increase the amount available for investment purposes.
- *Yield*: Earning the maximum return on cash invested.
- *Safety*: Protecting the assets of the school district against loss.[6]

To some extent, these three goals conflict with each other. For example, if cash is invested to increase the yield, it may not be available to pay current bills or meet payroll costs. Similarly, the higher the interest rate on an investment, the greater the risk. This concept must be tempered with the concept that the safety of principal is always the fundamental goal of public sector investment.[7] Since current obligations have to be met, and since risk must be minimized (generally, school districts are not permitted to invest money that is not fully insured by some agency), the yield goal is of necessity secondary.[8] While this limits the investment income available to the school district, there are ways to maximize cash inflow and minimize cash outflow to improve the investment yield. These techniques are described later in this chapter.

Concurrent financial goals of cash management are the minimization of the costs of the cash management process and minimization of borrowing. Clearly, the less money spent on personnel to manage the district's cash budget, the higher the yield on investments, although employment of a talented investment officer can often pay for itself in terms of improved yields. Moreover, it is important to have adequate staff to ensure that all cash actions are handled accurately and are not subject to embezzlement. While the minimization of borrowing is generally a good idea, there may be limited times when the school district can benefit from additional borrowing through arbitrage.[9] These topics are covered in more depth later in this chapter.

There are also a number of nonfinancial goals of a school district that can play a role in the school district's cash management policies. For example, the school district may wish to promote favorable business relations with local vendors and banks. Not only might this help to ensure the orderly conduct of the financial aspects of the school district's operations, it often helps to build trust and goodwill within the community.

Cash management can be identified as consisting of eight elements by which a sound cash management program can be classified.[10] The first is the administrative framework, which must be established to provide the policies and procedures for managing the district's cash resources. The next four elements are related to the availability of cash resources and include:

- Cash budgeting.
- Cash information and control systems.
- Collection and disbursement procedures for cash collection.
- Collection and disbursement procedures for borrowing.

These elements, when managed well, provide for maximum cash availability for the school district.

The remaining three elements of a cash management program are concerned with maximizing the yield on investments. They include:

- Knowledge of financial institutions.
- Knowledge of investment securities.
- Knowledge investment strategies.

Used together, these eight elements will assist a district business administrator to maximize the potential of his or her cash management program. The relationships among the eight elements are shown graphically in figure 13.1.

Cash Management: Administrative Framework

The day-to-day operation of the cash management program is conducted on the basis of the administrative framework established by the school district. The policies established for this work must encompass the legal and procedural requirements for cash management, as well as any school district specific needs, systems, and requirements. The achievement of a district's cash management goals is dependent on development of four actions:

Figure 13.1
The Elements and Goals of a Cash Management Program

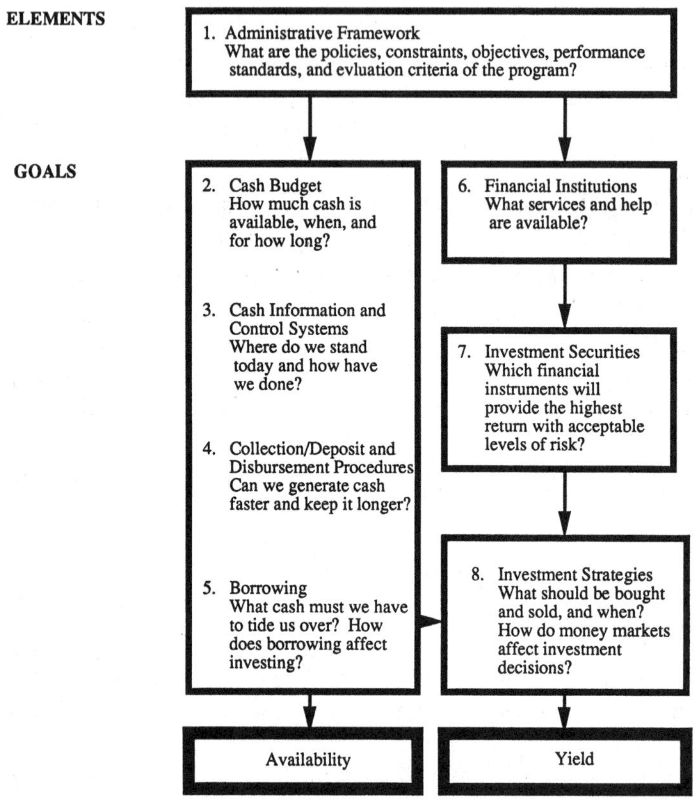

Source: Adapted from F.L. Dembowski. "Cash Management." in G. C. Hentschke. *School Business Administration: A Comparative Perspective*. Berkeley, CA: McCutchan. 1986. and F.L. Dembowski and R.D. Davey. "School District Financial Management and Banking." in R.C. Wood, ed. *Principles of School Business Management*. Reston, VA: ASBO International, 1986.

- An interpretation of relevant federal and state guidelines affecting cash management by the district's legal counsel.

- A board approved, written policy governing cash management operations.

- A process for maintaining a cash investment consolidation account that keeps track of the cash balance for different funds and accounts.

- An on-going evaluation process that reviews progress throughout the fiscal year and provides data to allow analysis of investment performance.[11]

Federal and Statutory Guidelines

All states have some sort of restrictions on the cash management activities of local school districts. While such rules and regulations vary from state to state, and sometimes on the basis of the size of a school district, there are three areas where such restrictions are most common:

- Stipulation of the dates and procedures for collecting taxes or fees.

- Restrictions on the kinds of securities and obligations for investment that are available to school districts.

- Limitations on the types of financial institutions with which districts may do business.

There are often accounting and procedural requirements that accompany the receipt of federal funds. Moreover, payment for these programs occurs at various times throughout the fiscal year. Consequently, the school district's cash management program must include plans for dealing with expenditure requirements in advance of the receipt of federal funds, and plans for investing federal funds that have been received but not expended that meet any requirements established by the federal government agency funding the program.

School District Policy on Cash Management

Drafting a policy statement about school district cash management requires special care. The policy needs to contain detailed specifications of the procedures, constraints, and limitations on actions of individuals. It is important to leave the specifics of the actual procedures to be used in handling and investing school district cash resources to the discretion of the school business official with the advice and consent of the superintendent along with the approval of district counsel.

A board of education policy should include the following points:

- Authority of cash management for the school district should be delegated to one official.

- The types of investment instruments in which the school district can invest should be delineated.

- Criteria for selection of depository banks for school district funds should be established in policy.

- The types and frequency of financial reports that must be submitted should be specified.

- Any other limitations on cash management.[12]

Consolidation of Cash Accounts

The size and wealth of a school district have a considerable effect on the rate of return to investment. Generally, the larger the school district, the greater the cash flow, and the more likely it is that there will be more cash available for investment at any point in time. Moreover, banks and other financial institutions often pay premiums for larger amounts of cash, and may offer a higher interest rate for a certificate of deposit (CD) for a higher amount. The benefit to the bank is that it has more cash for other purposes for the life of the CD.

A school district can take advantage of these economies of scale in investing by combining or pooling cash resources into one bank account. In most states, as discussed later in this chapter, it is possible for a number of districts to pool cash resources to create larger investment pools that can take further advantage of both larger sums to invest, and the ability to hire investment specialists who may also help improve the overall yield to the school district. In California, many counties have investment pools for all of the public jurisdictions within the legal boundaries (cities, school districts, and other special districts). In addition, the state maintains an investment pool for public agencies. Many school districts evaluate the performance of each investment pool, and shift excess moneys from one to another fund based on the current rate of return and the school district's own individual liquidity needs.

There are a number of advantages to pooling resources. In addition to the likelihood that there will be higher rates of return on larger amounts of cash invested.[13] Larger investment pools tend to reduce fluctuations in cash flow, simplify investment processes, and reduce paperwork.

Evaluation of Cash Management Performance

There are a number of factors that will affect the performance of a school district's cash management program. A number of external factors will affect the amount of money the school district is able to gain in return on its investments. These include current market conditions and the size and wealth of the school district. Factors internal to the school district include the types of investment instruments used and the skill of the school business administrator and/or the investment office.

It is difficult to determine the impact that a district's property wealth will have on the investment return it can expect. Wealthy districts typically receive more of their revenue from local property tax sources and less from the state. Pursuant to state statutes, property taxes are typically collected once or twice a year, meaning that there are specific periods of time when the school district is likely to have excess cash available for investment purposes. The greater the share of revenue derived from property taxes, the more excess cash the school district is likely to have available.

Each state uses a different formula and schedule for allocating state aid to school districts. If state funds are only apportioned two or three times a year, similar large sums of excess cash will be available to the school district. On the other hand, if apportionments occur more frequently, but in smaller amounts, there may be fewer opportunities to invest excess money to improve the school district's yield. Since the state has the same incentive for investment as do local school districts, wealthier school districts would appear to be at an advantage in this regard.

External factors, particularly the yield on investments and the types of investment instruments the school district is allowed to consider, will also affect the total yield from cash management activities. In addition to the general interest rate on investment securities in the economy, the amount to be invested, and the length of time for which it can be invested will affect the final return. In years when interest rates are uniformly high, districts can, and should, expect greater returns than in years when interest rates are lower.

Cash Budgeting

The purpose of cash budgeting is to determine when cash receipts are expected, how long those moneys will be available, and whether there is an opportunity for investment or a need to borrow funds.

A cash budget is different from a revenue and expenditure budget because it is concerned with the timing of receipts and expenditures, not with the overall revenue and expenditures to be expected during some time frame. Once a traditional budget has been developed, it is possible to estimate revenue and spending needs to establish the cash flow budget. Cash budgeting is thus a continual process with regular and frequent updates. With this knowledge, the school business administrator is able to design a strategy for investing and borrowing that maximizes the investment return and minimizes borrowing costs.

At the heart of cash budgeting is the cash flow schedule. A simplified cash flow schedule is displayed in table 13.1. This example simply lists the cash receipts and expenditures by month, the net flow of funds for the month, and the current cash balance. More detailed schedules can be developed, listing each individual receipt and expense. They can also be calculated daily or weekly rather than monthly. Most school districts will need to develop a cash flow schedule estimating daily surplus cash for investment. Regardless of the level of detail involved, the basic principle is identical to that displayed in table 13.1.

What is important to note from table 13.1 is that most months there is a positive cash balance that would be available for investment. In this example, the three months of July, August, and September, there is a negative cash balance indicating that the school district will have to borrow funds to meet its commitments. Also note that the cash balance differs from the monthly flow of funds. Table 13.1 reflects that there are only five months where revenue exceeds expenditures, and seven months where more money is expended than collected. This has important consequences for the fiscal management of the school district.

It is also important to note that in table 13.1 there is a beginning cash balance of $50,000, yet by the end of December the cash balance is zero. This indicates that during the twelve months represented in this example, the district's expenditures exceeded its revenue by $50,000. This can be confirmed by summing the estimated revenue and expenditure columns. Total revenue in table 13.1 amounts to $2,450,000, while expenditures are $50,000 higher at $2,500,000. Whether this is problematic depends on how the revenue and expenditure pattern appears to be evident in this table compared with the district's fiscal year performance. If the school district were to have a balanced budget, then in theory over the course of a fiscal year, revenue and expenditures should balance. The imbalance found during this twelve-month example could be the result of unanticipated expenditures, a decline in revenue, or simply the fact that the district's fiscal year runs from July 1 to June 30, and that at some point during the first six months of the next calendar year (the basis of this chart), revenue will be high enough to resolve the imbalance displayed.

Table 13.1
Sample Cash Flow Analysis for a School District

Month	Estimated Revenues ($)	Estimated Expenditures ($)	Monthly Net Flow ($)	Running Balance
Beginning Balance				50,000
January	400,000	200,000	200,000	250,000
February	250,000	200,000	50,000	300,000
March	400,000	200,000	200,000	500,000
April	150,000	200,000	-50,000	450,000
May	100,000	200,000	-100,000	350,000
June	50,000	400,000	-350,000	0
July	50,000	100,000	-50,000	-50,000
August	50,000	100,000	-50,000	-100,000
September	150,000	200,000	-50,000	-150,000
October	400,000	200,000	200,000	50,000
November	400,000	200,000	200,000	250,000
December	50,000	300,000	-250,000	0

Adapted From F.L. Dembowski and R.D. Davey. "School District Financial Management and Banking." in R.C. Wood. ed. Principles of School Business Management. Reston VA: ASBO International, 1986. pp. 241-43.

Making accurate cash flow projections is more straightforward for school districts than one might expect. Generally, revenue flows to the school district at specific times. For example, local property tax receipts are most likely received, pursuant to state statutes, once or twice a year on specific dates. Moreover, the schedule of payments for aid from the state is also known far in advance and can thus be predicted with relative ease. On the expenditure side, the largest single expense facing school districts is for salaries and benefits. Districts often allocate 80 to 90 percent of expenditures to personnel costs. These expenditures can be predicted with great accuracy. Other expenditures often occur in periodic ways that can also be predicted accurately, making it possible to estimate cash flow with considerable certainty.

Figure 13.2 displays a district's cash flow graphically. In this example, revenue is high in January, March, October, and November, reflecting either property tax collections or state aid payments. In other months, revenue is substantially lower. On the other hand, expenditures are more consistent at about $200,000 a month. They are higher in June, and lower in the summer, with an increase again in December.

As figure 13.2 shows, revenue exceeds expenditures in January through March, a time when the funds can be invested to improve school district revenue. In April through September, expenditures exceed revenue, and the school district will have to draw down the surpluses. Data developed through analyses similar to those displayed in table 13.1 and figure 13.2 will help the school business administrator determine what investment and borrowing decision should be made. Moreover, it will help the administrator determine the timing, amount, and maturity of each investment or loan.

Cash Information and Controls

Knowledge of the anticipated cash flow is only a first step. It is also important to have an effective cash information system if the school business administrator is to meet the cash management program's goals. A cash information system needs to maintain three types of records:

- A record of financial events such as investments made or loans incurred.
- Records to assist in investment decisions such as current cash balances in each account or fund.
- A permanent accounting of all cash by fund.[14]

A cash information system is basically an accounting process to keep track of daily receipts and disbursements, provide cash balances by fund and account, provide daily average-cost balances, provide a budgeted compared to actual report of receipts and disbursements, and provide information to update the cash flow schedule and evaluate the forecasting abilities of the cash manager.[15]

Figure 13.2
Sample Cash Flow for a School District

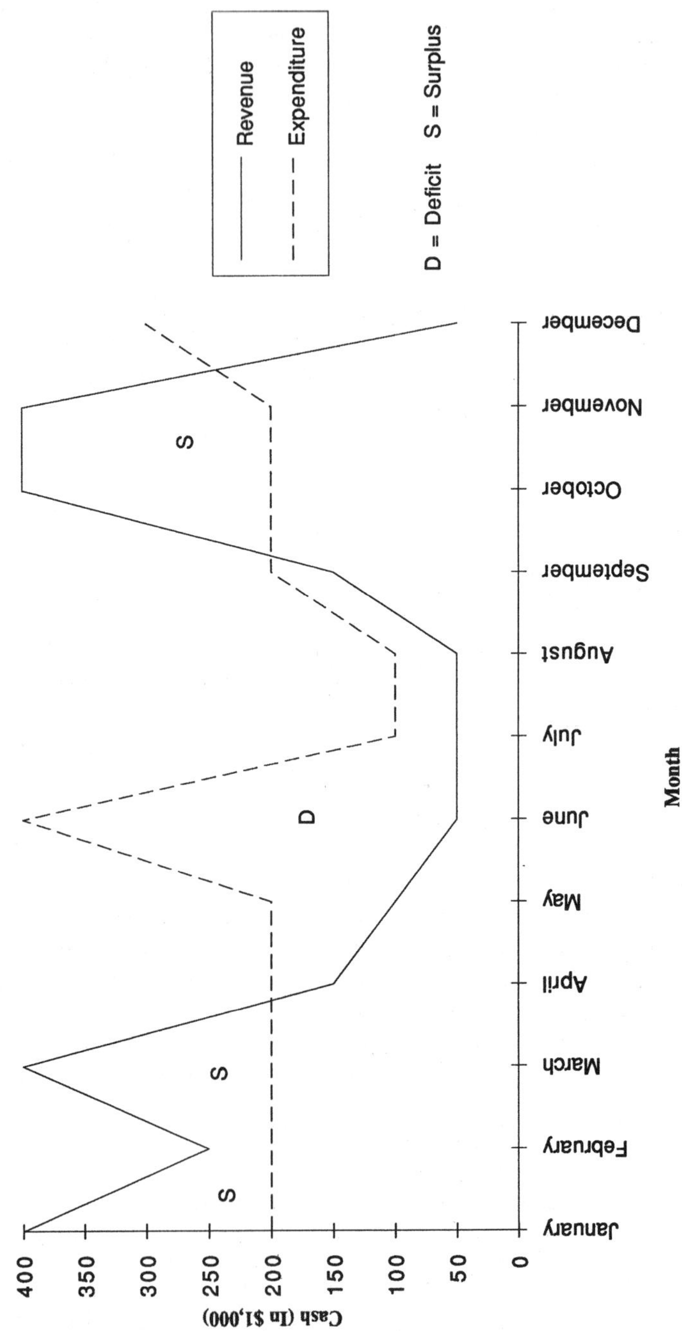

There are also a number of informal processes that are important to a cash management system. It is important for the school business administrator to maintain a close working relationship with other school district administrators, particularly those who receive and/or spend money. Often formal accounting systems do not provide the business office with information about commitments until they are actually made. In many instances it is helpful to know of anticipated spending, particularly for large or expensive items, well in advance. In this manner, the business officer can ensure that funds are available to pay for any expenses that are incurred.

Finally, it is important for the school business administrator to keep close track of the potential earnings of various investment instruments for those times when there is surplus cash to invest. In this manner, these investments can earn the maximum yield given the liquidity needs of the school district.

Cash Collection and Disbursement Procedures

The more cash a school district has available to invest, the greater return it can garner from investments. Collection and disbursement procedures should, therefore, be developed to bring anticipated revenue into the school district as soon as possible, and to keep the revenue in school district accounts as long as possible. To do this, procedures for revenue collection, bank deposits, and expenditures must be established. These procedures should ensure that funds are deposited in school district accounts as quickly as possible and maintained in interest-bearing accounts until needed to pay for approved expenditures.

Revenue Deposits

Presently, most major revenue sources can be transferred to district accounts electronically. State aid payments and transfers of property tax revenue from county tax authorities can be made available to the school district on the day and time the school district is eligible to receive the funds under state statues. This process is a vast improvement over earlier systems where districts either had to wait for the arrival of a check from the state via mail, or had to send a courier to the state capital to collect the check and return to the local bank in time to deposit the funds. The cost of wire transfers is minimal compared to the alternatives. Sending someone to the state capitol not only involves travel costs, but the value of the individual's time must also be considered as they are not available at the school district office to perform other functions. Moreover the interest lost over two or three days while waiting for a check to arrive in the mail can be substantial. For example, one day's interest on $1,000,000 at 5 percent amounts to $138.89. While this may appear to be a trivial sum, the loss would be substantial if a large school district were expecting to receive a substantial state aid payment.

In some states, notably Texas, local school districts are responsible for collection of property taxes. In these states, it is important to establish procedures to deposit tax receipts in school district accounts as soon as possible after they are received. Since the school district is likely to receive a large number of relatively small checks in the space of a few days, it is important that careful record keeping procedures be established to ensure that all tax receipts are credited to the proper accounts.

Most state aid systems are based on a school district's assessed valuation. There is generally an inverse relationship between the amount of state aid a school district receives and its per pupil assessed valuation. In most states, aid is based on a calculation that takes the local property tax rate and per pupil property wealth into account, and the actual aid payment based on that calculation. However, in some states, again Texas is an example, the portion of state aid funded through the guaranteed yield section of the state aid formula is based on actual tax receipts. Therefore, the school district has an incentive to ensure that all property taxes due are actually collected. The greater the state share of the yield guarantee, the more effort the school district will want to expend to collect delinquent tax payments. In all cases, districts may want to consider the employment of a collection agency whenever the costs of doing so are less than the anticipated revenue receipts.

Expenditure Procedures

Timing of expenditures is equally important. From the school district's perspective, it is desirable to make payments so that cash is removed from school district accounts at the last possible moment. This must be tempered, however, with the importance of paying bills in a timely manner. Vendors who are concerned over payment for services or materials may raise prices in the future, respond slowly to complaints, or perform poorly.

Virtually all states require that the school board approve the payment of bills.[16] Boards typically approve payments at regular meetings to ensure public accountability. Payments should be "aged" to improve control over disbursement of funds. Once an invoice is received, it should be processed and a payment date determined. The date clearly has to be after the next regularly scheduled board of education meeting so that approval can be obtained. Following approval, the bill can be paid on the predetermined date. Factors that affect decisions regarding payment date include the availability of discounts for prompt payment, previously agreed to vendor requirements for prompt payment, and method of payment. School districts generally seek aggressive discounts for prompt payment as these may lead to substantial savings, particularly if the discount exceeds the current rate of return on invested moneys.

The major expenditure item for a school district is its payroll. Today, electronic deposit of employee paychecks is a common practice. While this procedure has some disadvantages to school districts as the funds are transferred to the employees' accounts on pay day (ending the ability to earn interest on checks that do not clear for a few days), there are also substantial benefits to the school district in terms of reduced payroll costs and improved employee relations.

Offsetting these financial disadvantages is the fact that most school districts do not actually deposit 100 percent of the funds to the payroll account on the scheduled pay day. Generally, school districts do an historical analysis of the date on which the payroll account is cleared. Thus, money is then deposited into this clearing account on an as-needed basis. In this manner, maximum cash flow is obtained. If the school district must "borrow" money from the financial institution in order to clear the check, this amount has been prenegotiated in the contract with the institution. Again, it must be stressed that the school district's payroll account is the single largest expenditure for any school district. Thus, the maximization of the cash flow of this account is vital to the fiscal well-being of the school district.

Borrowing

It is possible that there will be times when a school district's expenditure needs exceed existing cash resources, and the school district has to borrow money. There are two basic borrowing strategies. Long-term loans provide districts with large sums of money, generally for a specific purpose such as construction of school facilities. Long-term borrowing is most often conducted through a bonding process, and is the topic of chapter 17.

When a school district has a short-term cash flow deficit, there are a number of instruments it can use to borrow funds for a short period of time. In short-term borrowing, money is usually borrowed in anticipation of expected revenues, i.e., state aid payments or receipt of property tax receipts, generally within a short period of time.

While many districts attempt to avoid short-term borrowing, often because of statutory restrictions or concern on the part of the school board regarding "going into debt," the use of short-term loans can be an important part of any cash management system, offering opportunities for net interest earnings if planned carefully. State statutes govern in a very strict fashion the manner and scope of school district borrowing. In a few states, such borrowing is directly from the state due to favorable rates and statutory restrictions. While statutes in each state vary, there are basically five types of short-term loan instruments available to school districts:

- *Revenue anticipation notes*: These are used for general purpose expenditures and are based on the anticipation of revenue, other than property taxes, to be collected by the school district in the near term.

- *Tax anticipation notes*: These are also used for general purpose expenditures, and are based on the expected collection of taxes levied or to be levied in the near term.

- *Budget notes*: These are used to provide revenue for unforeseeable expenditures in the current year and are generally paid off with taxes collected in the next budget year.

- *Capital notes*: These are short-term notes used to finance all or part of the cost of an item for which serial or sinking-fund bonds may be issued.

- *Bond anticipation notes*: These are issued whenever bonds have been authorized, but not yet sold. The proceeds from these notes may only be used for the same purpose as the long-term bonds, and they are paid off with proceeds from the bond issue.[17]

There are many state-specific restrictions on short-term borrowing. In many cases, the payback period is limited, and the reinvestment opportunities for borrowed funds tightly controlled. Typical payback period requirements include the fiscal year in which the funds were borrowed, or some period of time such as six or nine months. Some states place limits on the amount that may be borrowed on a short-term basis, typically something on the order of 75 percent of anticipated revenue, although this varies widely by state.

One way school districts can improve the yield of invested funds is to borrow money through one of the instruments previously identified, and reinvest it at a higher interest rate until it is needed to pay school district bills. This process is called arbitrage. Great caution should be exercised in the usage of such strategies in that there are severe Internal Revenue Service (IRS) penalties involved if certain procedures are not met. However, within severe guidelines this concept is possible because many financial institutions are willing to lend money to public agencies, including school districts, at below market rates of interest. Since interest earnings from loans to school districts are tax exempt, banks may be able to earn a higher yield by offering low interest loans to schools. School district officials can then invest the unexpended funds in other financial instruments with a higher yield, providing additional revenue to the school district.

Since the yield on the loan may be higher than for other instruments, and because loans to public agencies are generally very secure, banks are willing to participate in arbitrage. The only real loser in this process is the federal government, which loses tax revenue because the interest earnings of the bank are tax exempt. Because of this revenue loss to the federal government, the tax reform act of 1986 placed severe restrictions on the use of arbitrage by public agencies. The law establishes substantial penalties for "excessive" arbitrage earnings, making it difficult, if not impossible for school districts to earn money through this process.[18] Due to the severity of these regulations, the practical application of arbitrage is that it is no longer permitted for the vast majority of school districts in the country.

Selecting a Financial Institution

There are many options available to school districts in selecting financial institutions. Although there have not been any recently published studies, Dembowski found that in 1980 approximately 60 percent of school districts only used one bank, and more than half did not maintain savings accounts.[19] While maintaining cordial relations with a bank, particularly in a small community, is important, it is also important to maximize the return on invested funds and minimize banking costs to ensure that tax dollars are wisely spent. In many states the method and amount of funds that are invested in banks are strictly controlled by state statutes.

In the absence of highly specific and directive state statutes, school districts possess several options in placing its banking business within a given community and state. Again, the student of school business administration should always consult the applicable state statutes and competent legal counsel. Nonetheless, the concept is that the school district should always consider maximizing the best and most efficient use of taxpayer money.

Some school districts place all of the funds with a given bank based on the concept that they are more likely to be able to negotiate a better financial cost package to the school district. Other school districts use several different financial institutions within the boundaries of the school district so as to have a degree of political support and to benefit the community as a whole, rather than one financial institution. A variation of this concept is that some school districts place different funds with different financial institutions, e.g., the general fund, debt service, and auxiliary funds.

A variation of this concept is found when school districts rotate the financial services to local financial institutions based on board policy, which is a variation of the placement process. This variation does pool resources for a degree of efficiency but also eliminates the bidding or negotiations advantage as local financial institutions have no incentive to provide competitive rates.

Generally, the theory of school business management dictates the competitive bidding of such services. Where state statutes allow it, this process has proven to be the most cost-effective mechanism for the placement of school district funds. Board policy generally dictates the timeliness for bidding of such services. Typically, these policies allow for such rebidding every three to five years. All financial institutions that are insured and licensed to engage in complete banking services should receive either a request for proposal or a bid form. Great care and expertise must be exercised in the crafting of the request/bid document. Even greater care must be exercised in judging the lowest true cost to the taxpayers.

Many states regulate school district banking procedures. School districts are required to seek bids for banking services on a regular basis, and in some states, the school district must do business with a bank located in the county in which the school district is located, or strictly within the school district boundaries. Evaluation of bank depository bids can be a complex and difficult task given the range of services that might be offered,. It is important to understand what the school district's banking needs are and will be for the future.

Various guidelines for drafting requests for proposals for banking services have appeared within the school business management literature. Guidelines include information that the school district should make available to the banks interested in bidding to provide banking services for the school district as well as the services the school district should expect from the bank. Request for proposals generally should include the following concepts:

- Information concerning the financial status of the bank including its safety rankings from various agencies.

- The beginning and ending date of the contract with the bank including the terms by which the school board may withdraw from the arrangement.

- The cost by type of transactions with information provided as to minimums and maximums for cost projections. Financial institutions should not be prohibited from establishing minimum balances and other such requirements. The school district is concerned with the total overall cost, i.e., such requirements may prove to be less costly than a proposal that does not have a minimal balance requirement.

- A clear written understanding as part of the request for proposal that the school district may and will pursue all legally permissible investments if the receiving financial institution cannot provide competitive financial returns.

- The number as well as the complexity of accounts and services needed by the school district.

Generally, school districts should evaluate the bids based on the total overall cost. Additionally, similar to all other bid evaluations, the lowest responsible bid is one that combines the lowest cost with the best services needed by the school district. Most school districts examine the following issues when judging such proposals:

- A determination that the financial institution is fully insured by all applicable agencies and that it is in strong financial health as determined by state, federal, and private agencies.

- The ability to directly send and receive all wire transfers between all state, federal, and other agencies.

- The ability to engage in short term-loan arrangements against anticipated state aid payments as well as revenue from ad valorem taxes. A standard line of credit and the cost of such should be maintained and determined before the need arises.

- Availability and cost of a variety of services including safe deposit boxes for school district use and the availability and security of 24-hour depository services.

- Fees for every banking service including checks, stop payment orders, overdrawn accounts, and so on.

- Availability of ancillary services such as the cost of armored car services.

- Capacity, and fees if any, for handling bulk coins.[20]

It should be noted that the judgment of these issues is the prerogative of the school district. In some instances large school districts have hired financial consultants to make recommendations regarding these rather complex issues. Additionally, it is important to work closely with local banking officials, both to maintain a strong working relationship with members of the community, and to ensure that the school district receives the best possible service at the lowest possible cost. As one of the largest depositors in most communities, school districts should be in a position to negotiate agreements with favorable terms from local banks.

Investment Instruments

Like borrowing, school district investment strategies are regulated by state statute. In determining how to invest school district funds to obtain the maximum yield, the school business administrator must be aware of the risk involved in each investment and consider the liquidity of each in terms of the school district's future cash needs.

Risk

All investments should be considered in terms of financial risk. Financial losses should be avoided since it is taxpayer money that is at being used. Even in states that permit school districts to invest in high risk and speculative investments, such investments should be avoided. In many states, collateral on school district investments is required. The collateral requirement varies from state to state. Since the Federal Deposit Insurance Corporation (FDIC) only insures the first $100,000 of all deposits from any entity, other assets must be held as collateral. In general, banks bid lower rates of interest on investments that require collateral because it costs the bank to maintain the assets used as collateral instead of using them for other, more lucrative, purposes. The school business administrator should take possession of the collateral securities and require written documentation of the collateral.

Liquidity and Yield

Another important consideration in investing school district securities is the liquidity of the investment. Liquidity is the ability to convert the investment security into cash. The more quickly it can be converted to cash, the more liquid an investment. Generally, the more liquid an investment, the lower the yield. Thus, a school business administrator has to balance the higher yield of long-term investments against the school district's cash needs in deciding how much to invest in different types of securities. It is important to keep some highly liquid securities in a school district's investment portfolio to meet expenditure obligations on time, or to deal with unanticipated cash deficits.

The yield of an investment depends on the type of security purchased and the nature of the market at the time the investment is made. Depending on the need for cash for other purposes, banks may offer different rates of interest on comparable certificates of deposit. Thus, it is wise to seek competitive bids for investment of school district cash surpluses. There are a number of other factors that affect the yield of an investment. Generally, the following factors lead to higher interest rates on investments:

- Longer time to maturity.
- Greater risk.
- Lower liquidity.
- Larger sums invested.
- Lower or no collateral requirements.

Each security has a specifically defined method of calculating yield. Interest earnings are calculated on the basis of either a 360-day or 365-day year, causing small differences in actual earnings between bids with the same maturity. Thus, the school business administrators must consider not only the interest rate, but also total net earnings in dollars before making a decision.

Generally, there are four questions that a school business administrator should ask in analyzing each investment:[21]

- Does it conform to legal restrictions?
- What is the nature and amount of risk to be assumed?

- What is the yield?
- What is the degree of liquidity?

Types of Investments

There are a variety of investment instruments available to school districts. Among the most common types of securities and investment instruments used by school districts in most states are the following:

U.S. Treasury Bills, Bonds, and Notes

Backed by the U.S. Government, treasury issues are the lowest risk investment instrument available. They are sold at government auctions and reflect the interest rates in effect at the time of the sale. They can also be purchased on the open market, and most school districts purchase them through brokers. It is important to note that while the treasury security is perfectly safe, if it is purchased via a repurchase agreement through a broker or a financial institution, the broker or the financial institution is subject to a credit risk.

Once the issue has been sold at auction, the interest rate is fixed. Since that rate changes daily on new issues being sold on the open marketplace, fluctuations in the interest rate of existing treasury bills are reflected in the price on the face of the instrument. Thus, if the interest rate on new issues increases, the face value of existing bonds will be discounted to reflect the current interest rate.

Treasury *bills* have maturities of 13, 26, or 52 weeks and are sold in *$10,000* denominations. In contrast, treasury *notes* with a *$5,000* denomination mature in two or three years, while those with *$1,000* denominations mature in four to ten years.

Additional concerns are worthy of note. For example, in the event that the school district purchased such an issue via a weak financial institution, even thought it was perfectly legal, insured, as well as a collateralized issue, the school district would have to wait for the FDIC to settle the district's claim in the event of closure of the financial institution.[22] Thus, the issue of safety would be met while the issue of timeliness and liquidity would not be met.

Direct purchasing of treasury and governmental agency issues is not without peril. Many school districts purchase these instruments directly and do so very successfully. This type of investment is the safest, yet offers burdens in terms of the management of such issues. The school district must maintain liquidity. Thus, through the fiscal year, the maturities must be matched to the cash flow needs of the school district. To do this, the portfolio must be actively managed and traded.[23] Further, the school district business manager must control a sizeable portfolio in order to benefit from a satisfactory degree of economy of scale.

U.S. Government Agency Bonds

There are a number of other federal corporations that issue bonds for specific government-sponsored activities. Among the most familiar are the Federal National Mortgage Association (Fannie Mae), Government National Mortgage Association (Ginnie Mae) and the Federal Home Loan Mortgage Corporation (Freddie Mac). These are mostly secured by mortgages, but are not wholly insured by the full faith and credit of the government as are U.S. Treasury bills and notes.

Certificates of Deposit

Certificates of deposit (CDs) are used by school districts the same way that other investors use them. They provide a temporary place to invest money for varying lengths of time at varying interest rates. A major consideration in selecting a bank is the collateral used to back the CDs. In the past, some school districts have been caught with CDs that were secured by bad loans, and the districts have lost money. While a convenient way to manage surplus cash, most CDs assess a penalty for early withdrawal of funds. As a result, school business administrators should be confident of their projected cash flow before purchasing long-term CDs.

Repurchase Agreements

Repurchase agreements (REPOs) are used for short-term investments, often as short as overnight or through a weekend. The way these work is that a school district agrees to invest funds for a specific number of days and takes title to bank securities as collateral. At maturity, often the next morning, the bank repurchases the securities for the amount of the investment plus interest. These investments are very liquid. The level of risk is governed by the quality of the securities used as collateral. Although the yield is generally low, it is usually higher than simply keeping the funds in a checking account for the same period of time. Maynard and Wheatley's research suggest three cardinal rules regarding repurchase agreements:

- Know your counterparty.
- Take physical possession or independent book entry of the collateral.
- Establish and monitor a margin to protect the market value of the collateral from interest rate movements.

Sound investment strategy is specifically offered in the leading school business management journal, *School Business Affairs* as Maynard and Wheatley write:

> Most municipalities still execute "trust me" repurchase agreements. They do not take delivery of the repo collateral through an independent third party, but rather rely on a safekeeping receipt from the

> dealer or its agent, which is usually not sufficient to establish legal control over the securities. In some states, investments in repos without taking delivery of collateral may be illegal. Collateral should be delivered to an independent custodian under the [school district's] (not the dealer's) control, and there should be written agreements between the [school district] and the custodian. A [school district] might also choose to use a custodian which the two repo parties agree upon, but again, there should be separate written agreements between each of the parties to protect the [school district's] interest. Local [school districts] that have collateral delivery mechanisms have found that the costs tend to reduce net yields below those available from other equally liquid and safer investments.[24]

Money Market Certificates, Money Market Funds, and Passbook Savings

These are conventional banking instruments designed for use by the general public. However, they may be used as part of a school district investment program to deal with relatively small amounts of cash. Although the yield reflects the market interest rate, and is generally low, the accounts are highly liquid. The risk is related to the collateral provided by the bank and can be a problem if the district is not careful.

Commercial Paper

There are a several states that allow school districts to purchase corporate bonds and similar instruments. These are often referred to as commercial paper and carry a reasonably high rate of interest and risk. They possess greater risk because they are not backed by a governmental agency, but by the financial resources of the corporation that issued the paper. These bonds are subject to ratings by Moody's and Standard and Poor's rating services, which will help the school business administrator assess the level of risk of any individual bond offering.

State and Municipal Obligations

Many state and local governments issue a variety of securities to finance operations. School districts can invest surplus cash in these issues as well. Security of each is based on the credit quality of the agency issuing the bond, and repayment is made from anticipated revenue. In that these securities generally offer less returns due to tax issues, the school business administrator must weigh the relatively low rate of return versus a low risk factor.

Government Investment Pools

Many states have created mechanisms referred to as local government investment pools (LGIP).[25] These pools invest local governmental agencies' cash in order to maximize the efficiency through professional investment operations and economies of scale. Maynard and Wheatley write,

> When a local government invests in an LGIP it purchases 'shares' or 'units' of the pool. Monies from each participating entity are combined for purposes of investment but principal and earnings are accounted for separately. Therefore, each participants' fund maintains its own identity.
>
> These state investment pools are generally conducted by a state agency. . . . [M]ost allow deposits and withdrawals at any time, usually in any amount, and pay interest on a daily basis, net of pool expenses. . . . Yields are usually comparable to rates on large denomination money market instruments of $1 million and more and are normally in excess of what most units can earn individually.[27]

Many school districts use these government investment pools for a variety of reasons including:

- Portfolio diversification.
- Stringent credit analysis.
- Competitive rates of return.
- Daily liquidity.
- Economies of scale.
- Low overhead.[27]

In choosing where to invest school district funds, the school business administrator has a number of alternatives. The best strategy for any individual school district cannot be generalized, but is based on combining different securities to maximize gain, and minimize loses. The next section describes strategies that can be used to achieve that elusive goal.

Investment Strategies

The district's investment strategy governs the actions that are taken in the daily administration of the cash management program. The focus is on the three goals established at the beginning of this chapter, safety, availability, and yield. Three factors that must be considered are:

- The amount of money available for investment.
- Money market conditions.
- The mix of securities.[28]

As stated earlier, the goal is to maximize the yield to the school district, while ensuring that there is enough money to pay all of the district's bills. Thus, long-term, high-yield investments are to be preferred to short-term investments to the extent that the term of the investment does not interfere with the district's ability to meet its financial obligations in a timely manner.

Summary

This chapter provides an introduction to public education cash management for individual school districts. It is stressed within this chapter that individual cash management is highly predicated on applicable state statutes.

Several items should be remembered. For example, a cash budget differs from the revenue and expenditure budget, as it provides information on the amount of money available each day that can be used to pay the district's bills and meet other financial obligations. While it is essential to maintain enough cash to pay all of the bills on time, there are often periods of time when surplus cash can be invested in any one of a number of investment instruments, pursuant to state statutes, to provide income to a school district. The amount of income earned depends on the amount of money invested, the length of time for which money is invested, and the risk of the investment. Using the "prudent person rule," the cash manager's task is to minimize the risk of financial loss to the school district, ensure that funds are available when needed, and at the same time earn the maximum yield. These goals are often conflicting, and availability and security always must take precedence over yield in the management of public funds.

Endnotes

[1] Various portions of this chapter are based on and are adapted from F.L. Dembowski and R.D. Davey. "School District Financial Management and Banking." in R. Craig Wood, ed. *Principles of School Business Management.* Reston, VA: ASBO International. 1986.

[2] F.L. Dembowski and R.D. Davey. "School District Financial Management and Banking," 237.

[3] E. Maynard and P.M. Wheatley, "Private Sector Perspectives on Public Sector Investing," *School Business Affairs,* Dec. 1986, 29.

[4] F.L. Dembowski and R.D. Davey. "School District Financial Management and Banking," 237.

[5] D.E. Maynard and P.M. Wheatley, "Private Sector Perspectives on Public Sector Investing," 29.

[6] Adapted from F.L. Dembowski and R.D. Davey, "School District Financial Management and Banking," 237.

[7] D.E. Maynard and P.M. Wheatley, "Private Sector Perspectives on Public Sector Investing."

[8] F.L. Dembowski and R.D. Davey, "School District Financial Management and Banking," 237.

[9] Changes in the 1986 tax laws have limited the ability of local school districts to benefit from arbitrage. See e.g., W. Weldon. "Arbitrage Interest Rubs and School District Borrowing." *School Business Affairs,* Jan., 1989, 10-13.

[10] Adapted from F.L. Dembowski and R.D. Davey, "School District Financial Management and Banking," 238.

[11] F.L. Dembowski and R.D. Davey, "School District Financial Management and Banking," 238

[12] Ibid., 239.

[13] Higher rates of return are not guaranteed simply by larger investment pools. The rate of return is heavily dependent on the skill of the staff responsible for actually investing the funds it has available, as well as on the level of risk the investors are willing to assume, as well as the cash flow or liquidity needs of the district or districts that participate in the pool.

[14] F.L. Dembowski and R.D. Davey, "School District Financial Management and Banking," 245.

[15] F.L. Dembowski, "Cash Management.," in G.C. Hentschke. *School Business Administration: A Comparative Perspective,* Berkeley, CA: McCutchan, 1986. 224-225.

[16] While state statutes vary greatly, this statement is generally true. As an broad generalization, only the school board may constitute and pay an obligation. In many states, a subcommittee of the school board may examine all such requests for payments, yet it is important to remember that only the school board in a legally constituted meeting may approve actual payment. Several states allow certain exceptions to this general rule such as utility payments which can be adjusted at a later date if need be.

[17] F.L. Dembowski, "Cash Management," 249.

[18] W. Weldon, "Arbitrage Interest Rubs and School District Borrowing," *School Business Affairs,* Jan., 1989, 10-13.

[19] F.L. Dembowski, ed. *A Handbook of School Banking Relations.*, Albany: The New York State Ass'n of School Business Officials. 1981.

[20] This point is important as school district cafeterias, student stores, and activity programs often deal in large amounts of cash, often with considerable coinage.

[21] I.C. Candoli, W.G. Hack and J.R. Ray, *School Business Administration: A Planning Approach,* 4th ed. Boston, MA: Allyn and Bacon, 1992. 283.

[22] D.E. Maynard and P.M Wheatley, "Private Sector Perspectives on Public Sector Investing."

[23] Ibid.

[24] Ibid.

[25] States that have local governmental investment pools include such states as: California, Connecticut, Florida, Georgia, Illinois, Maryland, Massachusetts, Montana, New Jersey, North Carolina, Ohio, Oregon, Tennessee, Utah, Virginia, West Virginia, and Wisconsin. It should also be noted that many of these LGIPs are designed for local county or city governments in those states in which school districts are fiscally dependent as discussed in this chapter.

[26] Ibid.

[27] Ibid.

[28] F.L. Dembowski and R.D. Davey, "School District Financial Management and Banking." 255.

CHAPTER 14
Quantitative Decision Making

Introduction

One thing that becomes clear from reading the other chapters in this text is the complexity of managing the financial side of a school district's operations. Today's school business administrator is faced with a variety of responsibilities including: managing substantial cash investments, maintaining a comprehensive risk management program, operating a transportation system, overseeing a construction office, and planning for future district needs, all at the lowest possible cost. Doing this requires making important decisions, often on a daily basis. The purpose of this chapter is to describe a number of quantitative tools that the school business administrator can use to help make these important decisions. The approach taken follows Stokey and Zeckhauser's[1] model, offering tools that will help decisionmakers lay out their goals and then use logical processes to find the best alternatives for meeting those goals.

In recent years, schools have faced increasing scrutiny, particularly how they use the revenue they receive. This push for accountability has focused on ensuring that administrators maximize the resources available in the classroom. Yet despite this emphasis, recent research has shown that direct classroom instruction accounts for approximately 60 percent of total expenditures.[2] The remaining 40 percent includes many of the items directed by the school business administrator. By focusing attention on this 40 percent and ensuring that the services supporting the direct instruction program are operating in the most efficient way possible, it may be possible to direct additional resources to the classroom, even in a time of limited growth in real per pupil expenditures.

Although it has been demonstrated[3] that the tools described herein have been part of private sector decision making for a long time, and although there is nothing that guarantees that the usage will result in perfect solutions, these rational decision-making tools offer the opportunity for greater analysis of the alternatives. Moreover, the availability of tremendous computer power and high-quality software on individual desktops makes the use of these tool feasible in even the smallest and most remote school districts in the nation.

The tools described herein are based on techniques developed in the disciplines of economics, systems analysis, operations research, and mathematics. They offer ways to understand and analyze complex issues in a rational and straightforward manner. The emphasis of this chapter is on the tools themselves and not the school district data used as input to the process. Simplified examples will be provided to improve understanding of each methodology. Responsibility for the accuracy of data ultimately rests with the individual conducting the analysis.

Conceptual Framework

The first step in solving a problem is recognizing that there is a problem needing resolution. In the day-to-day world of school administration, there are often crises that need to be dealt with immediately, and decisions are often made too quickly to take advantage of the tools as described. However, as with any large and complex organization, there are many issues that can be considered and evaluated well in advance. Examples of such decisions include whether to build a new school, whether to provide a district operated transportation program or rely on a private contractor, how to find the lowest cost solution to adequately meet the long term maintenance needs of school facilities, or how to evaluate the costs and benefits of investment in a new computer system for administrative and/or educational purposes. These are difficult issues that lead to the expenditure of substantial sums of district funds. Ensuring that the school district is making the best decision possible is critical to the efficient operation of a district.

A Five-Step Process for Decision making

While there are a number of ways to structure complex problems like those posed, Stokey and Zeckhouser offer a five-step framework for structuring the problem and making choices among competing alternatives.[4] Their steps are:

Establish the Context

This first step is crucial to the analysis. Understanding the problem may well be the most difficult part of the process. Moreover, determining goals and objectives for dealing with the problem is essential at this early stage. For example, if a district is faced with a budget shortage and is seeking ways to reduce its pupil transportation costs, on the surface this appears to be a good way to minimize the impact of reduced funding on the classroom. However, it begs the question of whether reductions in transportation are the best approach to reducing expenditures. In other words, it is important to make sure what is the right question. Is it how to reduce transportation expenditures or how to reduce total expenditures for the district? What are the most likely or productive options available?

Determining the Alternatives

Once the problem is clear, it is important to determine all of the possible alternative solutions. Returning to the problem of reducing transportation costs, the possible solutions could include such things as increasing student walking distances before they are eligible for home to school transportation, staggering school starting times so that one bus and driver can handle multiple bus routes each morning and each afternoon, limiting the number of field trips each teacher may request for his or her class, increasing the length of time busses are operated before being replaced, charging students for home to school transportation (if this is legal in the particular state), or replacing a district owned operation with one provided by a private contractor. It is important to always include among the options the possibility of doing or changing nothing. That is, it may be that the status quo is in reality the best solution. In the example used here, that would mean finding an alternative way to reduce district expenditures and not changing pupil transportation services.

Predict the Consequences of Each Alternative

In this step, an estimate must be made for what the outcome of each alternative would be. Again using the transportation example, it would be necessary to estimate the savings possible from each option identified. The district already staggers school starts and end times to maximize the use of existing buses and drivers, the savings from this option may be minimal, whereas if all schools in the district begin and end at the same time, considerable savings may be possible through minor adjustments to the school times. Included in this analysis is the impact changes in school times would have on district students, employees, and parents. An important part of this step is the determination of which analytic tools are appropriate to the analysis, covered in the second, and most extensive, part of this chapter.

Placing a Value on the Outcomes

Here it is important to determine what criteria will be used to compare the analytical outcomes of the alternatives. If all of the options were simply monetary, such as how to provide school supplies to all schools for the lowest possible price, reliance on the costs of each alternative is adequate. However, most decisions made in a school district context must include the impact these decisions will have on students, teachers, and parents, as well as the community in general. If staggering school start and end times will lead to tremendous community opposition, another alternative may be more appropriate, even if the total savings to the district are lower.

Making a Choice

Once all of the alternatives have been carefully evaluated and comparisons made, it is possible to choose the option that appears to best meet the needs of the school district.

Following this process does not assure the best possible decision. Real problems rarely have neat, simple solutions. This framework merely provides a system that will help bring structure to the decision-making process. The final decision is only as good at these data and analysis the school business official brings to the process.

It is also important to recognize that this is only a framework. The process of making good decisions is an iterative one.[5] Often when moving from one step to another, it is discovered that all the alternatives were not considered, and it is necessary to return to an earlier step for more analysis. This is not a weakness, but rather a routine part of any decision-making process that deals with complex issues. The purpose of the structure outlined is not to force a rigid, lock-step process, but rather to offer guidance and a way to help keep track of the may loose ends and possible alternatives that are likely to develop as the analysis proceeds. In addition to making analysis more complete, this structure will help present the findings to others in a clear, concise, and organized manner.

The five-step framework presented is by no means the only way to approach problem solving. There are a number of alternative options, and modification of these five steps may suit the individual's management style better than the structure offered here. What is important is use of a systematic process for the identification of problems, determination of alternative solutions, and making a decision among those competing alternatives.

Models

In the process of conducting the analyses, school business officials will rely heavily on the use of models. The term model means many things to different people. Consequently, the next few pages offer a discussion of models in quantitative decision making. Models may be described as "a simplified representation of some aspect of the real world."[6] Models range from simple diagrams and maps to sophisticated computer simulations that attempt to estimate the performance of the United States' economy.

There are a variety of ways to look at models. This discussion begins with a description of three major types of models followed by a discussion of descriptive versus prescriptive models, consideration of deterministic versus probabilistic models, and advice about how to choose the right model for each problem.

Types of Models

Physical Models

Physical models include the plastic models of cars and airplanes that children build from kits. Similarly, architects often build physical models of new buildings to provide individuals with a better conceptualization of what the completed structure will look like. In both cases, the models are build on a much smaller scale than what they are representing. However, all physical models do not have to be smaller than reality. A model of a complex computer chip may be enlarged many times to help visualize all of its components and transistors.

Physical models are found in a number of places in school business administration. Many schools have physical models of new or planned schools located in a central district facility. Moreover, the person charged with developing the master schedule for a school may use a physical model of the classrooms available for class sessions. Even thought this may take the form of a large poster board with envelopes or boxes for each class, this is still a physical model of the school.

Diagrammatic Models

It is not always possible or necessary to create a physical model. For example, it would be impractical to build a physical model of a school district's geography, roads, and housing units to plot student transportation routes. A map or diagrammatic model of the district is not only an excellent substitute for such a physical model, but in this case might actually work better to describe the district's transportation needs and the bus routes developed to meet those needs. Moreover, as families move into and out of the district, a map is easier to update than would be a physical representation of the district.

Other examples of diagrammatic models include flow charts and blueprints. Each offers a visualization of the real thing, but offers less detail than a physical model. For example, the map used for pupil transportation purposes would not offer accurate information on how much traffic could be expected on each road along a bus route, nor would it be particularly helpful in ascertaining whether there were any major hills the buses would have to climb and descend on the routes. It also does a poor job of providing accurate information as to the safety for children at each of the bus stops. This information can only be gathered through a detailed survey of each bus route. The map only provides a guide to where the children live and where the buses that pick up and deliver them operate.

Diagrammatic models can also be used to clarify processes. Computer programmers frequently rely on flow charts to describe the steps that occur when a computer program is operated. For example, many school districts rely on automated student information systems to keep track of the pupils enrolled in the district and to complete required reports. Many individuals enter information or data into the systems, and through a series of steps, a variety of outputs are available to district officials. A flow chart can be used to describe what information is entered into the system, and by whom, as well as to show how that information is combined with other information to provide the required reports. Using the symbols, lines and arrows of a flow chart, it is possible to determine what information is available, and how it has been analyzed to produce a final report.

Conceptual Models

Conceptual models are more difficult to explain. They can be as simple as the visualization of some simple process like mowing a lawn or they can be complex mathematical formulations. Most people, faced with the task of mowing a large rectangular lawn with a fixed size lawn mower will seek some systematic process for mowing the entire lawn. It is unlikely they will randomly push the mower around until the task is complete. Whether they start on an outside edge and work their way in a concentric pattern, or simply go up and down, the approach they take is a simple conceptual model.

Mathematical models also fit this mode. When comparing a school district with neighboring districts, it is likely that required information includes how many students are enrolled and how big the district budget is. The budget will be divided by enrollment to determine the level of per pupil expenditures. That figure is used to compare districts. The act of simple division is a conceptual model. Other conceptual models include feedback mechanisms to describe the actions that take place.

Descriptive Versus Prescriptive Models

A descriptive model simply attempts to describe or explain a process or thing. It may also be used to predict how outcomes will change given a chance in some inputs to a system. A prescriptive model goes one step further and adds a set of procedures for choosing from among a set of alternatives derived from the descriptive model.

Deterministic Versus Probabilistic Models

In many models, the primary interest is knowing some outcome that is certain, or that will be treated as certain. Determining that outcome may require reliance on a number of assumptions. The real issue is ensuring that assumptions are as accurate as possible. Enrollment projections are an example of a deterministic model. Since most school districts receive funding on the basis of the number of pupils enrolled, accurate estimation of future enrollments is essential to good planning. The accuracy of future enrollment estimates is highly dependent on the quality of the data used and the assumptions made. Often it is helpful to conduct a sensitivity analysis to determine how changes in assumptions impact the analysis. The use of spreadsheets and microcomputers has greatly facilitated the ability of school officials to conduct sensitivity analyses quickly and efficiently.

Probability models describe situations where there is not one unique outcome and where you have to estimate the probability of any individual outcome occurring. Models of this sort are useful in estimating what programs are likely to be most helpful in improving student achievement. The probability of average student achievement increasing one grade level in one year can be estimated for a range of curriculum options and course delivery mechanisms. A decision tree can then be developed to estimate which curriculum methodology and delivery option is most likely to offer the greatest return in terms of student achievement.

Choosing the Best Model

Choosing the correct or best model for a problem is important in determining how well the problem is solved. The choice of models is highly dependent on the nature of the problem at hand and the consequences of making the wrong choice. In general, the model should provide enough information upon which an intelligent decision can be made. On the other hand, it should not exceed the parameters of the problem at hand. It is not productive to spend a great deal on a model predicting enrollment if the development of that model costs more than the expected cost of an error. For example, if current enrollment projection model is accurate to within ten students a year, and each student generates $5,000 in revenue, the potential error amounts to $50,000. If it would cost $100,000 to develop a model with a prediction accuracy of two students, the costs of the model would exceed the benefit of increased accuracy. On the other hand, if enhanced accuracy could be gained for $500, it would be cost effective to do so. Thus, it is important to match the scale of the model to the size of the problem.

The following provides an overview of a number of models that have been found to be helpful in school business settings. It is not an exhaustive list of analytic tools, but a description of those most likely to used.

Quantitative Tools for the School Business Administrator

There are a wide variety of analytical tools and possible models that school business administrators can use to help make important decisions in their school districts. Four of those tools are described in detail:

- Enrollment forecasting.
- Cost benefit analysis and a family of alternatives to cost benefit analyses including cost effectiveness, cost utility, and cost feasibility analysis.

- Linear programming.
- Decision analysis.

While this list is not comprehensive, it provides information on tools that are most likely to be of value to school business administrators on a regular basis.

Enrollment Forecasting

One of the school business administrator's most important tasks is to project future school district enrollments. Having an accurate estimate of the number of students expected to enroll in the district's schools during the next three to five years and beyond is essential good planning. Generally, state school finance formulas allocate revenue to school districts on the basis of pupil counts meaning that accurate revenue projections are highly dependent on accurate enrollment forecasts. Moreover, knowing how many students will arrive at school each fall is essential to knowing how much classroom space will be needed, how many teachers will have to be employed, how many textbooks must be purchased, and so on. In fact, the number of students drives almost all other school planning decisions. The importance of accurate enrollment projections cannot be underestimated.

Forecasting is generally defined as the projection of specified future events within a certain time frame, usually on the basis of incomplete information. The accuracy of a forecast depends on the recourses available and the validity of the methods used. The amount of effort that goes into a forecast should be a function of the stakes involved. In other words, how much error can the school district afford? Theoretically, the marginal investment in the forecasting effort should equal the marginal value of any improved accuracy in the forecast. Accurate forecasts depend on the following elements:[7]

- Obtaining accurate and reliable data.
- Preparing these data properly.
- Stating the basic assumptions of the model.
- Using a consistent calculus or mechanical means.
- Applying more than one kind of forecasting method with independent bases (sensitivity analysis is also indicated).
- Applying sound judgment and intuition.

The development of good assumptions is crucial to this process. The kinds and numbers of assumptions made in projecting populations depend on three factors as follows:

The Level of Detail Desired

If all that is sought is a measure of the aggregate change in total enrollment, then all that is needed is some measure of the aggregate change in the population. On the other hand, if more detailed data by age group is sought, or more likely, by grade level, additional assumptions must be made. These assumptions would deal with dropout rates, grade retention policies and actions, and in and out migration of families from the areas served by the school district.

The Span of Time Between the Present and the Last Projection

If the projection was for a relatively short period of time, many important variables can be held constant. A three-year projection can assume that the birth rate, for example, will show the same trend as recent years. However, if projecting changes over twenty-five years, changes in the birth rate could have a substantial impact on projected enrollments. This means that additional assumptions, particularly regarding birth rates, have to be made. In most school situations, it is probably only necessary to project enrollments forward three to five years, but given the relative ease with which microcomputers make such projections possible, the three- to five-year forecast should be updated at least once a year.

The Degree of Accuracy Desired

It is important to be as accurate as possible in developing enrollment forecasts, but they must also be conducted within the constraints of existing resources. Microcomputers have made it possible to dramatically increase accuracy of projections with little or no cost. Today, spreadsheet programs make it possible to use cohort survival and regression techniques to predict enrollments very easily. School officials can focus more attention on the accuracy of these underlying data and not on the time it used to take to hand calculate the enrollment projections.

Projection Methodologies

There is no single or best method for projecting student enrollments. In general, there are four types of demographic projection models and several non-demographic methods that can be used to project trends in any population.

Demographic Methods

Comparative Projection

The comparative or analogy methodology of population forecasting determines the population projections for the study area based on the growth pattern of an older area with similar social, economic, and political characteristics. The usefulness of this method for school districts is rather limited.

Projections by Extrapolation

There are two types of projections by extrapolations. Graphic extrapolation consists of plotting the population of past census years against time and sketching a line, which, in the judgment of the forecaster, appears to fit past data. This line is then extended into the future to obtain population estimates for future years. The graphic extrapolation techniques are relatively quick and simple, but tend to be inaccurate for all but very short time periods. In mathematical extrapolation, population is projected by fitting an equation to past data, rather than using a straight line as in graphic extrapolation. A straight line example is shown later.

Ratio and Correlation Methods

There are two ratio or correlation models that are used by individuals attempting population forecasts. The ratio method compares the ratio of the population of the study area to the population of a larger area in which it reside, and then applies that ratio to projections of the larger area. The second method compares the total population to a related variable such as employment or school enrollments. This ratio is applied to future estimates of this variable to determine future population. These methods only work if the relationship between the study area and the larger area or related variable can be shown to be historically accurate.

Correlation models rely on more complex methods such as multiple regression models to relate population to independent variables such as births, deaths, school enrollments, and auto registrations. This works well in the short run, but in the long run is only as good as the continuation of the relationships on which they are based.

Component Analysis

There are two principal variations of this approach. One compares births, deaths, and in-and out-migration. These are combined to estimate future population. The second method is cohort survival which forecasts future population by separate age, sex, and sometimes racial group. For each cohort, birth, death, and migration components are determined. This is believed to be the most reliable method for projecting populations, particularly for a three- to five-year time frame. An example of a cohort survival model is presented later in this section.

Non-Demographic Methods

Build-Out Rates

Build-out rates are used to project changes in population by looking at current and future housing developments in an area and multiply by an assumed household size. This is most accurate in the short term. It is an important tool in estimating enrollment in a school district as new housing developments can have a dramatic impact, often virtually overnight, on school enrollments. In many states and counties, developers are not required to meet with or discuss their plans with the local school district. This lack of communication makes it difficult for school officials to find ways to mitigate the influx of students the come with new housing.

Ultimate Population Analysis

This methodology is used to determine the saturation population at full development of an area. This method can be used to estimate the maximum number of students expected to reside within the district's boundaries. An upper limit on total population supposedly exists for any geographic area with fixed boundaries. However, this model does not help determine the time frame in which all the development will take place.

Enrollment Projection Examples

If the school administrator, for example, projects school enrollments for a five-year period, it would be necessary to assume that some things will remain in the future as they were in the immediate past, and that if they change, they will do so in a predictable manner. There are six general assumptions that need to be made (that is considered and tested in light of available data) when projecting populations of school children for a five-year period.

Mortality

Enrollment projections assume that there will be no major change in the death rate for any group of the population, particularly school age children. This is generally a relatively safe assumption to make as mortality rates among school age children are very stable over periods of time considerably longer than the five-year time frame considered here.

Fertility

Enrollment projections for school districts also that there will not be a dramatic change in fertility, or if changes are anticipated that they are predictable and can be factored into the analysis. Analysis of fertility is only important for projections that exceed five years since the kindergartners who will enroll in school in five years were born in the current year.

Local Economic Conditions

Trends in land use (industrial and commercial development) will impact enrollments. It is generally assumed that there will be no major changes in economic trends in the areas being studied. However, there are a number of situations where alternative assumptions may be important. For example, in areas impacted by the closure of a large military base that employed large numbers of civilian workers, there may be considerable immigration, leading to dramatic declines in student populations.

District Boundaries

It is important to consider the impact of potential changes in district boundaries. If annexations, unifications, or other changes in the areas served by the district are expected, the impact on enrollments must be factored into any future enrollment projections. Similarly, changes made in city or county comprehensive plans could have an impact on development, which could change predicted enrollments.

Administrative Policy

It must be assumed that policies on attendance, promotion, and pupil retention and acceleration will not change significantly or will change at a predictable rate. Implementation of a new attendance program that leads to increases in student attendance could improve enrollment, particularly in a state that uses average daily attendance as its measure for the distribution of state funds to school districts.

Acts of God

Most enrollment projections assume that no major catastrophe such as an earthquake, flood, or hurricane will occur, forcing families to move to other parts of the state or nation during the time frame of the analysis.

Cohort Survival Analysis

One of the most common, and accurate methods for estimating enrollments is the cohort survival method. This technique considers the "survival" rate of a cohort of students as they move through the grades. Relying on five years of actual enrollments, an average survival rate for each grade pairing can be determined. This rate is then used to estimate future enrollments. An example of this process is helpful.

The first step in a cohort survival analysis is to establish an accurate, grade-by-grade enrollment history for the past five years. Table 14.1 displays the enrollment history of a small K-8 school district for the past five years. As the table shows, the school district has experienced dramatic growth in student population in recent years. Total enrollment has grown from 1,882 students to 5,167 in this short period.

The second step is to calculate the average survival rate as students in each grade level advance to the next grade. The survival rates for each pair of years and grades are displayed in table 14.2, along with the average survival rate between each pair of grades. Determination of the survival rate in each cell of the table is straightforward. For example, the survival rate between

Table 14.1
Cohort Survival Analysis Enrollment History

Grade	Year 1	Year 2	Year 3	Year 4	Year 5
Grade K	270	394	481	601	786
Grade 1	285	345	526	600	787
Grade 2	247	345	403	541	686
Grade 3	222	299	404	466	610
Grade 4	199	233	326	441	552
Grade 5	219	232	304	355	529
Grade 6	164	223	264	342	425
Grade 7	149	198	262	316	431
Grade 8	127	168	212	279	361
TOTAL	1,882	2,437	3,182	3,941	5,167

years 1 and 2 for children advancing from kindergarten to first grade is 127.78 percent. This was calculated by dividing the first grade enrollment in year 2 by the kindergarten enrollment year 1. This yields the following calculation:

1st grade (year 2) / K (year 1) = Cohort Survival Rate

345 / 270 = 1.2778 or 127.78%

The last column of table 14.2 displays the average survival rate for the four pairs of years being considered for each of the grade pairings. In table 14.2,

Table 14.2
Cohort Survival Analysis Survival Rates

Grade	Year-to-Year Survival Rates (%)				
	Year 1 to Year 2	Year 2 to Year 3	Year 3 to Year 4	Year 4 to Year 5	Average Survival Rate
Grade K to 1	127.78	133.50	124.74	130.95	129.24
Grade 1 to 2	121.05	116.81	102.85	114.33	113.76
Grade 2 to 3	121.05	117.10	115.63	112.75	116.64
Grade 3 to 4	104.95	109.03	109.16	118.45	110.40
Grade 4 to 5	116.58	130.47	108.90	119.95	118.98
Grade 5 to 6	101.83	113.79	112.50	119.72	111.96
Grade 6 to 7	120.73	117.49	119.70	126.02	120.99
Grade 7 to 8	112.75	107.07	106.49	114.24	110.14

the simple mean was used to calculate the average survival rate. However, some analysts elect to weight recent years more heavily than earlier years. For example, the survival rate between years 3 and 4 and years 4 and 5 might be multiplied by 2 while the first two survival rates in each column would continue to receive a weight of 1. In this case, the average would be determined by dividing by the sum of the weights or 6. Another possible weighting scheme that is sometimes used is to weight the most recent pair of years by 4, the next most recent by 3, and so on. In this case, the weighting scale would look like this:

Year Pairing	Weight
Year 1 to Year 2	1
Year 2 to Year 3	2
Year 3 to Year 4	3
Year 4 to Year 5	4
Sum of Weights	**10**

There are a number of reasons to consider the use of weighting the survival rates in calculating the average survival rate. If the school administrator has reason to believe that recent trends in student enrollments are more representative of the future than past trends, the use of a weighting scheme allows for the increase for the importance of the more recent years. If the school district were to include a military base that is closing or has a number of new housing developments that have been completed and a similar number about to be completed, recent year-to-year changes in student enrollments may be more representative of what to expect in the future than changes that took place further in the past.

Because it is likely that different weighting schemes will result in different future projections, it is important to conduct a sensitivity analysis of the model's findings. This can be done by comparing enrollment projections from a number of different weighting schemes. If the results were to lead to similar projections, one can be fairly confident of the accuracy of the analysis. If, on the other hand, alternative weighting schemes were to lead to substantially different projections of future enrollment, then it is important to study these underlying data more carefully and determine the causes of the differences in cohort survival rates from one year to the next for each grade pairing. Once the reason for the differences has been determined, it is important to use sound judgment in determining whether future enrollments will follow early or later patterns of actual enrollment. Returning to the housing development example, if the district were "built-out," cohort survival rates for the future may more closely parallel those experienced before the housing development began. On the other hand, if the home building program in the district were only half complete, and there is a booming economy in the area, then more recent cohort trends may be more indicative of the future.

Using the average survival rates displayed in table 14.2, table 14.3 shows the projected enrollments for the next four years. Table 14.3 shows dramatic growth for the selected school district over this time frame. Table 14.3 provides a projection of both the total enrollment for the district, and grade by grade estimates of future enrollments. These grade level estimates are helpful in planning for classroom and teacher assignments each year.

Cohort survival analysis has proven to be a highly accurate technique for estimating changes in large populations. It is less accurate for smaller populations such as school districts.[8] The reason for this decline in the accuracy of this method is that economic or other changes in a small area are likely to have very little impact on the population of an entire state or of the country. However, if those economic changes take place in one district, or a small group of districts, the impact on those individual districts can be quite substantial. Because this projection tool, like others relies on historical data, it is unable to forecast the impact of major events. This is the reason the sixth step described is to apply sound professional judgment. The reader should also consult chapter 12 for a discussion of the cohort survival method.

Curve Fitting

As described, another approach to projecting student enrollments is to use curve fitting techniques. Essentially, this process looks at past enrollment trends and then estimates the path of future enrollment growth. While it is

Table 14.3
Cohort Survival Analysis Four-Year Enrollment Project

	Projected Enrollment (Future Years)			
Grade	Year 1	Year 2	Year 3	Year 4
K	870	788	760	750
1	1,016	1,124	1,018	982
2	895	1,156	1,279	1,159
3	800	1,044	1,348	1,492
4	673	883	1,153	1,488
5	657	801	1,051	1,372
6	592	735	897	1,177
7	514	717	890	1,085
8	475	566	789	980
Total	6,493	7,815	9,185	10,484

possible to make such estimates from a visual inspection of past trend lines, regression techniques offer a more precise way to estimate future enrollments by fitting a curve, or in the example below, a straight line, through past enrollment trends.

Regression is a statistical methodology that estimates the impact of one or more independent variables on some dependent variable. Using actual data from the previous five years, a simple linear regression can be estimated. The process is to fit a line to existing data as closely as possible, and then project that line into the future. The line of best fit is determined by finding a linear or other function that minimizes the square of the difference between the actual observations and the function being estimated. This can be visualized in figure 14.1 where these actual data are plotted along with the linear best fit function. The regression model estimated this line such that if the distance from each actual observation from that line were squared and the five squared distances summed, no other line could be drawn that would result in a smaller sum of the squares. This process is known as ordinary least squares, and is one of the most frequently used regression methodologies, although by no means the only one.[9]

The regression equation that results can be used to estimate future enrollments. In its simplest linear form, a regression equation for predicting future enrollments would look like this:

$$Y = C + Bx$$

Where:

Y = enrollment, C = a constant, x = the year, and B = a coefficient that shows the effect of each year on total enrollment.

Using the five-year total enrollments from table 14.1, the regression yields the following results:

$$Y = 899.6 + (807.4 \text{ x year})$$

Applying this equation to the previous five years and four future years yields the estimates of school district enrollments displayed in table 14.4.

Table 14.4

Enrollment Projections Based on Linear Regression

Historical			Future	
Year	Actual Enrollment	Predicted Enrollment	Year	Predicted Enrollment
1	1,882	4,707	6	5,744
2	2,437	2,514	7	6,551
3	3,182	3,322	8	7,359
4	3,941	4,129	9	8,166
5	5,167	4,937		

As displayed in figure 14.1, the equation's prediction for the first five years of data differs from the actual enrollments. This is the result of trying to fit a linear function to data that are not perfectly linear in nature. The estimated enrollments for the four future years were calculated by replacing the value for year in the equation with the future year values. Thus, the predicted enrollment for year 8 was calculated as follows:

$$Y = 899.6 + (807.4 \text{ x year})$$

Substituting the value 8 for year yields the following value for Y, the estimated enrollment:

$$Y = 899.6 + (807.4 \text{ x } 8) = 7{,}359$$

A review of table 14.4 shows that the estimated total enrollment for our school district appears to be lower each year using the regression methodology than using the cohort survival technique. The differences are displayed in table 14.5.

Figure 14.1
Actual Enrollments Compared to "Best Fit" Linear Projection

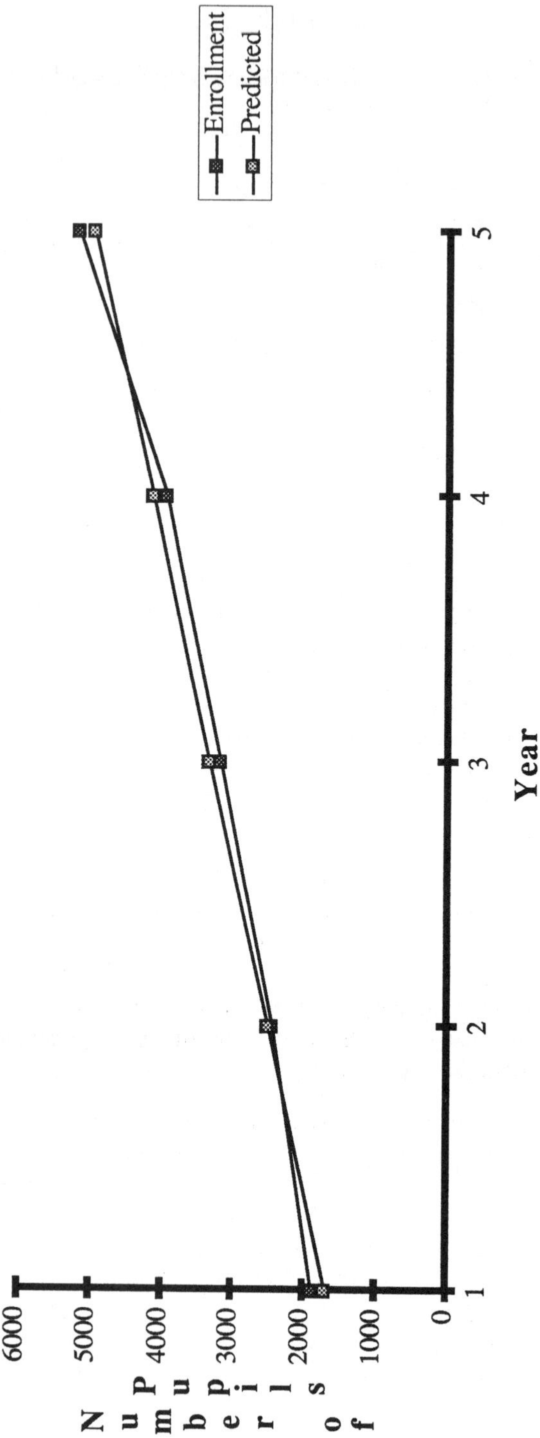

Table 14.5

Differences in Enrollment Projections Estimated Using Cohort Survival Analysis and Regression

	Method		
Year	Cohort Survival	Regression	Difference
6	6,493	5,744	749
7	7,805	6,551	1,264
8	9,185	7,359	1,826
9	9,734	8,166	1,568

Given the substantial differences that exist in the use of the two methodologies for the example presented in this chapter, what can be done to improve the precision of the estimates? After all, in year 9, the difference between the two methods amounts to 19.2 percent of the lower estimate and 16.1 percent of the respective totals. Figure 14.2 shows these differences graphically.

There are a number of sources to which the administrator can turn to gather additional evidence for predicting which of the estimates are more likely to be correct. This is important to consider when the statistical methodologies lead to relatively large differences in estimated future enrollments.

Among the sources of information concerning future growth in a school district are the major utilities including the gas company, electric company, and telephone service provider. Each of these organizations is responsible for ensuring that new housing units have the respective service as soon as people move in. As a result, they often have large planning staffs who can provide helpful information as to the pattern and timing of population growth in a given community.

The city or county planning department may also keep data on development activities, and in states and localities where zoning is the rule, comprehensive plans will help determine where different types of development will take place, further enhancing the ability to estimate not only the number of new students expected to enroll, but where they are most likely to reside. In addition, information about building permits issued will help in determining new dwelling units as well as where they will be located and how many individuals are likely to live in each household. It is likely that apartment and condominium units will have fewer children per unit than will single family houses.

Figure 14.2
Comparison of Cohort Survival and Curve Fitting Methods for Enrollment Projections

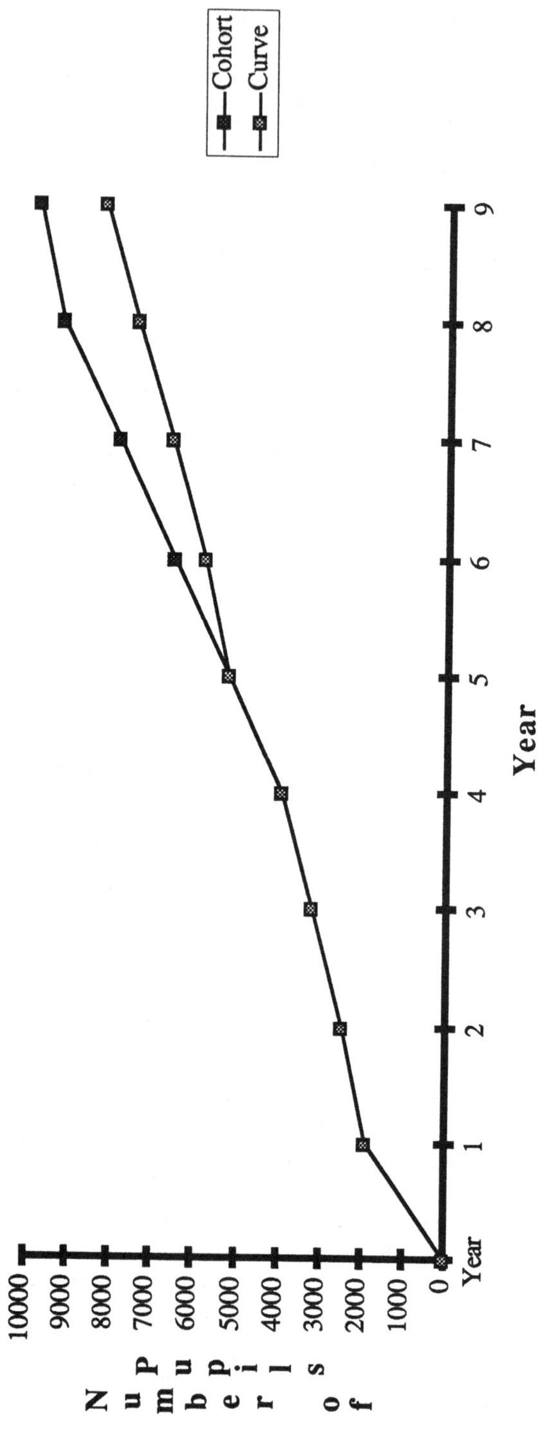

Finally, none of these techniques will work very well if there is a large influx of people into the district. A number of school districts in southern California and southern Florida have recently experienced a large influx of students due to immigration from Latin America. Although some of this migration is related to economic conditions, which can at least partially be estimated, some is the result of political factors in other countries, and nearly impossible to plan for.

This section has provided a discussion of one of the most important functions of the school business office, predicting future enrollments. This information is crucial to the long range planning of a school district. Another important business function is the efficient and economical use of the resources with which the business official is entrusted. This means ensuring that the district is able to purchase quality product, or undertake certain activities at the lowest possible cost, and to make sure that as the district chooses among competing alternatives for those resources, all of the opportunities are evaluated carefully.

Cost Analysis

School administrators always seem to be cutting budgets. Despite the fact that real per pupil expenditures increased by more than 200 percent between 1959-1960 and 1991-1992,[10] the demand for additional funds for education seems to have increased at an even faster pace. As a result, it is crucial that careful analyses of new programs be conducted before a school district commits to the expenditure of additional funds. In public agencies, when cost analyses are considered, they are generally cost benefit analyses. This chapter began with a brief discussion of cost benefit analyses, providing data concerning how to conduct a cost benefit analysis and how to interpret the results. Unfortunately, it is often difficult to place a monetary value on the benefits of an educational program. Thus, also described are some alternative tools that have been proposed (cost effectiveness, cost utility, and cost feasibility analysis).[11]

Why Conduct a Cost Analysis?

One of the primary responsibilities of a school business administrator is to manage the district's resources as efficiently and effectively as possible. Development of cost analyses to inform major expenditure decisions is critical to the success as an administrator, and, more importantly will ensure that the school district is able to maximize the use of the scarce resources.

By finding ways to use district resources efficiently, cost analyses can reduce the costs of reaching particular objectives or goals, and can expand what can be accomplished within a fixed budget or resource constraint. By helping the analyst compare both the costs and outcomes of a given project and choose the alternatives that provide the best results for any given resource outlay, cost analysis can help minimize the resource utilization necessary for any outcome. In effect, cost analysis allows the analyst to compare alternatives holding the outcome constant.

Cost Analysis Procedures: A Summary

There are five major steps to the conduct of a cost benefit analysis:

- Identify the project or projects that are to be analyzed.
- Determine all impacts of the proposed project. This includes both the positive effects of the project and any negative effects that may come from the project.
- Value each of the impacts. While the most common method is to estimate the dollar value of each impact, this is not always possible in education.
- Once the value of the impacts has been estimated, the net benefits of each must be identified by subtracting the value of the project from its cost.
- Choose the most cost effective alternative or alternatives.

The fundamental rule of cost analysis is that in any situation where one must choose among alternatives, choose the alternative that produces the greatest net benefit. This rule applies even if that means the choice is doing nothing because all of the other options have negative or lower net benefits.

A Simple Cost-Benefit Analysis

To fully understand cost-benefit analyses, it is helpful to begin with a simple straightforward example. For the moment, the issue of time is ignored. Time is a critical consideration and the tools to deal with it will be discussed later. Presently, the analysis will be simplified as it will focus only on the identification of costs and benefits.

Suppose that a district office is currently located in an old trailer in the parking lot of the high school. Because space inside the trailer is limited, many of the district administrative and management functions are scattered at a variety of sites throughout the district, some of them in rented commercial space. A proposal to build a new administrative services building for $350,000 has been proposed. This building will house all of the school district's central administration as well as other important management functions such as purchasing and warehousing, the transportation office (although not the bus garage), and the board meeting room. How would it be determined whether or not investing in such a facility would prove to be a cost-effective decision?

The first step would be to identify all of the costs involved in the project. Again, assume that the $350,000 includes all of the costs to purchase the land, design the building, construct the facility, and move all of the necessary staff to their new offices. Thus, the total cost would amount to $350,000.

The next step is to identify all of the benefits of moving to the new facility. These include all of the savings from not having to pay rent for commercial space, lowered travel reimbursements for district administrators who previously had to drive to different sites around the district to do their jobs, potential savings from purchasing supplies in larger quantity that can now be stored in the new central warehouse, energy savings that accrue because it cost more to heat and air condition the older district facilities that were being used for the various functions that would be centralized, and others that can be identified.

Obviously some of these savings will occur annually, but for the moment ignore the factor of time. If the sum of the total benefits for this project exceeds the $350,000 cost of its construction, then it would pay to build the facility. If the benefits are less than $350,000, it would not be cost-effective to allow construction to proceed. The measure used to determine whether this would be a sound investment is the "net benefits" of the project. If the net benefits were positive, then the project should proceed. If they were negative, the project should not be considered until such time as more information requires the benefits and costs be reevaluated.

Choosing from a Number of Alternatives

Another common situation faced by school officials is deciding among a number of alternative projects. Table 14.6 summarizes eight capital projects a school district is considering. The second column of table 14.6 provides information on the initial cost of the project, while the third column lists the total benefits to be derived from the project. The third summarizes the net benefits of the project. The question the district now faces is how to choose among these competing alternatives.

If the district were to elect to do only one of these projects, it would select project B since the fundamental guideline is to select the project producing the highest net benefit. There are two problems with this simple solution to the problem. The first is that it assumes moneys are no object, and that the school district will find the resources to undertake project B, even though it is the most expensive of the eight listed in table 14.6. On the other hand, if the district has more than the $500,000 needed to undertake project B, what other projects would it consider? Moreover, is Project B really going to offer the highest rate of return of all the projects, or are others really more cost effective? In summary, the identification of project B as the preferred alternative suffers from a lack of consideration regarding the relative scale of each project, the amount of moneys available to the district for construction

Table 14.6
Summary of Eight Possible Capital Projects for a School District

Project	Cost ($ 000)	Benefits ($ 000)	Net Benefits ($ 000)
A	100	600	500
B	500	1250	750
C	200	800	600
D	75	175	100
E	150	375	225
F	200	400	200
G	50	175	125
H	150	450	300

projects, and does not provide a ranking of the projects in terms of the rate of return.

To these issues, it is necessary to find a measure of each project that is not sensitive to scale. In this case, that measure is the ratio of net benefits to costs. By dividing each projects' net benefits by its costs, a benefit/cost ratio can be developed. If that ratio is greater than one, the benefits exceed the costs and the project should at least be considered for construction. If the benefit cost ratio is less than one, then the costs exceed the benefits and the project should not be undertaken. Moreover, the higher this benefit/cost ratio, the greater the district's return on its investment. The district should invest in the project with the highest ratio of benefits to costs. Table 14.7 adds two columns to those displayed in table 14.6. The fourth column in that table displays the benefit to cost ratio for each project and the projects have been ranked on the basis of this ratio. The fifth column of table 14.7 displays the cumulative costs the district would incur if it undertook each of the projects in the order in which they are listed.

Table 14.7 reflects that the highest ranked project is no longer project B, but rather project A. That is because the net benefit as a percentage of the initial cost of the project are the greatest. Project B, with a benefit cost ratio of 1.5 ranks sixth of the eight projects using this measure. Moreover, since it is likely that the district will have a fixed amount of moneys to invest in these capital projects, it can now be determined which projects will be considered. Suppose, for example, the district has $500,000 available for the eight projects. It could then afford to undertake the four highest ranked projects, A, C, G, and H. That would result in the expenditure of the full $500,000 available, as in the example. Only in this scenario, the school district would concentrate its activities on the four projects with the highest net benefit to cost ratio rather than spend it all on project B. The use of this benefit cost ratio will

Table 14.7
Eight Capital Projects Ranked by their Benefit/Cost Ratios

Project	Cost ($ 000)	Benefits ($ 000)	Net Benefits ($ 000)	Benefit/ Cost Ratio	Cumulative Cost ($ 000)
A	100	600	500	5.0	100
C	200	800	600	3.0	300
G	50	175	125	2.5	350
H	150	450	300	2.0	500
E	150	375	225	1.5	650
B	500	1,250	750	1.5	1,150
D	75	175	100	1.3	1,225
F	200	400	200	1.0	1,425

allow the school district to maximize its return on the moneys invested in capital projects.

Measuring Costs and Benefits

The most important, and difficult part of a cost analysis is the identification of costs and benefits. Unless the costs and benefits of a project are carefully identified and measured, the value of the analysis is very low. This section provides an in-depth discussion of the measurement of costs and benefits of educational programs.

The Distinction Between Costs and Expenditures

A distinction can be made between costs and expenditures as a great deal of attention to the complex issue of distinguishing between the two has been developed. [12] Costs can be defined as a measure of benefits that are foregone to realize some outcome or benefit, while expenditures are a measure of resources that flow regardless of the consequence. While this may seem an unnecessary distinction at first, analysis of expenditures to be made on a project, without a thorough consideration of its true costs may seriously under- or overstate the project's cost. Unfortunately, in educational settings, information concerning expenditures is more easily available than cost information, which has been more difficult to develop.

Defining Expenditures

A common approach to comparing the costs of alternative programs in educational institutions is to determine the monetary value of the resources necessary to implement each program, and compare the total expenditures across programs. This process implicitly assumes the two programs are intended to accomplish the same goals, and that each have identical inefficiencies in operation.[13] If these conditions do not hold (and there is little reason to expect that they do), comparisons of expenditures are invalid and can be misleading.

If, as is often the case in education, there are multiple goals established school district program, then estimation of the costs of that program must include all of the resources necessary to accomplish all of those goals. The difficulty is that a project's goals can be difficult to quantify or may even be contradictory. Determining the resources necessary to achieve the multiple goals of most educational programs is, at best, a difficult task. Because of this difficulty, many analysts stop short of estimating the true costs of a program, and instead focus on the expenditures required for its implementation.

In public elementary and secondary educational institutions, even determining the actual expenditures for a specific program can be difficult. Most state accounting systems require school districts to report spending by object (salaries, benefits, supplies, and so on), and sometimes by function (instruction, administration, instructional support, maintenance and operations, transportation, etc.).[14] Often these expenditure data are reported at the district level, and there is little or no information about how funds are used at the school or classroom level. Moreover, detailed information about specific programs within a district is often hard to discern from school district financial reports. In an object oriented system, estimating the expenditures for an individual program or project might require determining the salaries and benefits of staff members who work in that program, estimating what portion of their time is devoted to the project or program, and then determining which of the district's expenditures for supplies and materials should be attributed to that project or program.

Even in districts that are able to provide detailed information about the expenditures made for various programs, this information only provides a partial delineation of the full economic costs of the those programs. The other factors that must be considered when estimating the full costs of a program are described below.

Defining Costs

The textbook definition of the cost of a program is the benefits that are not realized through the best foregone alternative. Thus, if a resource were devoted to some use, the benefits associated with the best possible alternative use of that resource represent the "opportunity cost" of the program. Unfortunately, it is not always possible to determine what the best alternative use of those resources might be. Moreover, if that alternative were identified, determining its benefits may be a considerable problem. For example, if a district were considering the implementation of a new performance assessment program, the opportunity costs of that program would be equal to the benefits from any conceivable alternative reform that was not implemented.

In analyzing the costs of alternative program options, the range of alternative programs could be thought of as all the possible alternative programs the district could establish that address the same or similar issues. In this case, the benefits derived from the proposed program would be compared to the benefits derived from the best option facing the district other than the proposed program. The more beneficial the alternative given up, the more it will cost to devote resources to performance assessment.[15] However, before the benefits of a program can be measured, agreement must be reached as to the goals of the foregone activity. As discussed, reaching an agreement concerning the programs goals may be difficult to do.

In some cases, it may be appropriate to restrict the alternatives considered. For example, when analyzing the costs of upgrading school facilities, it may well be that the decision to be made is whether to retrofit the existing school or demolish the structure and build a new educational facility. In that case, the relative set of alternatives is limited to the remodeling program or the construction of a new school and the costs will be measured on the basis of the foregone benefits of the old facility.

It is also unlikely that a new program will require exactly the same level of resources consumed by the program it replaces, and it is also possible that a district would be reluctant to eliminate all of what it has done in the past and shift entirely to a new program overnight. In both cases, the total resources devoted to the program would need to be increased. In the first case, if the new program replaced the old program in its entirety but required more resources than the old program, the foregone benefits would include both the benefits of the old program plus all the benefits from other activities due to the transfer of resources to the new program.

Similarly, if the two programs were operated together, the costs of the program would include the benefits foregone by shifting resources into that programmatic area from other areas. If the old program were only continued partially, then the costs of the new program would include the portion of the benefits foregone from the old program along with any other benefits foregone through the resources that were shifted to the two programs.

To make a cost analysis useful to decisionmakers, cost analysts need to develop a common metric to measure the benefits of alternatives. Unfortunately, there is no simple way to compare the benefits of programs that have disparate goals. Since agreement on the spectrum of benefits may also be difficult to achieve, and since estimation of the benefits of a foregone alternative may require a great deal of time for what could be considered an activity with little value, many analysts simplify the issue by estimating the expenditures necessary to operate the alternative program. This is equivalent to estimating the monetary value of the resources directed toward the new program.[16] In effect, the analyst uses the dollar value of the actual or anticipated expenditures as a measure of the projects costs. Often called the ingredients method,[17] this approach relies exclusively on expenditures to measure costs and may lead to confusion about the difference between expenditures and costs.

If one believes that the benefits to be derived from an alternative assessment dramatically exceed the system being replaced, or if one anticipates that there will be improvements in student learning as a result of the new assessment system (an argument frequently made in favor of performance assessments and portfolios), then using the expenditures devoted to the alternative assessment program may in fact overstate the true costs of the program since the benefits derived exceed the benefits from the program or programs it replaces. Unfortunately, there is no way to estimate the size of this exaggeration. To resolve this problem, the analyst must make explicit assumptions about what factors could cause this overstatement, and the estimate costs with and without an adjustment for this issue. In the framework that is established, the ingredients approach to costs is used, and where necessary, adjustments for the potential overstatement of benefits that this could lead to is identified and possible adjustments considered.

Time as a Consideration

Expenditures for an asset (for example school bus, school building, major equipment or program), often do not occur at the same time as the receipts for the services of that asset or program. Moneys received sooner is more valuable than moneys received later. The difference in timing can affect whether acquiring the asset, or operating the program, is cost effective. Moneys received at different times are different commodities. Business officials, in both the public and private sectors, use interest calculations to make amounts of moneys to be paid or received at different times comparable.

When analyzing the costs and the benefits (or receipts) of a multi-year program, the present value of a stream of payments or benefits is a single amount of money that shows the economic equivalent of the entire stream at the present time. Interest rates, or discount rates as they are frequently called, when estimating the present value of future amounts are the measure used to equalize these amounts over time. Thus to compare a stream of benefits and costs over a fixed time period, it is essential to discount the streams of payments or receipts to a net present value. The discussion of present value begins with a review of how interest is calculated and accrued. This information is then applied to the use of discount rates for estimating the net present value of a stream of payments or benefits.

The concept of compound interest should be familiar, since earnings in a typical savings account accrue through compound interest. Basically, to compound interest means that the amount of interest earned during a period is added to the principal and the principal for the next interest period is larger. For example, if a deposit of $1,000 is made to a savings account that pays compound interest at the rate of 6 percent per year, it will earn \$60 by the end of one year:

$P \times r = I$
or
$1000 \times .06 = 60$
where:
P = principal

r = interest rate
I = interest payment.

During the second year, the principal of \$1,060 (the original \$1,000 plus the \$60 interest earned in year 1) will earn \$63.60 in interest (1060 x .06 = 63.6). Thus by the end of the year 2, the account contains \$1,123.60 (\$1,060 + \$63.60).

A common question then is how much will there be in n years if a certain amount is invested today at a given interest rate? Called the future value, this amount is calculated with the following formula:

$F_n = P(1+r)^n$
where:
F_n = the future accumulation value
P = one-time investment today (or principal)
r = the interest rate
n = the number of periods (years) from today.

Calculation of the present value is related to the calculation of future values. When calculating the present value, it must be determined how much principal (P) must be invested today in order to have a specified amount (F_n) at the end of n periods. The formula to calculate this is:

$P = F_n(1+r)^n$
where;
P = the present value of the future amount
F = the future value
r = the discount rate
n = the year

Over the life of a project, the flows of payments and benefits (income) may differ from year to year. Thus, in order to compare the costs and benefits of each program being evaluated, it is necessary to determine the net present value of the stream of payments and the stream of benefits. The formula for calculating the present value of this stream is:

$$NPV = \sum_{i=1}^{n} F_i /(1+r)^i$$

Note that this formula is the same as the present value formula previously presented, except now one is interested in the sum (the sigma) of the future values (F_i), and the n from the earlier formula has been replaced with individual n's for each year. These are represented by the i's. Table 14.8 demonstrates how to use this formula to calculate the present value of a stream of payments. The table assumes a discount rate of 5 percent.

Table 14.8

Calculation of the Present Value of a 5-Year Stream of Payments With a 5 Percent Discount Rate

	Year 1	Year 2	Year 3	Year 4	Year 5
Payment	10	4	6	1	8
NPV	$10/(1.05)$	$4/(1.05)^2$	$6/(1.05)^3$	$3/(1.05)^4$	$8/(1.05)^5$
Value	9.52	3.63	5.18	2.47	6.27

To determine the net present value of the stream of payments represented in the table, the last row is summed to give a net present value of 27.07. As this simple example shows, determination of the net present value is highly dependent on the choice of a discount rate.

There is considerable debate in the economic community as to how to select a discount rate for present value analyses. Essentially, a discount rate should reflect the return that would have been received had the moneys been invested in a safe investment instrument such as U.S. Treasury bonds. The higher the discount rate selected, the lower the net present value of a stream of payments. Moreover, the further into the future a payment takes place, the lower its net present value. As a result, the choice of a discount rate, along with determination of when costs and benefits are to be recognized can have a significant impact on the benefit-cost ratio of any project. In fact, it is entirely possible that a project with a benefit-cost ratio greater than one at one discount rate, will show a benefit-cost ratio less than one at another discount rate. This is likely to happen when the costs occur in the early years of the project and the benefits in later years. As the discount rate rises, its impact on payments further in the future is significantly different and could result in a different determination as to whether or not the project would pay off.

When some doubt exists as to the appropriate discount rate to use, analysts are advised to conduct a sensitivity analysis. If under all reasonable discount rates the ratio of benefits to costs remains above one, then the project will have net benefits. If on the other hand, some discount rates lead to a determination of positive net benefits and others to negative net benefits, it is important to reexamine all assumptions and give much greater consideration to the choice of a discount rate and the reasons for choosing it.

Alternatives to Cost-Benefit Analysis

One of the strengths of cost-benefit analysis is that it establishes an absolute condition for resource use. If benefits were not equal to or greater than costs, then a project should not be undertaken. In addition, the calculation of a benefit/cost ratio[18] makes it possible to compare different educational programs, and more importantly, compare those educational programs with other areas of public expenditure.

One of the difficulties with cost-benefit analyses is that it compares the net present value of a stream of estimated project costs to the net present value a stream of estimated project benefits. The process requires that the benefits be valued in monetary terms. Unfortunately, it is often difficult if not impossible to monetize the benefits of educational programs. This makes conduct of a cost-benefit analysis problematic. Consequently, two other tools that may be more appropriate, cost-effectiveness analysis and cost-utility analysis.[19] These tools attempt to compare the costs of a project or program with some measure of the program's effectiveness or its perceived utility.

Cost-Effectiveness Analysis

Cost-effectiveness analysis has been described as the evaluation of alternatives according to both costs and effects with regard to producing some outcome or set of outcomes.[20] As with cost-benefit analysis, the preferred alternative or program is the one that has the lowest cost for any given increase in the chosen measure of effectiveness. Thus, only programs with similar or identical goals can be compared, and that common measures of each program's effectiveness must be available to use a cost-effectiveness analysis.

There are a variety of effectiveness measures that can be used in educational settings. Among them are:

- The number of students completing the program.
- The number of potential dropouts who graduate.
- The number of graduates placed in appropriate jobs.
- Test scores in appropriate domains utilizing appropriate test instruments.
- Student assessment of the program on an instrument for measuring satisfaction.
- Evaluation of student physical condition and physical skills.
- The number of students placed in colleges of particular types.

- The number of courses and units received by students in advance placement, by subject.

Table 14.9 displays a hypothetical cost effectiveness analysis. It evaluates four different math programs that are being considered for implementation in

Table 14.9
A Hypothetical Cost Effectiveness Example

Program	Cost Per Student ($)	Effectiveness (Increase in Test Scores)	Cost Effectiveness (dollars per unit gain) ($)
A	550	25	22
B	100	5	20
C	150	15	10
D	65	5	13

a school. The second column of the table shows the per pupil cost of each of the four programs. These costs range from a low of $65 per pupil to a high of $550. The third column of table 14.9 contains the effectiveness measure. In this case, the measure is the estimated increase in test scores in math for each student enrolled in the program. The last column of table 14.9 computes the cost effectiveness of each option. The entries in that column represent the estimated cost of bringing one students performance up one point on the math test.

Looking at the last column of table 14.8, it is clear that program C is the most cost effective as it only costs $10 per unit increase in test scores in math. The other programs are all less cost effective than program C, with A and B being the least cost effective of the four. Under cost effectiveness analysis, the district would elect program C as its new math program if it wanted to maximize performance per dollar spent.

Cost-effectiveness analysis is a powerful tool as it allows the comparison of already available data on the effectiveness of a program with costs to help choose between alternatives. It works well for comparing alternatives that are considered for achievement of a similar goal, but is not useful if the goals of the programs to be evaluated differ. When the goals differ, cost utility analysis may be more appropriate.

Cost-Utility Analysis

Sometimes, it is difficult to get a useful measure of the effectiveness of a program. However, it may be possible to estimate (subjectively or otherwise) its potential effectiveness. It is also possible to estimate the utility or value of the improvement estimated to result from the program. Developing such an analysis is part of cost-utility analysis. In this method, many different types of outcomes can be considered in one analysis. However, the data requirements are less stringent, meaning school business administrators should have less confidence in the results.

Table 14.10
Hypothetical Cost-Utility Analysis

	Teaching Strategy	
	A	B
Probability of Raising Math Performance By One Grade Level Equivalent	0.5	0.3
Probability of Raising Reading Performance By One Grade Level Equivalent	0.5	0.8
Utility of Raising Math One Grade Level Equivalent	6	6
Utility of Raising Reading One Grade Level Equivalent	9	9
Expected Utility	(0.5) * (6) + (0.5) * (9) = 7.5	(0.3) * (6) + (0.8) * (9) = 9.0
Cost Per Pupil	$375	$400
Cost Utility Ratio	$50	$44.44

Suppose, as illustration, one wishes to compare two teaching strategies for both math and reading. It is not possible to combine the test results from each program in a cost effectiveness analysis. Table 14.10 offers one way to resolve this issue. The process entails seeking the opinion of faculty as to the effectiveness of each teaching strategy under consideration, and the desirability or utility of improving performance in each subject matter. Table 14.10 shows how the expected utility of two programs, A and B is estimated.

The first step is to estimate the probably that each strategy will raise math and/or reading performance by one grade level equivalent. The probabilities are shown in the first two rows of the table. The next step is to estimate the utility of raising performance by one grade level. These utilities are displayed in rows 3 and 4 of table 14.10. Row 5 shows the calculation of the expected utility of each teaching strategy which is determined by multiplying the probabilities by the utilities and summing them for the two subject areas. When combined with the per pupil cost of each program, a cost utility ratio can be calculated. In the sample displayed in table 14.10, the cost per unit of utility for teaching strategy B is $5.56 lower than for a unit of utility under teaching strategy A. As a result, this hypothetical example suggests that strategy B is more appropriate for this district.

Although the data requirements for cost-utility analysis are less stringent than for cost-effectiveness or cost-benefit analysis, the results cannot be reproduced on the basis of a standard methodology because most of the assessments of probably and utility are highly subjective. On the other hand, many different outcomes can be included in the evaluation, and cost-utility analysis offers a tool for systematically dealing with imperfect information and uncertainty.

Summary (Cost Analysis)

The three tools described, cost-benefit, cost-effectiveness, and cost-utility offer the school business administrator a set of important tools that he or she can use to compare alternative programs that are being considered in their districts. Comprehensive information on the costs and outcomes of different programs are critical to each of these tools, but the information gained through the conduct of such an analysis may help local school officials make stronger decisions as to how to improve school quality in a way that uses school funds efficiently.

Linear Programming

Linear programming is one of the most frequently used tools in the field of operations research. Operations research is concerned with the use of mathematical formulations to help choose among alternatives that are to be optimized. Although linear programming is a highly specialized field, it is used extensively to solve economic and industrial problems. Linear programming is used in the following places:[21]

- By commercial airlines to schedule aircraft and flight crews.
- To optimize the mix of ingredients in dog and cat food.
- To optimize product mix in hard goods manufacturing.
- By large commercial banks to manage their balance sheets.
- For long range capacity planning in production facilities.
- To minimize waste in the paper industry.
- By investment firms to optimize portfolio selections.

Used as a tool to allocate scarce resources among competing activities within given constraints in order to attain certain objectives, the linear program has many uses in education as well. Possible objectives in education include:

- Maximizing the number of students taught within a given budget constraint.
- Minimizing instructional expenses given the costs of various factors and a minimum number of students.

- Minimizing the distance students have to travel to school given the distances between their homes and schools, and the capacities of the schools, the school buses and any racial balance constraints the district must consider.
- Determining the best location for placement of a school bus maintenance facility to minimize daily distances busses must drive without children.
- Managing the district's cash flow.

Linear programming is a subset of mathematical programming that is concerned with the efficient allocation of limited resources to known activities with the objective of meeting a desired goal such as maximizing profit or minimizing costs. The linearity of some models can be justified on the basis of physical properties of the problem. Some nonlinear models can be linearized by the proper use of mathematical transformations.

The goal of a linear programming model is to optimize some function or objective, subject to one or more constraints. Education goals amenable to linear programming models include:

- Maximize:

  Student achievement

  Teacher experience

  Teacher training

  Time for instruction

  Availability of instructional materials

  Utilization of facilities

  Opportunity for extracurricular activities

  Subject offerings

  Nutritional value of school lunches

- Minimize:

  Cost of the educational program

  Cost of school lunches

  Cost of school facilities

  Transportation costs

  Cost of interest on school bonds

  Cost of equipment and supplies

Pupil-teacher ratio

Transportation time

Distance students must travel to school

Distance between student classes

Class size

As these examples show, linear programming can be a powerful tool for finding optimal solutions to complex educational problems.

The Simplex Method

Linear programming requires that the problem to be solved or the objective to be optimized be identified in mathematical terms that can be expressed as a linear equation known as the objective function. In general, there are also constraints that limit what can be done. The formulation of a linear programming solution requires that the objective function be established, and that the related constraints be identified. The simplex method provides an algorithm for solving such systems of equations. Most linear programming problems include many variables and a large number of constrains and are best solved using computers. Programs for solving linear and other mathematical programs are available for most mainframe and microcomputers. However, to fully understand how a linear programming model arrives at a solution to a problem, it is helpful to consider a simple graphical example.

The objective function establishes what is to be optimized. In addition, there are constraints that limit what is feasible or what can be done. For example, if attempting to minimize the number of portable classrooms needed at a school, the maximum class size permitted by either the teacher contract or state law would serve as a constraint. Thus, the first step is to determine the objective function and related constraints. For simplicity, the following solution provides a mathematical and graphical solution to a linear programming problem rather than attempt to include a real world example. The same methodology would apply to any optimization (minimization or maximization) problem that might be identified. The standard form of a linear programming model would be:

Maximize: $Z = X1 + X2$ (14.1)

Subject to: $X1 + 5X2 <= 5$ (14.2)

$2X1 + X2 <= 4$ (14.3)

$X1 >= 0$ (14.4)

$X2 > = 0$ (14.5)

Equation 14.1 is the objective function. In this case, we seek to optimize the combination of the two variables X1 and X2. Equations 14.2 and 14.3 are constraints which indicate the limits on the combinations of X1 and X2 that are possible. Note that the are expressed in terms of inequalities. It is possible that as the optimum solution is sought to the problem, there may be more of X1 or X2 available than we can use given the limitations of the problem. For example, if the maximum allowable class size is 25, and there are 65 students, the need is for three teachers. However, there is an excess capacity in teachers since those same three teachers could handle another 10 students. Thus, in equation 14.2, the sum of X1 and X2 can be no more than 5, but under optimal circumstances, it might be less than 5. The last two constraints which require that X1 and X2 not be negative (zero or greater) are necessary since it does not make sense to use negative amounts of any resource.

To solve this problem equations 14.2 and 14.3 are displayed graphically in figure 14.3.

Figure 14.3
Graphical Solution to a Hypothetical Linear Programming Problem

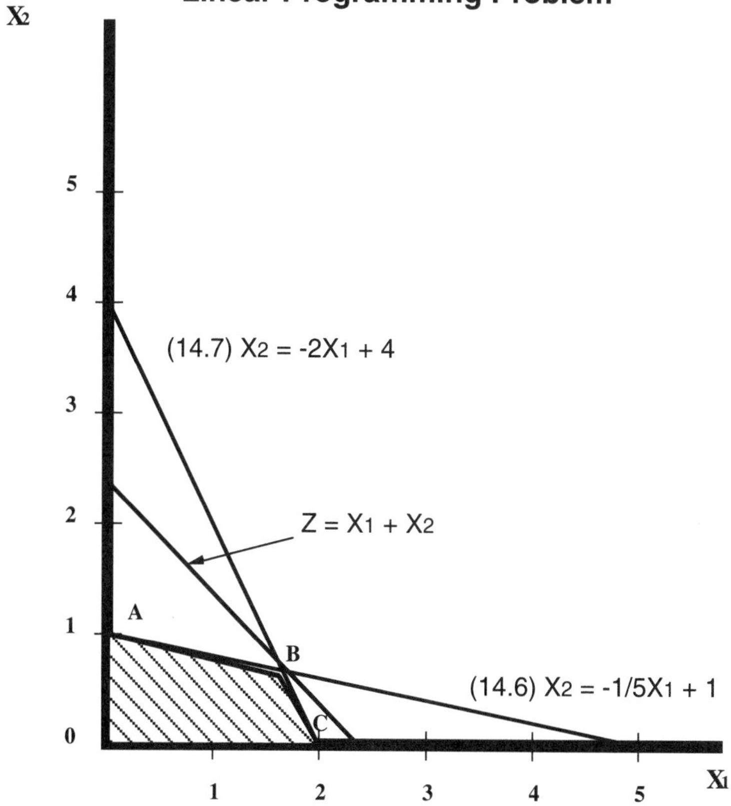

The standard mathematical form for the graph of a line is X2 = mX1 + b,

Where:

m = the slope of the line and b = the X2 intercept of the line.

Equation 14.2 can be put in this form as follows:

$$(14.6) X2 = -1/5X1 + 1$$

and equation 14.3 would be transformed into:

$$(14.7) X2 = -2X1 + 4$$

Note that both have been converted to equalities for this purpose. Each is displayed in figure 14.3. Since in fact, the constraints are inequalities, the solution to each is really the area below the two lines.[22] Constraints 14.4 and 14.5 require that X1 and X2 be nonnegative. As a result, the feasible solutions to this problem lie in the shaded region bounded by the origin and points A, B, and C.

The solution to this problem will fall on one of those three points, A, B, or C. To determine which one is the actual maximum, the values of X1 and X2 at each of the three points can be calculated by solving the pair of simultaneous equations and substituting them back into the objective function. Table 14.11 shows the solutions for Z at each of these three points. Point B represents the best combination of X1 and X2 that can be used to maximize Z. The graphical representation of the line represented by the objective function is shown in a dashed line in figure 14.3.

Table 14.11
Values of Z at Points A, B, and C

Point	X1	X2	Z
A	0	1	1
B	1.6	1.32	2.92
C	2	0	2

This mathematical example provides a conceptual example of the simplex method for solving linear programming problems. Most problems will have more variables and more constraints, making the determination of a solution more difficult. Computers can be used to find each of the intersection points of the constraints and test them in the objective function. Once the computer finds an intersection point that provides a higher value for the objective function than any of the adjacent intersections, the optimal solution will have been found.

The difficult part of linear programming is development of the objective function, the factor to be optimized (minimized or maximized), and making sure that all of the constraints have been considered and accurately described mathematically. The availability of computer programs makes the solution of these problems relatively straightforward.

Decision Case

The last quantitative tool is decision analysis that helps analysts sort out a complex set of decisions when the parts of the decision are both interrelated and the outcomes may depend, at least in part, on chance. The inability to establish a single solution to a problem brings in the element of risk into the decision-making process. Decision analysis offers a way to compare various alternatives and the risks associated with each to arrive at an expected value or outcome for each possible outcome.[23] Although a powerful analysis tool, the use of decision analysis in public policy has lagged behind other tools, mostly because it is difficult to estimate the probabilities and values of different outcomes.[24]

The most common approach to decision analysis is the use of a decision tree. In its basic form, the decision tree provides a road map for finding the way through a complex decision problem. The following six steps have been offered for development of a decision tree:[25]

- List the alternative decision possibilities for the resolution of the problem.
- Determine all alternative events that might result from the selection of each alternative.
- Review each event and assign the most likely probability to each event.
- Outline all outcomes and assign probabilities to the outcomes.
- Evaluate the consequences of each alternative according to criteria established. This is typically done in terms of monetary value, or the expected cost of not making the decision.
- Make a decision based on the estimated probabilities and values.

As illustration, the problem of evaluating four different math programs for use at the junior high school is revisited. Unlike the cost effectiveness discussion, in this instance, consider the relative probabilities that each program will meet the goals established for the mathematics program. Suppose that goal is that all students gain one grade level in math proficiency in the first year of the program. A decision tree reflects what the relative costs are of each program given the probability that each will succeed in meeting our goal.

Figure 14.4 shows a decision tree analysis of this problem and helps sort out which of the options appears to be the best under the current situation. The figure is a flow diagram that shows the structure of a decision problem. There are four elements in the diagram.

- Decision nodes indicating all possible courses of action open to the district (indicated by squares).
- Chance nodes showing the intervening uncertain events and all their possible outcomes (indicated by circles).
- Probabilities for each possible outcome (presented on parallel lines).
- Payoffs summarizing the consequences of each combination of choice and chance (the last column of the figure).

Figure 14.4
Sample Decision Tree for School Construction

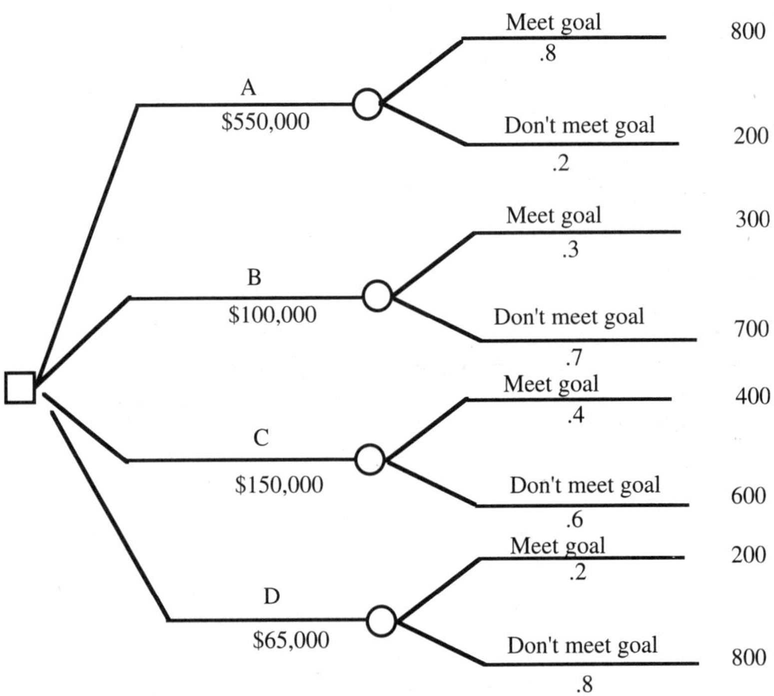

The first decision one has to make is which math program to select. The first set of decision are labeled A, B, C, and D, for the four program options. Also indicated is the total cost of each option assuming 1,000 students participate in the program. The second set of alternatives shows the probability of meeting the stated goal of all students gaining one grade level in math performance for each option chosen. The final column shows the expected number of students who will achieve the goal.

This information can be used to determine which of the options makes the most sense. Even with the decision tree, the example presented does not lend itself to a simple solution. The highest number of students expected to meet the goal is clearly through option A, but it is also the most expensive. On the other hand, option D is the least expensive, but the fewest students meet the goal. Choosing an alternative depends on the resources available and how this program relates to other district priorities.

Summary

This chapter provided an introduction to a number of analytical tools available to school business administrators in searching for solutions to many complex problems they face in their jobs. The descriptions of the tools were relatively brief and not designed to make the reader an expert in their use. Rather, the descriptions were designed to show how each of the methods could be used and provide some idea as to when each might be appropriate. Perhaps more important than the tools themselves are the modeling skills and techniques embedded within them. The demands of running the fiscal operations of a typical school district today require that officials understand the statistical and analytic tools available to help them maximize the use of their resources so that school services can be provided in a cost efficient manner throughout the district.

Endnotes

[1] E. Stokey and R. Zeckhauser, *A Primer for Policy Analysis.* (New York: W. W. Norton & Co., 1978.

[2] See e.g., B. Cooper, "Making Money Matter in Education: A Micro-Financial Model for Determining School-Level Allocations, Efficiency, and Productivity." J. of Education Finance. 20(1), 66-87, Fall 1994; and L.O. Picus, "The $300 Billion Question: How Do Public Elementary and Secondary Schools Spend Their Money?" Paper presented at the 1994 annual meeting of the American Education Research Ass'n, New Orleans, LA. April 1994.

[3] G. H. Baldwin, "Quantitative Decision Making." in R. C. Wood, *Principles of School Business Management*. (Reston VA: ASBO, International, 1986) 175-206.
[4] The material that follows is based loosely on Stokey and Zeckhauser's framework, *A Primer for Policy Analysis*, 5-6.
[5] Stokey and Zeckhauser, *A Primer for Policy Analysis* , 6.
[6] Ibid., 8.
[7] This list is adopted from G. Hentschke, Management Operations in Education, (Berkeley, CA: McCutchan Pub. Corp., 1975).
[8] For an in-depth discussion of this topic, see, D. Norval, Cohort Analysis. Sage University Paper series on Quantitative Applications in the Social Sciences, series no. 07-001. Beverly Hills, CA: Sage Pub., 1988.
[9] A more complete description of regression techniques is available in a number of college statistics texts.
[10] L.O. Picus, "The $300 Billion Question."
[11] H. Levin, *Cost Effectiveness Analysis: A Primer.* (Beverly Hills, CA: Sage Pub., 1983).
[12] D. Monk, The Cost of Systemic Education Reform: Conceptual Issues and Preliminary Estimates, prepared for the New Standards Project, The Cost of Systemic Education Reform;" for a discussion of the funding education based on the cost of education, see R. C. Wood and D. C. Thompson, "Funding Public Education Based on the Concept of Cost of Living," Educational Considerations, vol. 21, no. 2, Spring 1994, 60-72.
[13] Monk, The Cost of Systemic Education Reform.
[14] See Chapter 11 of this text.
[15] Monk, The Cost of Systemic Education Reform..
[16] Ibid.
[17] Levin, *Cost Effectiveness Analysis: A Primer.*
[18] Readers should note that it is possible to use a cost-benefit ratio as well. The cost-benefit ratio is calculated by dividing project cost by net benefits and is the inverse of the benefit-cost ratio. Analysis of cost-benefit ratios is the same as benefit-cost ratios, except that the most desirable projects are those with the smallest ratio of costs to benefits. The benefit-cost ratio was used in the text because the rankings it establishes are more straightforward to interpret and use.
[19] Levin, *Cost Effectiveness Analysis: A Primer*, includes a fourth analytical tool, cost feasibility analysis as well. In effect, a cost-feasibility analysis is making the determination as to whether sufficient funds exist to undertake a given project. While such an analysis is essential to the implementation of any new project,

simply having sufficient funds is not enough information upon which to undertake a new program. Consequently other tools such as cost-benefit, cost-effectiveness or cost-utility analysis must be used.

[20] Levin, *Cost Effectiveness Analysis.: A Primer.*

[21] L. Austin and J. Burns, Management Science: An Aid for Managerial Decision Making. (New York: Macmillian Pub. Co., 1985) 39.

[22] Note that if this were a minimization problem, equations 14.2 and 14.3 would use greater than or equal to signs, and the feasible solution would be above the two lines in figure 14.3.

[23] For a more complete treatment of decision analysis, see H. Raiffa, *Decision Analysis.* (Reading, MA: Addison-Wesley, 1968).

[24] Stokey and Zeckhauser, *A Primer for Policy Analysis,* 201.

[25] Baldwin, "Quantitative Decision Making," 190.

CHAPTER 15

Management Information Systems

Introduction

As society moves into the "information age," the ability of public schools to both provide students with adequate access to the vast amounts of information available and to manage the information needed to run and maintain a school system becomes more critical each day. Imagine the following scenario of a day in the life of an assistant superintendent for business administration in a medium-sized school district.

After a pleasant breakfast with your family, you turn on your home computer and log into an information service. The first thing you do is check the overnight electronic mail messages and respond to make sure there are no "unexpected" crises to deal with. You also answer a number of inquiries from staff and parents regarding district operations. In addition, there is a request from the local Boy Scout troop to use the junior high school multipurpose room on Monday nights for meetings. You respond to the scoutmaster by sending him, electronically, a form to fill out and an address to send it so that you staff can process it, get approval from the school board, and let them begin using the school as soon as possible.

When you have finished reading your e-mail, you log into the local bank's investment service to check on your district cash flow. You also check with your own internal financial management system to see how many checks were written, and based on past vendor performance, how long you expect it to take for the checks to clear. You run a series of quick calculations on a spreadsheet written specifically for this purpose and estimate the amount of funds to transfer from the investment pools to the district's checking account.

You then review the performance of the state and county investment pools in which you participate, and transfer the funds from the one that paid the lower interest rate overnight. Since the spread between the two is relatively small, and often fluctuates from day to day, you decide not to shift any funds between the two at this time. You also note the balance in both to ascertain whether sufficient funds are available to meet the payroll at the end of the week. This month, all seems fine because property taxes were due the previous month and the district has a surplus. Had funds been inadequate, you could have applied electronically for a tax revenue anticipation note (TRAN) from your local bank. In the past when you have filed such applications, the funds were available the following day.

Once the cash flow has been taken care of, you turn to the weather. You turn on your television to the local weather channel to discover that the storm predicted for today will not be as serious as first expected. However, the road over the pass to the North Valley section of your district may be closed later in the day as heavy snow is expected at elevations lower than the summit of the pass. This only poses a problem for the forty-five high school children who live in the North Valley. There is a K-8 elementary school for all the others, and one of the district's bus drivers lives in the valley and handles all busing for that segment of the district. She only drives the bus over the pass when maintenance and repairs are needed. Rather than risk driving a bus over the pass for the high school students, you call the district's technology office and alert them to set up the video link between the high school and the downlink at the elementary school. The bus driver is advised to pick up the high school students and deliver them to the elementary school, where a parent volunteer will monitor the class while instruction is presented via a video link.

Because of the various ages and class schedules of these forty-five students, the video link is not ideal, but it is better than having them absent from school completely. Moreover, the video room at the elementary school is also equipped with fifteen computers, all of which are connected to the district's local area network. Once the teachers at the high school learn that the students from the North Valley will not be able to make it over the pass, they can send assignments to them via electronic mail. Each of the students will be expected to log on to their account during the last hour of the day to make sure that they are aware of any assignments they might have missed. Since many of the students have computers and modems at home, they can continue to communicate with their teachers after school lets out.

You also remember that a pep rally is scheduled for the high school's basketball team. One of the star players will be "trapped" in the north valley. You send an e-mail message to the principal of the high school alerting her that the pass is closed and telling her to be sure the video system is hooked up so the North Valley students can participate in the pep rally, and that the player can give his brief talk to the rest of the students from his location.

Once these tasks are complete, you head for the office. On the way there, you get a call indicating that there is a leak in the roof of one of the elementary schools. You turn around and head for the school, calling the district's facility manager and asking him to meet you there. When you arrive, the custodian has already covered the leaky section of roof with a tarp, but it is clear that extensive repairs are needed, and that the sooner they can be completed the better. You and the facilities manager sit down at a table in the principal's office and type up a short request for proposals on your laptop PC. You then plug the PC into a network node in the school office and transmit the RFP to your secretary with a note to send it out right away to the list of approved vendors. With luck, you will have some bids in two days, and work can begin before the end of the week.

When you arrive at your office, you log into the district's information system to check the day's attendance. Since your state provides financial assistance on the basis of average daily attendance (ADA), which is monitored on a daily basis, the link between attendance and district revenue is critical. You realize that despite the closure of the pass and the absence of forty-five students, the pep rally at the high school has encouraged attendance, which is running at 99 percent today rather than the usual 95 percent. Even with the North Valley students missing, there will be more students on campus than usual. You call the districts food service provider and ask that an additional thirty-five meals be delivered to the high school today.

Your secretary walks in at this point to remind you that the Association of School Business Officials (ASBO) International is broadcasting a thirty-minute video on negotiating health care benefit costs with providers today. The broadcast is to start in five minutes, and your human resources staff and staff attorney are in the conference room ready to watch. She indicates that the VCR will tape the show, and the downlink is all set and working.

You go to the conference room and watch the broadcast with your staff, and then spend thirty minutes discussing what you have learned. You agree to use some of the strategies suggested when you meet later in the week with your insurance representatives. You also ask the director of human resources to prepare data on the impact of the "wellness" program implemented two years ago on health insurance rates. You ask that she provide that to you later in the day so you can share the information with the school board at its meeting later this evening. You also ask her to share the information with representatives of the district's bargaining units. This can be done via fax and electronic mail.

Following lunch, you sit down at your PC to prepare a presentation for tonight's board meeting. Specifically, the school board has asked you to prepare budget targets for the next fiscal year and to show where, if at all, additional funds will be available to implement some of their priorities. Using spreadsheet and graphing programs, you are able to put together a series of graphs and charts that show the current fiscal condition of the school district, expected revenue and expenditures for the next five years, and areas where resources for new programs might be found. Since they have provided you with six specific items they would like to see implemented, you prepare a graphic image showing the costs of each and indicating where funds can be found to implement each program. Since you know that the board will want to consider a number of alternative scenarios for these, you save these data on a spreadsheet program so that you can conduct "what-if" analyses with the board during the meeting.

When you finish at 2:30 p.m., you walk over to the district's "broadcast studio" where one of the high school teachers is providing instruction in Japanese to students in fifty surrounding high schools. This program, and others to teach Calculus, Latin, Korean, and Russian, is among the district's most successful. None of the high schools in the western half of your state are large enough to support any of these programs individually, nor could an adequate supply of trained teachers be found to teach the classes anyway. By relying on "distance learning" where one teacher's course is broadcast to high schools across the western portion of the state, students interested in these important math and language classes have access. Moreover, the nominal fees you charge other high schools for providing the instruction both pay for the program and allow your district to hire the very best teachers in each of these subject areas. In fact, the program is so successful, you are considering construction of a second studio to double your broadcast capability.

Initially, the satellite time to broadcast these programs was too expensive to justify. However, the installation of T1 land lines (via fiber optics allow for the transmission of broadcast signals over relatively short distances, similar to cable TV systems) has dramatically reduced the cost of distance learning. Before going back to your office you make a note to seek state construction funds for development of the second studio facility.

At 3:00 p.m. the superintendent's cabinet meets to discuss the board agenda. Since the pass is closed, the principal of North Valley elementary cannot attend nor can she link up via the video since it is being used to transmit instruction to the high school students. Consequently, you resort to a conference call. The superintendent reminds you that a number of people who have signed up to testify on the new curriculum standards for history, which are scheduled to be debated by the board tonight, live in the north valley area. You ask the North Valley principal to set up a video link from her multipurpose room and ask your secretary to notify the radio and TV stations in the area so that everyone knows that they can attend the meeting at the elementary school, and they don't have to risk driving over the snow covered pass. You suggest to the superintendent that if this is successful, similar video links should be made available at all future board meetings. Not only will it help residents of the North Valley participate, it will provide a great public relations boost when people realize that the district really is reaching out to them and wants to hear what they say.

Along those same lines, the superintendent directs his staff to try to call each of the individuals signed up to testify from the North Valley to make sure that they are aware that they only need to go the elementary school. A summary of the board agenda is faxed to each member, representatives of the bargaining units, and to representatives of the press. Full copies have already been sent out, but all have found the summary to be a helpful reminder.

At 5:00 p.m., as the cabinet meeting winds down, you turn on the television in the conference room to listen to the governor's State of the State message to be delivered from the state capital. You listen to his speech and then go to the superintendent's office to help her field calls from the press to discuss the impact of the proposed budget on your district. You also check your e-mail again and confirm that you will be available for a conference call with the other heads of the research committee of your state's school business officials' association the next day to consider lobbying strategy for next year's education budget.

After that you dash back to your office to modify your board presentation based on what the governor has said, copy it to a disk, and take it to the board room, where you load it into a computer and make sure that all of your charts and graphs display properly on the video projection equipment in that room. You also test to make sure that they can be transmitted to the site at North Valley elementary school. Since you still have a few minutes before the meeting, you draft a policy statement for the state school business officials association and send it via e-mail to all of the individuals who will be participating in the conference call tomorrow.

The school board meeting is successfully concluded. The parents from North Valley are able to testify over the video link, and questions directed to North Valley Elementary School's principal are answered immediately. Moreover, given the governor's proposal to eliminate regulations, the board asks you to prepare a report on potential savings. Since this is a topic that has concerned you for a long time, you are able to go to your office, retrieve a file from your computer and provide the board with specific saving estimates almost immediately. The board gives you a list of alternatives that should also be considered, and you agree to send them a preliminary analysis the following week.

At the conclusion of the meeting, you go back to your office to check your e-mail one more time and check the status of your cash flow and investments. That done, you go home for a well earned night's sleep.

If this scenario seems a little far fetched, it shouldn't. Everything our hypothetical school business administrator did relied on technology that is available and widely in use today. Unfortunately, these technological advances are often not available in schools, or if they exist, they cannot be integrated and coordinated as seamlessly as in the discussion. If school districts are to successfully manage the data and information needs they face, it is critical that they learn to use and to coordinate the application of appropriate technologies to facilitate instruction and management.

It is important to think of technology and management information systems as much more than simply the installation of computers in a school district. While computer technology is certainly at the heart of management information systems, other technologies like voice transmission and television transmission, either one way or both ways, are critical components of any well organized and effective management information system.

The purpose of this chapter is to provide a framework for implementing and evaluating management information systems and technologies. Rather than focus on specific systems, hardware and software, this chapter provides a framework that can be used across different types of technology to help you understand your information system needs, and establish systems that are fully integrated and coordinated to maximize the potential of your investment in these technologies and minimize the risk that your equipment or software will become obsolete too rapidly.

Most discussions of this sort focus on the machinery of technology, but as you will see throughout this chapter, the focus here is on the human element. Remember, a high tech machine is, in reality, only as good as the person making it work. Thus, we will stress over and over the importance of staff training. In fact, you will probably find that the single largest cost for implementation of new technology is not the computers, telephones, televisions, VCRs, projectors, and so on, nor is it the software to make them run. Rather, it is the expense of training the staff to use these tools effectively, maximizing their performance both in the management of your school system and for educating the students in your district.

A Framework for Analyzing Management Information System Needs

Regardless of the technology one seeks to implement in a district, it is important to analyze needs carefully to ensure that the installed systems, be they computers, interactive video systems, or simply telephones, are capable of doing the tasks envisioned now and for the foreseeable future. This last statement brings up the age old problem of obsolescence in technology. It is true that as soon as one purchases a computer personally or a group of computers for an office or for a school's computer laboratory, newer, faster, more powerful computers will be on the market. However, this should not be a problem if you remember a simple concept. *It is not obsolete as long as it is performing the tasks you want it to perform.* If a thorough needs analysis were conducted and extensively reviewed of the capabilities of the systems under consideration, there is no reason that a computer purchased today cannot last a minimum of five years. Moreover, even if it were necessary to replace some of the computers in one office to take advantage of higher processing speeds or new capabilities, the computers that are being replaced may still be useful in other areas of the district or in a secondary market.[1]

A seven-step systems analysis model for the selection, design and implementation of automated systems offers much assistance in this regard. The seven components of this systems approach include: [2]

- Preliminary analysis.
- Input and output analysis.
- Processing analysis.
- Alternatives and applications.
- Selection of equipment, software and vendors.
- Design.
- Implementation and management.

Before discussing each of these steps, a short discussion of what is meant by a systems analysis is presented.

The Systems Study

A system can be considered a group of interdependent elements working together to accomplish a predetermined purpose. In this context many things can be thought of as systems including a city, the environment, a human body, a business or industrial enterprise, a library, or a school. The term "system" is all-pervasive and can refer to biological, physical, and social organisms. For our purpose, the systems view has a management and technology orientation. A school is a system consisting of interdependent subsystems, including:

- An administrative subsystem.
- An instructional subsystem.
- A plant operation subsystem.
- A transportation service subsystem.
- A food service subsystem.
- A co-curricular subsystem.
- A health service subsystem.
- Possibly other subsystems depending on the individual circumstances of the school district.

Organizations have become increasingly complex over time. Just view the organizational structure, the data processing needs, the services provided, and the technologies used in today's schools compared to schools of one hundred, fifty or even twenty-five years ago. The complexity that has evolved in schools has occurred in all other organizations as well. To help managers deal with this complexity, a number of techniques, primarily quantitative and descriptive in nature, have been developed.[3] Many of these tools, such as time and motion studies and cost accounting have been adopted by business and industry and used to help solve operating problems. The introduction of computers into organizations made the use of these tools more efficient and increased individual work capacities and productivity. Computers also created additional complexity in the workplace. The computer is not always the answer to dealing with a problem. Other options such as a micrographics system or the redesign of a manual system may be more appropriate. Managers realized that systems analysis techniques could be used to integrate computers into an organization efficiently and effectively. Organizations, such as schools, have turned to the use of systems analysis to help them implement new computer applications.

In order to accomplish its purpose, the components of the system or subsystem and the activities associated with those components can be structured in a variety of ways. The role of systems analysis is to design the most feasible, suitable, and acceptable structure. This structure may be a manual system or a sophisticated computer-based system—the systems analyst is not limited to a specific design option. Instead, the entire spectrum of options is available for consideration. Systems analysis refers to the use of management techniques to accomplish the system design. A person using systems analysis needs to have the technical capability to analyze a problem. This is done by examining its parts. The analyst takes a system apart, examines it, and puts it back together again. Through this process the analyst understands what is going on, and through this understanding can design a new system if needed.

Preliminary Analysis

Problem Isolation: The first concern of a systems analysis is determining the administrative problem(s) which exist in a school. These problems can range from the simple to the complex, from inaccuracies in the names and addresses of students and staff to a failure of the accounting or grade-reporting system. Among the problems one should consider and be aware of are the following:

- The inability to quickly prepare reports, including state required financial, enrollment and attendance reports, district required grade reports, grade analyses, student directories, or enrollment counts.

- Inadequate statistical analysis of student demographic characteristics, including future enrollment trends.
- Scheduling of teachers, students, and classrooms,
- Scheduling of school buses for home-to school transportation, special education transportation, field trips, and athletic events.
- More timely budget projections and accounting records.
- Difficulty determining when library/media center items are overdue, or in ascertaining who has checked out which items.
- Generation of mailings to parents and special groups, and maintenance of accurate mailing lists.

The nature of the organization and the people within it can also have an effect on both the problem and its solution. It is helpful to understand the structure of each organization. In particular, one will want to know which units are interrelated within the district and which units have relationships with organizations outside of the district. One also needs to be familiar with the people, position classifications, and reporting procedures that affect the operation of the district and its component schools. An analyst may want to review annual reports, minutes of meetings, and personnel descriptions to obtain a sense of the relationships that exist. It is also helpful to review existing organization charts; or to draw such charts if none exist. Depending on the nature of the problems one is considering, knowledge of the organization can help define the issues that will be faced and will help you understand how those problems are affected by the organizational structures and personnel employed within the system. Without an understanding of the potential impact of a new system on the organizational climate of a school, it may be very difficult to use that system as an effective solution for the problem.

Other approaches to the isolation of problem areas include asking a set of basic questions concerning the school district's operation. Typical questions include the following:

- Are there areas with large information processing needs; that is, high volumes of inputs and/or outputs, and heavy processing requirements?
- Are there information retrieval problems such as reports not submitted on time, the inability to generate reports or information on important issues, inaccurate reports, or data quality problems?
- Are there areas with a large number of repetitive, clerical functions?

To determine which problems should receive highest priority, one could list all the activities, functions, tasks, or processes in the school and consider each one in light of these three questions. If a rating criteria were established for each of the three questions, one could then rate each activity, function, task, and process using those criteria. By summing the rating for each activity, one could determine which should receive attention first.

User Needs

In the concern with isolating and prioritizing problems the user must not be forgotten. It is important to perform a user needs assessment. The results of a user needs assessment should be matched against current district and school services and products to isolate problem areas. Training needs can also be identified. Questionnaires, interviews, and observations are a few of the techniques which can be used to gather information on user needs.

Organizational Goals and Objectives

Although they are not always formally stated, every organization has goals and objectives. They provide a plan for the future and offer benchmarks to measure how well the organization is succeeding in meeting that plan. In short, these goals and objectives provide criteria for evaluating the systems that make up the organization. Therefore, an understanding of the organization's goals and objectives is crucial to listing and prioritizing problems.

A goal is a long-range plan or a major end result. An objective is a shorter range task which must be successfully completed in order to attain a specified goal. Objectives should be thought of as the "means to an end." Another distinguishing factor between a goal and an objective is that goals tend to be broad statements of purpose, while objectives are narrower and frequently stated in quantitative terms. The following list of goals, if available to the analyst would help prioritize problems in need of a solution:

- Simplify student record keeping procedures and provide more timely information to school sites.
- Improve the timeliness of report generation for student demographic information and class scheduling.
- Reduce data processing costs while improving the level of service, or holding it constant.
- Increase awareness and use of library/media materials and services through better publicity regarding resources, programs, and services.
- Improve communications between the central office and the school sites.
- Develop greater awareness about school district programs and plans among the community.
- Provide school sites with the resources to successfully implement school site budgeting.

Problem Definition and List of Requirements

Once it is determined which problem(s) is to be consider further, it is important to describe that problem in more detail and outline the requirements of the system to alleviate the problem. System requirements are the specification of tasks associated with resolving the problem. One can think of these system requirements as functional requirements that outline which functions the system should perform. Figure 15.1 offers an example of an expanded problem definition including a list of system requirements for a student record keeping and scheduling system.

Figure 15.1
Sample Problem Statement:
Student Records and Scheduling

Problem Statement

A survey of administrators, staff, and teachers found genuine concern over the operation of the student record/scheduling system. At the present time, the system operates in batch mode, with limited interactive scheduling capacity. It is maintained by an outside data processing service bureau. Information used to update the student database is recorded on code sheets which must be sent to the service bureau. Frequent delays occur between the time the records are sent to the service bureau and the time they are actually input into the student scheduling database. There have also been substantial delays in receiving critical student record reports, such as the master student list for each school, and accurate class schedules. Users of these reports also complain about the difficulty of finding needed information quickly in the bulky printouts they receive.

Requirements List

- Interactive data entry and retrieval.
- Ability to query system for information on individual students, teachers, and class schedules.
- Generation of the following reports:
    - Master student lists
    - Master class schedules
    - Teacher lists
    - Homeroom enrollment and attendance lists

- Address labels for students, parents or teachers
- Immunization reports
- Other statistical summaries as required by the district or the state

In addition to the requirements listed in figure 15.1, specific information about the system itself will be required. This information specifies the unique characteristics of the school district and provides more details concerning the features desired in a new system. Among the specific pieces of information required for the student record keeping and scheduling system are the following:

- The number of schools expected to be part of the system.
- The number of faculty and number of students, specified by school level, in each separate application level. For example, data for the entire student body may be in the student record system, but only a portion of these data would be used in a scheduling system.
- The number of fields and data elements for each student or faculty record,
- The capability for handling personal data, such as locker number and language capabilities.
- The ability to enter additional data such as student enrollment in special education and bilingual programs.
- The number of courses and number of course requests per student.
- The number of school shifts, type of school cycle, and type of scheduling system (semester, trimester, quarter, or full-year).
- The number of periods in a day.
- Distinctions between required and optional classes.
- Ability to handle co-requisite courses,
- Study hall assignments and scheduling.
- The ability to create specific reports such as student lists by advisor, student name, courses, grade level, or homeroom.
- The frequency production for each report.

Objectives

Once the problem list is prioritized and a list of system requirements are determined, one can develop a set of objectives for each program area being considered. The purpose of these objectives is to provide a quantitative basis for judging the success of our application. The process is simply to take the problem statement and requirements list and rephrase them into standard objectives terminology.

A list of objectives for a student record keeping and scheduling system might look like the following:

Problem Objectives: Student Records and Scheduling

- To inaugurate an improved student record and scheduling system that will operate interactively, generate required reports on demand, and not increase operating costs

Benefits

There is no reason to develop a new system unless it will improve the district's ability to provide current services or to offer new services to the students, parents, or faculty of the school or school district. Often, there is a need to generate a list of benefits that will accrue to the school if a new system or application were developed. A list of benefits that might result from the development of a new student record keeping and scheduling system include:

- An easier to understand and more timely student scheduling procedure.
- More accurate reports on student attendance, enabling the district to maximize its per pupil funding from the state.
- Reduced teacher and student dissatisfaction because of the increased accuracy of the record keeping and scheduling system.
- Improved teacher and administrative productivity resulting from better, on-time data on students and class offerings.

Management Plan

Before proceeding to the next step of the systems analysis model, the analysis of inputs and outputs, it is helpful to develop a management plan. A management plan can help determine when specific tasks need to be completed. It can provide a check list of what should be done at what time, and it provides a good communications tool for management. Figure 15.2 lists the tasks associated with conducting a systems study.

Summary

The preliminary analysis phase of a systems study involves the definition of the problem(s). The analyst gathers all relevant data which might assist in defining the problem, including a determination of constraints and resources available, organizational structure and tasks, goals and objectives, and user needs. With the problem defined, the analyst outlines the system require-

Figure 15.2
Typical System Study Steps

1. Isolate the problem(s).
2. Understand user needs.
3. Determine the resource constraints.
4. Identify specific problems.
5. Outline the system requirements.
6. Develop a preliminary analysis.
7. Establish a management plan.
8. Analyze existing forms and files.
9. Analyze system documents.
10. Determine input requirements and define data elements.
11. Determine output requirements.
12. List system task requirements.
13. Flowchart the process.
14. Select software.
15. Select hardware.
16. Outline implementation tasks.
17. Perform implementation tasks.
18. Procure hardware, software, and personnel training.
19. Outline housekeeping procedures.
20. Develop system documentation.
21. Create a publicity program to inform system users of the changes.
22. Implement the publicity plan.
23. Test the system.
24. Train your staff.
25. Implement the system.
26. Establish feedback mechanisms.
27. Evaluate the results.

ments and develops a management plan of action. Once the preliminary analysis is complete, one can progress to the next step of a system study, the analysis of inputs and outputs.

Input and Output Analysis

A system consists of three elements: input, output, and process. A model of a simple system is shown in figure 15.3. The elements of a model include inputs, processes and outputs.

- An input is that which activates a system.
- A process is the activity which translates an input to an output.
- An output is a product of a system.

The success of a solution to the information management needs depends heavily on the ability to carefully define the inputs to the system, the manner in which they will be processed, and the design of the outputs.

Figure 15.3
A Simple System Model

Inputs

Typical inputs in a school district setting include pupil enrollment data, attendance registers, certificated and classified position counts, purchase orders, library charge cards, and student grade reports. Additionally, inputs may be in the form of files. Examples of files include student record files, equipment inventory lists/files, on-order files, accounting files, vehicle maintenance records, and physical plant repair files.

For a systems analyst to design a new system, he or she must understand the current systems. This includes detailed knowledge of the forms and files currently in use. Understanding these inputs will enable the analyst to determine the computer system size and the configuration that will meet the school or district's needs, and aid in the design of computer input formats, record structures, and outputs.

Output Analysis

An output is the product of a process. The same techniques used to describe inputs can be used to describe outputs, specifically, the use of form and file analysis. In simple terms, a form or a file could be either an input or an output, depending on the circumstances. For example, a pupil enrollment file would be considered an input if used to search for information on a specific student or group of students. It would be considered an output when it is the final product generated from filing the student enrollment forms, such as a student enrollment count by ethnicity sent to the state. Printouts and screen displays from computerized systems are also outputs and can be subjected to the same output review process as printed forms and manual files.

If the purpose of a system analysis were to help design a replacement system, existing outputs can be analyzed. If a new application were being designed, output may not exist. In this case, a description of the output, in addition to a listing of the specific data elements that are required, needs to be developed during the design phase of the systems analysis.

Processing Analysis

Task Analysis

Once the analysis of inputs and outputs is complete, one can direct attention to processing. In this context, processing refers to the routines associated with preparation of inputs and the use of the outputs. It is important to understand the processing of the current system in order to evaluate its efficiency, and to assure a complete understanding of the operation. Without this knowledge, it is not possible to revise the system processing to take advantage of technology-based designs.

There are a variety of methods which can be used to describe processing. One method is to use a narrative description. However, this technique is generally lengthy — and does not easily provide a 'picture' of the processing task(s) under review. To obtain an overall picture of processing that occurs, a more useful approach is to perform a task analysis. This analysis consists of generating a list of tasks associated with the process in sequential order.

A task analysis for the pupil enrollment portion of a student record system would include the following steps:

1. Parent or guardian requests a blank enrollment form.

2. Parent or guardian completes the form.

3. The form is submitted to the school clerk.

4. The clerk checks the form for completeness.

5. The clerk verifies data in the document.

6. The clerk indicates the data has been verified and notes how the verification was conducted.

7. The clerk initials the form.

8. The form is placed in the "files to be created" folder.

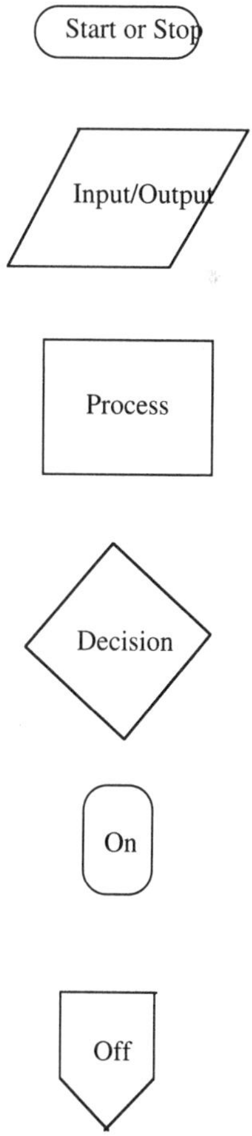

Figure 15.4 Standard Flowcharting Symbols

Flowcharting

A task analysis does not always provide a clear overall picture of the process or of the interrelationships among the tasks included in the process. A technique called flowcharting can help resolve these problems. Flowcharting is the process of successively dividing a complex problem into parts of progressively greater and greater detail, and representing this by graphic means. There are standard international symbols used for flowchart construction; and there are a basic set of charting conventions. The basic set of flowcharting symbols are displayed in figure 15.4.

The symbols displayed in figure 15.4 can be combined by using a set of generally accepted conventions that allows one to graphically describe a process. Figure 15.5 shows a flowchart of a task analysis for the pupil enrollment form process outlined. See how the flowchart makes it possible to quickly form an overview of the total process and follow the interrelationships of the tasks.

Alternatives and Applications

While the discussion has largely focused on computer applications, the same steps could apply to the acquisition of a variety of information technologies including video and voice. The options available can be as simple as the minor redesign of an existing system, or it could rely on a sophisticated computer-operated network linking microcomputers at user sites (such as schools, the maintenance facility or the bus garage) with a central mainframe computer, or even a video telecommunications network designed to deliver both instruction and offer a forum for public participation in school district activities.

In the past, computer options available to school business administrators for automating the functions of the business office were limited. The administrator could easily identify available systems in the marketplace to automate

Figure 15.5
Flowchart Example: Student Enrollment Form Processing Analysis

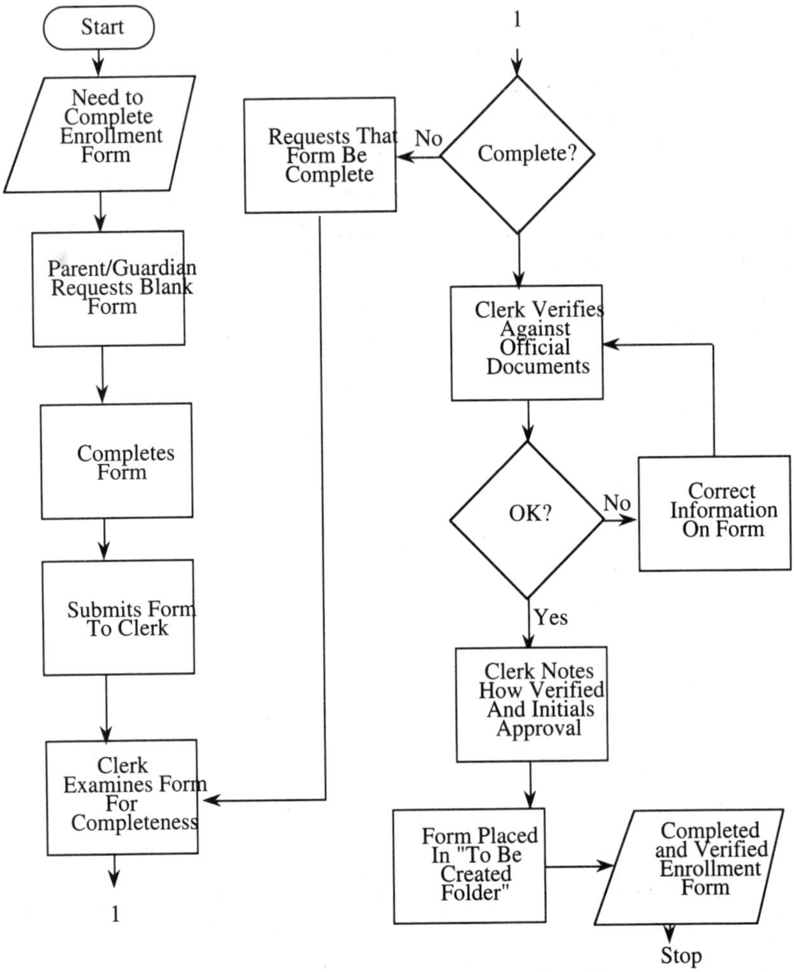

various administrative functions. This is no longer the case. Today's marketplace is extremely dynamic with a vast number of hardware and software options available to administrators. The issues which must be considered in selecting hardware and software are discussed.

Software

Educational administration software is software which facilitates the creation, manipulation, retrieval and reporting, and maintenance of data appropriate for the operation of a school or school district. This broad definition includes applications in many areas of educational administration, including student records, personnel records, finance and budgeting, management planning, cafeteria, pupil transportation routing and vehicle maintenance, and physical plant management.

It is possible to categorize the software marketplace by application area. But recent trends in integrated systems design sometimes makes this approach difficult. Integrated software packages are those that provide multiple computing functions in one program. For example, an integrated package may contain a word processing program, a spreadsheet, and a database manager. The advantage of an integrated package is that the various components are usually able to share files, making data transfer from one application to another easier. Integrated packages may also combine business functions like student enrollment, budgeting, and attendance accounting in one package. With today's emphasis on school accountability, such a package often has the capability of including student test scores along with pertinent demographic and family data. Similar data systems are being developed, often at the state level, for the tracking and reporting of teacher employment and qualification data.

The most sophisticated of these systems is in Florida where individual school sites use computers or terminals to report staff, student, and financial records to the state on a regular basis. These data not only provide a far more detailed understanding of how educational resources are being used in school systems, but they offer a wealth of opportunities for parties conducting research on how local schools and school districts use fiscal resources, and whether additional moneys matter in education.

Sometimes software is identified by the type of hardware on which it operates—mainframe, minicomputer, or microcomputer. However, presently, the migration of software from smaller to larger systems, and vice versa is commonplace. Thus, some software operates on a wide spectrum of hardware and also across hardware manufacturers. The biggest issue in this area today relates to the choice of platform for microcomputers. Currently Macintosh systems are not compatible with machines that operate in a DOS (Disk Operating System) or Windows environment. While some software is available for both platforms, many titles are only available on one of these systems. At present, if you select software for Windows for one application, you may foreclose some options for other software needs.

Fortunately, this problem is becoming less and less restrictive. In addition to the availability of many titles on both platforms, many new Macintosh computers offer software or hardware upgrades that allow them to run Windows and DOS software, although with some loss of performance. Moreover, the development of the Power PC chip may enable true cross-platform compatibility in the very near future.

Software may also be categorized by type, that is, the functions the software performs. Using this approach, educational administration software can be categorized as: work-enhancement software, database software, subsystem application software, and integrated software.

Work-enhancement software is that category of very popular software, mostly microcomputer-based, which facilitates office productivity. Included in this category are: word processing programs; spreadsheet packages; time management software; graphics program; communications software which allow access to other hardware, software, and data; and integrated work-enhancement programs which combine some or all of the above functions in one package.

Since so many educational applications are database dependent, this category is given special attention, even though in strictest terms database software could be considered work-enhancement software. Database software can be a list or file manager, which allows for the creation and manipulation of single, unrelated files. Text retrieval software, such as *Inmagic*, provides quick indexing of database records and individual words in those records. Database management software (dbms) is more complex in that it allows flexibility among fields and files. In a relational database manager, the relationships of data in separate files can be established as needed, rather than at the design step, as is the case of a file manager.

The subsystem application software available in the marketplace is designed to automate a single administrative function. Typically these applications are developed using a higher level programming language, such as BASIC, COBOL or C. Sometimes a general purpose dbms package is used to create standardized record structures, screen displays, and report formats, with programming provided so the user can take advantage of the full power of the dbms package, without learning all of its complexities.

Many schools would like to automate a number of application areas and/or have commonality (such as with user access, data files, and maintenance) across these application areas. Other districts elect to automate a single application leaving integration of additional application areas for consideration at a later date. There are a growing number of products in the marketplace which have this "modular" capability.

Computer Hardware

Hardware options can be divided into three categories: mainframe computers, minicomputers, and microcomputers. Mainframe computing is characterized by on-line processing typically distributed over a large area. Mainframe computers operate as multi-user and multi-application systems, they are expensive, and they need a controlled environment. Educational timesharing services, or educational computing service bureaus, typically operate on mainframe computers. Individual clients (schools) access and use the system through computer terminals which interface with the mainframe located at a remote site.

Minicomputers, like mainframes, operate in an on-line mode, are multi-user, and require a controlled environment for the hardware. But unlike a mainframe, minicomputers tend to be dedicated to a specific in-house application, such as school administration, and are housed on-site. The cost maybe within the realm of acceptability for large schools and school districts.

Microcomputers differ from the other two categories of computers because they can operate in a regular office or school environment, are relatively inexpensive, and until recently, have been single-user oriented. However, developments in microcomputer hardware and in telecommunications have allowed microcomputers to emulate the power and sophistication of larger systems, particularly the minicomputer. This has occurred through such developments as distributed processing where microcomputers are linked to minis and/or mainframes, and because of the development of local area networks (LANs) where individual microcomputers are linked to share software, data, and peripherals. Today, with increased memory, storage capacity, and processing speed available in microcomputers, it is difficult to make a clear distinction among these three categories of computers.

Often the best computer strategy is to link a series of microcomputers to a mainframe, or mini computer, for central processing. For example, a district with twelve schools could install attendance software on a microcomputer in each school office. Each day after the attendance was processed, the staff in the school office could "dial in" to the district's central minicomputer system and transmit the day's attendance figures to the central office. In the example at the beginning of the chapter, the school business officer would have used a system like this to understand the daily attendance and make the projection that he or she needed to order an additional thirty-five lunches for the high school.

Other Hardware and Equipment

Today, the management of information resources requires more than sophisticated computers. Much of the information received comes from television or even through the telephone. Moreover, there are many links between telephones and computers. The dramatic growth of the Internet in the last several years is evidence of the need to be aware of how these communication technologies work together.

The Internet is a series of large computers linked together through a series of protocols supported in large part by the National Science Foundation. Access to the Internet requires nothing more than a simple microcomputer (or even a "dumb" terminal), a modem and simple communications software. Using already existing phone lines, it is possible to use the Internet to search for and find information on almost any topic. Through a series of protocols, such as the text-based File Transfer Protocol and Gophers, to highly sophisticated software that actually allows the dissemination of graphic interface programs on the Internet, students, teachers, and school administrators can "surf" the Internet for a wide range of topics.

In addition to information retrieval services, many organizations offer subscriptions to so called listservs. A subscription is typically free, and subscribers receive regular postings from the organization providing information on specific topics. It is possible to subscribe to as many listservs as one desires, and subscriptions can be canceled. Many educational organizations are beginning to use these facilities for communicating quickly with their members.

Television also offers a powerful communications medium. The success of the controversial Channel One provides an example. Channel One was created by Whittle Communications to provide news programming to schools across the nation. Schools that signed up received a television set and VCR, the only stipulation being that students be shown Channel One's programming each day, either live or through tape. The daily program was fifteen-minutes long, and two of those minutes were devoted to commercials, which made the program controversial, but also paid for the broadcasts and the hardware donated to schools. Channel One provides an excellent example of both the power of television and video linkages between locations, as well as the potential dangers that must be considered as more curriculum content is provided via video connections.

Another powerful feature of television is the capability of interactive communication. Already some of the cable television firms across the nation have experimented with remote communications from subscribers. It may someday be possible to call into an audiovisual service and through a remote connected to the monitor in a classroom, select a particular program to view. The example at the beginning of the chapter where the students from North Valley High School received instruction and participated in a basketball pep rally from a remote location is possible today. Moreover, as the cost of satellite reception dishes continues to decline, it will be possible for schools and school offices to have access to the tremendous range of material available whenever they choose.

Selection

Up to this point, the systems analysis has included: problem definition, input/output and processing analysis, and a review of alternative software application designs. Careful review of available software to determine whether it meets the variety of needs is critical. Ideally, once a software selection has been made, appropriate hardware should be selected. Although this is the ideal approach, hardware is often already on-site. Therefore, software selections must be limited to those that operate on existing hardware systems. The specific software and hardware elements that should be considered while comparing alternatives are examined.

There are many software producers and hardware manufacturers competing for school districts' business. Selection is not always an easy process. However, there are some basic guidelines which can be used to review software and hardware options. The following points should be considered in software and hardware selection:

Characteristics of the Vendor/Producer

- Business reputation.
- Length of time in business.
- Knowledge of education and schools.
- Number of schools or school districts currently using the product.
- User references.
- Support services available from the vendor and from third parties.
- Geographic base of installations.
- Vendor accessibility by telephone.
- Other services provided.

Quality and Characteristics of Training

- Clear and Easy to Use Software and hardware documentation.
- Availability of training classes.
- Availability of self-paced computer-based training packages.
- Ease of learning the software.

Costs

- Understand all costs involved including Installation, Training, Maintenance, AND Upgrades. Conduct cost comparisons among available systems.

Software Selection Issues

- Loading time.

- Response time.
- Log-on procedures.
- User friendliness.
- Screen displays.
- Ease of procedures/commands.
- Ease of data entry.
- Ease of deletions/adds/changes.
- Flexibility of format.
- Searching ease.
- Quality of system output.
- Error messages.
- Security measures.
- Back-up procedures.
- Customizing features.
- Usage tracking.
- File and program size capabilities.
- Equipment compatibility.
- System/organizational compatibility.
- Expandability.
- Compatibility with future versions of the product.
- Integration with future applications.

Hardware Selection Issues

- Hardware reliability.
- Warranty.
- Servicing procedures.
- Response time.
- Software compatibility.
- System/organizational compatibility.

- Upgrading options.
- Ergonomic design features.

Selection Sources

Once the needs are understood, the marketplace can be explored. Most likely the first thing one will notice about the marketplace is the vast number of products available, and the confusing system requirements associated with each. Sorting through all that is available can be a difficult and time consuming process.

While there are many commercial sources of information, perhaps the most important source of information is other school administrators. One of the best ways to ascertain the strengths and weaknesses of any software/hardware combination is to ask someone who is already using it. Consult with one's colleagues in other school districts to see if any of them have found an application that meets their needs, and would be well-suited to meet the demands of the district as well. One should ask if others have considered the software program you are considering. If others have considered and rejected it, find out why. Was the program inadequate to meet their needs? Was support for the program weak? What didn't they like about the program, and why did they select a different one? If another district is using the program that is under consideration, questions should consider the program's strengths and weaknesses. Ask if you can visit the district and see the program in operation. One of the best ways to determine whether or not the program will work is to see how well it is working somewhere else. Armed with this knowledge and the selection criteria, the choice of specific hardware and software for the application can begin.

The previous analysis established that the student record and scheduling application has the following objectives:

- Input records and updates interactively.
- Retrieve immediate responses to information queries on individual students, teachers, and class schedules.
- Generate, on demand, student record and schedule reports including master student lists, master class schedules, teacher and homeroom lists, reports, address labels, and a variety of statistical summaries.
- Establish an improved student record and scheduling system which will not increase costs.

Further analysis of our needs revealed the following functional requirements for a student record system:[4]

- The system must handle a total district enrollment file of 20,000 records with 4,000 new records or updates per year.

- The system should connect each school site and district headquarters into a networked system.
- The system be able to handle an average of fifty new student records per school at registration time, and an average of five new records per school per month during the rest of the year,
- The system should be designed to handle an estimated 250 queries and 25 record updates per day.
- The system should have the ability to produce reports for individual schools.

The detailed input requirements for the application are generated from the results of the input analysis discussed. The forms input analysis provides the characteristics of these data that will be the system's inputs. In applications which involve the creation of databases, which many systems require, this step typically includes creation of the data dictionary. A data dictionary is a list of all these data elements in the system and each element's characteristics and special features.

The listing of requirements for this application indicates that there are a number of modules, such as student record and scheduling, that are integral to the automation effort. Consequently, any investigation of software should examine packages that provide all these components, and that offer them in an integrated approach.

A review of the educational administration software marketplace will most likely identify a number of packages worthy of consideration. An alternative would be to develop an in-house system designed specifically for the district's needs. This is often an expensive option, and in situations were a number of competing, stable programs are available, it is often a poor choice. This is because development costs would be high, the time needed to design, develop, and implement the application would be considerable, and there is no guarantee the end result would be better than existing, commercially available software.

Once the candidate systems have been compared, a final selection can be made. The reasons for choosing a particular system may vary from application to application, but might include the following:

- It received the highest rating through the district Software/System Evaluation.
- It best meets district functional requirements.
- It meets all of the district's technical and maintenance requirements.
- Its performance met the district's requirements by meeting or exceeding benchmark acceptance test levels.

- It is highly rated by other users.
- Site visits were positive; users at these sites rated the system highly in terms of functionality and performance.
- System costs for hardware and software are within the district's established cost framework.
- The system was rated as "one-of-the-best" by an independent evaluation team from a professional education association.
- The system provides an interactive, state-of-the-art user interface with the ability to meet the district's future growth needs.

There is a wide range of hardware and software in the marketplace. A systems analysis approach will provide an excellent basis for making the best choice for your school environment. There are a number of points that should be remembered when making these choices. These include the characteristics of the vendor and manufacturer, the quality and characteristics of the training that is available, costs, the technical capability of the system, and availability of support for both the hardware and the software. Many selection sources are available. Some are specific to schools and others are more general in nature. After careful consideration of the selection issues, a review of the selection sources, an analysis of specific needs (such as outlining the data dictionary specifications), and a careful review of options, one can make an appropriate choice.

Design

Systems design includes the design of a new input form as well as the design of input screen formats. As with the analysis phase, design involves detailed output descriptions. This involves outlining the characteristics of screen displays and computer-generated printed products. It is important to develop user-friendly interfaces. Additionally, the tasks which are to be performed in the new application are outlined. A number of systems techniques can be used to assist in this processing phase—flowcharting being one of the most useful. As appropriate, the issues associated with systems design may be used to fine-tune the requirements or selection criteria for commercially available packages.

Implementation and Management

System Housekeeping Tasks

In the analysis and design of information management systems, there are a number of other issues to be considered. First, one should develop general "housekeeping' procedures" These procedures include such obvious, yet important functions as preparation of software and establishing a system for file backup. Usually, the first instruction provided with a software package, whether this be an operating system or an application package, is how to make a backup copy of the software. If the software were damaged or misplaced, a backup will enable you to still run your application.

Similarly, any files that are created should be backed up on a regular basis. For example, when using a microcomputer application, it is a good idea to save your working files every fifteen to thirty minutes. If there is a power loss or if you remove the data disk incorrectly, your data loss will be minimized. It should be noted, however, that backup files consume substantial amounts of disk space. One should review backup files on a regular basis and delete files that are no longer needed. It is also good practice to store a copy of the software and important data files in another location. Thus, in the worst case situation (such as a fire in the office), all software and data files will not be destroyed. It is also important to store magnetic media, such as diskettes and tapes, in a location which will minimize environmental damage which might occur from dust or dampness.

Documentation

Once an information management system of any type has been analyzed and designed, it is essential to develop documentation about the application. This documentation includes: system/processing, programming, and user-based documentation. Documentation serves a variety of purposes and is used by a variety of different groups. Thus, documentation should be prepared with these varying purposes and users in mind. A commercially available system may have all these categories of documentation included as part of the procurement, or occasionally additional user-based documentation may need to be developed in-house. Points to consider in developing these different levels of documentation include:

- Knowledge concerning the level of user familiarity with the software/hardware.
- Use of clear, precise writing style with a minimum of jargon.
- Incorporation of a good table of contents and/or indexes into the documentation,
- Providing well-structured and organized documentation.
- Use of different type fonts, outlining, and blank spaces to highlight special or important points.
- Liberal use of illustrations and sample problems.
- Make certain that the documentation is current and that adequate procedures for updating it exist.

The following topics should be included in the documentation:

- **Overview**

  Introduction and background.

  Requirements of application.

  Hardware and software requirements.

  Overview of application (narrative).

  Overview of application (flowchart).

- **Data Preparation**

  Form examples with sample data.

  How to complete forms.

  How to prepare for data input.

- **Data Input**

  Computer operating procedures.

  Sample record.

  Input procedures.

  Screen displays.

  Problem solutions.

- **Processing and Products**

  Computer operating procedures.

  Sample screen displays.

  Sample products.

  Procedures for producing products.

  Problem solutions.

- **Housekeeping Routines**

  Creating backups and file deletions.

  Maintaining documentation.

- **Data and Programs**

  Sample record structures.

Program listings.

- **Index**

Implementation

Once a management information system has been designed, it must implemented, and mechanisms to monitor it and provide feedback must be in place. System implementation involves:

- Decisions as to selection of implementation procedures.
- Consideration of the various implementation tasks.

A system can be implemented in a variety of ways. For example, systems can be implemented all at one time, in parallel, or through a modular approach. Most small microcomputer systems are implemented at once. More complex applications, such as one involving all the components of a student record system, could be implemented in parallel. Parallel implementation requires maintaining the old system (and gradually phasing it out) while the new one is implemented. There are advantages to this approach, including the safeguard that if something goes wrong, there is a backup available; also such an implementation allows the opportunity for gathering effectiveness data on both systems.

As discussed, more complex systems consist of subsystems which interact. In a step-by-step implementation (modular) one takes small units to implement—and makes sure that the first unit works before progressing to the next unit. This approach has advantages including the fact that it is easier to implement smaller parts than a whole system at once; and staff can become accustomed to the new system more gradually. One of the problems associated with a modular design is that sometimes concern with perfecting individual components prevents the completion of the whole system.

Implementation Tasks

There are numerous tasks associated with the implementation of a computer system. The specific tasks vary by type of application. However, the following are among the most common tasks that will need to be completed to successfully install and operate a new computer application.

Management Planning. There are a number of management techniques, such as Gantt, PERT, and CPM, that can be used to organize and plan the implementation of any new information technology. What is important is to be sure that an implementation schedule is established and that all parties appear able to meet project goals and time lines.

Procurement. Purchasing the selected hardware and software is not always a straightforward matter. A formal procurement process may be required. In addition, a number of factors could cause a time delay in procuring the desired system. These factors could include: the need for administrative and school board approval, the district's purchasing department requirements may take a few days to complete, complex negotiations or the need to generate a complex contract with the vendor could delay the eventual purchase, or, delays may be caused by the temporary unavailability of hardware or software.

Conversion. In many applications it is necessary to convert existing manual files into machine readable form. For some applications, particularly those involving the creation of large databases, such as for student records, personnel, or library/media catalog records, the file conversion task may be considerable. For any file conversion, one should determine how the files will be converted, and who will perform that conversion. In some instances, it may be possible to interface with an existing database and download records into your own file. However, for most computer applications, it is far more common to convert files through manual keyboarding.

Site Preparation. The larger the on-site computer system, the more complex the site preparation. Both minicomputers and mainframes require exacting environmental controls for correct operation. These include the design of a facility for the hardware which has the following features:

- An uninterruptable power supply.
- Power panels and receptacles.
- Air conditioning, dust and humidity control, and adequate water supply for cooling.
- Static electricity control.
- Raised flooring (an/or false ceilings) for cabling.
- Adequate floor support.
- A telecommunications systems.
- Adequate space for hardware, storage, work space, and personnel.
- Paper waste disposal capabilities.
- Fire protection.

Lighting and power considerations are often forgotten in the crush of other, more immediate concerns. Lighting has an impact on both space and environment. Careful consideration should be given to the provision of adequate lighting and the problem of glare. To illustrate the dimensions of the problem, consider that the suggested illumination for accounting tasks (working with spreadsheets) is 150 footcandles, whereas viewing a video display unit requires only 30 footcandles Glare can be reduced by using low-luminous fluorescent lighting, drapes or shades on windows, screen hoods on terminal screens, and matte finishes on table tops. Individual, adjustable lights may be needed to illuminate the work area.[5]

The provision of adequate electrical power and telecommunications outlets is often a problem, particularly in older facilities. The cost of rewiring can be extraordinarily expensive. Modern methods such as flat wire under-carpet grids, and power poles to overhead plenum cabling hidden behind false ceilings offer less costly alternatives to traditional in-wall and under-floor systems.

Computers and data storage mediums such as floppy discs are subject to damage from a variety of electrical sources:

- *Surges and Spikes*: Of short duration, caused by utilities switching lines, nearby power lines being hit by lightning, and intermittent operation of motors such as in elevators and air conditioners.
- *Noise*: High frequency electrical impulses caused by auto ignition systems, transformers, fluorescent lights, static electricity, and lightning storms.
- *Voltage Fluctuations*: Caused by loss in power transmission, uneven regulation of voltage, and brownouts (intentional lowering of voltage by utilities during peak operating periods).

Equipment and data can be protected against surges and spikes by using powerline protectors that also double as multi-outlet junction boxes. In-line monitoring devices are available which warn of serious voltage fluctuations that can destroy equipment and/or data. Static electricity is often a problem in carpeted areas and during periods of low humidity. Care should be taken to select carpeting that is static-resistant such as olefin, nylon or acrylic.

Since microcomputers operate in the normal office or home environment, there is usually no need for a complex site redesign. However, it is necessary to provide new desk and work space for the microcomputer application. Likewise, it is important to investigate, if needed, provisions for adequate electrical outlets, static electricity safeguards, and disk storage devices.

Supporting a Microcomputer Facility

A look at one of the many mail-order computer supply house catalogs or a walk through a modern retail business supply store illustrate the variety of supplies, furnishings, and devices available for microcomputers. Just what the facility will require will depend upon a number of factors including:

- Type and variety of equipment available,

- Level of service provided.
- Budgetary limitations.

Even a bare-bones facility will need to consider of the following items:

- Dust covers to protect CPU's, monitors, and keyboards when not in use.
- Disk drive head cleaning kit.
- Floppy disk envelopes and binders or storage units.
- Surge protectors (see above).
- Blank diskettes for making backup and distribution copies of programs.

If line printers are provided, it is important to consider not only the kind of printer—dot matrix, daisy wheel, thermal, or laser—but also the costs of supplies, i.e., paper and ribbons.

Training: Training is the third component of any computer application (the other two being hardware and software). Without adequate training, the most elegantly designed computer application can, and probably will, fail. Thus, it is essential to develop appropriate training for staff who will use the system. Possible training methods include:

- Courses and workshops.
- Books and print materials.
- Documentation.
- Media, such as videocassettes.
- Computer-based training, e.g. diskette-based.

There are advantages and disadvantages to each of these training methods. However, the use of courses/workshops or computer-based training, followed-up with the use of printed materials is the most successful.

Public Relations

A public relations effort for a new system can be achieved through a variety of means, such as through new system publicity, adequate orientation or training procedures, and the design of easy-to-use, user-cordial systems. For example, the initial needs assessment indicated that administrators, staff, and teachers were dissatisfied with the student record/scheduling system. The new computer-based application resolves the problems associated with the old system — a publicity effort could be implemented to alert administrators, staff, and teachers concerning the implementation of a new, improved service.

Systems Monitoring: With the design and implementation of any system, it must be monitored to determine if it is achieving its objectives, within the constraints. Data are typically produced from the system and provided to management, who then can ascertain whether the system is operating correctly. In a school setting, systems monitoring and feedback refers to the generation of adequate reports, such as numerical compilations, including: tabulations of the number of items ordered; number of updates to various databases; vendor performance analysis; rate of production; and processing speeds. Additionally, the gathering of attitudinal responses from users also provides feedback on system performance.

Evaluation

In some instances it may be sufficient to provide feedback data on the operation of a new system through responses to attitudinal questions. In other instances, however, it may be more appropriate to describe the new system in terms of cost-effectiveness measures. Whereas feedback provide data in isolation to other systems, cost-effectiveness implies the comparison of the present system to alternatives, including the previous system.

It should be noted that in many systems studies, cost-effectiveness evaluation occurs prior to the design of a new system. One gathers cost and effectiveness measurements on the present system and compares these to the cost-effectiveness measurements of all possible alternatives — those systems that solve the problem and meet the objectives within the constraints. This type of comparison is known as a trade-off study. The utilization of systems analysis made the assumption that a computer-based application was the most cost-effective.

In a typical cost-effectiveness evaluation, data are gathered on the cost of the system, as well as on various performance measures. Performance measures to be used in the evaluation would vary by the type of system, but might include some of the following:

- User responses about a knowledge of products and services before and after the inauguration of the new computerized system.
- Overall user responses to products and services before and after the inauguration of a new computerized system.
- User evaluation of the system before and after the inauguration.
- Time to answer requests for information before and after the new system.
- Time to produce reports calculated before and after the new system.

- Cost to produce products and services before and after the inauguration of the new system.
- Overall cost and unit cost to produce the products and services versus the cost not to produce it, e.g. $0.

Conduct of cost effectiveness analyses is discussed in chapter 14 of this text.

Summary

In addition to the variety of tasks associated with the preliminary analysis, input and output analysis and design, processing concerns, and hardware and software selection, there are certain tasks important to the implementation and monitoring of a new system. The analyst should be concerned with developing documentation, implementation planning, procurement and site design, conversion efforts, and development of training and public relation programs. Upon completion of the implementation, the analyst should develop procedures whereby the system will be monitored and periodic feedback provided to management on the performance of the new system. In addition, a more formalized process, namely a cost-effectiveness evaluation study, can be conducted to provide additional evaluative data.

Managing the Design and Implementation of Information Systems

Up to this point, the discussion has focused on analyzing the needs and determining what software and hardware one should consider in designing management information systems. There is an additional, and critical, factor one must consider before beginning a computer implementation program — the human element. Computers do not solve problems, they are merely a tool that helps to solve a problem, to simplify a complex task, or to improve one's ability to complete repetitive tasks in an efficient manner. Automation of any task will only be successful to the extent that the people responsible for the task accept and use the computer. This section considers these human issues in more detail, offering suggestions on the roles of various staff members in the business office and throughout the school district, and offering strategies to assure successful implementation of a computing system in the district.

As outlined, there are numerous physical site design requirements associated with any type of computer system. Specific topics to be addressed in a computer site design include: power source and control; environment control through regulation of air, temperature, humidity, dust, and static electricity; appropriate flooring for cabling and support; water for cooling needs; telecommunications connections; waste control for large paper volumes; fire hazard reduction; and physical access control. To assist the school administrator, vendors of automated systems typically provide detailed requirements prior to installation. These specifications cover such topics as: space (system dimensions, system space requirements, storage space for disc packs, paper supplies, shipping cartons, documentation supplies, future expansion, and minimum room size); electrical (power lines, line conditioner, line labels, terminal power); telephone; environment (air conditioning, static conditions control); and area for receiving delivery of the system.[6]

An effective site design considers the interplay of the physiological, psychological, and the psychosocial needs of the individual, with the technology—the computer. Such a design should take the following into consideration:

- An effective work design includes the interplay of task, environment, and workplace.
- Attention to the micro environment, that is the design of individual space and to the mini environment such as the work desk, as well as to the macro environment including the building ceiling and flooring system.
- A design should allow for individual differences in VDT viewing distances and inclination, keyboard location, leg space, seating characteristics, foot support, writing and work surface requirements.
- The design should be based on the characteristics of the user group, e.g. school children have different physiological and psychological needs than working adults.
- There should be the provision for personal work space.
- It is useful to request feedback from employees to assist in developing staff satisfaction with technologies.
- A manager should incorporate adequate break/rest periods for staff who work with VDTs.
- Systems analysis techniques, such as project management, should be used in developing computer-based systems and this process should included attention to the "people" issue.
- Consider that there might be a change in the internal work processing flow due to use of a computerized systems—the work flow patterns used in a manual system may not longer be effective.

There are a number of issues that currently face the administrator of a computer system. These issues are the same whether the installation is a single PC-based system or a large mainframe. These issues include privacy and computer crime.

Privacy is the ability to control who uses personal information and to control the manner in which it is used. The creation and availability of databases of information concerning students and staff can make such personal data readily accessible to both those who "need-to-know" as well as to others. Security of such data is essential from those who should not have access. Additionally, other aspects of the privacy issue include the use of inaccurate data because of various human errors, such as data input. Unfortunately, sometimes such incorrect data has the authority of correctness since it is "in the computer." In addition, the fact exists that decisions can be made regarding individuals only using computerized data and not a complete body of information.

Various legislation has been enacted to assist in the privacy of individuals, which has relevance to schools. These include the Privacy Act of 1974 and the Family Educational Rights and Privacy Act (1974). This legislation indicates that individuals have the right to be informed as to the manner in which data are collected and to its use, and have the right to correct erroneous data. Also, data used for one purpose cannot be used for another without consent. More recently, additional legislation has been enacted in response to the illegal access of governmental computers by non-valid users and "computer-hackers" (The Comprehensive Crime Control Act of 1984).

Another major issue of concern is computer crime. Computer crime includes: the theft of property such as the stealing of hardware and software; the theft of data such as institutional mailing lists, theft of intellectual ownership through copyright infringement such as the unauthorized duplication of software and documentation (covered by the U.S. Copyright Act of 1978); and the theft of moneys, such as duplicate check writing, fraud, and embezzlement.

A number of security safeguards can be enacted to provide safeguards from computer crime and to make computer crime difficult. These include a variety of procedures, including the use of computer passwords to secure access to computers and the frequent updating of these passwords, the use of dial-back procedures for on-line access, and the separation of processing responsibilities. In must be emphasized that the computer facility, however small, represents an investment that should be protected. Security must be taken into account in selecting a location for the facility, and for securing access to and from the facility, such as through the use of card-keys and coded door keys. With the increasing emphasis on personal computers and LAN-based systems there is interest in the physical security of microcomputers. A number of devices are available to deter theft of these computers as well as theft of terminals which access larger systems. These include:

- Anchor pads: A system of adhesive pads and lockable rods attached to a slim steel housing that is bolted to the computer case making removal virtually impossible.
- Cable antitheft device: Uses small adhesive pads attached to a flexible cable that is in turn locked to the table top.

- Motion alarms: Movement of the equipment automatically triggers a piercing alarm.
- Lockable cabinets: Available in a wide range of styles and prices, from simple storage cabinets to full, lockable workstations.
- Power cord alarm: Disconnecting the power cord from this special socket activates an alarm.
- CPU theft covers: Metal enclosures that secure the CPU and monitor to the table top.

From a broader perspective, the school district can develop standards for use and ethical behavior regarding computer use, covering such topics as copyright infringement, personal use of computers, taking personal computers home, and monitoring employees' habits such as data input speed.

In the complex world of data processing today, microcomputers have become more powerful than many larger computers were just a few years ago. Moreover, the vast array of computers and peripherals is confusing to even the most sophisticated data processing manager. The concept of a "personal computer" has changed as more and more microcomputer users have established networks that link a number of users together, or link remote users to a larger computer located at a different site. Today, a microcomputer user can access information from other microcomputers or from a mainframe computer located at a central office.

The advantages of networks for school management should be obvious. For example, a network system would allow school attendance data to be sent to the central office on a daily basis. The central office could collect and total data from the individual schools, and prepare the required state reports accurately with a minimum of disruption to the local school staff. Or school principals could access a fiscal database and obtain an instant report on remaining budget balances for those funds over which they have control. Similar benefits from a shared data network would accrue to other financial matters, student achievement data, and information about personnel data throughout the school district.

Advances in computer software and hardware even make it possible for districts that have long relied on large mainframe systems to collect and share data with remote (school or office) sites with a minimum of modification to existing computers and software. Schools can be linked with computer cable, or more practically if there are substantial distances involved, through the use of modems and telephone lines connecting personal computers at school sites with larger host systems at the central office. Moreover, for smaller districts, advances in microcomputer capabilities may enable them to manage all of the data processing needs through a top-line personal computer. Finally, the availability of electronic mail networks can improve communications among school district staff. Often it is difficult for busy school principals to get in touch over the phone with equally busy central office administrators. Yet frequently, all that is needed is an answer to a question that does not require personal contact. The use of electronic mail, which allows people to send "mail" to one another can greatly facilitate most school district communications.

In response to the school reform movement of the 1980s, many districts are starting to implement school-based management programs. In addition to changing the basic authority relationships between the central office and local-school sites, school-based management has significant implications for administrative computing. It seems that one outcome of the school-based management approach is greater autonomy over the school's budget. While conceptually this is an important and straightforward issue, implementation of systems to make local control over larger segments of a district's budget are quite complex. For example, in most school districts today, the purchasing functions are handled by the business office. Most schools fill out purchase orders or requisitions when they want to purchase something, and the central office provides the requested materials. Schools also receive regular reports summarizing the expenditures and indicating the level of available resources for future needs. While similar systems will be needed under school-based management programs, if local schools were to have greater authority over budget and spending systems, the role of the central business office may change. In the past, the business office may have exercised substantial control over the flow of funds, now its role will be to facilitate school site decisions, and help sites avoid overspending the available fiscal resources.

While the complexities of these changes are substantial, networked computing systems seem to offer a powerful tool to help resolve these issues. It will become more important that school principals be aware, on a daily or weekly basis, of their account balances. Designing a system that allows this flexibility, within the constraints of existing district resources will challenge all business offices. The systems approach as outlined is one way to analyze these issues and make rational decisions that account for the individual characteristics and specific operational issues in a school district.

One of the most important considerations in designing a computer application to meet district needs is the human element. Unless the staff responsible for using and maintaining the system are "on board" the chances of system failure are substantial. The next section describes a number of important human considerations that need to be considered as one conducts a system analysis and begins implementation of an automated system.

Training

Perhaps the single most important human element in assuring the success of a new automated system is training. This step is often overlooked in the design and installation of new computing systems, yet unless the people responsible for using the system on a day-to-day basis are well versed in its use, they will not be able to benefit from all of the power it offers.

Training is an expensive and long term process. Individuals will require intensive training in the use of the new system once it is installed. It is important that employees be able to devote their full concentration to the training process. Little will be learned if the staff is trained at their desks while at the same time expected to answer the phone and fulfill their other job responsibilities. Training should take place in a quiet environment, with a skilled trainer in charge.

In addition to assuring adequate training at the beginning of the process, strong support is also necessary. There are three kinds of support to consider: (1) on site technical support from a vendor or an employee; (2) on site assistance from coworkers; and (3) documentation from manuals, tutorials, and on-screen help programs.

Once the initial training is complete and the staff begins using the new system, unanticipated problems are sure to arise. These result from workers not retaining all they learned during training, situations not covered in the basic training, or unanticipated complications. To resolve the first and second of these problems, someone will be needed on site for a few days of operation to refresh workers' memories. As for the third, if the problem were related to the system itself, help from the vendor to correct the situation may be necessary. If the problem were something that was considered in the system design, but simply an infrequent occurrence, then the district needs someone, either on staff or from the vendor, to be available to assist when the problem arises.

Once the system has been operating for some period of time and the workers have become comfortable with its operation, it is likely that some will be more proficient in its operation than others. These people should be encouraged to help their coworkers improve their skills and to continue learning about additional features of the system. For example, if a word processing program were introduced into the office setting, those workers who learn to use the basic system fastest should be encouraged to help their coworkers learn how to prepare documents using the new system. In addition, the district can help these people "stretch" their skills by relying on them to determine how to format more complicated documents, or to learn desktop publishing techniques and improve the presentation of materials.

Finally, one should be sure that each employee has available and knows how to use adequate documentation for the system. Often, solving a problem is simply a matter of looking up a simple command in the user manual. However, if finding the user manual requires leaving one's desk and searching out a copy, the manual is less likely to be used, and workers are less likely to find ways to solve problems on their own, potentially reducing the unit's productivity.

As this discussion shows, training is an expensive and continual process. However, this expense will pay off in the long run as all employees learn how to use the system, get used to helping each other solve problems, and find new and more powerful uses for the software they have been asked to use.

Endnotes

[1] There is a growing market for used computers. Many foundations, such as the New Life Computer Foundation in Los Angeles, California, have found considerable success in helping businesses donate used computers to schools and not-for-profit corporations. Similar efforts are possible within school systems. For example, if the district business office has a network of ten personal computers, and upgrades them to meet greater processing needs or to take advantage of new software, it is likely that the ten machines being replaced could form the basis of a computer lab at an elementary school or provide computers for ten classrooms in one school. If the computers being replaced were networked, then they might be ideal for a computer lab, where networking is often an expense not considered when computers are purchased.

[2] E.J. Kazlauskas and L.O. Picus, *Administrative Uses of Microcomputers in Schools: A Systems Analysis Approach to Selecting, Designing and Implementing Automated Systems.* (Reston, VA: Association of School Business Officials, Inter., 1990).

[3] See Chapter 14 of this text for a discussion of some of these quantitative tools.

[4] It should be noted that other modules in a student system, such as scheduling and grade reporting, would have separate additional requirements as these are not stated herein.

[5] A. Cohen and E. Cohen, *Planning the Electronic Office*, (New York: McGraw-Hill, 1983), 200-205.

[6] E.M. Coretz and E.J. Kazluaskas, *Managing Information Systems and Technologies,* (New York: Neal-Schuman, 1986), Chap. 4.

Chapter 16
Total Quality Management

Introduction

As school business officers strive to improve the management of public elementary and secondary school systems and to make the operation of the business office as efficient and service oriented as possible, one management tool that has found a large audience in recent years is Total Quality Management or TQM. The basic principles of TQM were developed by W. Edwards Demming in his work with Japanese industry during the 1940s and 1950s as those firms worked to rebuild following World War II. Although his views were shunned by most managers in the United States until recent years, Demming's concepts have had a considerable impact on the management of private industry in the United States and are making inroads in public management as well.

Today there are a variety of "models" of TQM. It has been implemented, with varying degrees of success in a wide range of business and government agencies including public school districts. The success of TQM in improving the quality of a concern's product has more to do with the way it encourages individuals to work together toward common goals than it does with the specific way it is implemented. TQM requires cooperation and communication among all the employees of the school district.[1] However, there are a number of basic themes and methods that are common to all TQM models no matter what they are designed to accomplish. This chapter begins with a definition of TQM, focusing on Demming's famous fourteen points. The next section of this chapter considers how well those fourteen points fit with the operation of public schools and school business offices. Following that discussion, there is a detailed description of how TQM can be put in place in a public school district. The final section of this chapter describes some common pitfalls that public school administrators have experienced when establishing TQM in school districts.

What is TQM?

Although TQM has been defined in a number of ways, Herman[2] has presented *in School Business Affairs* the definition of TQM as embracing the following concepts:

- *Philosophy* that believes that both internal and external customers are to be provided with products and services of ever-improving quality.
- *Goals* that state that each product or service and all activities or milestones leading to that final product or service shall be of the highest quality they can be.
- *Process* that accepts feedback from all stakeholders to define the products' and services' quality level and the specifications to be met. It utilizes empowered and well trained employees, within a systems structure, to develop products or services at each value-added juncture that will result in high quality delivery of those products and/or services and that will result in customer satisfaction.

Demming's fourteen points form the basis of TQM. They have been described by many authors over time. Milakovich summarizes the fourteen points as follows:[3]

1. Create constancy of purpose toward improvement of product and service.
2. Adopt a new philosophy.
3. Cease dependence on mass inspection.
4. End the practice of awarding business on the basis of price tag alone.
5. Improve constantly and forever the system of production and service.
6. Institute modern methods of training.
7. Institute leadership.
8. Drive out fear so that everyone may work effectively for the organization.
9. Break down barriers between staff areas.
10. Eliminate arbitrary numerical goals, slogans, and targets.
11. Eliminate work standards and numerical quotas.
12. Remove barriers that rob employees of their pride in workmanship.
13. Institute a vigorous program of education and retraining.
14. Take actions to accomplish the transformation.

Other definitions of TQM abound. Jablonsky's definition is "A cooperative form of doing business that relies on the talents and capabilities of both labor and management to continually improve quality and productivity using teams."[4] Vincoli defined TQM as ". . . a customer-focused, strategic and systematic approach to continuous performance improvement."[5] Other writers have defined additional components of TQM. Some of them include:

- TQM organizations should constantly improve the quality of their systems or processes.[6]
- TQM is about change, and change involves communications that cause changes in behavior.[7]
- The customer defines quality.[8]
- Employees are empowered through communication, training, and excellent leadership.[9]

Herman, in an article in *School Business Affairs*, adds a number of other important elements that are part of TQM. They include:[10]

- Customers are internal as well as external stakeholders.
- TQM has a value-added dimension. That is, quality is added at every step of the process which leads to an output in a business or service organization.
- Training is continuously required for management, employees and suppliers to make certain that they are constantly able to improve the quality of the product.
- Satisfaction of customers is gained through quality service.
- Extensive two-way communications with all stakeholders is required to define the quality specification to be used.
- TQM focuses on processes which are continuously improving to produce quality products at each step and to all stakeholders.
- Organizations which implement TQM will cause changes that will ultimately alter the culture of those organizations.
- An organizations' visions, mission, policies, and goals must focus on delivering quality to all customers.
- Problem-solving TQM teams are an integral structure within the TQM systems approach.

As this discussion shows, the general principles of TQM are to improve the quality of the product provided. Strategies for doing this include communication within the organization and communication with the suppliers and customers of the organization. In addition, TQM assumes that the individuals responsible for creating the product have good ideas regarding how product quality could be improved. There seem to be many places where such measures could be used to improve the performance of a school business office as well as a public school district in general. The next section of this chapter looks more closely at the fit between Demming's original fourteen principals and educational organizations, with specific emphasis on the school business office.

The "Fit" Between TQM and the School Business Office

This section relates Demming's fourteen points to the operation of a school system in general and to the school business office in particular. It is organized along the fourteen points as summarized previously, and follows the logic laid out in Herman's *School Business Affairs* article.[11]

Create Constancy of Purpose

This point suggests that an organization needs to focus on its purpose and work toward the qualitative improvement of its services or products. If the goal of the school system were high student performance (and there are other goals that are equally important), then the school district's leadership must focus the school district's employees, classified and certificated, parents, students, and all others with whom the school district has contact, toward improvements in student outcomes. Since many of a school district's outside contacts come through the business office in the form of vendors, tradespeople and service providers such as lawyers and health care professionals, constancy of purpose to improve quality within the business office is critical to the success of TQM and its ability to improve a school system.

While the impact of TQM principles in a business office may not have a direct bearing on the day-to-day learning that goes on in the public schools, to the extent that business employees work with, and not opposed to, the teaching staff, the more likely it is that teachers will have the supplies and materials they need in a timely manner. Having the tools to do the job, consigned from a helpful and supportive staff, may go a long way in improving teacher morale, and hence the quality of teaching. At a minimum, if high quality service makes getting supplies and resolving problems such as payroll changes easier, then teachers will have more time and effort to devote to teaching. While this is no guarantee that students will learn more, it seems logical that it will help. To assume anything else is counterproductive to the teaching-learning process on which public schools are based.

Adopt a New Philosophy

Herman[12] idenstifies three components of a "new philosophy" that are critical to the success of TQM. They are:

- Collaborative efforts among all staff toward the common objectives of the organization.
- Recognition for the contribution of all stakeholders. In a school setting this would include central school district management and staff, school district administrators and staff, teachers and other employees, parents, community members, individuals and companies doing business with the school district, and students.
- A dedication to delivering quality products and services to all customers.

These elements need to be present in the business office as well as other departments of a school district for TQM to be successful.

Cease Dependence on Mass Inspection

Movement away from large scale standardized testing of students is often unpopular and politically dangerous in today's educational environment. However, the principles of evaluation established through TQM, which focus on more careful assessment of the process used to produce a product or deliver a service with an eye to improving the process, can be utilized in management functions. TQMs approach to evaluation lead to monitoring the quality of each employee's work through assessment of the processes used at each step of the job assignment rather than basing the evaluation on some quantitative measure such as the number of batch entries completed in a day or the number of payroll errors requiring attention each pay period. By working with employees to improve job performance and monitoring the overall "quality" of the way they approach job tasks, long term improvements in service delivery are possible.

While such evaluations are possible in the classroom as well, regardless of whether standardized tests are used, a discussion of student assessment is beyond the parameters of a book on school business administration.

End Low Cost as the Sole Basis for Contracting

School districts are among the biggest consumers of goods and services in a local community. This standard suggests that price alone is not the best way to contract for goods and services. Rather, the standard that should be used is to seek the lowest price provided that meets the high quality standards established by the school district before the good or service is sought. If the quality specifications were not met, the lowest cost bid should be refused. This point has obvious importance to the operation of the business side of a school district. Whether it is ensuring that a new computing system does exactly what it is supposed to do, or that the paint on the school building lasts as long as it should, the issue here is that it is important to carefully define the quality standards necessary and then work with contractors and suppliers to make sure what is received meets those standards.

Constantly Improve the System of Production and Service

This means all employees of the school district should strive to improve the quality of the educational program of the school district. Even classified personnel can contribute to this endeavor. Assuring that the school buildings and the grounds and landscaping are well maintained may not seem directly related to student performance, but it will certainly add to student, parent, teacher, and community pride in the schools. Such pride is an important component of any successful school program. It shows the students that the community cares enough to provide a healthy and comfortable environment in which to go to school and that the schools are an important priority.

Institute Modern Methods of Training

Virtually all school districts have training programs in place for teachers, and many have them for other employees as well. TQM recognizes the importance of such training, which should include the following:

- Training should be conducted over time; quick fixes do not improve quality.
- Training should be provided for all categories of employees, not just teachers.
- Training should be offered to work teams as a unit, and not individually. That could mean principals, teachers, and aides all receive training at the same time. From the business perspective, training should be provided to custodians, clerks, secretaries, maintenance personnel, drivers, and food service staff in groups divided by schools served, rather than by job classification.
- Adequate resources be provided for continued staff development and training.
- New technologies should be integrated into the training process as appropriate.[13]

Institute Leadership

An effective school and effective school district requires effective leadership. Hiring strong leaders is critical to the success of any venture, and schools are no exception. Therefore, it is important to develop job descriptions that clearly specify the kinds of leadership skills that are desired and important to the various positions in the school district. Top managers need to spend time carefully selecting employees. In addition, high quality staff development programs are necessary to maintain strong leadership.

Eliminate Fear so Everyone Works for the Organization

Managing through fear, where employees are constantly concerned about loosing their jobs if they do not perform, is not the best way to develop a loyal and productive staff. It is more important to assess performance using techniques described in point three as previously cited. Further, it is critical to ensure two-way communication between supervisors and their employees, and empower all employees to do their job and make suggestions for improvements. These techniques will lead to a more productive and effective organization.

Break Down Barriers Between Staff Areas

The current trend toward site-based management for schools (see chapter 23 of this volume) is one example of a way to eliminate the barriers among staff charged with the responsibility of operating a school. The business office is often looked upon as one of the biggest barriers to change in school systems. It is frequently the business office that says there are not enough funds to institute a new program or purchase much needed materials and supplies. It is important that the business office work closely with all of the schools so that teachers, principals, and community members have a better understanding of the financial factors that must be considered in such decisions. Moreover, as site-based management gives more autonomy over budget and expenditure decisions to school sites, it will be important for school business officials and their staffs to find ways to work collaboratively with school sites to support and not hamper their efforts.

Eliminate Arbitrary Numerical Goals, Slogans, and Targets

It is important to have clear goals to guide school district decision makers. However, establishment of arbitrary standards will not necessarily help improve student performance or the efficiency of the business office. This becomes more critical as the role of the business office shifts from one of control to one of support under site-based management as discussed in chapter 23 of this text.

Eliminate Work Standards and Numerical Quotas

TQM calls for evaluating employees on the processes and products to be delivered on the basis of value-added at each step, rather than using rigid quotas and standards for assessing performance. Again, as business offices shift the focus to provide more service, this point becomes more important.

Remove Barriers that Rob Employees of Pride in Workmanship

This is achieved through empowerment of employees, two-way communications, collaboration among employees and teams of employees, and developing a culture of ownership and trust. Again, the school business administrator is the staff development leader for all of the support staffs of the school district.

Institute a Vigorous Program of Education and Retraining

It is important to help all employees upgrade their skills and knowledge so that they can become more productive over time. The types of training necessary for a school business office specifically and a school district generally are outlined as discussed in point 6.

Take Actions to Accomplish the Transformation

Change is difficult. TQM cannot be implemented overnight, and will not be truly effective unless fully supported by the employees. Until a change is internalized by the staff, it will require considerably effort on the part of the school business official to convince them that this is a better way to do things, and that overall performance of their unit will improve.

These fourteen points are critical to the success of TQM. The next section of the chapter describes in detail how TQM can be implemented in a school district and describes the models and systems that are part of successful TQM programs.

Implementation of TQM in a School District

Herman[14] identifies seven steps in the operation of a TQM model for school districts. He suggests that TQM requires a commitment to high quality and customer satisfaction. Once that has been achieved, the seven steps he describes help the school district's decisionmakers accomplish their goals. The seven steps are displayed graphically in figure 16.1 and discussed in detail as follows:

Step 1: Needs Assessment

The first step to successful implementation of TQM is to conduct a needs assessment of all "customer" groups of the school district or the school. In this case, the customers include students, employees, and external stakeholders, which include parents, community members, and local businesses. One might also want to include the state legislature as a stakeholder if the state were particularly centralized or provides a substantial portion of school district revenue.

The purpose of the needs assessment is to ascertain whether there are gaps in what the school district is providing each customer group and what is currently being done. Such assessments require a great deal of effort and attention to making sure that all interested parties in the community and school district have an opportunity to participate.

Step 2: Quality Specifications

It is important to develop quality specifications for both products and services provided by the school district. As shown in figure 16.1, they should be developed for each grade level, each functional area of the school district's operations, for each school building or site within the school district, and for each subject area offered by the school district. Functional areas most likely to be under the purview of the business office include personnel, purchasing, operations and maintenance, risk management, transportation, data processing, and so on. A separate set of quality specifications for each functional area should be developed based on the identified needs of the school district and its clients and customers.

Step 3: Strategic Plans

Strategic plans are an essential part of the TQM picture. Herman suggests that a strategic plan include specifications at three different levels:

- Mega specifications that relate to society and the external environment.
- Macro specifications that relate to the total school district.
- Micro specifications that relate to individual buildings, departments, grade levels and subjects.

These strategic plans should include the long term goals of the school district and describe what is needed to achieve the school district's vision of "what it should be." Strategic planning is discussed in detail in chapter 5 of this text. TQM can be successfully implemented within a well designed strategic plan for a school district.

Step 4: Tactical Plans

Tactical plans provide the details for how the school district will achieve its strategic goals. In short, the tactical plans provide the who, what, where, why, when, and how the school district will proceed. The tactical plans should provide for immediate actions to be taken to achieve the school district's strategic goals, and describe in detail the responsibilities of each individual for meeting the goals.

Step 5: Assessment or Evaluation

It is important to have an assessment component as part of TQM. This should include a formative assessment to make sure that the program is proceeding toward the established goals, and a summative evaluation to assess the degree to which the goal has been met.

Step 6: Two-way Communication

It is essential that all parties within the school district be involved throughout the process and that open communication be maintained throughout the process. The importance of this step is shown in figure 16.1 with the communication step linking to all of the previous identified steps. It is important that individuals communicate about the success or lack of success they are having in implementing the school district's plans.

Step 7: Recycling

Planning rarely is completely successful. This is particularly true when dealing with organizations as complex as a school district. Consequently, the TQM model provides for a recycling phase or step where an opportunity is provided to participants to help resolve any problems or conflicts that are found in matching quality goals with outcomes.

Sub-Systems to the TQM Model

Within the general framework of the model described in figure 16.1, Herman identifies a number of subsystems within the model. These include an employee subsystem, a student subsystem, an external environment subsystem, and a planning subsystem. Each of these is described in the following sections.

Employee Sub-System

There are four steps in the employee subsystem. The first is selection of employees who have a commitment to TQM, which includes a belief in:[15]

- The organization's culture.
- The value of collaborative efforts.
- The importance of quality products and services.
- The need to ensure customer satisfaction.

The second step is to provide extensive training in the knowledge and skills required for each position so that employees can assist with the third step which is the development of quality specifications and operational procedures to meet those specifications. The final step in the employee subsystem

Figure 16.1
A Generalized TQM Model for School Districts

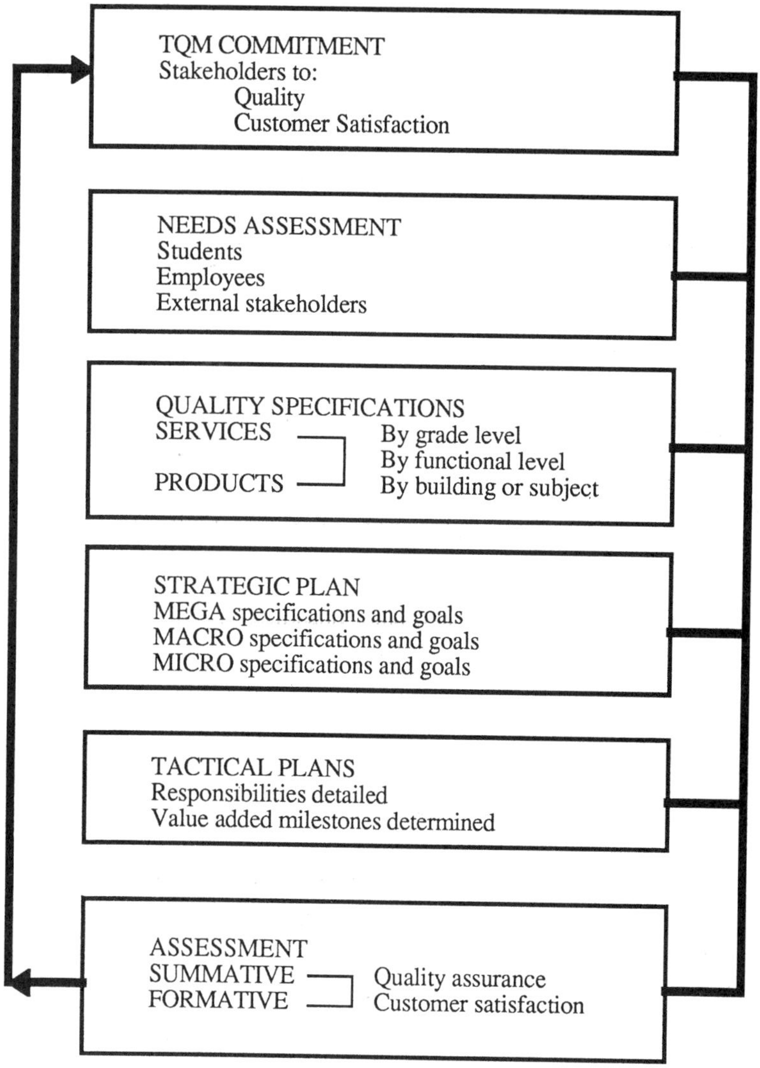

Source: J.J. Herman, "Total Quality Management Basics: TQM Comes to School." *School Business Affairs*, April, 1993.

is to assess the quality levels of the products produced and link that assessment into a recycle phase to ensure that all employees are providing the highest quality service to the organization's customers.

Student Sub-System

Herman defines four sub-steps for the student subsystem in addition to the ever present and important recycling phase. The four steps in this subsystem include:[16]

- Determination of outcome-based quality specifications for student achievement.
- Develop individual student educational quality achievement specifications.
- Provide instruction by teachers and appropriate technology to assist student learning.
- Assess the quality of learning achieved by each student.

As these statements indicate, achieving high student outcomes is not an easy task. In the vast majority of public school districts the establishment of such specifications and assessment of such standards is the most time consuming and important function of the school district's staff. Additionally, without careful planning and consideration, such topics may be so volatile as to cause great political and educational debate within a community. While on the surface, there appears to be little connection between this subsystem and the business office, as previously indicated, all are closely intertwined, and the smooth and effective functioning of the business office may have a more important impact on what goes on in the classroom than is readily apparent. At a minimum, the efficient operation of the school district's business function will help focus the maximum amount of resources toward student learning and minimize the level of resources needed for those things that have a less direct impact on student achievement.

External Environment Sub-System

This subsystem deals with the relationship between the school district and the stakeholders, generally the parents and community members living and working within the school district's boundaries. Herman suggests a four step model for dealing with this subsystem as well. The first step is a needs assessment to ascertain what the local community expects of the public schools. The second step is the establishment of the quality specifications for the services and products identified in step one. This may involve the business office more than one might suspect. Although it is likely that the community is most interested in student outcomes and preparing high school students for work and/or higher education, they may also be concerned with the appearance of the school district's schools, the way school buses are driven through their neighborhoods, and the quality of the food served in the school cafeterias. All of these functions are the purview of the business office. Consequently, there should be substantial involvement of the chief business officer and his or her staff in the design of quality specifications and the provision of these services.

The third step is a formative feedback stage where stakeholders provide input to the operation of the tactical plans. In other words, if the school buses do not run on time, this is when the chief school business official will hear about it. Finally, in step four, there is summative feedback from the customers to help determine if the strategic plan needs to be revised.

Strategic Planning Subsystem

The final subsystem in Herman's design is strategic planning. This subsystem involves the identification of needs, both from outside the school district and within the school district itself and in the subunits within the school district at the building and function level. Once the needs are identified at all three levels, the school district needs to develop a "vision." This vision should include the following components:[17]

- A high degree of customer satisfaction.
- An ever increasing quality of products and services.
- A system of continuous two way communication among all of the stakeholders.
- A collaborative and empowering environment for all customers, but especially for students and employees.

This vision forms the basis for the development of a school district's strategic plan. Ideally, the strategic plan includes both customer satisfaction goals and quality products and services. This is critical in the school business office. The strategic plan identifies how the school district will meet the needs of its clients, students, teachers, parents, and community members.

Implementation of the strategic plan is critical to the success of the school district in meeting its goals. Many of the services and products the business office provides to schools are essential to the educational program. Making sure that students have access to the instructional materials they need, that the materials are of high quality, and making them available when they are needed is important. Similarly, it is critical that the operations and maintenance staff ensure that facilities are open, clean, well kept, and accessible to all students. Finally, it is important to note that all of the functions of the personnel department are typically part of the responsibility of the chief business official. Making sure that teachers' paychecks arrive on time, and the benefits to which they are entitled are provided may have a considerable impact on morale.

An important component of the planning system is evaluation or assessment of the school district's ability to deliver on its plans. Assessments should be an integral and ongoing part of the planning subsystem. Both formative assessments designed to track the progress of the system and summative assessments designed to ascertain the degree to which the school district achieved its goals should be included.

The final part of the planning subsystem is tactical planning. It is in this stage that the detailed plans for implementing the school district's strategic plan and vision are mapped out. As previously described, these tactical plans guide the actions of staff and help direct the allocation of resources to functions in line with the school district's vision.

Summary

The last part of this discussion had provided a brief description of how TQM can be implemented in a local school district. As the discussion shows, there are many places where TQM intersects with the functions of the business office. The discussion makes it sound much easier than it is to establish a TQM management system in a school district. The last section of this chapter provides a discussion of seven common difficulties school districts can experience in establishing a TQM program.

Common Problems with TQM Implementation

James E. Law, in an article in the June, 1993, issue of *School Business Affairs*,[18] identifies seven common pitfalls or mistakes that school districts make when they try to use TQM methods. They are summarized as follows:

No Self Assessment

The first pitfall identified by Law is the failure to assess an organization's own strengths and weaknesses before setting up a TQM program. His point is that if an organization were to have internal or external problems, simply setting up a TQM program will not, by itself, solve those problems. Moreover, because TQM is a collaborative process, if all employees do not have confidence in their security, value, and ability, TQM will fail. Law suggests that if TQM were associated with a reduction in staff (a common occurrence in school districts), it will not be successful. He points out that, "People just will not contribute if doing so results in the loss of their own jobs."[19] and suggests that the solution to this problem is most likely attrition.

Top Management is Not Involved

It makes sense that without the support of top management, a new program is doomed to failure. Since TQM requires collaboration and communication at all levels of the organization, if the top management implementing TQM were not to participate, the program will fail.

Evaluation of Managers Does Not Change

If a TQM office were established in an organization and only that office is held responsible for the improvement in quality of the organization's products, then the collaborative backbone of the TQM model is negated, limiting or possible destroying the effectiveness of TQM principles. Unless all managers and departments are evaluated through TQM principles, it will not become a priority of managers.

Too Much Attention is Focused on TQM Processes Instead of Customers

While the process and tools of TQM are critical to its success, if an organization were to focus only on the internal processes of the organization, it will not succeed in improving the quality of its services and products. Law points out that these are tools to help an organization focus on what it is supposed to do and help it improve the quality of the product it provides. TQM should be thought of as an aid to improvement.

Too Much Training Too Fast

TQM is a long-term process. Not everyone in an organization is ready to commit to TQM at the same time. Individuals should only be trained in TQM processes when they will actually need it. Moreover, the training needs to include processes that both teach the tools of TQM and also motivate employees to actually use it. Training also needs to have follow-up components and refresher courses to make sure that employees continue to use the principles they have learned.

No Positive Environment

TQM assumes that a major part of the collaboration and communication will include suggestions for improving quality. If those suggestions are not taken seriously and rewarded by top management, people will not think their efforts are valued and the prognosis for TQM is poor. Management must have confidence in the employees' ability to solve problems and find better ways to do things within the organization. Moreover, if management is not responsive to these suggestions, employees will soon stop looking for ways to do things better.

Not Working to Improve with Customers and Suppliers

To work effectively, an organization must work with both its customers and suppliers if it is to ensure quality across the board. Working with suppliers (including teachers who supply teaching services to the school district) as well as with the customers (students, parents, community members, or local business firms), is critical to maintaining high quality.

In summary, as previously indicated, implementation of TQM is not simple or easy. TQM provides a method for working with the staff, students, and community to improve the quality of the school district's product—well educated children. The ultimate responsibility of every school district is to educate the next generation within our society. It takes a great deal of work and continuous monitoring to ensure that the school district is offering a high quality education, to all parts of the community.

Endnotes

[1] T.C. Valesky and J. Willis, "Workshop for Southern Association of Colleges and Schools,", Atlanta, 1993.

[2] J.J. Herman, "Total Quality Management Basics: TQM Comes to School" *School Business Affairs*, (April, 1993), 20-28 at 21.

[3] M. Milakovich, "Total Quality Management for Public Sector Productivity Improvement." *Public Productivity and Management Rev.*, (1992), 1, 19-32.

[4] J. Jablonski, *Implementing Total Quality Management —An Overview*. San Diego, CA: Pfeiffer and Co. 1991), 4.

[5] J. Vincoli, "Total Quality Management and the Safety and Health Professional." *Professional Safety*, 6, 28.

[6] M. Lynn and D. Osborn, "Demming's Quality Principles: A Health Care Application." *Hospital and Health Services Admin.* 1991, 1. 111-120.

[7] C. Quimby, et. al. "How Exactly Do You Communicate Quality?" *Quality Progress* (1991), 6. 52-54.

[8] C. Lee, "The Customer Within." *Training*. 7. 49-52.

[9] *Blueprints for Service Quality —The Federal Express Approach*. (AMA Membership Pub. Div., 1991).

[10] J.J. Herman, "Total Quality Management Basics," 20.

[11] Ibid., 24-27.

[12] Ibid., 24.

[13] See Chap. 15 of this text for specific discussion of this point.

[14] J.J. Herman, "Total Quality Management Basics," 21-22.

[15] Ibid., 21.

[16] Ibid., 22.

[17] Ibid.,23.

[18] J.E. Law, Leaping Over the Pitfalls of TQM," *School Business Affairs*, (June 1993), 21-25.

[19] Ibid., 24.

CHAPTER 17
Capital Outlay and Bonding

Introduction

The financing of educational facilities is an extremely broad and varied topic within the area of financing public education.[1] Nearly every state finances the construction of public school facilities from local sources of revenue. While the main source of revenue is the property tax, each state has its unique requirements concerning debt limits on borrowing, requirements for approval of bond issues, and numerous other administrative rules and regulations regarding procedural and technical specifications.

This chapter is intended to serve as a generic overview for school business administration students concerning the major tasks associated with financing large capital programs. There are numerous loan and grant programs, state purchases of local school district bonds, and combinations of these programs that are not addressed by this chapter. The chapter presents numerous examples to acquaint the student of educational administration with the tasks involved. It is vitally important that the examples not be used as legal or financial advice. Discussions and examples are for illustration only and do not imply the offering of any legal or financial advice by the authors.

Historical Setting

Financing capital outlay was of little concern in the early years of our country, when schools were constructed by volunteers with no thought or consideration for such things as tax rates, bonds, and bond referenda. The National Education Finance Project Study reported that 100 years ago when legislatures began enacting laws to allow the issuance of bonds for public school districts, they were primarily concerned with three things: protection of the bond purchasers, limitation of the amounts spent on construction, and limitation of the public debt. Through the early 1900s, the major responsibility of financing the public school building rested with local communities. As late as 1941, it was reported that only twelve states made some financial provisions for capital outlay and debt service. These states had "crude and inequitable distribution systems in which the principles of equalization were neither recognized or applied . . .." Some states did loan money to local districts with low interest rates.[2] After World War II, however, the number of states that contributed to capital outlay and debt service increased dramatically.[3] Possible reasons for this increase include: the fact that few school buildings were built during the Depression, population growth[4], a surplus of state revenue[5] and a growth of minimum state standards for facilities.[6]

Recent Trends

Since the 1960s, most school finance reform has involved state equalization plans. While reformers have tended to concentrate on the basic general aid distribution programs, relatively little examination has been made of the area of capital outlay. Despite what are in many instances pressing needs, school districts with low property values have great difficulty financing educational facilities in many states.[7] If, in fact, the distribution of state aid for capital outlay is distributed inappropriately, such negative aspects may occur as a direct result. On the other hand, it is critical to note that if a state were to have a reasonably equitable capital outlay formula, or include it in the overall general fund distribution scheme, such inequitable features may not necessarily exist.[8]

State aid for capital outlay and debt service grew from $78 million in 1951, to $633 million in 1970,[9] to an estimated $1.4 billion in 1979.[10] During 1990-91, state aid for capital outlay had reached approximately $4.9 billion.[11] Yet, even after this tremendous expansion of state aid, it is estimated by the authors that 80 percent of all funds for public school capital outlay and debt service will originate as local property tax. In perspective, the total capital outlay costs in 1979-80 amounted to approximately 7.5 percent of all funds expended for public elementary and secondary education in the United States.[12] For the 1990-91 school year, state aid for local school construction, as specifically funded separate from the general fund, ranged from zero in several states to a high of 11.2 percent in California.[13]

Bond Referendums

Local communities generally have strong input into the financing of school facilities. Well over 75 percent of all capital outlay costs are funded through bond sales that must be approved by a public referendum. In general, the approval rate of such referendums has declined since the late 1950s. While the rate of passage has steadily declined, the number of offerings held for public approval has also declined. Offsetting this decline has been a steady and rapid climb of par values[14] due to the escalating costs of construction and the decline of purchasing power by school districts due to inflation. During the 1990s the rate of inflation and the cost of construction has declined in significant proportions.

The successful selling of a bond issue to the public is an involved and lengthy local political process. Herman has offered the following overall suggestions regarding the selling of a local issue to a given community:

- Develop an informational videotape[15] and script that will be presented by the citizen committee members who will be accompanied by a board member and/or an administrator as they attend meetings or go to numerous coffee klatches sponsored by supporters.
- Allow time at the meetings for a question-and-answer period, or request that community members phone or write in the question they wish answered. If the same questions are asked numerous times, answers should be placed in writing and distributed throughout the community.
- Distribute information continuously by mail, by meetings, by phone call-in sessions, and, when possible, by radio and TV call-in programs. This information should be different for the different publics, but the entire community should be provided with the basic information.
- Take a straw vote poll, but make certain that the community will not see this as a negative pressure tactic.
- Make certain that everyone knows that taxpayers funds are not used to print or distribute information, or to incur any market costs related to the bond issue. Clearly state that all materials are paid for by the citizen's committee to promote passage of the issue. Also, always emphasize how the new facilities will help students.
- Obtain and publicize endorsements from the media, civic groups, and "key communicators" in the district.[16]

It should be noted that not all states require a public referendum for the sale of long-term indebtedness. Further, more and more states are allowing school districts to engage in various forms of certificates of participation, i.e., technically short-term debt obligations that are renewable by school boards at the end of each fiscal year at a given bid rate of interest over the life of the issue for the purposes of constructing school facilities. This procedure is discussed later in this chapter.

Construction and Interest Costs

The number of classrooms constructed rose from 30,900 in 1949 to 75,000 in 1967[17] and then decreased dramatically. During this period, the average expenditure per classroom rose from $32,815 in 1949 to well over $60,000 in 1980. Historically, inflation has been a major factor in the steadily increasing capital expenditures. The Cost of Commercial and Factory Building Index has increased from 100 in the base year 1972 to 201.8 in 1980, as estimated by the U.S. Department of Commerce. The Composite Construction Cost Index, which rose from 64.8 in 1963 to 2,340.2 in December 1980, represents a 255 percent decrease of purchasing power in seventeen years. The Composite Construction Cost Index rose nearly 40 percent from 163.7 in November 1977 to 228.5 in November 1980.[18] During the early 1990s these rates have significantly declined.

During approximately this same period of time, interest rates increased from 3.11 percent in the early 1960s to a high of 10.56 percent in 1981 for all general obligation bonds. In the early 1990s the interest rates for municipal bonds declined to almost 5 percent and have stabilized at approximately 7 percent at the time of this writing. Consistently, general obligation bonds have shown the highest bond ratings and, conversely, the lowest interest rates. Nonrated bonds had the next highest interest rates, while revenue bonds had the highest overall interest rate reaching 11.41 percent in 1981. In 1993 the rate was 7.5 percent. The net increase costs of bonds for school purposes sold by states have tended to be lower than those sold by minor government divisions (civil or school).[19] Over the past few years, not only has the cost of construction increased substantially, but also the cost of obtaining the money.[20]

Population Projections

In many communities there is a backlog of needed classroom construction. This backlog has continued to grow and add to the facilities needed to house new programs such as early childhood education. For example, special education will continue to need new and replacement facilities. The popular literature is full of overviews of the deteriorating conditions of the American public education infrastructure. There is perhaps no area of our society that does not have specific and significant needs in this area.[21]

Since it is projected that the school-age population will decrease, or at least remain stable during the 1990s, it may be assumed simplistically that public school districts will need fewer classrooms and thus need to build less in each succeeding year of the 1990s and beyond. The public is beginning to examine more critically the issuance of new debt because of rising construction costs dramatically juxtaposed with declining enrollments. On closer examination, however, local taxpayers may be called on to assume a greater tax burden for any of the following reasons:

- Population shifts within districts.
- Enrollment pattern among grade levels.
- Buildings that do not comply with safety and health standards.

- New educational programs.
- Changes necessary because of energy consumption.[22]

Additionally, on a state and regional level there are substantial variations in student population projections ranging from large increases to large decreases depending on migration patterns, economic conditions of the areas, and fertility rates. Some areas will need more classrooms while others will have a surplus.

Bonds

The vast majority of all money for capital outlay are raised through general obligation municipal bonds. School districts issue a bond where the payment is guaranteed by the full faith and credit of the issuer. A school district bond is a debt obligation of the school district. The general rule is that higher quality bonds pay lower yields (interest) to the investors. Generally, the bonds are payable by the ad valorem (property) taxes of the school district and in some cases other general revenue.

As a generalization, a municipal bond is any bond issued by a state or local governmental unit including a school district. Since many variations exist due to state statutes, some fiscally dependent school systems issue bonds in the name to the county commissioners, city, or town. In some instances, the state issues all school bonds. There are numerous bonds that may be issued by governmental agencies ranging from lease-rental bonds to industrial aid revenue bonds. Brief mention is made of revenue bonds, certificates of participation, and holding corporations but this is mostly for informational purposes as presently few states permit school districts to engage in such activities. While it is stressed for the purposes of this introduction that such plans are quite limited, it should be noted that more and more states are allowing such plans.

Generally, municipal bonds receive federal tax exempt status on the interest earnings.[23] They are not exempt from federal inheritance taxes. Most states and localities also exempt municipal bond interest payments from taxes for qualified issues within the respective state of the resident. Municipal bonds may be subject to state taxes in some states while other states exempt them only in the state of issuance. Such federal tax benefits are very attractive for many individuals. Presently, 75 percent of all municipal issues are owned by individuals either directly or through municipal funds.[24] If an individual were to invest in school district bonds returning a tax free interest rate, then the net interest rate would be actually higher, depending upon the person's tax bracket. For example, an individual with a 6 percent school bond and in the 31 percent tax bracket would receive a tax-equivalent yield of 8.7 percent. This same individual with a 36 percent tax bracket would receive a yield of 9.38

percent. Thus, dependent upon the individual's circumstances, one can readily see the attractiveness of such investments.

It can be argued that only relatively wealthy individuals pursue such strategies and thus the federal government actions are in fact a subsidy to the rich as well as suppression of the true cost of capital. The opposing view allows that this tax shelter is available to any person willing to invest and the suppression of capital ultimately lowers the cost of borrowing. This allows property tax rates to be significantly lower in all communities, even the ones that are relatively poorer. Needless to say, this debate will continue for some period of time.

The price of municipal bonds will vary depending on their attractiveness to investors. Often investors will gauge the cost of a twenty-year government obligation bond to that of a twenty-year treasury bond. Many professional market analysts judge the strength of municipal bonds in terms of investment to the cost of taxable treasury bonds. In this manner, it may be determined whether government obligation bonds are relatively undervalued or overvalued. Generally, if there is a decline in the supply of government obligation bonds, coupled with rising tax rates, this would tend to drive the price of governmental obligation upward. Thus, while this area is highly contingent on local needs, it is also a function of the overall laws of investment supply and demand functions.

The security of most highly rated municipal bonds has been expressed by the New York Stock Exchange as follows:

> They are considered second only to obligations to the Federal Government. During their severest test—The Great Depression— only about 1/5 of 1 percent suffered any loss of principal and less than 2 percent defaulted, that is, failed to pay interest or principal on time. In most of these cases, bondholders eventually received their interest and principal.[25]

Nearly all school bonds are serial bonds, as proscribed by state statutes, as a result of the mismanagement of term bonds during the early 1900s. Serial bonds are numbered and are payable, in accordance with state statute, typically semiannually or annually over the life of the issue. The lower numbered bonds are retired first, paying greater shares of interest during the first stages of the issue. As the interest is paid, greater portions of the principal are paid. A sample amortization schedule is included in this chapter illustrating this concept.

In a few rare instances, some states permit the issuance of term bonds. A term bond permits payment on the interest only with the principal repayment due at the end of the issue. Again, adequate bond counsel is necessary to answer specific questions concerning the permissibility of term issues.

Sinking Funds

Allowed in a few states, sinking funds permit the school district to levy a specific tax, invest the funds over a period of time, and use them for the construction of an educational facility. Critics of this process cite the erosion of the savings through inflation, numerous instances of mismanagement, poor investment practices, and changed facility priorities over the life of the tax. Proponents cite the savings of the interest rates as well as the ease of not facing a referendum. Again, it must be stressed that the use of these cumulative building funds is limited to a few states.

Current Revenues

Financing capital outlay from current revenue is often suggested by those who cite the interest savings and ease of administration. The use of money for capital outlay projects does not provide for leverage within the budget. Assuming maximization of the school budget, most money should be allocated to the direct learning activities of the schools. It is rare that the private sector finances its own internal growth from earnings. In like manner, school districts must issue bonds due to:

- Lack of cash flow or resources to tax.
- The attempt to maintain the tax rates on a reasonably consistent basis.
- The principle of intergeneration states that the school facility will be used by several generations of residents, and, therefore, it is improper for one generation to assume the entire burden.
- If the assumption is made that there will be some inflation within our society, the value of today's money will be less in the future in purchasing power; hence, simple rate of return on investment schedules may be determined reflecting that, over time, the building is being paid off with "cheaper" money.

Holding Corporations/Certificates of Participation

The term holding corporation includes all arrangements by which a public agency leases a turnkey project and escapes the requirements for public bond referenda. In a few states, school districts are allowed to form holding corporations which, in actuality, are lease-rental arrangements for the construction and purchase of educational facilities. Essentially, states allow school districts to enter into lease-rental agreements with a public or private corporation in accordance with specific statutes. Such agreements usually specify that the school becomes the property of the school district when it is paid for in full. In the few states where such arrangements are allowed, the indebtedness under these arrangements is not considered against the constitutional or statutory limitations of bonded indebtedness. The general rule of law, while not specifically applicable to any state,[26] is that three overall conditions must be met for a lease obligation to be acceptable in terms of not incurring long-term bonded indebtedness:

- The lease or other agreement must be entered into in good faith.
- It must not create an immediate indebtedness for the aggregate installments provided for.
- It must confine liability to each installment as it falls due that year in consideration for the actual benefit of the facilities or equipment made available that year.[27]

Generally, these holding corporations consist of a group of local citizens who form a corporation for the sole purpose of building a new facility. The holding corporation acquires the revenue under the auspices of the school district as a nonprofit activity. Again it must be stressed that the employment of legal and financial counsel is absolutely essential, as such activities are limited to a few states.

A more recent variation of this is the concept of certificates of participation. This arrangement is similar to those previously described. Bonds are sold to the public, without a public referenda, with the technical provision that the debt must be renewed by the school board every year of the issue. Thus, it is not considered to be a long-term debt and is not considered against the overall percentage of allowable debt in relation to the assessed valuation of the school district. The major drawback is that the year-to-year feature of the arrangement must be accounted for in the bidding of the bonds and such arrangements carry higher interest rates for this increased risk.[28] However, this negative feature may be somewhat offset by the issuance of municipal bond insurance against this potential occurrence.

As a broad generalization, certificates of participation represent fractional, undivided interests in basic rent payments. The school district has the option to purchase the facility, generally for a nominal fee of one dollar, once these payments are made in their entirety. The interests of the investors are secured through a Certificate Trust Agreement. The issue may be sold in order for the school district to, among other things, acquire land, construct and equip school facilities, fund reserve funds, and pay the expenses for the overall issue. Generally, the investors are registered in book entry form meaning that the fiscal agent registers the owners of the bond for payment and tax purposes. Thus, the owners do not receive the bonds directly in a book entry format. Further, as a general rule, such lease obligations are not considered to be a general obligation or a pledge of the faith and credit of the school district or the state in repaying this debt. Notwithstanding these limitations, many school districts do purchase private insurance to insure that the basic rent payments will be made to the investors.

Bond Ratings

The investment community, which requires some form of uniform standard by which to judge the attractiveness of investments, has established a ratings system. Bond rating agencies are extremely important in protecting the investor's money as well as for the overall organization and comparability of issues. General obligation issues of any substantial size are rated generally by a private agency.

The primary national bond ratings agencies are Moody's Investors, Standard and Poor's, and Fitch Investors. Higher ratings represent better issues. It is important to note that the listing does not make the issue, but rather the issue makes the rating. With very few exceptions, the higher the rating, the lower the interest rate. Since lower ratings reflect greater risk, school districts are willing to pay higher interest rates to those who assume the risk.

Not all school districts' bonds are rated. For example, if the issue were to be sold to local investors, i.e., a local bank, rather than being rated, the bank officials may request certain financial information directly from the school district. Additionally, below certain dollar amounts, the rating agencies do not issue the ratings (e.g., Moody's does not issue ratings for general obligation bonds below $1.5 million).

While each agency is different with respect to rating structure, the ratings are based on the soundness of the investment to the lender. The following are explanations provided by Moody for its municipal bond rating system:

Aaa

Bonds rated Aaa are judged to be of the best quality. They carry the smallest degree of investment risk and are generally referred to as "gilt edge." Interest payments are protected by a large or exceptionally stable margin, and principal is secure. While the various protective elements are likely to change, such changes as can be visualized are most likely not to impair the fundamentally strong position of such issues.

Aa

Bonds rated Aa are judged to be of high quality by all standards. Together with the Aaa group they comprise what are generally known as high grade bonds. They are rated lower than the best bonds because margins of protection may not be as large as in Aaa securities, fluctuation of protective elements may be of greater amplitude, or there may be other elements present that make the long-term risks appear somewhat larger than in Aaa securities.

A

Bonds rated A possess many favorable investment attributes and are to be considered as upper-medium-grade obligations. Factors giving security to principal and interest are considered adequate, but elements may be present that suggest a susceptibility to impairment sometime in the future.

Baa

Bonds rated Baa are considered as medium-grade obligation; e.g., they are neither highly protected nor poorly secured. Interest payments and principal security appear adequate for the present but certain protective elements may be lacking or may be characteristically unreliable over any great length of time. Such bonds lack outstanding investment characteristics and, in fact, have speculative characteristics as well.

Ba

Bonds rated Ba are judged to have speculative elements; their future cannot be considered well-assured. Often the protection of interest and principal payments may be very moderate, and thereby not well safeguarded during both good and bad times. Uncertainty of position characterizes bonds in this class.

B

Bonds rated B generally lack characteristics of the desirable investment. Assurance of interest and principal payments or of maintenance of other terms of the contract over any long period of time may be small.

Caa

Bonds rated Caa are of poor standing. Such issues may be in default or there may be present elements of danger with respect to principal or interest.

Ca

Bonds rated Ca represent obligations that are speculative in a high degree. Such issues are often in default or have other marked shortcomings.

C

Bonds rated C are the lowest rated class of bonds, and issues so rated can be regarded as having poor prospects of ever attaining any real investment standing.

Con. . .

Bonds for which the security depends upon the completion of some act or the fulfillment of some condition are rated conditionally. These are bonds secured by:

- Earnings of projects under construction.
- Earnings of projects unseasoned in operating experience.
- Rentals that begin when facilities are completed.
- Payments to which some other limited condition attaches. Parenthetical rating denotes probable credit stature upon completion of construction or elimination of basis of condition.

Those bonds in the A and Baa groups that Moody's believes possess the strongest investment attributes are designated by the symbols A1 and Baa1. The vast majority of school district bonds are in the "investment-grade" categories, which include AAA, AA, A, and BBB.[29]

Standard and Poor's system of ratings have different symbols that reflect the soundness of investments. Briefly, Standard and Poor's system ratings are as follows:

AAA	Prime
AA	High Grade
A	Upper Medium Grade
BBB	Medium Grade
BB	Lower Medium Grade
B	Some Degree of Speculation
CCC	Greater Degree of Speculation
CC	Greatest Degree of Speculation
C, DDD, DD	Various Degrees of Default

Each rating agency charges a fee for rating an issue. Charges vary from a few thousand dollars for a small and simple issue to a high of several thousand for large and complex issues. The rating for a particular issue is normally in effect until the issue is retired. Each firm releases the rating to the *Daily Bond Buyer,* in order for the issue to be examined by the investment community. Additionally, each firm publishes this information and makes it available to the investment community.

Additionally, many states place the full faith and credit of the state behind all school district general obligation bonds. These bonds require state approval and must meet all state prescribed standards. Generally, these issues take on the rating of the state itself which is generally higher than the individual school district would have obtained.[30]

It should also be noted that not all issues are rated. Generally, small issues are not rated. Many issues are sold, pursuant to state statutes, in public sales without a rating. Small school districts issuing notes for rather small amounts often find it advantageous to sell the notes directly to a local financial institution. Usually the financial institution will resell the issue to local investors.

An unrated bond may not be a poor investment. However, it should be noted that unrated bonds issues are growing at a rapid rate. For example, in 1980 there were 1,626 unrated issues. By 1991 there were 3,100 such issues. During 1991, the latest available year, 28 percent of all municipal issues were unrated. The default rate for unrated municipal issues during the first two years of issuance averages 2 percent. However, the average time of an unrated municipal obligation default is nearly six years. Thus, unrated bonds tend to carry significantly higher risk and commensurate interest rates.

Legal Services

Every school district must employ competent legal counsel in issuing a bond. The school business administrator must, without exception, consult closely with a law firm that has extensive experience with the issuance of public bonds. The bond attorney functions as the legal adviser to the board of education and ultimately to the investor. The attorney is concerned with the mechanics of the issue as well as the overall thrust of the issue in relationship to the legal thinking of the courts as well as the Internal Revenue Service (IRS) regarding the issue's appropriateness.

Most experienced bond firms maintain the necessary forms and documents normally via software. Fees vary from state to state based on the complexity of the state statutes and the time and costs incurred in examining all the necessary information. Bond counsels usually employ a quality school bond consultant to assist with the transaction. A small, privately sold issue may result in very modest legal fees. A large, complex undertaking, such as a formation of a holding corporation for a special education cooperative involving many school districts, might result in fees of several hundred thousand dollars.

Moody's has offered the following advise regarding the selection of a legal adviser:

- The importance of competent legal guidance from the very start of your issue's preparation cannot be overemphasized. Once you decide on an improvement program and choose to issue bonds instead of paying as you go, your first step should be the selection of an attorney skilled in handling municipal bonds. In general, the more experienced the attorney, the more money he [or she] will save you as a bond issuer and the less trouble you will have in bringing your bond issue to market.

- Your attorney drafts or approves the original bond resolutions and ordinances, the notice of the bond election, and the form of the ballot if an election is required under local law. Once voters have approved the issue, he [or she] reviews the preparation of a possible prospectus, financial advertisements, and the entire "package" presented to underwriters for bidding. After the bonds have been sold, printed, paid for and delivered, he [or she] renders a final legal opinion. Bond buyers today rarely accept delivery of bonds if such an opinion is not furnished.
- Perhaps your public unit's own counsel or another local attorney may be able to do as fine a job as a specialist, but it is better not to take the chance. Just a tiny technical error may require that you re-advertise the offering. Also, a bond accompanied by what underwriters call a "local" opinion generally sells at a lower rate (i.e., higher interest cost) than a bond of the same public unit that carries a "marketable opinion." A marketable opinion is simply an opinion by a legal specialist whom underwriters and investors recognize as of proven competence.
- Dealers want to know before they bid that a final legal opinion will be furnished with the bonds when delivered.
- If, for some reason, your community decides not to secure a legal opinion, it must allow the bidders to name their own attorneys. This, quite naturally, will be at the community's expense. Bidders will subtract it from the price they offer for the issue.[31]

Moody's offers a specific checklist for the issuance of municipal bonds that is applicable to school business administrators involved in a school bond issue. Some of the items on the checklists include:

- Retained an attorney with extensive experience in the laws of your state that affect municipal bonds.
- Retained an attorney who has a record of "marketable opinions."
- Turned over to the attorney all proceedings preliminary to those in which he, . . .[or she] . . ., participated.
    - Consulted the attorney in all phases of issue, including:
    - Election (if any) notices.
    - Election (if any) procedures.
    - Method of determining the best legally accepted bid.

Financial Advisement

Increasingly over the last few years, governmental units have been turning to financial advisement firms for expertise in issuing bonds. School districts have often found that the independent financial adviser is an extremely cost-effective means to increase the return on investments. School districts rarely retain on staff a person skilled in bond market analysis, preparing the prospectus, advanced refunding, and reinvestment of proceeds. A financial adviser may be particularly helpful in answering the following questions for a school district:

- Is the financial undertaking feasible?
- Are the economic resources sufficient to support a marketable financial plan?
- What are the various financial options?
- What is the most marketable, least costly debt retirement program?
- What are the most efficient issue sizes, maturity schedules, prepayment options, and bidding provisions for the proposed project?
- What covenants will protect the interests of the investor and the school district?
- What is the optimum timing for the new issue?
- What should be done to make an accurate presentation to the rating agency?
- How can the school district create maximum demand for the issue?
- How can investments of the proceeds maximize the use of temporarily idle funds?
- What are the future needs, based on the economy and growth of the school district?[32]

The independent financial adviser often times has the expertise to examine the school district's financial position and to represent the district in the bond market as well as to handle the reinvestment schedules for maximum efficiency to the district. These services are often necessary for the small school district that has rarely sold bonds and does not expect to sell additional bonds within the near future. The financial adviser should be willing to provide certain services such as:

- Consult with architects, engineers, and bond attorneys in the preparation of the financing plan, resolutions, ordinances, and certificates.
- Assist in the timely publication of official documents.

- Coordinate the formation of an appropriate holding corporation/building authority (where legally permissible).
- Answer any questions, if necessary, of various governmental authorities or school community groups.
- Prepare and distribute the official prospectus to bidders in the major money markets.
- Prepare and present the appropriate materials for optimum credit ratings.
- Arrange and conduct informational meetings for the investment community.
- Conduct the sale
- Coordinate the printing of bonds and closing documents for timely delivery to the purchaser.
- Analyze cash flow and manage the investments of bond proceeds.[33]

The financial adviser should tailor the services to fit the needs of the school district. For example, the needs and desires of a small school district that may face the issuing of general obligation bonds once every thirty years are obviously different than that of a major urban school district that is constantly entering into long-term debt as well as retiring debt issues.

The worthiness of hiring a firm to serve as the financial adviser in issuing general obligation bonds or entering into any form of debt cannot be looked upon lightly. Employing such a firm may prove to be quite cost effective for many smaller school districts. Such firms must possess adequate professional errors and omissions insurance in order to protect local property taxpayers. In most states a bond issue may be sold through a process of negotiated sales, i.e., no public bid process is used. The underwriter simply sells the issue at the best price the firm thinks appropriate.

The Prospectus

The prospectus, or official statement, is the document that communicates all the needed information concerning the school district to the financial community. It is, by necessity, a complete description of the community, i.e., the school district and the factors that bear on the future economic outlook of the community.

An outline of a Prospectus follows each section. The details will change according to each situation and the applicable state law. Under no circumstances should a school business administrator or student of educational administration view these examples as being applicable to his or her situation. Nor, is it to be interpreted as legal or financial advice.

Statement

The official statement is usually found at the beginning of the prospectus. It summarizes and highlights the issue. An example of a twenty-year general obligation bond statement appears as follows:

OFFICIAL STATEMENT

(NEW ISSUE)

SCHOOL BUILDING BONDS

SCHOOL DISTRICT OF________________

Bonds Dated __________
Principal Due__________

These Bonds are being issued pursuant to ______________ for the purpose of paying for construction, erecting and equipping ___________.

The Bonds will be general obligation of the District for which the full faith and credit and unlimited taxing powers are pledged.

The bonds are being issued in registered for, $_______ denominations. Interest is payable _______, ______ and semiannually thereafter.

Bids must be for not less than _________ and accrued interest on the total principal amount of the issue. Bids must be accompanied by a certified or cashier's check in the amount of ___________ payable to the order of the School District Treasurer.

The Bonds will mature __________________ in the amounts and years as follows:

1995	____________	2002	____________	2009	__________
1996	____________	2003	____________	2010	__________
1997	____________	2004	____________	2011	__________
1998	____________	2005	____________	2012	__________
1999	____________	2006	____________	2013	__________
2000	____________	2007	____________	2014	__________
2001	____________	2008	____________		

> Bonds maturing in the years _____________ are subject to redemption on ________,______ and any interest payment date thereafter at par plus accrued interest.
>
> The Bonds are offered subject to the legal opinion of
>
> ____.
>
> In the opinion of Bond Counsel, interest on the Bonds is exempt from present Federal Income Taxes.
>
> Bid Opening & Award Time ______________________________
>
> Rating: An application has been made to ____________________
>
> The Date of this Official Statement is_____________________

Purpose

The statement of purpose would usually follow the Official Statement. Depending on the complexity and various state legal requirements, it may be simple or extremely detailed. An example of a simple statement might be, "The bonds will be issued to finance the cost of improvements to school buildings." However, more complex circumstances and legal requirements may warrant a more detailed explanation, for example:

> These Bonds are being issued pursuant to ____________ for the purposes of building and equipping an elementary school complex and a swimming pool addition. The new school will replace the _____________ Elementary School that was gutted by fire on _____________. The School Board unanimously voted on ___________ to accept an insurance settlement of _______. These funds, in addition to the proceeds of this issue, comprise the building program.
>
> Students displaced from the burnt facility were housed in various rural structures, churches ,and the City's Armory. Additional cost to the District is per month for rental of _________ structures.
>
> On _______ the electors of the District authorized by a vote of _____ "yes" to _______ "no" to borrow an amount not to exceed ______ of its general obligation bonds for the purpose of building and equipping an elementary school complex; and by a vote of ______ "yes" to _____ "no" to borrow an amount not to exceed _____ of its general obligation bonds for the purpose of building and equipping a swimming pool addition to the elementary school complex.
>
> The new school will be built on land already owned by the District. An architectural firm has been retained. At the present time, the School District intends to sell the old site and fire damaged building

> and use the proceeds to reduce outstanding indebtedness. Occupancy of the new school is anticipated to occur in the fall of _______.

Date, Type, Denomination, and Maturities

The dates, types, and denominations are shown in the following example. This example reflects a non-callable issue.

> The Notes will be dated as of _______ will be issued as registered negotiable investment securities, will be in the denomination of ________ each unless designated otherwise by the purchase within 48 hours of award of sale, and will mature serially, without option of prior payment, on ________ in the following years and amount:

PAYMENT	PERIOD	PRINCIPAL	INTEREST	TOTAL
1.	12/2/94 12/1/95		$593,750	$593,750
2.	12/2/95 12/1/96	$115,000	$475,000	$590,000
3.	12/2/96 12/1/97	$125,000	$464,075	$589,000
4.	12/2/97 12/1/98	$140,000	$452,200	$592,200
5.	12/2/98 12/1/99	$150,000	$438,900	$588,000
6.	12/2/99 12/1/00	$165,000	$424,650	$589,650
7.	12/2/00 12/1/01	$180,000	$408,975	$588,975
8.	12/2/01 12/1/02	$200,000	$391,875	$591,875
9.	12/2/02 12/1/03	$215,000	$372,875	$587,875
10.	12/2/03 12/1/04	$240,000	$352,450	$592,450
11.	12/2/04 12/1/05	$260,000	$329,650	$589,650
12.	12/2/05 12/1/06	$285,000	$304,950	$589,950

13.	12/2/06	12/1/07	$315,000	$277,875	$592,875
14.	12/2/07	12/1/08	$345,000	$247,950	$592,950
15.	12/2/08	12/1/09	$375,000	$215,175	$590,175
16.	12/2/09	12/1/10	$410,000	$179,550	$589,550
17.	12/2/10	12/1/11	$450,000	$140,600	$590,600
18.	12/2/11	12/1/12	$490,000	$ 97,850	$587,850
19.	12/2/12	12/1/13	$540,000	$ 51,300	$591,300
TOTAL			$5,000,000	$6,219,650	$11,219,650

Previous data are for a twenty-year general obligation bond at 9.5 percent over the life of the issue. Note the delay of the payment of the principal. This example reflects an even payment period over the life of the issue. In many states this even payment schedule is required by statute.

Paying Agent

It is customary that the successful bidder be allowed to reasonably designate the depositor to which the future payments shall be made. The following statement reflects this practice and the conditions of specificity:

> Principal and interest will be made payable to any suitable bank designated by the successful bidder within 48 hours after award of sale, subject to approval by the School Board, and the District will pay the reasonable and customary paying agency charges. If the paying agent is not selected within 48 hours after the award of sale, or if it is not satisfactory to the District, the District will select the paying agent.

Interest Payment Dates and Rates

The prospectus clearly outlines the method of bidding the fractions and the maximum amount that the district will entertain in the cost of borrowing funds. In order to make the bidding even more competitive and ease ascertaining the lowest bid, it is common to see a specified range for the bids (e.g., 1.5 percent). It is customarily stated that the bonds be bid on face value, i.e., for the full amount plus interest as opposed to a discount, which is for less than the face amount, the difference is the discount reflecting the interest. A statement to that effect is as follows:

> Bidders are requested to name the interest rate or rates the bonds are to bear in multiples of one-twentieth of one percent, but not exceeding _______ percent per annum. The difference between the highest and the lowest rate bid shall not exceed _______ percent. There will be no limit on the number of rates of interest that may be specified. One rate of interest shall apply to all bonds of the same maturity. Interest payable on any bond on any interest payment date shall be represented by payment schedule, and no bid will be considered for less than par and accrued interest for all of the bonds.

Delivery

Delivery deadlines and language often vary; however, a time line, cost (if any), and other dates as necessary may be included. Three examples of a delivery statement follow:

Example 1. The bonds are expected to be ready for delivery within 30 days after the sale. Delivery anywhere in the continental United States will be made at the expense of the school district. Payment for the bonds must be made in federal funds or other immediately available funds.

Example 2. The district will furnish without cost to the purchaser and within 60 days after award of sale the printed and executed Notes, the legal opinion of ______, and the usual no-litigation certificate. Delivery will be made without cost to the purchaser at the place of the purchaser's choice. Payment for the Notes must be made in funds immediately available to the District on the date of delivery.

Example 3. Within 40 days of the date of the sale, and without cost to the purchaser, the District will furnish and deliver at any point in the continental United States, the printed and executed certificates, and the opinion thereon of _______, to the effect that the certificates of indebtedness are valid and legally binding obligations of the District and, unless paid from other sources, are payable from ad valorem taxes levied upon all of the taxable property within the District without limitation as to rate or amount.

Legal Opinion

Legal opinions will vary dependent on the state and firm offering the opinion. If an unfavorable preliminary opinion were to be issued, the sale should be canceled or delayed by the school business administrator until all legal qualifications have been fully met. Generally, the final legal opinion is not issued until after the sale. These examples are given to familiarize the school business administration student with legal opinions. Under no circumstances should these examples be considered legal advice. Additionally, legal opinions will vary depending on the legal firm and the applicable state statutes.

EXAMPLE 1. We have acted as Bond Counsel in connection with the issuance by School District of ____, State of_______, of its Certificates of Indebtedness, dated as of ______, in the aggregate principal amount of ______.. The certificates bear interest at the rate of _____ percent (_____%) per annum and are payable as to both principal and interest on _____, without option of prior payment. The Certificates bear serial numbers and are in denominations as follows:

Serial Numbers	Denominations, Each
#############	________________

Both principal and interest are payable at ________, in__________, ______.

For the purpose of giving this opinion, we examined certified copies of certain proceedings taken by the School District in the Authorization, sale and issuance of the Certificates, including the form of the Certificates, and certain other proceedings and documents furnished by the School District. From our examination of such proceedings and other documents, assuming the genuineness of the signatures thereon, and the accuracy of the facts stated therein, and based upon laws, regulations, rulings and decisions in effect as of the date thereon, in our opinion:

1. The Certificates are in due form, have been executed and delivered and are valid and binding general obligations of the School District in accordance with their terms to which the full faith and receipts of the School District are pledged.

2. The Certificates are issued in anticipation of the receipt of taxes levied for its _____ fund for collection in _______, of which an amount equal to the principal and the interest on the Certificates has been irrevocably appropriated to a special debt service fund for the payment of such principal and interest. All of the taxable property within the School District is subject to the levy of additional taxes if needed for the payment of the Certificates, without limitation as to rate or amount.

3. The Certificates are exempt from taxation as property by the State of _______ and its subdivisions and municipalities and the interest to be paid on the Certificates is not included in gross income of the recipient for United States or State of _______ income tax purposes, except ________ corporate franchise taxes measured by net income.

Dated at _______,_______,_______,_______.

EXAMPLE 2. The bonds are dated _______ are of $_______ denomination; are numbered 1 to ______, inclusive; mature serially on _______ of each of the years and in the amounts as set forth below and bear interest payable ________ and semiannually thereafter on and of each year at the rates per annum as follows:

Year	Amount	Interest Rate
####	$$$$$$	%%%%%%%%%%

Bonds maturing on and after _____,_____ are subject to redemption prior to maturity as a whole, or in part in inverse order of maturity and within each maturity by lot, on _____,_____ and on any interest payment date thereafter at the principal amount thereof and accrued interest to the date of redemption.

We have examined the documents which we deem pertinent to the validity of said bonds, including the certified record evidencing the authorization of said bonds by the City Council of said city. On the basis of such examination, we are of the opinion that said bonds have been lawfully authorized and issued under the laws of the State of _____; that they are the lawful and enforceable obligations of said city in accordance with their terms; that they are payable from taxes to be levied on all taxable property in said city, without limitation as to rate or amount; and that executed bond number one of said issue, which we have examined, is in proper form.

An example of a tax exemption statement from legal counsel, pursuant to the current IRS regulations generally entails quite detailed information with certain caveats that will be noted by the investment community. An example of such is as follows:

In the opinion of ___________, Bond Counsel, under existing law, the Bonds and the income thereon are exempt from taxation under the laws of the state of _____, except estate taxes and taxes imposed by §§ ___________ on interest, income or profits on debt obligations owned by corporations, banks and savings associations,: and the interest on the Bonds is excluded from gross income for federal income tax purposes under Section 103(a) of the Internal Revenue Code of 1986, as amended (the "Code"). and is not treated as an item of tax preference under Section 57 of the Code for purposes of the federal alternative minimum tax imposed on individuals and corporations. An opinion to those effects will be included in the legal opinion of Bond Counsel. Bond Counsel will express no opinion as to any other tax consequences regarding the Bonds.

The opinion on the tax matters will be based on and will assume the accuracy of certain representations and certifications and compliance with certain covenants of the District to be contained in the transcript of proceedings and which are intended to evidence and assure the foregoing, including that the Bonds are and will remain obligations the interest on which is excluded from gross income for federal income tax purposes. Bond Counsel will not independently verify the accuracy of the certifications and representations made by the District.

The Code prescribes a number of qualifications and conditions for the interest on state and local government obligations to be made and to remain excluded from gross income for federal income tax purposes, some of which, including provisions for potential payments by the District to the federal government, require future or continued compliance after issuance in order for the interest to be and to continue to be so excluded from the date or issuance. Noncompliance with these requirements could cause the interest on the Bonds to be included in gross income for federal income tax purposes and to be subject to federal income taxes retroactively to the date of their issuance. The District will covenant to take actions required of it for the interest on the Bonds to be and to remain excluded from gross income for federal income tax purposes, and not to take any actions that would adversely affect that exclusion.

Code provisions applicable only to corporations (as defined for federal income tax purposes), which impose an alternative minimal tax on 75 percent of adjusted current earnings (for taxable years that began before 1990, 50 percent of the excess of adjusted net book income) over other alternative minimum taxable income, may subject a portion of the interest on the Bonds earned by certain corporations to that corporate alternative minimum tax. In addition, under the Code interest on the Bonds may be subject to an environmental tax imposed on certain corporations for taxable years which begin before 1992, a branch profits tax imposed on certain foreign corporations doing business in the United States, and a tax imposed on excess net passive income of certain S corporations.

Under the Code the exclusions of interest from gross income for federal income tax purposes can have certain adverse federal income tax consequences on items of income or deductions for certain taxpayers, including among the financial institutions certain insurance companies, recipients of Social Security and Railroad Retirement benefits, and those that are deemed to incur continued indebtedness to acquire or carry tax-exempt obligations. The applicability and extent of these or other tax consequences will depend upon the particular tax status or other tax items of the Registered Owners of the Bonds. Bond Counsel expresses no opinion regarding those consequences.

From time to time, there are legislative proposals in Congress, which, if enacted, could alter or amend the federal tax matters referred to or adversely affect the market value of the Bonds. It cannot be predicted whether or in what form any such proposal might be enacted or whether, if enacted, it could apply to bonds (such as the Bonds) issued prior to enactment.

Types of Bid and Award

Certain states have a prescribed manner in which bids must be let. A few states specify that all bids must reflect one interest rate for the entire issue. Generally, the bidder may specify any combination of rates. The comparison of multiple rates is relatively simple. Good faith checks, based on statute and custom vary from state to state. A good faith check generally is required for 1 to 5 percent of the par value of the issue and is included in the bid documents. The following reflects a multi-interest rate bid, the method of calculation, and the manner of award.

Sealed bids may be mailed or delivered to the undersigned and must be received prior to the time of said meeting. Each bid must be unconditional except as to legality, which may be conditioned upon the legal opinion specified previously, and must be accompanied by a cashier's or certified check or bank draft in the amount of _____ payable to the School District Treasurer, to be retained by the School District as liquidated damages if the bid is accepted and the bidder fails to comply therewith. The bid authorizing the lowest net interest cost (total interest from date of certificates to maturity less any cash premium) will be deemed the most favorable. No oral bid, and no bid of less than par will be considered, and the School Board reserves the right to reject any and all bids and to waive any informality in any bids.

Since the purpose of the prospectus is to attract the highest possible bond rating and the lowest interest rate for the school district, full and honest disclosure of all pertinent financial information is absolutely necessary. Failure to provide full and complete disclosure may drive up the interest rate or preclude any bids. Specifically, the school district is legally required to disclose certain facts as would all private issues.[34] The following data are representative of the types of information normally found in a prospectus:

VALUATIONS

	1995/96 Equalized	1995/96 Assessed
"A" County	$150,092,698	$66,838,503
"B" County	29,321,180	4,754,233
Prior Years:		
1993/94	$154,815,000	
1992/93	131,405,800	
1991/92	116,830,900	
1990/91	99,564,700	

Analysis of Current Equalized Valuation of a School District

In School District	1995 Equalitzed	% of District Valuation
City of "A"	$43,960,940	24.50%
Township of "B"	58,628,070	32.68
Township of "C"	14,873,790	8.29
Township of "D"	11,766,652	6.56
Township of "E"	1,627,396	.91
Township of "F"	6,168,500	3.44
Township of "G"	6,561,750	3.66
Township of "H"	6,505,580	3.62
Township of "I"	5,550,700	3.09
Township of "J"	23,770,480	13.25
	$179,413,878	100.00%

General Obligation Direct Debt—As of January 1, 199-

Purpose	Dated	Rates	Maturities	Principal Amount Outstanding
School Bldg. Bonds	4/1/77	3.25%	4/1/1997	$165,000

Various states have limitations on the amount of debt that may be incurred by school districts within those states. This limitation varies among the states and must be carefully ascertained. Generally, the total outstanding principal of all pure general obligations bonds, as opposed to certificates of participation, is limited to a percentage of the assessed valuation. The prospectus should reflect such a condition, for example:

DEBT LIMITATIONS

Pursuant to ________ Constitution and Statutes, the School District of ___________ is subject to a debt limitation of percent of the value of taxable property located herein as assessed for state purposes.

Legal Debt Limit for District

(___% of Assessed Value)$$$$$$$$$$

Debt Outstanding as of_______ $$$$$$$$$$

Current Debt Capacity
$$$$$$$$$$

It is also important to realize that bond ratings reflect total outstanding debt from all forms of governmental indebtedness. That is, the taxpayers who reside within the district are also subject to additional taxes. Despite the fact that a given school district may have little, if any, debt, the outstanding debt of other governmental units may reflect adversely on the school district. In those states where school systems are part of and co-mingled with county/city forms of government, the problem of overlapping debt is not reflected. The debt of the school district is reflected within the overall governmental structure. A brief example of an overlapping debt schedule is reflected as an example of this display:

OVERLAPPING DEBT

Valuation in District Net General Obligations

Total 95/96

	Assessed Value	Amount
"A" County	$4,588,800,295	$472,295,567

"B" County	163,280,902	1,292,263
City of "C"	475,172,286	472,295,567
City of "D"	20,947,286	1,292,263
I.S.D. No. 1	2,854,868,356	473,577,830
Metro Council	8,174,627,264	473,577,830
Metro Transit	7,580,857,145	473,577,830
Metro Airport	8,174,627,264	473,577,830

	Percentage	Debt	Net G.O. Debt to District
"A" County	10.29%	$41,450,000	$4,265,205
"B" County	.79	780,000	6,162
City of "C"	99.39	29,454,984(1)	29,275,309
City of "D"	6.16	6,020,000	370,832
I.S.D. No. 1	16.58	16,300,000	2,702,870
Metro Council	5.79	36,300,000(2)	2,101,770
Metro Transit	6.24	9,850,000	614,640
Metro Airport	5.79	0	0
			$39,336,458

I. Excludes $4,015,000 Utility bonds and $525,000 State-Aid Street Bonds which are G.O. backed but supported primarily by systems revenues and state aids, and $1,080,000 recreational Facility Bonds payable from gross revenues. Also less $16,000,000 sinking funds and $5,500,000 utility revenue supported portion.

2. Excludes general obligation where bonds of the Metropolitan Council refunded January 1, 19—, and supported by an escrow fund established from refunding bond proceeds. Also excludes $945,000 Tax Anticipation Certificates due January 1, 19—. Excludes 487,000,000 of sewer bonds issued by the Metropolitan Council and metropolitan sewer system, which debt has been assumed by the Metropolitan Council. These bonds are supported from sewer service charges to the governmental units within the metropolitan

sewer system. Includes Metropolitan Council bonds for parks and open space.

3. Excludes $175,000,000 current debt service of the Metropolitan Airports Commission, which is required to be fully supported by revenues from charges for the use of airports by airlines and others. Any tax levies for deficiencies must be reimbursed from increase charges.

The taxes per $1,000 of assessed value is usually found within the prospectus for the past several years. In many states the taxes are reflected on the basis of dollars and cents per $100 of assessed valuation. An abbreviated example of an assessed valuation table with the necessary explanatory statements are shown:

TAXES PER $1,000 OF ASSESSED VALUE (1)

	1998/99	1997/98	1996/96	1995/96
Special Sch. Dist. No. 1	$49.529	$49.426	$50.220	$50.463
City of "A"	43.421	48.564	48.571	43.711
Misc. (2) "B"	3.493	3.052	4.416	3.508
County	29.352	29.050	27.787	26.436
	$125.795	$130.029	$129.994	$124.118

1. Before property tax relief credit on homestead property at 45 % of taxes (except County and City debt service taxes) to a maximum of $325.

2. Miscellaneous levies include those for the Metropolitan Council, Metropolitan Transit Commission, Housing and Redevelopment Authority, Mosquito Control Board, Park Museum, and Watershed District No. 1. In many states the mill rate is computed on the basis of several funds that make up the total school district.

In those states that employ the use of multiple levies, a detailed sub-fund mill tax rate is shown such as the following example:

DETAIL OF 199_/9_ SCHOOL DISTRICT MILL RATE

General Educational Program	31.002
Maintenance	3.808
Transportation	1.444

Additional School Services	1.772
Community Service	.570
Municipal Employees Retirement	2.596
Capital Outlay	3.398
Umemployment Compensation	.342
Debt Service	6.909
Statutory Operating Debt	1.690
	53.531

Normally, a brief explanation of the method of school tax collection is included in the prospectus. The following is an example:

> The tax year in ________ is January 1 to December 31 and taxes are collected by the county treasurer. The sequence of events in the taxation of property begins with the certification of the property tax budget levy to the county auditor in the fall. The county auditor then calculates the mill rates and spreads the taxes designed to meet these budget requirements. The resulting taxes on property are payable the following year. The due dates for payment of taxes on real property are: one-half on or before May 31 and the remaining one-half on or before October 31. The due dates for payment of personal property taxes are: one-half on or before February 28 and one-half on or before June 1. On June 1 a penalty of 3 % is charged against all unpaid taxes on property that has been accorded the homestead classification, and a penalty is charged on the first day of each month up to and including November 1 for both homestead and non-homestead property. On the first Monday in January following the year in which taxes are due, the penalty increase to a maximum of 10% on homestead and 14% on non-homestead property. Interest commences the second Monday in May following the year in which the taxes became due at the rate of 6 % per annum. Personal property taxes are penalized at a rate of 8 %.

Another example from another state is as follows:

> All real and tangible personal property taxes are due and payable on November 1 of each year, or as soon thereafter as the tax roll is certified and delivered to the Tax Collector. A notice is mailed to each property owner on the tax roll for taxes levied by the county, school district, and other taxing authorities. Taxes may be paid upon receipt of such notice, with discounts at the rate of 4 percent if paid in the month of November, 3 percent if paid in the month of December; 2 percent if paid in the month of January: and 1 percent if paid

> in the month of February. Taxes paid during the month of March are without discount. All unpaid real and tangible personal property taxes become delinquent of April 1 of the year following the year in which the taxes are levied.
>
> The delinquent real property taxes bear interest at the rate of 18 percent per year from April 1 until a certificate is sold at auction, from which time the interest rate shall be as bid by the buyer of the certificate. Delinquent tangible personal property taxes also bear interest at a rate of 18 percent per year from April 1 until paid. Delinquent personal property taxes must be advertised within 45 days after delinquency, and after May 1 the property is subject to levy, seizure and sale.
>
> _____ law provides that tax liens are superior to all other liens, except prior United States Internal Revenue Service liens. The Tax Collector advertises once each week for four consecutive weeks and sells tax certificates on or before June 1 for unpaid tax bills. Tax certificates not sold at auction become the property of the county.

Often the prospectus reflects the total tax levy that is not the amount of money that the school district actually receives. In these instances, a tax collection table is necessary. The following illustrates an example of a tax collection table.

	199_	**199_**	**199_**
(a) Total Tax Levied	$57,118,863	$64,283,788	$67,243,572
Total Tax Collected	54,995,058	61,524,630	64,579,954
Discounts Allowed	1,708,552	1,947,579	2,019,1112
(b) Total Credits on Collections	56,703,610	63,472,209	66,617,066
(b) as a % of (a)	99.27%	98.74%	99.06%

1. Tax collection figures are as of June 30 of the year following the year indicated. Tax collections begin November 1 of the year indicated, and taxes become delinquent April 1 of the following year. Tax certificates are sold mid-May to June, and the collections are certified in the tax collector's report to the county commission on the third Monday in July.

2. Taxes levied and collected include overlapping taxing districts of the county, but exclude taxes collected for municipalities with the county.

In order to satisfy the requirements of full disclosure, the prospectus reflects an overall budgetary picture. Each state has different reporting and accounting requirements. Regardless of the state requirements, the prospectus must reflect generally accepted accounting practices. Usually, it is necessary for the school district to meet the requirements set forth by the Governmental, Accounting and Financial Reporting (GAFR) Audits of State and Local Governmental Units Standards.[35]

The school business administrator must employ a qualified accounting firm to examine, audit, and provide a financial overview acceptable to the investment community. In many states this is done routinely by a state agency. The school district must meet all the rules and regulations of the GAFR and all applicable accounting procedures above and beyond what the state or local requirements may require. The following is an example of notes to financial statements pertaining to accounting procedures.

SCHOOL DISTRICT "A"

NOTES TO STATEMENTS OF RECEIPTS AND DISBURSEMENTS

Year Ended June 30, 199_

Note 1. *Summary of Significant Accounting Policies*

a) Accounting Method: the accounts of the school district are maintained, and the accompanying statements have been prepared on the cash basis of accounting in conformity with accounting practices prescribed by the _______________ Department of Public Instruction. Accordingly, income and expenses are recognized only as cash as received or disbursed and receivable, inventories, prepaid expenses, accrued expenses and vouchers payable are not reflected in the financial statements. Purchases of property and equipment are recorded as disbursements of various funds when paid. Checks for teachers' salaries and related benefits for July and August are written prior to June 30, and are therefore reflected in the bank balance.

b) Uncollected Local Property taxes: At June 20, 19__, uncollected local property taxes aggregating $ _____ were due the School District for the year then ended. This represents an increase of $ ____ over the amount receivable one year earlier. These property taxes receivable have not been reflected in the accompanying statements.

Note 2. *Cash And Investments*

The debt service fund, building fund, and pupil activity fund each account for transactions through a separate and distinct bank account or investment fund. Transaction of the other funds are all accounted for through a common bank account or investment fund.

June 30, 19__ cash and investment balance requirements of funds that have common bank accounts and investments are as follows:

General Fund $$$$,$$$.$$

Special Project Fund $$$$,$$$.$$

Total $$$$,$$$.$$

The common cash and investments of the District at June 30, 19__ were as follows:

"A" State Bank*

Checking Account "#l" $($$$,$$$.$$)

Investments $$$$,$$$.$$

$$$$,$$$.$$

Note 3 . *General Fund Advances Due From Other Funds*

As of June 30, 19__, the following fund is obligated to the general fund for previous cash advances:

Community Service Fund $ $$$.$$

Note 4 . *Long-Term Debt (See Long Term Debt Section of the Official Statement.*

Note 5. *Employee Retirement Plan*

The School District substantially provides all of its contracted teachers and other professional staff with a retirement plan funded through contributions to the _______ Retirement System. The retirement plan requires an employee contribution equal to __ % of the teaching salary plus an annually determined employer contribution rate (set at ___ % for the year ended 6-30-__). The total required contributions to the retirement system for the year ended June 30, 19__, were as follows:

School District's Share $ $$$,$$$.$$

The annual employer's contribution rate, which is actualized by the State of _____, provides for funding of prior service costs, including interest, over a period of __ years. Information concerning the amount, if any, of the excess of the actuated computed value of vested benefits over the total available in the pension fund is not maintained on a district-by-district basis.

In states where school districts make payments to the state/teachers' retirement plan such a statement may be illustrated as follows:

> The majority of the employees of the School District are covered by the Teachers' Retirement System of __________, which is a contributory defined benefit plan. During the fiscal year ended June 30, 19__ the _____ Board of Education contributed $ __________ (_____ percent of current covered payroll) and the employees contributed $ __________ (___ percent of current covered payroll) to the plan. The Board has no further liability under the plan other than the future contributions to the plan which are established each year.

Cash and Investments

A detailed account or an aggregated statement should reflect the cash and investments on hand. While often reflected in the prospectus as "Cash on Hand," the reality of sound school business management practices dictates that all "Cash on Hand" be in various investments as discussed in the chapter on cash flow. An example of an aggregated statement follows:

> Cash and Investments—For all funds of the District as of June 30, 19__ were $________
>
> A detailed finite statement is illustrated:
>
> CASH AND INVESTMENTS—June 30, 19__

General Fund	$$,$$$,$$$
Board of Education	$$$,$$$
Library	$$$
City-County Health	$,$$$
Sinking Fund	$$$,$$$
Central Equipment & Stores	$$$
Combined Capital Improvements	$$$
Miscellaneous	$$$
	$,$$$,$$$

Future Financing

The investment community is obviously interested in ascertaining if additional bonds are anticipated within the near future by the district or through an overlapping debt structure. Such a statement might read:

Future Financing

None of the municipalities within the District anticipate the issuance of bonds/notes within the next six months for capital improvements.

Another example of debt may reflect the presence of Tax Anticipation Notes, which pursuant to state statutes may or may not serve as a binding obligation on the ad valorem taxes of the school district. For example:

The District has outstanding $__________ in Tax Anticipation Notes, issued pursuant to §§ __________ to provide interim funds for the payment of operating expenses of the District for its fiscal year commencing July 1, 19__. The Tax Anticipation Notes are payable from the receipts of certain ad valorem taxes levied for operating purposes. However, the Tax Anticipation Notes do not constitute a pledge of the full faith and credit of the District and the holders thereof will not have the right to compel the ad valorem taxing power of the Board. The funds collected from the applicable tax receipts have been set aside to meet the current debt obligation. The District does not expect to issue any tax anticipation notes during the fiscal period ending June 30, 19__.

Additionally, other debt may be present, for example:

The District had one short-term obligation outstanding as of __________, incurred under the authority of §§ __________ . The borrowing occurred in _____ in the amount of __________ at an interest rate of _____ percent to be repaid in _____ annual installments. As of __________, there was an unpaid balance of $ __________

§§ __________ defines this type of loan as one which anticipates receipt of future revenues. Neither a pledge of credit of the District nor a requirement to levy future taxes to secure this type of loan is allowed by law.

Pending Litigation

The prospectus normally specifies whether there is any pending legal action that may have an effect on the sale, issuance, or ability of the district to repay the obligation. It is important that bond counsel draft this section. A statement written by bond counsel is illustrated:

There is no controversy or litigation of any nature now pending or threatened restraining or enjoining the issuance, sale, execution, or delivery of the bonds, or in any way contesting or affecting the validity of the bonds. There are no suits for damages currently pend-

> ing against the District that are significant or that are not fully covered by liability insurance. Neither the District nor its attorney anticipate any litigation that would contain a significant demand for financial relief against the District or its properties that would not be fully covered by liability insurance.

A somewhat simpler statement bond counsel might draft in some states is illustrated:

> There is no litigation now pending or, to the knowledge of District officials, threatened that questions the validity of the Bonds or of any proceedings of the District taken with respect to the issuance or sale thereof.

In the event there is anticipated or pending litigation, bond counsel must reflect the nature of the circumstances surrounding the possibility. Such statements be drafted by legal counsel.

Certification Statement

Often a certification statement is expected by the investment community. The certification statement is an assurance by the school district that all the information contained in the prospectus is accurate. An example of a certification statement commonly found in a prospectus follows:

> CERTIFICATION
>
> The District will furnish, upon request, a statement to the effect that the Official Statement to the best of their knowledge and belief, as of the date of sale and the date of delivery of the bonds, is true and correct in all material respects, and does not contain any untrue statement of a material fact or omit to state a material fact necessary in order to make the statements made therein, in light of the circumstances under which they were made, not misleading.

General Information

General economic and demographic information is included. The smaller the district the more acute the need for this type of information. There are few absolutes in what must be included or how it is organized and displayed. Often pictures are included of the present educational facilities, major industrial employers, banks, and so on. Underscoring the need for this type of data, Moody's Investors Services explains:

> There is a large, comprehensive body of data available to an analyst on economic resources, industry and employment, and population characteristics for the states, the counties, the cities of over 10,000 population. More limited data are available for places as small as

> 2,500 population. For districts that cut across boundaries, these standard data are lacking, and these kinds of districts include the most numerous single class of borrowers—school districts.[36]

General information for a relatively small community should be as thorough and complete as reasonably possible. This section would normally include such things as:

- Geographic and demographic information.
- Number, type, age, capacity, and enrollments of the present educational facilities, etc.
- Prior enrollment patterns by grades.
- Number and classifications of personnel.
- Private school enrollments and status.
- University, college, and technical schools within the community.
- Agricultural information (if the community is agricultural), including such items as size, cropland values, market value of agricultural products sold, and data available from the U.S. Census of Agriculture.
- Information on the form of city government and municipal services.
- The unemployment rate available from the state.
- Amount and value of building permits for the past few years.
- A listing of the largest property taxpayers within the district.
- An overview of the financial institutions of the district.

Sale of Issue

The actual sale of the issue is usually held at a time and place printed in the official prospectus. The opening of the bid should not take place at the same time as the awarding of the bid. The actual award should be held at a later date. The following outline is offered:

- Schedule the actual bid opening for the day of a board meeting. At the time of the actual opening have all parties present to witness the opening and reading of the sealed bids.
- Analysis of the bids would then occur over the next several hours (days). If multiple interest rates were received, they must be examined carefully.

- The final awarding of the bids would then occur at the board meeting pursuant to the appropriate statutory guidelines and based on the analysis and advice of financial and legal counsel.

The bids must always be sealed, delivered in an appropriate manner, and received before the scheduled opening. The good faith check of the winning bid is deposited to the school district accounts. The uncashed good faith checks of the other bidders are normally returned via registered mail. It is generally customary for the winning bidder to be informed immediately by telephone and by certified mail.

In determining the best bid, the district tries to ascertain which one carries the lowest cost In a single rate bid, the lowest rate determines the best bid, i. e., the total cost minus all premiums. In a split rate bid, the financial adviser or school business administrator must determine the net dollar cost to the district, i.e., the total cost minus all premiums. If the total cost is the same for the bids, the higher premium will determine the lowest net cost. It is important to remember that the total net cost must be determined by subtracting any premium charge.

Bond Registration

Presently, all bonds must be registered. Typically, the financial agent, i.e., the bank, is the registrar. Generally, municipal bonds are exempt from federal and state taxes but they are not exempt from estate taxes. The previous system of bearer bonds was disallowed under federal law effective July 1, 1983, when the federal government mandated the registered format.[37] The reader should note that in 1994 the municipal bond market was estimated at $1.2 trillion for new issues. At the time of this writing, the 80,000 municipalities are not required to register these issues with the Securities and Exchange Commission (SEC). However, the SEC has proposed that such registrations be required for the future.

Both the SEC and the Municipal Securities Rulemaking Board are in the process of proposing and implementing new rules and regulations concerning the advertisement, sale, and monitoring of municipal bonds. SThe financial advisor and underwriter of any school bond issue should know all of the applicable rules and regulations in order to protect the interests of the school district as well as their own corporate interests.

Investment of the Proceeds

Although the total amount of a school district's bond issue may be obligated to a project, portions are only payable on certain dates over the life of the project. As a result, when most school districts undertake sizable capital outlay projects, they have various amounts of excess money for different periods of time. These surplus funds must be invested pursuant to state and federal statutes. The investment of the cash flow—or more properly stated, the reinvestment of the issue—can be a complex task.

Great care and diligence should be exercised with the architectural/construction management team to set forth a viable construction payment schedule. With a construction payment schedule, the school district can then set forth an aggressive investment schedule for the maximum legal returns. The use of such money may be dictated by statute in some states, and the statutes must be examined closely by legal and financial counsel. The reinvestment schedule must be fully examined by legal counsel in order to meet all applicable IRS rules so they may not be considered arbitrage bonds. Arbitrage is making money due to the difference between the interest rates that school districts pay on borrowed money and the interest rates they could earn by investing the same funds.[38]

Olson has noted the following general arbitrage concerns:

> • Whether a debt obligation is an "arbitrage bond" is generally determined by the issuer's "reasonable" expectations as to the use of the borrowed funds on the date of issue. . . . [T]he issuer must "certify" as to its expectations. Nevertheless, the New Regulations now specifically provide that a "No-Arbitrage Certificate" is only "evidence" of the issuer's expectations, and that a Certificate does not establish any conclusions of law or presumptions of reasonableness.
>
> • *Small issues*: There is no need for a Certificate if the debt does not exceed $250,000. (Note that if the debt is really part of a larger issue, e.g., it is incurred within 15 days of similar debt (same terms, interest rate, security, etc.), this exception will not apply.)
>
> • *Leasing transactions and other "non-cash" borrowings*: Not infrequently, the "debt" of the borrower is incurred in connection with a "non-cash" transaction. For example, a school district may "lease" computers or office equipment under a financing lease that is in effect at the time of purchase or may purchase land or other proper such as vehicles, with seller-providing financing. No certificate is necessary in these situations because their are no actual "proceeds" that could be invested other than in the "non-investment" property being purchased.

The current IRS regulations require that the school district must certify, "within six months from the date funds are borrowed, that is reasonably expects to incur a substantial binding commitment to expend at least 5 percent of the 'sale proceeds' of the debt obligations." Olson further cautions:

> "Sale proceeds" are all amounts received including underwriter's discount and capitalized interest. Therefore, it is now more important than ever to make sure that a project being financed with tax

> exempt debt be well on the way to reality when money is borrowed. Otherwise, if at least 5 percent of the proceeds are not expected to be "committed" within six months, investment of the borrowed funds must be yield-restricted from the outset.[39]

Olson, writing in *School Business Affairs*, the professional journal of the Association of School Business Officials International, has succinctly stated the requirements and exceptions regarding arbitrage penalties:

> Rebate:
>
> Unless one of the exceptions described below is applicable, school districts, . . . are required to "rebate" or pay, to the U.S. Treasury any "arbitrage profits." The amount that must be rebated is the difference between the present value of all payments made to acquire investments and the present value of all amounts received from such investments, each discounted at the "yield" on the borrower's debt. It is important to remember that this comparison is yield to yield, not interest earned to interest paid. Yield must be determined under the economic accrual method.
>
> [T]he borrower can pick any convenient date during the first five years after the debt was incurred (or the date the debt is retired, if sooner) to make the first computation of required rebate. (It is not necessary to use the fifth anniversary of the issue date.) Thereafter, the computation is made on the fifth anniversary of the last computation date or the date all the debt is retired.
>
> While the method for determining the rebate amount is essentially the same as under the prior regulations, there have been a few minor changes to the computations required, which may make some of the bond documents read a little differently than they did previously. For example, under the prior regulations, the borrower was allowed a $3,000 "Computation credit" per computation period and there was a $100 "round down" provision. The latter provision has been deleted and now there is a $1,000 per year payment credit . . ..
>
> The new regulations permit refunds of overpayment of amounts that were previously rebated to the Treasury. It does not matter whether the overpayment occurred because an amount was erroneously overpaid, ...[40]

There are certain exceptions to rebates regarding smaller issues below an amount of $5 million per year and where the earnings do not exceed $100,000 per year. This "safe harbor" is met if the average annual debt service on the debt does not exceed $2.5 million. Other exceptions included the transfer of dollars involved in an advanced refunding issue. Additional exceptions apply if all money from all proceeds was expended within six months. Further, after June 30., 1993, if the six-month rule cannot be met, the district may have up to 18 months if the project is not a construction project. Additionally, there are numerous reimbursement regulations with a required expression of official intent. There are exceptions to these regulations that the legal and financial counsel should be knowledgeable of and able to complete.

The IRS regulations also cover tax anticipation notes (TAN) that school districts may issue. Again, Olson's overview is a most helpful introduction to this area.

> . . . Generally, the proceeds of a TAN may be invested without regard to yield for a temporary period of 13 months. The New Regulations also permit a longer period, up to the earlier of two years from the date of issuance or 60 days after the last date for payment of the taxes that will be used to redeem the TANs, for issuers that have budget cycles exceeding a year.
>
> The New Regulations do not prescribe how much money "small" issuers not subject to rebate can borrow for working capital purposes.[41]

However, as Olson points out, the regulations do control the amount of the working capital reserve:

Under the New Regulations, the amount of the working capital reserve has been substantially limited—now, only 5 percent of the prior fiscal year's actual operating fund (which can include capital expenses paid out to the issuer's operating revenues, if any) may be treated as a reserve that is unavailable.[42]

It is critical that the chief financial officer along with the superintendent and board of education have competent legal counsel in these matters.

Advanced Refunding

Where permissible, school districts may engage in advanced refunding to lower the total cost to the taxpayers. Advanced refunding is the issuance of a new bond issue to pay off the old bond one. If interest rates have declined, the new issue is less expensive and, therefore, sold at the lower rates.

The bond markets can fluctuate for several reasons. Generally, advance refunding is used when the outstanding bond has not reached the first call date or is not callable under the conditions of the sale. The proceeds of the new issue are invested in the highest return investments (all fully insured) with the maturities due before the payment schedule due date. The investment is normally in an escrow account established exclusively for the purpose of establishing the purchase of refunding bonds for the purchase of U. S. Treasury Securities. The escrow account then pays the refunded bonds to the maturity or the call date. Normally, a 1 to 1.5 percent difference in interest rates is necessary for such activity to be cost effective. In periods of stable interest rates, the number of refinancing issues generally declines. In recent years, 1993 was the highest year for refinancing municipal obligations with over $90 billion refinanced during the first half of 1993 alone.

Most states allow for such activity while regulating the procedures and terms of such sales. The refunding process is restricted as to the investment yield based on Treasury Department's arbitrage rules. Generally, the new issue will have a lower net interest cost due to a lower bond market/higher reinvestment schedule. It is important to note that the old issue is fully secured by the new issue's escrow investments.

Final interpretation of the present tax code can only be engaged in by competent legal and financial counsel.

Bond Insurance

Insurance is available for municipalities to assure bidders that they are purchasing guaranteed obligations. The largest such insurer, the Municipal Bond Insurance Association (MBIA), and the Financial Guaranty Insurance Company (FGIC), as well as other large insurance corporations, such as AMBAC, offer municipalities an AAA rating from Standard and Poor's and Aaa from Moody's. MBIA coverage is available to new issues that qualify. The municipality may purchase the coverage as a part of its offering, or it can qualify its issue as rated so as to have the bidder pay for the coverage. In theory, the purchase of insurance is cost effective because the school district pays less interest due to its higher rating as a result of the no-risk element to investors.

A brief example of municipal bond insurance language from a prospectus is shown. It is important that this is only the introduction to this section given the typical length of the display.

> Concurrently with the issuance of the Bonds, __________ will issue its Municipal Bond Insurance Policy for the Bonds (the "Policy'). The Policy unconditionally guarantees the payment of that portion, of the principal and interest on the Bonds, by the District. On the day on which such principal and interest is due or on the business day next following the business day on which __________ shall have received notice by telephone or telecopy, subsequently confirmed in a signed writing, or written notice by registered or certified mail, from an Owner of Bonds, for the benefit of the Owners, or, at the election of __________, directly to each Owner, in ether case upon receipt by __________ in form reasonably satisfactory to it or (a) that is due for payment and (b) evidence, including any appropriate instruments of assignment, that all of such Owner's rights to payment of such principal and interest shall be vested in __________. The term "nonpayment" in respect of a Bond, include any payment of principal or interest made to an Owner of a Bond,

> that has been recovered from such Owner pursuant to the United States Bankruptcy Code by a trustee in bankruptcy in accordance with a final, nonappealable order of a court having competent jurisdiction.
>
> The Policy is non-cancelable and the premium will be fully paid at the time of delivery of the Bonds. The Policy covers failure to pay principal of the Bonds, or their respective stated maturity dates, or dates on which the same shall have been duly called for mandatory sinking fund redemption, and not on any other date on which the Bonds, may have been called for redemption, acceleration or their advancement of maturity, unless __________ shall elect, in its sole discretion, to pay such principal due upon acceleration together with any interest accrued to the date of acceleration, and covers the failure to pay an installment of interest on the stated date for its payment.

The school district bond prospectus may contain an appendix that may include such items as the official school board resolution, the auditors' financial statement for the district, the actual legal opinion, and a specimen of the municipal bond insurance policy. While the appendix will vary from issue to issue, a complete financial and legal representation should be allowed for within the prospectus.

Overall, Moody's has offered the following checklist for issuing ratings:

PRIMARY DOCUMENTS

- Official statement or prospectus.
- Notice of sale.
- Annual reports or audits for the last three years.
- Most recent budget for operations.
- Capital budget or planning document.
- Copy of local charter or document that describes governmental structure.
- Ten-year enrollment trend and projection, school plant description, including facilities and capacity.
- Legal opinion.

SUPPORT DOCUMENTS (FOR INFORMATION NOT CONTAINED IN ANY OF THE AFOREMENTIONED DOCUMENTS)

- Assessed valuation for the last five years.
- Equalized ratios for the last five years.
- List of ten largest taxpayers, their assessed valuation, and type of business.
- Current population and latest census estimates.
- Tax rates and levies for the last five years, current and total collections, tax rates of overlapping jurisdictions.
- Tax assessment and collection procedures, including due dates and penalty rates.
- Tax and levy limitations.
- Debt or interest rate limits.
- Statement of direct debt and debt of overlapping debt issuers, including allocatable share.
- Principal and interest payments of outstanding debt, each year-by-year, bonds segregated by security.
- Future borrowing plans.
- Number of building permits over the last five years, plus the dollar amount.
- Number of governmental employees and whether unionized or contract status.
- Statement regarding status of employee pension funds, latest actuarial study, if available.
- Local and area unemployment rate.
- List of ten major employers, number of employees, and type of business.
- Area of the issuer (in square miles) and percentage of land that is developed.[43]

Summary

The area of capital outlay and bonding for educational facilities is an extremely broad and complex topic. Careful planning must be undertaken to ensure the efficient use of the facilities and taxpayers' money. Financial and legal counsel must be secured at the outset of the planning. Thus, the financial as well as legal interests of the school district as well as the taxpayers may be ensured. The school business administrator is the person in most districts who represents the district and the board of education to the investment community. The school business administrator should be prepared to gather and display basic information regarding the school district. Ultimately, it is the school business administrator who is the key person in the capital outlay programs of any school district.

Endnotes

[1] Portions of this chapter are based on the previous edition written by R. C. Wood.

[2] G. D. Webber, *State Equalization of Capital Outlays for Public School Buildings* (Los Angeles, Univ. of California Press, 1941), 5.

[3] W. M. Barr and K. F. Jordan, "Financing Public Elementary and Secondary School Facilities," *Planning to Finance Education*, ed. R. L. Johns, *et al.*, (Gainesville, FL: National Educ. Finance Project, 1971), 240-241.

[4] R. L. Johns and E. L. Morphet, *The Economics and Financing of Education*, 3d ed. (Englewood Cliffs, N.J.: Prentice-Hall, 1975), 279.

[5] W. M. Barr, *Financing Public Elementary and Secondary School Facilities in the United States* (Bloomington, IN: Bureau of Surveys and Administrative Studies, 1970), 141.

[6] Barr, 25.

[7] See, D. C. Thompson, and R. C. Wood, *Analysis for Plaintiffs: Roosevelt Elementary School District et al v. C. Diane Bishop et al.*, A technical report to the plaintiffs in action before the courts in Arizona, (Gainesville, FL Wood, Thompson & Associates, Inc.), Sup. Ct. No. CV-93-0168-T/AP, 1994.

[8] See, D. C. Thompson, R. C. Wood, and M. D. Miller, *Findings of Fact and Opinion on the Equity and Fiscal Neutrality of South Dakota's State Aid Formula to Public Schools: Expert Analysis on Behalf of the State in Bezdichek et al. v. State of South Dakota*, Technical Report to the Attorney General of the State of South Dakota, 1993, (Gainesville, FL: Wood, Thompson & Associates, Inc.). See also, R. C. Wood, M. D. Miller, and D. C. Thompson, *Response By Defendant Experts to Various Reports by Plaintiffs Regarding the Equity of South Dakota's Public School Finance Plan,* Technical Report to the Attorney General of the State of South Dakota, 1994, (Gainesville, FL: Wood, Thompson & Associates, Inc., see also "Planning for Capital Outlay, Maintenance, and Operations," in D. C. Thompson, R. C. Wood, and D. S. Honeyman, *Fiscal Leadership for Schools: Concepts and Practices*, (White Plains: Longman Inc.)

[9] Barr, 143-144.

[10] Estimate based on data appearing in *State Public School Finance Programs, 1978-79*, ed. E. O. Tron (Washington, D.C.: U.S. Office of Education, 1980).

[11] Estimate based on data appearing in S. D. Gold, D. M. Smith, S. B. Lawton, and A. C. Hyary, *Public School Finance Programs of the United States and Canada, 1990-91*, (American Education Finance Ass'n and the Center for the Study of the States, Albany, NY).

[12] Estimate based on data appearing in *Finances of Public School Systems in 1979-80* (Washington, D.C.: U.S. Bureau of Census, CF-80, No. 10).

[13] S. D. Gold, D. M. Smith, S. B. Lawton, and A. C. Hyary, *Public School Finance Programs of the United States and Canada, 1990-91.*

[14] Par values refer to the amount of principal.

[15] See also, D. Lindstrom, "Bond Tip: Let's Go to the Videotape," *The American School Board J.*, July 1994, 33-34.

[16] J. J. Herman, "Control Points in Planning a New or Remodeled School Building," *School Business Affairs*, Jan., 1991 at 23.

[17] Barr, 241.

[18] The Composite Construction Cost Index reflects the Cost of Building Index and the previously issued Cost of Materials Index.

[19] Based on major general obligation issues appearing in the *Bond Directory*.

[20] By design, this chapter does not discuss relatively rare municipal bond investments such as inverse floater municipal issues. Suffice to say that such issues do exist where permitted.

[21] See, "Planning for Capital Outlay, Maintenance, and Operations," in D. C. Thompson, R. C. Wood, and D. S. Honeyman, *Fiscal leadership for Schools: Concepts and Practices*, (White Plains: Longman Inc.).

[22] W. R. Wilkerson, *Problems and Issues of Fiscal Neutrality in Financing Construction* (Washington, D.C.: Office of Health Education and Welfare, Bureau of Elementary and Secondary Education, 1973), 9.

[23] It should be understood by the reader that under certain circumstances issues bought in the secondary market purchased at a discount may be taxed, dependent upon the date of purchase, either as capital gains or as income.

[24] It should be understood that recently some municipal bonds are listed for trading on the New York Stock Exchange, however, the vast majority are not listed for offer and bid prices.

[25] *Understanding Bonds and Preferred Stocks* (New York: New York Stock Exchange, Inc. 1978).

[26] Lease Purchase Agreements are not allowed in New Mexico for the purposes of escaping applicable debt to assessed valuation limitations.

[27] P. A. Covert, ""Lease Financing: Advantages and Risks," *School Business Affairs*, Jan., 1993 at 46.

[28] For an article concerning the risk of such issues, see, D. Vrana, "Some Government Bonds are no Longer Good Investments," *California J.*, Aug., 1992, 403-406.

[29] D. Hitchcock, "How School District Bonds are Rated," *School Business Affairs*, Jan., 1992 at 15.

[30] It should be noted by the reader that such guarantees generally do not cost the state money if a district were to default. Such payment would be made by the state as a deduct from any state aid payment.

[31] *Pitfalls in Issuing Municipal Bonds* (New York: Moody's Investors Ser., Inc.), 24.

[32] Adapted from *Public Finance and Management*, (Minneapolis: Evenson-Dodge, Inc.), 5.

[33] Ibid., 7.

[34] SEC Rule 15c 2-12.

[35] Further discussion of these points are contained in S. B. Tidwell, "Educational Accounting Procedures," Chap. 5 in R. C. Wood, *Principles of School Business Management*, (Reston, VA: Association of School Business Officials, 1986). The reader may also wish to consult S. B. Tidwell, *Financial and Managerial Accounting for Elementary and Secondary School Systems,* 3d., (Reston, VA: Association of School Business Officials, 1985).

[36] *Pitfalls in Issuing Municipal Bonds,* 36.

[37] For a complete discussion of the requirements of registration see, J. E. Petersen and M. P. Buckley, A *Guide to Registered Municipal Securities* (Chicago: Municipal Finance Officers Ass'n, 1983).

[38] C. D. Olson, "New Regulations Affect Debt Financing," *School Business Affairs*, Sept., 1993, at 18

[39] Ibid., at 19.

[40] Ibid., at 18-19.

[41] Ibid., at 22.

[42] Ibid., at 22.

[43] *Moody's Bond Record, Special Edition*, "School District Ratings," (Moody's Investors Ser., Oct., 1984).

Chapter 18 Facility Planning and Management

Introduction

The range and scope of educational facility needs in our society is almost overwhelming. Urban school districts as well as very rural school districts typically share in the need to construct, renovate, and update educational facilities. The planning of educational facilities calls for a series of interrelated activities, conducted by many professionally and technically trained personnel and involving many community people.[1] Some of the activities described herein can be completed independently of other activities. The activities may be discrete, yet interrelated, as each activity fits into a chain of events that results in an educational facility for the school district.

Planning activities associated with the construction and renovation of educational facilities are technical and fall generally within the purview of the school district finance office. With the normally large expenditure of funds for planning and building, these expenditures constitute the most expensive event or planning activity done by a school district. Thus, it is important that the planning activities in the area of capital outlay be done expeditiously and exactingly. Poor school facility planning is costly and the consequences almost always remain for the life of the building. Therefore, no effort should be spared in augmenting planning for capital projects.

The general phrase "school facility planning" encompasses all activities concerning educational facilities. These activities are supervised and directed by school personnel trained in educational administration. The school district must see that all of these activities are completed in accordance with applicable school policy and controlling state statutes. Thus, this chapter offers a generic overview of school facility planning. The student of education finance management should be aware that controlling state statutes and related rules and regulations may dictate a different course of action in certain instances.

Large school districts need large organizational staffs, as they must provide new or different housing for the student population. In small school districts, where the facility need is not as great, the staff is smaller. Someone, however, in every school district has responsibility for facility management and planning. In smaller systems, the school business manager is usually responsible for facility planning and maintenance and operations. The general rule is that the smaller the organization, the more outside expertise the school district needs in order to successfully complete a major educational facility. This outside help will ensure the efficient use of taxpayer money while meeting the educational needs of the students.

Generally, in large urban school districts, an office for facilities is responsible for planning and overseeing the construction or renovation of all educational facilities. This office may be headed by an assistant superintendent for school facilities. Under this office directors may supervise a large number of people, each with specific responsibilities and duties. This office is generally responsible for the following activities:

- Demographic planning.
- Capital program planning.
- Site selection on acquisition.
- Facility planning and programming.
- Architectural review.
- Construction.

In large urban school districts each of these areas is headed by a professional or technical expert. The site selection and acquisition office may be directed by someone who has a law degree with specialization in real estate. Likewise, a person with an architectural registration might direct the architectural review office. All of these programs work together to complete each project contained within the capital improvement program (CIP) and budget.

Each project for planning a new school or a major renovation, should be directed and guided by one person who is assisted by appropriate school district personnel. Titles for such a position include: supervisor of planning, facility planner, educational planner, project manager, or planning director. Characteristics of the person holding this position might include experience as a teacher and principal, planning knowledge and experience working with governmental agencies outside of the school district, some knowledge of the design work of architects or of construction, and the ability to work with large and diverse groups. These skills may be hard to find in one person, but such qualifications are needed in the course of planning a school facility. At minimum, the person should know the school district and its planning process; the person can then gain experience by engaging in the planning process. It is of utmost importance that one person be charged with the responsibility of carrying the project through to completion.

Planning, designing, and constructing an educational facility is arduous and time-consuming. Therefore, it is necessary to plan a facility long before it will be needed to house pupils. Historically, many school districts, for example those in Florida, have not been able to stay ahead of the rapid pupil population growth. Given proper planning expertise and financial support, school districts should be able to avoid this problem.

Long-Range Planning

Long-range planning in school districts is paramount to facility planning. The limited resources available to public schools and the increasing demand for services from the public make it mandatory that every school district engage in substantive long-range planning. Additionally, to secure a high level of cooperation and coordination among the total staff in working toward the common goals of the school district, the board of education needs this type of long-range assessment to serve as a guide for the future.

The Council of Educational Facility Planners International (CEFPI) lists a seven-step methodology for working with community members in the planning process. The emphasis in this planning model is an involvement of community members in a decision-making process. This methodology lists the following steps:

1. *Get organized/define problems and goals*: First, community groups and individuals must define the problem, decide what is to be accomplished, create time lines, outline responsibilities, and determine how the job will be accomplished.

2. *Gather information*: The information needed to answer the above questions is identified and collected. Both hard and soft data are gathered. Those involved in the planning process must locate the source and compile data in appropriate form for use in subsequent steps.

3. *Develop priorities*: The list of needs created from step two will need to be prioritized. The needs with the highest priority will be identified.

4. *State program and facility needs:* The planning group then assess which priority needs can be met by existing resources and what changes are needed to make that happen. It will also determine program and facility needs that cannot be met by existing resources.

5. *Explore options*: The options identified need to be shown to the community, and community reactions need to be gathered. The options need to be evaluated so that one preferred option can be selected.

6. *Refine preferred option*: Once community agreement has been achieved on a preferred option, it is necessary to put the plan into action. The kinds of

details that will need to be addressed will depend on the nature of the alternative that best meets the needs of the community.

7. *Follow through*: Execution of the plan of action follows. The implementation of the plan begins by hiring staff, initiating capital projects, signing agreements, or getting out publicity.[2]

The planning group also addresses other concerns after it has completed the final step, such as its continuing role in the planning process, periodic review of the plan, and communication of the preferred option and its implementation.

The CEFPI model is based on wide community and organizational involvement.[3] Because of the nature of the school district it is important to involve both the clients and the operators of the organization. Each school district, however, must decide the methodology to be used in developing a plan so that local conditions, available expertise, indigenous processes, and state requirements are fully observed.

Long-Range Plan Elements

The specific topics to be covered in any particular long-range plan may be a function of controlling statutes, but they generally address the following issues:

- *Description of the community*: The particular community and its inhabitants are described in detail including demographic statistics. Community resources such as agencies, organizations, physical features, and even individuals may be described here.

- *Description of the educational program* : The goals and purposes of the organization should be delineated and refined into objectives, sometimes referred to as the mission statement. Following this, a description of the educational program that stems from the goals or mission statement should be specified in detail. Proposed changes to the program should also be stated. Scope and sequence needs to be enunciated, and the human resources needed to staff the educational program should be identified.

- *Identification and projection of clientele to be served* : This section of the report looks at the number and kinds of students to be educated now and in the future. The pupil population of the schools should be projected for five years so that educators can anticipate the kinds and amounts of materials, textbooks, staff, furniture, and facilities that will be needed. The projections should not be limited to the age groups traditionally served by the school, but should also project numbers of pre-schoolers and adults that will be served by programs sponsored by the public schools.

- *Appraisal of existing facilities*: The existing facilities must be appraised to determine their adequacy in terms of the kinds of educational programs to be implemented and the number of students to be housed. The results of this evaluation, on a school-building-by-school-building basis, are listed. Buildings to be abandoned, renovated, and improved, and where applicable, the number and kinds of new facilities needed should be stated. From the appraisal of the existing facilities, items for the maintenance section are identified and written directly into the proposed capital improvement program.[4]

- *Physical changes and new construction:* In this section of the document the necessary changes, improvements, and new construction need to be identified specifically. This section differs from the previous section in that physical changes, maintenance items, and new construction are identified and prioritized. The previous section contains the entire evaluation of each facility. This section includes evaluative statements plus physical changes necessary to bring the particular facility up to standards. For example, each facility has to be judged as to whether it has to be modified to meet current Americans with Disabilities Act (ADA) standards[5] as well as Section 504 standards for special education students, and what the associated costs would be.

- *Financial plan*: The financial resources necessary to fund the type of program need to be detailed. The plan should identify additional staff, increases in instructional materials, and equipment to implement the new educational program. The capital needs for the new program are also specified. Both the operation and capital costs need to be projected over the life of the long-range plan.

After the long-range plan is developed, it is submitted to the board of education for approval. When approved, the document serves as the guide for all planning activities in the school district. The life of the plan, however, may be only one year; the staff and board of education should review the long-range plan each year to determine what changes and revisions need to be made. Often the changes are such that a new long-range plan must be developed. Generally, revisions are made annually; at the end of five years, a new long-range plan is developed.

Data contained in the long-range plan regarding facilities become data for the CIP and the capital budget. Changes in the long-range plan would produce subsequent changes in the CIP. The same sort of relationship exists between other segments of the long-range plan and various offices of the school district. For example, the type of staff needed to implement the educational program becomes data for the human resource plan of the school district. With each subsequent change in the long-range plan, all segments of the school district may need to alter their plans.

Site Selection and Acquisition

Selection and acquisition of suitable sites for school facilities are a very important part of the school planning process. Acquiring the proper site for a school facility has a tremendous impact on the potential for service to the community. Often such a site is unattainable. In fact, obtaining good sites for schools continues to grow in complexity in many communities. Reasons that account for this phenomenon include: competition for sites by the commercial sector; the increase in site size to accommodate an expanded educational program; the inflationary rise in land prices; and, in urban areas, the movement of people into heavily populated areas where there is little open land.

Often, the selection of a school site is one of the most controversial issues educators face in planning a new school. The process is so controversial that some school systems conscientiously do not involve or notify community members that a site for a new school is to be selected. This is particularly true in large school systems where the political process is volatile. However, many school districts, as a matter of policy, involve community members in this phase of school planning.

The process of selecting a site for a school facility normally hinges on the long-range development plan of the school district. As stated earlier, the plan should identify those areas of the school district where growth and development will take place and then identify those neighborhoods affected by site selection. These data from the long-range plan are then entered into the CIP for the appropriate funding year. When the particular project becomes funded, the appropriate office or agency in the school district is then authorized to initiate the site selection process.

Selection Process

The selection of a site, unless controlling statutes state otherwise, generally entails three basic tasks:

- Developing criteria to govern the selection of a site.
- Applying selection criteria to each prospective site,.
- Approving the selection by the board of education and other appropriate agencies.

The actual selection of a school site involves many organizations and people with a variety of skills. The site team may include any and all of the following: educational administrators, educational consultants, architects, urban and regional planners, landscape architects, recreation planners, and legal counsel. Each person or organization contributes to the process as the need demands. The primary share of responsibility falls with the school district administrators, facility planners, and architects because they generate much of the data that are needed to develop and apply criteria used in making recommendations to the board of education. The team is headed by a school district administrator who assists the group in completing the task.

After the criteria have been identified and approved, the team identifies potential sites for consideration. Several sources of data can be used by the team to obtain data such as:

- The comprehensive urban plan as well as all other state and local planning documents.
- The comprehensive regional, urban, or community land use maps.
- Aerial photographs.
- Redevelopment authority maps and official minutes.
- An actual tour of the proposed area to be served by the new school.

Site Selection Criteria

The site team develops a set of criteria based on the educational requirements of the proposed school. They then use these criteria to evaluate each potential site. The criteria may vary one geographic area to another, but certain similarities exist in all criteria used by school districts. The CEFPI suggests that the criteria address the following questions:

- Will the site support the educational program?
- Is the site's location convenient for the majority of pupils?
- Is the site the right size and shape?
- Is the topography conducive to desired site development?
- Is the general environment aesthetically pleasing?
- Is the site safe?
- Is the air quality healthful?
- Is the site free of industrial and traffic noise (both air and ground)?
- Does the land drain properly and are other soil conditions good?
- Does the site have desired trees and other natural vegetation?
- Is water available?
- Is the site near community services—library, parks, museums?
- What is the relation of the site to existing educational facilities?

- How is the surrounding land zoned? Will its development enhance the school site?
- Are utility services available?
- Is the site served by public agencies—police, fire department, etc.?
- Is the site easily accessible to service vehicles?
- Is the site available?
- Is the price affordable?
- Can the land be shared with other community facilities and organizations, especially parks?
- Will the site provide desirable open space for the community where it is needed?
- Is the site expandable in the future?

In addition, the criteria should address important community concerns including the impact the acquisition of a site has on existing housing and the district's desegregation plan. Especially in urban areas where adequate, existing housing is so valuable, displacement could induce a community response that could adversely affect the purchase of a site. Of course, the final criterion for any site is whether it can pass political muster. The selected site must be politically acceptable or, at least, not unacceptable to the power structure of the community before the board of education can take any action.

Testing the political acceptability of any potential site is usually done informally; however, in most areas the selected site is approved by the regional or urban planning commission as a matter of formality. In so doing the political power structure can make its desires known. In populated areas, a school site is usually designated long in advance of need by the local governing body after consideration of the long-range development plan of the political subdivision. In this manner the judicious placement of such community resources as schools, recreation areas, parks, libraries, and other facilities is determined on a broader basis that is designed to serve the whole community.

Application of Criteria

In developing a list of potential sites, certain general criteria apply before a site is placed on the list. The team looks for sites that meet basic predetermined standards. Once the list of sites has been developed, the team evaluates each site both objectively and subjectively. Objective comparisons will be based on such data as: size of parcel, price of land, distance of utilities to the site, soil percolation, quality of the air, location, and distance of the clientele to be served. Subjective judgments may be based on: the aesthetic environment, impact of the proposed school on the neighborhood, use of other community services, marketability of the site and adaptability to future grade organization changes.

After applying the criteria to each site, the team narrows the list of potential sites to one or two. If the school district has several promising sites to consider, it may want to compare the two best sites in the final stage. Often, however, school systems do not have the luxury of choosing among several sites, but are limited to selecting from two possible sites. In this case, the site team may want to conduct a physical examination of each site before formally applying the criteria. Additionally, some school systems have soil testing and inspection completed before making a decision. After evaluating all available data, the site team makes its recommendation to the board of education for approval or rejection. If approval is given, the board of education then authorizes the legal counsel of the school district to obtain the site.

Site Acquisition

Pursuant to state statutes, generally, a school district may obtain a site in any of four ways.

- Negotiations with the owner and subsequent purchase.
- Condemnation procedures and subsequent purchase.
- Acceptance of a gift of land.
- Receipt of surplus governmental property.

The most common method of obtaining a site is through negotiation with the owner and subsequent purchase. The vast majority of sites are obtained through this process, with the proper board of education representative conducting all the negotiations. There are, however, instances when the owner and the board of education cannot resolve the purchase price for a site. Pursuant to state statutes, when an impasse in negotiations for the purchase price of a site occurs, the board of education has recourse through the exercise of the right of eminent domain. This right stems from the state; by virtue of the fact that the local school district in every state is a subdivision of that state, therefore, it possesses the right of eminent domain.[6] This is the right to directly obtain a parcel of land when the owner refuses to sell, regardless of the reason.

The process of obtaining a site through this method is called "condemnation." The board of education passes a resolution that a particular site is needed for school purposes, and a proper disposition is made to the designated local court. The court then conducts proceedings to determine the equitable price of the parcel of land. The court either appoints appraisers or receives appraisals from the school district and the owner. In either event, the court determines the equitable price of the land, and the school district obtains clear title to the site on payment of the determined price. Condemnation is rarely used because it is very time consuming and can result in a huge public relations problem. Nevertheless, in special situations a board of education may choose to use this method under controlling state statutes.

In cases where patrons or organizations donate a parcel of land to a school district, the board of education can use the gift of land for a school site or sell the land and use the proceeds of the sale for any important purpose. The board can also exchange the land for either a different site or other physical facilities. The general rule is that the board of education can accept gifts and may use the gift in whatever manner deemed prudent for the benefit of the school district. The deed to the land, however, may have a codicil to the effect that the land can be used only as a school site, in which case the board of education must use the gift only in that manner. If the site is not used for a school facility, the title to the land may then revert, if contested in court, to the original owners.

In some areas of the country, state and local jurisdictions have enacted ordinances stipulating that developers of large tracts of land for housing purposes must deed a certain amount of land for public services such as a school, fire station, library, or other functions. In lieu of the dedication of land, the developer may pay a lump sum of money to the local government equal to the cost of land that would normally be dedicated.[7] Where such ordinance or restrictions exist, the board of education must work closely with the appropriate planning department of the local government to ensure that the dedicated tract of land will indeed serve the functions of the school district.

The fourth method of obtaining land for educational purposes is through the surplus property programs of either the federal government or in some cases the particular state. The instances where local school systems have obtained useful sites for schools through this means is very limited. But the federal government and selected states do on occasion declare real property surplus and available for the local governmental use. Although the needs of the school district and the availability of property from other governmental units seldom mesh to the benefit of both parties, in some cases arsenals, parks, camps, and other types of real property have passed onto school districts for good use in housing the educational program.

After the board of education has authorized the purchase of a certain site for school use, the appropriate representative of the school district negotiates the purchase price. The board of education then must agree to that price and give the negotiator the authority to draw up the contractual documents to obtain the site. In all states the local school district must use the services of legal counsel to inspect the title and to draw up the documents for the purchase. It is important that the school district obtain and hold a clear and free simple title to the property to protect their rights to any facility that is constructed on the site. The local board of education holds the title to the land for all sites owned and controlled by the school district. In a few states, the legal title is in the name of the city, county, or other governmental unit as specified in statute.

The Design Process

When an architect is employed by a school district, certain standard services can be expected to be performed in designing a capital project. Five basic services include:

- Schematic design.
- Design development.
- Preparation of contract documents.
- Monitoring the bidding or negotiations.
- Supervising the construction.[8]

Architectural services are specifically covered in the contract document that the school district and architect negotiate and sign. The specific provisions of the relationship between the school district and the architect must be identified and explained in that document. Occasionally, there are predesign planning activities in which the architect is engaged by the school district. Such activities are usually covered in the standard contract document along with other services. Some architects consider predesign activities separate from basic services.

Predesign Planning

The tasks to be completed in this phase of development are the selection of the site and the preparation of educational specifications. Often, school systems have the architect serve as an adviser or observer to this process, but the actual work of selecting the site and preparing the educational specifications is reserved for an educator. Local practice will usually determine the extent of involvement of the architect in the predesign planning process. The most beneficial use of the talents of an architect at this stage may well be in providing advisory-technical assistance to the site selection team.

Schematic Design

This phase of project development begins with the educator translating the educational needs of the school district, as expressed in the educational specifications, to the architect. This phase is followed by a period of intense interaction and communication between the educator and the architect in which several concepts or designs are reviewed. The result of this phase is a set of line drawings of the entire facility. At this stage, the board of education reviews the work of the architect (as represented by the schematic drawings) and approves the work to date. This is an important milestone of the project, for it is at this point that the basic building design is set and little deviation in the exterior structure of the building can take place. The interior of the building, however, will change during the next stage of development. The school district must review the design during this period to ensure the proper development.

Design Development

After the schematic design is approved by the board of education and any other necessary governmental agency, the architect develops and refines the drawings. There is an intense dialogue between educator and architect while the design is being refined. The plans and other construction documents are then prepared. This may take from six to thirty months depending on the complexity of the project, and the administrators in the school district must constantly monitor the progress and work of the architect and the architectural staff in developing the design of the facility.

Preparation of Construction Documents

The architect is responsible for the preparation and production of the formal construction documents. These documents consist of a complete set of working drawings and a set of technical specifications. Some variation may exist between states and regions, but the working drawings generally consist of a series of drawings that are prepared along the lines of the various disciplines: architectural, structural, heating/ventilation/air conditioning (HVAC), electrical, plumbing, fire protection, interior design, and landscaping. Depending on the complexity of the project, the documents may include all of these, or they may simply contain the architectural, electrical, plumbing, and HVAC drawings. The working drawings are the documents used by contractors and builders to prepare bids for submission to the school district.

The second part of the contract documents is a book generally called "technical specifications." This document contains:

- Bidding documents that contain the bidding instructions and other legal conditions.
- Conditions of the contract including general and specific provisions that will govern the relationship between the contractor and the school district.
- Technical specifications that provide a detailed description of all material and workmanship to be used in the facility.[9]

Monitoring the Bidding Process

Depending on the size of the school district, the architect provides several services in this stage of the project such as: preparing the bid advertisement, issuing bid documents, answering questions and inquiries regarding construction, evaluating the bids, and assisting in the preparation of the contracts.

Usually the legal counsel of a school district completes all legal requirements of the bidding process such as preparing the advertisement and the contract, but the architect may assist the legal counsel by providing data regarding the project to be included in the legal documents. The architect may assist the school district by identifying potential bidders in the construction project, providing lively competition that can result in lower bids.

In cases where it is necessary to negotiate with contractors regarding a bid, the architect can present the needs of the school district. In some states, school districts negotiate with the lowest responsible and responsive bidder when that bid is over the budgeted amount. In such circumstances, the architect can identify options that may assist the bidder in meeting the needs of the school district. The school district may, therefore, proceed with construction rather than reject all bids, redesign the facility, and rebid the project. The latter may only produce higher bids because of inflation or delay the project beyond the date desired.

Some states allow school districts, under specific controlling statutes, to negotiate with construction management firms to complete a building project rather than submit the project to competitive bidding. Where this is allowed, the school district that already has an architect under contract may use that person in the negotiation process with the construction management firm.[10]

Construction Supervision

The services offered by the architect during construction under the basic services section of the contract generally include:

- Representing the owner.
- Making periodic visits to the construction site.
- Approving payments to the contractor.
- Interpreting the requirements of the contract and rejecting work that does not conform to contract documents.
- Approving shop drawings.
- Preparing change orders.
- Conducting inspections to determine dates of completion.
- Issuing the final certificate for payment of the contractor.[11]

The assistance of the architect during the construction stages in interpreting the requirements of the construction contract is vitally important to the success of the project. When the project is completed, certain activities are required of the architect, including securing bonds and guarantees, and issuing the final certificate for payment of the contractor. The latter officially completes the work of the architect.

Additional Services

The school district can reasonably expect the architect to provide the described basic services, but should additional services be required, the school district must contract separately with the architect for the work. This contracting is typically done by an addendum to the architect's contract. Services that are outside the realm of what is normally considered basic architectural services include:

- Assisting in site selection.
- Constructing special analyses and studies of the owner's needs beyond those called for in the schematic design.
- Making measured drawings of existing buildings.
- Providing detailed cost estimates beyond the statements of probable costs called for in the basic services.
- Providing interior design and services for special selection of furnishings.
- Planning surveys and site administration.
- Reviewing previously approved drawings to make changes not initiated by the architect.
- Furnishing additional services resulting from default of contractors or other similar events over which the architect has no control yet which involve significant additional time and cost.
- Providing post-occupancy evaluation services.[12]

One area that causes confusion is the supervision of construction. The usual American Institute of Architects (AIA) contract does not provide for daily supervision of the construction project; in fact, it calls only for periodic visits by the architect or the architect's staff. Should the school district desire such supervision from the architect, an addendum to the contract would be needed to account for the extra cost of this service.

A school district may handle the day-by-day supervision of a construction project by employing its own supervisor. This method offers the potential advantage of reducing costs and having a school district employee, rather than an employee of the architect, on the scene every day. School districts need to have such daily supervision to protect their investment and to ensure the timely completion of the project with the desired types and kinds of workmanship.

Fee Schedule

Several methods exist for compensating an architect. The most prevalent is to use the construction cost of a facility as a basis for determining architectural fees.[13] Under this procedure, architects are paid a certain percentage of the construction cost. The percentage varies with the amount of construction to be done and with the negotiating process. Some states have published guidelines for percentages of construction cost to be used in employing an architect. The state chapter of the AIA also can supply a list of suggested percentages to be applied in negotiating a contract.

The construction cost consists of the cost of completed construction for the project, less cost of the land, right-of-way, or similar costs. When the school district negotiates a contract with an architect, the method of compensation is determined and becomes a part of the contract. When services outside of the basic architectural contract are desired by the school district, architects are compensated separately from the basic fee. The usual fee for such additional services is the cost of direct personnel services plus reimbursable expenses, where the personnel costs are the salaries of professional, technical, and clerical employees and their benefits.

The architect will expect to be paid periodically based on the progress of the work. Although the school district and architect may set any mutually agreeable payment schedule, an initial payment of 5 percent of the total fee is frequently made to the architect when the agreement is signed. Payments are usually made on a monthly basis, in proportion to the services rendered. The following schedule of payments is taken from the AIA agreement, Document B 141, and defines the percentage of the total fee that should be paid at the completion of each of the basic service phases:[14]

Architectural Contract

The school district may use a contractual document that is prepared by its legal counsel or they may use the standard AIA Document. Should the school district decide to write its own contract, the assistance of a competent architect will be needed by the legal counsel to advise on the needed services and to make provision for all conditions. Inasmuch as the contract represents a considerable amount of money and delineates the exact relationship between the board of education and the architect, care must be exercised in writing the document to ensure that all exigencies are covered.

Architect Selection

A licensed architect has passed a state qualifying examination and is normally a member of the AIA, which mandates a code of professional ethics. Not every architect, however, can work successfully in the atmosphere of a public school district where there are many levels of approval. Therefore, in selecting an architect, the board of education must make certain that the successful firm understands the problems and complexities inherent to school districts.

There are three methods of selecting an architect: direct employment, design competition, and comparative selection. Direct employment is where the school district contacts one architectural firm and enters into a contractual arrangement to design the capital project. This method is often used when the architectural firm has done work previously for the school district or the school district is situated in a small community.

With the design competition method, the school district selects or invites architects to submit designs for the capital project. The contract is awarded to the architect who submits the "best" designs. Most architects do not favor this method, since it requires each firm to expend considerable resources without any way to recoup their money if not selected.

Under the comparative method, architects are evaluated on the evidence they submit to the school district and on data collected by school district personnel through interviews and visits to former clients and facilities. The successful candidate is selected based on the evaluation done by the school staff and board of education. The steps in the comparative selection method are:

- Identify the pool of interested architects. Obtain names from the local AIA chapter, colleagues in education, and the telephone directory.

- Send out information about the project and a questionnaire seeking information about the firm.

- Review the architect firm's capability statement and completed questionnaire.

- Identify five to ten firms that seem capable and interested. Send an additional questionnaire seeking further information, including names and addresses of former clients and facilities that can be visited.

- Evaluate the material from the above firms and narrow the list to three firms.

- Conduct an intensive background investigation on the three firms by contacting former clients and colleagues and visiting facilities designed by the prospective firms.

- Hold interviews with the three architects and their staffs to gather first-hand impressions and additional data.

- Visit the offices of each of the firms in order to gain an impression of the working atmosphere.

- Select one firm from the list of three based on the visits, interviews of the candidates, and interviews of former clientele.

- Present the candidate firm to the board of education.[15]

Most of the work, if not all, is done by school district's staff and not by members of the board of education. This is because administrators have more experience in interviewing and selecting personnel than most board of education members. School administrators should be given the responsibility of making the investigation and making the recommendation to the board of education.

Criteria for Selection

Criteria for evaluating an architect should be developed by the school district staff members, with input from all segments of the district and the board of education. The CEFPI has recommended certain points to be considered when establishing a selection criteria:

- *Registration and professional activities*: The architect must be licensed by the state and should be a member of the American Institute of Architects.

- *Experience*: The architect should have experience in designing the type, kind, and size of project that the school system needs. Experience in designing a school might be beneficial, but more important is whether the architect has worked with a school organization.

- *Quality of work*: The school system will want to interview clients of the architect to ascertain how they feel about the work performed by the architect. Contracting firms can also be a source of information regarding the preciseness of the documents produced by the firm as well as the relationship between the contractor and the architect. Observing and evaluating buildings designed by the architect will give further insight into quality of work.

- *Staffing and facilities*: The size and quality of the architect's office and staff may be of concern to the school system. The complexity of the school project may demand considerable expertise that the architect should be able to document and produce. School system personnel have to make a judgment as to whether the firm can handle a large project. Judgments need to be made about the capability of the staff based on information submitted by the architect and gathered during a visit by the school system personnel to the architect's office.

- *Method of operation*: Working with a school system is significantly different from other typical architectural clients in two respects:

Obtaining approval from various people and groups inside and outside the organization.

Presenting ideas to a large number of people and groups.

The school system should contract with an architect who feels comfortable working within this type of environment. Architects who have had previous experience working with school systems can be expected to work efficiently. Architects who have not had such experience but who, in the determination of the school personnel and board of education, possess the ability to work within such an organization should be considered for employment and then oriented properly to the planning mode of the school system.

Obtaining a competent architect who will meet the needs of the school district is one of the most critical tasks in the planning for a school facility. Much of the success of the whole project rests with this one decision. Having the appropriate architect will enhance all of the other planning efforts of the school personnel and contribute greatly to the success of the capital project.

Even with the best architect under contract, however, the success of the project is assured only if the proper architect/owner relationship is observed. The architect's job is always to be in an advisory capacity. The architect does not make decisions for the school district; the architect suggests solutions to problems raised by the school district. The educator must exercise the leadership necessary to present the problem correctly and in sufficient detail to the architect so that the architect can offer effective solutions. This means that the educator must carefully detail in writing the needs of the school district and then take the necessary time to translate these needs and evaluate the solutions offered by the architect.

Developing Educational Specifications

The process of planning for school facilities, whether new construction or renovation of existing structures, is a complicated series of interrelated processes, starting with identification of need and proceeding through the final phase of occupying the new facility. One of these processes is the development of a set of educational specifications.[16] Educational specifications have been used for many years by a variety of school districts very successfully. For many reasons, this process is often the weakest link in planning facilities. Reasons why educational specifications are not more widely developed include:

- The lack of leadership in education in assuming responsibility for developing educational specifications.

- The belief that some educators do not know how to direct the process of developing clear and concise educational specifications.

- The vague demarcation of responsibility between the educator and the architect.

- The desire on the part of the architect to provide extra services.

- The misguided belief that the school system does not have the capacity and/or the staff to do the task and cannot afford to purchase outside assistance.[17]

Roles and Responsibilities

For most educators, planning and constructing a new facility or remodeling existing facilities are not common occurrences. Planning a new school facility in the average school district is perhaps a once-in-a-generation happening. Consequently, in many school districts, educators are often unfamiliar with the separation of responsibility and relationships among participants in the planning process for a capital construction project. This is especially true in the design phase.

The School District

The board of education is the owner of all facilities in the school district and acts in that role.[18] The superintendent is the chief administrative officer of the board of education and in this role makes decisions regarding every capital project within broad guidelines. The board of education does make certain major decisions in the process of planning, designing, and constructing facilities, but the superintendent and staff augment these major decisions with daily decisions that enable the project to proceed in an orderly and timely fashion.

The Architect and the Design Team

Following the procedures outlined in the preceding section, the board of education contracts with a firm, corporation, or individual to provide the architectural expertise necessary to complete the design of the proposed capital project. In turn, the architect organizes a team of individuals that will work on the project is usually called the "design team." It is headed by an architect charged with the major responsibility of the design work and includes the engineers who will design the electrical, mechanical, structural, and HVAC systems.

During the course of the project, the design team may be augmented by other individuals such as acoustical engineers, color consultants, soil experts, educational consultants, landscape architects, and energy consultants. This group, along with draftsmen and other technicians, will design a facility and produce a set of documents and working drawings that will be used to tender a bid.

The Design Review Team

On the other side of the design process, there should be designated a group of individuals who serve as the reviewing agent for the board of education. This group of school district employees is usually called the "design review team," and it reviews and approves the work of the design team in preparation for board of education approval.

The composition of the design review team may vary from state to state and from school district to district, but it is imperative that one person be given the authority to act as representative for the local school district throughout the design and construction phase of the project. This person should be the same one who heads the entire school planning effort. The major duties of this person, usually designated as the "educational planner," include: heading the design review team, interpreting the educational program to the architect, coordinating the local reviews, submitting documents for state approval, and most importantly making the necessary decisions, in cooperation with the rest of the design review team, in a timely fashion to ensure completion of the project.

Working Relationships

Figure 18.1 illustrates the members of the design team and the design review team and the relationship between the two groups during the design phase.

The interface of these two groups is an extremely important one for the success of the project. During a very intensive design process, the interface may be on a weekly basis and in certain instances on a daily basis.

During the design phase of any capital project, members of the design review team should be able to take a broad view of the project in terms of time and space. They must recognize the need for flexibility to meet changing program needs and yet provide adequately for the present. Members of the design review team who are responsible for the review of the total concept must consciously keep priorities and programs before them and assume the initiative in requesting information that will enable them to make intelligent decisions concerning development of the design. The design review team cannot abrogate this leadership responsibility and expect the design team to anticipate all of their conditions.

It is also extremely important to establish early, close working relations with all members of the design team. The most propitious time to employ an architect is long before any design work is needed. On almost every capital improvement project, both architect and engineer can provide data in assisting educators to make early decisions regarding the facility. There are fewer

Figure 18.1
Design Team/Design Review Relationship

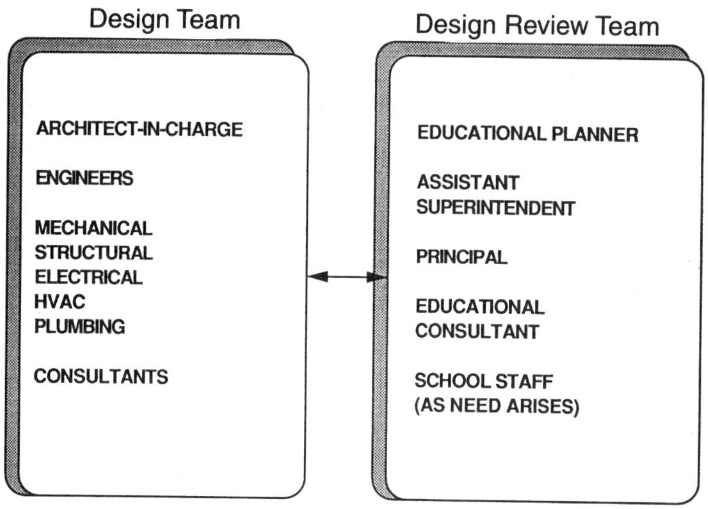

problems in projects where members of both the design team and the design review team meet early and establish good working relations.

Evaluating Proposed Designs

In order to properly evaluate any proposed design work of an architect, it is necessary to have firmly in mind what the intended facility is to accommodate in terms of activities, programs, and personnel. The basis for that knowledge is the approved set of educational specifications. This document will serve as the guide in evaluating the architect's suggested designs

The design stage of the school planning process, however, should be viewed as a period of time where the architect develops, from a clear set of educational specifications, the design of the proposed facility beginning with the very initial concept through the schematic stage to the working drawings and bid documents. The design review team works closely with the design team during this stage to facilitate the process, to assist in translating educational needs, and to systematically review the work of the design team. The design stage may take as little as six months for a rather small capital project, to as long as thirty months, for a more complex project such as a large high school.

Plan review should occur regularly during this period. The intensity and frequency of the review will depend to a great extent on the type of project designed and also on the stage of development. The design review team should not allow any significant amount of time to elapse without review and approval of segments of the plans. Close monitoring of the project needs to be maintained at all stages of development.

The Construction Process

After the architect has completed the design of the facility and completed the contract documents consisting of the working drawings and technical specifications, the school district is ready to enter into the bidding stage. Before any bidding, the board of education must decide whether it desires to enter into a contract with a single contractor or with several contractors. In some states, controlling state statutes do not permit school districts to enter into one contract, and the board of education must sign contracts with contractors representing the different divisions of labor.[19] After the decision has been made, the final bidding documents are prepared by the architect.

The Bidding Process

Generally, there are five steps involved in the bidding process.

Locating Bidders

School systems must advertise the bidding of the project; this is the formal method of notifying potential bidders. Often the architect can be instrumental in securing potential bidders.

Issuing and Retrieving Documents

In the advertisement, the school district identifies the location of the bidding documents (board of education office, architect's office, or other locations). The procedures for obtaining the documents are also set forth. Usually a potential bidder must deposit a certain amount of money in order to obtain the documents. The deposit is normally refunded when the documents are returned. This checkout system allows the architect to have some control of the documents and to recover the cost of the sets of plans.

Receiving and Tabulating the Bids

Bids must be received at the predetermined time and place set forth in the bid documents. The bids that have been received are opened by the representative of the board of education and tabulated. The bidders usually attend this public opening of the bids. The bids must be submitted, dependent upon controlling state statutes, with a surety bond indicating that the bid is genuine and that the contractor is able to accept the offer of a contract should that bid be the one selected.

Analyzing the Bids

Analyzing the bids should be done carefully by both the school personnel and the architect. The architect can provide a great deal of assistance during this stage by evaluating the bids and the firms making the bid. School districts are by law required to award the contract to the lowest responsible and responsive bidder. Thus, not only is the amount of the bid evaluated, but also the technical specifications. Helping to determine the adequacy of the latter is where the architect can have the greatest input into the decision. When alternate bidder arrangements are permitted, the architect must help the school personnel make the difficult decisions regarding the various alternative bids and the consequences of each alternative.

It is possible for school districts to reject any and all bids if they are above the budgeted amount for the project. Dependent upon controlling state statute, school districts may negotiate with the lowest bidder on the type of the work and subsequent price to arrive at an acceptable bid. The architect must then reduce the scope of the project while giving the school district what it needs and be able to negotiate with a bidder.

Awarding the Bid

After the bids are analyzed, the architect makes a recommendation to the school staff, which then considers the bids and recommendation and in turn makes a recommendation to the board of education. In a formal resolution of that body, the bid is awarded and the legal counsel of the school system draws up a contract to be signed by the successful bidder and the board of education. This contract is then the basis of the construction work performed to erect the facility. When the contract is signed, the successful bidder gives a performance bond to the school district. This is a guarantee that the contractor will do the job it has contracted to do, including all plans and provisions. Usually this bond, again dependent upon controlling state statutes, is not less than 100 percent of the contract price.

Following the formalities of the contract award and signing, the contractor begins to mobilize the work force in anticipation of executing the contract.

The Construction Contract

There are several contractual arrangements a board of education can enter into regarding a capital construction project. In all cases, dependent upon controlling state statutes, the board of education must procure services through a competitive bidding or negotiation process. While the end result of any procurement arrangement is basically the same, the relationship between the board of education and the firm responsible for constructing the building is somewhat different in each arrangement. Some of the standard contractual arrangements are as follows:

Single Contract

The single contract arrangement is the most conventional approach and is used by the vast majority of public jurisdictions in securing services. Traditionally, a school district enters into a capital construction project after completing the educational planning stage and having the facility designed by an architect. When the bid process is completed, a contract for construction is let to the lowest responsive and responsible bidder. Figure 18.2 illustrates the traditional relationship between the board of education, the architect, and the contractor.

The prime contractor secures the services of necessary subcontractors to provide for the mechanical, structural, electrical, and plumbing systems. Under this arrangement, the prime contractor is directly responsible to the school board.

Figure 18.2
Traditional School Board/Contractor Relationship

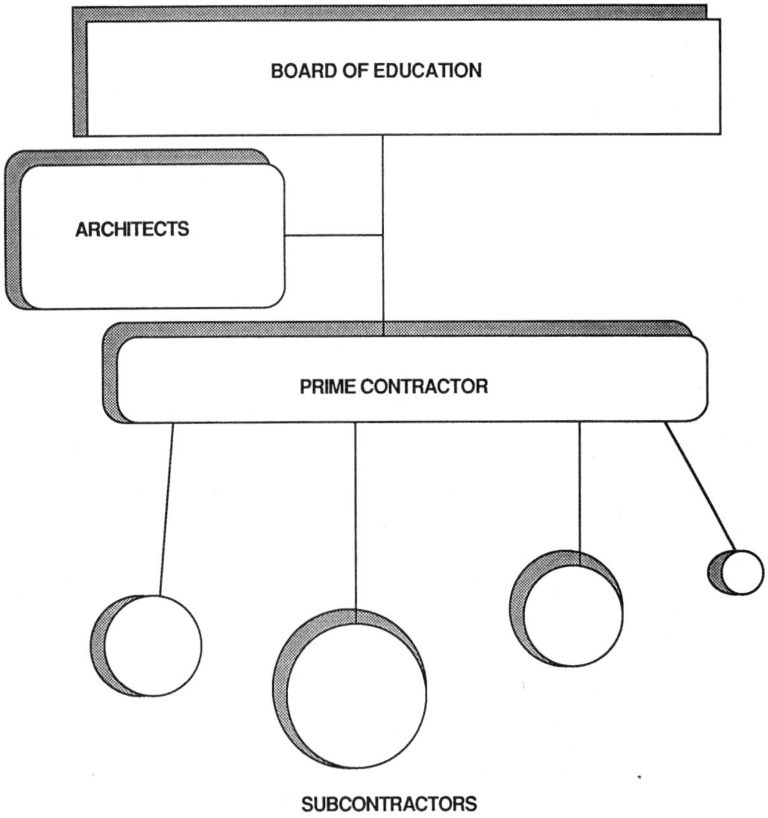

In some states, school districts are required to enter into separate contracts along the lines of the disciplines of the subcontractors. With this type of arrangement, the public jurisdiction must then provide for the coordination between the contractors either through their own personnel or through outside assistance. Figure 18.3 shows the relationship between the school district, architect, and contractors under this arrangement.

In large school districts, the type of coordination necessary to successfully complete the project is not a serious problem because personnel are readily available. In small school districts, it is often necessary to look outside for personnel to coordinate the work. The situation with the small schools seems

Figure 18.3
Multi-Contract Relationship

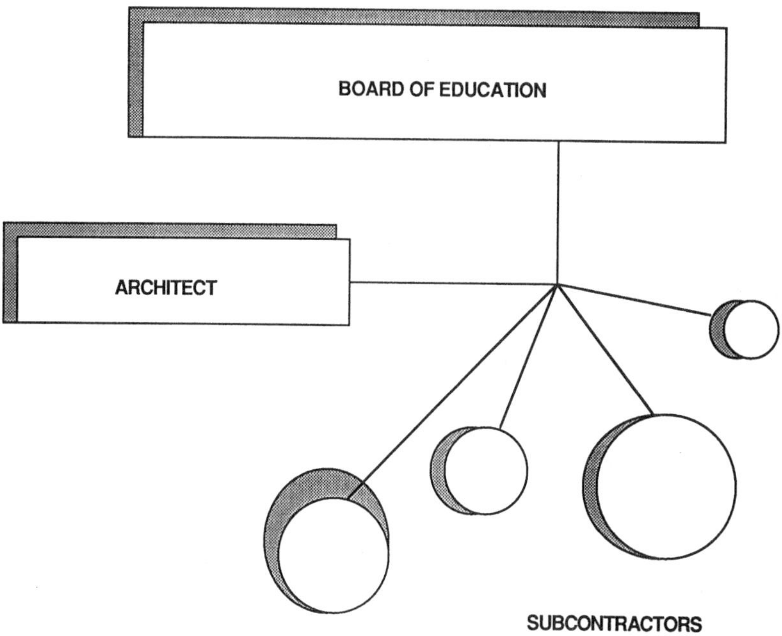

ready-made for a type of service relationship generally known as construction management.

Construction Management

The construction management contractual arrangement is shown in figure 18.4.

While the construction management arrangement may seem similar to the conventional approach of using a prime contractor, it differs in the area of responsibility.[20] As commonly practiced, the construction management firm

Figure 18.4
Construction Management Relationship

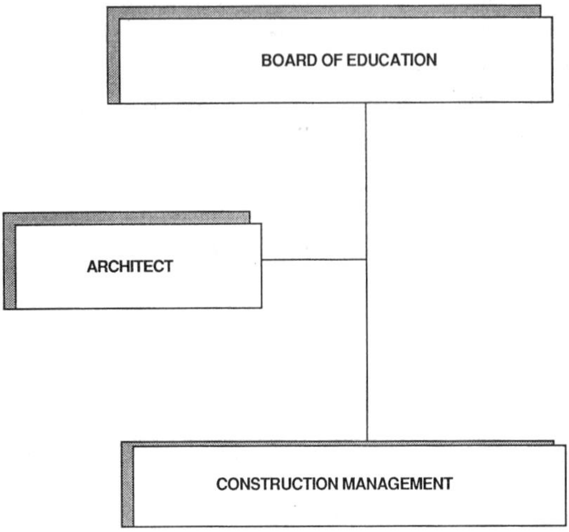

is employed by the school board to provide continuity of leadership from the design phase through occupancy. The construction management firm makes significant inputs to the architect relating to cost, quality of materials, and time. In this manner the architect can then adjust the bid package to take advantage of the market. The actual bidding of the project is left in the hands of the construction management firm. If cost overruns become apparent, the bidding specifications may be adjusted to stay within the budget. The duties of the construction management firm end with the occupancy of the building. The adjustments and guarantees made by the construction management firm become the responsibility of the board of education at the time of occupancy. Such conditions can add significantly to the total cost of the facility.

Construction management contracts vary and should be reviewed by legal counsel before the contract is signed. This type of contractual relationship is used widely in the private sector where there are few governmental restrictions regarding this relationship. Before a school district contemplates this type of contractual relationship, either state or local legal counsel should be consulted to determine its legality.

Design/Build

Another contractual arrangement that school districts often use, again within controlling state statutes, is referred to as design/build.[21] Under this arrangement, one firm does the design work and the construction work. An advantage of this arrangement is that the school district has only one firm or team to deal with through the design and construction phase, as opposed to the usual two or more under the conventional contractual method. Figure 18.5 shows this contractual relationship.

Figure 18.5
Design/Build Relationship

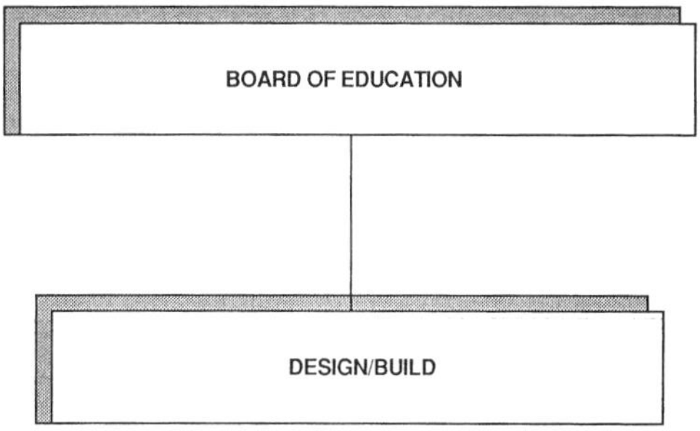

The design/build contractual method is defined as a situation where a group of professionals such as architects, engineers, and others join the builder in the formation of a team that will provide to a school district all of the services from design to construction. Each team interested in a proposed capital project is invited to submit a cost bid to design and construct a facility based on the educational program and criteria. Each team then submits to the board of education a preliminary design and construction cost estimate to complete the project. The board of education then selects one team for the contract based on both design and costs.

Initially this contractual method may seem to offer school districts several advantages, but caution must be exercised before such an arrangement is completed. First, such arrangements must conform to controlling state statutes. Some school districts are required by law to award a construction contract to the lowest responsible and responsive bidder following a period of competitive bidding. Such a provision may preclude a school district's entering into this type of contractual relationship.

Some school districts have the ability to secure such services through a process called competitive negotiations. Competitive negotiation allows a school district to choose a design/build team based on successive competitive negotiations with several qualified firms. Obtaining the services of a design/build team may well fall into that category of service procurement; however, the school district should obtain the opinions of its legal counsel and from the legal counsel of the department of education and the Office of the Attorney General.

Construction Supervision

When a capital project of any scale is under construction, the school board needs to have the project proceed in an orderly and timely fashion with the assurance that all contractual documents and building specifications are met. The periodic supervision and interpretation of the contract documents, usually furnished by the architect, does not ordinarily provide the daily inspection of construction that fully protects the owner's interest. A clerk-of-the-works or owner's supervisor, well qualified to inspect the work in progress, is often selected prior to signing the construction contract.

Continuous inspection is the only way to observe the quality of work in a construction project and assist in correction of inevitable problems. Without such inspection, no one can verify that the contract documents are being fulfilled.

Under the standard AIA contract, the architect is not responsible for exhaustive continuous inspection unless provided by agreement in the contract.[22] The amount of the fee allocated to construction administration is not sufficient to provide this service. Therefore, the board of education is required to pay the salary of a full-time on-site owner's representative.

The primary function of the clerk-of-the-works is to understand the contract documents, be aware of the construction schedule, observe the work in progress, ensure that the architect's representative is on the premises at the proper time, and see that the necessary inspections are conducted. This person should maintain records of working documents and specifications, correspondence, conferences, original contract documents, addenda, change orders, supplementary drawings, and keep a log book recording all daily activities including weather and personnel on the job. Any discrepancies observed in work performed should immediately be reported to the architect and entered in the log. It is important that the owner's representative understand the limitations of the authority of this position and that this supervision is only a supplement to the work of the architect's employees. The clerk-of-the-works is not authorized to make any decisions concerning the design or construction of the building.

Planning for Occupancy

Numerous groups and individuals must do a great deal of work to plan, design, and construct a school facility, and yet it is the staff of the school district that makes the facility educationally viable. Therefore, it is imperative that the teachers and classified staff of a new or remodeled building be oriented to how it operates.[23] Not only does an orientation serve to introduce the staff to the physical features of the building, it brings a closure to the teacher and staff involvement in the planning process. Ideally, teachers and administrators are involved in helping to formulate the scope of the program that will be conducted in the new or remodeled facility. In fact, some of those who were involved in the initial planning may be assigned to the new facility. Thus, some individuals involved in the first part of the planning process are then involved in the orientation program for moving into the facility. People who are expected to work in a new facility should know how it operates (where certain facilities are, how the new technology functions, how the structure is expected to respond to the educational program, and what the health and safety features are).

The educational planner who took the lead in the school planning effort should also exert leadership in orienting the staff. The principal of the school and the architect should also be involved in the planning and execution of the orientation program, inasmuch as the orientation needs of the teaching staff differ from that of the classified staff. The orientation effort should have several focuses. The CEFPI suggests that orientation programs consist of a user orientation program and a public information program. The former is devoted to helping those who will work within the facility become familiar with how it operates. The latter is designed to promote the idea of the school building as a community resource.

Staff Orientation

Orientation programs for teachers may take a variety of forms designed to enable everyone to understand the functioning of the building and how it supports the educational program. These activities may include: building tours, in-service sessions with fellow educators, mock fire drills, discussions centering on floor plans, and question-and-answer sessions with the architect. Orientation sessions for the classified staff may take a similar form, but the content will be directed more to how these service providers perform their job functions within the new facility.

It is very important that the custodial staff and those responsible for the HVAC system thoroughly understand the workings of the system. Many times the firm that supplied the HVAC system will conduct orientation sessions with the staff.

Public Information Programs

Public information programs are designed to let community members know what they have taxed themselves for, tell them how they may use the facility, and to encourage them to use it. Whatever the format, this program should serve to bring closure in the minds of the community with respect to the planning and construction process. Activities to be considered include: a building dedication, an open house, a student orientation, building tours for the community, media publicity, and distribution of printed materials.

Post-Occupancy Evaluation

Following the successful occupation of the completed facility, a post-occupancy evaluation is highly recommended. Data obtained at this time can be critical to future capital planning efforts. If it were not collected in a timely and organized fashion, it may never become available.

There are two things that need to be evaluated at this time—the product and the process. The product of the school planning effort is the building itself, and the process includes all of the activities of the many people and groups involved in planning and constructing the facility. Both product and process are legitimate subjects of evaluation.

Process Evaluation

The process used in planning the facility should be evaluated while it is still fresh in the minds of the individuals who participated. Effective evaluation helps to correct problems and ensures a better process for the next capital project. People and agencies involved should be surveyed to determine their perception of the success of the process and how they feel it should be improved.

There is no standard form or method for obtaining data regarding involvement. The educational planner can prepare an instrument tailored to the area that reflects the planning process used in the school district. A questionnaire can be completed by the participant, or interviews can be used to collect data, if there are sufficient resources. Data desired deals with what people feel their contribution was to the project, how other groups and individuals contributed, and whether the process was conducive to good planning as they understand it.

Product Evaluation

Evaluating the building involves determining how it operates, how it supports the educational program, how much it costs to operate the facility, and whether the school district received what it purchased. The evaluation can vary according to the desired data. There should, however, be an attempt to collect more than just the quantifiable data associated with square foot ratios, costs per square foot, cost per pupil, square feet per pupil, life-cycle comparisons, construction time, and the number and amount of change orders. The evaluation should also gather data to assess such things as architectural and educational innovations and the impact of various design features on the educational program and student movement.

The school district will need to develop and finalize the evaluation, develop proper instruments to gather data, identify people to gather data, and identify people or agencies to receive the data. Of course, the purpose of the evaluation must be identified and approved as the first step because the purpose will determine the data needed and how it will be used. The school district also needs to decide how often future evaluations will occur.

Maintenance and Operations

Operations Procedures

Regardless of how well conceived and constructed the facility might be, unless it is kept clean and in good working order the facility will not effectively support the educational program. A clean and functioning school building is extremely important to the success of the educational program. All of the successes in planning and designing a school facility can be overshadowed by poor housekeeping.

The maintenance and operations staff of an entire school district is usually headed by one individual. Depending on the size and complexity of the particular school district, this person might be either an executive director, director, or supervisor of maintenance and operations reporting to an assistant superintendent for operations or support services. In small school districts, the school business administrator may have responsibility for supervising this department along with others.

Regardless of the organizational structure, the purpose of maintenance and operations services is the same, i.e., keeping the facilities clean and in good repair. The operations staff of the school district seeks to keep the facilities as clean as possible. Efforts of maintenance staff, on the other hand, are directed at keeping the building functioning. These services can include routine and emergency repairs, scheduled preventive maintenance, painting, and other work that provide the teaching staff a healthy environment that is in working order. [24]

While most school building administrators do not directly manage the building's custodial program, they are ultimately responsible for the cleanliness and well-being of the facility they administer. However, the administrator may not be knowledgeable about the technical details of how the custodial staff should do the work to clean the building.

Roles and Responsibilities

The line of authority and responsibility for the custodial staff of an individual school building is usually shared between the principal and the school business administrator of the school district.[25] The dual nature of authority/responsibility of the custodian lies in the fact that he or she is hired and trained by one office of the district to do a service for the staff of an individual school building that has a principal who is responsible for all activities within the physical structure. The principal of a school is normally in charge of all programs and personnel within the building, or, as is commonly stated, the principal is responsible for "everyone under the roof of the school." Because of this authority, all personnel who work in that building are under the direct supervision of the principal. Thus, the principal supervises the work of the custodial staff to the extent that all custodial work must be acceptable to the principal. At the same time, the custodian is directly responsible to the school business administrator's staff. The custodian is in direct line of authority from the supervisor or director of custodial services. This relationship is illustrated in figure 18.6.

Custodians are responsible to the local school building principal for the following:

- Ensuring cleanliness of all aspects of the building and grounds.

Figure 18.6
Principal/Custodian Relationship

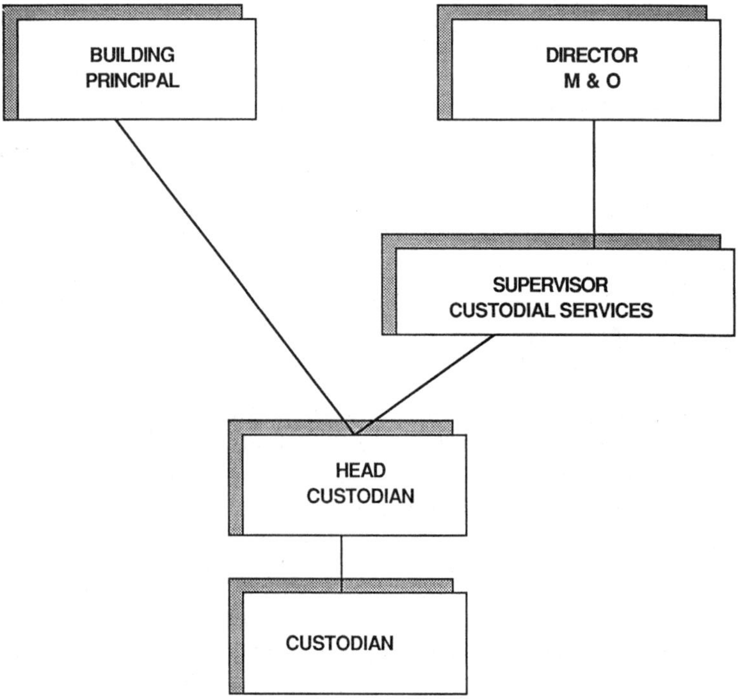

- Maintaining the proper environment, which includes temperature control, ventilation, building safety, and healthful conditions.

- Scheduling all custodial activities so as not to interfere with educational and related activities, and making the school and its facilities available as the principal directs.

- Cooperating with the staff to achieve the purposes of the school.[26]

It is important that a good cooperative working relationship exist between the custodial and instructional staffs for the successful operation of the school organization. Each staff must respect and support the work and responsibility of the other, and recognize that both staffs are working toward the same goal i.e., effective education of students.

Custodians and members of the cleaning staff are employed by the school district and are assigned to various locations. They are trained by a central administrative officer or staff in a prescribed method of cleaning and operating the school building, and their technical expertise is supervised by this staff. Therefore, the local school building custodial staff members are responsible to the central staff for the following activities:

- Carrying out established, system-wide procedures and standards of performance.

- Operating equipment and machinery of the capital plant of the school.

- Taking the necessary steps to protect the building and equipment of the school system in case of emergency.[27]

The school custodian's position is a demanding one and should be occupied by a person of the highest possible qualifications. By demanding more of applicants, the school district will obtain better employees.

Although a person meeting all qualifications might command a salary higher than what a school district may be able to pay, there are other benefits that a school district can offer prospective employees that more than compensate for the differences in salary. The retirement system of most school districts is much better and more reliable than what is available in industry. Additionally, other employee benefits are at least equal to what the private sector offers. A capital improvement program. An intangible benefit, but one that perhaps far outweighs the rest, is the pleasant surroundings in which the custodian works and the prospect of building good working relationships with a variety of employees over a period of years.

Training and Orientation

Once a person has been employed by the school district to perform custodial duties, sufficient training and orientation should take place so that the person can perform efficiently. No matter how well qualified a person might be, that individual needs to be trained in the methods and procedures of the particular school district. This training is usually conducted on a district-wide basis by central administration staff.

The type of training offered new custodial workers varies from school district to school district. Such training may be as simple as a few verbal instructions to a more elaborate preservice training program of several days conducted by larger school districts. In some instances, state departments of education have conducted custodial training in areas of the state that might benefit from such services. Because of the indigenous nature of the custodial work, the most effective programs are those that are directed to the needs of the local school district. The new custodial worker needs to learn the methods of cleaning preferred by the school district, the schedule of work at a particular station, the responsibilities that will be expected of the worker, and the level or quality of work that will be expected. Such minimal preservice training should at least enable the worker to begin work immediately and effectively. Custodial workers should never be allowed to learn on their own or to find out how to do the job themselves. To let his happen invites unmet performance standards and a sense of failure on the part of the employee.

If a school district does not have a preservice training program for new custodial workers, one needs to be established. Several avenues of assistance are open to a system that desires to establish such a program. Commercial concerns that manufacture custodial products are a good source of assistance in training to use their products. Neighboring school districts, especially larger ones, are an excellent option. In some instances, the person responsible for training custodial workers in a large school district is permitted by the superintendent to conduct preservice and inservice activities in neighboring school districts, thereby establishing a uniform training program for the region.

Preservice training is never enough to ensure competency over a long period of time. The school district should plan for a systematic inservice training program for all custodial workers. New methods and products for cleaning should be introduced at these sessions, which should be mandatory for all staff members. These inservice training sessions should not be limited to cleaning methods and products, but also cover a variety of topics and concerns to help the custodial worker become an effective member of the organization. The results of such training will improve service and overall building operations as well as boost the morale of the entire school staff.

Oversight and Evaluation

In large school buildings where there are several custodial workers, one person is normally named as the head custodian. This person usually organizes the work force and schedules the activities. The head custodian makes certain that all the tasks necessary for a clean school are discharged in accordance with the staff available and that all of the activities that take place in the building follow school district policy. Although each school district organizes the custodial program around the needs of the particular organization and available resources, local custom and union agreements often are controlling factors in specific work agreements.

The building principal should periodically review the work schedule of the custodian(s) who are assigned to the building to make certain that all tasks are covered and assignments are completed. If changes are needed in the schedule to improve the routine, the principal should confer with the director or supervisor of custodial services for the school district.

The evaluation of custodial workers is usually done by the head custodian and by the director of custodial services. The principal of the building to which the custodian is assigned has input into the evaluation process, but the final evaluation is completed by the people who hired and trained the custodial worker. Custodial staff evaluations are usually informal, but in some larger school districts, systematic evaluation programs are used.

The Maintenance Function

Every school organization has some system for completing repairs and minor improvements to school buildings. Such activities are usually the responsibility of the central maintenance staff. This staff provides necessary services to the local building principal beyond the normal scope of work of custodians. These services include: routine and emergency repairs, scheduled preventive maintenance, major and minor renovations, and updating of facilities for teaching needs such as installation of chalkboards, shelving, new lighting, and plumbing.[28]

The maintenance program is administered by the central administration under the supervision of a person usually titled director or supervisor of maintenance. This person is augmented by a trained staff representing various crafts and trades. The size of the staff and the number of crafts and trades represented vary greatly from one school district to another. Obviously, the larger the school district, the larger and more diversified is the maintenance crew. Small school districts on the other hand, usually have limited maintenance staffs and contract outside of the school district for difficult repairs. Many school districts find it more economical to contract larger maintenance tasks to outside agencies and employ a small maintenance crew to perform smaller services. Each school district, however, determines the best process and staffing configuration to take care of both large and small maintenance needs.

Requests for repair work or building improvements are channeled through the principal's office and forwarded to the central maintenance department. Each school district has forms to be used for requesting such service. These requests are prioritized, screened for budget resources, and assigned to a crew for completion. Each crew documents to the central maintenance department that the work has been satisfactorily completed. Through a feedback system, the principal of the building is notified of the disposition of the particular item requested.

In addition to requests from principals, the maintenance department initiates requests following inspection of the facilities. Such requests typically focus on technical details that an educator might overlook, such as the condition of the boiler or the roof. These inspections are usually conducted on an annual basis, although some school districts conduct inspections more frequently. Occasionally, special inspections for internal maintenance items are conducted at certain times of the year. Additionally, the daily inspection of the building by the school custodian should also serve to identify needed repairs and building improvements.

The maintenance items identified through such inspections fall into two categories—small items that the school district personnel can complete and large items for which the school district does not have the in-house expertise. Examples of the latter category include boiler replacement, new roof installation, or upgrading rest rooms. These items are then included in the CIP of the school district, funded under the capital budget, prioritized according to local needs, and completed through outside contracting.

Preventive Maintenance Programs

The preventive maintenance program is a regularly scheduled plan to extend the life of each piece of major mechanical equipment and to preserve the physical properties and function of the building itself.[29] Under this type of program, custodians and maintenance personnel inspect and service all mechanical equipment on a regularly scheduled basis to prevent any malfunctions. Likewise, periodic inspections are made of the building structure to determine wear or faults that need to be corrected. A preventive maintenance program is developed by the central administration staff for every facility in the school district in conjunction with architects and engineers. Naturally, the best time to develop such a program is during the design and construction stage of a new facility. Undoubtedly, most school districts need outside assistance to develop such a program that would encompass all buildings. Cost savings on future repairs and shutdowns more than compensate for the cost of developing such programs.

Energy Management Monitoring Systems

Many school districts have developed energy management monitoring systems that use a computer to centrally obtain data on various aspects of building conditions and energy use. Such systems monitor the building temperature, outside air, intake air temperature, output of the heating plant, speed of fans, and other similar components in order to provide the central of office with a composite look at energy consumption at any particular facility.[30] By centrally monitoring the HVAC system, appropriate measures of control can be exercised for a system that is not functioning properly. In some instances, corrections can be made to the local building heating and cooling system by the central computer, eliminating the need to send a maintenance crew to the building. Of course, such corrections are limited to adjustments to those components under surveillance, but the computer can immediately call attention to any condition that cannot be corrected through computer intervention and that needs personnel to go to the scene to make repairs. As the availability of such technology increases and the cost decreases, even small school districts will use such systems to control energy use and cost.

Role of Maintenance and Operating Staff in Energy Conservation

The duties of custodians may overlap with maintenance personnel as the first line of defense in energy conservation and in preventive maintenance. Custodians thus may make observations concerning the wear, adjustment, and regulatory levels of equipment, make minor adjustments; and report the need for further services. Regardless of the division of work between the custodial and maintenance staffs, the energy conservation efforts of both groups involve four functions: observing, cleaning, repairing, and measuring.[31]

Custodial personnel should be trained to constantly observe the workings of the facility. The custodian should observe the operation of thermostats, water faucets, toilets, and closing mechanisms. The custodian serves as the eyes and ears for the principal and maintenance staff for the safe and efficient operation of the building. The custodian can determine whether the condition can be remedied by the school custodial staff or whether the maintenance crew must be notified. In order to conserve energy, the custodian must be trained to notice energy wasting conditions. In the daily rounds of cleaning the building, the custodian should observe and check each part of the system to make certain that it works properly. Custodians should recognize that by keeping the building, lighting system, equipment, and filtration system clean they are helping to conserve energy.

Normally custodians do not assume responsibility for measuring such things as the consumption of energy, but in the present environment, custodians should be trained to take measurements in order to provide a database for energy use decisions.

Future Issues

The planning of educational facilities addresses new and changing needs of society. For example, the growing trend of site-based management as discussed earlier in this text will require additional space for conferences and meetings with the various constituencies of the school. Curriculum integration in the sciences will require greater multipurpose laboratories. More active roles for students will require greater flexible seating arrangements.[32] Greater participation of teachers will also require additional space.

Most public schools are continuing to meet the needs of special education students via Section 504 guidelines and state regulations. Additionally, all public schools must meet the Americans with Disabilities Act requiring full and open access to all individuals in our society.

The growth of technology will continue for the life of the school facilities. School planning should take into account, as much as possible, today's technology as well as allow for growth in the use of technology in the public schools.

Schools are being used as community centers in many districts. The use of these facilities for more than just formal K-12 education is projected to grow for a variety of reasons. Most school districts have determined that this is a wise use of facilities in that it fosters community support for public education.

Schools of the future must be designed with built-in flexibility. Brubaker has described this best:

> School design has learned a lesson from the modern office building, the model of flexible, adaptable space. Just as office tenants change, school programs, educational philosophies, and enrollment requirements change. So, also like offices, interior spaces in schools will be designed to meet and accommodate the new needs that are sure to arise.[33]

In many communities, the designing of educational environments should be guided, to a large extent, for safer schools. Research has indicated that the design of safe schools may have a direct relationship to criminal behavior.[34]

The future of designing, constructing, and utilizing public educational facilities continues to open a wide variety of issues and opportunities for public school leaders to enhance the learning environment and equal educational opportunities for all children in our society.

Endnotes

[1] Much of this chapter is based on the previous edition written by Glen Earthman, a noted educational facilities professor and consultant from Virginian Polytechnic Institute and State University.

[2] *Guide for Planning Educational Facilities*, (Columbus, OH: Council of Educational Facilities Planners, Intern., 1976).

[3] *Guide for Planning Educational Facilities.*

[4] For a discussion of appraisals see, M. MaRous, "What Real Estate Appraisals Can Do for School Districts," *School Business Affairs*, Oct., 1990, 48-49.

[5] See, L. Davis, "Get Your Building ADA OK: Implementing a Compliance Plan," *School Business Affairs*, Jan. 1994, 21-25, see also, J. Whisenant, "Access, Public Schools, and the Americans with Disabilities Act," *CEFPI's Educational Facility Planner*, vol. 31, no. 1, 1993, 11-13; J. Cherry, "Get on the ADA Bandwagon: Complying with the American with Disabilities Act," *School Business Affairs*, Oct., 1992, 44-46.

[6] *Schoolhouse Planning*, (Reston VA: Ass'n of School Business Officials, Inter., 1980).

[7] See specifically, California State Allocation Board, *Handbook: Options to School Districts* (Sacramento, CA: 1983); other examples of such procedures can be found in Florida and Texas.

[8] S.A. Kliment, "A-4 Architecture: Professional Options," *Architects: Handbooks of Professional Practices,* (Washington DC: American Inst. of Architects, 1982).

[9] American Inst. of Architects, *Standards Form of Agreement Between Owner and Architect*, Doc. B141 (Washington DC: AIA, 1974).

[10] For a discussion of construction management versus that of utilizing a general contractor, see, J. McKinley, "Construction Manager or General Contractor?" *School Business Affairs*, Jan., 1991 12-15.

[11] AIA Doc. B141.

[12] AIA Doc. B141.

[13] *Schoolhouse Planning* .

[14] AIA Doc. B141.

[15] B. Castaldi, *Educational Facilities: Planning, Modeling and Management*, (Boston: Allyn & Bacon, Inc.),1989.

[16] For a general overview of educational specifications see, J. Herman, "Educational Specifications for Building or Remodeling Schools," *School Business Affairs*, Oct., 1990, 44-47.

[17] G. I. Earthman, *The Process of Developing Educational Specifications,* (Monograph, College of Education, Virginia Polytechnic Inst. and State Univ.,)1-8.

[18] Again, it must be realized that in several states the legal owner is the city, county, or other governmental unit as required by statute. Generally, these school districts are financially dependent on other governmental agencies.

[19] *See, e.g., Va Code* § 11-17-1. *See also*, A.M. Goldblatt and R.C. Wood, *Construction Management for Educational Facilities: Professional Services Procurement and Competitive Bid Statues* (Topeka: NOLPE, 1985).

[20] *Schoolhouse Planning.*

[21] *Schoolhouse Planning.*

[22] AIA Doc. B141.

[23] J.D. MacConnell, *Planning for School Buildings*, (Englewood Cliffs: NJ: Prentice-Hall, Inc., 1957).

[24] *Custodial Methods and Procedures Manual*, (Reston, VA: Ass'n of School Business Officials).

[25] *Custodial Methods and Procedures Manual.*

[26] Castaldi, *Educational Facilities.*

[27] E. Stoops, M. Rafferty, R.E. Johnson, *Handbook of Educational Administration* , (Boston: Allyn & Bacon, 1975).

[28] *Custodial Methods and Procedures Manual.*

[29] *School Facilities Maintenance and Operations*, (Reston, VA: Ass'n of School Business Officials).

[30] *School Facilities Maintenance and Operations.*

[31] *School Facilities Maintenance and Operations.*

[32] R.E. Babineau, "Restructuring Education-Restructuring Facilities," *CEFPI's Educational Facility Planner*, vol. 29, no. 1, 1991, 5-9.

[33] C. Brubaker, "These 21 Trends will Shape the Future of School Design," *American Sch. Board J.*, April, 1988, 31-33.

[34] See, T.D. Crowe, "Safer Schools By Design," American Society for Industrial Security, 1991; see also, L.V. Mione and J. Robertson, "Safe School Environment Through Design," *CEFPI's Educational Facility Planner*, Vol. 20, no. 1, 1992, 53; S. Sabo, "Security By Design," *American School Board J.*, Jan., 1993, 37-39.

CHAPTER 19
Risk Management

Introduction

Throughout the earlier chapters in this text, a great emphasis has been placed on issues having the potential to create significant leadership opportunities or, alternatively, significant problems for public school administrators who have been placed in positions of high trust in school districts.[1] Although these topics have been seemingly endless in scope, the list is not yet complete because it is not enough to make good educational and fiscal decisions—at times it is necessary to anticipate disaster and to plan for its prevention or its amelioration. Although every task in this text is the most important at different moments in the life of schools, an overshadowing and constant concern in districts is the topic of protecting against harm and liability of many types. Known as risk management, this arena is vitally important to school business administrators and to all other persons involved in the educational enterprise.

Although it is possible to spend many years in schools with no evident reason to be concerned for risk management, data dispute any unlikelihood of need to prepare in advance for risks. In the general population, in 1993 alone 72,000 people were killed by accidental injury. Just for vehicular accidents, more than 2 million people were affected by temporary or permanent total disability. These disasters have taken on an air of commonality via the matter-of-factness of media reporting, but the totals are startling. In 1993, there were 42,000 motor vehicle deaths, 13,500 deaths from falls, 7,200 deaths from various poisonings, 4,800 deaths from drowning, 4,000 deaths from fires, 2,900 deaths from choking, and 1,600 accidental deaths from firearms.[2] Many of these deaths occurred in the workplace or in places of public accommodation, including schools. When millions of disabling injuries are added to the picture, news reports of yet another accident or shooting do not do justice to the complacency that surrounds the average American's worklife and homelife.

For school personnel, the risks have become greater due to changing social attitudes that provoke more violent behavior and because of changing attitudes that more aggressively confront harm. Whether such attitudes stem from opportunism or other motivation is not clear, but it is clear that schools are not exempt from these risks. For schools, anticipating and controlling for risk is an extremely important aspect of school business management since there is an increased standard of care owed to children who are viewed as unable to know and anticipate harm. As schools have increasingly become full service agencies providing more services during a longer day, the legal responsibilities have been increased as well, particularly as moral issues about unsuitable adult behavior have become more widely known. Every school is at risk for injury or death from vehicles, falls, poisonings, drowning, choking, firearms, and other risks such as natural disaster.[3] Since there is no way to completely prevent such events, schools must establish good risk management policies and plans in order to protect the public treasury and to protect the children and adults who work in schools.

Risk management for public school districts in America is thus a highly complex responsibility for local administrators and school boards given the varied and multifaceted tasks of public education. The purchase of insurance is a major component of risk management activities in a school district. However, insurance is not the only component of an overall risk management program. Other issues must be considered, and options may be available to help the school district simultaneously reduce both risks and costs. Because risk is a constant companion of the administrator, it is important for school professionals to have a thorough grasp of the issues and best available solutions. In this chapter risk management is explored from the view that if wisely structured, the district's risk management plan can serve patrons, taxpayers, staff, and children effectively and economically.

This chapter is meant to both introduce and update knowledge of risk management issues for school business administrators. From this chapter, school business administrators can gain a new appreciation of the risks faced by districts and employees in safeguarding the human and fiscal assets making up the context of modern education.

Qualifications of a School District Risk Manager

Although boards of education are abundantly aware of the enormous financial investment in property and personnel making up every school district in the United States, boards nonetheless often fail to establish sound risk management policies simply because they are unaware of the benefits of 'managing the unpredictable.' In the same vein, boards are also reluctant in many instances to recognize the importance of risk management leadership until unexpected events capture their immediate and complete attention in distressful ways. The anxiety and tenseness surrounding a rude awakening may result from many different sources such a revised state code section exposing the district to previously unanticipated risks or from the filing of a lawsuit against the district and the board individually and severally.

In other instances, however, districts have responded either in advance of need or in answer to a crisis by establishing the position of risk manager for the school district or some position that includes such responsibilities. Such appointment should be made carefully and should be based on wise discernment of school district needs and the qualifications of the risk manager. The lack of uniform response, however, has caused qualifications of successful school district risk managers to be almost as varied as the districts in which they practice. But under ideal circumstances, the risk manager will be a senior member of the business office staff with particularly important traits and qualifications. These skills encompass at a minimum:

- A trained understanding of the entire risk management concept.
- Some insurance experience.
- Budgetary knowledge.
- Knowledge of existing codes and statutes relating to risk.
- Knowledge of legal contracts relating to risk.
- Knowledge of building design requirements.

The breadth of these areas is imposing in its weightiness. While it is beyond the scope of most people to be expert in each of these areas, the competent risk manager will have some exposure and general knowledge and competence in each of these areas. In larger districts, the risk manager will often have previous experience in one or more technical insurance fields, as well as the ability to have subordinates with special skills that round out the entire complement. At the minimum, however, the person designated as risk manager should be empowered to enlist the services of other staff members or outside persons as technical resources to permit functioning at an acceptable level of efficiency. Beyond this point, it is critical that the risk manager be aware that risk management is an integral part of the overall management of the district and requires constant attention to planning and evaluating risk management, rather than consideration that arises only in response to a particular problem that has occurred.

As noted elsewhere in this text regarding many different issues in school districts, policy forms a fundamental element of risk management. The basic concept of risk management should be of primary concern to all persons in the district, and policy addressing the operation of the district should be set out in writing by the school board. Risk management should be part of the district's policy development activities and in the broad oversight context should include:

- Board policy statements addressing the importance of risk management and setting out the district's attitude toward careful planning.
- Specific procedures to implement these board policies.

- Objectives of the risk manager's position, leaving no doubt about the importance and centrality of this person's work.

- Statements about the responsibility and authority of the risk manager to accomplish the task of managing the district's risk program.

These elements of policy can be further understood to form what might be viewed as a series of job and task descriptions that should be built into board policy and operational policy manuals. Development of a good risk management program at the school district and building levels requires at least five broad areas, with specific subparts and accompanying policy manuals, all having direct implications for the qualifications and tasks of the risk manager. According to one very recent source, an appropriate elemental list would include at least:

Broad Areas

- Identification of components.
- Identification of program goals and objectives.
- Development of school board policies.
- Responsibility identification.
- Organization of site committees.

Subparts

- Organizational chart of responsibility.
- School district policies, emergency procedures, and administrative procedures.
- Safety committees.
- Student safety instruction.
- Employee safety training.
- Site risk analysis.
- Risk inspection procedures.
- Safety devices.
- Accident reporting and investigation.
- Accident analysis.
- Costs of claims analysis.
- Safety promotion and awards programs.

- Hazardous waste program.
- Radon monitoring program.
- Asbestos management program.

Policy manual

- Employee accident, injury, investigation, and prevention:
  - Definitions.
  - District manager responsibilities.
  - Site manager responsibilities.
  - Notice of injury forms.
  - Accident investigation and prevention.
  - Disposition of claims correspondence.
- Extended leave for injury or illness in line of duty:
  - Definition.
  - General information.
  - Forms and applications.
  - Student or visitor injury: reporting, investigation, and preventions.
  - Definition of a student/visitor injury.
  - Site manager responsibilities.
  - Student/visitor injury form.
  - Accident investigation and prevention.
  - Disposition of claims correspondence.
- Student injury insurance claims:
  - Definitions of student injury insurance claims.
  - Site manager responsibilities.
  - Student injury claim forms.
  - Student injury claims processing.
  - Disposition of claims correspondence.
- Vehicular accidents:

- Responsibilities for district owned/leased vehicular accidents.
- Private vehicles used for official business.
- Reporting forms.
- Vehicle accident report.
- Driver education accident report form.
- Disposition of claims correspondence.

- Claims for damages to personal property other than vehicles:
  - Definition of personal property.
  - Site manager responsibilities.
  - Disposition of claims correspondence.
- Protective clothing and equipment program:
  - Definition and references.
  - Application.
  - Responsibilities.
  - Safety self-inspection program.
  - Inspection forms.
  - Frequency of self-inspection.
  - Inspection checklist retention.
- Unsafe equipment:
  - Definitions.
  - Responsibilities.
  - Procedures.[4]

Although many aspects of risk management are internal to a district as seen by the early sections of this chapter, much of the success of the risk manager will depend on the ability of the individual to work well with the various segments of the insurance community. This success will depend in large part on the ability to evaluate the advice, counsel, and marketing techniques of insurance brokers. This ability rests largely on experience in that it is necessary to distinguish between sound advice and marketing efforts. The competent risk manager must not succumb to either extreme in regard to dealing with insurance brokers. On the one extreme is the risk manager who believes everything represented by the broker. Equally bad is complete distrust of any information originating from the insurance industry. Qualifications for risk manager, then, rest in high technical skills and the ability to be discerning by understanding that competing interests are at stake in the commercial marketplace that must be offset against the superior skill of insurance professionals in that field and against the needs of the district to provide a comprehensive risk management strategy.

Fundamentals of Risk Management

The ultimate goal of risk management is to minimize the risks of doing business and to do so at an acceptable cost. While a for-profit business which faces prohibitive costs of risk can decide to discontinue its operations, schools usually cannot take such definitive action because they are established for a set of reasons not related to a profit-loss statement. Rather, schools must continue to operate and manage risk and the attendant costs, regardless of whether continued operation is fiscally efficient. In such situations, the risk management function is a vital aspect of district operations in that its management can be critical to the financial well-being of the district. In a corporate situation, managers are responsible to stockholders for the responsible management of company assets; managers of school assets, however, are responsible not only for management of fiscal assets, but also for how effectively assets are used in the educational process.

Certain fundamental aspects of risk management need to be identified. The fundamentals of risk management may be broken down into four basic elements making up the process of managing risk:

- Identification and measurement.
- Risk reduction or elimination.
- Transfer of risks incapable of assumption.
- Assumption of remaining risks.

Identification and Measurement

The first element of identification and measurement states that it is impossible to manage risks if they are not known. Thus, one of the most important roles of risk managers is to establish an information gathering system that will inform and advise concerning all aspects of the operations of the district. This process should not be limited merely to the physical risks of district operations, but should also extend to many areas of events transpiring in schools such as curriculum development, parent clubs, community use of facilities, break-time activities of the staff, classroom activities, and field trips. The successful risk manager should take strong steps to be viewed throughout the school system as an advisor to whom other administrators or personnel can come for help in evaluating all types of risk. For example, participation in curriculum development committees will enable the risk manager to provide input during the planning stages of new programs, and participation in recreation planning will assist in knowing and assessing risks relating to playgrounds, athletic fields, and so forth. Products stemming from such involvement might include various flow charts, checklists, inventory forms, and questionnaires to gather information and to assess the degree of risk associated with the many activities sanctioned by the district. It is important that such aids form part of an overall information gathering system and not merely be completed as a matter of form.

The role of risk management relates to measuring the amount of risk associated with each operation in the district. Three dimensions of risk deserve particularly careful consideration when seeking to measure risk and are known as *probability*, *frequency*, and *severity*. While each needs to be considered individually, the relationship between them is critical in that these parts interact to form the element of risk exposure.

A detailed discussion of hypothetical probability distribution is beyond the scope of this chapter. However, a general understanding of this subject is essential to the risk manager, and it must be recognized that probabilities form the basis of actuarial underwriting by casualty insurance companies. By definition, probability is the likelihood that some event will occur. For example, insurance premiums are based on the probability that some loss will occur wherein the higher risks require higher premiums since insurance companies are not in business to lose money. It is helpful to divide probability into categories indicating the potential for occurrence, such as *none*, *low, moderate, high*, and *definite*.

A corollary concept is frequency, in that the probability of an event may be augmented by a greater degree of frequency. For example, a risk that is relatively low in probability might also be expected to occur with less frequency, while a high probability can also be expected to occur with more frequency. Various combinations of these elements are possible, making it important for insurers or risk managers to anticipate the likelihood of a given occurrence and its potential for reoccurrence.

While probability and frequency are important considerations, the severity of loss is actually the more important consideration. For example, glass breakage may be highly probable and frequent, but its severity in terms of loss may be fairly low. In contrast, bus accidents carry far more severity, while probability and frequency may be low. Management of catastrophic losses is therefore the single most important function of the risk manager.

Risk Reduction or Elimination

The second element of risk reduction or elimination forms a critical aspect of risk management and states that a primary task of the risk manager is to reduce or eliminate risk at every opportunity. Unfortunately, only in rare instances will the risk manager be able to eliminate risks completely. Generally, the genuine contribution of a risk manager is to educate other decision-makers concerning the hazards of doing business and to direct energy and expertise toward reducing risks. For example, one major area of concern is the handling of moneys in the district. Eliminating risk is impossible, but reducing risk is possible by setting up a system of checks and balances that is adequate to meet all legal and insurance requirements. In this instance, this would include proper accounting for grants and bequests, as well as all other funds. Effective controls over financial matters thus provide a way to reduce risk and lead to lower premium costs when purchasing fidelity coverages because frequency and probability have been reduced.

Transfer of Risk

The third element of good risk management is transferring risk when economically efficient to do so. Risk transfer can be divided into two basic areas. The first is the non-insurance transfer. Often a risk can be transferred to another party through contract or other agreement. An example of transferring risk in this manner is requiring contractors to accept liability for all losses resulting from their work. Another example is a "hold harmless" clause written into a contractual agreement. In this case, one party to the agreement agrees to hold the other party harmless for certain acts or conditions. Hold harmless agreements are often included in standard contracts or sales agreements. Caution must be observed, however, in that legal precedent is not clearly established in all instances since some liability cannot be transferred by such agreement.

The remaining method to transfer risk is the insurance transfer. In theory, the purchase of insurance should be the last step taken because it costs money to purchase insurance, while other means such as reducing risks or requiring other parties to assume liability carry no direct payments. However, when the cost of insurance is not prohibitive, the purchase of insurance is often the only way to effectively transfer risk. As the cost of insurance rises and/or the insurance market refuses to insure certain types of hazards, insurance purchasers become more creative and resourceful.

Assumption of Risks Incapable of Transfer (Retention)

The fourth element of good risk management may include a decision to retain a risk. Such a decision is not to be undertaken lightly and will depend on many serious factors. Identifying those factors in a prescriptive sense is difficult because the contingent factors vary as market conditions change. As a general rule, however, risks should be retained when the frequency is high and the severity is low. In such cases, the purchase of insurance is not necessarily efficient since it basically amounts to "trading dollars." For example, if the district were to have a high frequency of glass breakage, it may be more effective to have glass replacement handled by maintenance personnel than to purchase insurance against such breakage because the district could trade dollars with outside contractors through insurance premiums equal to or greater than the cost of internal repair. Use of insurance deductibles is another form of risk retention. High deductibles on relatively low risks are not a wise use of money in that a higher deductible amounts to assuming risk for an event of low probability. The calculated use of deductibles based on an analysis of frequency and severity can thus reduce the combined cost of insurance and retention. More problematic, however, is that the district may occasionally be forced to retain risks when no market exists for the placement of insurance or the cost is prohibitive due to unfavorable probability, frequency, and severity.

Legal Aspects

An especially important aspect of risk management is understanding that risk is an aspect of district operations with tremendous legal burden for liability and statutory form. A major part of wise risk management therefore includes a realization that a risk management system is not complete until legal counsel has reviewed and approved the district's plans. Insurance companies hire attorneys to review the operations, and school districts should do the same since many areas of risk management involve legal research and analysis as well as other legal skills such as letting bids, conducting negotiations, drawing contracts, negotiating settlements, and so forth. Further, applicable laws may be either state or federal, requiring interpretation and defense. As a result, the risk manager needs frequent access to competent legal counsel. If such counsel is not readily available, it is imperative for the risk manager to negotiate some arrangement to provide convenient access to legal opinions, as proper use of legal counsel avoids many problems. Too often, attorneys are called upon to defend actions that would never have been taken had counsel been given earlier.

Contracts

A legal area of particular import for risk management involves contracts. Generally underestimated by most administrators, school districts daily assume liability by contract without being fully aware of the exposure. Examples abound, including purchase orders, sales contracts and agreements, construction contracts, maintenance agreements, building and equipment leases, and employment contracts. They all contain hidden liabilities in the sense that clever opposing counsel can show greater liability for the district than was ever intended from the district's view. As a matter of extreme principle, all contracts entered into by a district should be reviewed for insurance considerations, as well as budgetary and legal ramifications. Many districts have an agreement to send copies of all contracts to the insurance broker of record for review and also to staff attorneys or outside attorneys under retainer.

As established in chapter 7 concerning legal liabilities of administrators, districts and employees may not assume they are protected by sovereign immunity. By implication, then, risk is assumed in all instances where immunity is not available. This includes risk for buildings, as well as vehicles, and other general liability. Many districts lease buildings from other owners, making the district a tenant. Using the same logic stated earlier in which the district seeks to transfer risk, it is reasonable to assume that lessors would seek to transfer liability to the lessee. The result is a strong need for competent legal review of any and all leases because the district as tenant may be required to purchase building insurance. With the advent of vehicle leasing, the same requirements apply. Leases for equipment and so forth are becoming more common. Under these conditions, special attention to legalities of liability takes on a new dimension. The lease must be specific as to:

- Amount of coverage.
- Source of valuation and frequency of update.
- Additional insured requirement.

The issue of indemnity is a particularly difficult one in that many contracts seek to demand assumption of liability by one party to the exclusion of the other party. Under such an agreement, one party assumes some or all of the other party's liability. While such agreements are subject to negotiation and reflect current practice in a particular area, the party in the weaker negotiating position is often forced to agree to such an arrangement. The district must be aware of the consequences of agreeing to accept liability through a hold harmless agreement, especially since insurance contracts in force may not permit such agreement. As a general rule, the district should attempt to require the other party to hold the district harmless. Such a clause should stipulate that the indemnitor will not only pay any damage award, but also will defend or pay the cost of defense on behalf of the district. The latter is preferable because it would allow the district to choose its own counsel.

If the district were to be held harmless by another party to a contract, legal counsel is nonetheless critical, as it is important to verify that the other party will be able to pay an award for damages. In addition to requiring the other party to show proof of insurance, the district should also require that the district be named as an additional insured. Sample statements to be reviewed by legal counsel include language such as:

SAMPLE WORDING FOR AN ADDITIONAL INSURED CLAUSE

The named district is to be included as an additional insured as respects liability arising out of the use of _____. It is further agreed that ______ liability policy should be endorsed to be primary insurance as respects the interests of the district and that any other insurance maintained by the district is considered excess and not contributing insurance.

SAMPLE WORDING FOR A HOLD HARMLESS AGREEMENT

Contractor shall take all responsibility for the work, shall bear all losses and damages directly or indirectly resulting to the contractor, to any subcontractor, to the district, or to district employees on account of the performance or character of the work, unforeseen difficulties, accidents, occurrences or other causes. Contractor shall assume the defense cost of and save harmless the district and its officers and employees from all claims, loss, damage, injury and liability of every kind, nature and description directly or indirectly arising from the performance of the work. Approval of the insurance contracts does not relieve the contractor or subcontractors from liability under this hold harmless clause.

As noted, use of legal counsel and appreciation for the legal nature of risk management is also found in relationships between statutes and the district's liability exposure. As discussed in chapter 7, liability is a serious matter for administrators and districts, and the source of such liability rests in statutory code that sets out immunity or exceptions to immunity. A major aspect of the risk management program is knowing the controlling state statutes. Such knowledge should be the basis of any decision to purchase insurance or to retain a risk. Some states allow the district to elect certain options in insurance and self-insurance (see later section on self-insuring). When code requirements for insurance do exist, the district should seek assistance in assuring that the code section numbers are listed in the insurance policy or endorsements, as it is a principle of insurance that any loss not technically covered will not be paid.

Another critical aspect of statutory code relates to how insurance is to be obtained if it is legal to do so.[5] In some states, requirements to bid insurance can limit the options available to the risk manager. On the other hand, many districts assume that such a requirement to bid coverage exists even though the bidding requirements for their state may actually apply only to the purchase of materials and supplies. State requirements should be reviewed with district legal counsel because bidding insurance may not always produce the best result. Putting the district's insurance requirements out to bid every year does not provide continuity between the district and its insurance carriers.

While this may not seem important in a soft insurance market, such continuity can provide a vital loyalty edge when market conditions tighten. An alternate method with many variations involves the selection of an insurance broker through some type of competitive selection process based upon specifications of services desired. The broker is then charged with obtaining competitive quotes from various insurance carriers based on specifications developed jointly with the district. Since total risk management needs much more than just the purchase of insurance, selection of a competent broker is critical to the risk management program of the district. The broker should have staff qualified to advise the district on the condition of the insurance market, as well as assisting the school district in developing the best overall package of insurance coverage at the best price available.

Loss Control

Although the focus of many districts' risk management programs is on insurance, there are steps that should be taken before insurance is relied upon. As stated earlier, this derives from the perspective that loss to districts should be prevented rather than indemnified. Although insurance admittedly forms a major element of a risk management program, district-initiated loss control measures should be an integral aspect of a total risk management plan.

Whether highly sophisticated or structurally simple, the goal of any loss control program is to conserve the district's many and varied assets. These assets can be divided into three general categories of people (students, staff and public), property, and cash. To successfully achieve the goal of asset conservation, a variety of strategies should be implemented, including:

- Loss identification and measurement.
- Loss avoidance or prevention.
- Loss reduction.
- Post-incident claims control.

Without these measures, school districts are likely to have insurance claims denied and may become largely uninsurable for failure to take ordinary and prudent precautions to stop avoidable losses.

Loss Identification and Measurement

The most basic steps in identification and measurement were discussed in the beginning of this chapter. This section goes further by emphasizing the importance of adequate record keeping to promote effective loss control.

Loss avoidance or prevention.

Obviously, the most effective loss control activity is the avoidance or elimination of any activity or circumstance that may produce a loss. It is at this level of loss control that the greatest impact on district finances can be accomplished. If the risk administrator is involved in the planning stages for all new activities of the district, many ill-advised activities can simply be avoided. Board policy should require that all new curriculum programs be reviewed by the proper panel of district personnel prior to implementation. Such review helps ensure that new programs conform to established district policy and would include loss control considerations. Examples of such policy requiring review can be seen in figure 19.1 and figure 19.2.

Figure 19.1
Sample Policy Statement

It shall be the policy of the governing board of the ____ School District to maintain a risk management program consistent with all legal requirements pertaining thereto and within the financial capability of the district, providing a program of protection and/or insurance against:

> Loss, damage, or liability as mandated or permitted by the education, government, labor, insurance and other pertinent codes of the state of _______; to include, but not limited to, loss or damage to the real and personal property of the district, liability of the district and/or the personal liability of members of the governing board and employees of the district for damages to persons or property, including injury or death, losses due to employee dishonesty, worker's compensation liability, and the provision of a program of health and welfare benefits for employees to the limits established from time to time by the governing board.

Under the provisions of this policy, the superintendent shall develop and maintain rules and regulations necessary to implement and administer all aspects of this policy.

Figure 19.2
Rules and Regulations for Administration and Supervision of the District's Risk Management Program

- **Management and Control.**
- Management and control of the District's Risk Management Program and any and all insurance authorized by the Governing Board shall be a function of the Business Division.
- The _______ shall be the Risk Manager for the District.
- The Risk Manager shall have authority to establish rules and procedures within Board policy to insure the safety and wellbeing of pupils, employees and the public while on or in district property. Such rules and procedures shall be designed to maintain the District's liability at a minimum and insurance premiums as low as possible consistent with insurance requirements.

- The latest available professional risk management techniques shall be utilized to identify, analyze and minimize risks inherent in the operation of District programs.
- Risk considered to be unretainable shall be transferred. Such transference shall normally be through the purchase of insurance; however, the use of a pooling agreement is permissible if considered in the best interest of the District.
- **Program/Site Manager Responsibility**
- It shall be the responsibility of each Program/Site Manager to be constantly alert to the risks inherent in its operation and to make every effort to minimize such risks.
- The Program/Site Manager shall educate employees concerning their responsibilities under the Risk Management Program and enlist their support in minimizing such risks.
- Each employee shall be charged not only with operating in the safest manner possible, but with alerting others, especially management, to any hazard and to safer methods of operating.
- Each Program/Site Manager shall be charged with the safekeeping of all assets charged to the program, and with the security of the facility which houses the program, unless that duty has been specifically charged to another person.
- Program/Site Activities. Each Program/Site Manager shall be responsible for the following:
- Completion of questionnaires and survey forms on existing activities.
- Notification of new or proposed activities to the Risk Manager.
- Cooperation with the Risk Manager for onsite inspections and analyses of physical operations.
- Program/Site Safety Committee
- Each Program/Site Manager shall form a safety committee which shall meet at least monthly to inspect facilities and equipment.
- Copies of safety committee reports shall be forwarded to the Risk Manager.

Placement of Insurance

- The Risk Manager shall make every effort to obtain insurance coverage at the most economical cost, consistent with required service, by obtaining quotations, by competitive bidding, or through negotiation, using whichever method is considered most advantageous to the District.
- The Risk Manager shall report insurance requirements, together with all relevant information including premium quotations, and recommendations for placing such insurance to the Superintendent, who shall present such recommendations to the Governing Board.
- Upon authorization by the Governing Board, the Risk Manager shall place such insurance in accordance with said authorization and the relevant provisions of these rules and regulations.
- In an emergency, the Risk Manager shall place insurance; and the Superintendent shall immediately report such placement for ratification by the Governing Board.
- The Risk Manager shall be responsible for maintaining continuous coverage of all insurance programs mandated by law or Board Policy.
- **Types of Exposures to be Protected Against**
- The following insurance shall be carried in accordance with applicable rules and regulations and/or such self-insurance program as the Governing Board may authorize:
- Fire and extended coverage insurance covering all buildings at replacement cost which are owned or occupied by the District, in such amounts as are authorized by the Governing Board.

- Comprehensive liability insurance or a self-funded program covering the District, members of the Governing Board, District officer and employees while acting in the discharge of their duties within the scope of their employment and/or under the directions of the Governing Board. Errors and Omissions coverage should be included in the comprehensive liability package or as a separate policy.
- Fidelity bonds protecting the District against loss occasioned by fraud or dishonesty of officers, employees and agents of the District.
- Worker's Compensation insurance or a self-funded program covering all employees and volunteers of the District, pursuant to the Labor Code of the State of and sufficient to provide benefits as prescribed by law.
- Casualty, fire, theft and extended coverage insurance on selected vehicles owned or operated by the District to the limits as specified by the Governing Board.
- Burglary and robbery insurance covering moneys and property as specified by Board action.
- Boiler and pressure vessel property damage insurance, covering boilers and such pressure vessels and property to such limits as authorized by the Governing Board.
- Sprinkler leakage insurance on property where designated by the Risk Manager and authorized by the Governing Board.
- Umbrella accident insurance for students participating in field trips and athletic events and who are not otherwise covered.
- Student accident insurance will be made available and premium paid by the parent or guardian.
- Programs of insurance in the area of employee fringe benefits such as: hospital and medical insurance, dental insurance, life insurance, long-term disability insurance; but only in the kinds and to the limits authorized by the Governing Board.
- Such other insurance coverage as the Governing Board may authorize.
- **Insurable Value—Buildings and Contents**
- The insurable value of a building shall be defined as the replacement cost of such building less the non-insurable items set forth in the District's fire insurance forms.
- The term "contents," as used in connection with insurance, shall be defined to include all personal property not specifically excluded by terms of standard form for fire insurance policy.
- **Settlement of Losses**
- When any property covered by insurance is lost, damaged or destroyed, a notice concerning the loss shall be sent immediately to all affected carriers or their representatives. The Risk Manager shall act as adjuster for the Governing Board in the settlement of losses and shall sign the Proof of Loss as authorized. The Superintendent shall recommend to the Governing Board, as soon as practicable, a basis for settlement and, upon adoption by the Governing Board, the agreed amounts shall be collected from the insurance company(ies) and

deposited to the credit of the District in accordance with procedures governing such collection.

- **Liability Claims Procedure**
- All Claims for Damages filed with the District pursuant to the Government Code of the State of ______ shall be referred to the appropriate insurance carrier or self-insurance administrator for recommendations regarding acceptance or rejection of the claim. Such recommendations shall be presented to the Governing Board for Approval. For all claims rejected by the Board, a written notice of rejection shall be sent to the claimant. Such notice shall contain a warning advising the claimant of the statutes for filing court action pursuant to Section ____ of the Government Code of the State of

Many decisions regarding new programs, materials, personnel, sites, and equipment are often made or directly influenced by staff members having no guidelines regarding loss control or loss control activities. Although not all losses are avoidable, the frequency of losses can be reduced through proper loss control activities. Information concerning losses that have occurred previously is essential to the prevention of reoccurrences.

Loss Reduction

Two different activities need to be implemented to effectively reduce the severity of loss. First, as much preventive work as possible must be undertaken to minimize the severity of a loss. Second, a great amount of planning needs to be undertaken to provide necessary training, procedures, and materials for post-incident claims control. The list of preventive activities seen earlier in this chapter can assist greatly in loss reduction.

Post-Incident Claims Control

Post-incident claims control involves those activities that occur in the critical period immediately following the discovery of a loss. For example, the point at which a fire is extinguished is the starting point for post-incident claims control. These activities include recovery of salvageable materials, security to prevent further loss, and locating and checking inventory lists.

School Board Commitment

Without doubt, implementation of a loss control program must begin at the top of the organization. Without support of the school board and the superintendent, a loss control program is merely a literary exercise. The board of education should adopt a simple and clear policy statement expressing commitment to controlling losses and should instruct the superintendent and staff to develop the necessary regulations needed to implement such a policy. This stated commitment by the school board will make it known to the staff and community that loss control is a district way of life and not a paper project. A policy similar to the one as reflected in figure 19.2 is illustrative of the convincing detail such policy should entail.

General Issues

Of equal importance to loss control programs is assignment of staff responsibility for risk management. If this assignment is made too far down the organizational ladder, the program will suffer from lack of attention and apparent unimportance. The importance of the loss control administrator to the system was apparent earlier in the chapter when it was argued that this person needs specific skills in the areas of local codes, insurance principles, claims adjusting procedures, building codes, management techniques, and finance and accounting operations. The development of loss control programs for the district is not as difficult as is the communication of these programs to the staff. Many sources of model programs exist for the novice loss control administrator, including professional organizations, large corporations, nearby districts, and any number of professional publications from the Association of School Business Officials, International.

Communication of loss control programs and the importance to the district are the most difficult steps. It should be obvious that one administrator or even the entire business office staff cannot implement complex programs without the cooperation of the entire district staff. No loss control administrator could ever find the time or expertise to develop and implement effective and efficient loss control programs alone. An organizational chart such as in figure 19.3 detailing risk management should be developed to assist in securing the school district's interests.

Pre-employment and Ongoing Measures

The complexity and wide-ranging impact of risk management seen in figure 19.3 indicates that risk policies are not onetime measures that can be considered complete at any point in time. Rather, an important part of a loss control program is the district's pre-employment measures and ongoing processes that take advantage of preventing a high probability of claim loss. Although space prohibits a full discussion of all these steps, a series of important measures could include most or all of the following:

- Pre-employment physical exams.[6]
- Job-related safety training programs.
- Safety communications system.
- Accident investigation programs.
- Safety committees.

Figure 19.3
District Risk Management Diagram

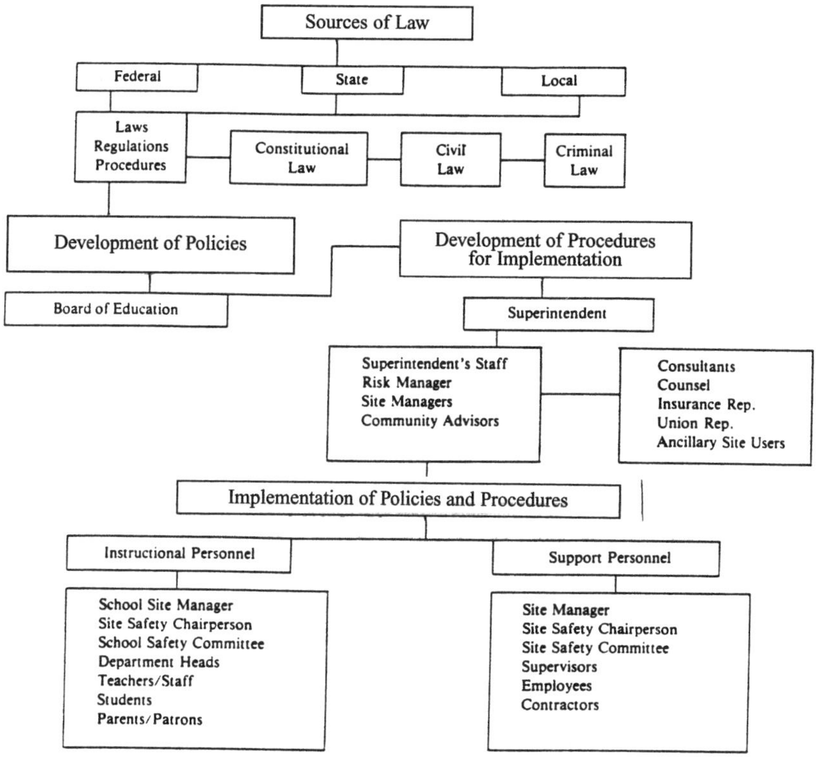

- Safety inspections.
- Safety materials, equipment, and training.
- Recordkeeping system.
- Safety performance as part of employee performance evaluation.
- Adequate inventory procedures.
- Participation with other agencies, companies, or coalitions to share ideas and resources on managing risk.

A complete risk management program is likely to contain nearly all of these items. Yet despite the obvious benefits of these steps, care should be taken to observe all requirements of law relating to special problems that could arise when implementing some of these items. For example, establishment of an effective pre-employment physical can be difficult due to a complex maze of federal, state, and local regulations. In particular, under the ADA federal antidiscrimination laws can be very intimidating. However, exemplary effective programs are in place in both the public and private sectors. One critical point is that such a program to screen out high risk through physical examinations must be based upon the actual physical requirements of each job. If a particular physical condition has no bearing on performance in the position in question, any consideration of that condition will almost certainly be considered discriminatory. Determination of whether an applicant can physically perform a particular job can only be made by a qualified health professional. Most importantly, the health professional cannot make a defensible determination without a recent and detailed job description of the subject's position. If properly structured, the pre-employment physical examination as a legitimate part of the employment process is of vital interest to the loss control administrator. It must be treated as part of a web of legally sensitive areas relating to:

- Permissible questions on job applications.
- Statement attesting to truthfulness of statements on the application, signature, and consequences for falsification.
- Use of medical information to dismiss or not hire.
- Confidentiality of medical information.

As a general rule, districts should pursue any means to reach their goals provided that they observe the intent of the laws. In the case of physical exams, such measures have value only in the losses they prevent. Further, no discrimination may occur due to a disability as defined under the Americans with Disabilities Act.

Other measures, however, should be pursued aggressively. For example, accident investigation is an accepted method of loss control, since the purpose of investigating accidents is to prevent future occurrences, rather than assessing blame. Establishment of a closed loop accident reporting-investigation-prevention program cannot be accomplished without the cooperation of the employees. Statistics indicate that about 85 percent of all worker's compensation claims are caused by improper work habits, leaving only 15 percent caused by defective "things" such as equipment or stairs. As a result, loss prevention through accident investigation involves data collection, data analysis, plan formulation, and on-the-job safety measures.

Data Collection

Many of the most effective loss control programs are those which are developed in-house to correct some known problem area. Therefore, a principal element in loss prevention is identification of a method of early detection that will catch problems before costs mount or before work habits become too firmly established to reverse easily. In larger organizations, computerized databases are often used to sort and classify large quantities of data on risks experienced. Regardless of the size of an organization, simple and concise procedures for gathering data concerning causes and cures for various loss-generating incidents must be established.

To assist with data analysis, a loss matrix can be developed. For example, in the case of worker's compensation claims, the first step in loss analysis should be to categorize actual loss experience data, i.e., type of employee by type of injury, part of body by type of injury, type of accident by type of injury, part of body by type of accident, and type of employee by type of accident. Such a system can be useful across a district's curriculum and its classified operations as well. When these data have been arrayed in matrix format, it is relatively simple to decide where the frequency and severity of losses warrant attention and the priority order for implementing a loss prevention program. When the categories have been prioritized, data from accident investigations can be used to develop plans to prevent future recurrences. Once the closed loop of accident investigation—>plan—>corrective action—>evaluation is established, refinements can be made in the system until it is truly responsive to changes in the patterns of accidents. Such a system is also favorably rated by casualty insurers who are understandably reluctant to cover organizations that do not methodically analyze their risk profile.

As the first line supervisor becomes involved in investigating accidents and completing the proper reports, the result will be awareness of which factors cause the most accidents. When management has factual information on the actual causes of accidents, it can make adjustments in work routines, equipment, facilities, and training to reduce or eliminate the causes of many accidents.

On-the-Job Safety

Finally, a risk management program should take into account issues relating to on-the-job safety. While this aspect is built into the previous discussion, there are additional steps that can be taken to enhance this aspect of loss prevention. Most of these steps relate to worker awareness and participation in loss prevention since about 85 percent of employee compensation claims are due to employee work habits.

An important and effective element of safety programs is the formation of safety committees. Experience shows that safety committees provide significant feedback on how employees view their jobs, as well as on potential problems. It is only reasonable since employees are the people who do the work and who are injured, to make them well informed concerning the probability and causes of injury. However, forming safety committees means the employer must be prepared to correct all deficiencies identified by the safety committee; otherwise, the committee can become a negative factor, and the district will create additional liability for itself if it fails to act on known safety problems. Safety committees must receive direction from the employer if they are to serve the intended purpose.

To maximize the effectiveness of loss control programs, it is important to include as many employees as possible. Extensive and meaningful employee involvement can be aided by provisions that favorably impress employees and make them want to cooperate. These may include:

- Agreements to share verifiable savings with employees.
- Including safety programs in contract negotiations.
- Making safety performance part of the employment and performance review process.
- Education programs for employees to explain and "sell" safety.
- Incentive programs to create peer pressure for those who might abuse the workers' compensation system.

OSHA—Occupational Safety and Health Act

In addition to general liability arising from state statute and the general benefits of risk management programs, the Federal Occupational Safety Health Administration (OSHA) laws provide an incentive for employers to provide a safe work environment. Under federal legislation, OSHA enables individual states to adopt statutes regulating safety at the workplace, but OSHA provides that state laws may not be less stringent than federal law. Since many states have exercised this option under OSHA, it is difficult to provide an overview of state occupational safety laws. It is more useful to advise administrators to obtain a copy of the *General Industry Safety Orders* or the equivalent for the appropriate state from either federal or state OSHA offices. The goal of all federal and state OSHA laws is the protection of workers from the preventable and/or negligent hazards in the workplace. Importantly to administrators, civil and criminal penalties were included in the programs in order to ensure compliance. The following are general procedural requirements of OSHA.

Posting Requirements

The district is required to display in a prominent location the appropriate poster explaining the intent and coverage of the act in each location where there are employees. Annual summaries from the log of occupational injuries and illnesses also must be posted where they are visible to all employees. Likewise, citations received from federal or state OSHA must be posted in a prominent place where the violation occurred.

Recordkeeping Responsibilities

A daily log of reportable occupational injuries and illnesses must be maintained. The definition of reportable injuries may differ from state to state. Detailed records of each reportable injury or illness must be maintained and available for inspection by OSHA inspectors. At the end of each year, a summary of the log of occupational injuries and illnesses also must be prepared. Administrators are required to immediately report to the proper state or federal department all serious or fatal injuries. Definitions may vary. If toxic substances or harmful physical agents were present, records must be maintained and made available to employees.

OSHA Inspections

The district must allow a compliance engineer to enter the workplace without delay to inspect or investigate during regular working hours. The compliance engineer may privately question any employee or administrator. The district must allow an employee or a representative to accompany the compliance engineer. All required OSHA records must be made available to the engineer. If the investigation reveals a violation, the district will be given a written citation describing the violation and the length of time to correct the situation.

Complaints

Complaints of unsafe conditions by employees must be investigated within three days, with priority given to the most serious complaints. The name of the person submitting a complaint is confidential. Any action against employees submitting complaints is forbidden. No action may be taken against any employee(s) refusing to work where any safety or health standard is violated when such violation creates a hazard to any employee.

Appeals and Penalties

If the district receives a citation, it has fifteen working days to notify OSHA of the desire to appeal. The appeals board can affirm, modify, or cancel the citation and proposed penalty. Generally, public employers are exempt from civil penalties; however, criminal penalties may be levied against specific managers. Any officer, manager, or supervisor may be guilty of a misdemeanor for knowingly violating any standard or safety order of OSHA or for causing another to do so. Further, being charged under OSHA does not relieve the manager from being charged under other state or local statutes as may be applicable. OSHA is a serious matter, as company officials have been charged and convicted for the deaths of employees where the hazards were known to officials of the organization.

Fundamentals of Insurance

The realities of modern life indicate that there is little that can be done to fully prevent risk beyond the measures as described to this point. Because no risk can be fully eliminated and because schools are required by law to exist and to engage in certain activities, the remainder of risk management strategy rests with the concept of insuring against loss.

The Nature of Insurance

The concept of insurance is familiar to nearly everyone since life in the modern world demands that businesses and private citizens alike have a need for insurance. Insurance is a simple concept in that an insurance company agrees to pay for losses incurred by the insured in exchange for payment of insurance premiums. There should be no mistaking that insurance companies are profit enterprises that expect to collect premiums in excess of loss payments to their clients. As such, insurance is exchanging a known willing loss (payment of premium amounts) in return for reducing an unknown and possibly severe loss. Insurance carriers thus provide a service or product in the form of risk assumption for sale. The product, however, is different from manufacturers of tangible products in that the carrier's obligation to provide its product (claims payments) depends on an actual occasion demanding performance and is at some point in the future. In this respect, insurers are more like banks than like product manufacturers. The nature of insurance is such that government has felt obligated to provide some level of assurance that insurance carriers will be able to meet obligations because modern life would be nearly impossible without the security that various insurance products bring to both personal and business activities.

Government Regulation

The McCurran-Ferguson Act (P.L.15) was passed in 1945 to reduce the confusion following several court decisions involving insurance. This act stipulated that regulation of the insurance industry shall remain with the states except when Congress specifically enacts legislation dealing with that industry. Examples of federal legislation overriding state control would include E.R.I.S.A., minimum wage laws, civil rights laws, and other laws of a general or specific nature that could impact state control over the insurance industry. The primary purposes of both federal and state regulation of the industry are the financial stability of insurance carriers and an assurance of fair trade practices. Regulations promote a balance between financial stability and still allowing for competition to be the main regulator.

The effect of regulation is the same in each state. Each state government has an insurance commissioner who is usually appointed by the governor, although in some states insurance commissioners are elected. The insurance commissioner is responsible for enforcing the insurance code of the given state. In practice, this amounts to examining the financial security of companies petitioning to operate in the state, granting licenses to companies and agents based on security and standards, investigating complaints about practice, and acting upon requests for rate changes from insurers. Most commissioners belong to the National Association of Insurance Commissioners (N.A.I.C.), which has its own accrediting standards.

Rating Bureaus

Of considerable importance to school business administrators seeking to know more about insurance and insurers are organizations known as rating bureaus. Rating bureaus exist for the primary purpose of pooling the experience of many insurers to provide comprehensive rating information. Insurers are not required to participate in rating bureaus; however, participation is an indicator of stability and willingness to submit to examination. Participation is not an automatic stamp of approval, however, as some insurers participate in one or more bureaus to obtain rating information and then choose to deviate from those rates for certain classes of business. The operation of bureaus will differ by geographic area, as will the requirements for participation. Insurers not participating in any bureau are usually referred to as independent insurers. Care must be taken in choosing insurance companies, as good companies exist inside and outside of rating bureaus.

In some areas of the nation, the term "bureau" has been replaced by the term "insurance service office" (ISO). Indirectly, the ISO exerts some regulatory pressure on its participating members by establishing rates and by classifying risks in a more uniform and comparable manner. Schools may be forced to go outside an ISO, however, as an unfavorable classification of a risk by an ISO may force the insurance purchaser to seek an independent insurer in order to obtain coverage. Many independent insurers will use bureau rates as a guideline, however, especially in areas where they have inadequate rating information on which to base an amount of exposure when underwriting some liability.

Carrier Status

The McCurran-Ferguson Act establishing state control of insurance industries has resulted in a system which identifies companies by the home office. Insurance carriers are considered to be domestic companies in the state in which they are incorporated, while they are foreign companies in any other state. Such terminology has nothing to do with international issues, and it has nothing to do with the quality of an insurer—it is merely a designation indicating whether the insurer is operating in home territory. An insurance company licensed to conduct business and sell insurance is referred to in many states as an admitted carrier. Due to limited staffing of state insurance departments or commissions, active regulation is generally limited to domestic carriers. Thus, purchasing insurance from a foreign or nonadmitted carrier in a state may carry some additional risk compared to purchasing coverage from a domestic carrier. In most states, the insurance commissioner will only assist in collecting from admitted carriers. Insurance placed with a foreign carrier is usually placed through a broker with a surplus lines license. This usually involves at least one additional broker and, therefore, another level of commissions. Sound professional advice is needed in order to make the best decision because rates are not the only element of economy in risk management.

The Role of Agents and Brokers

An array of terms apply to the individuals who represent insurance companies. These terms are not generally synonymous, as each term indicates differences in the role these people may play. Private individuals seeking insurance usually work with an agent who is licensed by the state and the company to sell certain lines of insurance such as life, health, or casualty coverage. Agents may also sell commercial lines of insurance. The distinction of an agent is that the person works exclusively for a particular insurer. In contrast, an insurance broker does not represent one particular insurance company. A broker may place coverage with any carrier that will accept the risk. Differences also exist in how agents and brokers are paid. Agents generally receive a percentage of the premium as commission for the sale. An agent may also participate in bonuses for the amount of business placed with a particular company or on that company's good experience with the agent's insureds. A broker's arrangement is similar, although the broker hopes to make more money by shopping the policy in order to please the client, and additionally, the broker may receive incentives from the insurer in return for placing clients in that insurer's portfolio. These distinctions do not exist in all states, and some states treat brokers and agents exactly the same. Likewise, some states issue different licenses to brokers and agents, while others issue only one category of license.

For practical purposes, it is important to understand the status of a broker in a given state. Most school districts deal with brokers due to the unique needs of the school for broad insurance coverages and because brokers often have more experience with commercial lines of coverage. School district risk managers are often concerned with the commission arrangements between their brokers and carriers. If questions arise as to whether commissions are excessive or vary greatly between carriers, other arrangements can be made to compensate the broker for services. Insurance may be billed net of commissions so that the risk manager may see the exact cost of insurance. A commission percentage may be agreed upon for each line of business between the school district and the broker. The broker may also be compensated on a fee-for-services basis. While this may require some extra recordkeeping, it may be useful to monitor commissions to note whether commissions are excessive on some insurance placements.

Insurance Contracts

The basic element of insurance is the contract. Insurance contracts are the vehicle by which the terms and conditions of coverage are expressed, stating covered losses and exclusions to coverage. The insurance contract is subject to the same basic contract laws as any other contract as discussed earlier in chapter 7. However, the insurance contract is usually much longer and more detailed than most other contracts simply because the insurance industry is actuarially based, meaning that the risks have been carefully calculated by the insurer, with the practical effect that highly unfavorable risks are likely to be excluded from coverage. In addition, the contract is lengthy because many of the terms used in the insurance contract are unique to the insurance industry, making it necessary for disclosure purposes to state terms and definitions in the body of the contract. The wisdom of professional counsel is particularly applicable to the insurance contract, as evaluating insurance agreements is often beyond the capabilities of even a seasoned risk manager.

Although highly complex, much language in insurance contracts or policies is standardized by policy type and by the regulation of the insurance industry through state insurance commissions. Thus, a standard fire policy will have nearly the same language regardless of the company. There are, however, important exceptions to this generalization that must be noted when applicable.

District Responsibility

As implied earlier regarding reasonable and prudent efforts on the part of the insured to prevent and minimize loss, the insurance contract does not create a one-sided agreement that protects the district unilaterally. Rather, the district has responsibilities that are built into any insurance contract. The following is a list of basic responsibilities of a school district in the insurance transaction:

- *Prompt reporting of claims and potential claims.* Some policies may contain a provision that the carrier is not liable unless the insured reports the claim within a certain period of their first knowledge of the claim.

- *Cooperation with the defense of claims.* Various adjusters, attorneys, and investigators may need assistance with anything from photos and records to aid in locating witnesses.

- *Employee education.* District employees must be educated not to make prejudicial remarks or statements to anyone regarding an accident or other incident. Even a careless statement made immediately after an accident may turn out to be very damaging by establishing liability.

- *Good faith negotiations.* If a claim could have been settled below the insured's retention or deductible but is eventually settled for much more, the carrier may refuse to pay claiming "bad faith."

- *On-going contact.* Notify the insurance carrier promptly when additional exposures are created such as adding vehicles, buildings, employees, etc.

Types of Insurance

Most insurance bought by school districts falls into one of three classifications of property, automobile, or general casualty. An additional class of insurance relating to employee benefits also may be provided by the district.

Property Insurance

Property insurance is designed to cover perils relating to loss or damage to physical items such as buildings, automobiles, and attendant items such as possessions, contents, and so forth.

Fire and Extended Coverage

Of all major perils, fire is one of the most deadly to life and one of the most damaging to property. Because this peril is so devastating, the fire insurance policy has been developed and is a mainstay of the insurance industry. Several levels of fire coverage are available. The basic policy is known as standard fire coverage. In this instance, only loss from fire is covered. Other forms of extended basic coverage are also available so that a policy can be customized to include protection against loss from fire and lightning and other optional coverages by endorsement relating to contents of the building. The basic fire insurance policy is probably the most standardized policy in the entire insurance industry. The major sections of the fire policy are identifiable as:

- *The declarations page.* This section contains the names and addresses of the insured, the name of the insurance carrier, the name of the agent or broker, the name of the mortgage company holding a lien for loss-payable purposes, the premium rate, and the property description and location.

- *The insuring clause.* This section indicates that the insurance company will indemnify the insured for loss to the property described when it is caused by the perils stated, provided that loss occurs during the policy period.

- *Conditions.* This section covers collecting when there is other insurance, changes in conditions of the property, and concealment or fraud.

- *Loss provisions.* This section deals with required notice, details of the loss, inventory lists, protection of property, loss assessment, subrogation, arbitration, and limitations on the insured's right to bring legal action.

- *Exclusions*. Coverage for certain occurrences such as acts of war, military operations, insurrection and rebellion, and nuclear accident are commonly excluded under most fire policies.

- *Fire insurance forms*. Coverage under standard fire insurance policies is expanded by means of adding the following forms and endorsements: dwelling and contents form; building and contents form; blanket form; floater form; schedule form; reporting form.

- *Coinsurance (average clause)*. Fire insurance carriers offer lower premiums in exchange for the requirement to carry insurance equal to a certain percentage of property value. If insurance equal to that percentage were not carried, then the carrier will not pay for the full loss, instead depreciating it from replacement cost. This potential penalty places a burden on the insured to maintain insurance to valuation or to increase the insured's exposure.

- *Agreed amount clause*. This clause will, for an additional premium, remove the fear of being under-insured. An annual valuation is supplied to the carrier by the insured. The acceptance by the carrier eliminates any penalties under the coinsurance clause by automatically adjusting the replacement cost.

- *Extended coverage*. This is probably the most used endorsement as indicated by the common reference to "fire and extended coverage insurance." The additional coverages include windstorm, hail, explosion, riot and civil commotion, damage from vehicles and aircraft, and smoke damage.

- *Additional coverages*. Depending on the circumstances, districts may choose to opt for additional coverage for such items as: business interruption insurance (based upon earnings or value); rental value insurance; valuable papers insurance; demolition insurance; personal property of employees; personal property off premises; trees, plants, and shrubs.

- *Deductibles*. Premium reductions are available in return for accepting a deductible. Sometimes called "first dollar loss," a deductible shares the loss between both parties. Care should be taken in determining the deductibles for each line of coverage. Schools should almost always have some deductible on property coverages. The size of the deductible would be influenced by the size, financial strength, loss history, and management philosophy of the school district, as well as market conditions.

If large deductibles are chosen, it is often advisable to purchase aggregate stop loss coverage to limit the exposure of the district to a certain number of deductibles per year. For example, with a $10,000 deductible per occurrence, a $50,000 aggregate might be chosen. Another decision is whether to opt for actual cash value or replacement cost. Actual cash value equals replacement cost less depreciation. This type of coverage should not be purchased unless the district can set aside the funds necessary to offset depreciation because the additional premium for replacement cost coverage is generally more cost effective.

Allied Lines

In addition to property coverages, other coverages which are closely associated with fire coverage are referred to as allied lines. These lines continue the extension of coverage beyond basic and extended coverage to include more specialized losses that might be incurred by the insured. Broadly, allied lines may cover:

- Crop insurance (hail, wind, insects, disease, freezing).
- Earthquake coverage.
- Water damage coverage.
- Sprinkler leakage coverage.
- Boiler and machinery coverage.

Allied lines offer benefits for educational organizations in that school districts are responsible for a wide variety of property for which special insurance coverage should be considered. Although most property is owned by the district, some property in the care of the district or on the school premises is owned by other parties, including teachers and students. The district is required to protect such property from a number of potential types of loss. A variety of different insurance vehicles are available, including:

- **Personal property floater.**

May be written with either a property schedule or a stipulated amount of coverage. A deductible is almost always used to reduce premium costs. Recovery is usually limited to theft or vandalism. Mysterious disappearance is usually not covered.

- **Messenger robbery.**

Covers loss from robbery with threat or actual force. This coverage is primarily for large school districts where fairly large amounts of cash or negotiable papers are carried by district messengers.

- **Inside robbery.**

Covers robbery on the premises. Its use is normally restricted to locations with large amounts of cash such as bookstores, cafeterias, and sporting events.

- **Safe burglary.**

Covers loss from burglary of a locked safe or vault on the premises. Recovery is contingent on proof of forced entry.

- **Fidelity bond.**

Insures the district against loss from the dishonest acts of its employees. Coverage may be by position or by name. Coverage is often required by state code.

- **"3-D" bond.**

Coverage is for destruction, disappearance, or dishonesty. Blanket type coverage against multiple types of loss may be written in broad form providing very good, but expensive, coverage.

While purchase of the previously mentioned property coverage would ensure comprehensive protection for a school district, the costs would be prohibitive without carefully selected deductibles to control costs.

Boiler and Machinery Coverage

In addition to the specialized coverages, still another set of special protections are available and are known as boiler and machinery coverages. This is probably the least understood area of property insurance. The unusual aspect of boiler and machinery coverage is that the insurance coverage is not as important as the highly technical and specialized inspection service provided by all carriers. Coverages under this category may include:

- Actual property loss.
- Expenses of repair.
- Property damage liability.
- Bodily injury liability.
- Defense settlement and supplementary payment.
- Automatic coverage for newly acquired property.
- Use and occupancy loss.

Boiler and machinery insurances may cover the following types of property:

- Boilers, fired vessels, electric steam generators.
- Unfired vessels.
- Air conditioning and refrigeration systems.
- Electrical transformers.
- Pumps and compressors.
- Engines and machines of various types.
- Fans and blowers.

These items represent considerable loss to the district if damaged or destroyed, but premiums mount up quickly as additional coverages are purchased. Coverage on actual cash value basis will be less expensive to purchase, but this may leave the district with a large amount at risk in an era of rising equipment costs. Replacement cost coverage with a reasonable deductible will probably produce more favorable results.

Automobiles

The second major kind of insurance purchased by school districts covers vehicles of various types owned or operated in connection with the educational enterprise. Similar to property insurance, automobile insurance comes in many different types. Automobile insurance is particularly confusing due to the wide variation in state laws regarding compulsory coverages. Many states have compulsory insurance laws, while others have enacted no-fault laws or uninsured motorist laws. While the laws among the individual states vary, almost every state has some form of financial responsibility law. These laws require the suspension of drivers' licenses and/or vehicle registrations if the owner or operator fails to furnish insurance coverage or to furnish proof of financial security sufficient to cover payment of a judgment arising from an accident in which liability is fixed.

Liability and Property Damage

Automobile insurance policies are typically made up of coverage for automobile liability, collision, and comprehensive physical damage. Automobile liability coverage is similar to that of general liability in which the insurer agrees to defend claims for liability brought against the insured. Physical damage (property damage) coverage or collision coverage pays, after the deductible is charged back to the insured, for loss to owned automobiles and to non-owned autos operated by an insured driver. Comprehensive physical damage coverage pays for physical damage other than collision and may be written with or without a deductible. Comprehensive claims could stem from fire, windstorm, flood, and theft or proximate cause, such as glass breakage resulting from a collision. Coverage for personal effects in the owned auto and transportation expense reimbursement are also included with comprehensive coverage. Towing costs may also be included or may be a separate premium.

As with all insurance policies, the contract is written and contains sections similar to those discussed earlier under property insurance. Sections of the automobile insurance policy include a declarations page and definitions of terms and conditions. The declarations section identifies the insured, the insurance company, coverages and limits, forms and endorsements attached, and the policy number. The conditions section specifies which vehicles are covered, where and when coverage applies, the insurer's obligations, exclusions, who is insured, and territories covered.

An important element of automobile insurance is coverage for damage caused by uninsured motorists. Although the laws of most states require liability coverage as a condition of vehicular privilege, lawbreakers exist and can cause untold damage while having no financial security against which to seek recovery. Uninsured motorist coverage is contained in many automobile package policies, and any school district should be certain that its coverages include protection against such disaster. One concern for the insurance buyer is to require sufficiently high limits on this coverage. Another concern is to be sure that coverage for under-insured motorists is also included, as many drivers carry only the lowest limits of liability available by law. Such limits are often too low to cover the losses incurred in a major accident. As a general rule, the highest limits of all coverages should be selected, while choosing deductibles wisely according to the financial stability of the district.[7]

General Casualty

As seen earlier in chapter 7, liability represents a sizable threat to all individuals and organizations in the modern age. School districts are no exception as the rule of sovereign immunity has generally been stripped away or severely limited. As such, liability is an insurable hazard. Liability is a creation of law unlike the natural hazards to which property is exposed. This can best be understood by saying that, unless legally determined to be responsible, there is no hazard named liability. Such determination may be by a court of law or by an administrative body such as a worker's compensation appeals board. Liability is often called the "law of torts." Liability can also be created by accepting liability through some form of agreement or contract as discussed earlier in this chapter.

Liability

As discussed earlier in chapter 7, three areas of torts are generally recognized: intentional torts, strict liability, and negligence. An intentional tort is one where the person committing the act intends to commit the act. For example, battery qualifies as a tort and is defined as the intentional touching without permission of another person. Other examples of tortious acts are slander, libel, and trespass. Strict or absolute liability is assigned to persons dealing in certain hazardous situations such as the use of explosives. Any damage caused by use of explosives will be the responsibility of the person using the explosives. Negligence is defined as the omission to do something that a reasonable person would do or the doing of something that a reasonable and prudent person would not do. In order for negligence to be found, actual loss or damage must arise from the failure to conform to a standard of conduct when there exists an actual obligation to conform to a measurable standard.

Sources of Liability

In the instance of school districts, liability may result from negligence in three areas:

- Injury to pupils resulting from failure to provide adequate supervision. The district has a statutory obligation to provide adequate supervision for students. Failure to provide such supervision may be construed as negligence. A serious injury will usually result in such a charge.

- Injury to persons or property resulting from the failure to maintain a safe place or to conduct operations in a safe manner.

- Injury to outside persons or property resulting from the negligence of officers, employees, or agents of the district.

Anyone authorized to act for the district or apparently authorized to act will be held to the standard of a "prudent person." That is, a prudent standard means to act as would be expected of a reasonable and prudent person. Obviously, what constitutes prudence is subject to wide interpretation. Some states or other jurisdictions may impose standards which are more strict than ordinary prudence. Statutes often place stricter requirements on school districts than on other public agencies due to the nature of the school's mission and the inexperience of a young population in knowing and avoiding risk.

The demise of sovereign immunity has meant that liability insurance is mandatory for school districts in most states. Some states also allow the district to self-fund all or part of the liability exposure to the district or its trustees. Coverage against personal liability of officers and employees while acting within the scope of employment may be required or allowed, depending on state statutes. Certain acts by officers or employees may be considered outside the scope of employment, such as libel or slander, and thus may not be insurable. Liability insurance coverage includes investigation of liability claims and defense against liability suits. Under these conditions, the liability carrier is obligated to defend the district without regard to the merit of the claim. The carrier, however, will usually settle any case that its legal counsel does not expect to win. Settlement by the carrier may occasionally be made even though the claim has no merit, if winning is very unlikely. Districts should always try to participate in any decision regarding whether to settle or defend because reputations and principles are at stake. Settlements may also be costly to the district because they are always subject to policy limits and deductibles or to self-insured retentions. Other costs may also be incurred if the district loses a case, as court costs are usually assigned against the loser in a liability suit. In most states, interest begins to accrue from the date of judgment. The posting of a bond will usually be required if an appeal is filed.

The sources of liability are wide as discussed in chapter 7. As discussed at length, use of school facilities is one of the areas where considerable risk is involved. In the modern school age, school buildings are used by many people and many organizations every day. Any of these uses can lead to suits alleging a variety of negligent acts or unsafe conditions. To protect against significant loss, liability coverage should be as broad as possible to protect the district against the wide variety of facility uses. In addition to insurance, the district can do much to protect itself against this type of claim. The district should require user organizations to provide certificates of insurance and to name the district as an additional insured on the liability policy. In addition, every school board should adopt a policy concerning the use of facilities. Such use may be partially governed by statute or community expectations. Regardless, properly drawn administrative procedures will allow the district to better control its liability exposure by eliminating improper uses and controlling other uses. Organizations desiring to use school facilities should have a clear understanding of these rules and requirements.

Defenses Against Liability

As also discussed at length in chapter 7, some legal doctrines known as defenses can bar a plaintiff from recovering fully from the defendant. The three most important are assumption of risk, contributory negligence, and comparative negligence.

- *Assumption of risk.* The injured party must have voluntarily, and with knowledge, decided to do something which placed the claimant at risk, e.g., the plaintiff voluntarily was a passenger in a car of a driver who was obviously drunk.

- *Contributory negligence.* The injured party must, by personal action, have contributed to the harm being claimed.

- *Comparative negligence.* In those states that have adopted comparative negligence statutes, the preceding defenses would not be a bar to recovery of damages, but would bring about a reduction in the amount of the recovery. In the past, governmental units such as school districts often were immune to tort actions. Court decisions, and legislatures, have removed or eroded such immunities to the point where schools are sued with great frequency in most states. Since the doctrines of governmental immunity vary widely from state to state, it is important to thoroughly understand the applicable laws of each state.

Liability Coverage

Fortunately, broad comprehensive liability coverage is available and should be sought by every school district unless state statutes are precise in granting complete immunity to liability claims. Special school policies, having the advantage of including the coverages needed by most districts without numerous endorsements, may also be available.

The elements of a standard liability policy are common to all general liability coverages, setting out:

- *Insuring agreement*. This section contains the specific responsibilities of the insurance carrier.

- *Conditions*. This section contains the responsibilities of the school district and outlines the procedures to be followed in case of a claim.

- *Definitions and special features*. This section defines the covered hazards and special features of the policy.

- *Exclusions*. Exclusions are of two types: standard exclusions such as nuclear war, and manuscript exclusions that are appended to the contract rather than being part of the preprinted contract. The district should be very careful about accepting any manuscript exclusions. These can lead to unpleasant results when claims are filed. Manuscript exclusions are much more common when the insurance market tightens.

- *Umbrella coverage*. It is usually more economical to secure basic liability coverage limits of $500,000 to $1,000,000 and add increased coverage through an umbrella policy. Another advantage of an umbrella policy is that it usually contains few exclusions and is therefore broader than the underlying liability policy. Any loss covered under the umbrella but excluded under the primary policy is subject to a self-insured retention, usually $25,000.

- *School board liability (errors and omissions*). Also known as trustees liability, school leader's liability, or educator's liability, this coverage has become almost a necessity for all school districts. This type of policy provides protection for negligent acts, errors, omission, or breach of duty by the institution, trustees, officers, administrators, employees, student teachers, and volunteers. The policy pays after a deductible is applied. Trustee's liability supplements general liability coverage and does not replace it. Court awards are usually apportioned between trustee's liability and general liability policies.

- *Awards for damages*. Damages awarded by a judge or jury fall into three categories: special, general, or punitive. Special damages are usually out-of-pocket expenses incurred by the plaintiff such as medical bills, loss of earnings, property repairs, and legal fees. General damages may be allowed for losses which are not directly measurable such as pain and suffering. Punitive damages may be allowed where the conduct of the defendant is grossly negligent, reckless, and done without regard to life and property. Such awards are not insurable in most states.

Worker Compensation

A final type of liability insurance is known as worker compensation coverage. Worker's compensation laws were enacted to provide a satisfactory method of handling the volume of industrial accidents. Prior to 1900, common law defenses of contributory negligence, assumption of risk, and negligent acts of fellow servants made the handling of claims for injury or death slow and the outcome highly uncertain. Between 1900 and 1910, employer liability laws were adopted by some states. In 1911, the first worker compensation law was adopted.[8] Today each of the states has such laws, in addition to laws at the federal level. Since state laws vary so greatly, it is incumbent on the school district risk manager to become familiar with the requirements of the appropriate state.

Historically, a basic tenet of worker compensation law was that such law was the only source of recourse available to an injured employee. Thus, an employer gave up the presumption of fault in return for a single course of action by the employee. Although courts have eroded this doctrine in some states, it is still basically true. A primary goal of worker compensation laws was the sure, prompt, and reasonable payment of benefits and medical care to injured employees. This goal has been met more successfully than any of the others. Another goal of worker compensation law was the reduction of payments to attorneys for handling personal injury claims. This goal has been less successful for a variety of reasons.

Rates for this form of insurance are established by a worker compensation rating bureau for a particular geographical area. Rates are established for classes of employees, i.e., firefighter, general clerical, and so on. In some states, school districts may be considered as single class occupations, that is, the rate per $100 of payroll would be the same for all employees. To encourage safety, worker compensation rates are modified by a formula established by the bureau, the state fund, or the commercial carrier. Districts below a certain size may be subject to minimum premium requirement and be charged a higher modification factor than the actual losses would indicate. The general formula for calculating a district's worker compensation premium is expressed as:

Payroll x Rate x E.M.F. = Premium

Where E.M.F. = Experience Modification Factor

All states now allow unlimited medical care for industrial injuries or illnesses. Most states also provide indemnity benefits (temporary disability) at 66.66 percent or higher, with some allowing 100 percent. Permanent disability benefits are available according to schedules for the permanent impairment from an industrial injury. Death benefits are paid to the beneficiary of the victim of an industrial accident or illness. Most states allow qualifying employers to self-fund worker compensation exposure. However, some of these states do not extend this privilege to school districts. Where self-funding is available, there may be requirements as to minimum size, posting of bonds, or other requirements. Excess worker compensation insurance is almost always purchased to stop the loss from any one incident at a predetermined level such as $100,000. Some states also allow the purchase of aggregate stop-loss insurance to protect against combined losses.

Employee Benefit Coverage

In addition to the types of insurance described, employers often provide other insurance coverages on a voluntary or negotiated agreement basis known as employee benefit coverages. Employee benefits have traditionally been handled by the personnel department; however, this situation has changed in many districts due to spiraling premiums that began to accelerate during the late 1970s and have continued until today. The decisions concerning placement of benefit coverages require more background in risk management than is often found in personnel departments. Indeed, in most districts the cost of benefits is second only to the cost of salaries.

A thorough analysis of various benefit plans is beyond the average risk manager; in fact, it is beyond many brokers and consultants. This is an area in which it is often wise to consult with an expert to ensure that the benefit plans are designed effectively, understandably, and affordably. For descriptive purposes, however, employee benefit coverages are most usually provided in the areas of medical, disability, and other coverages, taking in a range of options.

Medical Coverage

Medical plans are comprised of two basic types, although there are many variations of each. They are Health Maintenance Organizations (HMO) and indemnity plans.

The HMO Act of 1973 required employers with over twenty-five employees to offer employees a choice between the district's indemnity plan and a federally qualified HMO. The HMO must be offered if it approaches the district, or if employees request an HMO. More than one HMO may be offered if the district so chooses and has enough employees to meet any minimum enrollment standards of the plans. Health Maintenance Organizations are often able to offer lower premiums than is true for indemnity plans. This is due to the fact that the HMO can better control costs and can emphasize preventive care. Drawbacks exist, however, as some employees will not choose an HMO because they desire to choose their own doctor.

In contrast, indemnity plans include those offered by associations such as Blue Cross and Blue Shield and those offered by commercial insurance carriers. The similarities between these plans far outweigh their differences. Generally speaking, the "Blues" are often more generous with respect to minimum group size than are commercial insurers. Indemnity plans can be categorized by the types of coverage provided, such as major medical, basic medical, and hospitalization, or a combination of several different coverages.

The great majority of policies written for school districts are the combination form that provide basic coverage after application of a deductible, with major medical costs usually picked up on an 80:20 cost-share ratio. In a few areas, 100 percent plans have been negotiated. These plans are attractive to employees, but they are very expensive and tend to encourage over-treatment since almost everything is paid after the deductible.

In recent years, medical benefit costs have soared almost beyond reach for districts and employees. Most 100 percent plans have disappeared, and carriers have dramatically increased deductibles and reduced benefits because of the high incidence of malpractice claims and astronomical jury awards for damages. Lack of employee motivation to control costs and lack of understanding of where benefit dollars go have also aided the cost spiral. Under traditional health plans, there was little incentive for employees to control costs or even to pay attention to costs. From the perspective of school districts, little can be done about increased costs until tort reform to cap medical awards is achieved. In the interim, school districts must continue reviewing existing plans and attendant costs, seeking alternative funding arrangements, negotiating hard deals with preferred provider organizations, pressuring professional service review organizations, and demanding audits of provider billings. These areas can be enhanced by school districts, although the best available talent should be consulted for this activity.

Disability Coverage

In addition to regular medical plans, districts usually offer disability coverage of some type. Disability insurance may be for temporary disability, for long-term disability, or for a combination of coverages. The most important consideration in the purchase of disability insurance is coordination with existing pension plans and other programs. District purchasing programs not designed with disability insurance specifically in mind will end up paying for coverage never used due to overlap of benefits. An elimination period is generally present, that is, benefits do not begin until the person is disabled for a specified period of time, and then benefits are usually paid at a set percentage of salary. Benefits may be limited to a set number of weeks, months, or until retirement or until some other event occurs such as eligibility for state disability programs.

Other Coverage

Still other types of coverage may be provided by the district on a cost share basis, employee elective premium, or district assumption of costs. In some instances, these coverages may be riders to other types of policies, such as basic medical coverage. Often these coverages entail areas such as dental or vision care or group life insurance.

The inclusion of dental insurance in employee benefit plans has increased greatly since 1970 until it is now nearly standard, either as a rider to basic medical plans or as a separately purchased policy. Three types of plans are generally found in school districts: incentive care plans, prepaid plans, and scheduled plans. With incentive care plans, dental costs are reimbursed up to a specified percentage until the plan's cap is reached. General benefits such as preventive care are covered at a higher rate than other types of more major coverage, and a deductible may apply. Orthodontia benefits are often minimal. With prepaid programs, participants are usually required to use the services of listed dental offices, and deductibles are often a significant cost to the employee. In the case of scheduled programs, payments up to the annual limit are based on a fee schedule, with additional costs borne by the employee. Premiums to the employer are lower since increased fees do not result in increased reimbursement. Obviously, plans have varying degrees of popularity, with many variations existing locally.

One of the more popular forms of "other" insurance through districts is life insurance. The availability of life plans varies with current tax laws regarding pre-tax premium payments and with the ability of the district to help fund premium costs. Although changing tax laws often make a discussion obsolete, the enduring feature of life plans can be found in the types of coverage provided. Four types of coverage are usually offered by school districts: level term life, decreasing term life, accidental death and dismemberment, and travel accident. Not all types are offered by every district. Level term allows each eligible district employee to receive the same amount of life insurance for a level premium amount for the life of the contract without regard to age or sex. Additional optional coverage may be made available through payroll deduction. Decreasing term varies the amount of coverage for a premium that depends on the age of the employee, with coverage decreasing with age, although premiums are level. Accidental death and dismemberment is a more specialized form, generally as additional coverage against accidental injury that can be added to one of the other term life programs for a modest premium. Additionally, travel accident coverage may be available, as some states permit the purchase of this coverage for employees traveling on school district business, with costs usually very low for $100,000 or more of coverage.

Self-Insuring Alternatives

The risk management process also allows for self-insuring plans when permitted by the laws of individual states. Although the terms self-insuring and self-funding are often used interchangeably, some technical distinctions between the two concepts may be drawn. Self-insurance implies that the self-insurer is retaining the full risk and funding losses on a "pay as you go" or cashflow basis. Self-funding implies that funds have been set aside and allocated to pay for anticipated losses. Authority to engage in either form is always statutory and does not exist in every state. As used here, the terms are synonymous unless differentiated in text.

Self-insurance was originally considered only after traditional insurance became unavailable or prohibitively expensive. As the success of self-insurance programs has become more widely known, however, it has become more of an option than a last resort. Nonetheless, self-insurance is not a venture to be entered into lightly. Since most administrators having the responsibility for risk management or insurance purchasing have little background in the technical areas of probability and distribution theory or actuarial accounting, it is not difficult to underestimate loss potential.

Evaluating Self-Insurance

Self-insurance is not entirely different from ordinary insurance. That is, the basic requirements of spread of risk, frequency of occurrence, potential for catastrophic loss, and determination of risk are all elements that must be taken into account. Consideration of self-insuring any risk must be based upon objective analysis of these factors. If the necessary skills to accomplish such an analysis were not extant in the district, they should be obtained elsewhere before proceeding. As a general rule, the following concepts apply when considering self-insurance:

- *Minimum size.* It is obviously impractical to self-insure the liability exposure for one automobile. However, it may be quite prudent to self-insure a fleet of 1,000 automobiles simply because of the size of the budget that supports the initial purchase. For school district administrators, however, there is no magic formula for computing the minimum size at which the laws of large numbers come into play for any given risk.

- *Spread of risk.* Using the example of a fleet of 1,000 autos, self-insuring for the theft and physical damage portions would be more practical if those vehicles were garaged in several facilities in several cities. It might not make economic sense to self-insure them if they were all housed in one large facility.

- *Potential for catastrophic loss.* Again using the same fleet of vehicles, the potential for catastrophic loss is much greater if all vehicles were housed together, rather than spread across a larger geographical area. Housing all vehicles together would make the purchase of some form of stop-loss insurance almost mandatory. Unless insurance is economically unfeasible, self-insurance should be considered only as an alternative to purchasing insurance. That is, the decision to self-insure should be made only after comparing the attendant costs, benefits, and drawbacks of all available options. Self-insurance should not be a goal, but rather one tool to be considered in the search for ongoing solutions to the problems of managing school district risk. To summarize, the district should consider retaining those losses that may reasonably be anticipated, are recurring, and limited in severity. The threshold for inclusion in these categories will vary as the insurance market fluctuates.

- *Funding.* Items which qualify for self-insurance are usually funded in one of two ways. Minor items with high frequency such as glass breakage may be treated as expense items. If these items are to be treated as an expense, then funds should be allocated on a line item basis. For larger items with less frequency or higher potential, reserve funds should be set aside. Funds may be reserved in several ways. In those states which allow joint pooling, the funds may be held and invested by a joint powers authority.

- *Joint exercise of powers.* Some states statutorily allow public agencies to jointly exercise any power that they individually possess. This device will allow self-insurance to become an option for many small districts that could not meet the minimum qualifications for self-insuring. The authority to exercise joint powers may include a number of restrictions including how funds can be invested, what risks can be treated, minimum size, state regulation, deposits or bonds to be posted, licensing, etc. The restrictions imposed are usually a small price to pay for the greatly expanded opportunities available to districts under a self-insurance program. The spread of joint powers authorities has been phenomenal in some states. It has been estimated that in some states 90 percent of all school districts participate in a joint powers authority for one or more coverages.

Summary

In the modern context of liability and litigation, the school business administrator must be aware of the opportunities and practices making up the risk management function in school districts. As insurance premiums continue to grow and as coverage becomes more complex, increased awareness and skill are becoming absolutely critical to survival.

Endnotes

[1] Portions of this chapter are based on the previous edition written by L. N. Randal.

[2] Excerpted from *The World Almanac and Book of Facts 1995.* (Mahwah, NJ: Funk & Wagnalls Corporation 1994), 964.

[3] D. C. Thompson, R. C. Wood, D. S. Honeyman. *Fiscal Leadership for Schools: Concepts and Practices.* (New York: Longman 1994), 478, see also Chap. 10 of this text for a differently focused discussion of risk management.

[4] Adapted and modified from D. R. Dunklee and R. J. Shoop, *A Primer for School Risk Management.* (Boston: Allyn & Bacon, 1993), 39-40.

[5] See chapter 7 in this text for a discussion of immunity and assumption of risk following from insurance purchasing. The implication is that purchasing insurance may waive immunity or, alternatively, it may not be statutorily permissible to buy insurance when immunity is secure.

[6] Great caution should be exercised with legal counsel to engage in these within ADA guidelines.

[7] In some states, school district liability does not extend to vehicles, regardless of usage, unless the employee is using a commercial driver's license in the course of his or her work. Thus, classroom teachers who may be engaged in work-related activities, but without specific job descriptions and the required CDL, assume all automotive liability coverage in that controlling statutes specifically exempt this from public liability.

[8] Chamber of Commerce of the United States, *Analysis of Worker's Compensation Laws.* (Washington, D.C.: Chamber of Commerce of the United States, 1979) 3.

CHAPTER 20
School District Purchasing

Introduction

Public school districts engage in the purchasing of goods and services every day.[1] Purchasing is a highly specialized activity that is related to receiving, storing, maintaining inventories, and distributing. Purchasing seeks to maximize the value of limited public resources in order to maximize the educational programs of school districts and individual schools.

The Procurement Function

Students of educational administration need to acknowledge that no one model of purchasing is appropriate for all school districts. Private sector purchasing models are often inapplicable to school districts due to controlling state statutes. Further, purchasing models are not generalizable to both large urban school districts and small rural districts as the needs of each type of school district are significantly different.

Principles of Purchasing

Purchasing is a major function in educational resource management related to providing the materials necessary to operate the educational programs and facilities of the school district. Purchasing involves expenditure of public moneys necessary for the efficient operation of public schools.

Management of the purchasing operation requires working closely with other departments and administrative units in developing a positive relationship with other school functions. Supply management is the ordering, bidding, purchasing, warehousing, distribution, and accounting within the purchasing process.[2] Purchasing services the vital operations of the school district.[3] I

In general, the objectives of purchasing are to buy materials, supplies, and equipment of the right quality in the right quantity from the right source at the right price at the proper time. School administrators must strike a balance between the quality of a product and service with the economics of that product and service. Quality may be defined as " . . . the sum or composite of the properties inherent in a material or product."[4] "Much of the theory in supply management has stemmed from trying to answer two basic questions: how much to order at one time, and when to order it.[5]

Once the quality of a product has been determine, it is necessary to decide on the quantity to be purchased. The procurement program of a school district must provide adequate quantities of material of acceptable quality in a timely manner that will provide for the educational needs of the various programs.

Ethics of Purchasing

School personnel authorized to make purchases for the school district as well as other staff engaged in the purchasing operation must maintain high ethical standards of conduct. The reputation of the school district is reflected through the practices of the school district's purchasing official. A code of ethics should be developed to guide the acts of purchasing and the school district's methods of doing business.

The importance of a municipality's ethical conduct in managing its purchasing program is emphasized by the development of a "Model Procurement Code for States and Local Governments" by the American Bar Association. The Association of School Business Officials International (ASBO) developed a code of ethics that provides guidance for the purchasing function. Briefly, the ASBO Code of Ethics pledges the individual's actions to be:

- Best for the pupil.
- Fair to all concerned.
- Designed to build good will and better understanding.,
- Based upon respect for the past, knowledge of the present and concern for the future.
- Legally and morally right.
- Loyal to my associates, to my school district, to my state or province and country, and to my conscience.
- The best that I can perform.[6]

The New York State Association of School business Officials adopted the following code of ethics for school purchasing officials:

- To consider first the interests of the school district and the betterment of its educational program.
- To endeavor to obtain the greatest value for every tax dollar expended.
- To be receptive to advice and suggestions from colleagues, both in the educational field and in other departments of business administration, inso-

far as such advice and suggestions are not in conflict with legal or moral restrictions in purchasing procedures.

- To strive for knowledge of school equipment and supplies in order to recommend items that may either reduce cost or increase the efficiency of the means of education.

- To insist on and expect honesty in sales representation whether offered verbally or in writing, through the medium of advertising or in the sample of a product submitted.

- To give all responsible bidders equal consideration and the assurance of unbiased judgment in determining whether their product meets specifications and the educational needs of the district.

- To discourage the offer of, and to decline, gifts which in any way might influence the purchase of school equipment and supplies.

- To accord a prompt and courteous reception, insofar as conditions permit, to all who call on legitimate business missions,

- To counsel and asset fellow school purchasing officials in the performance of their duties whenever occasion permits.

- To cooperate with educational, governmental, and trade associations in the promotion and development of sound business methods in the procurement of school equipment and supplies.

A code of ethics combined with school regulations and adherence to various state statutes will provide the groundwork for a respectable and reputable purchasing operation. A highly ethical standard of conduct results in a reputation for the school district and purchasing official above any suspicion of unfair dealings.

Methods of Purchasing

There are several methods of purchasing available to school districts. Variations will be necessary to accommodate differences due to the controlling laws of the state and policies of the local school district.

Before engaging in the purchasing of goods and services for a school district, the purchasing official must become familiar with the state laws, regulations of the State Department of Education, applicable state procurement regulations, and all local policies and regulations governing purchasing within the public sector, particularly educational institutions. Many states and local school districts have rules and regulations requiring competitive bids for purchases beyond a stipulated amount. Also, some states have available state purchase contracts with various vendors and state agencies at given prices for use by local school districts.

Contracts and Purchasing

With any discussion of school district purchasing the nature of contracts and the surrounding issues must be fully understood. As arms of the state, school districts have been granted permission to enter into contracts for hiring of staff, purchasing supplies and equipment, and contracting for capital facility needs. Although contract law is a broad and highly specific field, some major highlights are relevant for this discussion.

Schools are contracting organizations in that they buy goods and services. As such, they may be liable for contractual performance requirements and for various forms of liability whenever a contract is breached by nonperformance or whenever a contract is improperly executed. Thus, it is important to consider how contracts are properly constructed to contain all essential elements, whether school districts in fact have the authority to make contracts, how board members and administrators relate to the power to contract, how recovery generally operates under a breached contract, and other issues associated with the concept of contract law.

Elements of Contracts

Regardless of the purpose of a contract, there are five key elements that must be observed in order to create a legally enforceable contract. The first element of mutual assent simply means that both parties to a contract must agree to the terms and conditions of the contract, whereupon mutual assent is said to include offer and acceptance or such terms. The second element of consideration means that there is an exchange of something valuable given by both parties as inducement to enter into the contract, such as money, services, or merchandise. The third element of a contract requires that both parties be legally competent, meaning that each party entering into the contract must have the legal capacity to do so, both as a proper party to the contract and of intellect and age to make a valid contract. The fourth element of subject matter not prohibited by law is a requirement that simply precludes the enforceability of any contract that is based on some illegal activity. Finally, the fifth element of agreement in form required by law simply means that there are legal requirements in some instances as to the actual form of the contract itself, such as written agreement for the purpose of real estate. Each of these elements is both common to contracts generally and governed in specific by varying individual state statutes.

Authority to Contract

Legal authority to enter into a valid contract depends entirely on the laws of a given state. Since school districts have been held to mere statutory creations or extensions of the state, it follow that the power to contract derives from that same statutory source so that school districts are in fact unable to contract outside the parameters and limits of law as intended by the respective legislatures. Further, since the board of education is in all instances charged with the operation of the district, only the board is authorized to engage in contractual relationships. As a general rule, the powers of a board of education are described completely in the controlling state statutes. Generally, chief among those is the power to contract, in that a board of education is recognized as a legally competent entity to enter into a contract.[7] It is important to note, however, that in many circumstances, courts have acknowledged the power of agency to administrators to enter into contracts on behalf of the board of education, although the board may not have delegated its discretionary duties, which include the expenditure of moneys. Further, all contracts, including those by agency, must be ratified by the board of education in order for funds to be legally expended.

Many of the challenges involving contracts have centered on whether a contract is valid under such conditions because a board of education may wish to nullify contracts extended on its behalf.[8] For example, the court in *Community Projects* [9] held that only the board of education had contractual powers and that an agreement to purchase goods that had been approved by a principal was invalid. In contrast, an opposite ruling was reached in *Hebert*[10] as the court held that because the board had ultimate power to contract and had given power to a building principal over extracurricular activities that led to a contract, it has granted an implied power to contract. Similarly, in *Smith*[11] the court found that a superintendent had the authority to modify a board contract.

The concept of agency can be one of the most troublesome issues faced in school districts. The basis for agents acting on behalf of a board of education generally lies in controlling state statutes. There is no question that a school board may appoint agents, but there are limits on the act of agency, particularly relating to the power to contract. Although interpretation of such limits is complex, the scope of agency can be better appreciated utilizing a general concept discussed earlier, whereby all board of education acts may be classified into the two concepts of ministerial and discretionary duties of boards. Ministerial duties are those responsibilities expected of a board of education, which it may choose not to perform for itself. Most daily functions of a school district are ministerial duties, such as daily control of playground supervision or developing course content. It is generally agreed that boards may delegate ministerial duties, but it is still true that the board of education has final responsibility as the overseer of the total educational program.

In contrast, discretionary duties are those responsibilities that the board of education has chosen to perform or are specifically required by controlling state statute. Many discretionary duties fall to the board, such as evaluation of the superintendent or adopting a budget. It is generally held that these discretionary duties may not be delegated in any manner. These distinctions apply directly to contracts through agency where expenditure of funds is a discretionary power that a board of education may not delegate. Practically, however, boards usually appoint an agent whose task is to deal with contractual matters on behalf of the district. The risk arises by appointing someone to enter into contract agreements without either statutory power to do so or the legal authority to create a valid contract.

It is well established that a school district is not bound to a contract when an unauthorized employee enters into an agreement in the name of the school district. The contract can be voided by the courts, and the district may refuse to ratify the contract. While this may create problems for vendors or others who relied in good faith on the agent as representing proper authority to enter into the contract, courts will require the person relying on such representation to ascertain whether authorization of an agent to contract on behalf of the district was accurately presented and allowed for under controlling state statute.[13] A board of education may choose, however, to ratify an invalid contract. The burden always falls heavily on anyone dealing with the public treasury that the outcome of a suit against a school district is perhaps somewhat unusual.

For example, in *Joske*[14] a construction company that was resurfacing an athletic track failed to complete the job in a timely manner, whereupon the district locked the construction company out of the construction site. The construction company sued for breach of contract and the school district also counter sued for breach. The district court's ruling against the school district was twice reversed on successive appeals, with the U.S. Court of Appeals, Eleventh Circuit, holding that the district could not be liable because construction and maintenance of school facilities were governmental functions and thus immune from liability.

Similarly, the power of a board of education to contract has been repeatedly upheld, even in the face of higher governing authority. For example, in *Anderson*[15] a school board had applied for capital construction funds, whereupon state board of education funds were disbursed to the county treasurer. The county board of education, however, refused to countersign the fund warrants to release the funds to the local school district. Upon suit, the appeal courts found that the district had acted within the scope of its authority and that the county board had no discretion in releasing the moneys because the legislature had placed the application process on the district, thereby empowering the district to award contracts.

Although the facts of each case are different and controlling state statutes on contracts vary among the states, the net sum of contract disputes finally relates to the discretionary functions of government, wherein it is well settled that a board of education cannot contract away its governmental obligations.

In *Coalition*[16], a contract between the board of education and a neighborhood group to establish an experimental school for three years was invalidated because under the terms of the contract the board could not close the school at its discretion, which effectively resulted in the board impermissibly contracting away one of its discretionary duties. Boards, therefore, have general and broad authority to enter into contracts by virtue of statutory enablement, and they may appoint agents to act on behalf of the school district in most states, although generally no contract is final until it receives board ratification according to the requirements of the controlling state statutes. As a result, knowledge of individual state laws is the only way to avoid serious problems relating to the authority to contract.

Competitive Bid Laws

In addition to general complexity of contracts, an area of potential trouble for school districts may arise protect the public treasury through the use of contract restrictions known as competitive bid laws. This is a result of the enormous pressure on government to seek the greatest utility of each tax dollar so that competitive bidding has been in use for many years. As the Mississippi Supreme Court in *Beall v. Board of Supervisors* [17] noted in 1941, " . . . the objective is to prevent 'private' and secret machinations." The purpose of such statutory bid procedures is to demand strict accounting of public expenditures under sunshine laws requiring that all public business be conducted under the rules governing open meetings. Without exception, school districts in all states are subject to various sunshine laws and further subject to competitive bid statues in most states. Because most states require all units of government to conform to bid laws, school districts are generally included under the local governmental unit provisions of applicable state statutes. For school districts, the effect is that no funds can be spent except by formal approval within a legally constituted board of education open meeting and in agreement with statutory provisions relating to the applicable threshold amounts requiring competitive bids. This returns to the earlier concept that only the board of education can expend funds, either by initiation or ratification of contracts recommended by its agent, and such expenditures must be for a legal purpose and approved in a duly advertised open board of education meeting.

Purchasing requires observance of special statutes that usually include competitive bidding requirements. Such laws usually have a minimum threshold below which competitive bidding does not apply due to the expense of bidding or because vendors might not bother to bid on low amounts. But as the purchase price increases to where large sums of moneys are involved or when the price is great enough that the number of competitors is small, states usually require bidding because large purchases may be discounted or because a single vendor's monopoly might increase costs. Bid laws are usually written to cover a broad area of supplies, goods, machinery, and equipment.

Although bid laws vary by individual states, they contain several common elements. A typical bid law includes statements about threshold expenditure amounts, splitting bids, unfair bid restrictions, method of bid openings, awarding or rejection of bids, bid bonds, and exceptions under emergency conditions. Threshold expenditure amounts are intended to trigger bid laws only when large purchases are involved. Bid statutes often prohibit dividing any purchase into smaller separate amounts in order to prevent subvention of the bid law intent. Unfair bid restriction prohibits writing specifications so as to exclude all vendors except a preferred supplier. Methods of bid opening and awarding of bids attempt to develop standard procedures that leave no questions concerning ethics and openness. The law, therefore, usually requires bids to be opened in public, acceptance of the lowest responsible bid with the reasons stated for rejecting all other bids, and reasons for not accepting the absolute lowest bid given in writing. Emergency exceptions to bid laws recognize that good reasons may occasionally arise where time is of the essence wherein there is not sufficient time to seek bids in a deliberate fashion. Finally, such laws usually require financial commitment by the bidder in the form of a bid bond in case the contractor does not deliver as required. The complexity of these laws is illustrated by excerpting from one state's bid law affecting public schools:

> 1. When the total amount of any purchase does not exceed $15,000 or the total annual rental payments do not exceed $10,000, the purchasing unit may buy on the open market without the giving of notice or the receiving of bids, in one of two ways: (a) The purchasing unit may invite quotes from at least three persons known to deal in the classes of materials by mailing them a copy of the specifications for the materials not less than seven days before the time fixed for receiving quotes. If satisfactory quotes are received, the unit shall award the contract to the lowest responsible and responsive quota for each class of materials required. The unit may reject all quotes and if no valid quotes are received for an item, purchase or lease that item on the open market without further invitations for quotes, (b) The purchasing unit has the option of purchasing or leasing materials in the open market without inviting or receiving quotes.
>
> 2. The total purchase of similar materials by a purchasing unit, from one person, during a six-month period under paragraph 1 may not exceed $15,000 and the total annual rental payments to one person under all leases for similar materials may not exceed $10,000.
>
> 3. When a purchase or lease exceeds the $15,000 or $10,000 respective amount, the purchasing or leasing unit shall prepare specifications describing the kind and quantity of the materials needed, but they shall avoid specifications that might unduly limit competitive bids. Under certain conditions, specifications may be solicited by way of publishing a legal notice… .When specifications have been

completed, a notice of the time and place for receiving bids shall be given by publication.... All meetings for receiving bids must be open to the public. All plans and specifications shall be kept in a place available for public inspections ,which shall be specified in the notice.

4. The bids shall be opened publicly and read aloud at the time and place designated in the notice.

5. After a satisfactory bid is received, the purchasing unit shall award a contract to the lowest responsible and responsive bidder for each class of materials required. If a contract is awarded, but not to the lowest bidder, the factors used to justify that award must be stated in writing at the time the award is made, and a copy of that statement must be kept available for public inspections.

6. The purchasing unit may reject all bids and ask for new bids. If no valid bids are received for an item, the unit may purchase or lease that item on the open market without further advertisement for bids.

7. A certified check shall be filed with each bid by a bidder not having his principal place of business in the state in the amount determined and specified by the purchasing unit in the notice of the letting. For the purpose of this statute, it is lawful to submit a draft, cashier's check, or money order issued by a financial institution insured by an agency of the United States in all cases where it is required that a certified check be submitted with a bid. A bond shall be filed with each bid by a bidder having his principal place of business in the state in the amount determined and specified by the purchasing unit in the notice of the letting. The amount of the bond or certified check may not be set at more than 10 percent of the contract price. This does not preclude the use of a performance bond in addition to the bid bond or certified check in the amount of the performance bond is listed in the specifications and in the notice of the letting.

8. The purchasing unit shall allow a bidder not having his principal place of business in the state to submit a bid bond if the bond is secured by a surety having his principal place of business in the state.

9. All checks of unsuccessful bidders shall be returned to them by the purchasing unit upon selection of successful bidders. Checks of successful bidders shall be held until delivery or until completion of the contract.

10. Each bid for a purchase or lease of materials must be executed on forms prescribed by the appropriate state agency.

11. As an alternative to advertising for new bids, a purchasing unit may extend beyond the contractual period, a contract let under the provisions of this law.

12. Within thirty days after the acceptance of a bid, quote, offer, estimate, or contract for the purchase or lease of materials, the purchasing unit shall issue an original purchase order to the successful bidder, retain a copy for the unit's records, and file a copy for public record.

13. A purchasing unit, upon declaration of an emergency, may make any purchase or lease materials without giving notice or receiving bids if quotes are invited from the giving of notice or receiving of bids.

14. Materials may be purchased or leased from an agency of the federal government without the giving of notice or receiving of bids.[18]

The seriousness of competitive bidding is evident by the substantial requirements of law. The law must be followed closely by school districts to avoid allegations of impropriety by both unsuccessful bidders as well as the public.

The error prone and complex nature of contracts and bid laws is well illustrated by several cases involving disputes over terms and technical minutiae. For example, the requirement to accept the lowest responsible bid usually turns on whether a board of education can withstand a challenge to its decision to reject the absolute lowest bid, as in *Baron*[19] where the board was sued by a construction company for disqualifying the low bid. On review, it was revealed that the company bringing the action had been rejected by the board because it did not have the proper state licenses. The court held that the board's disqualification was not arbitrary and that the requirement to accept the low bid was waived in such instances. Similar technicality held for the defendant board of education in *Great Lakes*[20] as suit was brought when a bid for construction work was filed five minutes after the deadline for bid specifications. The court held that the board of education could exercise discretion to not consider late bids without interference unless allegations of fraud or abuse of discretion were made. Broad leeway in board of education discretion regarding bidders was present in *Crest*[21], as a disappointed bidder for a school building contract sued, claiming that it should have been given the contract since it held the low bid and had been prequalified. The state supreme court held, however, that the board could award the contract to the contractor of its choice despite the merits of the low bidder. Similar leeway was seen in *Corner*[22] as a defendant board was relieved of any duty in not awarding a school renovation contract to the low bidder. The court held that even under the state's bid law the contractor had no protected property interest by virtue of being the low bidder and that the school district's reason for not accepting the low bid—dissatisfaction with the plaintiff's predecessor corporation on an earlier project—was not stigmatizing so as to implicate a liberty interest demanding due process.

While board of education leeway exists as seen in the case discussed, the seriousness of observing bid laws and the technicalities is evident as typified by many other cases. A board of education was held to have violated state bid laws in *Byrd*[23] where taxpayers challenged irregularities in bid procedures. The appeals court held that a construction contractor's bid using an unbonded subcontractor's quotation, when bidders were required by law to select from an approved list of bonded subcontractors, was a material variance in bid requirements. Similarly, a board's inability to defend a decision not to accept the low bid for a bus purchase contract in *Ohio Bus Sales*[24] led the court to order award to the low bidder since the board of education never actually reached a determination on what constituted the lowest responsible bid. Likewise in *Borden*[25], the appeals court held for the contractor in a dispute when a board rejected the lowest bid for milk supplies. The board had defended its decision, stating its displeasure with the fact that the contractor was only self-insured, rather than commercially insured, with a $3 million deductible for general liability exposure. The court held that bid specifications issued by the board did not exclude self-insurers and that the contract should be awarded to the low bidder.

Similarly, a board's defense was overturned in *1st Westco*[26] where a bid law operated to exclude non Pennsylvania construction contractors from any work in that state. The court held that the statute violated the Privileges and Immunities Clause of the United States Constitution because although the statue was enacted to combat state unemployment, the defendant was unable to show that non-Pennsylvanians were a particular source of evil or that the statute was closely able to aid the unemployment of those for whom the statute was enacted. Likewise, the court held for the contractor in *Bloomfield*[27] by requiring a board of education to enter into arbitration although the contractor's claims would result in expenditures by the board in excess of legally appropriated amounts for the construction project in question.

Recovery Under Contracts

When a contract is alleged to have been breached, the question of recovery usually arises either by virtue of damages to the aggrieved party or for recovery of products themselves. Under these conditions, the likely scenario is that at least one party acted in the belief that a valid contract existed and thereby provided products or services, while the other party subsequently failed to make payment for those goods or services. The issues of greatest concern in such situations are whether there is a contract either expressly or implied or whether it can be established that the contract itself was invalid for any of several reasons. While the topic is considerably more complex than presented herein, it is worthwhile to touch briefly on the concept of recovery in that school leaders undoubtedly will be faced with such a claim at some point.

An express contract is a contract in which the duties and rights thereunder are expressly agreed to by all parties to the contract. An implied contract is one in which the parties have acted in such a manner as to read a contract into the actions of the parties, even though both parties have not expressly agreed to such terms. The difference is not merely semantic; in that the elements of contracts are used to establish whether a contract expressly or implied exists. For example, accepting goods or services on behalf of a school where there is no firm evidence suggesting that such services were intentionally *pro bono* or that the goods were a deliberate gift may result in obligation to both parties. Under these conditions, the right of recovery is a reasonable question that can be raised. In the case of an implied contract, the question must consider the value of goods or services rendered so that a fair value for the disputed issue can be established. If recovery were permitted on an implied contract, recovery is both bounded by and limited to the reasonable value of goods or services. In the case of an express contract, however, the general rule holds that recovery is controlled by the value expressly agreed to without consideration of whether the amount is reasonable. In the latter instance, it is particularly prudent to avoid contracts that are unreasonable as courts will generally bind both parties to an improvident contract if the contract were determined to be express.

In some instances, however, the contract itself may be invalid for any of several reasons. The first clear instance of an invalid contract is one that fails to observe all the elements of contracts as discussed earlier. For example, no contract could exist if mutual acceptance were missing or if one party was legally incompetent to make a contract, such as lack of authority to enter into contracts or when the contract is *ultra vires*. Similarly, absence of consideration would void a contract, as promises to make gifts or to perform free services fail to meet the consideration test. Clearly, no contract can exist if it calls for or results in an illegal act. Still other contracts, however, may be invalid because of failure to follow a prescribed form.

For example, a contract required to be in writing but only agreed upon orally is invalid because it fails to meet the statutory requirements. These instances create particular problems for courts under recovery claims in that services or products may have already been rendered or consumed before the dispute arises, making an equitable claim more difficult by placing the question before the court as to whether to relax the statutory requirement or to allow an uncompensated benefit to one party. While each case is unique, the general rule has been that if all other elements of a contract were observed and if the district were to have the authority to expressly contract, then it will be liable under an implied contract as long as the form or requirements of the contract did not violate other statutory provisions. To some extent, the outcome of such a dispute rests in long-standing evidence that a court's disposition toward the question of whether equity[28] or strict statutory application[29] is more important when examining issues in the public policy arena.

One of the areas of most frequent contract dispute involves extra work by one contracting party beyond the expectations of the other party. A review of recent claims before courts suggest that this situation most often arises in construction projects. For example, the court held for the plaintiff contractor in *Winchester* [30], as the contractor sued the board of education and its individual members to recover additional costs incurred in a building project. The court held that the board of education was not a legal entity subject to suit under state law, but that the individual board members were subject to suit in their official capacities and that the contractor was entitled to benefit of the board's waiver of sovereign immunity.

Likewise, the court held for the contractor in *Flower City*[31] where a contractor, who was hired to remove asbestos from a school, also cleaned up and repaired fire damage whereupon the district rejected a claim for payment based on extra services rendered by the contractor. Following denial of both parties' request for summary judgment, an appeals court determined there were genuine issues of fact regarding the claim for retainage under the contract.

In contrast, however, the court held for the board in *Owners Realty*[32] where a contractor had performed extra work when removing asbestos. The court reasoned that no other recourse was available, in that the written contract explicitly barred claims for additional compensation and that the contractor had further failed to present notice in a timely fashion.

As a general rule in disputed public contract cases, courts will examine three principal issues in recovery claims. If just cause exists for a claim against a public entity, the first issue is whether the goods are capable of being returned. If the goods are returnable, courts will generally permit recovery. The second issue is that no implied contract will be found where it is determined that no express contract authority originally existed, as such contracts will be held *ultra vires*. The third issue is that *caveat emptor* will apply, except where fraud or abuse of discretion is found. None of these instances, however, defeat the intent of contracts by making it possible for a board of education to misuse its preferential position regarding the public treasury. For example, a claim against a school district for unpaid fees for legal services was upheld, as the state supreme court held that payment could not be refused under existing contracts.[33] Likewise, a claim for contractor services under an oral contract held that the alleged damages could not have reasonably been foreseen or contemplated by the contractor at the time the contract was entered into, making equitable claim a valid issue against the board.[34]

The net sum of this brief discussion of contracts yields several useful points. First, school districts are generally held to possess contractual authority. Second, the elements of contracts apply equally to schools, and the further operation of statutory requirements protecting the public treasury also apply. Third, requirements of statutes concerning contracts differ via controlling state statutes. Fourth, the purpose of sunshine and bid laws is to scrutinize acts involving the public purse, and these requirements will not be easily set aside as violation of these state statutes are counter to the serious open intent of most legislation. Fifth, boards may appoint agents, but they may not delegate their discretionary duties, of which contracts and expenditure of funds qualify as nondelegatable. Sixth, there is a predisposition of courts to guard the public treasury at the expense of outside parties. Seventh, boards may not take advantage of such presumption so as to engage in intentional misconduct. Eighth, contracts are generally unassailable when executed in proper form and written to be so specific as to avoid vagueness and to bar unforeseen problems such as disputes over additional work. As a result of this complex arena known as contract law, knowledge of controlling state statutes and tightly drawn terms of agreements are the only methods by which schools and school professionals to avoid serious contractual problems in the purchasing of goods and services.

Additional Purchasing Arrangements

State Contracts/State Agencies

When state contracts or agencies are utilized, purchase orders are issued directly to the contractors without the necessity of the district securing bids or quotations, since the state has already taken competitive bids or made and/or supplied the items through various agencies such as the state corrections department. In a few states, this procedure is utilized as the primary method of purchasing by many school districts.

Cooperative Bidding

State statutes sometimes authorize two or more school districts to join together in an agreement to make purchases through competitive bidding on a cooperative basis.[35] This has been done for some period of time in certain states through boards of cooperative educational services agencies. Cooperative purchasing agreements generally must have the approval of the board of education of each of the school districts and operate specifically under permissible state statutes.

Government Surplus

State laws permit the purchase of surplus and secondhand supplies, material, or equipment without competitive bidding from the federal government, the state government, or from any other political subdivision or district. Such purchases from a private person or corporation are subject to competitive bidding if the cost were beyond a certain amount.

Professional and Personal Services

Generally, state statutes have been judicially construed as having no application to contracts for professional services or for services requiring special skill or training. For example, contracts for legal, medical, dental, nursing, or engineering services are not subject to competitive bidding requirements in most states.[36]

Monopolies

Where a subject of a contract is controlled by a monopoly and there is no possibility of competition, the courts will not require the empty formality of inviting proposals. Examples of such monopolies may include utilities such as telephone and electric services.

Petty Cash Purchases

Such funds shall be used for the payment of properly itemized bills of nominal amounts and under conditions filing for immediate payment to the vendor upon delivery of any such materials or supplies or the rendering of any such services. Petty cash accounts should be reconciled on standard forms with the business office prior to reimbursement. Requirements of reconciliation should include, at a minimum signature of person making expenditure, receipt for amount spent, and approval of the person responsible for the fund. Petty cash funds must be operated in accordance with the regulations established by the statutes, the state accounting code, and school district policies.

Standardized Lists

Standardized lists of supplies and equipment can be helpful to the school district by facilitating the competitive bidding process and avoiding special order type items. The benefits of standardization may be listed as follows:

- Allows lower costs from bids on large quantities of one item.
- Reduces and facilitates repairs and replacements.
- Reduces inventories, thereby reducing storage costs, and at the same time increases the amount of school funds available for other purposes.
- Speeds delivery of materials or equipment.
- Reduces the number of materials and equipment for which specifications must be written.

- Reduces the work of the purchasing department, including that of business office record-keeping.[37]

When developing standardized lists, the personnel concerned must be involved in determining the standards and specifications of the items.

Some states also allow for school district standardization for a specific kind of equipment or supply by having the board of education of that district adopt a resolution indicating the need for standardization. This does not, however, eliminate the need for competitive bidding, but permits the purchase of a specific make and model. Some reasons for such standardization may include the ability to interchange parts or match existing furniture. Examples may include school buses and microcomputer equipment.

Procurement Procedures

The purchasing process is not an isolated management function of the business office but involves close working relationships with other departments of the school district. The needs of teachers in various departments at various grade levels must be taken into consideration. Communication must be open among the users of the materials, school officials, and vendors.

Fundamental Concepts

A purchasing manual developed by the Kentucky School Business Officials Association and Kentucky State Department of Education lists four basic principles of procurement.

- Buy the quality standard for the purpose required.
- Have the product or service available at the time and place requested.
- Purchase on the most feasible and economical basis.
- Purchase from the lowest responsible bidder meeting the required specifications.[38]

The California Association of Public School Business Officials lists the following procedural principles:

- Accomplishment of definitive objectives in the shortest possible time and the easiest manner consistent with accuracy and efficiency.
- Simplification of tasks to speed operations and reduce potential for error.
- Establishment of procedures that are definitive and understandable, and will eliminate friction, duplication, and confusion.

- Placement of responsibility for each step of performance.

- Establishment of procedures that are flexible enough to allow for expansion as the district grows.

- Provision of a purchasing system that is economical and consistent with the job to be done.

- Assurance that the procedures are adequate to carry out the tasks for which they were developed.

Steps in the Purchasing Function

Differences in the size of a school system, various state laws and regulations, and administrative organizational structures may necessitate variations in purchasing procedures. It is not possible to establish an "absolute" system of purchasing procedures for all public school districts. Procedures that work well for one school district may be inadequate or too complex for another school district.

The procurement function is generally considered to cover the complete supply transaction beginning with the recognition of a need and ending with the approval of a claim voucher for payment. The following ten steps describe in detail the school procurement process:

- *Recognition of a Need.* Someone in the district recognizes a specific need for a supply, equipment item, or type of service. This could be anyone in the district—teacher, principal, aide, clerk, or custodian.

- *Processing of Requisitions.* Requisitions to supply a specific need are initialed by the user (teacher, principal, custodian, or stockroom clerk) and forwarded to their supervisor for review and approval.

- *Preparation of Specifications.* Before an order is placed or a quotation requested, the purchasing official must describe what is desired in order that prospective vendors may intelligently quote prices and fill orders.

- *Request for Bids or Quotation.* The purpose of obtaining bids or quotations is to encourage competition in the procurement of supplies, equipment, and services. It also allows the discovery of what a given item will cost, where it can be obtained most economically, and when delivery will be made. Recommendations will be made to the board of education on formal bids.

- *Issuance of Purchase Order or Award of Contract.* Purchase orders are signed by the purchasing official and issued from the purchasing office, subject to the purchasing policies of the district. All contracts that require the public advertising of the competitive bidding process must be awarded or rejected by resolution from the board of education. Certified checks and bid bonds received from unsuccessful bidders will be returned. A regular purchase order is then issued after award by the board of education.

- *Follow-up of Order for Delivery.* The purchasing official has the responsibility for following up contracts and orders to expedite delivery. The requisitioner will be kept advised of the status of the requisition and purchase order. Contact with vendors should be limited to the purchasing official only.

- *Invoice from Vendor.* When the invoice and/or claim form is received from the vendor, the purchasing official will review it to verify that the price agrees with quotations or contract figures and the terms of the order, including f.o.b. shipping or delivery point.

- *Receipt of Goods.* Upon receipt, goods should be checked for condition and for quantity and quality as stated in the purchase order or contract. This is accomplished by having the requisitioner send the receiving copy of the purchase order and the packing slip to the purchasing office as evidence of receipt of goods and materials. Each of the items must be signed by the administrator and have indicated on the document "received and accepted." In the case of partial orders, a signed packing slip will suffice. If the order is found to be unacceptable, notify the purchasing office.

- *Quality Control.* Items should be tested for conformity with specifications based on needs or use.

- *Approval for Payment of Invoice or Claim.* The final step in the purchasing procedure is the approval of the invoice or claim for payment, either by the board of education or the internal auditor. Prompt payment of claims could mean taking advantage of discounts which resulting in substantial savings.

In order to properly execute these basic steps in the purchasing process, it may be necessary to develop as a part of the purchasing system some general operating procedures. The following are basic procedural statements, which could appear in a purchasing operations manual:

- Only the person designated by the board as purchasing agent may commit the district to a purchase.

- The materials, equipment, supplies, and services to be purchased shall be of the quality required to serve the function in a satisfactory manner, as determined by the requisitioner and the purchasing agent.

- It is the responsibility of the requisitioner to provide an adequate description as required by the purchasing agent, so that he or she may be able to prepare the specifications and to procure most expeditiously and economically the desired commodity and/or service.

- It is the responsibility of the purchasing agent to make alternate suggestions to the requisitioner if, in the judgment of the purchasing agent, the specifications would restrict competition or otherwise preclude the most economical purchase of the required items. In the case of disagreement, either party may refer the matter in accordance with established procedure.

- When a low bidder proposes an alternate as "an equal" to that specified, it is the responsibility of the purchasing agent to determine whether the proposed substitution is, in fact, an equal. Such decision shall be based on his or her evaluation and that of the requisitioner. In the case of disagreement between the requisitioner and the purchasing agent, either party may refer the matter in accordance with established procedure.

- Each building should designate a person to act as building purchasing contact to whom all correspondence and inquiries should be made from both within the building and from the purchasing office.

- Once items are delivered to the building, the security of those items becomes a building responsibility.

- All correspondence with vendors regarding placement and status of orders will be handled by the purchasing office. No orders are to be picked up by individuals without prior authorization of the purchasing office.

- To the extent possible, the purchasing department will try to adhere to a bid calendar in which various categories of purchases will be planned at designated times throughout the school year. This will provide for cyclical purchasing in which certain items can be purchased off season when prices may be lower and deliveries more complete and accurate.

The steps and procedures are a part of a total purchasing system developed to accomplish the necessary transactions to obtain the needed supplies, services and equipment as quickly, accurately, economically, and efficiently as possible.

Administration of the Procurement Process

The purchasing official for the school district must deal with all aspects of the district, both educational and otherwise. This requires close work and coordination with various levels in the school system's administrative organization and requires the purchasing function to be integrated with all phases of the school operations. Throughout the procurement process, decisions must be made on such matters as quality, quantity, price, and source.

Establishment of Authority

The organizational relationship of the purchasing official to all other administrators must be clearly delineated and the scope of authority of the person clearly defined and understood. The local school board should designate by formal resolution the purchasing agent of the district. In many cases, this is the same individual responsible for the administration of the business affairs of the district, such as the school business administrator. In large districts, it could be a separate designated full-time position responsible to the school business administrator. The designated individual is the only employee with authority to contract for supplies, equipment, and services of the school district. Only orders for goods and services that have been approved by the purchasing agent become obligations of the school district and give rise to a claim against the school district. It is upon the authority of his or her signature that a purchase order becomes approved and issued, thereby creating a school district obligation.

The purchasing transactions of a school district must be centralized to provide for control and efficiency of operation. Systems and procedures must be developed that provide an efficient means for being responsive to the various needs of the different departments in the organization, and at the same time provide the necessary control for managing the funds of the district.

The Act of Purchasing

In order to provide for an efficient procurement operation with the proper authority and delineation of responsibility, it is necessary to have written policies establishing the procedures used in purchasing of goods and services for the district. Written policies provide for a systematic process built around established procedures. Local policies in conjunction with state and federal law and various regulations of state departments provide the framework for an operation in which the purchasing agent can exercise professional judgment in administering his or her duties and responsibilities for the procurement of goods and services of the school district. These policies may then be developed into district-wide procurement regulations and guidelines. Some of the areas for which guidelines should be provided within a school district purchasing systems manual are listed as follows:

Purchasing Systems Manual Table of Contents

I. THE PURCHASING FUNCTION

- Introduction
- Purchasing Guidelines
- Methods of Purchasing
- Steps in the Procurement Process
- Purchasing Practice
- Purchasing Procedures

II. OPERATIONAL PROCEDURES

Preparation of Purchase Orders

Cancellation of Purchase Orders

Outstanding Purchase Orders

Special Item Requisitions

End of Year Transactions

Purchasing Guidelines

- The purchasing agent is appointed by the board of education and is responsible for developing and administering the purchasing program of the school district.

- Competitive bids or quotations shall be solicited in connection with all purchasing whenever possible. Contracts shall be awarded to the lowest responsible bidder complying with specifications and with other stipulated bidding conditions.

- The purchasing agent is authorized to issue purchase orders without prior approval of the board of education where formal bidding procedures are not required by law and when budget appropriations are adequate to cover such obligations.

- All purchase contracts for materials, equipment, or supplies involving an annual expenditure over $_______ and all public work contracts involving expenditures over $_______ shall be awarded on the basis of public advertising and competitive bidding. The purchasing agent (or other designated person) is authorized to open bids and record the same pursuant to law.

- All contracts that require public advertising and competitive bidding shall be awarded by resolution of the board of education. Recommendations for the award of all such contracts shall be submitted to the board of education by the purchasing agent.

- Residence or place of business of local bidders may be a consideration only in cases where identical bids have been submitted.

- Purchases shall be made through available state contracts of the Division of Standards and Purchasing through the State Office of General Services whenever such purchases are in the best interest of the school district. Also, purchases of prison-made and blind-made products may be purchased through the appropriate agencies without competitive bidding.

- Items commonly used in the various schools or units thereof shall be standardized whenever consistent with educational goals and in the interest of efficiency or economy.

- Petty cash funds shall be established annually in accordance with the regulations of the board of education and the Department of Education.

- The purchasing procedures employed shall comply with all applicable laws and regulations of the state and Department of Education.

- A statement of "General Conditions" shall be included with all specifications submitted to suppliers for their bids. These general conditions shall be incorporated in all contracts awarded for the purchase of materials, equipment, and supplies.

- Opportunity shall be provided to all responsible suppliers to do business with the school district. To this end, the purchasing agent shall develop and maintain lists of potential bidders for the various types of materials, equipment, and supplies. Such bid lists shall be used in the development of a mailing list for distribution of specifications and invitations to bid. Any supplier may be included in the list upon request.

- No board member, officer, or employee of this school district or their families shall be interested financially in any contract entered into by the board of education.

In these procedural manuals it is important that appropriate statutory authority, state board of education rules and regulations, and board policies be cited.

Management of Purchasing

In order for the purchasing process to be initiated, an appropriation for the requested goods and services must be available. This requires the integration of the budgeting function with the purchasing function. The budget development process must provide for educators to plan their program needs in such a manner that they are reflected in the various budgetary appropriations. The budget becomes the financial plan for the education program and should strike a balance between the educational needs of the school system and its ability to support those needs and the quality of goods and services and the economics of those goods and services.

Once the budget is approved by the board of education, the purchasing of supplies, equipment, and services reflected in the budget appropriations is initiated. The purchasing agent will be responsible for the coordination of the implementation phase of the budgeting process, which includes the purchasing function. The budget becomes a management tool for planning, organizing, and controlling district operations, which includes the purchasing process. In order to achieve the necessary controls the budgeting system must be integrated with the district accounting system.

As discussed in chapter 11, encumbrance accounting is a procedure that provides for control of budget expenditures by reserving funds for various district purchase orders. It obligates a portion of an approved budget appropriation for purchase commitments made by the school district. Once payment is made for the goods, the encumbrance is adjusted to reflect the actual expenditure. This provides the purchasing agent with current information on the status of various budget appropriations.

In the Association of School Business Officials International's *Purchasing and Supply Management Handbook for School Business Officials,* outlines the following procedures for maintaining adequate budgetary and accounting control:

- All purchases should be based on purchase orders initiated by those having appropriate authority.
- Purchase orders should be approved through one central office by those having appropriate authority.
- Purchasing practices should be reviewed periodically in order to disclose inequities.
- Incoming merchandise should be controlled by the preparation of a receiving report.
- Vendor invoices should be mailed directly to the account department for verification with the purchase order and receiving report.
- Invoices should not be recorded as a liability until properly approved.
- Adequate separation of accounting, purchasing, and receiving duties should be provided.
- Prompt payment of vendors' claims is necessary.[40]

The purchasing procedures of a school district and the approved budgetary appropriations must be coordinated and monitored as part of the budget administration process.

The size of the school district, the controlling state statutes, its administrative organizational structure, and various state and local policies and regulations are some of the factors that should be taken into consideration when designing a specific purchasing system. The system should also allow for cost analyses and other management reports.

Warehousing and Distribution

Once the ordered supplies have been received by the school district, the purchasing agent must manage the supply storage and distribution processes. Whether a centralized or decentralized supply management operation is utilized depends largely upon the size of the school system.

Supply Storage

One of the first decisions a school district must make is whether to use centralized storage facilities or decentralized site-based storage facilities. Decentralized storage and warehousing allows for the receiving and storing in each individual school of all materials needed by that school. Centralized storage provides a warehouse for receiving and storing of all goods purchased by the school district. Factors include the annual consumption of various supplies and the extent of standardization of supplies. There are no definitive factors that clearly indicate when one storage facility is more effective than another.

A summary of the advantages are listed in the Association of School Business Officials International's *Purchasing and Supply Management Handbook for School Business Officials*:

Advantages of Centralized Storage

- Flexibility of purchasing allows large volume orders to be placed at any time of year, thereby providing opportunity for savings in supply expenditures.
- An adequate and periodic supply of materials can be provided with minimal storage in schools
- Purchases can be scheduled to avoid ordering during suppliers' peak seasons, thereby allowing more timely deliveries and minimization of back orders.
- Adequate inspection and quality control can be maintained in a systematized manner.
- Centrally controlled supply inventory records can facilitate studies of consumption, price trending, usefulness of materials, standardization, delivery times, demand pattern, and basis for scheduling purchases.
- Centralized storage of frequently used supply items combined with a reliable distribution system can eliminate the need and temptation to hoard and stockpile materials thus reducing loss resulting from obsolescence and waste.

Advantages of Decentralized Storage

- Reduces school district supply costs by reducing or eliminating intra-district deliveries and operating costs of a central warehouse.
- Reduces administrative and clerical costs for a central warehouse; however, results in more school unit duties for administrative and clerical staff related to processing of orders, inventory, and control,

- Due to limited storage space in school districts, storing large volumes of supplies may not be feasible and may actually be less expensive to provide limited school storage space than to build a warehouse.

- Where statute permit, standard purchasing contracts with vendors providing periodic deliveries throughout the year can eliminate the need for a central warehouse. This, in effect, uses the vendors' storage as an extension of the school system's storage facilities.

- Obsolescence is reduced by making frequent purchases of supplies delivered directly to the user.

Distribution and Inventory

Once materials and supplies are received at a central warehouse or building sites, a system of distribution is needed to assure the various materials reach the source of need. Plans for the distribution of educational materials should include use of requisitions, flow charts, records of distribution, and inventory records. A central store catalog is useful for requisitioning items from a central warehouse. The catalog can be broken into various categories and incorporate a coding system for use by staff and for purposes of inventory control. When filling orders, items should be properly packaged, identified, and accounted for from point of distribution to point of use.

An adequate inventory control system for maintaining records of supplies and equipment is a necessity. Such a system should provide for the maximum utilization of school materials and supplies, as well as avoid delaying or hampering the educational process due to lack of needed supplies.

There are two basic types of inventory systems: perpetual inventory and periodic inventory. The perpetual inventory system provides an up-to-date, continuous tabulation of materials on hand, thereby reflecting the current stock and distribution status. It is most suited for the centralized storage and warehousing concept. Periodic inventory provides an accurate accounting of stock only after an actual physical inventory count and is most suited for use with annual requisitioning or once-a-year large orders.

The purchasing official must be responsible for safeguarding public funds and hence materials and supplies. Adequate records of supply inventories is strategic to maintaining needed safeguards. Various computer models for inventory control are available and may be used to expedite the process and provide greater efficiency and economy in inventory control, thereby enhancing the educational program.

Purchasing: A Systems Approach

The purchasing function can be treated as various systematic techniques applied to the procedures and processes involved in the procurement and distribution of educational goods. The techniques may involve computer applications as well as other related mathematical principles and models.

Functional Analysis

The purchasing function of the school district, which begins with the recognition of a need and ends with paying of an invoice, must be governed by a system of procedures needed to accomplish the tasks necessary to procure the goods and services in order to operate efficiently and accurately.

In order to evaluate the efficiency and adequacy of the purchasing operation, school business officials should analyze the processes of each phase of the operation and examine what changes, if any, in procedures would allow the department to function more effectively. The procedures must be designed to fit the needs and organizational structure of the school district, recognizing that there is no one specific procedure for all school districts.

Organizational Analysis

The organizational structure of purchasing departments vary with the size of the school system. The efficiency and effectiveness of the purchasing function will depend somewhat on the organization of the operation. The work flow is affected by the organization of tasks and assignment of responsibility. As changes occur in various rules and regulations governing the purchasing operation and as procedures are altered to conform with these changes, the organization of the purchasing operation becomes subject to modification and improvement.

Once the awards are complete, the total amount of requests for various buildings can be checked against the approved budget appropriations for sufficiency of funds. This links the purchasing function to the budgetary process. When insufficient funds are noted, specific orders can be rejected and the requisitioner directed to adjust the order or secure additional funding. Once the appropriations are justified to meet the purchasing obligations, the computer can print out the purchase orders for processing.

Systems application utilizing operations research theory can be applied to the purchasing function for developing optimum ordering strategies. Once determinations of the quality and types of goods has been made, the basic supply management decision of how much to order and when to order or reorder must be decided. Decision rules can be formulated such as "when the inventory level of supply item A is at X units, order a new shipment of Y units." The optimum economic order quantity and reorder point must be determined.

It is a basic tenant of purchasing that low ordering cost is accomplished at the expense of high inventory holding cost and low inventory holding cost is accomplished at the expense of high ordering cost. The objective is to find a minimum cost point between the two extremes referred to as the "optimum order quantity."

Various models are based on a variety of assumptions regarding demand and lead time, thereby enabling decisions to be made with those certainties. However, these assumptions are not always fixed. Demand rate may vary from one day to the next. Lead time between orders often varies and may result in a stock depletion. As a result, a model must be developed dealing with uncertainty of demand and lead time. Again, commercial software is available to assist in developing a viable model that can be used to simulate the operation under the conditions of uncertainty and determine the optimum order quantity.

Purchasing Systems Manual

The purchasing procedures, rules, and regulations must be communicated to all levels of the organizational structure of the district. Employees must be informed of the process and given the information needed to effectively procure their various needs and successfully perform their part in the educational organization.

A purchasing systems manual is an ideal way to inform all management staff and other employees of the school district concerning the district's procurement process. As previously indicated in the sample table of contents, the manual should include information concerning state and local laws and regulations and policies concerning procurement; basic guidelines to follow when using the system; the steps of the procurement process; and purchasing practices and procedures. In addition, it should provide a detailed account of how to use the system and what to do in special situations. Persons responsible for various aspects of the process should also be indicated. The appendices should include a sample of the forms to be used (along with instructions on how to complete them and route them), and a copy of state and/or local written policies.

In addition to providing information and a means of communication, a written manual provides standardization and consistency in policies and procedures, establishes uniformity within the process, and serves as an authoritative reference for management.[41]

In short, the manual represents the promulgation of the district's purchasing function in a single source document. As such, it becomes the written communications link among the school business office, central office administrators and supervisors, building administrators, teachers, vendors, the board of education, and the community with respect to purchasing.

The development of a purchasing systems manual is a major task. Once completed and adopted, it must be maintained and updated to reflect changes in statutes, regulations, policies, procedures, and organizational structure. After some use, the manual may need alterations in format and content. It is important to remember that the manual is a dynamic and authoritative management tool requiring commitment from all sectors in its maintenance, application, and enforcement.

Endnotes

[1]Portions of this chapter are based on the previous edition written by J. L. Natale.

[2]G. C. Hentschke, *Management Operations in Education*, Berkeley,: McCutchan Pub. Corp., 1975, 90.

[3]C. I. Candoli, et al, *School Business Admin.* 3d, (Boston: Allyn and Bacon, 1982) 208.

[4]Candoli, 205.

[5]Hentske, 90.

[6]R.E. Munsterman, *Purchasing and Supply Management Handbook for School Business Officials,* Reston: Assn of School Business Officials, 1978, 49-50.

[7]NEA Wichita v. Unified School. Dist., 234 Kan 512, 74 P.2d 478 (1983); Allen v. Town of Sterling, 329 N.E.2d 756 (Mass. 1975).

[8]See e.g., Spradlin v. Board of Trustees, 515 So. 2d 893 (Miss. 1987); Responsive Environments v. Pulaski County, 366 F.Supp. 241 (AK, 1973).

[9]Community Projects v. Wilder, 298 S.E.2d 434 (N.C. App. 1982).

[10]Hebert v. Linvingston Parish School Board, 438 So. 2d 1141.

[11]Smith v. Fort Madison, 334 N.W.2d 701 (Iowa, 1983).

[12]See, Cado Business Systems v. Board of Educ., 8 Ohio App.3d 385, 457 N.E.2d 939 (1983).

[13]See, R. W. Anderson Housewrecking and Escavating v. Board of Trustees, 681 P.2d 1326 (Wyo. 1984).

[14]Joske Corp. v. Kirkwood School Dist., 903 F.2d 1199 (8th Cir. 1990).

[15]Anderson County School Dist. v. Board f Educ., 371 S.W.2d 807 (S.C. App. 1988).

[16]Coalition v. School District Kansas City, 649 S.W.2d 533.

[17]Beall v. Board of Supervisors, 191 Miss. 470, 3 So. 2d 839 (1941).

[18]Excerpted from Indiana Code § 35-1-9.

[19]Baron Construction v. Jefferson Parish School Board, 606 So. 2d 27 (La. Ct. App. 1992).

[20]Great Lakes Heating v. Troy School District, 494 N.W.2d 863 (Mich. Ct. App. 1992).

[21]Crest Construction v. Shelby County Board of Educ., 612 So. 2d 425 (Ala. 1992).

[22]Corner Construction v. Rapid City School Dist., 845 F.Supp. 1354 (D.S.D. 1994).

[23]Byrd v. Independent School Dist., 495 N.W.2d 226 (Minn. Ct. App.) 1993).

[24]Ohio Bus Sales v. Toledo Board of Educ., 610 N.E.2d 1164 (Ohio Ct. App. 1992).

[25]Borden v. Jefferson Parish School Board, 609 So. 2d 1180 (La. Ct. App. 1992).

[26]1st Westco Corp. v. School Dist. of Philadelphia, 811 F.Supp. 204, (E.D. Pa. 1993).

[27]Board of Educ. v. Christa Const., 593 N.Y.S.2d 178 (N.Y. 1992).

[28]Burk v. Livingston Parish School Board, 190 La. 504, 182 So. 656 (1938) stated that to leave a claimant remediless to the exclusive benefit of the other part is an offense to justice.

[29]Oberworth v. McCreary County Board of Educ., 275 KY. 319, 121 S.W.2d 716 (1938) stated that uncompensated benefit at the expense of the claimant may be a lesser injustice than endangering public resources so as to indemnify claimant at public expense.

[30]Winchester Construction v. Miller Count Board of Educ., 821 F.Supp. 697 (M.D. Ga. 1993).

[31]Flower City Insulations v. Board of Educ., 594 N.Y.S.2d 473 (N.Y. App. Div. 1993).

[32]Owners Realty Management v. Board of Educ. of New York, 596 N.Y.S.2d 416 (N.Y. App. Div. 1993).

[33]School Comm. v. City of Chicopee, 590 N.E.2d 1158 (Mass. 1992).

[34]Wasilewski v. Dalton Sch., 580 N.Y.S.2d 265 (App. Div. 1992).

[35]For an example of this process see Florida State Board of Educ. Adm. Rules Rev. 6A-1.013.

[36]For a more complete discussion of these issues, see, R. C. Wood, "Competitive Bidding of Architectural Services," School Business Affairs, July 1985, 49-51.

[37]P. E. Burrup and V. Brimley, Financing Education in a Climate of Change, 3d, Boston: Allyn and Bacon, 1982, 365.

[38]Kentucky School Business Officials Assn and Kentucky Dept. of Educ., Purchasing Manual , 1975, 9-10.

[39]For a more complete discussion of federal law, specifically the Uniform Commercial Code and its application to public school districts, see the chapter "Purchasing," in D.C. Thompson, R. C. Wood, and D. S. Honeyman, *Fiscal Leadership for*

Schools: Concepts and Practices, (White Plains, NY: Longman Publishing Co.) 1994.

[40]Munsterman, 40.

[41]See, A. J. Andrews, "Preparing and Using Administrative Manuals," J. of Systems Management, March 1980, 15.

Chapter 21 Pupil Transportation

Introduction

Perhaps the most frequent interaction most members of the public have with their public schools is the sighting of a yellow school bus transporting children to and from school, on a field trip, or to an athletic event. Since over 22 million children ride school buses to and from school each day, many parents' most visible reference as to how well their schools are doing is related to the operation of the pupil transportation system. If school buses do not run on time, or if children were not picked up and dropped off in a safe manner at the same location every day, parents may suspect that the school district is not doing other things well either.

Despite the importance of pupil transportation, both as a means for getting children to school, and as a necessary educational device, most educators have little training or knowledge concerning the operation of school buses and the systems that support the transportation operation. This chapter describes the important factors that school business administrators must know to operate their transportation system, and discusses a number of current issues in pupil transportation funding and service provisions. It begins with a discussion of the scope of pupil transportation in the United States.

The Scope of Pupil Transportation

The first thing one notices concerning pupil transportation services in public school districts in the United States is the large size of the system, and the dramatic growth that has occurred in pupil transportation in the last fifty or so years. Thompson, Wood, and Honeyman point out that in 1929, there were approximately 25.7 million children enrolled in the public schools in the United States. Of these, just over 1.9 million or 7.4 percent of the total were provided transportation at public expense at a total cost of $54.8 million.[1] By 1992-93, 22.3 million, or 52 percent of the nation's 43.3 million public school children were transported to and from school each day at public expense at a cost of nearly $10 billion. That represented something on the order of 4 percent of the total K-12 educational expenditures that year.

The transportation of public school children varies dramatically across the states. Table 21.1 shows the total public school enrollment and number and percent of students transported to and from school for each of the fifty states for the 1992-93 school year. As the table shows, in some states the vast majority of students receive pupil transportation, while in others a relatively small proportion are provided this service. Minnesota, for example transports some 90 percent of its public school children, while at the other extreme California only provides transportation for approximately 18 percent of its children.

As discussed, the factors that affect the number of students transported vary greatly among school districts as well as states. Most students are transported to and from school because they live too far from the nearest school to walk. However, in some areas, safety hazards such as busy streets, railroad tracks, or the lack of sidewalks demand that public school systems provide transportation to students living closer to their school. Other reasons for transporting students to school include the need to bus them to different schools to meet desegregation requirements, or to shift students to different schools in an attempt to alleviate overcrowding at their neighborhood school.

Approximately 70 percent of school buses used to transport public school children are owned and operated by school districts. The remaining 30 percent are operated by private bus contractors hired by local school districts. Table 21.2 shows the number of district-owned and contractor-owned buses, as well as the percentage of buses owned and operated by the public schools in each of the fifty states for which data were available. These data also point to the dramatic size of the public school bus fleet, with over 300,000 buses in operation in 1992-93. Since data are missing for a number of states, it is likely that this figure is substantially higher than 300,000.

Another measure of the extent and size of the pupil transportation system is the total miles buses are driven each year. Table 21.3 provides these data for each of the fifty states for which data were available in 1992-93. The table shows that school buses were driven over 4.2 billion miles that year.

A state-by-state data analysis concerning pupil transportation expenditures for 1992-93 reveals that the total expenditures for public transportation were $9.8 billion. This ranges from a high of nearly $1.3 billion in New York to a low of $17.8 million in Nevada. Total transportation costs is a relatively limited way to compare transportation costs across states given the tremendous diversity in the size and population density of the fifty states.

Expenditures per pupil transported averaged $486 in 1992-93.[2] This figure ranged from a high of $2,678 in Vermont to a low of $158 per pupil in South Carolina. Expenditures per bus averaged $23,068 ranging from a high of over $42,000 in New York to a low of $11,523 in Iowa. Finally, when analyzed in terms of expenditures per mile traveled, including miles driven for service purposes, the costs averaged one dollar ninety-nine cents per mile, and ranged from over four dollars in New Hampshire, to 62 cents per mile in Louisiana.

Table 21.1
Students Transported by State 1992-93

State	Enrollment	Number of Pupils Transported	Percent of Pupils Transported (%)
Alabama	714,916	443,248	62
Alaska	123,161	45,570	37
Arizona	649,320	253,235	39
Arkansas	433,314	285,987	66
California	5,195,777	935,240	18
Colorado	612,635	214,422	35
Connecticut	478,879	335,215	70
Delaware	105,547	88,659	84
Florida	1,980,758	871,534	44
Georgia	1,225,729	919,297	75
Hawaii	177,000	35,400	20
Idaho	217,933	106,787	49
Illinois	1,893,077	1,003,331	53
Indiana	964,352	694,333	72
Iowa	489,670	249,732	51
Kansas	451,536	189,645	42
Kentucky	658,406	454,300	69
Louisiana	795,690	572,897	72
Maine	214,658	176,020	82
Maryland	772,638	517,667	67
Massachusetts	864,500	561,925	65
Michigan	1,639,372	819,686	50
Minnesota	786,413	707,772	90
Mississippi	473,262	397,540	84
Missouri	765,419	443,943	58
Montana	163,020	66,838	41
Nebraska	284,459	76,804	27
Nevada	222,846	75,768	34
New Hampshire	185,360	133,459	72
New Jersey	1,143,387	365,884	32
New Mexico	309,352	148,489	48
New York	3,200,000	1,984,000	62
North Carolina	1,398,072	671,075	48
North Dakota	118,756	36,814	31
Ohio	1,790,871	1,307,336	73
Oklahoma	640,000	320,000	50
Oregon	516,000	232,200	45
Pennsylvania	1,717,613	1,202,329	70
Rhode Island	92,000	65,320	71
South Carolina	646,985	407,601	63
South Dakota	148,804	41,665	28
Tennessee	906,975	562,325	62
Texas	3,535,742	1,131,437	32
Utah	461,259	152,215	33
Vermont	9,803	7,842	80
Virginia	1,013,967	790,894	78
Washington	915,694	448,690	49
West Virginia	313,997	241,778	77
Wisconsin	829,415	530,826	64
Wyoming	100,899	38,342	38
Total	43,349,238	22,363,315	52

Source: California Department of Education

Table 21.2
Comparison of District and Contractor Owned Buses for Pupil Transportation by State: 1992-93

State	District Owned Buses	Contractor Owned Buses	Total Buses	Percent of Buses Owned By Districts (%)
Alabama	7,183	292	7,475	96%
Alaska	n/a	n/a	n/a	n/a
Arizona	n/a	n/a	n/a	n/a
Arkansas	5,823	0	5,823	100%
California	14,649	6,953	21,602	68%
Colorado	4,017	244	4,261	94%
Connecticut	574	4,514	5,088	11%
Deleware	440	940	1,380	32%
Florida	13,810	910	14,720	94%
Georgia	10,211	51	10,262	100%
Hawaii	16	857	873	2%
Idaho	1,751	561	2,312	76%
Illinois	8,790	13,201	21,991	40%
Indiana	8,234	2,783	11,017	75%
Iowa	5,675	240	5,915	96%
Kansas	4,513	1,251	5,764	78%
Kentucky	8,942	207	9,149	98%
Louisiana	3,937	3,614	7,551	52%
Maine	2,084	519	2,603	80%
Maryland	2,881	2,844	5,725	50%
Massachusetts	2,362	2,840	5,202	45%
Michigan	n/a	n/a	n/a	n/a
Minnesota	4,268	5,491	9,759	44%
Mississippi	5,330	0	5,330	100%
Missouri	6,370	5,431	11,801	54%
Montana	1,046	867	1,913	55%
Nebraska	2,083	625	2,708	77%
Nevada	1,305	0	1,305	100%
Nw Hampshire	347	1,600	1,947	18%
New Jersey	6,292	8,700	14,992	42%
New Mexico	624	1,604	2,228	28%
New York	15,725	15,000	30,725	51%
North Carolina	12,836	0	12,836	100%
North Dakota	1,113	415	1,528	73%
Ohio	13,800	1,000	14,800	93%
Oklahoma	n/a	n/a	n/a	n/a
Oregon	2,523	1,693	4,216	60%
Pennsylvania	n/a	n/a	n/a	n/a
Rhode Island	909	420	1,329	68%
South Carolina	n/a	n/a	n/a	n/a
South Dakota	1,146	493	1,639	70%
Tennessee	4,455	1,943	6,398	70%
Texas	n/a	n/a	n/a	n/a
Utah	1,783	82	1,865	96%
Vermont	1,080	720	1,800	60%
Virginia	10,305	20	10,325	100%
Washington	6,057	1,013	7,070	86%
West Virginia	3,385	0	3,385	100%
Wisconsin	n/a	n/a	n/a	n/a
Wyoming	1,553	15	1,568	99%
Total	210,227	89,953	300,180	70%

Table 21.3
Total Number of Miles Driven for Pupil Transportation in the United States: 1992-93

State	Total Miles Driven
Alabama	63,179,155
Alaska	7,278,600
Arizona	53,166,741
Arkansas	43,090,826
California	250,595,738
Colorado	64,988,074
Connecticut	n/a
Delaware	18,441,776
Florida	230,218,623
Georgia	111,360,240
Hawaii	7,272,000
Idaho	24,015,909
Illinois	245,509,813
Indiana	81,407,525
Iowa	60,273,137
Kansas	72,684,633
Kentucky	92,882,562
Louisiana	310,147,200
Maine	31,700,487
Maryland	99,084,755
Massachusetts	n/a
Michigan	194,781,240
Minnesota	136,670,610
Mississippi	48,708,345
Missouri	107,842,894
Montana	18,300,152
Nebraska	37,319,060
Nevada	19,182,172
New Hampshire	11,000,000
New Jersey	n/a
New Mexico	30,375,085
New York	359,001,358
North Carolina	128,848,000
North Dakota	24,869,717
Ohio	175,263,660
Oklahoma	76,686,346
Oregon	44,889,929
Pennsylvania	228,708,784
Rhode Island	n/a
South Carolina	59,243,032
South Dakota	20,000
Tennessee	74,088,680
Texas	301,223,426
Utah	26,874,326
Vermont	n/a
Virginia	109,611,985
Washington	87,681,985
West Virginia	41,454,200
Wisconsin	n/a
Wyoming	15,684,437
Total	4,225,627,217

These data show tremendous variation in costs across the fifty states. Many of these differences are accounted for by differences in the way school buses are used to transport pupils. In states where school buses must travel long distances over relatively good roads through sparsely populated areas, the per pupil costs might be relatively high and the costs per mile relatively low. Similarly, in states with severe weather, transportation costs could be considerably higher due to greater maintenance needs.

As this brief description shows, providing transportation for public school children is a big business. In recent years it has become more and more complex and expensive as schools not only have to provide traditional transportation services, but also need to ensure that children with disabilities are offered appropriate transportation services so they too can attend public schools. Additionally, in many states, the transportation of pupils who attend religiously affiliated schools is the statutory responsibility of local public school districts. More importantly, the transportation of some 22 million children a day requires extensive training and effort to ensure that school buses are safe. Fortunately, today's public school buses are among the safest forms of transportation available to the public.

The remaining sections of this chapter describe the history of pupil transportation, provide information on the maintenance and operation of pupil transportation fleets, and discuss the financing of pupil transportation. The remaining sections also cover a number of critical issues facing school business administrators including decisions regarding district provision of pupil transportation services versus hiring a contractor, and the issue of seat belts in school buses.

An Historical Perspective on Pupil Transportation

The rapid growth in the number of students transported to and from school since the beginning of this century is closely linked with the rapid consolidation of school districts. Odden and Picus show that the number of school districts in the United States declined from more than 117,000 in 1940 to less than 16,000 in 1990.[3] With this consolidation came fewer schools and longer traveling distances for children.

The first state law authorizing the use of public tax dollars for pupil transportation was passed in 1869 in Massachusetts.[4] By 1900 eighteen states had enacted pupil transportation laws allocating governmental money for the purpose of transporting children to and from school. In 1900, schools spent $14.5 million to transport 356,401 students.[5] Even with the advent of increased pupil transportation programs, it was impossible to transport students long distances such as are common today. There have also been tremendous advances in vehicle building technology making it possible for buses to travel longer distances and at greater speed with increased safety. This has also facilitated the growth of pupil transportation systems and the consolidation of school systems.

The growth in pupil transportation during the twentieth century resulted in the need for administrative regulations to assure uniformity and safety. School districts established minimum distances for which busing would be provided, school bus safety standards were adopted and strengthened, and more efficient bus routing algorithms were developed to reduce costs.

In recent years, the role of pupil transportation has grown for a number of reasons. Intradistrict busing of minority and majority students to achieve racial integration in schools became not only a major demand on transportation systems but a very controversial political issue as well. Buses are also used more frequently today to provide transportation to children with disabilities who cannot walk to their local school, or who must be transported longer distances to specialized programs designed to meet their needs. As the number of field trips and athletic events have increased, additional uses of the school buses have been found. As a result, it is rare not to see school buses on the road today.

One of the important things to keep in mind throughout this discussion is that transportation differs from other educational services. Pupil transportation is not confined to school grounds and school buildings, and school buses must travel on city, county, and state highways where tremendous liability exists.[6] As a result, many of the rules, regulations, and laws affecting the operation of a pupil transportation system are enforced by agencies other than the school district or the state education agency.

Funding Pupil Transportation

The $9.8 billion a year spent on pupil transportation represents 4 to 5 percent of total K-12 educational expenditures. However, these expenses do not fall uniformly on school districts across the country or within individual states. There are obvious differences in terrain, population density, and safety factors among school districts that lead to different transportation needs, and hence different levels of expenditure. In addition to factors over which school districts have little or no control, there are other transportation policies that also affect the costs of pupil transportation. For example, establishing minimum walking distances has a tremendous impact on the number of students who need to be transported to school each day. As the number of students transported increases, so too do the costs of providing that transportation.

These differences, some controllable by the school district, others not, have important implications for education finance equity. In many states, court rulings or legislative actions have established standards to reduce differences in per pupil expenditures across school districts. If pupil transportation were included in total equalized expenditures, then school districts with particularly expensive transportation programs will not have the resources to provide the same level of educational programming as other school districts. Consequently, forty-nine states provide transportation aid to local school districts. Presently, the only state that does not do so is New Hampshire.

State transportation aid programs vary in scope and distribution method. Zeitlin's analysis of state aid formulas found that state goals for state transportation aid programs could be categorized into three types as follows:[7]

- Ensure that all districts provide a minimum level of transportation.
- Encourage districts to provide efficient and effective transportation service.
- Distribute state funds equitably to equalize the educational opportunities in each district.

She also found that state funding formulas could be divided into four categories as described as follows:[8]

Funding Method	Number of States
Funding the actual cost	13
Flat rate per unit	4
Multivariate calculations and factors	29
No distinct transportation funds	4

As these data show, the majority of states rely on a number of variables or factors in determining local district transportation aid. Thompson, Wood, and Honeyman demonstrate that the factors most often included in a pupil transportation aid system include the following:[9]

- Expenditures per pupil.
- Population density.
- Bus capacities.
- Flat grants.
- Eligible miles.
- Eligible hours of operation.
- Number of pupils transported.
- Distance traveled.
- Hazardous walking factors.

- Cost of bus replacement.
- Desegregation costs.

With the exception of Hawaii, which has one state-operated school district, no state fully funds local district pupil transportation systems. While this may seem onerous to many, in reality it makes for sound education finance policy from the state perspective. Requiring local school districts to fund a portion of the costs helps ensure that pupil transportation funds are spent efficiently and wisely. Moreover, since most states provide for equalization of general fund revenue, the burden of the local share is placed more heavily on those districts with the best ability to pay those costs. Some states establish separate categorical programs for funding pupil transportation while others provide funding through the state general funding formula.

Table 21.4 provides a state-by-state summary of the transportation formulas used by the fifty states. It shows the diversity of methods used to help local school districts fund pupil transportation services, but as a review of the table shows, the basic components as identified herein are used in most of the state systems. In addition to the diversity of funding methods, it should be noted that the proportion of transportation costs for which districts receive reimbursement varies from as much as 90 percent of total costs in New York and Colorado to as little as 53 percent in Mississippi. Maryland reflects 100 percent reimbursement, but this fiscal aid is based on 1981-82 expenditures, and does not reflect increases in costs since that time. In some states, the proportion varies depending on a variety of factors, generally inversely related to district wealth.

California, for example, has a unique funding formula. State transportation aid to school districts amounts to 95 percent of the previous year's state aid allocation. Allocations were initially based on 90 percent of approved home-to-school expenditures in 1982-83. This reimbursement is also supposed to be modified by a cost of living adjustment (COLA), but the legislature does not appropriate funds for the COLA every year. This declining reimbursement has resulted in current state funding levels that average approximately 50 percent of total home-to-school transportation costs.[10] Perhaps more importantly, since the funding is based on expenditures in 1982-83, the formula makes no adjustment for differences in the rate of growth in spending. As a result, some school districts receive reimbursement for as much as 90 percent of current transportation costs and other school districts as little as 10 percent.[11] In a state where general education revenues is highly equalized, these differences can have substantial unequalizing implications for general education funds for public education.

Transportation funding generally creates problems in equalization formulas because of its differential impact on school districts. District pupil transportation costs are often a function of things out of the school district's control

Table 21.4
State Aid Formulas for Pupil Transportation

State	Funding Method
Alabama	Allowable ADA cost through general aid
Alaska	Fully funded categorical
Arizona	Dollar amount per mile in each district
Arkansas	Per pupil amount and sparsity factor
California	Percentage of prior year aid plus COLA
Colorado	Percentile reimbursement up to 90%
Connecticut	Equalized between 10% and 60%
Delaware	Fully state funded within guidelines
Florida	Density index
Georgia	Combination of standard and variable costs
Hawaii	Full state funding
Idaho	85% of allowable costs through general aid
Illinois	Per pupil amount and/or allowable costs
Indiana	Equalized with density factor
Iowa	Funded through general aid
Kansas	82% of actual costs on density index
Kentucky	Density and interschool cost comparisons
Louisiana	Funded through general aid
Maine	Funded through general aid
Maryland	100% reimbursement based on 1981-82 amounts
Massachusetts	Reimbursement at legislative rate of 100% in 1.5 miles
Michigan	Formula based on costs averaging one third of costs
Minnesota	Complex formula with equalized features
Mississippi	Density and fixed rate table averaging 53% of costs
Missouri	80% reimbursed subject to efficiency audit
Montana	Statutory schedule of costs per bus mile
Nebraska	Weighted equalization formula
Nevada	General aid equaling 85% of allowable costs
New Hampshire	No aid to general transportation
New Jersey	90 % equalized formula
New Mexico	Categorical with equalization features
New York	90% equalized formula
North Carolina	Pays bus, staff per pupil/mile on prior year basis
North Dakota	Mileage and per-pupil reimbursement
Ohio	Partially funded through general aid
Oklahoma	Density formula
Oregon	70% reimbursement
Pennsylvania	Equalization formula
Rhode Island	Reimbursed through general aid
South Carolina	Fully state funded and monitored
South Dakota	Reimbursed through general aid
Tennessee	60% per pupil and density formula based on prior year
Texas	Density formula through general aid
Utah	Linear regression formula
Vermont	Equalized categorical aid
Virginia	Formula on pupils, mileage, and bus counts
Washington	Reimbursement on bus fleet size
West Virginia	80% of all costs
Wisconsin	Flat amount per pupil based on distance
Wyoming	75% through general aid formula

Source: D.C. Thompson, R. C. Wood and D. Honeyman, *Fiscal Leadeship for Schools: Concepts and Practices*, New York NY: Longman, 1994, 524. Adapted with permission.

such as distance, population density, and road conditions. Consequently, any program that reimburses a school district for a fixed proportion of its costs runs into a problem with the fact that the fixed proportion could be substantially more money per pupil for some school districts than for other school districts. The typical solution is to include a sparsity factor, a school district size factor, or both in the formula, rewarding school districts with difficult or expensive transportation needs with more funds than those with less difficult and expensive needs. Determining what the appropriate factors are is a difficult and often politically charged process.

Beyond the issue of revenue for pupil transportation, school districts are still faced with the task of operating an efficient, safe transportation system that meets the needs of the school district's students for both home-to-school transportation and other activities as approved by the board of education. The following section provides a discussion of the important components of pupil transportation operations.

Operation of a Pupil Transportation System

Regardless of whether a school district operates its own buses or contracts with a private firm for transportation services, understanding the basics of pupil transportation systems is important. This section focuses on a number of factors related to the operation of pupil transportation programs in public school districts. These include:

- Personnel, especially bus drivers.
- Bus purchasing and replacement.
- Bus maintenance.
- Bus routing.
- Safety.

Following this section, three additional current topics in pupil transportation are discussed. These include the transportation of children with disabilities, the use of seat belts in school buses, and contracting versus owning of the transportation fleet for public school districts.

Personnel Issues in Pupil Transportation

A school bus driver is responsible for the safety of all of the students on his or her bus at any time. This is no small task as the driver has to maintain discipline and drive safely. In addition, since the school bus is both the most frequent contact the public has with the school system, and because it is the first and last contact the district's children have with school, maintaining discipline and driving in a courteous manner are also critical parts of a school bus driver's job. The fact is, driving a forty-foot-long bus with fifty to ninety energetic children over windy mountain roads, or along busy city streets is a very difficult job. It requires an individual with both excellent driving skills and the ability to deal effectively with student discipline issues.

These multiple responsibilities mean it is critical that the school district must hire competent drivers. This is often a difficult task because driving a school bus is generally a part-time job. People are expected to drive a bus early in the morning for two or three hours, and again in the mid-afternoon for another two or three hours. It is expensive to hire different individuals to work only one shift due to training and personnel administration costs, but it often is difficult for drivers who work both shifts to fill the hours between shifts with other jobs.

Once the supervisor of transportation is able to find people willing to deal with the hours and responsibility necessary for successfully driving a school bus, the transportation supervisor will still have to find qualified individuals. Transportation supervisors must seek individuals who exhibit sound judgment based on their maturity. Potential bus drivers must have the necessary motor skills and physical characteristics, as well as the willingness to be trained to drive a public school bus. Additionally, under federal law, all school bus drivers must possess a commercial driver's license and be subject to drug testing procedures.

Once hired, driver training is a critical part of any bus driver's employment. States have certification programs for bus drivers and offer training for new hires. Many large school districts maintain their own training staff due to the large number of drivers they employ and the frequent turnover the school district experiences.

In addition to providing training in the operation and handling of a school bus, driver training should also focus on student discipline. Maintaining discipline on the bus is critical to the safety of the students riding to and from school. There are many techniques for dealing with discipline in a positive manner, and the importance of the task is twofold. First, and most importantly, it is necessary to ensure that the school bus driver can operate his or her vehicle safely. Second, enforcing discipline on the school bus before students arrive at school helps create a better school climate once the students are at school. If they arrive acting unruly, their teachers will have a more difficult time getting their attention and getting them to learn effectively.

In addition to the initial training, it is important to provide refresher courses in both driving and student discipline techniques. No matter how good a school bus driver may be, constant reminders of safety policies and practice in handling a school bus will only help them remain safe, courteous drivers.

Other Transportation Personnel

Regardless of the size of a school district, and no matter whether the school district contracts for transportation or provides this service itself, it is critical to employ someone who is responsible for the operation of the pupil transportation program. In small districts, particularly ones that rely on contractors for transportation, this may be one portion of the responsibilities of the assistant superintendent for business. In larger districts, there may be a transportation supervisor. This individual needs to understand both how to operate a bus fleet and how to train and work with the bus drivers, mechanics, and other personnel to maintain safety. Specifically, the transportation supervisor is also responsible for hiring and managing the mechanics who keep the district's buses operational and safe. The transportation supervisor is ultimately responsible for the total transportation program. He or she must be able to implement school board policy within severely restrictive budgetary constraints.

Bus Purchasing and Replacement

School buses are used on a daily basis, and every pupil transportation program needs to have a program for replacing buses as they wear out. It is important to replace older school buses to take advantage of improvements in safety standards and safety design. In 1977, the National Highway Traffic Safety Administration (NHTSA) issued a set of safety standards for school bus design known as the Post-DOT bus safety standards. These standards, which became effective on April 1, 1977, were designed to make buses stronger and better able to withstand an accident. The standards also included requirements for padding the backs and sides of seats in the bus to minimize injury to students. One common measure of the quality of a school district's or a state's school bus fleet is the number of pre-1977 buses still in use. Since pre-1977 school buses would be nearly 20 years old or more today, one would expect to see few of them still on the road.

In fact, in a number of states, all pre-1977 school buses have been taken out of service. States that have achieved this goal include Arizona, New Mexico, Vermont, Massachusetts, North and South Carolina, Iowa, Hawaii, Virginia, West Virginia, Kentucky, Delaware, and the District of Columbia. There are a number of states with a negligible number of pre-1977 buses in use as well. At the other extreme, approximately 21 percent of the buses in use in California and 22 percent of the buses in Washington are pre-1977 buses.

While some state aid formulas provide funds for school bus replacement, other states leave this capital expenditure to the local school district. In either case, planning for school bus replacement is an important component of the transportation budget since the purchase of a school bus is a considerable investment. Today the price of a new bus is around $35,000, but costs escalate quickly if the bus needs to be outfitted with special equipment such as a wheelchair lift. Bus prices of $70,000 to $100,000 are not uncommon in today's market.

The purchase of large items such as buses is usually governed by school district bidding requirements. In many instances, the state will bid all school buses. In those instances where the state does not actually bid the buses, it promulgates very specific state specifications and bid procedures for the purchase of school buses. Detailed written specifications for buses are prepared and manufacturers, or local dealers, submit sealed bids. There are two bid components for the purchase of a school bus — one for the body and one for the chassis. Districts seek separate bids on each, and the body manufacturer then builds the body to the specifications of the chassis selected.

There are an infinite combination of options for a school bus. Typically, the most important decisions to be made have to do with the engine type (gasoline or diesel), the size of the bus, and such large and important factors as transmissions. Decisions on engine type are generally based on the terrain to be driven, the type of use that will be made of the bus, and availability of diesel fuel. Buses that will be used in heavy traffic for long periods each day make considerably different demands on the engine than do buses that drive relatively open highways at high speed for short periods of time each day.

The size of the bus is also an important consideration. While bus bodies that accommodate as many as ninety students are available, configurations seating sixty-five or seventy-seven are more common. It is not unusual for particular school buses to cost upwards of $100,000 to meet particular needs. Smaller school buses are also readily available. Many school districts like to purchase large buses because they offer more flexibility. However, under certain circumstances, such as long travel distances with few children and crowded inner-city streets, there may be considerable efficiencies to be gained in the use of smaller buses. Smaller, relatively specialized school buses have become more common in recent years as the demands for transportation of children with disabilities increases.

Similarly, transmission decisions should be based on the type of driving that will occur and the relative skill of the drivers. While manual transmissions may offer some savings in operational costs, the convenience of automatic transmissions may outweigh the savings in many instances.

Bus Maintenance

Maintaining the school district's bus fleet is as important as ensuring that there are qualified drivers. A good maintenance program is essential to assuring adequate school bus safety. In addition, good maintenance will extend the useful life of a school bus, reducing replacement costs significantly.

Proper school bus maintenance begins with drivers' daily inspections of their vehicles. Each bus driver should do a walk-around for his or her bus before starting the engine to make sure the tires are inflated and that no leaks or other problems are apparent. Once the bus engine is started, drivers should check to make sure that headlights, rear lights, brake lights, running lights, flashers, blinkers, crossing lights, stop arms, and other safety features are working properly. Often drivers can do this in teams with one driver watching each function while the other engages the various features. Once complete for one school bus, drivers can change roles and check the other school bus. In many pupil transportation systems, drivers are required to fill out a daily inspection sheet indicating whether each component is working.

Drivers are also the front line for noting maintenance needs. They know when the engine sounds different or when the brakes are showing signs of wear. Drivers should have a simple system for reporting maintenance needs to the mechanics, and there should be a provision for telling drivers what actions have been taken to repair noted problems on their school buses.

Equally important is a preventive maintenance program. A routine schedule for all maintenance functions from oil changes to brake inspections to engine overhauls should be established. The bus manufacturer's guidelines provide an excellent place to start in designing a preventive maintenance program, but services should be adjusted to meet the individual conditions of the school district. However, at no time should the manufacturer's maintenance recommendations be exceeded. The time between services may be shortened if conditions warrant, but lengthening them is ill advised.

No matter how skilled the transportation supervisor is at overseeing the preventive maintenance program for the school bus fleet, at some point a bus will break down, require a major repair, or find itself with a flat tire. The school district generally maintains an adequate number of spare buses to deal with these situations, as well as to provide relief so that other routine maintenance can be performed. If the transportation fleet is large enough to have a number of mechanics employed, much of the preventive maintenance can be scheduled at night and during other off-use times.

The transportation department generally maintains detailed records on each school bus in the fleet. These should include copies of work orders for all maintenance and repair actions, as well as a record of the vehicle's daily operations. There are a number of commercially available computer programs that will help manage bus maintenance records and will remind the shop mechanics when a school bus is due for preventive maintenance servicing. These tools not only help ensure that the school buses are operating at peak efficiency and safety, but can also be used to track operating costs of each school bus over virtually any period of time. For larger fleets with a variety of types of school buses, these data can be quite valuable in analyzing which school bus and chassis manufacturer's equipment is the most cost efficient for that particular operation. The data can help transportation department staffs measure the costs and benefits of various options purchased for the transportation fleet.

Bus Routing

Efficient routing of school buses is another important component of pupil transportation. Since it is essential that school buses run on time and provide a seat for all of the children entitled to ride to school, developing an accurate data base of where all the children live, where they go to school, and what is the most efficient way to transport them to their respective school buildings is essential. In addition to ensuring that all children receive a safe and timely ride to school, the person responsible for routing must make sure that children do not spend too much time on the bus, that buses are not filled beyond capacity, and that the distance buses are operated empty (known as deadheading) is minimized.

The options for scheduling buses are as varied as the number of districts attempting to find an efficient routing structure. There are a number of factors that school districts need to consider, however. Perhaps the most important in terms of transportation efficiency is staggered school starting times. If all schools start and end at exactly the same time, each school bus can only operate on one bus route before and after school. However, if school starting times are varied by a few minutes across the district, it may be possible for the school district to use each bus on two, three, or even four routes in both the morning and afternoon. This dramatically reduces the number of school buses and drivers needed and provides longer hours of operation for buses. This not only makes the buses more efficient, but provides better employment for school bus drivers. The question every school district needs to examine as a local policy issue is the question of how much school starting times may be staggered within a local community. In many cases, there is considerable community opposition to staggered hours of operation. In these cases, it may be good policy to show the district residents how much money can be saved with reasonable, minor adjustments in school times.

Regardless of the number of routes a school district operates each day, another important goal is to minimize deadhead mileage and time. Driving a school bus empty is necessary before picking up the first child and after dropping off the last child each day. However, if routes are designed in a circular rather than linear fashion, it may be possible to reduce some of the deadhead time. Unfortunately, there are often geographical constraints to doing this, forcing school districts to rely on relatively linear bus route patterns.

Another factor to consider is minimum busing distances. Many districts have a policy of busing all children outside a given radius of a school. These distances are usually in the range of a mile or a mile-and-a-half at the elementary level, and somewhat greater at the secondary level. Increasing the school district's mandatory walking distance can often save considerable money as a number of children previously able to ride the school bus are no longer able to do so and are expected to walk to school each day. This discussion may not be applicable in those states where such issues are dictated by state statutes or regulations. Moreover, safety factors involving students must be taken into consideration within every such instance.

Another important function for school bus routing is accurate information concerning the location of each of the children to be bused to and from school. Fortunately, there are a number of computer programs currently available that allow the transportation office to minimize school bus routes once the basic information concerning each child has been accurately placed into the system. Once the computer software is able to determine where each child lives and where he or she goes to school, the software program can begin to match students with buses and bus capacities to find the minimum daily mileage needed to transport children to and from individual schools. Advances in these geocoding systems have reached the point where they can be operated by school officials from a simple desktop personal computer.

Safety

The overall safety of the pupil transportation program is critical. All of the issues discussed herein contribute to the safety of the entire transportation system. In addition, it is critical that safety be uppermost in everyone's mind when school buses are operating. Safety includes making sure that all school bus stops are adequately protected from traffic, and that students have ample time to cross busy streets when they are picked up and dropped off. Safety also means ensuring that there is a place at each school where buses drop off and pick up students. If possible, it should be on the other side of the school from the entrance and exit for children who walk, and it should be well separated from the area used by parents relying on private vehicles for transporting their children to and from school. The critical integration of transportation needs with facility planning is necessary for these safe loading zones within each school facility.

Safety on the school bus should be part of the curriculum in each class at the beginning of each school year, and bus drivers and mechanics should have frequent refresher courses on school bus safety as well. While all of this may seem burdensome, this diligence in transportation safety will ensure school buses continue to be one of the safest modes of transportation in the United States for years to come.

Other Issues in Pupil Transportation

In addition to the important components of successfully operating a safe and efficient pupil transportation system, there are a number of other important issues that pupil transportation system directors must deal with on a regular basis. They include whether to contract for pupil transportation or to have the school district operate its own school bus fleet, transportation of children with disabilities, and the use of seat belts in school buses.

Contract Versus Owning

Data presented in table 21.2 at the beginning of this chapter indicate that some 30 percent of the school buses in use today are owned and operated by private contractors and not public school districts. One of the most hotly debated issues in pupil transportation today is over which system, contract or district ownership, provides for better transportation. The answer, not surprisingly, is not straightforward, and in fact the right choice for one school district may not be the right choice for another school district.

Major advantages attributed to relying on bus contractors for the provision of pupil transportation services include:

- It relieves school officials of a major management responsibility for maintenance and operation, one for which they typically have received little or no formal training.

- Capital outlay expenditures for new buses are not necessary.

- Some of the complaints and criticisms about bus service can be directed to the contractor.

- There may be more flexibility for adding new bus routes or obtaining an additional bus for some extracurricular activity, particularly if the contractor serves a number of districts in the vicinity.

- The contractor is responsible for driver hiring and training.

On the other hand, there are a number of important advantages to school district ownership of the buses. These include:

- Possibly lower costs.

- It may be easier to use school buses for other programs, like field trips, since it is possible that a contractor will use the school buses for other purposes when not engaged in school district contract activities.

- The ability to integrate transportation services into the entire instructional program.

- More control over the hiring and training of drivers.

- Possible reduction in legal complexities should litigation occur.

What is notable about these two lists is the fact that some of the benefits of one can be thought of as benefits of the other option as well.

There are alternatives to simple school district ownership or contracting. Many large school districts employ both contractors and district owned and operated school buses. Operation of a dual fleet of school buses helps the school district ensure that the contractor is keeping prices competitive since the school district knows what it costs to operate its own school buses. Similarly, if the school district cannot provide transportation services for the same or less than the contractor, then it is clear that inefficiencies may exist in the transportation department. Other large school districts rely on a number of contractors for the provision of transpiration services. Doing so provides competition among the contractors and helps keep contract costs lower.

Other options include contractor ownership of buses that are operated and maintained by the school district, and school district owned buses that are maintained and operated by a contractor. All of these options are viable ways for school districts to ensure that transportation services are provided safely and efficiently.

Transportation of Children with Disabilities

There is no question that provision of transportation services for children with disabilities is more expensive than providing pupil transportation for all other children. Because the transportation of these children is often provided through the regular pupil transportation department, it is often difficult to separate these costs. However, Miller found that in one county-wide service unit in Southern California, transportation costs were as much as $800 per pupil higher than regular education costs.[12]

There are a number of reasons for these higher expenditures. In the case of children with severe disabilities, educational services are generally provided through special programs located at a few schools. Thus, the students need to be transported greater distances than most regular education home-to-school students. In addition, many children with severe disabilities require specialized services and equipment, both in school and on the school bus. School buses must be able to transport wheelchairs and other special equipment, often requiring modifications at considerable cost. Finally, given the importance of minimizing the time children spend on the bus, the capacity of any one bus is dramatically reduced. Another factor is the length of time it may take to assist a disabled child board the school bus. All of these factors result in the use of a relatively large number of small, highly specialized school buses.

In addition to the physical requirements of the school buses, drivers often require special training to deal with both the discipline and health needs of children with disabilities.

Many children with disabilities do not require such extensive services and facilities. Moreover, the practice of placing children with disabilities in the least restrictive environment means that to the extent possible they should be placed in regular classrooms, and ride the school bus with other children. The current movement toward inclusion in special education has placed additional pressure to provide services for children with disabilities in regular environments.

This movement for inclusion has considerable implications for pupil transportation. In many instances, school buses need to be modified to transport children with disabilities. Wheelchair lifts, kneeling buses, and specialized seats and restraints (seat belts) may be required to transport many of these children safely. In addition to the expense of modifying school buses with this specialized equipment, the bus routing and scheduling tasks become much more complex as it is now important to have a properly equipped school bus on each route. Where at one time the dispatcher only needed to be sure that the school bus capacity was adequate for the number of children on the route, he or she now has to ensure that the proper support equipment is also available. If a regular route bus were to break down and a suitable replacement were not available, other transportation arrangements would have to be made for the children requiring specialized services for some period of time. Such flexibility in the pupil transportation system requires the availability of other school buses or vehicles and qualified drivers, at very short notice. Such flexibility is obviously expensive.

In the case of many of these specialized facilities for transporting children with disabilities, it takes longer for children to board the school bus. In some instances, the driver has to leave his or her seat to operate a wheelchair lift or other piece of equipment. The time this takes must be factored into the routing structure to ensure that school buses arrive at school on time each day. If one route were to have a number of children with disabilities requiring specialized equipment, the time factor can be substantial. In addition, adequate safety precautions need to be taken to ensure that the school bus cannot be moved by others on the bus while the driver is not seated. This either requires turning off the motor, or finding some other way to keep the bus from being moved. In either case, it may take time for the driver to implement these safety precautions.

In addition, the transportation of children with disabilities on regular bus routes also requires additional safety and first aid training for the bus drivers. Coping with discipline problems can be more difficult when children with disabilities are part of the student mix. In addition, drivers require training in the specialized equipment necessary for transporting children with disabilities.

Seat Belts

One of the most controversial issues in pupil transportation has been and continues to be the use of seat belts on school buses. Seat belts have been required in passenger automobiles since the 1960s, yet a similar requirement, on the national level, does not exist for school buses. On the surface, it seems that requiring seat belts on school buses would help improve their already excellent record in terms of the number of fatal accidents per miles driven. However, the issue is complex and reveals a debate among pupil transportation specialists and the general public.

The arguments in favor of seat belts should be familiar to most readers. Seat belts in cars have been proven over and over again to save lives in automobile accidents. Yet, studies of seat belt use on buses found that lap belts offered no additional protection to children on those buses unless they were used in conjunction with high back seats.[13] The 1977 bus safety requirements issues by NHTSA did not require seat belts in school buses, but did require high back seats and additional padding on both the front, and back of each seat.

Since that time, the National Transportation Safety Board (NTSB) has found that the Post-DOT bus standards have satisfactorily protected children in school bus accidents. NTSB has argued that there is little justification for extending seat belts to school buses.

Opponents of mandatory seat belts argue that they are only effective if they are used by students, which they fear will not be the case. Moreover, they contend that there is a risk of injury to students from the buckles of seat belts that become projectiles in accidents or in incidents of disruption. Finally, opponents of seat belts argue that the cost of fitting all buses with seat belts is very high, and that those funds could be better spent to improve school bus safety through better bus and seating design. In summary, the seat belt issue remains unresolved.

Despite the fact that a few states and several school districts require seat belts for school buses, the usage of seat belts on school buses is minimal across the nation. The controversy continues after many years of debate regarding this volatile topic. More than ten years have passes since *School Business Affairs* published the arguments made on both sides of this issue. As cited, the advantages and disadvantages are as follows:

Pro

- Data prove that seat belts save lives.
- Kids who learn to buckle up on school buses will carry the habit over to automobiles.

- Studies on compartmentalization are not complete and need to be restructured.
- Studies that say seat belts hurt young bodies are not true.
- Younger parents grew up with seat belts and demand additional protection.

Con

- School buses have the best safety record in transportation.
- Other measures, including compartmentalization and driver training, are more cost effective than seat belts.
- Children will not wear their seat belts.
- Belts can harm young bodies.
- Belts are too costly to install.
- Buses are so large that they do not suffer damages in the same way an automobile does in a collision.[14]

Summary

Pupil transportation is one of the most important and least understood functions of school operations. On a daily basis, more people have daily contact with school buses than with any other facet of the public education system. The goal of transporting students is to do so in a safe and cost-effective manner. Various state statutes control the manner and operation of student transportation. As a nation, pupil transportation consumes some 4 to 5 percent of educational resources, and is responsible for delivering over half the nation's students to and from school each day.

Endnotes

[1] D.C. Thompson, R.C. Wood, and D. Honeyman, *Fiscal Leadership for Schools: Concepts and Practices*. New York, NY: Longman, 1994. 512.

[2] Note that this figure is a simple average of the values for the states for which data are available. Since not all figures were available for all states, it was impossible to provide a weighted average expenditure that might have provided a more accurate measure of the mean costs of pupil transportation per pupil across the United States.

[3] A.R. Odden and L.O. Picus, *School Finance: A Policy Perspective.* New York, NY: McGraw Hill. 1992, 4

[4] *Acts and Resolves*, Commonwealth of Mass., Chap. XCXXII, § 1, (1869).

[5] D.H. Stollar and K. Tanner, *Student Transportation Study for the State of Indiana* (Knoxville, Tenn: Univ. of Tennessee, 1978).

[6] R.C. Wood and R.W. Ruch, "Negligence in Transporting Students," *School Business Affairs*, Nov., 1986, 48, 50-51.

[7] L.S. Zeitlin, "State Pupil Transportation Funding: Equity and Efficiency." *School Business Affairs*. 56(4), 1990, 23.

[8] Zeitlin. "State Pupil Transportation Funding," 22-25.

[9] Thompson, Wood, and Honeyman, *Fiscal Leadership for Schools*, 520-523.

[10] P. Goldfinger, Chap. 5 in *Revenues and Limits.* Sacramento, CA: School Services of California, 1994,

[11] J. Murphy, *Measuring the Encroachment of Special Education and Transportation Programs on the General Fund of California's Unified School Districts in Nine Southern Counties: Implications for State School Finance Equity and Adequacy.* (Univ. of Southern California, Dissertation, 1994).

[12] L.O. Picus and C. Miller. "Cost and Service Delivery Trade-Offs in Providing Educational Services for Students With Severe Disabilities." *Educ. Admin. Quart.* 31(2).

[13] *Issues in Pupil Transportation*, (ASBO: Reston VA: 1987) 71.

[14] F.J. Rex, "Seat Belts in School Buses: A Study of Current Protective and Potential Action," *School Business Affairs*, April, 1985, 31.

Chapter 22
Foodservice

Introduction

The many faces of school business administration include a wide variety of opportunities for leadership and frontline drama in the context of modern public education. Many areas of school business administration, however, have suffered from relegation to a mundane status because they are not directly involved in the teaching and administrative roles. Yet, other areas of school district management are absolutely vital to successful district operation, in that frontline instruction could not be carried out well absent support services of many kinds. Many of these areas command large budgets, such as transportation and maintenance, where millions of dollars are needed to ensure that children safely reach school and are served in a clean and hospitable environment.

The importance of foodservice in the scheme of school business management is no less vital, in that much data show that hungry children make extremely poor learners. In the modern context, foodservice commands enormous budgets and human resources, playing a role that has increased as schools have moved into providing hot lunch and breakfast programs, free and reduced meals, federal and state lunch price supports, and many other elements of attempting to provide children with better learning skills through health and nutrition. The importance of these programs has always been obvious, but the imperative to do more was best stated in the observation that " . . . even yet, children come to school hungry, and stay that way, even when federal and state programs exist to feed them."[1]

The broad area often known as support service or auxiliary service is thus vital to how schools carry out the total mission of education. These operations are larger and more expensive than most people realize. In the case of foodservice, people must eat before can they do anything else. It makes no difference whether a district is rural or urban—these needs are the same, with the only differences resting in the manner of delivery and economies of scale that help set costs in such programs. For example, foodservice requires employment of many laborers, with urban schools employing hundreds of workers in mass production operations and rural districts employing many people relative to smaller numbers of students and sites served. The result is a need for effective and efficient organization in order to help students perform better in the classroom by virtue of satisfying the basic human need for nourishment.

Given such reality, this chapter is intended to introduce and update general knowledge of foodservice concerns affecting school business administrators. Although not exhaustive of the many foodservice issues faced by districts, this overview of concerns lays a foundation for informed and productive work on the part of administrators by granting respect to still another important element of school business operations.

History of the Foodservice Function

General Issues

As stated at the outset of this chapter, the importance of the school foodservice function cannot be overemphasized because the role of good nutrition in the mission of schools is only fundamental. Data from earlier portions of this text and other sources[2] indicate that schools face tremendous problems beginning in the home that are laid on the school as barriers to effective teaching and learning. For example, data show that unwed births have reached epidemic levels, with 50 percent of all illegitimate children born to teenage mothers. These same data point to the fact that every day forty teenage girls bear their third child, a shocking fact worsened by a confounding reality that these infants are often malnourished by virtue of being born to child mothers who are themselves poorly fed.[3]

Many children come to school from disadvantaged homes. Data on American families show that many school children are at economic and social risk. These data indicate that of every 100 infants, twenty are illegitimate, twelve have parents who divorce before the child is eighteen, another five have parents who will separate, six will lose a parent to death, forty will live in a female-headed household, thirteen are born to teen mothers, fifteen are born to homes where no parent is employed, fifteen are born to homes with poverty wages, and 25 will eventually receive welfare.[4] Many homes are ill-prepared to provide good nutrition, thereby worsening the effect of other disadvantages.

These data are accompanied by other indications that poverty is steadily growing. In 1992, almost 13.9 million children (21.1 percent) were designated below the federal poverty definition, a change from 1970 when 10.2 million (14.9 percent) were in poverty.[5] As Hodgkinson noted over a decade ago, a child under age six has a 600 times greater chance of living in poverty than a person over age sixty five, in part because federal spending for children's programs has tended to decline over recent years, while federal spending for aging populations, until recently, has tended to increase or at least exhibit no shrinkage.[6] While poverty declined throughout the first seventy-five years of the twentieth century, by 1979 that trend had clearly reversed with significant levels of poverty still evident today on the threshold of a new century.

These and other data linking poor school performance to nutrition and social issues have sparked historic interest at all levels of federal, state, and local government concerning what can and should be done to alleviate the problems of poorly nourished children. The debate has raged for many decades, with all levels of government aiding health and nutrition in various ways, ranging from cash subsidies to surplus commodity distributions designed to simultaneously benefit different segments of the social and economic climate of the nation. While these aids have been useful, it has not been sufficient to assure that school foodservice programs have provided solutions to the problems children bring to school.

The role of foodservice in education has only increased with the passage of time, as poignantly illustrated by the fact that most schools have now moved from only hot lunch programs into full service breakfast programs designed to aid student learning through good nutrition. This was prompted by the fact that so many children come to school poorly nourished—a fact that is nearly always accompanied by severe limitations to educational readiness. While foodservice cannot possibly solve all the problems cited here, it aims to make school a better experience for tomorrow's income-producing citizens.

Federal Support

As stated earlier, all three levels of government have attempted to respond to nutritional needs in schools. As noted in the first edition of this text, beginning in the early 1900s teachers were trying to find ways to feed poor children. Farmers would donate food, and the smell of food cooking on a wood stove was a common aspect of schooling in the early twentieth century. These realities marked the beginning of a concerted effort to provide children with better educational opportunities through recognition that nutrition and achievement are both part of a good school experience.

The first federal legislation involving aid to nutrition, however, actually grew out of an economic interest in agriculture resulting from the Great Depression. In 1935, the Bankhead-Jones Act (P.L. 74-182) authorized federal grants to states to create agricultural experiment stations, and in the same year Congress approved the Agricultural Adjustment Act (PL 74-320) authorizing 30 percent of customs receipts for encouragement of exportation and domestic consumption of agricultural surplus commodities. The effect of this law was that surplus commodities began to be used in school lunch programs in 1936, wherein both agriculture and schools were the primary beneficiaries in that a nutrition agenda was linked with a desperately low market for excess commodities.

Not long afterward, Congress passed the National School Lunch Act (PL 79-396) authorizing assistance through grants-in-aid to states for the purpose of helping provide food and facilities for nonprofit school lunch programs. By 1954, Congress had passed the School Milk Program Act (PL 83-597), which appropriated federal funds to help buy milk for school lunches. As a general rule, the interest of Congress in school nutrition has not slowed, particularly as it has pursued an equality agenda focusing on federal programs for low income families.

As seen in table 22.1, federal involvement in education has been substantial and broad based, covering a wide array of social and economic national interests. Table 22.1 indicates that federal support for education was estimated

Table 22.1
Federal On-Budget Funds for Education—1965-1994

Level or other educational purpose, by agency and program	1965	1970	1975	1980	1985	1990	1992	1993	1994[1]
1	2	3	4	5	6	7	8	9	10
Total, all programs	**$5,331,016**	**$12,526,499**	**$23,288,084**	**$34,493,502**	**$39,027,876**	**$51,624,342**	**$60,479,844**	**$67,740,618**	**$68,364,174**
Elementary/secondary education programs	**$1,942,577**	**5,830,442**	**10,617,195**	**16,027,686**	**16,901,334**	**21,984,361**	**27,926,888**	**30,834,326**	**34,318,808**
Department of Education[2]	567,343	2,719,204	4,132,742	6,629,095	7,296,702	9,681,313	12,057,746	13,058,974	14,825,817
Grants for the disadvantaged (includes Chapter 1)	—	1,339,014	1,874,353	3,204,664	4,206,754	4,494,111	6,158,813	6,615,047	6,900,113
Impact aid program	349,671	656,372	618,711	690,170	647,402	816,366	794,794	432,153	982,705
School improvement programs (includes Chapter 2)	72,298	288,304	700,470	788,918	526,401	1,189,158	1,514,892	2,032,552	1,663,417
Indian education	—	—	40,036	93,365	82,328	69,451	68,523	99,925	82,170
Bilingual education	—	21,250	92,693	169,540	157,539	188,919	198,332	124,778	257,449
Education for the handicapped	13,849	79,090	151,244	821,777	1,017,964	1,616,623	2,243,338	2,564,070	3,603,794
Vocational and adult education	131,525	335,174	655,235	860,661	658,314	1,306,685	1,079,054	1,190,449	1,317,242
Education Reform - Goals 2000[3]	—	—	—	—	—	—	—	—	18,927
Department of Agriculture	623,014	760,477	1,884,345	4,064,497	4,134,906	5,528,950	6,714,082	7,154,483	7,800,902
Child nutrition programs	178,580	299,131	1,452,267	3,377,056	3,664,561	4,977,075	6,126,983	6,596,588	7,236,654
Agricultural Marketing Service—commodities[4]	340,073	341,597	248,839	388,000	336,502	350,441	400,000	389,900	400,000
Special milk program	86,609	83,800	122,858	159,293	15,993	18,707	19,178	15,535	21,730
Estimated education share of Forest Service permanent appropriations	17,752	35,949	60,381	140,148	117,850	182,727	167,921	152,460	142,518
Department of Commerce	—	—	—	54,816	—	—	—	—	—
Local public works program—school facilities[5]	—	—	—	54,816	—	—	—	—	—
Department of Defense	73,000	143,100	264,500	370,846	831,625	1,097,876	1,197,318	1,259,374	1,206,145
Junior ROTC	—	12,100	12,500	32,000	55,600	39,300	54,746	84,100	108,800
Overseas dependents schools	73,000	131,000	252,000	338,846	613,437	864,958	912,916	895,674	841,845
Section VI schools[6]	—	—	—	—	162,588	193,618	229,656	279,600	255,500
Department of Energy[7]	100	200	300	77,633	23,031	15,563	15,236	6,254	6,254
Energy conservation for school buildings[8]	—	—	—	77,240	22,731	15,213	12,586	5,054	5,054
Pre-engineering program	100	200	300	393	300	350	2,650	1,200	1,200
Department of Health and Human Services[9]	79,999	167,333	683,885	1,077,000	1,531,059	2,396,793	3,310,200	4,114,498	4,771,327
Head Start[10]	—	—	403,900	735,000	1,075,059	1,447,758	2,201,800	2,776,286	3,326,285
Payments to states for AFDC work programs[11]	—	—	—	—	—	459,221	594,184	736,474	859,913
Social Security student benefits[12]	79,999	167,333	279,985	342,000	456,000	489,814	514,216	601,738	585,129
Department of the Interior	130,096	140,705	220,392	318,170	389,810	445,267	517,666	536,483	587,817
Mineral Leasing Act and other funds:									
Payments to states—estimated education share	11,075	12,294	27,389	62,636	127,369	123,811	122,045	108,924	123,991
Payments to counties—estimated education share	10,731	16,359	29,494	48,953	59,016	102,522	45,805	34,903	39,536
Indian Education:									
Bureau of Indian Affairs schools	92,603	95,850	141,056	178,112	177,265	192,841	325,582	368,817	399,234
Johnson-O'Malley assistance[13]	15,534	16,080	22,251	28,081	25,675	25,556	23,590	22,980	24,326
Education expenses for children of employees, Yellowstone National Park	153	122	202	388	485	538	644	859	730
Department of Justice	6,402	8,237	9,822	23,890	36,117	65,997	94,724	107,857	125,228
Vocational training expenses for prisoners in federal prisons	1,466	2,720	3,039	4,966	8,292	2,066	1,944	1,725	3,395
Inmate programs[14]	4,936	5,517	6,783	18,924	27,825	63,931	92,780	106,132	121,833
Department of Labor	230,041	420,927	1,097,811	1,849,800	1,945,268	2,505,487	3,708,362	4,240,990	4,535,859
Job Corps[15]	—	—	175,000	469,800	604,748	739,376	925,826	949,287	1,027,939
Training programs—estimated funds for education programs[16]	230,041	420,927	922,811	1,380,000	1,340,520	1,766,111	2,782,536	3,291,703	3,507,920
Department of Transportation[17]	—	45	50	60	60	46	60	60	40
Tuition assistance for educational accreditation—Coast Guard personnel[18]	—	45	50	60	60	46	60	60	40
Department of the Treasury	32	—	847,139	935,903	273,728	—	—	—	—
Estimated education share of general revenue sharing—[19]									
State[20]	—	—	475,224	525,019	—	—	—	—	—
Local	—	—	371,915	410,884	273,728	—	—	—	—
Tuition assistance for educational accreditation—Coast Guard personnel[18]	32	—	—	—	—	—	—	—	—
Department of Veterans Affairs[21]	41,250	338,910	1,371,500	545,786	344,758	155,351	190,608	222,567	320,210
Noncollegiate and job training programs[22]	14,550	281,640	1,249,410	439,993	224,035	12,848	—	—	—
Vocational rehabilitation for disabled veterans[23]	17,400	41,700	73,100	87,980	107,480	136,780	184,500	216,276	252,960
Dependents' education[24]	9,300	15,570	48,990	17,813	13,243	5,723	6,108	5,840	5,737
Service members occupational conversion and training act of 1992[25]	—	—	—	—	—	—	—	451	61,513

—Federal on-budget funds for education, by level or other educational purpose or activity, by agency and program: Fiscal years 1965 to 1994—Continued

[In thousands of current dollars]

Level or other educational purpose, by agency and program	1965	1970	1975	1980	1985	1990	1992	1993	1994[1]
1	2	3	4	5	6	7	8	9	10
Elementary/secondary education programs (continued)									
Other agencies									
Appalachian Regional Commission[26]	—	33,161	41,667	9,157	4,632	93	5,182	5,382	5,300
National Endowment for the Arts[27]	—	—	3,686	4,989	4,399	4,641	5,000	5,000	5,000
Arts in education	—	—	3,686	4,989	4,399	4,641	5,000	5,000	5,000
National Endowment for the Humanities[28]	—	20	149	330	321	404	809	1,645	1,623
Office of Economic Opportunity[29]	182,793	1,072,375	16,619	—	—	—	—	—	—
Head Start[30]	96,400	325,700	—	—	—	—	—	—	—
Other elementary and secondary programs[31]	20,000	42,809	16,612	—	—	—	—	—	—
Job Corps[32]	34,000	144,000	—	—	—	—	—	—	—
Youth Corps and other training programs[33]	31,000	553,368	7	—	—	—	—	—	—
Volunteers in Service to America (VISTA)[34]	1,393	6,498	—	—	—	—	—	—	—
Other programs									
Estimated education share of federal aid to the District of Columbia	8,507	25,748	42,588	65,714	84,918	86,579	109,894	120,759	127,286
Postsecondary education programs	**1,197,511**	**3,447,697**	**7,644,037**	**11,115,882**	**11,174,379**	**13,650,915**	**14,384,138**	**17,844,015**	**14,128,963**
Department of Education[2]	237,955	1,187,962	2,089,184	5,682,242	8,202,499	11,175,978	11,323,584	14,660,704	10,698,931
Student financial assistance[35]	—	—	—	3,682,789	4,162,695	5,920,328	7,071,440	7,678,293	7,420,623
Federal Direct Student Loan Program[36]	—	—	—	—	—	—	—	10,000	153,905
Federal Family Education Loan Program[37]	—	2,323	111,087	1,407,977	3,534,795	4,372,446	3,253,648	5,554,920	1,911,525
Higher education	218,264	1,029,131	1,838,066	399,787	404,511	659,492	718,406	1,041,583	880,673
Facilities—loans and insurance[38]	3,588	114,199	16,292	−19,031	5,307	19,219	25,984	−2,976	−5,720
College housing loans[38,39]	—	—	—	14,082	−164,061	−57,167	−39,907	−28,355	3,027
Educational activities overseas	129	774	1,881	3,561	1,838	82	—	—	—
Historically Black Colleges and Universities Capital Financing, Program Account[40]	—	—	—	—	—	—	—	—	151
Gallaudet College and Howard University	15,974	38,559	111,971	176,829	229,938	230,327	263,497	353,390	286,468
National Technical Institute for the Deaf[41]	—	2,976	9,887	16,248	27,476	31,251	30,516	53,849	42,279
Department of Agriculture	—	—	6,450	10,453	17,741	31,273	34,238	32,730	33,373
Agriculture Extension Service, Second Morrill Act payments to agricultural and mechanical colleges and Tuskegee Institute[42]	—	—	6,450	10,453	17,741	31,273	34,238	32,730	33,373
Department of Commerce	5,081	8,277	14,973	29,971	2,163	3,312	3,270	3,549	3,600
Sea Grant Program[43]	—	—	1,886	3,123	2,163	3,312	3,270	3,549	3,600
Merchant Marine Academy[44]	3,570	6,160	10,152	14,809	—	—	—	—	—
State marine schools[44]	1,511	2,117	2,935	12,039	—	—	—	—	—
Department of Defense[45]	77,500	322,100	379,800	545,000	1,041,700	635,769	680,194	696,800	684,500
Tuition assistance for military personnel	—	57,500	86,800	(46)	77,100	95,300	102,400	123,400	132,600
Service academies[47]	77,500	78,700	86,200	106,100	196,400	120,613	125,146	128,000	143,700
Senior ROTC	—	108,100	116,500	(46)	354,000	193,056	193,348	184,100	190,100
Professional development education[48]	—	77,800	90,300	(46)	414,200	226,800	259,300	261,300	218,100
Department of Energy[7]	3,000	3,000	3,000	57,701	19,475	25,502	34,373	17,654	17,654
University laboratory cooperative program	3,000	3,000	3,000	2,800	6,500	9,402	19,100	4,000	4,000
Teacher development projects[49]	—	—	—	1,400	—	—	—	—	—
Graduate traineeship programs[50]	—	—	—	—	—	—	—	—	—
Energy conservation for buildings—higher education[8]	—	—	—	53,501	12,705	7,459	9,573	6,654	6,654
Minority honors vocational training[51]	—	—	—	—	150	—	—	—	—
Honors research program[51]	—	—	—	—	120	6,472	1,000	1,000	1,000
Students and teachers[52]	—	—	—	—	—	2,169	4,700	6,000	6,000
Department of Health and Human Services[9]	469,223	981,483	1,686,650	2,412,058	516,088	578,542	743,456	720,911	759,516
Health professions training programs[53]	139,795	353,029	599,350	460,736	212,200	230,600	305,829	299,785	305,583
Indian health manpower[54]	—	—	—	7,187	5,577	9,508	19,460	26,100	27,400
National Health Service Corps scholarships	—	—	1,206	70,667	2,268	4,759	58,706	33,323	42,270
National Institutes of Health training grants[55]	—	—	154,875	176,388	217,927	241,356	348,034	350,804	373,394
National Institute of Occupational Safety and Health training grants	4,327	8,088	7,182	12,899	8,760	10,461	10,972	10,472	10,472
Alcohol, drug abuse, and mental health training programs[56]	85,101	118,366	83,727	122,103	43,617	81,353	—	—	—
Health teaching facilities	—	—	353	3,078	739	505	455	427	397
Social Security postsecondary students' benefits[57]	240,000	502,000	839,957	1,559,000	25,000	—	—	—	—
Department of Housing and Urban Development[38]	220,744	114,199	−55,418	—	—	—	—	—	—
College housing loans[38,39]	220,744	114,199	−55,418	—	—	—	—	—	—

at $68.4 billion for fiscal year (FY) 1994, an increase of nearly twofold since 1980. More than 43 percent of the $34.3 billion spent by the federal government in FY '94 for elementary and secondary education came from the Department of Education, an office created in 1979 to oversee federal education programs. The range of the Department of Education has indeed been broad, overseeing major programs in the areas of Chapter 1, grants to local education agencies to improve education for the disadvantaged, education for individuals with disabilities, Chapter 2, school improvement programs, vocational and adult education, impact aid, and the new program known as *Goals 2000,* which is now enjoying major national reform attention. According to the Department of Education, *Goals 2000* will create a national structure for reform that supports state and local efforts to improve teaching and will work

Federal on-budget funds for education, by level or other educational purpose or activity, by agency and program: Fiscal years 1965 to 1994—Continued

[In thousands of current dollars]

Level or other educational purpose, by agency and program	1965	1970	1975	1980	1985	1990	1992	1993	1994[1]
1	2	3	4	5	6	7	8	9	10
Postsecondary education programs (continued)									
Department of the Interior	30,153	31,749	50,844	80,202	125,247	135,480	140,266	132,916	147,001
Shared revenues, Mineral Leasing Act and other receipts—estimated education share	6,260	6,949	15,480	35,403	71,991	69,980	68,982	61,566	70,082
Indian programs:									
Continuing education[58]	8,993	9,380	13,311	16,909	24,338	34,911	38,970	39,840	43,184
Higher education scholarships	14,900	15,420	22,053	27,890	28,918	30,589	32,315	31,510	33,735
Department of State	53,420	30,850	50,347	—	—	2,167	9,057	10,211	9,496
Educational exchange[59]	53,420	30,850	50,347	—	—	—	—	—	—
Mutual educational and cultural exchange activities	47,025	30,454	50,300	—	—	—	—	—	—
International educational exchange activities	6,395	396	47	—	—	—	—	—	—
Soviet-East European Research and Training[60]	—	—	—	—	—	2,167	9,057	10,211	9,496
Department of Transportation[17]	—	11,197	11,885	12,530	55,569	46,025	53,991	57,576	61,014
Merchant Marine Academy[44]	—	—	—	—	19,898	20,926	27,007	26,788	29,934
State marine schools[61]	—	—	—	—	19,777	8,269	11,072	10,320	11,714
Coast Guard Academy[18]	—	9,342	9,780	10,000	11,857	12,074	13,071	13,602	13,032
Postgraduate training for Coast Guard officers[62]	—	1,655	1,855	2,230	3,499	4,173	2,540	6,306	5,883
Tuition assistance to Coast Guard military personnel[18]	—	200	250	300	538	582	301	560	451
Department of the Treasury	8,208	—	268,605	296,750	—	—	—	—	—
General revenue sharing—estimated state share to higher education[19,20]	—	—	268,605	296,750	—	—	—	—	—
Coast Guard Academy[18]	6,815	—	—	—	—	—	—	—	—
Postgraduate training for Coast Guard officers[62]	1,293	—	—	—	—	—	—	—	—
Tuition assistance to Coast Guard military personnel[18]	100	—	—	—	—	—	—	—	—
Department of Veterans Affairs[21]	55,650	693,490	3,029,600	1,803,847	944,091	599,825	854,480	919,991	1,070,421
Vietnam-era veterans[63]	33,950	638,260	2,840,600	1,579,974	694,217	46,998	—	—	—
College student support	—	—	—	1,560,081	679,953	39,458	—	—	—
Work-study	—	—	—	19,893	14,264	7,540	—	—	—
Service persons college support[64]	—	18,900	74,690	46,617	35,630	8,911	—	—	—
Post-Vietnam veterans[65]	—	—	—	922	82,554	161,475	88,500	65,894	51,953
All-volunteer-force educational assistance[66]	—	—	—	—	196	269,947	650,540	745,786	912,292
Veterans[67]	—	—	—	—	—	183,765	530,820	626,669	806,592
Reservists[68]	—	—	—	—	196	86,182	119,720	119,117	105,700
Veteran dependents' education[69]	21,700	36,330	114,310	176,334	131,494	100,494	103,440	96,311	94,176
Payments to state education agencies[70]	—	—	—	—	—	12,000	12,000	12,000	12,000
Other agencies									
Appalachian Regional Commission[26]	—	4,105	2,545	1,751	—	—	1,487	1,587	1,500
National Endowment for the Humanities[28]	—	3,349	25,320	56,451	49,098	50,938	58,512	57,804	56,825
National Science Foundation	27,170	42,000	60,283	64,583	60,069	161,884	210,375	246,591	283,015
Science and engineering education programs	27,170	37,000	60,283	64,583	60,069	161,884	210,375	246,591	283,015
Sea Grant Program[48]	—	5,000	—	—	—	—	—	—	—
U.S. Information Agency[71]	7,512	8,423	9,405	51,095	124,041	181,172	207,676	256,068	266,292
Educational and cultural affairs[59]	—	—	—	49,546	21,079	35,862	38,858	44,183	28,961
Educational and cultural exchange programs[72]	—	—	—	—	101,529	145,307	168,818	211,885	237,331
Educational exchange activities, international	—	—	—	1,549	1,433	3	—	—	—
Information center and library activities[73]	7,512	8,423	9,405	—	—	—	—	—	—
Other programs									
Barry Goldwater Scholarship and Excellence in Education Foundation[74]	—	—	—	—	—	1,033	2,900	3,023	3,054
Estimated education share of federal aid to the District of Columbia	1,895	5,513	10,564	13,143	15,266	14,637	16,382	14,247	14,294
Harry S Truman scholarship fund[58,75]	—	—	—	-1,895	1,332	2,883	2,401	2,894	3,221
Institute of American Indian and Alaska Native Culture and Arts Development[76]	—	—	—	—	—	4,305	6,612	7,462	12,913
James Madison Memorial Fellowship Foundation[77]	—	—	—	—	—	191	885	1,298	2,344

conjointly with a new vocational opportunity program known as *School to Work* that seeks to help youth develop skills, knowledge, and abilities in order to better address national concerns about economic unpreparedness of the next generation of taxpayers.[7]

All federal interest in education, however, has not come through the Department of Education. While federal interest through that office totaled $34.3 billion in 1994, a second federal agency also provided massive funding in the form of child nutrition programs. Harking back to programs begun early in this century, the Department of Agriculture has become the second largest provider of federal resources to K-12 education by funding nutrition programs of remarkable size. As also seen in table 22.1, the Department of Ag-

—Federal on-budget funds for education, by level or other educational purpose or activity, by agency and program: Fiscal years 1965 to 1994—Continued

[In thousands of current dollars]

Level or other educational purpose, by agency and program	1965	1970	1975	1980	1985	1990	1992	1993	1994[1]
1	2	3	4	5	6	7	8	9	10
Other education programs	**374,652**	**964,719**	**1,608,478**	**1,548,730**	**2,107,588**	**3,383,031**	**3,991,955**	**4,107,193**	**4,804,970**
Department of Education[2]	182,021	630,235	1,045,659	747,706	1,173,055	2,251,801	2,579,883	2,526,372	3,038,576
Administration	17,732	47,456	108,372	187,317	284,900	328,293	368,420	353,545	469,157
Libraries	26,111	108,284	225,810	129,127	85,650	137,264	214,928	181,219	166,104
Rehabilitative services and disability research	137,313	473,091	709,483	426,886	798,298	1,780,360	1,991,875	1,983,848	2,396,903
American Printing House for the Blind	865	1,404	1,994	4,349	4,230	5,736	4,587	7,774	6,414
Trust funds[56]	—	—	—	27	−23	148	73	−14	−2
Department of Agriculture	87,551	135,637	220,395	271,112	336,375	352,511	400,442	443,650	452,745
Extension Service	85,924	131,734	215,523	263,584	325,986	337,907	385,087	424,928	434,582
National Agricultural Library	1,627	3,903	4,872	7,528	10,389	14,604	15,355	18,722	18,163
Department of Commerce	251	1,226	2,317	2,479	—	—	—	—	—
Maritime Administration:									
Training for private sector employees[44]	251	1,226	2,317	2,479	—	—	—	—	—
Department of Health and Human Services[9]	3,953	24,273	31,653	37,819	47,195	77,962	97,643	96,860	110,060
National Library of Medicine	3,953	24,273	31,653	37,819	47,195	77,962	97,643	96,860	110,060
Department of Housing and Urban Development	512	—	—	—	—	—	—	—	—
Urban mass transportation—managerial training grants[78]	512	—	—	—	—	—	—	—	—
Department of Justice	3,850	5,546	42,818	27,642	25,517	26,920	34,526	37,028	27,236
FBI National Academy	1,850	2,066	5,100	7,234	4,189	6,028	10,631	14,388	12,004
FBI—Field Police Academy	1,450	2,500	5,254	7,715	10,220	10,548	12,578	10,563	9,015
Narcotics and dangerous drug training	550	980	1,152	2,416	83	850	695	230	230
National Institute of Corrections[79]	—	—	31,312	10,277	11,025	9,494	10,621	11,847	5,987
Department of State	10,780	20,672	28,113	25,000	23,791	47,539	44,086	58,840	38,534
Foreign Service Institute	6,395	15,857	20,750	25,000	23,791	47,539	44,086	58,840	38,534
Center for Cultural and Technical Interchange[59]	4,385	4,815	7,363	—	—	—	—	—	—
Department of Transportation[17]	—	3,964	11,877	10,212	3,785	1,507	2,419	500	467
Highways training and education grants[80]	—	2,418	3,250	3,412	1,500	—	1,945	—	—
Maritime Administration:									
Training for private sector employees[44]	—	—	—	—	1,135	1,507	474	500	467
Urban mass transportation—managerial training grants[78]	—	1,546	2,627	500	1,150	—	—	—	—
Federal Aviation Administration[81]									
Air traffic controllers second career program[82]	—	—	6,000	6,300	—	—	—	—	—
Department of the Treasury	—	18	3,096	14,584	16,160	41,488	51,694	56,779	58,276
Federal Law Enforcement Training Center[83]	—	18	3,096	14,584	16,160	41,488	51,694	56,779	58,276
Other agencies									
Agency for International Development	63,329	88,034	78,896	99,707	141,847	170,371	212,220	215,248	231,180
Education and human resources	53,968	61,570	58,349	80,518	115,104	142,801	195,570	190,020	201,980
American schools and hospitals abroad	9,361	26,464	20,547	19,189	26,743	27,570	16,650	25,228	29,200
Appalachian Regional Commission[26]	—	572	1,574	8,124	113	—	938	1,005	975
Corporation for National and Community Service[84]	—	—	7,045	2,833	1,761	8,472	8,600	8,500	180,000
Estimated education funds[85]	—	—	7,045	2,833	1,761	8,472	8,600	8,500	180,000
Federal Emergency Management Agency[86]	—	290	290	281	405	215	261	76,467	65,660
Estimated architect/engineer student development program[87]	—	40	40	31	155	200	250	375	102
Estimated other training programs[88]	—	250	250	250	250	15	11	92	108
Estimated disaster relief[89]	—	—	—	—	—	—	—	76,000	65,450
General Services Administration[90]									
Libraries and other archival activities	4,013	14,775	22,532	34,800	—	—	—	—	—
Japanese-United States Friendship Commission[91]	—	—	—	2,294	2,236	2,299	1,610	1,503	1,833
Library of Congress	15,111	29,478	63,766	151,871	169,310	189,827	296,044	311,453	320,325
Salaries and expenses	11,421	20,700	48,798	102,364	130,354	148,985	252,623	260,918	264,882
Books for the blind and the physically handicapped	2,317	6,195	11,908	31,436	32,954	37,473	38,688	45,261	49,013
Special foreign currency program	1,187	2,273	2,333	3,492	4,621	10	10	30	—
Furniture and furnishings	186	310	727	14,579	1,381	3,359	4,723	5,244	6,430
National Aeronautics and Space Administration									
Aerospace education services project	100	350	600	882	1,800	3,300	6,100	6,500	6,300
National Archives and Records Administration[92]									
Libraries and other archival activities	—	—	—	—	52,118	77,397	99,412	106,975	108,231

riculture provided over $8 billion in total aid to elementary and secondary education nutrition programs in FY '94. Of all its school support programs, by far the largest are the child nutrition programs, which in FY '94 sent nearly $7.34 billion to public school lunchrooms. Likewise, the Agricultural Marketing Service's commodity program provided over $400 million to schools in 1994 as a result of a 1935 program that has carried forward to the present time.[8]

The Department of Agriculture also provided another $21.7 million through the special milk program, and an estimated $142.5 million came from education's share of Forest Service permanent appropriations. While data in table 22.1 clearly show that levels of support from all federal sources have

Federal on-budget funds for education, by level or other educational purpose or activity, by agency and program: Fiscal years 1965 to 1994—Continued
[In thousands of current dollars]

Level or other educational purpose, by agency and program	1965	1970	1975	1980	1985	1990	1992	1993	1994[1]
1	2	3	4	5	6	7	8	9	10
Other education programs (continued)									
National Commission on Libraries and Information Science[93]	—	—	449	2,090	723	3,281	1,437	867	954
National Endowment for the Arts[27]	—	340	1,068	231	1,137	936	3,286	2,784	2,250
National Endowment for the Humanities[28]	—	5,090	38,486	85,805	76,252	89,706	99,782	100,826	99,093
Smithsonian Institution	2,233	2,461	5,509	5,153	7,886	5,779	6,578	7,828	7,678
Museum programs and related research	2,133	2,261	4,203	3,254	4,665	690	93	1,165	1,000
National Gallery of Art extension service	100	200	300	426	675	474	793	763	810
Woodrow Wilson International Center for Scholars	—	—	1,006	1,473	2,546	4,615	5,692	5,900	5,868
U.S. Information Agency—Center for Cultural and Technical Interchange[59]	—	—	—	15,115	18,966	20,375	29,550	31,991	38,283
U.S. Institute of Peace[94]	—	—	—	—	—	7,621	11,350	10,468	11,549
Other programs									
Estimated education share of federal aid for the District of Columbia	948	1,758	2,335	2,990	7,156	3,724	4,095	4,749	4,765
Research programs at universities and related institutions[95]	1,816,276	2,283,641	3,418,374	5,801,204	8,844,575	12,606,035	14,176,863	14,955,083	15,111,432
Department of Education[96]	13,248	87,823	82,770	78,742	28,809	89,483	154,800	232,165	316,417
Department of Agriculture	58,362	64,796	108,162	216,405	293,252	348,109	437,967	436,187	441,483
Department of Commerce	4,015	4,487	21,677	48,295	52,951	50,523	77,240	70,805	87,798
Department of Defense	436,912	356,188	364,929	644,455	1,245,888	1,871,864	2,070,959	2,002,572	1,857,274
Department of Energy	439,334	548,327	761,376	1,470,224	2,205,316	2,520,885	2,867,528	2,763,515	2,623,337
Department of Health and Human Services	474,362	623,765	1,273,037	2,087,053	3,228,014	4,902,714	5,210,711	5,952,976	6,065,854
Department of Housing and Urban Development	—	510	2,650	5,314	438	118	203	401	374
Department of the Interior	9,839	18,521	28,955	42,175	34,422	49,790	57,449	54,049	31,023
Department of Justice	—	1,945	8,902	9,189	5,168	6,858	4,986	3,496	3,400
Department of Labor	—	3,567	6,124	12,938	3,417	5,893	1,169	600	600
Department of State	—	8,220	10,973	188	29	1,519	200	—	—
Department of Transportation	—	12,328	28,478	31,910	22,621	28,608	35,015	57,789	48,372
Department of the Treasury	—	—	—	226	388	227	85	133	167
Department of Veterans Affairs	337	518	1,112	1,600	1,000	2,300	2,491	2,550	2,550
ACTION	—	—	36	—	—	—	—	—	—
Agency for International Development	—	—	—	77,063	56,960	79,415	32,979	27,659	25,850
Environmental Protection Agency	—	19,446	33,875	41,083	60,521	87,481	152,012	124,500	169,500
Federal Emergency Management Agency	—	—	—	1,665	1,423	—	—	—	—
National Aeronautics and Space Administration	208,688	258,016	197,301	254,629	485,824	1,090,003	1,377,322	1,367,542	1,451,185
National Science Foundation	154,046	253,628	475,011	743,809	1,087,046	1,427,007	1,664,697	1,833,563	1,958,329
Nuclear Regulatory Commission	—	—	7,093	32,590	30,261	42,328	27,418	22,934	26,344
Office of Economic Opportunity	7,078	20,035	—	—	—	—	—	—	—
U.S. Arms Control and Disarmament Agency	—	100	—	661	395	25	100	25	25
U.S. Information Agency	—	—	—	—	—	—	—	—	—
Other agencies	10,055	1,421	5,913	990	432	885	1,532	1,622	1,550

[1] Estimated.
[2] The U.S. Department of Education was created in May 1980. It formerly was the Office of Education in the U.S. Department of Health, Education, and Welfare.
[3] This program creates a national framework for education reform and meeting the National Education Goals. This program also has the School-To-Work Opportunities program which will be used to initiate a national system to be administered jointly by the U.S. Departments of Education and Labor. Both departments are to establish a national framework within which all states can create statewide systems to help youth acquire the knowledge, skills, abilities, and labor market information they need to make an effective transition from school to work or to further their education or training.
[4] These commodities are purchased under Section 32 of the Act of August 24, 1935, for use in the child nutrition programs.
[5] This program assisted in the construction of public facilities, such as vocational schools, through grants or loans. No funds have been appropriated for this account since FY 77, and it was completely phased out in FY 84 after the monitoring of closeouts of projects was completed. Data are not available for previous years.
[6] This program was funded by the U.S. Department of Education in FYs 65–81 in the Impact Aid program. This program provides for education of dependents of federal employees residing on federal property in cases where free public education is unavailable in the nearby community.
[7] The U.S. Department of Energy was created in 1977. It formerly was the Energy Research and Development Administration and before that the Atomic Energy Commission.
[8] This program was established in 1979. Funds were appropriated for this program in FY 80.
[9] The U.S. Department of Health and Human Services was part of the U.S. Department of Health, Education, and Welfare until May 1980.
[10] The Head Start program was formerly in the Office of Economic Opportunity, and funds were appropriated to the U.S. Department of Health, Education, and Welfare, Office of Child Development, beginning in 1972.
[11] This program was created by the Family Support Act of 1988. It provides funds for the Job Opportunities and Basic Skills Training Program.
[12] After age 18, benefits terminate at the end of the school term or in 3 months, whichever is less.
[13] This program provides funding for supplemental programs for eligible Indian students in public schools.
[14] This program finances the cost of academic, social, and occupational education courses for inmates in federal prisons.
[15] The Job Corps program was formerly in the Office of Economic Opportunity, and funds were appropriated to the U.S. Department of Labor beginning in 1971 and 1972.
[16] Some of the work and training programs included in this program were in the Office of Economic Opportunity and were transferred to the U.S. Department of Labor in 1971 and 1972.
[17] The U.S. Department of Transportation was created in 1967.
[18] This program was transferred from the U.S. Department of the Treasury to the U.S. Department of Transportation in 1967.
[19] This program was established in FY 72 and closed in FY 86.
[20] The states' share of revenue-sharing funds was not spent on education in FYs 81–86.

Source: U.S. Department of Education, National Center for Education Statistics, *Federal Support for Education.* (Washington, D.C.: National Center for Education Statistics, 1994).

varied over the years from 1965 to 1994, considerable sums of money have been directed toward education generally and specifically toward nutrition assistance—a fact well illustrated by total support increasing from just over $623 million in 1965 to nearly $7.81 billion in the Department of Agriculture alone. In the larger context of the Department of Education, data show funding increased more than seventeen-fold in roughly a thirty-year period.

A graphic summary of federal aid to education is seen in figure 22.1 showing total federal support for education and related programs by granting agency for FY '94. Figure 22.1 shows that of the $68.4 billion on-budget funds going to education, 42.2 percent came from the Department of Education. The second largest direct share totaled 12.8 percent, coming from the Depart-

Figure 22.1
Federal On-Budget Funds for Education FY 1994

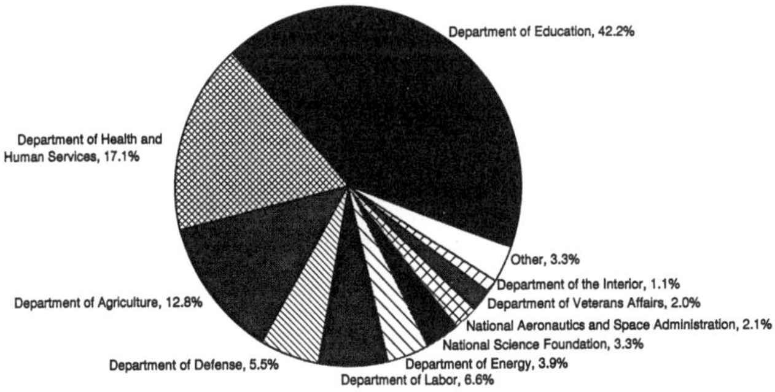

Total=$68.4 billion

Source: U.S. Department of Education, National Center for Education Statistics, Federal Support for Education. (Washington, DC: National Center for Education Statistics, 1994).

ment of Agriculture as just discussed. Although Health and Human Services (HHS) appears to have provided a larger percent, the key is direct aid since HHS also funds other services benefiting education such as Head Start, provides financial support to students under age nineteen who are covered by social security benefit programs, funds Aid for Dependent Children work programs created by the Family Support Act of 1988, and other programs as well. Other massive federal programs also appear in table 22.2, showing in summary form the largest providers of federal on-budget education funding and also comparing changes from 1980 to 1994. When other off-budget moneys and nonfederal funds generated by federal legislation are considered, federal programs provide valuable aid to public education.

State Support

The second unit of government that has participated in foodservice programs is the states. Historically, in most instances state participation has been a function of three choices that states have had to make. The first choice is whether the state even wishes to participate in the National School Lunch Act of 1946, a federal law that placed a matching funds requirement on states in order to qualify for the sizable federal funding discussed earlier. If a state chose to accept available federal food aid, federal funding would be determined in a participation agreement reflecting the current federal reimbursement rate at that given point in time, and the state would be required to match federal money with a certain percentage from its own funds. These funds

Table 22.2
Largest Providers of Federal On-Budget Education Program Funding: 1980, 1985, 1990, and 1994

[Amounts in billions of constant FY 94 dollars]

Agency	FY 80		FY 85		FY 90		FY 94*		Percent change, FY 80 to FY 94*	Percent change, FY 90 to FY 94*
	Amount	Percent of total	Amount	Percent of total	Amount	Percent of total	Amount	Percent of total		
Total	**$62.7**	**100.0**	**$53.3**	**100.0**	**$59.1**	**100.0**	**$68.4**	**100.0**	**9.0**	**15.7**
Dept. of Education (ED)	23.9	38.1	22.8	42.8	26.5	44.9	28.9	42.2	20.9	8.8
Dept. of Health and Human Services (HHS)	10.2	16.3	7.3	13.6	9.1	15.4	11.7	17.1	14.7	28.6
Dept. of Agriculture (USDA)	8.3	13.2	6.5	12.3	7.2	12.1	8.7	12.8	5.2	21.8
Dept. of Labor (DOL)	3.4	5.4	2.7	5.0	2.9	4.9	4.5	6.6	33.9	57.8
Dept. of Defense (DOD)	2.8	4.5	4.3	8.0	4.1	7.0	3.7	5.5	32.1	−9.2
Dept. of Energy (DOE)	2.9	4.7	3.1	5.8	2.9	5.0	2.6	3.9	−9.3	−9.7
National Science Foundation (NSF)	1.5	2.3	1.6	2.9	1.8	3.1	2.2	3.3	52.5	23.3
National Aeronautics and Space Administration (NASA)	0.5	0.7	0.7	1.3	1.3	2.1	1.5	2.1	213.7	16.5
Dept. of Veterans Affairs (VA)	4.3	6.8	1.8	3.3	0.9	1.5	1.4	2.0	−67.4	60.7
Dept. of the Interior (INT)	0.8	1.3	0.8	1.4	0.7	1.2	0.8	1.1	−4.4	6.1
All other federal agencies	4.2	6.7	2.0	3.7	1.7	2.8	2.3	3.3	−45.9	35.3

* Estimated.

NOTE: Percentages based on unrounded numbers. Because of rounding, details may not add up to totals. See table B in appendix for current dollars.

Source: U.S. Department of Education, National Center for Education Statistics, *Federal Support for Education.* (Washington, D.C.: National Center for Education Statistics, 1994).

then flowed to the individual school districts operating approved school foodservice operations.

The second choice followed from whatever the state had concluded in its first decision—i.e., states clearly had the option to participate or to refuse to participate, but failure to meet federal participation requirements would result in denial of important federal funding, forcing a third choice of whether the state would participate at all in foodservice programs. State support, then, can be seen as both voluntary and compulsory in the sense that options were in fact available—however, states have had a generally difficult time refusing federal aid because of the political attraction of these funds.

In contrast to only one federal government, the history of individual state participation is more difficult to summarize because there are fifty different states that have been free to make independent decisions about the extent of state involvement. It is clear, however, that most states have chosen to participate in federal food aid.

Based on available data, states are remarkable for both their differing approaches and for their commonalities, as some features are both surprising and expected. Not surprising was a common feature in all states reporting receipt of federal moneys. All such states complied with the extant federal matching requirement of 30 percent state funding in order to qualify for assistance under the National School Lunch Act. Intriguingly, no reporting state contributed to foodservice in excess of 1 percent of total general fund budget for education. Most states reported the federal matching requirement and indicated how many state dollars went to foodservice specifically, while other states did not break out food aid from other categories for reporting purposes, rolling the totals together into a lump sum going to "other programs,"

The state of Washington has taken the position that it was not statutorily or constitutionally obligated to aid foodservice, although federal requirements for qualifying state participation nonetheless caused the state to contribute about $3 million.[9] In Wisconsin, the state contributed foodservice aid for a broader purpose than child nutrition, granting $3.6 million to federal matching requirements and opting to add another $155,000 for elderly meal assistance and an additional $290,000 for low income kindergarten milk. While state participation is broader than outlined here, a genuine sense of the nature of foodservice is well captured by the data in table 22.3.

Table 22.3
Overview of State Food Participation

State	Method and Extent of Food Aid
Alabama	None reported.
Alaska	None reported.
Arizona	None reported.
Arkansas	$1.7 million. <1% of total state aid. Meets federal matching.
California	$53.2 million. <1% of total state aid.
Colorado	None reported.
Connecticut	$6.4 million. <1% of total state aid. Meets federal matching.
Delaware	None reported.
Florida	Not distinguishable; rolled into $865 million taking in 28 categorical aids.
Georgia	$24. million. <1% of total state aid. Formula on efficient and meal count.
Hawaii	$43.5 million operating fund. Investment capital fund $10.6 million.
Idaho	None reported.
Illinois	$13.1 million. <1% of total state aid. Follows National School Lunch Act.
Indiana	None reported.
Iowa	None reported.
Kansas	$2.44 million. <1% of total state aid. Meets federal matching.
Kentucky	None reported.
Louisiana	None reported.
Maine	None reported.
Maryland	$6.6 million. <1% of total state aid. State pays all cost above federal aid.
Massachusetts	$6.2 million. <1% of total state aid. Meets federal matching.
Michigan	None reported.
Minnesota	$4.6 million. <1% of total state aid. Meets federal matching.
Mississippi	None reported.
Missouri	None reported.
Montana	None reported.
Nebraska	$810,000. Rolled into 'miscellaneous, each providing less than $1 million'
Nevada	None reported.
New Hampshire	None reported.
New Jersey	$6.6 million. <1% of total state aid. Meets federal matching.
New Mexico	None reported.
New York	None reported.
North Carolina	$11.9 million. <1% of total state aid. Districts receive $38,242 per director.
North Dakota	None reported.
Ohio	$8.9 million. <1% of total state aid. Districts pays costs above federal match.
Oklahoma	$5.4 million. <1% of total state aid. Meets federal matching.
Oregon	None reported.
Pennsylvania	Not distinguishable. Rolled into $90 million supporting 11 categoricals.
Rhode Island	None reported.
South Carolina	Provides $623,000 to supervisor salaries.
South Dakota	None reported.
Tennessee	None reported.
Texas	None reported.
Utah	None reported.
Vermont	None reported.
Virginia	$5.8 million. <1% of total state aid. Meets federal matching.
Washington	$3 million. <1% of total state aid. No constitutional duty to aid food.
Wisconsin	$4.1 million. <1% of total state aid. Matching; elderly; kindergarten milk.
Wyoming	None reported.

Source: excerpted and modified from American Education Finance Association, *Public School Finance Programs of the United States and Canada 1990-91*, ed. Steven Gold. (Albany: American Education Finance Association and the Rockefeller Institute of Government, 1992).

The commonalities and differences in state participation in school foodservice programs are summarized in table 22.3. These data illustrate both the importance of foodservice aid plans and the relationship of amounts of aid to total educational funding. While on one hand the dollars are large, state funding is relatively small compared to both the federal contribution and to the proportion of spending for school purposes that occurs in the United States. The large dollar amounts contributed to foodservice still represent only pennies per child, as the population served is large enough to render aid totals small when considered on a per-child basis. But as a result of considering the relationship between federal and state governments to foodservice, the federal government can be said to be the largest contributor to the nutrition support arena.

The Local Role

In comparison to state and federal funding for foodservices, the local role may seem insignificant. Perspective, however, causes all ideas to take on different meanings—an observation particularly important in the foodservice function of school business operations since foodservice programs are generally intended to be a self-funding cash operation. From one perspective the local role is great while from another view it has been relatively minor.

The perspective of the local role as minor in foodservice operations flows from the observation that if the federal government were the largest contributor and states must match contributions in meaningful proportion, any unfunded portion of foodservice costs are divided between the local school district and its cash patrons. This perspective is fundamentally correct, although it understates the decisions that must be made when dividing remaining costs among the parties. For example, if federal reimbursement for regular paid meals was currently at $.3050 and if the state rate was $.0500, then the unfunded portion of a meal would be $1.30 if the actual price charged was $1.65. Similar calculations occur for free and reduced meal prices. The local role is then one of determining how much, if any, the district wishes to subsidize meal costs through shifting local tax revenue to foodservice operations—the corollary to which is how much each child must pay for a meal to cover any remaining unfunded portion after federal, state, and local shares have been subtracted. The perspective of a minor local role is accurate only if "minor" is defined as dealing with the leftover costs after federal and state revenue is applied.

On the other hand, leaving the unfunded portion of meal costs to the local district creates a tremendous burden for the school administrator and the board because such decisions carry weighty moral and financial implications. If the district was to choose to pass on all unfunded costs, some children might not receive the intended benefit of the foodservice program. If the district was to subsidize meals from tax revenue, hard economic choices would have to be made since all revenue is finite and every expenditure decision results in reducing revenue and flexibility from other educational and support areas. Fortunately, some of this burden is lifted by virtue of the fact that federal reimbursement for children qualifying for free and reduced lunch, breakfast, or milk programs is sharply higher than for regular paid meals.

For example, in FY '95 federal reimbursement for regular paid lunches was the $.3050, but the reduced price lunch reimbursement was $1.4650, and free lunch was reimbursed at $1.8650. Using a $1.90 actual meal price, the district would not have a tremendously difficult decision to make in the case of a free lunch since $1.90 - $1.8650 = $.035 unfunded; however, if the board were to fully fund such a difference in highly impoverished areas, the costs could be more significant. Often the biggest problem, however, is rising costs associated with purchasing and preparing food such as supplier price increases, increasing labor costs, and the cost of equipment and facilities. Creating a self-funding program is difficult under these conditions, a fact that is not lost on school business administrators who are charged with overseeing the local role in foodservice operations.

Proposed Federal Funding Changes

School foodservice programs, however, have not gone uncriticized for a variety of complex reasons. Although 92,000 of the nation's schools offered free or subsidized lunches to 26 million students in 1995, of whom 13 million such children came from low income families receiving one-third to one-half of their daily nutrient intake from school food programs, there have been proposals in Congress to significantly change how federal funding of school foodservice programs is structured. These changes, if approved, would represent the first major shift in direction for federal food subsidies since the National School Lunch Act of 1946 was approved in response to a perceived national crisis stemming from failure by many young recruits to pass their draft physicals in the world wars of this century as a result of nutrition-related diseases.

The political revolution that swept the United States in 1994 generated new legislation aimed at restructuring some of America's entrenched social programs that consume vast resources. In the broader context, the Republican *Contract with America* was a promise to reform the welfare state and to return the nation to its hardworking roots by applauding a meritocracy. A bill known as the *Personal Responsibility Act of 1995*[10] was introduced in the Congress that would, according to its title, restore the American family, reduce illegitimacy, control welfare spending, and reduce welfare dependence. This measure, if enacted, would seek to reduce welfare payments to illegitimate children and aliens, curb growth in welfare spending, increase states' flexibility in administering such programs, and achieve other related provisions intended to restore national integrity and the work ethic. The contents of the bill were far-reaching, addressing not only welfare payments and illegitimacy, but also measures designed to require work in exchange for income and to test welfare recipients for drug use in order to better control how welfare payments are spent.

The impact of the bill on school foodservice programs would be of major significance, as the Act would consolidate food assistance programs by repealing the Food Stamp Act of 1977,[11] the Child Nutrition Act of 1966,[12] the Emergency Food Assistance Act of 1983,[13] the Hunger Prevention Act of 1988,[14] the National School Lunch Act,[15] the Commodity Distribution Reform Act and WIC Amendments of 1987, and other lesser federal laws addressing similar issues. Repeal of these federal laws would not mean termination of all such assistance, but rather these laws would be replaced with state food assistance block grant programs and a new food coupon program. In addition, the sale of surplus commodities to the states would be authorized to provide food assistance to needy groups. The intent of consolidation and turning programs back to the states would be to bring such issues closer to their roots, removing the federal government from a caregiver role in matters that could be better and more efficiently managed by the states.

Block grants to states under this proposal would be strictly guarded to preserve the intent of legislation. The bill would permit the Secretary of Agriculture to make grants to the individual states in response to needy populations as federally defined, with authorization to fund the bill with $35.6 billion beginning in 1996 and adjusted annually thereafter. Eligibility of a state to receive a block grant would be based on:

- Assurance that such a grant would be expended to provide food aid to people meeting the description of economic disadvantage.
- Assurance that not more than 5 percent of the grant's funds would be used for administration.
- Assurance that not less than 12 percent of each grant from 1996-2000 would be spent for food aid and nutrition education for pregnant women, postpartum women, lactating women, infants, and young children.
- Assurance that not less than 20 percent of each grant from 1996-2000 would be spent to provide:
    - Nonprofit school breakfast programs.
    - Milk in nonprofit schools and nonprofit nursery schools, child care centers, settlement houses, summer camps, and similar institutions devoted to economically disadvantaged families.
    - Nonprofit school lunch programs.

• Summer foodservice programs in nonprofit food authorities, local governments, and nonprofit higher education institutions participating in the National Youth Sports Program.

• Assurance that the amount of food assistance provided to any person will be proportionately reduced if the person were otherwise eligible for aid but has not worked at least thirty-two hours in a program on behalf of the state or other political subdivision during the month prior to receiving assistance.

• Assurance that no government worker will be replaced by people who qualify for food assistance that would result in wrongful treatment of the displaced worker.

Although the Personal Responsibility Act and related subparts are not yet law and may fail to become law, these provisions are highly representative of the debate surrounding the role of government in welfare-based programs. These issues raise questions about how far government should become involved in the lives of people, how much responsibility people should take for themselves, how much responsibility citizens have toward each other, and whether the nation has been on the right course during this century. Indeed, such issues raise questions about the meaning of a capitalist democracy, suggesting either that the nation's economic structure has drifted too far from its roots or, alternatively, that the nation's economic structure itself is wrong.

While this text cannot either resolve those questions or take a position on which is the more accurate view, it can be noted that such questions are deeply rooted in political philosophy and that the answers are mutually exclusive because there will be winners and losers regardless of which position is finally adopted. It can also be observed that changes in federal foodservice laws will have a large impact on how schools operate the foodservice function. For example, if reform laws pass, one midwestern urban center expects to lose at least $1.3 million in cash and commodities that presently serve a student population where 73 percent of all pupils qualify for free or reduced lunches.[16] Other districts fear that rising lunch prices could force indirect problems, such as more brown-bagging that would reduce the number of meals served, thereby raising costs in an economy of scale equation since cafeteria facilities would be serving fewer meals. Under these conditions, school districts might withdraw from federal programs because the cost of federal standards would exceed the benefits received.

While no laws have yet changed, the future of foodservice is in a state of flux. America has always been about change, about reform, and about independence. Americans have always chafed about taxes, and taxes paid for the purpose of charity have always been a sore point for the American people. Political pendulums swing, and many people now believe that government is too big, too generous, and too undemanding. Programs resembling welfare

are currently a target, and food assistance programs at the federal level will be significantly shaped by this debate on the threshold of the twenty-first century. Regardless of the outcome of presently proposed changes, the debate will not abate because the losing side will continue to lobby for its cause.

Nature of the Current Basic Program

Although change may be eminent, basic foodservice programs are currently in place that represent the interrelationships between federal, state, and local governments. While sessions of Congress have brought some adjustment to the program, the basic nature of food assistance has not changed greatly since its inception.

Broad Compliance Requirements

The issue of compliance is a common one whenever outside aid is provided. The granting agency generally wants assurances in return for aid, and those assurances align with the purpose of the program's intent. In the case of foodservice, assurances are generally directed toward the federal government in return for the sizable dollars and commodities that were described earlier in this chapter. At the state level, districts must assure states that compliance is observed so that the state can then provide assurance to the federal government. Likewise, many states also aid foodservice for which the states themselves want a separate compliance monitoring system.

Historically, eligibility for federal foodservice funds has been based on compliance with thirteen broad policy statements issued in federal rules and regulations. Specifically, schools and states must agree that the foodservice program in each school will:

- *Operate a nonprofit program*: This means that only up to a three-month operating balance may be kept on hand and still be considered nonprofit. In contrast, management companies (discussed later) operating for a profit motive are exempt.

- *Serve meals that meet nutrition requirements*: This may be accomplished by one or more types of meals. An approved program can offer a Type A lunch (single menu), a fast-food choice menu, or an a la carte menu. Each must be planned to meet minimum dietary recommendations.

- *Price meals as a unit*: This limitation does not prevent foods from being sold as separate items. In order for a lunch to be counted as a reimbursable lunch, however, it must be priced as a unit. Sales of seconds are additional income.

- *Supply free and reduced price meals to eligible needy children*: Uniform policy requirements must be followed.

- *Agree to avoid discrimination*: No child may be refused a meal under the program because of inability to pay or due to race, sex, color, or national origin.

- *Keep accurate records of income and expenditures*: Records are subject to intensive state and federal audits. Records must be maintained for three prior school years plus the current year.

- *Complete a formal reimbursement claim*: The written claim must be sent on a timely basis each month to the state.

- *Distribute applications for free and reduced price meals*. At the beginning of each school year, districts must actively inform each student enrolled in an eligible school and must receive any submitted application to participate in the program.

- *Review and act on free and reduced price meal applications*: In accordance with state materials, parents/guardians must be notified with respect to any action taken on their application.

- *Develop and implement a verification procedure*. To confirm eligibility for free and reduced price meals, a systematic :method of verifying accuracy of applications must be developed, disseminated, and followed.

- *Maintain accurate participation records*: The district must establish procedures for obtaining accurate meal counts by reimbursable category (paid, free, reduced, and adult meals on the serving line at the point of service).

- *Establish and implement purchasing procedures*: Purchasing procedures must be in compliance with state and federal regulations.

- *Use federally donated foods or commodities*: This includes the ability to store commodities properly without spoilage.

State eligibility requirements beyond the minimum federal mandate can vary, making it impossible to fully describe, although a flavor of different state programs can be gathered from table 22.3 seen earlier in this chapter.

Regulations Regarding Foods Served

In addition to broad compliance requirements, federal law requires participating foodservice programs to comply with other, more specific regulations regarding foods served. The goal of these regulations is to assure that foodservice programs are providing healthful meals that assist in daily nutritional requirements of children. These requirements are part of the National School Lunch and Child Nutrition programs and can be better understood by carefully examining the school lunch patterns provided in table 22.4, which are taken from federal food requirements.

The nutritional goal for school lunches is to provide approximately one-third of the Recommended Dietary Allowance (RDA) for children of various ages as specified by the National Research Council. The nutrition standards include nineteen vitamins and minerals, plus protein and calorie requirements. Accordingly, the school lunch patterns shown in table 22.4 provide a framework for planning nutritionally adequate school lunches. According to federal regulations, in order to receive reimbursement a school lunch must contain four components and five menu items. The four components are meat or meat alternate, vegetable and/or fruit, bread or bread alternate, and milk. As table 22.4 indicates, two vegetable/fruit components must be included on the menu. The lunch patterns provide five portion sizes based on ages and grades. Group IV meets all age group requirements. It also most resembles the Type A lunch that was used as the standard guide until 1980. Similar regulations apply to schools offering breakfast programs, with comparable nutritional requirements. Group IV serving sizes form the basis of the following list of menu requirements:

Table 22.4
School Lunch Pattern for Various Age/Grade Groups

U.S. Department of Agriculture, National School Lunch Program

USDA recommends, but does not require, that you adjust portions by age/grade group to better meet the food and nutritional needs of children according to their ages. If you adjust portions, Groups I-IV are minimum requirements for the age/grade groups specified. If you do not adjust portions, the Group IV portions are the portions to serve all children.

COMPONENTS		MINIMUM QUANTITIES Preschool ages 1-2 (Group I)	ages 3-4 (Group II)	Grades K-3 ages 5-8 (Group III)	Grades 4-12 age 9 & over (Group IV)	RECOMMENDED QUANTITIES[2] Grades 7-12 age 12 & over (Group V)	SPECIFIC REQUIREMENTS
Meat Or Meat Alternative	A serving of one of the following or a combination to give an equivalent quantity:						• Must be served in the main dish or the main dish and one other menu item. • Vegetable protein products, cheese alternate products, and enriched macaroni with fortified protein may be used to meet part of the meat/meat alternate requirement. Fact sheets on each of these alternate foods give detailed instructions for use.
	Lean meat, poultry, or fish (edible portion as served)	1 oz	1½ oz	1½ oz	2 oz	3 oz	
	Cheese	1 oz	1½ oz	1½ oz	2 oz	3 oz	
	Large egg(s)	½	¾	¾	1	1½	
	Cooked dry beans or peas	¼ cup	⅜ cup	⅜ cup	½ cup	¾ cup	
	Peanut butter	2 Tbsp	3 Tbsp	3 Tbsp	4 Tbsp	6 Tbsp	
Vegetable And/Or Fruit	Two or more servings of vegetable or fruit or both to total	½ cup	½ cup	½ cup	¾ cup	¾ cup	• No more than one-half of the total requirement may be met with full-strength fruit or vegetable juice. • Cooked dry beans or peas may be used as a meat alternate or as a vegetable but not as both in the same meal.
Bread Or Bread Alternate	Servings of bread or bread alternate A serving is: • 1 slice of whole-grain or enriched bread roll, muffin, etc. • A whole-grain or enriched biscuit, roll, muffin, etc. • ½ cup of cooked whole-grain or enriched rice, macaroni, noodles, whole-grain or enriched pasta products, or other cereal grains such as bulgur or corn grits • A combination of any of the above	5 per week	8 per week	8 per week	8 per week	10 per week	• At least ½ serving of bread or an equivalent quantity of bread alternate for Group I, and 1 serving for Groups II-V, must be served daily. • Enriched macaroni with fortified protein may be used as a meat alternate or as a bread alternate but not as both in the same meal. NOTE: *Food Buying Guide for Child Nutrition Programs*, PA-1331 (1983) provides the information for the minimum weight of a serving.
Milk	A serving of fluid milk	¾ cup (6 fl oz)	¾ cup (6 fl oz)	½ pint (8 fl oz)	½ pint (8 fl oz)	½ pint (8 fl oz)	At least one of the following forms of milk must be offered: • Unflavored lowfat milk • Unflavored skim milk • Unflavored buttermilk NOTE: This requirement does not prohibit offering other milks, such as whole milk or flavored milk, along with one or more of the above.

NOTES: 1. Group IV is highlighted because it is the one meal pattern which will satisfy all requirements if no portion size adjustments are made.

2. Group V specifics recommended, not required quantities for students 12 years and older. These students may request smaller portions, but not smaller than those specified in Group IV.

Meat or Meat Alternate

One serving is equal to two ounces of edible lean meat, poultry, fish, or meat alternate, such as two ounces of cheese, one egg, 1/2 cup of cooked dried peas or beans, four tablespoons of peanut butter, or any combination of these items. The meat or meat alternative must be served in the main dish with one other item. No one form of meat or meat alternate should be served more than three times per week if no choice is offered.

Vegetable and/or Fruit

The required 3/4 cup must consist of two or more servings of vegetables or fruit, or both. Full-strength vegetable or fruit juice cannot be counted for more than half of the total requirement.

Bread or Bread Alternate

The minimum quantities for the bread or bread alternate component are eight servings per week. Regulations specify that a minimum of one bread or bread alternate must be served each day. Only breads or bread alternates that are whole grain or enriched may be used to meet the required amounts. One serving of bread or bread alternate is equivalent to: (a) one slice of whole grain or enriched bread; (b) an enriched biscuit, roll or muffin; (c) 1/2 cup of cooked enriched rice; (d) 1/2 cup of cooked macaroni, noodles or other whole grain or enriched pasta; or (e) 1/2 cup serving of cereal grains, such as corn [or] grits . . ..

Milk

The milk component is met by serving 1/2 pint of fluid milk. Schools must offer unflavored lowfat milk, unflavored skim milk, or unflavored buttermilk. Flavored milk may also be offered as a choice.

Other Foods

Other foods may be added to the menu as needed to complete lunches, to help improve acceptability, and to provide additional food energy (calories) and other nutrients. To help ensure that all school lunches meet the nutritional goal, the following recommendations have been provided:

- A choice of foods should be offered.

- The amount of fats, salt, and sugar in the lunch should be kept to a minimum.

- Vegetables or fruits rich in Vitamin A should be served at least twice per week.

- Fruits or vegetables rich in Vitamin C should be served several times per week.
- Several foods rich in iron should be served each day.

As may be seen by this brief description of foodservice regulations, the goal is a well balanced diet for growing children to assist in meeting their basic nutritional needs. Nutritional standards for school lunches were established on the basis of knowledge of food needs for school-age children, and devotion to this goal has not wavered since the program's inception in the 1940s. The goals of the original law remain intact, stating that school foodservice should serve nutritionally adequate, attractive, and moderately priced meals, should help children grow socially and emotionally, should extend educational influences to the homes of children, and should improve children's food habits with the ultimate goal of developing physically fit adults. Likewise, foodservice operations in schools have been extensively organized for these same purposes.

Organizing for Foodservice Operations

Discussion in this chapter thus far has led to the observation that foodservice, like other aspects of school business management, is big business. As such, organization for foodservice operations is extremely important to effectiveness and efficiency in that a well-intentioned program that runs poorly is not achieving its aims and is, in fact, wasteful. How a school district organizes foodservices depends on a variety of factors, including the relative importance placed on nutrition and foodservice as a contributor to educational goals, the size and number of schools in the district, the size of the school-age population, the economic background of students, the plant facilities and equipment, and the district's budget. These considerations affect the foodservice function in terms of goals, type of management system operated, food systems selected, and the staffing and structure of operations.

Goals

All effective administrators are familiar with the role of goals in organizational activity. Goals generally drive the overarching aspects of any enterprise in that goals represent the ends that the organization seeks to achieve. Goals also apply to the foodservice function in that goals may drive the structure and implementation of a foodservice operation. For example, the school district must determine the relative importance that will be placed on foodservice at school. Thoughtful establishment of foodservice goals shapes whether the district views foodservices as a necessary evil or a co-contributor to the educational program. As with all other areas of school business administration, the district should think through its goals and objectives for foodservice, building policy at the board level that helps shape attitudes and outcomes. Examples of useful policy and goal statements include:

- The goal of the foodservice operation is to provide nutritional food that students enjoy.
- The goal of the foodservice program is to run a fiscally successful operation while keeping within the funds available.
- The goal of the foodservice operation is to provide affordable nutritious food to children.

Clearly articulated program goals form a sound basis for shaping the nature and functioning of any aspect of school operations. As goals become articulated, they spread throughout the constituency, communicating the district's philosophy toward nutrition and education and assisting with the public relations of good school management. In fact, the communication and support goals of a school district can be readily served by the foodservice function. For instance, offering services to senior citizens and day care centers connect people with the public schools who might otherwise form opinions from data uncontrolled by the district. The result may be a higher level of support and greater accuracy of information throughout the community.

Type of Management System

There are essentially three configurations for management that affect the organization and implementation of foodservice in a school district. One configuration calls for contracting with an outside management company to meet the district's foodservice goals. A second configuration calls for a central administration role in coordinating and administering the food program for all schools in the district. The third configuration calls for having each school, or some clustering of schools conjointly, coordinate and administer the foodservice program through a decentralized system.

Management Company

The nature of contracting with an outside management company is implicit to its name. Under this arrangement, the district opts to not undertake foodservice operations, choosing instead to reach an agreement with an independent provider who guarantees satisfactory performance in keeping with the district's goals and fiscal limitations. A foodservice management company generally charges the district a fee based on the number of meals served, although some arrangements provide a fixed fee agreement in which the outside contractor accepts a flat amount for the total operations. Depending on the individual state, bid laws may apply to districts desiring to contract out the foodservice operations, making accurate and detailed bid specifications an extremely important task in order to match program delivery with program goals.

In most cases, the primary reason for contracting foodservices is the anticipated benefits relating to fiscal and managerial efficiency. School districts that have used outside food contracts have generally been satisfied with the decision. Benefits have included: increased and expanded menus, elimination of the deficits marking many foodservice operations, and improved communication due to the market responsiveness of contractors who have the profit incentive to please their clients. The literature reports many school districts arguing that foodservice contracts have proved more efficient and less costly than in-house service providers.[17] Specifically, the arguments favoring external foodservice operations have focused on the following concepts:

- School administrators have more time to spend on school curriculum in an era when noninstructional responsibilities seriously reduce instructional contact.

- Because districts are relieved of the personnel function of foodservice operations, districts have been able to reduce wage and benefits costs, disputes, and grievances.

- Because foodservice contractors are professionals in a specialized field (as contrasted to school districts, which operate foodservice as an ancillary or auxiliary service), districts have experienced improved menu planning and purchase prices.

- Because the district is no longer performing the food function, there has been a significant reduction of in-house record keeping requirements.

- Because contractors are commercial suppliers, there is strong incentive for foodservices to become a profit centers if the contractors are to survive financially.[18]

Although there are significant benefits to contracting for school foodservices, the district should undertake a careful review of potential outcomes of opting for outside contractors. First, the district should carefully screen available providers to secure only the most reputable and experienced firms. This can be done in various ways including attending trade conventions and reviewing the experiences of other contracting districts for advantages, disadvantages, and recommended contractors. The district should also prepare a set of written goals and objectives for the foodservice system well in advance of the time when a contracting decision will be made. This will facilitate the development of bid specifications and assist prospective bidders in their decision about whether to bid, as well as clearly setting out the criteria for evaluating proposals by bidders. Evaluation of bidders' proposals by knowledgeable administrators with the help of outside consultants is also highly useful. Because a contractual relationship should last for quite some time and because public relations can be damaged by unwise choices, careful pursuit of the best bid for foodservice contractors is time well invested.[19]

Centralized/Decentralized Management

The remaining options call for either a central administration role in coordinating and administering the food program for all schools in the district, or for each school or a group of schools in the district to coordinate and administer a foodservice program through a decentralized system. Each of these options has benefits and drawbacks.

For districts opting not to contract foodservices, the only option is to provide this function internally. If districts were to choose to provide internal foodservice operations, the full list of federal and state regulations discussed earlier must be observed. For example, the National School Lunch Program requirements outlined earlier must be met, taking in the scope of requirements including:

- Operating a nonprofit program in which only a three-month operating balance may be maintained to preserve nonprofit status.
- Serving meals meeting nutrition requirements under Type A single menu, fast-food choice menu, or an la carte menu.
- Pricing meals as a unit.
- Supplying free and reduced price meals to eligible children.
- Agreeing to avoid discrimination.
- Keeping accurate records of income and expenditures for state and federal audits.
- Completing formal reimbursement claims.
- Distributing applications for free and reduced price meals.
- Reviewing and acting on free and reduced meal applications.
- Developing and implementing a verification procedure.
- Maintaining accurate participation.
- Establishing and implementing purchasing procedures.
- Using federally donated foods or commodities, including the ability to store commodities properly without spoilage.

These are complex tasks for school foodservice personnel in that there is so much reporting and supervising that districts operating foodservice functions must generally hire a full-time director of foodservices. This person should be a certified dietician who is highly trained to provide efficient and effective operation of a modern complex support service. This person should work with building principals and the central business office and should have direct responsibility for supervising foodservice workers. The position has been described as one of "planning, doing, and comparing results"[20] because the manager is ultimately responsible for a range of skills taking in menu planning, food purchasing, and hiring and dismissal of foodservice personnel.

The foregoing considerations apply regardless of whether foodservices are centralized or decentralized in that coordination must occur on a district-wide basis. Districts opting for decentralized services need to provide these skills at each location, a formidable task that has led many districts to modify the concept to more closely resemble quasi-decentralized systems operating under a centralized oversight position—a reality that has evolved even further into a concept discussed in the next section regarding central food kitchens that send food to decentralized satellites. Districts opting for fully centralized operations can achieve certain benefits, however, in that centralization nearly always results in efficiency in use of personnel and cost-effectiveness. General benefits assigned to a centralized system are usually cited as:

- Reducing the work load of principals by assigning foodservice supervision outside the building.
- Increasing expertise by enabling the district to employ someone with training and experience in foodservice management at a central level.
- Improved staff training due to a centralized position devoted only to foodservice.
- Reducing duplication of effort by having central office personnel handle regular activities such as menus, state reports, and purchasing.
- Securing lower prices through competitive bids and volume buying.
- Helping ensure compliance with federal and state regulations and, thus, reducing the opportunity for errors in record keeping that could jeopardize outside funding.

Food Systems

The type of management system adopted by a district corresponds closely to the type of food system it chooses. On-site preparation is the most common form of food system used in the nearly 90,000 schools in the United States. But as labor costs have increased, school districts have increasingly examined alternative food delivery systems since on-site preparation is also the most expensive system—a fundamentally decentralized method even in systems with central oversight management structures.

As discussed earlier, in some instances districts have determined that the best effectiveness and efficiency can be obtained from outside contracting of foodservices. In other instances, districts have chosen to stay with on-site preparation due to local preferences, the high cost of constructing central preparation facilities, or the impracticability of off-site preparation in the case of remotely located buildings in a geographically large district. Still others, however, have opted for the satellite concept as a solution that compromises the strengths and weaknesses of both on-site and fully centralized facilities.

A satellite foodservice system prepares food at one location for serving at another location. Satellites first gained popularity in the 1960s when increased funding became available under child nutrition laws. A school receiving centrally prepared food may have a finishing kitchen or merely a serving kitchen. Basically there are three types of satellite systems. The first type is a bulk system where food is transported ready to serve hot or cold, depending on the type of food. The second type is a cook/chill or freeze method where the food is transported precooked and ready for reconstituting in the ovens at the satellite's finishing kitchen. The third type is a pre-plated system where meals are centrally prepared and pre-portioned onto plates that are immediately ready to serve or, alternatively, ready to be reheated on site in a serving kitchen.

There are several advantages and disadvantages to any of these systems. Benefits of satellite systems include reducing expensive on-site preparation equipment, elimination of some personnel, and the economy of bulk buying. In many cases, satellite schools may only need some warming space, freezer/refrigerator storage space, and other minimal equipment requirements. However, any amount of decentralization still requires some duplication of personnel and facilities, larger and newer facilities, and specialized transportation. Further, duplication of facilities is not completely avoided by full centralization since at least a serving kitchen is required at each satellite site, and the method of centralized preparation may also require significant satellite kitchen equipment as well. In all instances, multiple overlapping personnel is a fact of life. Similarly, other issues become important. A centralized system is predicated on an assembly line production model, requiring sufficient volume to be cost effective. Political considerations also apply, as converting from an on-site program to a satellite operation may be met by community resistance. In all instances, the rule is to balance people considerations against the cost-benefit ratio.

Staffing and Structure

As stated earlier, decisions concerning how foodservice operations are to be organized have implications for staffing and structure. As implied previously, districts running internal foodservice programs will likely differ in structure depending on the relative degree of centralization or decentralization. Similarly, the type of facilities and equipment and the number and expertise of personnel assigned to the foodservice function will differ considerably. These elements can be illustrated hypothetically using a variety of charts and diagrams.

Organizational Charts

An example of a well designed centralized foodservice organization chart can be seen in figure 22.2. The interrelationships among the various aspects of a sound operation can be seen clearly in that information flows are multiple and appropriate. As with all good systems, policy formation rests first with the board of education, affected by outside rules and regulations from organizations such as the state and county health service regulators and the state school foodservice organization. Implicit to the chart is the effect of federal regulations, in that states act as pass-through organizations for federal aid to foodservice operations. Similarly, effectuation of board policy flows through the superintendent and through the organization, providing multiple feedback loops for monitoring the system and assuring appropriate communication. Although only administrative ranks are shown in figure 22.2, continuation of coordination and communication can be envisioned to the extent that children and parents are reached as the ultimate consumers of foodservice operations.

Figure 22.2
Example of Centralized Organizational Structure

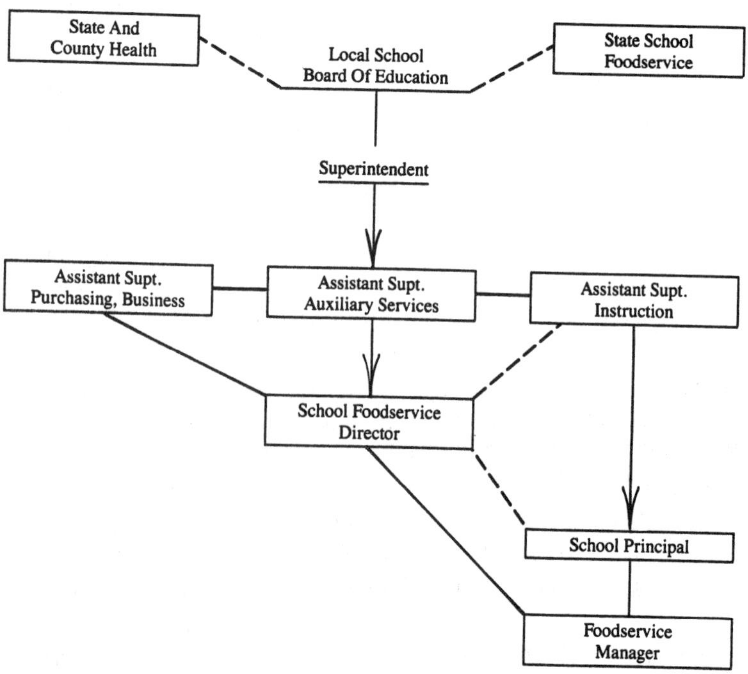

A variation on the same theme can be seen in figure 22.3, which depicts a possible organizational chart for a large, centralized foodservice operation that includes satellites of some type. In this instance, the chart presumes the same vertical hierarchy beginning with the board of education and superintendent, but focuses on how the system would extend beyond the position of director of foodservices. In this case, the system is large enough to include area superintendents for the district. Likewise, area supervisors form a part of this organizational chart, as do various clerks and managers, stemming from a concept of specialization and span of control.[21]

Figure 22.3
Example of Large Centralized Structure

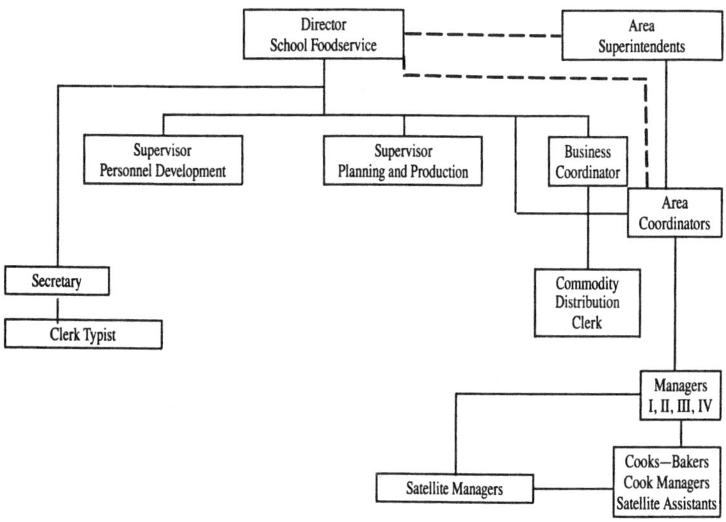

Still a third organizational chart can be observed in figure 22.4 which shows a large secondary school with a la carte sales. In this instance, the chart is exploded to focus on the internal operation at the school level, again presuming all higher vertical organizational features remain in place since no individual school can have a fully autonomous foodservice operation because of the oversight needs of coordination, purchasing, and paperwork, all of which are ultimately driven by law and local board policy.

Staffing Numbers and Configurations

In addition to organizational structure considerations, the number of staff and possible configurations of personnel are real concerns in foodservice operations. These elements drive much of the cost of operations, and they

Figure 22.4
Example of Large, Centralized Secondary Organizational Structure

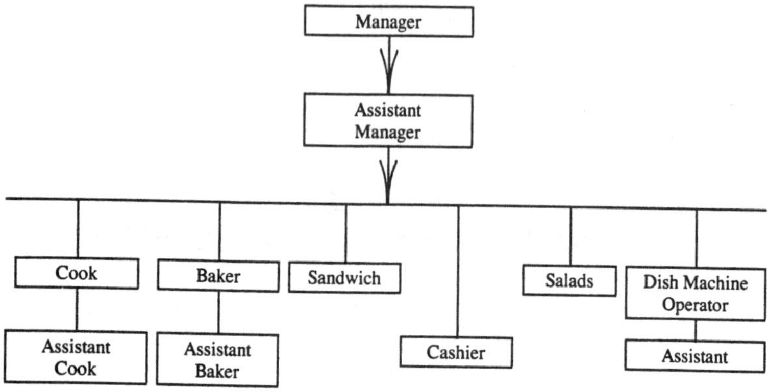

depend in large part on the expertise available and on the size of population that must be served.

The number of staff needed is usually based on the number of people needed for serving purposes at any one time. By drawing out a symbolic kitchen/ serving area, it may be easier to establish how many people may be needed. A number of different staffing formulas are available that are based on labor hours involved in a production operation. A fairly typical formula is provided in table 22.5 and can be considered as reasonably representative of common state and school district standards.

Table 22.5
Staffing Guidelines for School Lunch Program

Number Lunches Served	Meals per Labor Hour	Total Hours
Up to 100	9½	9-11
101-150	10	10-15
151-200	11	15-7
201-250	12	17-20
251-300	13	20-22
301-350	14	22-25
351-400	14	25-29
401-450	14	29-32
451-500	14	32-35
501-550	15	35-36
551-600	15	36-40
601-700	16	40-43
701-800	16	43-50
801+	18	50+

[1]These recommendations do not include hours for breakfast, a la carte, or dinner.

Figure 22.5
Staffing Guidelines for Elementary Schools

Serving 400 Lunches on a Line

(Lunch Served from 11:00-12:45)

Position	Arrival/ Departure	Total Hours**	Assignment At Serving Time
Manager	8:00-3:00*	6½	Supervision Back Up Serving Line
Cook	8:00-2:30*	6	Cooking
Baker	8:30-2:00*	5	Serving
Salad	9:00-2:30*	5	Backing Serving Line
Staff	10:30-2:00	3	Serving
Cashier	10:00-2:00*	3½	Cashier
		29 (staffed at 14 meals per labor hour using disposables)	
Dish Machine Operator	10:30-2:00	3	Dishwasher
		32 (staffed at 13 meals per labor hour using dish machine)	

*Includes 30 minutes for lunch (on their own time).

**Hours paid.

Calculations to derive the necessary staffing numbers are intuitive to table 22.5. After the number of hours and numbers of people have been established, the next step is dividing the labor hours among the number of people. Meals per labor hour are calculated by dividing the total number of meals served by the total labor hours at the location. An equivalency formula can also be applied for breakfast, a la carte, or dinner by using the national average cost of producing and serving a lunch adjusted for current price levels.

Staffing configuration should also pay close attention to the types of work that must be done and the necessary skills that must be available. As stated earlier, the foodservice supervisor or manager should have a nutrition, foodservice, and administrative background. As district size increases, span of control and specialization re-enter the equation, calling for multiple supervisors, specialists, accountants, buyers, and clerical staff. The organizational charts earlier in this chapter address such issues in larger schools. At the school level, the manager may have assistant managers, heads of departments such as salads, bakery, etc., and a large number of staff people as required by the number of serving lines and types of food preparations. Sample staffing guides for elementary and secondary schools are seen in figure 22.5 and figure 22.6 and can be adjusted according to specific needs, such as in the secondary school where self-serving can be used to reduce staff requirements. For smaller schools, however, the task may be even more difficult in that small programs require fewer people but require people with multiple skills.

Figure 22.6
Staffing Guidelines for Secondary Schools

Serving 1500 on 3 Serving Lines, Salad Bar and Snack Bar

(Lunch Shifts: 11:00, 11:30, 12:00, 12:30)

Position	Arrival/ Departure*	Total Hours**	At Serving Time
Manager	7:30-3:00	7	Supervision of 3 Serving Lines and Kitchen
Asst/Manager	7:30-3:00	7	Supervision of Snack Bar and Salad (backing)
Cook	8:00-2:30	6	Cooking
Asst/Cook	8:00-2:30	6	Cooking and Backing Line 1
Baker	8:00-2:00	5½	Backing Line 1
Asst/Baker	8:30-2:00	5	Backing Line 3
Salad	8:00-2:00	5½	Backing Salad Bar
Asst/Salad	9:00-2:00	4½	Backing Lines for Fruits and Salads
Snack Bar	8:30-2:00	5	Serving Snack Bar
Asst/Snack Bar	9:00-1:30	4	Serving Snack Bar
4 Cashiers	10:00-2:00	15 (3½@)	Cashiering Line 1, 2, 3 and Salad Bar
Head Cashiers	9:30-2:30	4½	Cashiering/Head
3 Staff	9:30-2:00	12 (3 each)	Serving Lines 1, 2, and 3
		87 hours (staffed at 17 meals per labor hour using disposables)	
2 Staff	10:30-2:30	7	Dish Machine Operators
		94 (staffed at 16 meals per labor hour using dish machine)	

*Includes 30 minutes for lunch (on their own time).
** Hours paid.

Administration and Financial Management

The final area of concern in this chapter relates to the administration and financial management of foodservice programs. While this topic is complex and cannot be treated fully here, some general discussion in the areas of managing pupils, personnel, purchasing, plant and equipment, and general financial management is appropriate.

Pupil Management

While the issue of pupil management is a far-reaching concern taking in all school personnel, a major concern for foodservice operations is a set of decisions relating to duration of the lunch period. This decision is based on such factors as service population, number of staff, and plant layout and facilities.

Development of general operating procedures has shown that most students can eat without rushing in twenty minutes after receiving the food. The traditional thirty-minute lunch period is thus adequate for most situations if no student were to stand in line for more than approximately ten minutes. Studies have shown that an average cafeteria line with trained servers and cashiers can expect to serve ten meals per minute, making it possible to serve 100 meals every ten minutes. If lines are not moving at this rate of speed, one or more of the following problems may be the cause:

- Running out of food.
- The line is not set up in a logical or flowing manner.
- Poorly organized employees.
- More self-service items are needed.
- Inefficient procedures for collecting money or tickets.

In setting lunch periods, then, limitations to this rule of thumb should be taken into account, as well as the number of seats in the dining room and the number of serving lines available. All these factors will affect pupil management and can be altered to increase efficiency.

Personnel Management

As with all other areas of administration, foodservice operations are vitally concerned with personnel management on many levels. Of particular interest are the areas of:

- Development of clearly written job descriptions.
- Clear communication of division of responsibilities,
- Regular, ongoing, and periodic summative evaluations.
- Regular staff training and development.
- Determination of pay scales and fringe benefits.
- Development of effective work schedules.

As any experienced educational administrator knows, many problems can be avoided by the use of clear job descriptions. Job descriptions are meant to convey to supervisors and workers the expectations and duties of a given position so that no mistake can be made about responsibilities when employees are evaluated. Figure 22.7 gives a brief series of possible job description statements, although these must be adapted to the needs of each organization according to local structure and accountability. A related benefit to clear job descriptions is the communication of division of responsibility among all the

The bid process may take several different forms, depending on custom and statute. For smaller school districts, open-market price quotes from two to four companies may be the most effective method in terms of time required to solicit, examine, and award bids. Under this system, quotes are compared and the bid is awarded to the lowest responsible bidder. An alternative informal bid format is also often used where the bid has certain specifications and is based on a written quote, but without the format of an actual bid contract binding both parties to the result. As a result, either party may renege on the agreement—as a benefit, however, informal bidding may result in lower prices because the supplier's overhead costs of carefully calculating a firm bid price are substantially avoided. A third format of formal bidding is a system that includes detailed written specifications for all products, with the quotes legally binding on both parties.

In many instances, and under enabling state statutory guidelines, significant price breaks can be gained by using collective purchasing procedures. This system is especially useful to enhance purchasing power when individual district purchasing occurs on a relatively small volume basis. By joining a collective purchasing group, small districts are able to purchase at bulk supply rates normally open only to larger districts that consume quantities qualifying for discount. Likewise, competition among vendors is greater for large contracts, making volume buying even more attractive for all types of districts. Many sources of information concerning bidding and collective purchasing exist, most of which alert new buyers to pitfalls and suggest cost-saving techniques that have been proven by experience. Among the recommended procedures are such cautions and tips as:

- Start collective purchasing on a small scale until pitfalls can be identified and a code of ethics between supplier and buyer is worked out.
- Write specifications precisely to avoid unpleasant outcomes:
- Name of the product (generic, not brand, e.g., gelatin, not Jell-O)
- Federal grade, brand, or other quality designation.
- Unit of purchase (pound, case, carton, bag, etc.)
- Geographical area of production (such as Blue Lake green beans or domestically-grown pineapple)
- Variety, type, style, size, etc.
- Condition of merchandise upon receipt.
- Quantity to be purchased.

As with all other areas of school business management, purchasing for foodservice operations should be done with a high standard of ethics established by the administration or school board, rather than by division managers. Foodservice directors, supervisors, managers, and purchasing agents should be informed and aware of those standards. At the root of all such standards should be obtaining the best price for the best product. If the quality of products are not acceptable and exactly according to bid specifications, committees can be organized to determine better ways to reach quality standards.

Plant and Equipment Issues

Planning foodservice facilities and equipment is a complex task that should involve experts from several different fields such as architects, educational consultants, and the food personnel who will work in the facility. It is extremely important to engage in thoughtful and expert planning because, to some extent, the layout of the dining room and preparation facilities will affect staff configuration and the organization of the entire foodservice function. At a minimum, good planning will result in making serving and dining more pleasant, while poor planning may cause severe disruption or slowdown in otherwise well-laid plans.

For quite some years, there have been significant efforts to avoid an "institutional" look and to develop dining as a more aesthetic experience. Varied table sizes, comfortable seating, room dividers, good acoustics, bright colors, adequate lighting, and pleasing decor help to achieve a more pleasant dining experience. Good storage facilities—both dry and refrigerated—help

Table 22.6
Storage for Food by Type

Type	Temperture	Use	Characteristics
Dry-Cool	50°-70°	Semi-perishable and Staple Items	Well Lighted Dry Well Ventilated Easy to Clean Well Constructed Shelves Floor Racks Thermometer
Dry		Detergents Paper Goods	Same as above
Refrigerated Reach-in or Walk-in	32°-40°	Perishables Semi-Perishable Raw and Cooked Foods for Short Periods of Time	Thermometer Adjustable Shelves
Frozen Food Storage	0°F. or below	Frozen Foods	Thermometer Shelves or Racks Properly Wrapped Packages

keep foods safe, wholesome, and appetizing. Careful consideration must be given to providing needed storage facilities to ensure that all foods, whether U. S. Department of Agriculture commodities or purchased items, will have adequate storage to maintain quality and nutritive value until consumed. Temperature, humidity, ventilation, rodents, and insects must be controlled throughout the entire storage area. Professional advice on achieving these goals is imperative and will generally meet or exceed the standards shown in table 22.6.

In addition to the foregoing, equipment replacement represents a major cost factor for school districts. Architects and engineers suggest that all major equipment be replaced every twenty-five to thirty years. Since many schools were built in the 1950s and 1960s, equipment replacement is a looming necessity for many schools. Added to this reality are rising labor costs, environmental issues such as nonchlorofluorocarbon (CFC) compressors in major air-moving systems, and increased technological costs associated with sophisticated, electronic and computer-controlled equipment.

General Financial Management

The entire thrust of this chapter has centered around foodservice management in the context of wise and efficient use of fiscal resources. The financial management aspect of the foodservice function has great implications for productivity and efficiency because large amounts of money are devoted to it. Financial management of foodservices, however, is similar to that of other areas of school management with respect to securing funds, establishing a budget, cost-effectiveness, and accountability. Foodservice is nearly unique, however, in that it must establish prices for services in a nonprofit organization, and it also "earns" revenue. Further, foodservice is subject to unique federal and state reimbursement procedures as discussed earlier. As a result, foodservices budget procedures have an added complexity: the need to combine a largely federally funded program with other issues relating to pricing and cost accounting.

Fundamentally, the school district receives funding on the basis of its contract with the state if it participates in the national school lunch and child nutrition programs that were applicable at a given point in time. Table 22.7 illustrates the available revenue to the typical district using FY '95 as the year of record, with each successive year subject to the normal variations that follow from changes in federal reimbursement as Congress deliberates aid to nutrition services. Claims are submitted for reimbursement by local school food authorities, and federal funds and commodities are distributed to state education agencies for redistribution to authorities at the local level. In return, participating states must follow the federal matching formula, although states are individually free to exceed minimum federal requirements. In addition, local units may subsidize meal costs to keep prices low, or local units may choose to pass on the unfunded costs to clients.

Table 22.7
State Worksheet on Meal Subsidies

		TOTAL ANNUAL MEALS	FEDERAL RATE	FEDERAL Reimbursement	STATE RATE	STATE Reimbursement	DISTRICT LOCAL PRICE	DISTRICT LOCAL REVENUE	TOTAL 7-1-94 to 6-30-95
LUNCHES									
Paid Elem	1.	65,000	.3050	$19,825	.0500	$3,250	1.10	$71,500	$94,575
Jr. High	2.	36,000	.3050	$10,980	.0500	$1,800	1.20	$43,200	$55,980
Sr. High	3.	34,000	.3050	$10,370	.0500	$1,700	1.20	$40,800	$52,870
Free	4.	20,700	1.8650	$38,606	.0500	$1,035			$39,641
Reduced	5.	12,100	1.4650	$17,727	.0500	$605	0.40	$4,840	$23,172
Adult	6.	7,700					1.50	$11,550	$11,550
TOTAL	7.	175,500		$97,508		$8,390		$171,890	$277,788
BREAKFAST									
Paid Elem	8.	6,000	.1900	$1,140			0.95	$5,700	$6,840
Jr. High	9.		.1900	$0				$0	$0
Sr. High	10.		.1900	$0				$0	$0
Free	11.	1,800	.9600	$1,728					$1,728
Reduced	12.	900	.6600	$594			0.30	$270	$864
Adult	13.							$0	$0
TOTAL	14.	8,700		$3,462				$5,970	$9,432
KINDERGARTEN									
MILK									
Paid	15.	4,000	.1100	$440				$0	$440
Free	16.	1,000		$0					$0
TOTAL	17.	5,000		$440				$0	$440
OTHER CASH									
Sales/Income	18.								$0
12 Months Total Income	19.			$101,410		$8,390		$177,860	$287,660

Although budgeting for foodservice must follow state accounting procedures, making it difficult to engage in generalizable discussion about detailed budget procedures, part of every financial management decision involves the development of a budget. While there are sizable similarities between the budget forms for all the different funds operated by the district in most states, budget behavior in foodservice must be different simply because other funds anticipate regular and predictable revenue flows. For foodservices, however, revenue flow is much more volatile in that it is based on anticipated income that may not be realized or, alternatively, may exceed projections. This has real implications for monitoring the budget carefully, particularly because spoilable supplies are involved through the purchase of foodstuffs. Likewise,

Figure 22.8
Food Costs for the Month Worksheet

A. Beginning Purchased Food Inventory First of Month $__________
B. Food Purchases during Month $__________
C. Total A & B $__________
D. Ending Purchased Food Inventory for Month $__________
E. Total Purchased Food Cost for Month (- D from C) $__________
F. Beginning USDA Commodity Food Inventory for Month $__________
G. Commodities Received during Month $__________
H. Total F & G $__________
I. Ending Commodity Inventory for Month &__________
J. Total Commodity Food Cost for Month (- I from H) $__________
K. Total Food Cost for Month
Add total on lines E ($__________)
and J ($__________) = $__________

the earlier discussion concerning changing supply prices makes fixing prices a tenuous task. Further, the expectation that foodservice will be largely self-supporting complicates this problem.

These issues make it important to place the foodservice budget with someone having considerable budget knowledge and experience. Fortunately, however, budgeting can be greatly assisted by an effective cost accounting system that addresses the four main concerns of food purchases, supplies, equipment, and labor. Inventorying supplies is essential to budgeting, along with good cost accounting activities, such as tracking food costs as illustrated in figure 22.8.

The most critical aspect of budget management rests with pricing meals at the point of delivery. As stated earlier, difficult decisions have to be made regarding shifting costs among parties. Prices charged to students and adults for complete lunch and breakfast meals are usually established at the board level based on recommendations by the superintendent and staff. The superintendent will usually base the recommendation on the advice of the foodservice director or the business manager after taking into consideration all costs and projections. Costs attributable to foodservice operations generally include all or part of the food, labor, employee benefits, repairs, replacement of equipment, cleaning and paper supplies, telephone, travel for banking and training meetings, and salary costs for administrative, clerical, and supervisory staff. Meal prices charged to adults should fully cover the cost of the meal, including all preparation and service, since no financial aid can be channeled to adults. Only student meals are subsidized by federal and state aid to foodservice.

To a large extent, pricing is a function of marketing, just as it is with most businesses in the for-profit sector. Schools generally raise meal prices in response to market changes, as in 1981 when reduced federal funding forced higher meal prices—a situation that will widely reoccur if current legislation before Congress under the Personal Responsibility Act of 1995 is approved. Increased prices create new problems, however, as price hikes can affect participation, creating a new spiral of cost problems as cafeteria production goes down. Other problems also multiply, as in the case of credit extended to clients. Many school districts extend credit to students through charge slips, although there is actually no good accounting provision permitted for such activity because foodservice accounting is on a cash basis. In this case, many school districts set up a charge slip system wherein each day the foodservice manager exchanges the charge slips for cash from the principal's office, permitting the manager to record the meal served on that day. Collections on unpaid charge slips then become the task of the building principal's office, creating new problems for charges that go unpaid. There are few good alternatives to such issues, as children are involved in ways that make adults reluctant to pursue collection aggressively.

In the most elementary terms, then, financial management in foodservice operations is a function of multiple complex tasks. These may be identified broadly as accounting for and using the following nine aspects of foodservice budgeting:

- Establishing historical data and interpreting such information.
- Following demographic changes on school openings or closings.
- Projecting enrollments.
- Observing and anticipating effects of menu changes.
- Planning and coordinating changes in operating procedures.
- Setting goals and carrying out effective and efficient plans.
- Anticipating and budgeting for changes in food and labor costs.
- Establishing meal price and equitable charges.
- Observing state and federal guidelines.[22]

Conclusion

The challenges associated with running a school foodservice program are many and differ widely, depending on specific and changing circumstances of federal, state, and local origin. Ensuring compliance with state and federal regulations forms a major part of the foodservice function. Cost control and accounting are vital to the operation, as these elements determine a district's foodservice budget and are tied directly to federal and state reporting requirements that come together through reimbursement and audit activities. Still another challenge stems from viewing school foodservices as a quasi-business from the viewpoint that such programs are expected to be self-sustaining and must follow cash accounting procedures. The business aspect continues into another role of foodservice, in that management must be particularly concerned with client tastes and preferences, as well as pricing and packaging. Ultimately, the goal of foodservice as an aid to instructional programming cannot be forgotten.

As stated at the outset of this chapter, no area of school business management is disconnected from the educational goals of schools. The foodservice operation provides an opportunity to significantly enhance the equality of the educational opportunity. The reverse is also true—failure to apply great care and skill to auxiliary services such as the foodservice function is certain to have negative consequences for all school district operations. The role of school business administrators is to provide wise management in all aspects of school district operations in order to best enhance the total school mission.

Endnotes

[1] Quoted in D. C. Thompson, R. C. Wood, and D. Honeyman, *Fiscal Leadership for Schools: Concepts and Practices.* (New York: Longman, 1994) 512.

[2] *See* generally Chap. 1 discussion of demographics in D. C. Thompson, R. C. Wood, and D. Honeyman. *Fiscal Leadership for Schools:,* 3-72.

[3] H. Hodgkinson, *All One System: Demographics of Education, Kindergarten through Graduate School.* (Washington, D.C.: Inst. for Educ. Leadership, 1985) 3.

[4] Children's Defense Fund, *A Vision for America's Future: An Agenda for the 1990's: A Children's Defense Budget* (Washington, DC: Children's Defense Fund, 1989), cited in D. C. Thompson, R. C. Wood; D, Honeyman. *Fiscal Leadership for Schools:,* 13.

[5] U.S. Dept. of Educ., National Center for Educ. Statistics, *Digest of Education Statistics 1994* (Washington, D.C.: National Center for Educ. Statistics, 1995) 29.

[6] Hodgkinson, *All One System*, 5.

[7] U.S. Dept. of Education, National Center for Educ. Statistics, *Federal Support for Education.* (Washington, D.C.: National Center for Educ. Statistics, 1994), 9.

[8] Commodities are purchased under §32 of the Act of August 24, 1935, for use in child nutrition programs.

[9] *Seattle School Dist. v State of Washington,* (citation omitted, 1983).

[10] H.R. 4 Personal Responsibility Act of 1995 (Contract with America), *A Bill to Restore the American Family, Reduce Illegitimacy, Control Welfare Spending, and Reduce Welfare Dependence.* Introduced January 4, 1995 in the House of Representatives by Rep. E. Clay Shaw, Jr. (R-Fl) with 121 co-sponsors. Sections of the bill are as follows: Title I, Reducing Illegitimacy; Title II, Requiring Work; Title III, Capping the Aggregate Growth of Welfare Spending; Title IV, Restricting Welfare Growth for Aliens; Title V, Consolidating Food Assistance Programs; Title VI, Expanding Statutory Flexibility of States Title; Title VII, Drug Testing for Welfare Recipients; and Title VIII, Effective Date.

[11] The Food Stamp Act of 1977 (7 U.S.C. 2011 et seq.).

[12] The Child Nutrition Act of 1966 (42. U.S.C. 1771 et seq.).

[13] The Emergency Food Assistance Act of 1983 (7 U.S.C. 612c note).

[14] The Hunger Prevention Act of 1988 (P.L. 100-435; 102 Stat. 1645).

[15] The National School Lunch Act (42.U.S.C. 1751 et seq.).

[16] "School Lunches at Risk," (Kansas City, Mo: Kansas City Star, February 13, 1995), A-9.

Chapter 23
Site-Based Management

Introduction

One of the most prominent educational reforms of recent years has been the movement toward decentralized management. Most frequently referred to as site-based management (SBM), or site-based decision making (SBDM), this trend in the organization of schools and school districts has important implications for school district business administrators. SBM represents a fundamental change in the way schools, and consequently school districts, are organized. By shifting the focus of power and decision making to individuals at the local school, the role of the central office changes from one of regulation to one of support. Additionally, as schools are given more authority over budgets and expenditures, the functions of a business office change as well.

This chapter provides a definition of site-based management and provides school business officials with information concerning how SBM has been implemented in selected school districts. This will enable them to think through the complex set of changes that are required to support the implementation of educational decision making at the school level. While the focus of this discussion is on the implications of SBM for school business administration, this chapter will also describe current research on the management and organizational changes in school systems that are needed to implement SBM successfully.

Defining Site-Based Management

There is general agreement concerning the definition of site-based management. One author stated that school-based management " . . . means simply that schools within a district are allotted money to purchase supplies, equipment, personnel, utilities, maintenance, and perhaps other services according to their own assessment of what is appropriate."[1] Others offer a similar definition, stating that "[s]chool-based management is a system designed to improve education by increasing the authority of actors at the school site"[2] It is further argued that the rationale behind site-based management is a belief that the closer a decision is made to a student, the better that decision is likely to serve the student. The concept of site-based management is not new; certain school districts in Florida first gained national notice in the 1970s. In fact, one district has been using site-based management for over thirty-four years[3] — nonetheless, SBM has become more prominent in the last ten years for three reasons:

- The importance of site-based management in the effective schools literature.[4]
- The importance ascribed to decentralization in the so-called "second-wave" of educational reform.[5]
- The availability of inexpensive, yet powerful computing and networking tools that make it possible to transfer budget and other management information between school sites and central offices in a timely fashion.

Because site-based management places greater authority in the hands of school-site personnel, it is not surprising to find wide variety in the way site-based management plans have been developed and implemented. In its various forms, site-based management provides greater school site autonomy over some combination of budget, personnel, and program decisions.[6] Most site-based management plans also include a formal governance structure at the school site to involve local parents and teachers in the decision-making process. Due to enabling statutory guidelines, as well as local school board policy, the actual role and authority of these site councils varies widely from an advisory role in many districts, to authority to hire and fire the school's principal under Chicago's school-based management program begun in 1989.

The early literature on site-based management was dominated by project descriptions, status reports, and position pieces, with few systematic, empirical examinations of the topic.[7] Most of this literature describes exemplary programs and provides little insight into the impact SBM might have on both school/district relationships, and district/state or school/state relationships.

It has been suggested by many that today's education reforms require a different management structure that decentralizes decision making to the school level. The goal of this change in school organization is to have all students achieve at a high level and achieve mastery over all subjects including math, science, writing, history, geography, and physics.

Odden, Wohlstetter, and Odden argue that experiences in other public and private sector enterprises show that higher levels of performance can be achieved through decentralized, high involvement management, and suggest that this model is the most promising new management strategy for education as well. They suggest that SBM provides the methods for reorganizing schools to make them more effective by doing the following:[8]

- Restructuring site/district relationships to provide more power, authority, and accountability to the school site.

- Establishing conditions to restructure school and classroom organization.
- Allowing school faculties to reallocate money currently in the system and to target new resources into productive uses as defined at that site.

Complex, collegial, and uncertain work best lends itself to decentralized strategies that rely on employee involvement in the production process. [9] Teaching is a complex task, and it is best done collegially as it faces uncertainty in its day-to-day tasks. Thus, there is a strong rationale for implementing a high involvement, decentralized, or school-based management structure in public education [10]

Implementation of a SBM structure in a school district will not guarantee that student performance will improve, however. Research[11] has found that prior to 1990, most of the SBM programs implemented in school districts focused on the decentralization of budgeting and often only on a small portion of the budget at that. Teams of researchers found few examples of decentralization of authority over personnel, nor did they find many examples of improved student achievement as a result of the decentralization of authority.

Other recent studies of SBM found that districts often did not use teachers to make curriculum decisions and/or changes,[12] and that few SBM programs relied on student achievement or performance data to assess the effectiveness of such programs.[13] Such studies lend credence to the argument that SBM is not necessarily being used as a tool to improve student achievement.

It has been suggested by some that SBM requires the redesign of the whole school organization. They argue that for SBM to work effectively, people at the school site must have real authority over budget, personnel, and curriculum. In addition, if SBM were to have an impact on student performance, this power must be used effectively. It is also argued that in addition to this power, individuals at school sites must have the knowledge and information they need to make decisions, and they must be rewarded for their accomplishments.[14]

The next section of this chapter provides a more detailed discussion of how these four concepts—power, knowledge, information and rewards—are linked to effective implementation of SBM, and how they might affect the operation of a school district's business office. Following this discussion, detailed descriptions of budgeting procedures, authority changes for SBM, the role of school-site councils ,and implications for school business management are provided.

Framework for Successful Implementation

Although SBM may not, by itself, lead to improved student outcomes, it does appear to have considerable value in creating or establishing improved school culture that may lead to higher quality decision making at the school-site level. Some suggest that replacing traditional, top-down structures by involving more of a school's stakeholders can lead to higher levels of enthusiasm among staff. It is claimed that such involvement "holds the promise of enabling schools to better address students' needs."[15]

Similar to Lawler's conclusions in *High Involvement Management*,[16] which dealt with private companies, the successful implementation of SBM in public schools requires more than changing the power relationships within a school district. Schools also need support through knowledge, information, and rewards as well. Each of these four cornerstones of successful SBM implementation are described in the following sections.

Power

Historically, it has been the responsibility of the business office to establish annual budget targets. The business office makes sure that schools expend within the restrictions established and monitors the school spending patterns to assure that the budget totals are not exceeded.

SBM, if fully implemented, fundamentally changes this relationship by giving the power to make budgetary decisions to the school site. Changing the relationship between the school district and the individual schools does not mean that the business office must close. On the contrary, the business office may have to work harder than ever before. Where once all schools had to maintain budget and expenditure patterns based on dictates from the central office, now it is the site that determines spending priorities. This forces the district office to support the wide variation in decisions that are made by school site personnel, and see to it that all of the staff at the schools are adequately prepared to handle the added budget responsibility.

If this power shifts to the school site, who is responsible for making decisions that were once the responsibility of the central office? While models of site-based power vary, Wohlstetter and Buffett identified three general models of site-based authority.[17] The first is principal dominated. In school districts relying on this model, authority rests with the school principal. In successfully restructuring schools, principals in school districts using this model of SBM typically rely on committees of teachers and/or community members to give them advice.[18]

The other two models for distributing power to the school site rely on more powerful site-based councils. In one model, the council is dominated by teachers and in the other by community members. While the composition of a site-based council is often hotly debated, research has indicated that the composition of the council itself did not matter much in the success of a school's implementation of SBM, but rather the relationships among the members determined a school's ability to take advantage of its new power.[19]

Successful SBM also requires the active participation and support of the central office. Research on SBM has found that central offices that are able to shift the focus from one of control to one of support and who view the schools as the customers tend to be most successful in implementing decentralized decision making. Moreover, it is clear that changes of this magnitude do not take place overnight, but rather they require considerable time and effort to implement successfully.[20]

As suggested, simply changing the power relationships within a school district will not, by itself, lead to successful site-based decision making, nor will it lead directly to improved student performance. To have a truly high involvement model operate well, three other components— knowledge, information, and rewards—are also necessary.

Knowledge

The literature from both the private and public sectors shows that knowledge and skills are required for employees to develop their new roles in ways that lead to improved performance.[21] The literature suggests four areas where new knowledge and skills are needed:[22]

- Interpersonal or team skills for working together effectively in a group setting.
- Technical knowledge and skills for providing the service.
- Breadth skills for engaging in multiple tasks, in particular those tasks decentralized to the work team as a result of a flattened organizational structure.
- Business knowledge and skills for managing the fiscal aspects of the school.

The development of these skills among staff at the school will require considerable investment in professional development, considerably more than is generally spent by all but the highest spending districts in the United States. To meet these considerable staff development needs, either additional revenue would be needed, or spending priorities would have to be revised. The research on SBM has consistently shown that without the knowledge and skills as outlined, successful restructuring of a school or school district is much more difficult, if not impossible.[23]

SBM will require a new set of skills and attitudes by many public school administrators. Odden and others writing in *School Business Affairs* have succinctly outlined the following skills and attributes that will enhance the success of a SBM plan. As a part of SBM, building principals need to:

- Become expert in budgeting as the site develops its own budget from a base zero.
- Involve teachers in the recruitment, selection, development, and evaluation of all school staff.
- Become brokers of information, professional knowledge, and programs that could help teachers accomplish school goals.
- Help find professional development and training opportunities,
- Become entrepreneurs for additional resources.
- Buffer teachers from distracting forces either at the site or from the central office or community.
- Help create school visions.
- Foster a culture that supports shared decision making and collegial approaches to school actions.
- Develop an awards structure that provides meaningful intrinsic and extrinsic rewards for the new, hard efforts of teachers.
- Become an effective manager of a complex change process as the school restructures its organization, curriculum and instructional program, and teacher jobs.
- Be a liaison between the school and the school's "customers" including parents, community members, the central office, and the school board.

Additional skills and expertise are need by central office personnel including:

- Devolve budget and personnel decisions to the site.
- Ensure, perhaps through a formal budget set aside (2-4 percent of a site's budget), that schools invest in ongoing professional development and training for faculty and staff.
- Create a fiscal accounting system that provides revenue and expenditure information to sites through an interactive computer system as well as provide other data on student demographics and achievement.
- Begin developing an awards and compensation structure that rewards teachers individually for expanding their professional expertise and col-

lectively for producing significant improvements in student performance, particularly achievement.

School boards will need to be involved in:

- Setting goals and directions.
- Engaging in strategic planning to implement them.
- Stimulating curriculum development aligned with national, state, and professional curriculum standards and creating staff development structures to enable teachers to teach them.
- Designing mechanisms for restructuring education systems and changing school management.
- Developing specific local education goals, consistent with state and federal educational goals.
- Creating policy coherence around local curriculum, categorical program requirements, student testing, and teacher professional development that are linked to the goals.
- Designing site-based management and organizational structure consistent with the need to create high performance schools across the entire district.
- Implementing an assessment and indicator system to track school and district progress toward accomplishing the goals.
- Develop a process for periodically reviewing each school's performance.
- Design a phased-in assistance/intervention program for schools where performance lags.[24]

Information

If a school is to make good decisions, it is important that the individuals at the school have access to information about organizational goals, objectives and levels of performance. They also need data on the work processes. This would include information concerning district and site revenue, expenditures, cost structures, customer satisfaction, comparisons with other schools, and data on the work environment. More importantly, it is critical to share this information with everyone engaged in the school community, including school site administrators, teachers, parents, and as appropriate, even students.

Detailed databases on school revenue and expenditures are essential to the development of annual budgets, and are therefore a critical management tool in determining a school's fiscal performance. Data concerning student performance and outcomes are also essential if the site-based decision-making structure were to help find ways to improve student learning. Much of these data can be collected and analyzed using today's microcomputers and school-wide and district-wide local area networks. Having access to these data in a timely manner is necessary if a school is to be held accountable for its performance.

Rewards

It is argued that by some that successful decentralization of authority requires that employee compensation systems be redesigned so that they align the self-interest of the faculty with the organization's objectives.[25] This implies a shift from seniority based pay systems to salary structures based on assessments of knowledge and skills. Another possibility is to provide compensation based on group or team performance rather than individual performance.[26]

Typical of the types of financial rewards used by SBM schools are money for differentiated staffing positions, additional compensation for administrative responsibilities, funds to participate in staff development activities, and small grants for teachers who put in extra time on a project important to the school's goals and/or mission. Since changing compensation systems in schools is often a difficult task, and despite the fact that such changes need not cost more money, the combined resistance of teachers and the lack of new sources of funding often leads school districts developing SBM programs to rely on non-monetary compensation schemes. These might include mentoring or grant writing assignments, reduced teaching loads for other responsibilities, or even notes of appreciation from the principal.[27]

These components have been found to improve the success of SBM in school districts. However, within this framework, it is important to look at how SBM actually performs in the operation of a school district. The following sections consider that issue with a focus on the implications for the school business office.

School District Budgeting Procedures

Budgeting, or the allocation of resources to achieve institutional or organizational goals, is one of the most important functions of school district management. In most school districts in the United States, this function is carried out centrally with input from individual school sites. As described in chapter 12 of this text, discussions of budget theory are filled with high expectations. It has been stated that the school district budget " . . . is an important tool for school administrators to understand and utilize in achieving their basic mission — educating children in the most effective and cost-efficient manner."[28] Unfortunately, as others have argued, "actual budgeting practices fall far short of these high aspirations."[29]

Typically, a school district's budget is a process of balancing expenditures with revenue. After estimating revenue for the next year, budget officials ascertain the costs of providing the existing program in the following year. After providing for current programs, if additional funds are available, new programs may be funded. On the other hand, if resources were not adequate to fully fund the current program, reductions are required. As a result, school district budgets are typically incremental in nature. That is, changes in spending patterns, if they are made at all, occur at the margins, with little change in the year-to-year allocation of resources among schools and between the central office and school sites.[30]

To allocate funds across sites and programs, school districts typically use a set of allocation rules.[31] These allocation rules focus on three components:[32]

- Formulaic ratios.
- Centrally controlled budget categories.
- Pay scales.

Enrollment projections are crucial to the allocation of funds since they drive both the revenue received from the state and the allocation of resources to school sites. School districts are generally known to use two formulaic ratios:[33]

- A staffing formula.
- Funding formula for supplies, materials, and textbooks.

Staffing formulas for teachers are usually established on the basis of a predetermined average class size. For example, a school district may provide one teacher for every twenty-three elementary school students, or one teacher for every eighteen high school students. Administrative staff are often assigned on the basis of school size. Almost all schools have a principal.[34] Larger schools may require additional administrative staff, and a district may have a policy of providing an assistant principal for any school over a certain size. Counselors and guidance personnel are typically provided on a ratio basis, such as one counselor for every 350 or 400 students. Clerical positions are filled on the basis of school enrollment or number of teaching positions. Other classified staff such as custodians, groundskeepers, and maintenance personnel are frequently assigned on the basis of the some factor such as building square footage, number of classrooms, or site acreage. It should be remembered that these allocation rules are determined by individual districts and may vary considerably across districts.

An important consideration in budgeting for personnel is how to charge sites for the people working at each school. Salary schedules for both certificated and classified staff typically include a considerable range in the salary of individuals. If funds were allocated to schools on the basis of a formula based on students, for example, then schools with highly paid (usually more senior) teachers could have trouble balancing the budget, while schools with less experienced (and thus less costly) teachers could have a surplus of funds. Since this could also lead to unanticipated incentives in personnel decisions unrelated to teacher quality, many school districts with SBM plans allow school sites to purchase classroom teachers at the district's average cost, minimizing this problem.

Funding formulas for supplies, materials, and textbooks are typically based on enrollments or staff positions as well. Some school districts have very detailed allocation procedures, providing separate formulas for each item or group of items, while other school districts provide a lump sum payment, usually based on enrollment, for schools to purchase these items as they see fit. Central control over supplies, materials, and textbooks often extends beyond these allocation formulas as well, with schools being required to purchase items through a central purchasing system.

These allocation rules assume that there is a single best production process for instruction and that that process is imbedded in those rules. Moreover, these allocation rules greatly simplify the work of the central administration. In addition to providing a metric for distribution of resources, they are politically useful because they are objective and treat schools in an equal manner.[35]

Hentschke concludes that current budgeting practice "rules out all but the most marginal decisions about improved instructional programming."[36] He points out that two proposed reforms of the 1960s and 1970s, planning, programming, budgeting systems (PPBS) and zero-based budgeting (ZBB), did little to change school district budgeting practice. Hentschke suggests that the initial popularity of PPBS and ZBB was based on the assumption that improved budgeting technology would lead to better decisions concerning the allocation of school resources.

PPBS is a top-down budgeting system that requires agreement as to the goals of the institution and an analysis of those programs that will best achieve those goals. Alternatives are analyzed, and those found to be most cost effective are implemented. Enactment of a full PPBS system in any jurisdiction requires considerable changes in the structure of the budget process, and in many cases requires evaluations of alternatives that are difficult to make since it is hard to measure the output of educational processes.

ZBB, on the other hand, is a bottom-up process that requires production units (for example schools) to develop and cost out a set of decision packages. These packages are then ranked, and funded on the basis of those rankings until available resources are consumed.

The failure of these budgetary reforms in schools were the result of several pressures including the overall concept that these reforms did not change the authority relationships over the distribution of resources within a school district. [37] It is often argued in the professional literature that to successfully implement a SBM system, certain changes in these authority relationships are essential. These are described in the next section.

Authority Changes for Site-Based Management

Because site-based management can take a number of forms, it is difficult to ascertain which provisions of a site-based management plan actually alter decision-making arrangements in schools and school districts.[38] Yet, it is not surprising to expect that decentralization of school systems results in some type of changes in authority relationships between schools and school districts.

Authority over Utilities and Substitute Teachers

Substitute teachers and utilities are two examples of school resources that are typically allocated centrally. Expenditures for these items are controlled through the central office, and budgeted on the basis of past experience. Under this arrangement, local school sites have no incentive to minimize expenditures on these items since any savings that accrue are realized by the central office and not by the school site.

It has long been argued that an alternative would be to provide each school with the resources for these items. Many school districts, throughout the nation, have long since budgeted these issues in a manner where each school receives incentives to stay within budgetary constraints. If utilization levels were less than budgeted, the school could then share in the savings, using the funds for resources that more directly work to achieve educational goals. To make this system work, the school site would also have to share in at least part of the risk of over-expenditures, and the central office would not be able to penalize school site savings by budgeting less in future years.

Authority over Staff Development, Curriculum Development, and Other Central Office Support

Historically, such issues as staff development, curriculum development, and other central office support functions were typically provided to schools by the central office. Often the support service needs of the schools are determined centrally, with consultation with the school sites. With the advent of SBM such decision making is becoming relatively open in the management of public education.

Authority over the Mix of Professionals

As described, school districts typically established allocation formulas for personnel. Under SBM, changes in authority relationships are generally granted to individual schools concerning the order of personnel units. For example, one school might choose to eliminate an assistant principal position in favor of an additional teacher to reduce class size, while another school might choose slightly larger classes, but provide a part-time aid in each classroom in the school, while a third might trade a teaching position for a counselor to deal with problem or at-risk students. Providing this type of staffing flexibility would allow school sites to develop educational programs that more closely met the established goals.

Authority over the Source of Supply

Most school districts operate central purchasing programs to minimize the costs of materials and supplies. In addition to these attributes, the school district is the corporate entity that must meet all state imposed bid statutory requirements. Having individual schools engage in non-bidding practices when purchasing goods and services may collectively pose statutory complications. This concern, regardless of manner, must be resolved by the school district business manager. Some SBM plans have no restrictions in terms of where schools are able to purchase resources from district or non-district suppliers. This policy decision has obvious advantages and disadvantages as discussed in the purchasing chapter of this text.

Authority to "Carry Over" Resources to the Next Fiscal Year

Where permissible under statutory control, and independent of other fiscal dependency to other governmental units, and where permitted by local board policy, most school units may become more efficient if they are able to carry over the unexpended funds to the next fiscal year, without penalty. This would allow them to consider larger equipment purchases, such as computer systems, which would not be possible under a system that takes back any unspent balance in an appropriation account.

Relief from Regulation

Virtually every state allows for a school district, or individual school, with local board of education support, to petition the state board of education to be released from any state mandate or rule/regulation. A common approach to site-based management is to release schools from certain requirements of school district policy or from certain portions of employee collective bargaining agreements. Often, district policy, or employee contracts, restrict the actions of a school. A number of recent site-based management programs have provided for relaxation of these restrictions in schools that develop individual educational plans.

For example, once a school develops its own plan, the school board may provide exemptions to district policies that would prevent some of the actions desired on the part of the school. Schools seeking to undertake actions prevented by the district contract with the teachers, must also receive approval from the bargaining representatives. Generally, programs that release a school from district or contractual rules require the submission and approval of a school plan prior to approving the relaxation of the standards.

There is one other authority issue that must be considered in looking at site-based management programs: the treatment of local or site decision making. Is the principal responsible for making all site decisions? What role do teachers and parents play in the management of a local school? And how are the sometimes conflicting interests of these parties resolved? This important issue is the topic of the next section of this chapter.

The Role of School-Site Councils

Although school districts are managed by a professional staff headed by a superintendent of schools, final authority over district operations lies with the board of education. As more authority is vested at the school-site, it seems logical that interested parties should have a voice in determining both the goals and objectives of the school's educational program and the methods used to meet those goals and objectives. Consequently, virtually all site-based management programs include provisions for school-site councils. The role and function of those councils is often a matter of considerable debate and acrimony. This section considers three issues regarding the composition and operation of school-site councils:

- How is membership on the site council determined?
- What matters should the site council consider?
- Who has formal decision-making authority—the principal or the council?

These matters have been resolved differently in different locations. While local variation in the role of the site council is to be expected, it is important that site councils operate cooperatively if the goals of site-based management are to be met.

Council Membership

School-site councils are not a new phenomena. Many federal and state categorical programs have for years required the establishment of a site advisory council. For example, schools receiving Chapter 1 funds or bilingual money from the federal government are required to establish parent advisory councils to provide input into the operation of the program. In California, the School Improvement Program requires that schools receiving funds establish an advisory council to provide guidance to the principal in the development and implementation of improvement plans. In addition, many states and/or local districts have specific requirements for parent advisory councils at school-sites, regardless of the acceptance of categorical program funds.

Consequently, most schools, and the principals, have had some experience working with advisory councils. Two of the major differences between most school site councils established under site-based management programs are:

- The role of teachers on the site council.
- The level of decision-making authority.

The mission of most school site councils, regardless of the final decision-making authority, includes the overall operation of the school as opposed to advising school staff on one individual program. Thus, the potential power of the council is considerable. As a result, teachers have demanded that they play an important role on the site council. In fact, many teacher organizations argue that teachers should constitute a majority of the council's membership. In most cases, the argument boils down to whether the teacher representatives should constitute one half of the site council or a majority of the council.

Each district has resolved this issue itself. Typical models include a council with 50 percent public members and 50 percent teachers, with the principal representing the tie-breaker, while others provide a one-member majority for either the public or for the teachers. Of course, a simple majority in no way guarantees control of the council, as teachers may have disagreements over the school's goals, and not all community members will be automatically opposed to everything the teachers think is important.

Another issue that must be resolved is the public membership. Should site council members be required to have children attending the school or is it beneficial to allow non-parents to participate in school decisions? Moreover, how should public membership be established? For example, public membership could be constituted through a community election as in Chicago, through appointment by the principal, or some combination of the two? Similarly, who appoints the teacher representatives? For example, the teacher representatives could be appointed through the school's union representative, the bargaining unit, or teachers' elections at each school.

Finally, once a site council has been selected, what rules guide its decision making? Many attempt to operate through consensus. If a consensus were not to be reached, majority voting policies seem like a logical alternative for formulating council policies and goals. However, this raises again the issues of membership and representation of different groups on the council. One option might be to give each group one vote and require that they reach consensus among themselves first. For example, regardless of the council membership, in decision making, the teachers would have one vote, the public one vote, and the principal one vote. This alternative might reduce some of the difficulties associated with determining the membership composition of the site council.

School-Site Council Authority

Once matters of membership and organization have been decided, it is still critical to determine what issues the council can consider. Is the council's role one of fundraising and "cheerleading" or will it be expected to play a substantive role in the educational program of the school? In a true site-based management program, one would expect the council to be able to consider whatever matters they deemed important. However, many school districts, either through policy or as a result of the terms of the various collective bargaining agreements, have limited the authority of site councils. Assuming that the site council is charged with establishing the educational goals of a school, what role should the site council play in:

- Determining how progress toward meeting the goals should be measured?
- How instruction should take place?
- What materials should be used?
- What role should the site council play in hiring teachers and other staff?
- Determining teaching assignments?
- Establishing discipline policy?
- Allocating school resources across program areas?

These are important considerations if the council is to operate successfully. Moreover, they are closely related to the final issue to be considered, the relative authority of the council and the school principal.

Location of Formal Decision-making Authority

Perhaps the most controversial issue of all is the formal decision-making responsibility vested in the site council. What is the role of the principal in relation to the site council. Who has final authority and thus responsibility for major policy decisions and for the day-to-day operations of the school? How these issues are resolved plays a crucial role in the way the site council operates. For example, does the principal have the authority to override a decision of the site council, or must he or she carry out the decisions of that council. Who has authority to hire and fire school principals—the district's central administration or the site council.

For the most part, site council powers seem to be limited to an advisory role, with the principal having final authority over most matters. This is not universally the case, however. In Chicago, the recent school-based management reforms have placed final authority for the operation of a school in the hands of the locally elected site-based council. That council has the authority to employ the principal, and by the end of the 1990-91 academic year, all of the site-based councils in the district had made a decision as to whether to renew the principals' contract. Half of the site councils made that decision during the 1989-90 school year. While most chose to renew the principal's contract, 49 of 272 school councils did not do so. In some instances, the sitting principal was allowed to reapply for his or her position although no guarantees were provided.

The relative authority of the site council and the school principal will have a strong influence on the motivation of teachers and community members to serve on the site council. Principals and administrator organizations are universally opposed to placing final decision-making authority in the hands of site councils, arguing that it is hard to hold a committee accountable and that it is unreasonable to hold the principal accountable for actions he or she is required to carry out by order of a site council. Since this is a relatively new issue, there is little empirical evidence concerning the success or lack of success of different authority structures. Over time, as more school districts struggle with this issue and make decisions regarding the role of the site council, more evidence on the merits of various options will emerge.

The remaining question then is, of what benefit is site-based management, and how can it help improve schools. The next section of this chapter describes how site-based management may help improve school accountability, productivity, and flexibility.

Site-Based Management as a Tool for Improving School Accountability, Productivity, and Flexibility

The chief outcomes of site-based management or organizational decentralization are said to include more accountability for schools, increased productivity, and flexibility of decision making. Each of these benefits is discussed.[39]

Accountability

One of the most important tenets of the school reform movement of the 1980s was the desire to hold schools accountable for the educational outcomes of students. Site-based management per se does not improve accountability, but it does provide two ways to make schools more accountable for their performance:

- By making the community part of the management process.
- By making school officials more accountable to the community.

An important role for any site council is to help establish the educational goals of the individual school it represents. In addition to establishing goals, it must decide upon measurement standards to use in judging how well the school is performing in relation to its stated goals. By participating in this process, the site council becomes part of the accountability process. If the school is successful, the members of the council can share in that success. If the school fails to meet its established goals, the public can look both to the professionals in the school and the site council for explanations and for improvements. Instead of simply criticizing the school, community members will be challenged to help resolve the problems that keep the local school from achieving its goals.

The school becomes more accountable to the community as the site council becomes involved in the school's operation. Parents, through their involvement, will gain more confidence in school budgeting and financial practices. Moreover, by reporting to the parents, school officials will be more aware of, and responsive to, community concerns.

Productivity

Although scant, the literature on site-based management indicates three areas where productivity might be enhanced:

- Increased services or reduced costs.
- Increases in student access.
- Lower costs of administering decentralized schools.

Studies of decentralized schools have shown greater variation in the ways schools spend funds for supplies. If one assumes local personnel know best how much should be spent on supplies relative to other educational priorities, then spending, at least for supplies, in decentralized school districts would be more efficient. [40]

It has also been suggested that site-based management could improve equality of educational opportunity since the school could tailor programs to meet the needs of the students enrolled at that particular school. One way to achieve greater equality among students, particularly disadvantaged students, is to permit school personnel the discretion to use resources for students as they see fit, rather than based on a set of requirements handed down from the state, federal government, or the district's central office.[41]

Finally, there is at least one study that estimates that site-based management will result in additional administrative tasks at the school site, something that might require additional resources. While this may be the case, the analysis does not consider the impact of decentralization on the central office. If central office positions were eliminated, then economies in the cost of administration may be available, further enhancing the productivity of site-based management. Moreover, the advent of relatively inexpensive microcomputers and networks make it possible to develop management systems that minimize the additional work load at the school site and provide better information for making critical decisions. These factors, which may be accompanied with initial investment costs, would appear to have long-term potential for reducing the costs of administration in school systems.[42]

Flexibility

While it seems obvious that increasing local authority would result in increased flexibility for schools, such may not be the case. Even under site-based management, many schools must still meet certain district and state requirements. For example, California has strict pupil-teacher ratios that must be adhered to by grade level. These restrictions could eliminate a certain amount of flexibility within a school. Similarly, resource constraints, a common ailment for school districts, could limit the flexibility a local site has to establish an educational program.

To the extent that site-based management programs allow waivers from state and/or district requirements, flexibility is enhanced. Obviously, such waivers and decentralization do not have to go hand-in-hand, but to the extent that the two are related in state or district policy, the greater potential flexibility schools will have as they strive to initiate individual educational programs.

Summary

Site-based management is a procedure whereby some portion of a school district's management authority is transferred to the school site. The rationale for doing this stems from the argument that it is the individuals at the local school who best understand the needs of the children enrolled in the school and what programs will best serve them.

A number of authors have argued that for site-based management programs to succeed, authority relationships between the school site and the central office need to change. They claim that school sites should have more autonomy in allocating budget resources, and should be given the ability to carry over unspent funds however they accrue.

One of the major issues in site-based management is the establishment of a school site council. Many issues regarding the membership, voting procedures, and authority of the council need to be established if site-based programs are to be successfully implemented. Site-based management also appears to offer several potential advantages in terms of school accountability, productivity, and flexibility, although the literature on these subjects to date is rather limited.

Finally, if states wish to encourage school district decentralization, they can either offer incentives for site-based management programs or provide large portions of state aid in the form of general grants that will not unduly restrict school districts from implementing such programs.

Endnotes

[1] D. J. Brown, *Decentralization and School-Based Management.* London: The Falmer Press., 1990, vii.

[2] W. Clune and P. White, *School-Based Management: Institutional Variation, Implementation and Issues for Further Research.* New Brunswick, NJ: Rutgers University, Center for Policy Research in Education, 1988, 1.

[3] *Id.* at 3.

[4] See, e.g., S.C. Purkey, and M.S. Smith, "The District Policy Implications of the Effective Schools Literature," *Elementary School J.*, 1985, 85, 353-389.

[5] See, e.g., *A Nation Prepared: Teachers for the 21st Century.* New York, NY: Carnegie Corporation, 1986.

[6] B. Malen, R. Ogawa, and J. Kranz, "What Do We Know About School Based Management? A Case Study of the Literature - A Call for Research," in W. Clune and J. Witte, eds. *Choice and Control in American Education: Vol. 2, The Practice of Choice, Decentralization and School Restructuring.* Bristol, PA: The Falmer Press, 1990.

[7] *Id.*

[8] A. Odden, P. Wohlstetter, and E. Odden, "Key Issues in Effective School Based Management," *School Business Affairs*, May, 1995, 4-16.

[9] E. Lawler, *High Involvement Management.* San Francisco, CA: Jossey-Bass, 1986.

[10] S. Mohrman, E. Lawler, and A. Mohrman, "Applying Employee Involvement in Schools," *Educational Evaluation and Policy Analysis.* 1992, 14(4), 347-360.

[11] See, e.g., B. Malen, R. Ogawa, and J. Kranz, "What Do We Know About School Based Management;" P. Wohlstetter and A. Odden, "Rethinking School-Based Management Policy and Research," *Education Administration Quart.*, 1992, 28(4), 529-549.

[12] See M.A. Smyle, "Redesigning Teachers' Work: Connections to the Classroom," *Rev. of Research in Educ.* Washington D.C.: American Educ. Research Ass'n., 1994, 127-177.

[13] A.A. Summers and A.W. Johnson, *A Review of the Evidence on the Effects of School Based Management Plans.* Paper presented at the Conf. on Improving the Performance of America's Schools: Economic Choices. Board of Science, Technology and Economic Policy, National Research Council, National Academy of Sciences, Washington, D.C. 1994.

[14] See e.g., P. Wohlstetter and S. Mohrman, *School Based Management: Promise and Process.* New Brunswick, NJ. CPRE Brief, FB-05-12/94.

[15] *Id.* 1.

[16] E. Lawler, *High Involvement Management.*

[17] P. Wohlstetter and T. Buffett, "Decentralizing Dollars Under School-Based Management: Have Policies Changed?" *Educational Policy*, 1992, 6(1), 35-54.

[18] Wohlstetter and Mohrman, "School Based Management."

[19] Wohlstetter and Mohrman, "School Based Management," 2.

[20] P. Wohlstetter, R. Smyer, and S. Mohrman, "New Boundaries for School-Based Management: The High Involvement Model." *Educ. Evaluation and Policy Analysis* 1994, (16)3, 268-286.

[21] For a discussion of the literature on decentralization in the private sector, see S. Mohrman, "High Involvement Management in the Private Sector," in S. Mohrman and P. Wohlstetter (eds.), *School-Based Management: Organizing for High Performance.* San Francisco, CA: Jossey-Bass, 1994, 25-52.

[22] A. Odden, P. Wohlstetter, and E. Odden, "Key Issues in Effective School Based Management."

[23] See, e.g., P. Wohlstetter, R. Smyer, and S. Mohrman, "New Boundaries for School-Based Management."

[24] A. Odden, P. Wohlstetter and E. Odden , "Key Issues in Effective School Based Management," 14.

[25] A. Odden, P. Wohlstetter, and E. Odden, "Key Issues in Effective School Based Management."

[26] For a description of how such systems work in the private sector, see E. Lawler, "Strategic Pay: Aligning Organizational Strategies and Pay Systems," San Francisco, CA: Jossey-Bass, 1990.

[27] A. Odden, P. Wohlstetter, and E. Odden, "Key Issues in Effective School Based Management."

[28] W.T. Hartman, *School District Budgeting*, Englewood Cliffs, NJ: Prentice Hall, 1988, 1.

[29] G. Hentschke, "Budgetary Theory and Reality: A Microview," in D. H. Monk, (ed.) *Microlevel School Finance: Issues and Implications for Policy,* Cambridge, MA: Ballinger Pub. Co. 1988, 311.

[30] The reader may wish to refer to Chapter Twelve of this text for a more detailed description of budgetary theory and practice.

[31] G. Hentschke, *School Business Administration: A Comparative Perspective,* Berkeley, CA: McCutchan Pub. Co., 1986.

[32] J. Guthrie, *Understanding School Budgets.* Washington D.C.: U.S. Dept. of Educ., 1988.

[33] M. Kirst, "The Internal Allocation of Resources within U.S. School Districts: Implications for Policymakers and Practitioners." in D. H. Monk, (ed.) *Microlevel School Finance.: Issues and Implications for Policy,* Cambridge, MA.: Ballinger Pub. Co. 1988.

[34] In small, one school districts, the principal may serve in the dual role of superintendent/principal, or the school may be managed by a head teacher/ principal. In addition, districts with very small schools may assign one principal to two schools. But in general, most schools in the United States have a full-time principal.

[35] G. Hentschke, "Budgetary Theory and Reality."

[36] G. Hentschke, "Budgetary Theory and Reality," 312.

[37] G. Hentschke, "Budgetary Theory and Reality;" see also J. Brackett, J. Chambers and T. Parrish, *The Legacy of Rational Budgeting Models in Education.*, Rpt. 83-A21. Stanford, CA: Inst. for Educ. Finance and Gov. 1983.

[38] See, e.g., B. Malen, R. Ogawa, and J. Kranz, "What Do We Know About School Based Management."

[39] D. J. Brown, "*Decentralization and School-Based Management..*"

[40] Ibid.

[41] Ibid.

[42] Ibid.

Chapter 24

School Business Management and Future Issues

Introduction

This book concentrates on expenditure side of education finance rather than the revenue portion of the discipline. The authors have attempted to call on their years of research as well as public education administrative careers. The text is not capable of fully answering every potential question that is placed before school districts in every state during every legislative session. Nonetheless, it is hoped that a broad introduction to this arena will suffice in leading to further study and practice by students as well as practitioners of the profession of school business management.

Public education, in our society, is the one public endeavor that perhaps touches the lives of every person. The role of public education in our society encompasses a wide variety of social, economic, and educational goals. Public education is clearly one of the most important societal functions in order to ensure the well being and the future of any nation. Thus, the study and practice of school business management draws from a variety of activities and is complex in its application to a wide array of tasks and duties within every public school district.

In a very brief overview, chapters 1, 2, and 3 respectively presented an introduction to the field, the legal context, and taxation of public education, all of which contribute to an introduction to the basic factual and conceptual elements to the profession. Chapter 1 introduced to school business management. In this regard, much has changed in the field of school business management over the years, and yet the practice of school business management remains fundamentally the same. The reader is certainly encouraged to read the works of the early pioneers of the field of the practice of education finance, such as Fred Hill, Schuyler Joyner, and Charles Foster, to name a few of the early professionals who were visionaries in that they engaged in practice while writing seminal works for the profession. Their writings dating back to the early 1950s serve as a guide and inspiration for beginning school business managers throughout the nation. The Association of School Business Officials International, and its monthly professional journal, *School Business Affairs*, as well as the association's annual meeting and its numerous publications will continue to point the way toward future issues. For the beginning chief finance officer, as well as the seasoned veteran, this professional association is indispensable for continuing education. While the future cannot be predicted with certainly, the history of the profession clearly reveals that the association will continue to offer guidance and vision for the practice of education finance.

Chapter 2 presents the legal context of school business management. Despite the fact that much of school business management is largely state statutory dependent, the overriding major public policy issues continue to be established by the Supreme Court. Given the nature of our litigious system of resolving public policy issues, additional public policy controversies will continue to be adjudicated and will, in many instances, eventually reach the Supreme Court. These issues, encompassing the great social and economic controversies of the day, will surely affect public schools as nearly all such decisions. Thus, the public schools will continue at the forefront of public discussions and debate.

Chapter 3 on taxation establishes the framework of society paying for public education. It is notable that the future will reflect even greater attacks on public education and the related taxes committed to its existence. Public education has never been financed based on the benefits to the individual. It has always been financed based on the benefits to that of society as a whole. This fundamental concept appears to have been lost in contemporary times. It is as if segments of society wish to withdraw to their enclaves and separate societies. These social, political, religious, and racial issues are interwoven with economic ones, specifically who pays for what within our society. These debates will most likely continue and even escalate over the next several years.

Chapters 4 and 5 concerning management theories and strategic planning introduced areas that school business managers must fully understand within school district activities. The basis of many administrative actions are time honored, well proven administrative theories. Often, public education presents such a complex picture that school board members and even public school administrators have difficulty understanding and grasping the overall direction of the organization. The chapter on strategic planning offers a view that certain school districts have found to be productive and useful. Surely, other such models will emerge with varying degrees of success. Nonetheless, a current model is always helpful to business offices in public education. Again, it is highlighted that the business office is never removed from the educational process of the school district.

Chapter 6 presents the major policy discussion of fiscal inequity among school districts. This discussion represents the most intense area of education finance having an affect on all of public education in America. As the lead authors of this text have served as expert witnesses in more than a dozen education finance equity suits affecting more than 1 million children, it is an area to which they bring a unique perspective. As witnesses for both plaintiffs and defendants, the two lead authors have served in more education finance suits than perhaps all others combined over the past decade. It is their view that this arena will continue to be the major policy arena in all of public education for many years.

Chapter 7 on legal liability introduced an area that is omnipresent in many school districts. Similar to other such legal areas, it will grow in complexity in most states. State legislators must continue to examine the public policies of protecting the public while at the same time protecting the public treasury.

Chapters 8, 9, and 10 present discussions on personnel, collective bargaining, and reduction-in-force, areas that are highly related to the major expenditure of all school districts—people. Public education is essentially a highly intense labor organization. The vast majority of school districts expend nearly 90 percent of general fund budgets on salaries and fringe benefits. Thus, these discussions as presented regarding personnel, collective bargaining, as well as reduction-in-force are reflective of the costs of public education. These issues will continue to be time consuming and, on a local school district level, some of the most important areas of school business management regarding the proper functioning of the school district in offering educational services to the next generation.

Chapters 11 and 12 are interwoven in the discussions of accounting and budgeting, vital activities for the business office. While it is not necessary that the chief fiscal officer be an accountant, the chief fiscal officer needs to be quite conversant with accounting principles, practices, and procedures. The budgeting chapter is one that reflects a task area of the business office and the decisions therein. Each budgetary year reflects a shared vision of the local community and the state legislature as to the priorities of each local school district.

Chapters 13, 14, 15, and 16 covered quantitative management of a school district, as well as long-range planning activities. These chapters on financial management, quantitative decision making, and management information systems reflect the technical and quantitative nature of public school administration. With the advent of more sophisticated software and research appearing almost daily, these tasks may become even more sophisticated and available to all school districts. On the other hand, the chapter concerning total quality management reflects both a quantitative and qualitative view of reforming public education.

Chapters 17 introduced capital outlay and bonding, while chapter 18 operationalized these issues with facility planning and management. As the public becomes more reluctant to fund capital outlay and the need for fiscal plants continue to grow, almost every school district will face demands within very limited resources. The authors' research and publications have consistently pointed out the need for capital outlay concerns for more than a decade. Unfortunately, while inconsistent in its application, this need will not abate through the next several generations.

Chapter 19 operationalized risk management for school districts. A sound understanding of risk management elements will continue in its applicability for school business managers.

Chapters 20, 21, and 22 presented task areas that answer to the business office. Purchasing, transportation, and food services, similar to all other task areas, will grow in terms of complexity as well as expectations.

Lastly, chapter 23 introduced site base management as a growing school district option in administrating public school districts. Because this movement toward more control at the building level will continue, school business managers will in many instances operate in a somewhat different manner than previously within many school districts.

In an overall sense, students of educational administration, and certainly chief fiscal officers, are well aware that public education is facing unprecedented issues. These issues vary greatly in form and impact. These issues dictate that the student of school business management must truly understand how the role of the chief finance officer is not one of simply "keeping the books" while ignoring the educational issues of the school district. Every successful school district reflects sound fiscal management while maximizing the efficiency of the limited financial resources at its disposal. This can only be done with comprehensive understanding and active involvement by the chief fiscal officer with the entire educational program of the school district.

The role of the school business manager is to build the finest public school district and to expend the moneys in the most efficient manner. The person is employed by the local board of education to engage in these tasks. His or her school district may in fact be educationally superior to another school district by virtue of being more efficient or by pursuing more grants, which is not the burden of the individual finance officer. Thus, these skills and expertise are largely up to local school districts to employ.

For the reader of this text, it should be clear that each chapter presents a wealth of material that exists in the research and professional literature. In fact, the text is deeply rooted in the professional literature and practice as typified by the Association of School Business Officials International and the authors' own research and practice.

It must be noted that nothing in public education occurs vacuously. It is difficult to grasp that the education codes of the various states need the complexity and detail that they represent. However, state statutes, as well as rules and regulations, represent what different legislatures think to be wise public policy and practice that must be engaged in by public school officials. Further, the reader must assume that in some manner, the legislature does represent the collective public will. On many occasions the reader may note how the legislature has chosen process over efficiency and procedure over the final cost in order to ensure the public treasury.

Public education continues to be called upon by society to meet a wide variety of socioeconomic ills in the most efficient manner. Efficiency is often defined as doing more with fewer fiscal resources. While it may be argued that the fiscal management of every school district grows more complex every year, this concept is probably true for most professions and most public agencies.

Unfortunately, what has historically separated public education from other issues in the minds of the public is that professional educators, as well as social reformers, have seized public education as the vehicle of virtually every social reform issue in America. When the general public perceives that public education and the related fiscal support must cover everything from reading, writing, and arithmetic to driver's education and sex education, the public grows less supportive as a whole.

Contrasting this miscast reliance on public education to solve the social ills of society is the abandonment of state policymakers' support of public education. Public schools reflect the complex socioeconomic mosaic of society while at the same time reveal how society as a whole has yet to accept and agree upon the most basic values that public education must transmit to the next generation. Such evidence is clearly found in the movements toward vouchers, tax credits, and charter schools. While these options certainly having limited attractive features, they nonetheless will fiscally degrade the limited resources to which the vast majority of public schools have access.

The lessons of history are unforgiving. Public education has provided an unparalleled overall success for American society over the past dozen generations. It has experienced successes and certainly failures, of untold magnitude.

In the future, changing demographics in America dictate that more and more school-aged children will enter school from single- parent families, unprepared to learn, with English being the non-primary language of the home. These issues will have a profound impact on school staffing and needs.

Changing educational needs and school finance formulas lead to challenges in courts due to even greater perceived obligations. Further, greater fiscal equity needs will exist by those who perceive that the educational finance distribution formula is the root of all fiscal problems within certain school districts. The fact that the judiciary will settle the major education finance policy issues will most likely continue for some period of time.

With the globalization of our nation, schools will continue to be thrust into computerization with access to information worldwide. Schools will face even more pressure to have all students computer literate. In the first edition of this text it was noted that many public schools of the nation were essentially fighting tomorrow's war with yesterday's technology. This new globalization is largely a result of the shift from industrial society to an informational society. Thus, without great forethought and resources, the poorest of our society will grow even poorer.

In summation, the state and quality of public education in America is largely interwoven with the overall state of the economy and public education is directly related to the ability of the nation to compete in the global economy. This is overshadowed by the reality that the next decade will offer bleak fiscal resources to public education. While the relationship between public education and the economy is highly interrelated, public education will have great difficulty meeting the expectations of the public while being almost precluded from assisting the nation in the competitive world of a global economy. This is due to ever-increasing numbers of students entering the public schools because of rising birth rates in segments of the population, increasing immigration numbers, and the general baby boom.

The public has been not been impressed with the results of public education and its current returns to society. With the current (1995) $270 billion being spent toward public elementary and secondary education, the public perceives it is not receiving a reasonable return on its investment. Thus, public education must, in the future, consider educational research, which has indicated that lowering class sizes in the elementary grades is quite cost effective in terms of return of investment, while in the higher grades, such actions produce marginal results. School districts must examine where the local school district perceives to gain the greatest educational benefit in relation to the investment.

Also, public education cannot continue to build public schools without examining alternatives. All public school districts must examine year-round schools as well as the availability of surplus spaces that might be converted to educational programs within the community. They desperately need to engage in competitive concepts as well by breaking down rigid contracts and assignments of all individuals within the school district.

School districts must collectively and individually pursue the rapid deregulation of public education. Every time an opportunity presents itself to free public school districts from needles rules and regulations, school districts must collectively pursue these possibilities.

Lastly, public education needs to understand, accept, and fully participate in the public policy debate regarding the role of public education in our society. Too many school board associations are overly concerned with minutiae, state finance officers with minor public issues, school board members with narrow parochial interests, and school superintendents with distrust of the public policy debate that has become the issue within every state in every legislative session. In the meantime, the opponents of public education have become well organized, well financed, and well grounded in the political reality and arguments in the state capital. Public educators, and in particular chief fiscal officers, need to join this dialogue and bring their individual and collective expertise on behalf of public education to the forefront of these public policy discussions.

Index

Symbols

A

D

E

F

G

H

I

J

K

L

M

N

O

P

Q

R

S

T

U

V

W

X

Z

Cases

A

B

C

T

U

V

W